Ralf-Dietrich Kahlke (Ed.)

The Pleistocene of Untermassfeld near Meiningen (Thüringen, Germany) Part 5

MONOGRAPHIEN

des Römisch-Germanischen Zentralmuseums

Band 40, 5

Römisch-Germanisches Zentralmuseum
Leibniz-Forschungsinstitut für Archäologie

RGZM

SENCKENBERG
Leibniz Institution for Biodiversity and Earth System Research

Römisch-Germanisches Zentralmuseum
Leibniz-Forschungsinstitut für Archäologie
und
Senckenberg
Research Station of Quaternary Palaeontology Weimar

Ralf-Dietrich Kahlke (Ed.)

THE PLEISTOCENE OF UNTERMASSFELD NEAR MEININGEN (THÜRINGEN, GERMANY)

PART 5

Mit Beiträgen von
Jessica Arnold · Stefan Flohr · Nuria García · Helmut Hemmer · Alessio Iannucci
Ralf-Dietrich Kahlke · Uwe Kierdorf · Lutz Christian Maul · Paul P. A. Mazza
Beniamino Mecozzi · Jelle W. F. Reumer · Raffaele Sardella · Hans van Essen

Verlag des Römisch-Germanischen Zentralmuseums **Mainz 2022**

Redaktion: Bärbel Fiedler | Senckenberg Research Station of Quaternary Palaeontology Weimar
Bildbearbeitung: Evelin Haase | Senckenberg Research Station of Quaternary Palaeontology Weimar
Umschlaggestaltung: Evelin Haase | Senckenberg Research Station of Quaternary Palaeontology Weimar – Right upper canine of *Homotherium latidens ultimum* (natural size) from Untermassfeld and detail of the excavation area (1979–2015 field seasons)
Satz: Claudia Nickel (RGZM)

Bibliografische Information der Deutschen Nationalbibliothek

Die Deutsche Nationalbibliothek verzeichnet diese Publikation in der Deutschen Nationalbibliografie; detaillierte bibliografische Daten sind im Internet über **http://dnb.d-nb.de** abrufbar.

ISBN 978-3-88467-347-8
ISSN 0171-1474

Druck: Beltz Grafische Betriebe GmbH, Bad Langensalza
Printed in Germany.

CONTENTS

Dedicated to the former Senckenberg director

Fritz F. Steininger

with our thanks for many years of support

FOREWORD

More than four decades of continuous excavation and research have been devoted to the Pleistocene deposit of Untermassfeld and its extraordinarily rich and diverse fossil content. The prepared and preserved finds comprise a total of 18,250 catalogued specimens or series and fill about 1000 drawers at the Senckenberg Research Station of Quaternary Palaeontology in Weimar. The corresponding research results produced by an international multidisciplinary team of scientists are available in five ongoing monograph volumes. The long-term Untermassfeld project was, in many ways, influenced by changing constellations in scientific policy. The character of the Weimar Quaternary palaeontology department, which was in charge of the overall project, changed from being an independent institute into a working group in a geoscientific institute in Jena, before becoming integrated into the Senckenberg Research Institute in Frankfurt a. M. Our special thanks go to the then director Fritz F. Steiniger, who strongly supported the interests of Quaternary palaeontological research in Weimar and the progress of the excavations in Southern Thuringia.

Over the years, the field, conservation, and evaluation work was carried out with the assistance of different funding models. For regular support we thank the Free State of Thuringia, represented by the Thuringian State Office for Heritage and Archaeology with its president Sven Ostritz (Erfurt, Weimar), the City of Meiningen, and the District of Schmalkalden-Meiningen. Assistance in setting up storage rooms in two newly occupied institute buildings during the years 2005–2009 was received from the Senckenberg Society for Nature Research, which was always generous in its support of the work on Untermassfeld. The palaeontological expert of a renowned national organisation for research funding, on the other hand, recommended as early as the early 1990s that the excavation work at Untermassfeld should be discontinued in the near future. If this advice had been followed, almost 9,000 Early Pleistocene vertebrate finds from a reliable stratigraphic context would not be available to researchers today.

Over the course of this long-term project, the technologies used for the recovery and documentation of the fossil material developed considerably. In addition to the use of efficient laser surveying technology in the field, find documentation was converted to digitally created excavation plans, allowing selective access of each individual specimen. The use of digital photography and, occasionally, applied computer tomography also improved the visualisation possibilities of the examined fossil material. A new impregnation method specifically for the Untermassfeld vertebrate fossils was also developed during the project in order to preserve the osteological material as permanently as possible. The following staff members of the Weimar Institute have been involved in the overall project since the publication of Part 4 of the Untermassfeld monograph (2020): Jessica Arnold (preparation), Susann Döring (photography), Bärbel Fiedler (editorial manuscript management), Evelin Haase (cataloguing, documentation, graphics), John-Albrecht Keiler (conservation technology, field management, funding), Frank Krause (technical support), Dennis Rössler (preparation), Sabine Schneider (finances), Gerald Utschig (collection management), Rebecca Michalik (preparation). Our heartfelt thanks go to all of them. It goes without saying that for the whole of the team the challenging work on the Untermassfeld project and the resulting collection, as well as the numerous collaborative relationships, was an enriching experience.

After the then Director General of the Römisch-Germanisches Zentralmuseum Mainz Konrad Weidemann † included Parts 1–3 of the Untermassfeld monograph in the RGZM's publication series (1997, 2001), it was Sabine Gaudzinski-Windheuser (Mainz, Neuwied) who later made the continuation of the Untermassfeld series possible. The printing of Parts 4 and 5 was financially supported by the Senckenberg Research Institute in Frankfurt a. M. within the framework of a cooperation agreement. We sincerely thank all those involved.

In addition to the description of a pathological finding, the present volume contains new research results on suids and elephantids and in particular on new findings from ursids and hyaenids, including an analysis of the bone modifications they caused. New results are also presented for all six felid species, micromammals, and cercopithecids. A synopsis on site origin, palaeobiodiversity, taphonomy, palaeoenvironment, chronostratigraphy and significance of the Untermassfeld fossil assemblage for western Palaearctic faunal history concludes the monograph. Since the treatises on the research results on Untermassfeld are now quite extensive, cross-references to further and more precise explanations are unavoidable. Corresponding citations are mostly provided with page and illustration references to enable the quick location of corresponding passages.

The manuscripts for Part 5 of the Untermassfeld monograph were kindly reviewed by Diego Jaime Álvarez-Lao (Oviedo), László Bartosiewicz (Stockholm), Marco Cherin (Perugia), Eric Delson (New York City), Helmut Hemmer (Mainz), Uwe Kierdorf (Hildesheim), Adrian Lister (London), Simon Parfitt (London), and one anonymous colleague. The majority of the texts were linguistically edited with expertise and much attention to detail by Christina Nielsen-Marsh (Leipzig). Claudia Nickel (Mainz) took care of the final editing and layout of the individual contributions in her usual meticulous manner.

Since the excavation site of Untermassfeld has now been secured against theft and associated destructive interventions, it would make sense to present the findings and current research results to a wider public. Perhaps in the future there will be the possibility of a permanent exhibition in Thuringia that focuses on the discoveries of Untermassfeld in an ecological context. In so doing, unique and irretrievable material, witnesses of life a million years ago, could be made accessible as tangible 3D reproductions. Other exhibits displaying the original specimens could also be made available to the viewer, and which might occasionally even permit handling. It is to be hoped that the site, its fossil content, and the knowledge gained from its study will form the basis, starting point, and inspiration for future palaeobiological research and museum experiences.

Weimar, January 2022 *Ralf-Dietrich Kahlke*

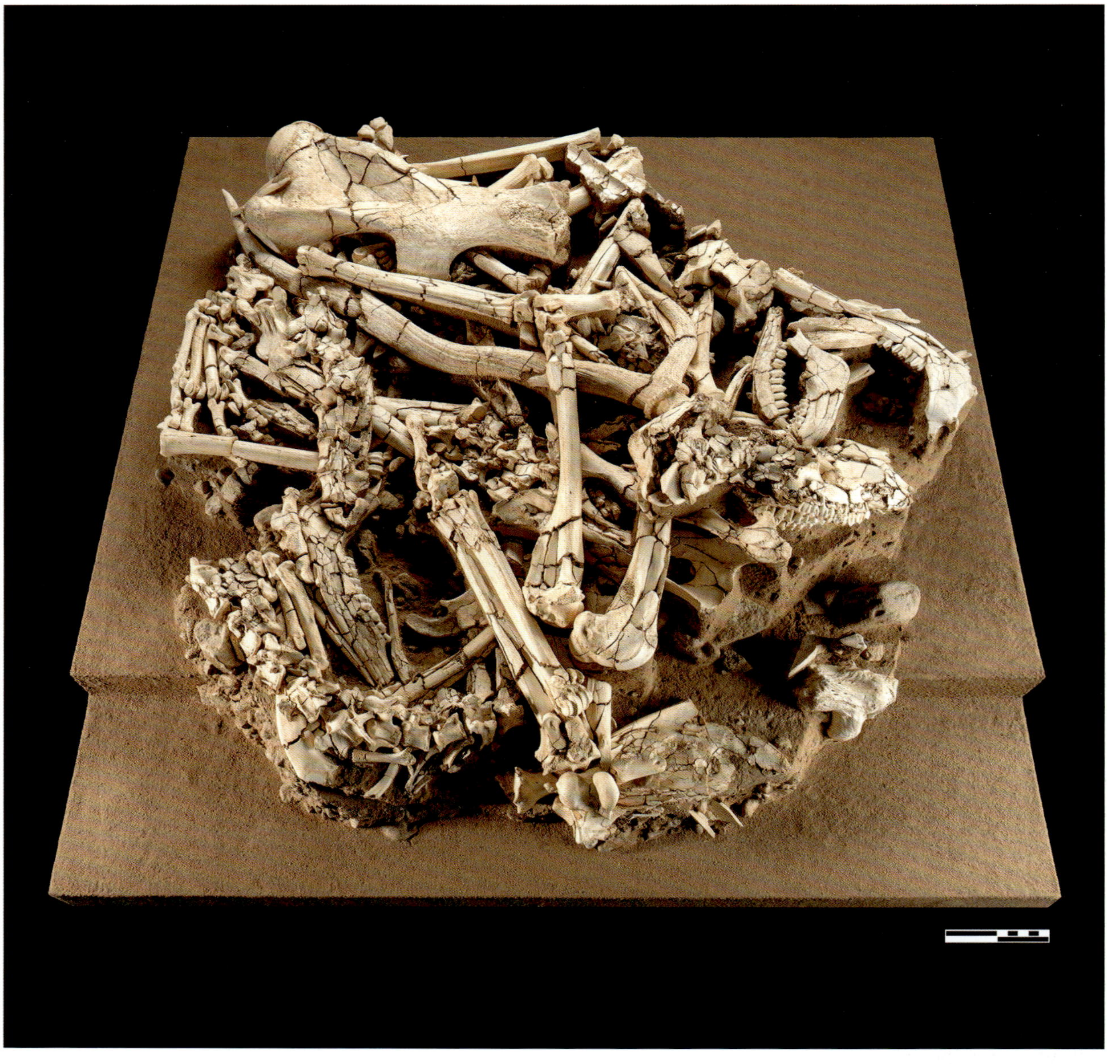

Untermassfeld research excavation, in-situ block preparation IQW 2014/41 900 (Mei. 47 209) with a polyspecific bone concentration embedded in channel lag deposits of Upper Fluviatile Sands [UFS(l)] from grid squares Q 21, 22, and 49 (Kahlke in this volume, Fig. 15, **Foldout III**), 2.14–1.90 m below site 0-level (see Kahlke 2020, Fig. 5a–b for images showing the block during excavation and its recovery by crane), 255 finds exposed (a few more than are visible in the excavation plan) including a skull of *Pseudodama vallonnetensis* (de Lumley, Kahlke, Moigne et Moullé) with its left antler (Breda et al. 2020) and a substantial part of the forequarter skeleton from a female *Acinonyx pardinensis pleistocaenicus* (Zdansky) (Hemmer and Kahlke in this volume), original find position faced in a south-westerly direction. – Dimensions of the preparation: 1.0 × 1.0 × 0.46 m.

References

Breda, M., Kahlke, R.-D., Lister, A., 2020. New results on cervids from the Early Pleistocene site of Untermassfeld. In: Kahlke, R.-D. (Ed.), The Pleistocene of Untermassfeld near Meiningen (Thüringen, Germany). Part 4. Monographien des Römisch-Germanischen Zentralmuseums 40, 4, 1197–1249.

Hemmer, H., Kahlke, R.-D., in this volume. New results on felids from the Early Pleistocene site of Untermassfeld, 1465–1566.

Kahlke, R.-D., 2020. Excavation progress and research at the Early Pleistocene site of Untermassfeld during the years 1997–2015. In: Kahlke, R.-D. (Ed.), The Pleistocene of Untermassfeld near Meiningen (Thüringen, Germany). Part 4. Monographien des Römisch-Germanischen Zentralmuseums 40, 4, 1031–1078.

in this volume. The Pleistocene of Untermassfeld – A synopsis of site origin, palaeobiodiversity, taphonomic characteristics, palaeoenvironment, chronostratigraphy, and significance in western Palaearctic faunal history, 1635–1734.

UWE KIERDORF · STEFAN FLOHR · RALF-DIETRICH KAHLKE

POST-TRAUMATIC OSTEOMYELITIS IN AN IMMATURE *BISON MENNERI* FROM THE EARLY PLEISTOCENE SITE OF UNTERMASSFELD

Abstract

Palaeopathological studies can contribute to the reconstruction of the biology of extinct species. Here we present the case of a pathological radius of a *Bison menneri* calf from the late Early Pleistocene fluviatile site of Untermassfeld (Central Germany). Based on macroscopic and radiographic analyses, a diagnosis of post-traumatic osteomyelitis is put forward. According to our reconstruction, a comminuted open fracture of the distal radius was accompanied by invasion of the fracture site by microorganisms. This infection caused a massive inflammation that spread from the diaphysis via the growth plate to the distal epiphysis and led to the formation of an extended, multilocular Brodie's abscess. Accompanying inflammation of the periosteum (periostitis) caused the formation of a thick cloak of periosteal new bone in the distal radius and an associated thickening and abnormal contour of this bone portion. The radius furthermore shows premature onset of epiphyseal fusion at both ends. Based on the pathological findings it is suggested that the fracture and associated inflammation of the radius was survived for some weeks. It is possible that due to lameness resulting from the condition and despite the likely protection by its mother and the herd, the bison calf was killed by a predator. It is, however, also conceivable that the disabled animal drowned during one of the high flood events that led to the formation of the site, or that spreading of the infection via the vascular system had caused sepsis and death from organ failure. The presented case expands the spectrum of skeletal pathology known in large mammals from the Untermassfeld site and in fossil species of the genus *Bison*.

1. Introduction

Studies on the palaeopathology of extinct mammals and other vertebrates have a relatively long history (Moodie 1923; Tasnádi-Kubacska 1962; Baker and Brothwell 1980; Rothschild and Martin 1993; Bartosiewicz 2013) and can contribute to our understanding of the lifestyle, behaviour, and ecology of animals from the past (Garutt 1999; Iurino et al. 2013; 2015; Bishop et al. 2015; Foth et al. 2015; Haynes and Klimowicz 2015; Kierdorf et al. 2016; Palombo and Zedda 2016; Woolley et al. 2021). Incorporation of palaeopathological data into the study of extinct species is therefore recommended when attempting to reconstruct their biology.

More recently, the need for increased scientific rigor and terminological precision in palaeopathological reporting and diagnosis has been emphasised by different authors (Ortner 2012; Appleby et al. 2015; Buikstra et al. 2017; Klaus and Lynnerup 2019). While these requirements were expressed mainly with respect to human palaeopathology, the same also applies to investigations in animals. A further necessity in palaeopathological studies on both human and animal remains is the distinction between pathological changes and alterations formed postmortem due to taphonomic processes (Bartosiewicz 2008; 2013; Haridy et al. 2019; Klaus and Lynnerup 2019).
The late Early Pleistocene site of Untermassfeld (Thuringia, Germany) has been investigated for more than four decades, and the recovered fossil remains constitute the most complete mammal assemblage from the Epivillafranchian biochron (1.2–0.9 Ma BP) of the western Palaearctic (Kahlke 1997; 2001; 2006; 2020). Some palaeopathological findings on large mammals from this locality have previously been reported (Kahlke and Kierdorf 1997; Kierdorf et al. 2012; Kierdorf and Kahlke 2020). In continuation of this research, the present contribution describes a radius of an immature *Bison menneri* from Untermassfeld showing evidence of post-traumatic osteomyelitis.

2. Material and methods

2.1. The Untermassfeld fossil assemblage

The site of Untermassfeld is situated on the eastern valley slope of the river Werra 2 km south of the town of Meiningen in Southern Thuringia, Central Germany. Continuous excavations at Untermassfeld were performed between 1979 and 2019 (Kahlke in this volume and references therein). They yielded about 18,250 palaeontological specimens or series (status as of 1 July 2020), of which nearly 14,300 are large mammal finds that were mostly identified at the species level (Kahlke 2001; 2006; in this volume). The Untermassfeld finds are stored and curated at the Senckenberg Research Station of Quaternary Palaeontology in Weimar, Germany (IQW).
Accumulation of the unusually high amount of vertebrate remains is attributed to a succession of high flood events that carried complete or partial carcasses in different stages of decomposition as well as isolated skeletal elements to the site. The material accumulated under the lee of a clastic mudflow fan close to the right bank of the river (Ellenberg and Kahlke 1997; Kahlke et al. 2020; Kahlke in this volume, Fig. 2). It has been assumed that the finds include both direct victims of the floods as well as leftovers of predator and scavenger activities, swept away from areas upstream of the site (Hemmer 2001; Kahlke and Gaudzinski 2005; Kahlke 2006; in this volume).
The Untermassfeld faunal assemblage is assigned to Marine Isotope Stage (MIS) 31 and has an absolute age of slightly more than one million years (Kahlke 2006; 2007; in this volume; Gerdes et al. 2020). The composition of the fauna indicates a temperate climate with warmer summers than occur today in the region and mild winters (Kahlke 2001; 2006; Böhme 2020). In addition to *Bison menneri*, the Untermassfeld ungulate fauna includes *Capreolus cusanoides*, *Cervalces carnutorum*, *Eucladoceros giulii*, *Pseudodama vallonnetensis*, *Hippopotamus antiquus*, *Stephanorhinus hundsheimensis* and *Equus wuesti*. Carnivora are represented by *Panthera onca gombaszoegensis*, *Acinonyx pardinensis pleistocaenicus*, *Puma pardoides brevipes* (Hemmer and Kahlke in this volume), *Megantereon cultridens adroveri*, *Homotherium crenatidens*, *Pachycrocuta brevirostris*, *Ursus* cf. *dolinensis*, *Canis* (*Xenocyon*) *lycaonoides*, *Canis mosbachensis* and others (complete faunal list in Kahlke in this volume). The climatic conditions indicated by the composition of the Untermass-

feld faunal record are supported by oxygen isotope data obtained on the phosphate in the mineral phase of *Equus* bones from the site (Stephan et al. 2001).
According to the palaeoenvironmental reconstruction, at the time of formation of the site most of the Werra river valley was covered by deciduous riparian forests, while the area outside the valley was a park-like landscape with sparsely wooded to open biotopes (Kahlke 2006; Van Asperen and Kahlke 2017). No hominin remains have been recovered at Untermassfeld, and there is no evidence for the presence of hominins at the site during its formation (Kahlke 2006; Roebroeks et al. 2018).

2.2. The Untermassfeld bison material

The vast majority of bovid finds from Untermassfeld are considered to represent a single taxon that was described as a new species, *Bison menneri*, with the Untermassfeld site as the type locality (Sher 1997; Bukhsianidze 2020). *B. menneri* is the most common mammal species in the Untermassfeld fossil record with almost 2,200 bone samples recovered during the excavations (detailed compilation in Arnold in this volume). These finds represent an MNI of 69, including 51 adults of both sexes and 18 immature individuals. The mortality profile of the Untermassfeld bison material corresponds to that of a natural population. Therefore, it seems likely that larger bison groups with herd-like age structures fell victim to the recorded high flood events (Kahlke and Gaudzinski 2005; Kahlke 2006).
B. menneri was a very tall and slender species with long legs, a high hump and a rather small head with short horns (Sher 1997; Kahlke 2006; skeletal reconstruction in Van Asperen and Kahlke 2017; Kahlke in this volume, Fig. 5). Based on dental mesowear data, the species has been characterised as a mixed feeder similar to the extant North American wood bison (*B. bison athabascae*) (Van Asperen and Kahlke 2017). In more recent taxonomic works, extant bison is often assigned to the genus *Bos* (e.g., Groves and Leslie 2011). Contrary to this usage, in the present contribution the generic name *Bison* will be retained for both extant and extinct bison species.

2.3. Imaging methods

The pathological bison radius was photographed with a digital camera (Nikon D 300). Subsequently, the specimen was X-rayed (70 kV, 16 mAs) and CT scanned (Phillips Brilliance 64 CT scanner; 120 kV, 250 mAs, 1 mm slice thickness). The radiograph and the initial axial and reformatted CT images were viewed and processed using the RadiAnt DICOM viewer and Adobe Photoshop CS4.

3. Case description

The pathological specimen IQW 1980/16 656 (Mei. 16 177) is a right radius of *B. menneri* of undetermined sex (**Fig. 1**). The ulna, which in bisons is initially attached to the radius by a syndesmosis and later united synostotically, is not fused to the pathological radius. This demonstrates that the two elements of the antebrachium had still been connected only syndesmotically. However, despite the individual's young age, both epiphyses are already attached to the diaphysis. Proximally and distally, the epiphyseal lines are still completely visible (**Fig. 1**), indicating a very early stage of epiphyseal fusion. Data on the normal age at epiphyseal fusion are not available for *B. menneri*. For extant *Bison* species, fusion of the proximal radial epiphysis

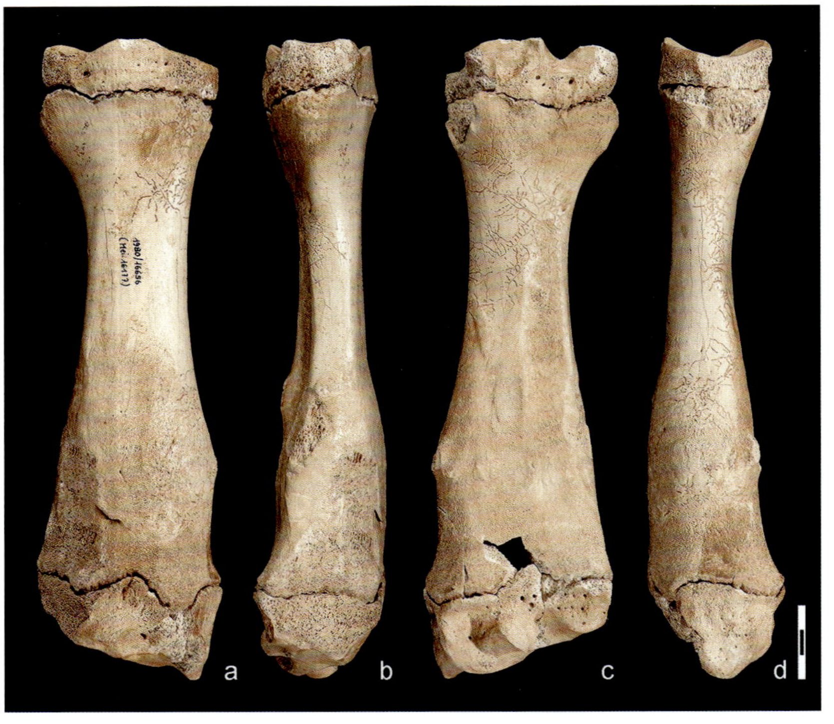

Fig. 1 *Bison menneri* Sher, Untermassfeld. – **a–d** Right immature radius IQW 1980/16656 (Mei. 16177) with angulation of its distal portion towards lateral, loss of cortical bone caudodistally, and signs of root etching, cranial, lateral, caudal, medial views. – Scale: 30mm.

has been reported to start at the beginning of the fourth year of life, while fusion of the distal epiphysis occurs early in the sixth year (Duffield 1973).

Greatest length of the pathological radius is 259mm. The greatest length of radii of adult *B. menneri* ranges between 350 and 370mm in females (n = 5) and between 375 and 418mm in males (n = 8) (Sher 1997). Thus, on the basis of its length it can be confirmed that the pathological radius is that of an immature individual. The fossil assemblage from Untermassfeld contains only one further radius of an immature *B. menneri* (also from the right side of the body). This specimen IQW 2011/38035 (Mei. 35009) is represented by the diaphysis only, which has a greatest length of 235mm and shows no pathological changes. Greatest length of the diaphysis of the pathological radius is 204mm, suggesting that it belonged to a considerably younger individual than the mentioned non-pathological radius diaphysis.

Data for radius length in the European bison or wisent (*B. bonasus*) were reported by Koch (1932). The highest value recorded for the radius of an adult individual was 318mm. This is considerably less than the values reported for *B. menneri*, which is a particularly long-legged species (Sher 1997; Kahlke 2006). According to Koch (1932), greatest radius length in 8-month-old *B. bonasus* calves (n = 4, both sexes represented) ranges between 244 and 246mm, while a value of 261mm is given for the radius of a 1-year-old individual of undetermined sex. Comparing the length of the pathological *B. menneri* radius (259mm) with the values given above, it is concluded that the individual to which it belonged probably died in its first year of life, i.e., was a calf (Fuller 1959). The fact that both epiphyses are already initially fused to the diaphysis therefore clearly denotes premature onset of epiphyseal fusion in the pathological radius.

The material state of preservation of the pathological radius is good, as is the case for most of the Untermassfeld vertebrate remains. There is some postmortem loss of cortical bone (exposing the spongiosa) at the epiphyses, as well as at the proximal (caudomedially) and distal ends (caudally) of the diaphysis (**Fig. 1**). In the latter location only a thin sheet of cortical bone is present, a portion of which has been fractured away (**Fig. 1c**). As is denoted by their light colour, some of the breakage surfaces are fresh, indicating that the damage occurred during or after recovery. Signs of root etching in the form of a dendritic pattern of shallow

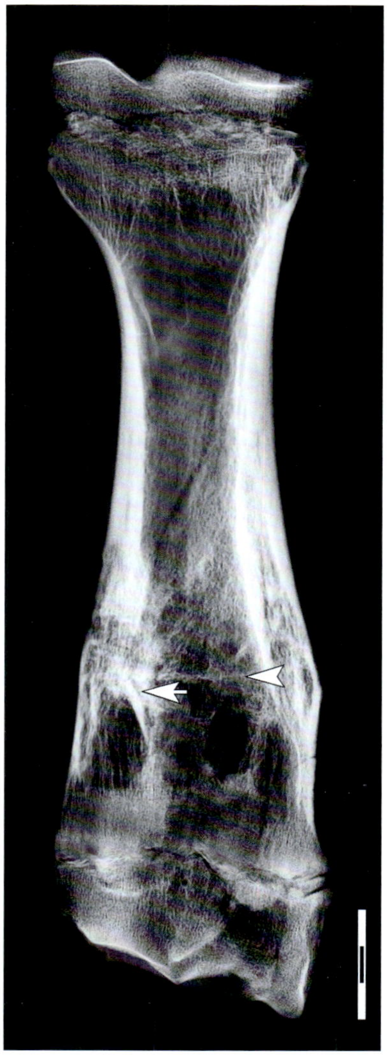

Fig. 2 *Bison menneri* Sher, Untermassfeld. – X-ray image of right immature radius IQW 1980/16656 (Mei. 16177) with radiolucent (osteolytic) areas in the distal metaphysis and epiphysis, multilayered periosteal new bone formation in the distal half of the diaphysis, fragments of the original cortex (arrow) and transverse radiodense line (arrowhead), craniocaudal projection, proximal to top of figure. – Scale: 30 mm.

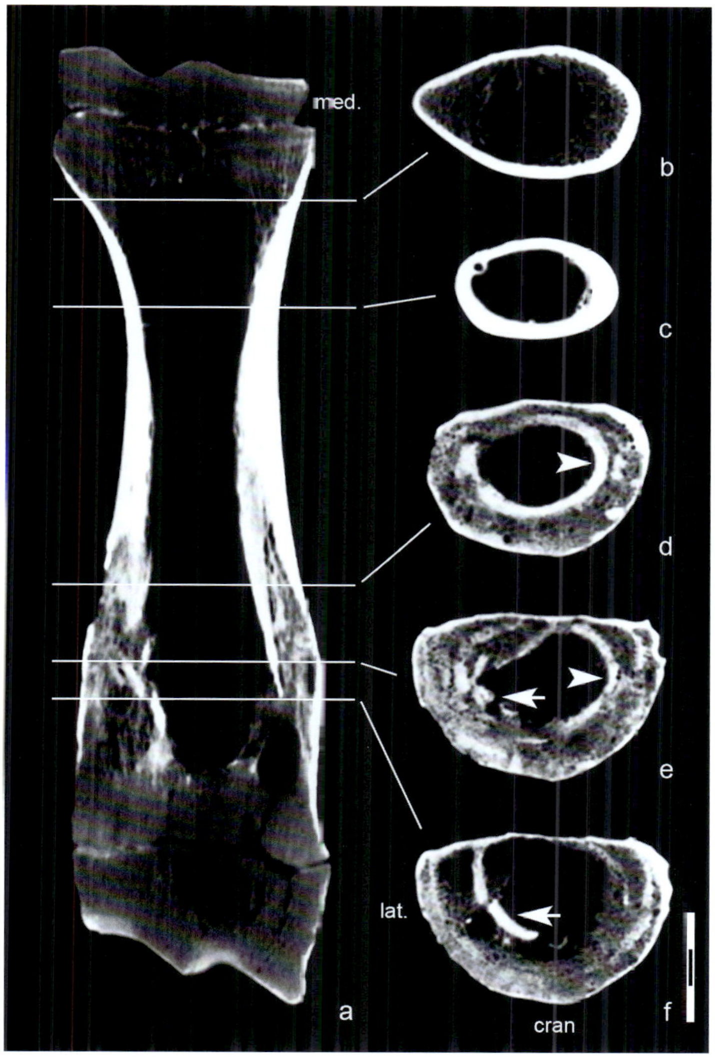

Fig. 3 *Bison menneri* Sher, Untermassfeld. – CT images of immature radius IQW 1980/16656 (Mei. 16177). – **a** Coronal section. Note lytic lesions in the distal metaphysis and epiphysis and local widening of the physeal space. – **b–f** Axial sections at different levels along the proximodistal axis. Proximally, the diaphyseal cortex shows a normal structure (**b–c**). In the distal half of the diaphysis, the original cortex is partly (**d–e**) or completely (**f**) destroyed, and pronounced deposition of periosteal new bone has occurred. – Arrows: fragments of lateral cortex. – Arrowheads: endosteal border of original cortex. – Scale: 30 mm.

grooves are present in various regions of the radius (**Fig. 1**), demonstrating that it had been in contact with plant roots for some time after its deposition. Corresponding traces have been observed also in numerous other finds from Untermassfeld (e. g., Kahlke 2001, 935, 976; 2006, 81; Kotowski et al. 2020, 1274 f.). Bite, chew or gnaw marks are missing in the pathological radius, suggesting that the specimen had been rapidly covered by fluviatile sediments after its final deposition.

The proximal half of the diaphysis of the pathological radius is of a normal shape. In contrast, the distal half of the bone is thickened, has an irregular contour and exhibits an abnormally coarse surface (**Fig. 1**). Conventional X-ray (**Fig. 2**) and CT images (**Figs. 3–5**) reveal a normal cortex in the proximal part of the diaphysis as well as normal bone architecture in the proximal metaphysis and epiphysis. They further demonstrate a

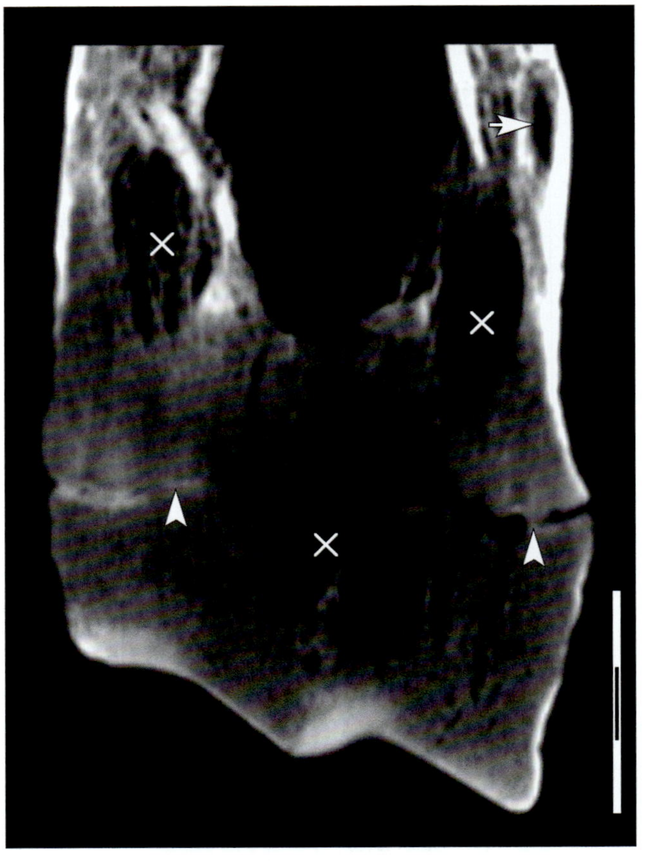

Fig. 4 *Bison menneri* Sher, Untermassfeld. – Higher magnification of a coronal CT section of the distal diaphysis and epiphysis of immature radius IQW 1980/16656 (Mei. 16177) showing metaphyseal and epiphyseal lytic lesions (×), lytic lesion in periosteal new bone (arrow), bony bridging between epiphysis and metaphysis (arrowheads), tracts between metaphyseal lesions and physis, and fragments of the original lateral diaphyseal cortex. – Scale: 30 mm.

fracture in the distal half of the diaphysis. Several fragments of the lateral cortex are visible on the X-ray and CT images, and the fracture is accordingly classified as comminuted (Resnick et al. 1995a; Lovell et al. 1997). The fracture is undisplaced, but the radius exhibits a slight deformity as its distal portion is angulated towards lateral by about five degrees (**Fig. 1a, c**). The X-ray image (**Fig. 2**) shows an increased radiodensity extending along the fracture line that is oriented more or less at a right angle to the shaft. This indicates that some fracture consolidation had already occurred at the time of death (de Jong et al. 2014).

The X-ray image further shows several radiolucent areas in the distal portion of the radius that denote marked bone destruction in the course of an inflammatory process (**Fig. 2**). CT imaging (**Figs. 3–5**) demonstrates that the osteolytic areas in the metaphysis are confluent, indicating that they represent a large multilocular abscess cavity. A rim of sclerotic bone surrounding the lesion is not apparent on the images. The metaphyseal lytic lesions are connected to the growth plate region by tracts of varying widths (**Fig. 4**), indicating spreading of the inflammation from the metaphysis to the physis. CT imaging further demonstrates a local widening of the physeal space (**Fig. 4**). From the growth plate, the bone destructive process further spread into the distal epiphysis, as is indicated by a large osteolytic lesion representing an extended medullary abscess cavity (**Figs. 2–4**). Coronal and sagittal CT sections also show localised bony bridging between the metaphysis and epiphysis both proximally and distally (**Figs. 4–5**).

The X ray and CT images moreover demonstrate pronounced periosteal new bone formation in the distal diaphysis of the pathological radius (**Figs. 3–4**). Except for a dense outer layer, the periosteal new bone, which partly appears multilayered, is of a loose structure (**Fig. 3d–f**). Its deposition has led to a marked

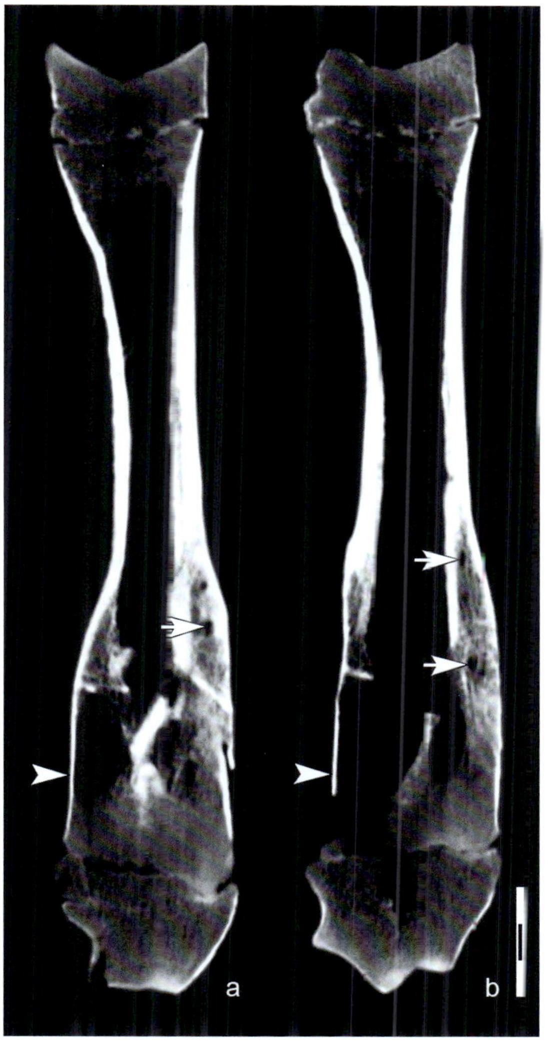

Fig. 5 *Bison menneri* Sher, Untermassfeld. – Sagittal CT sections of immature radius IQW 1980/16656 (Mei. 16177) with small lytic lesions in periosteal new bone (arrows), bony bridging between epiphysis and metaphysis at both ends, bone loss except for a thin sheet of cortical bone caudally (arrowheads), and fragments of the original diaphyseal cortex. – **a** lateral half of radius. – **b** medial half of radius. – Scale: 30 mm.

thickening of the distal diaphysis. Small intracortical abscess cavities are also present in the periosteal new bone (**Figs. 3–5**). The underlying original diaphyseal cortex has partly been destroyed (**Fig. 3d–f**).

4. Discussion

The macroscopic and radiological findings in the pathological radius demonstrate a transverse fracture in the distal half of its diaphysis with fragmentation of the lateral cortex. The fracture pattern suggests a strong traumatic impact from lateral as the likely cause of the fracture.

The radius moreover shows distinct signs of bone destruction in the course of an inflammatory process, indicating that the fracture was complicated by infection. It is concluded that the fracture was an open one and that the fracture site was invaded by infectious agents, most likely bacteria, causing post-traumatic osteomyelitis. Direct implantation of infectious material, as is thought to have occurred in the present case, is one possible route of bone infection; others being haematogenous spread and spread from a contiguous source (Resnick and Niwayama 1995a; Craig et al. 2016; Roberts 2019).

The cavitary lesions in the metaphysis and epiphysis are evidence of pronounced bone destruction in the course of an extended inflammatory process. The lack of sclerotic rims in the cancellous bone surrounding the lytic lesions suggests that the destructive process was ongoing at the time of death of the calf. The cortical fragments present in the distal diaphysis of the radius are considered to be fracture-related rather than formed by sequestration. However, like sequestra the fracture fragments represent pieces of necrotic bone that can harbour bacteria and thereby sustain an inflammatory process (Resnick and Niwayama 1995a; Freyschmidt 2016). In the course of the bone inflammation, the original diaphyseal cortex was partly destroyed.

Our findings indicate that the infection spread from the metaphysis to the growth plate region and further to the epiphysis. Spreading of the infection across the growth plate is considered to have occurred via transphyseal blood vessels, which in humans and other mammals are present in young individuals (Resnick and Niwayama 1995a; Blickman et al. 2004; Craig et al. 2016). The widened space in the physeal area denotes local destruction of the physis in the inflammatory process, which is a common finding in cases of transphyseal spreading of infection (Craig et al. 2016).

Periosteal new bone formation in the distal diaphysis of the pathological radius, which led to the thickening and abnormal contour of this bone region, denotes affection of the periosteum by the inflammatory process (periostitis). Formation of an extensive cloak of periosteal new bone is often seen in cases of osteomyelitis in young individuals (Resnick and Niwayama 1995a). In these individuals, the periosteum is thick, elastic, richly vascularised, and only relatively loosely attached to the subjacent bone from which it can easily be lifted (Resnick and Niwayama 1995a; Bisseret et al. 2015). The cambium layer of this active periosteum has a high osteogenic potential, causing the deposition of large amounts of bone upon stimulation (Bisseret et al. 2015). The presence of lytic lesions (intracortical abscesses) within the periosteal new bone shows that soon after its formation the newly deposited bone became affected by the inflammatory process.

Based on their radiological appearance, we diagnose the lesions in the distal metaphysis and epiphysis of the pathological bison radius IQW 1980/16656 (Mei. 16177) as indicative of a subacute to chronic form of osteomyelitis known as Brodie's abscess (Resnick and Niwayama 1995a). This type of abscess shows a predilection for the ends of tubular bones, but can also occur in other locations (Resnick and Niwayama 1995a; Kornaat et al. 2010; Craig et al. 2016; Freyschmidt 2016). It has been suggested that a Brodie's abscess develops when the infective organism has a relatively low virulence and/or the host shows an increased resistance to the infection (Resnick and Niwayama 1995a; Freyschmidt 2016).

The typical presentation of a Brodie's abscess is that of a well-demarcated pyogenic lesion of oval shape surrounded by a rim of sclerotic bone (Resnick and Niwayama 1995a; Kornaat et al. 2010; Craig et al. 2016). This abscess type was first described in the amputated leg of a 24-year-old male patient who complained about leg pain for more than twelve years. In the man's distal tibia, Brodie (1832, 242) observed »a cavity of the size of an ordinary walnut filled with dark-coloured pus« and further noticed that the »bone immediately surrounding this cavity was distinguished from its neighbourhood by being of a whiter colour, and of a still harder texture, and the inner surface of the cavity presented an appearance of high vascularity«. He further observed the deposition of periosteal new bone in the tibia that he regarded as »the result of inflammation of the periosteum at some former period« (Brodie 1832, 241).

Since the original description by Brodie (1832), the spectrum of bone abscesses classified as Brodie's abscess has widened considerably (Miller et al. 1979; Kornaat et al. 2010). It has been argued that, notwithstanding the marked differences in their radiological appearance, all these lesions should be considered as a single entity with variable expression (Miller et al. 1979). Thus, in humans, cases described as Brodie's abscesses include multilocular lytic lesions, abscesses traversing the growth plate, and abscesses surrounded by an only barely perceptible cancellous sclerosis (Miller et al. 1979). Moreover, although Brodie's abscesses are not normally associated with the formation of sequestra (Freyschmidt 2016), such instances have occasionally been reported (Weaver 1972; Miller et al.1979; Kornaat et al. 2010).

The observations in the studied pathological bison radius suggest a rather vigorous inflammatory process. However, in line with the view on the variability of the lesions in the case of Brodie's abscesses (Miller et al. 1979), interpretation of the present case as representing a variant of this condition is considered appropriate. It has been suggested that Brodie's abscess has a natural history that includes a destructive phase followed by a containment phase (Miller et al. 1979). The apparent lack of sclerosis in the cancellous bone bordering on the abscess cavities of the pathological bison radius could indicate that a transition to the containment phase had not yet occurred when the individual died. However, the possibility of some taphonomic loss of bone material along the lesion margins must also be considered.

The clinical stages of osteomyelitis are frequently referred to as acute, subacute and chronic (Resnick and Niwayama 1995a), with some variation in definition among authors. Osteomyelitis is usually characterised as subacute when the condition is present for more than 10 to 14 days but less than six to twelve weeks, while it is referred to as chronic when it has existed for longer (Riise et al. 2008; Freyschmidt 2016; Mandell et al. 2018). In humans, lytic bone lesions and apposition of periosteal new bone become detectable by plain radiography only after a delay period of about one to two weeks (Resnick and Niwayama 1995a; Riise et al. 2008). In post-traumatic osteomyelitis in dogs, radiological signs of a periosteal reaction were first detected between two and four weeks after the trauma (Pascalev 1999). In the human and veterinary medical literature, extended deposition of periosteal new bone onto the altered cortex is therefore considered evidence of subacute and chronic osteomyelitis (Resnick and Niwayama 1995a; Pascalev 1999). The recorded spectrum of bone lesions in the pathological bison radius from Untermassfeld suggests that the fracture was probably survived for some weeks, allowing the post-traumatic osteomyelitis to progress to the subacute or chronic stage.

Localised bony bridging between the metaphysis and epiphysis was observed at both ends of the pathological bison radius. Premature epiphyseal fusion is a known sequel of osteomyelitis of long bones in humans and other mammals, which can result in angular deformity of limbs, limb shortening, or both (Resnick et al. 1995a; Piddo et al. 2000; Bigham and Shafie 2008). In the case described here, the angular deformity of the bone is, however, mainly attributed to the fracture rather than premature epiphyseal fusion.

In conclusion, the lesions recorded in the pathological radius of the *B. menneri* calf from Untermassfeld are regarded as indicative of post-traumatic osteomyelitis with formation of a Brodie's abscess and accompanying ossifying periostitis. According to our interpretation, the condition was caused by a comminuted open fracture of the distal radius with bacterial invasion of the fracture site. The fact that the fracture is undisplaced and shows only minor angulation is attributed to stabilisation by the attached ulna and by the thick and elastic nature of the periosteum in the young individual. Our findings suggest that the infection started in the distal diaphysis and from there spread to the physis and further to the epiphysis. The inflammation also affected the newly deposited periosteal bone, causing intracortical abscesses, and furthermore led to premature fusion of both epiphyses.

In the palaeopathological case discussed here, the diagnosis is based solely on gross pathologic and radiological findings, while bacteriological and haematological data as well as clinical signs, which are of major importance for diagnosing osteomyelitis in human and veterinary medicine, are missing. Despite these inherent limitations, we conclude that on the basis of the documented spectrum of bone lesions a well-founded diagnosis of post-traumatic osteomyelitis following a comminuted fracture of the distal radius can be put forward.

Regarding the differential diagnosis of the condition, primarily bone destruction in the case of malignant bone tumours (osteosarcoma, Ewing's sarcoma) must be considered (Resnick and Niwayama 1995a; Resnick et al. 1995b; Chhem et al. 2008; Freyschmidt 2016). Multi-layered periosteal bone apposition can likewise be evident in the case of malignant bone tumours (Resnick and Niwayama 1995a; Resnick et al. 1995b; Freyschmidt 2016), but is also present in other conditions, including hypertrophic osteoarthropathy (Resnick and Niwayama 1995b; Flohr et al. 2018). In the latter disease, the cortex of the affected bones is, however, not altered (Flohr et al. 2018). As detailed above, in the present case, the available evidence in our view clearly supports the diagnosis of a fracture-related pyogenic inflammation in the radius of the bison calf.

Whether the radius fracture and its sequels were causally related to the death of the bison calf remains speculative. The individual's locomotion was certainly impaired by the condition. Due to lameness, the animal may have fallen victim to one of the large predators present in the Untermassfeld fauna (see section 2.1.), despite the protection provided by its mother and the herd. It is, however, also conceivable that the disabled animal drowned when large parts of the valley floor were inundated during a torrential high flood event. A further possibility to be considered is that spreading of the infection via the vascular system had caused sepsis and death from organ failure.

This is the second report of a limb bone fracture in *B. menneri* from Untermassfeld. Previously, a displaced tibial fracture in an adult *B. menneri* bull from the site has been described (Kierdorf et al. 2012). This fracture had also been comminuted and had healed with massive callus formation and shortening of the tibia by about 6 cm. Contrary to the present case, the tibia did, however, not show signs of osteomyelitis. Limb bone fracture has occasionally also been reported in other bison species. Thus, Kirillova et al. (2019) observed the fracture of a metatarsal in a Late Pleistocene steppe bison (*B. priscus*) from NE-Russia. Also in this case, the fracture had healed with pronounced callus formation and shortening of the bone. In an extant American bison (*B. bison*), a displaced femoral fracture healing with massive callus formation was reported by Moodie (1923).

In human archaeological skeletons, post-traumatic osteomyelitis is most commonly observed in the cranium and long bones (Lovell 1997). Cases of post-traumatic osteomyelitis affecting skull bones or postcranial skeletal elements have previously also been described in different species of Pleistocene mammals (Tasnádi-Kubacska 1962; Capasso 1998; Garutt 1999; Iurino et al. 2015; Nowakowski and Stefaniak 2015). However, to the best of our knowledge, the case presented here is the first report of post-traumatic osteomyelitis in a Pleistocene bison. Our observations thus expand the spectrum of skeletal pathology known in large mammals from the Untermassfeld site and in fossil species of the genus *Bison*.

Acknowledgements

We thank J. Arnold (Weimar) for her support in updating the MNI of bison found in the Untermassfeld fossil assemblage. S. Döring (Weimar) kindly took the photographs for **Fig. 1** and E. Haase (Weimar) arranged the figures. We are grateful to B. Holland and A. Wisotzki and their team at the Clinic for Diagnostic and Interventional Radiology and Neuroradiology of St. Bernward's Hospital, Hildesheim, for taking the conventional radiograph and CT scan of the pathological *Bison menneri* radius.

References

Appleby, J., Thomas, R., Buikstra, J., 2015. Increasing confidence in paleopathological diagnosis – Application of the Istanbul terminological framework. International Journal of Paleopathology 8, 19–21.

Arnold, J., in this volume. Bovid bone modifications by *Pachycrocuta* from the Early Pleistocene site of Untermassfeld. 1423–1464.

Baker, J., Brothwell, D., 1980 Animal diseases in archaeology. London.

Bartosiewicz, L., 2008. Taphonomy and palaeopathology in archaeozoology. Geobios 41, 69–77.

2013. Shuffling nags, lame ducks. The archaeology of animal disease. Oxford.

Bigham, A.S., Shafiei, Z., 2008. Unilateral premature closure of distal radial growth plate due to hematogenous osteomyelitis in a German shepherd dog. Comparative Clinical Pathology 17, 275–277.

Bishop, P.J., Walmsley, C.W., Phillips, M.J., Quayle, M.R., Boisvert, C.A., McHenry, C.R., 2015. Oldest pathology in a tetrapod bone illuminates the origin of terrestrial vertebrates. PLoS ONE 10, e0125723.

Bisseret, D., Kaci, R., Lafage-Proust, M.-H., Alison, M., Parlier-Cuau, C., Laredo, J.-D., Bousson, V., 2015. Periosteum: Characteristic imaging findings with emphasis on radiologic-pathologic comparisons. Skeletal Radiology 44, 321–338.

Blickman, J.G., van Die, C.E., de Rooy, J.W.J., 2004. Current imaging concepts in pediatric osteomyelitis. European Radiology 14, L55–L64.

Böhme, M., 2020. New results on amphibians and reptiles from the Early Pleistocene site of Untermassfeld – Bioclimatic analysis of the herpetofauna. In: Kahlke, R.-D. (Ed.), The Pleistocene of Untermassfeld near Meiningen (Thüringen, Germany). Part 4. Monographien des Römisch-Germanischen Zentralmuseums 40, 4, 1141–1158.

Brodie, B.C., 1832. An account of some cases of chronic abscess of the tibia. Medico-Chirurgical Transactions 17, 239–249.

Buikstra, J.E., Cook, D.G., Bolhofner, K.L., 2017. Introduction: Scientific rigor in paleopathology. International Journal of Paleopathology 19, 80–87.

Bukhsianidze, M., 2020. New results on bovids from the Early Pleistocene site of Untermassfeld. In: Kahlke, R.-D. (Ed.), The Pleistocene of Untermassfeld near Meiningen (Thüringen, Germany). Part 4. Monographien des Römisch-Germanischen Zentralmuseums 40, 4, 1169–1195.

Capasso, L., 1998. Cranial pathology of *Ursus spelaeus* Rosenmüller & Heinroth from Chateau Pignon, Basque territories (Spain). International Journal of Osteoarchaeology 8, 107–115.

Chhem, R.K., Saab, G., Brothwell, D.R., 2008. Diagnostic paleoradiology for paleopathologists. In: Chhem, R.K., Brothwell, D.R. (Eds), Paleoradiology. Berlin, 73–118.

Craig, L.E., Dittmer, K.E., Thompson, K.G., 2016. Bones and joints. In: Maxie, M.G. (Ed.), Jubb, Kennedy, and Palmer's pathology of domestic animals. Vol.1. St. Louis, 16–163. (6th ed.)

de Jong, J.J.A., Willems, P.C., Arts, J.J., Bours, S.G.P., Brink, P.R.G., van Geel, T.A.C.M., Poeze, M., Geusens, P.P., van Rietbergen, B., van den Bergh, J.F.W., 2014. Assessment of the healing process in distal radius fractures by high resolution peripheral quantitative computed tomography. Bone 64, 65–74.

Duffield, L.F., 1973. Aging and sexing the post-cranial skeleton of bison. Plains Anthropologist 18, 132–139.

Ellenberg, J., Kahlke, R.-D., 1997. Die quartärgeologische Entwicklung des mittleren Werratals und der Bau der unterpleistozänen Komplexfundstelle Untermaßfeld. In: Kahlke, R.-D. (Ed.), Das Pleistozän von Untermaßfeld bei Meiningen (Thüringen). Teil 1. Monographien des Römisch-Germanischen Zentralmuseums 40, 1, 29–62, pls. 1–18.

Flohr, S., Jasch, I., Langer, A., Riesenberg, M., Hahn, J., Wisotzki, A., Kierdorf, H., Kierdorf, U., Wahl, J., 2018. Secondary hypertrophic osteoarthropathy in a male from the Early Medieval settlement of Lauchheim, Germany. International Journal of Paleopathology 20, 72–79.

Foth, C., Evers, S.W., Pabst, B., Mateus, O., Flisch, A., Patthey, M., Rauhut, O.W.M., 2015. New insights into the lifestyle of *Allosaurus* (Dinosauria:Theropoda) based on another specimen with multiple pathologies. PeerJ 3, e940.

Freyschmidt, J., 2016. Skeletterkrankungen. Klinisch-radiologische Diagnose und Differentialdiagnose. Berlin. (4th ed.)

Fuller, W.A., 1959. The horns and teeth as indicators of age in bison. Journal of Wildlife Management 23, 342–344.

Garutt, N., 1999. Skull pathologies in *Coelodonta antiquitatis*: implications about social behaviour and ecology. Deinsea 6, 175–186.

Gerdes, A., Gärtner, A., Linnemann, U., Kahlke, R.-D., 2020: In situ U-Pb geochronology of re-deposited travertine from the Early Pleistocene site of Untermassfeld. In: Kahlke, R.-D. (Ed.), The Pleistocene of Untermassfeld near Meiningen (Thüringen, Germany). Part 4. Monographien des Römisch-Germanischen Zentralmuseums 40, 4, 1105–1115.

Groves, C.P., Leslie, D.M., 2011. Family Bovidae (hollow-horned ruminants). In: Wilson, D.E., Mittermeier, R.A. (Eds.), Handbook of the Mammals of the World. Volume 2, Hoofed Mammals. Barcelona, 444–774.

Guermazi, A., Mohr, A., Genant, H.K., 2003. Brodie abscess: another type of chronic posttraumatic osteomyelitis. European Radiology 13, 1750–1752.

Haridy, Y., Witzmann, F., Asbach, P., Reisz, R.R., 2019. Permian metabolic bone disease revealed by microCT: Paget's disease-like pathology in vertebrae of an early amniote. PLoS ONE 14, e0219662.

Haynes, G., Klimowicz. J., 2015. A preliminary review of bone and teeth abnormalities seen in recent *Loxodonta* and extinct *Mammuthus* and *Mammut*, and suggested implications. Quaternary International 379, 135–146.

Hemmer, H., 2001. Die Feliden aus dem Epivillafranchium von Untermaßfeld. In: Kahlke, R.-D. (Ed.), Das Pleistozän von Untermaßfeld bei Meiningen (Thüringen). Teil 3. Monographien des Römisch-Germanischen Zentralmuseums 40, 3, 699–782, pls. 132–143.

Hemmer, H., Kahlke, R.-D., in this volume. New results on felids from the Early Pleistocene site of Untermassfeld. 1465–1566.

Iurino, D.A., Fico, R., Sardella, R., 2015. A pathological Late Pleistocene badger from San Sidero (Apulia, Southern Italy): Implications for developmental pathology and feeding behaviour. Quaternary International 366, 96–101.

Iurino, D.A., Fico, R., Petrucci, M., Sardella, R., 2013. A pathological Late Pleistocene canid from San Sidero (Italy): implications for social- and feeding-behaviour. Naturwissenschaften 100, 235–243.

Kahlke, R.-D., 1997. Bisheriger Gesamtbefund zur Geologie, Paläozoologie, Taphonomie, Ökologie und Stratigraphie der unterpleistozänen Komplexfundstelle Untermaßfeld. In: Kahlke, R.-D. (Ed.), Das Pleistozän von Untermaßfeld bei Meiningen (Thüringen). Teil 1. Monographien des Römisch-Germanischen Zentralmuseums 40, 1, 385–418.

2001. Die unterpleistozäne Komplexfundstelle Untermaßfeld – Zusammenfassung des Kenntnisstandes sowie synthetische Betrachtungen zu Genesemodell, Paläoökologie und Stratigraphie. In: Kahlke, R.-D. (Ed.), Das Pleistozän von Untermaßfeld bei Meiningen (Thüringen). Teil 3. Monographien des Römisch-Germanischen Zentralmuseums 40, 3, 931–1030, suppl. I–XV.

2006. Untermassfeld. A late Early Pleistocene (Epivillafranchian) fossil site near Meiningen (Thuringia, Germany) and its position in the development of the European mammal fauna. British Archaeological Reports, International Series 1578. Oxford.

2007. Late Early Pleistocene European large mammals and the concept of an Epivillafranchian biochron. In: Kahlke, R.-D., Maul, L.C., Mazza, P.P.A. (Eds.), Late Neogene and Quaternary biodiversity and evolution: Regional developments and interregional correlations. Proceedings of the 18th International Senckenberg Conference (VI International Palaeontological Colloquium in Weimar). Volume II. Courier Forschungsinstitut Senckenberg 259, 265–278.

2020. Excavation progress and research at the Early Pleistocene site of Untermassfeld during the years 1997–2015. In: Kahlke, R.-D. (Ed.), The Pleistocene of Untermassfeld near Meiningen (Thüringen, Germany). Part 4. Monographien des Römisch-Germanischen Zentralmuseums 40, 4, 1031–1078.

in this volume. The Pleistocene of Untermassfeld – A synopsis of site origin, palaeobiodiversity, taphonomic characteristics, palaeoenvironment, chronostratigraphy, and significance in western Palaearctic faunal history. 1635–1734.

Kahlke, R.-D., Gaudzinski, S., 2005. The blessing of a great flood: differentiation of mortality patterns in the large mammal record of the Lower Pleistocene fluvial site of Untermassfeld (Germany) and its relevance for the interpretation of faunal assemblages from archaeological sites. Journal of Archaeological Science 32, 1202–1222.

Kahlke, R.-D., Kierdorf, U., 1997. Diffuse idiopathische skelettale Hyperostose bei einem Bären (*Ursus* sp.) aus dem Unterpleistozän von Untermaßfeld bei Meiningen (Südthüringen, Deutschland). Quartär 47/48, 159–175.

Kahlke, R.-D., Linnemann, U., Gärtner, A., 2020. New results on the origin and geological history of the Early Pleistocene site of Untermassfeld. In: Kahlke, R.-D. (Ed.), The Pleistocene of Untermassfeld near Meiningen (Thüringen, Germany). Part 4. Monographien des Römisch-Germanischen Zentralmuseums 40, 4, 1079–1104.

Kierdorf, U., Kahlke, R.-D., 2020. Pathological findings on remains of hippopotamids from the Early Pleistocene site of Untermassfeld. In: Kahlke, R.-D. (Ed.), The Pleistocene of Untermassfeld near Meiningen (Thüringen, Germany). Part 4. Monographien des Römisch-Germanischen Zentralmuseums 40, 4, 1251–1272.

Kierdorf, U., Kahlke, R.-D., Flohr, S., 2012. Healed fracture of the tibia in a bison (*Bison menneri* Sher, 1997) from the late Early Pleistocene site of Untermassfeld (Thuringia, Germany). International Journal of Paleopathology 2, 19–24.

Kierdorf, U., Meng, S., Kahlke, R.-D., 2016. Resorptive depressions on a horn core of Late Pleistocene (MIS 3) *Bison priscus* (Bovidae, Mammalia) from northeastern Germany. International Journal of Paleopathology 15, 76–82.

Kirillova, I.V., Shidlovskiy, F.K., Zinoviev, A.V., 2019. Metapodials of ancient bison (*Bison priscus*) of northeast Russia: »stress markers«, sex and withers height. Integrative Zoology 14, 270–279.

Klaus, H.D., Lynnerup, N., 2019. Abnormal bone: Considerations for documentation, disease process identification, and differential diagnosis. In: Buikstra, J.E. (Ed.), Ortner's identification of pathological conditions in human skeletal remains. London, 59–89.

Koch, W., 1932. Über Wachstums- und Altersveränderungen am Skelett des Wisents. Abhandlungen der mathematisch-naturwissenschaftlichen Abteilung der Bayerischen Akademie der Wissenschaften, Supplement-Band. 15. Abhandlung. München, 552–681.

Kornaat, P.R., Camerlinck, M., Vanhoenacker, F.M., De Praeter, G., Kroon, H.M., 2010. Brodie's abscess revisited. Journal of the Belgian Society of Radiology 93, 81–86.

Kotowski, A., Stefaniak, K., Kahlke, R.-D., 2020. A rhinocerotid skull from the Early Pleistocene site of Untermassfeld. In: Kahlke, R.-D. (Ed.), The Pleistocene of Untermassfeld near Meiningen (Thüringen, Germany). Part 4. Monographien des Römisch-Germanischen Zentralmuseums 40, 4, 1273–1293.

Lovell, N.C., 1997. Trauma analysis in paleopathology. Yearbook of Physical Anthropology 40, 139–170.

Mandell, J.C., Khurana, B., Smith, J.T., Czuczman, G.J., Ghazikhanian, V., Smith, S.E., 2018. Osteomyelitis of the lower extremity: pathophysiology, imaging, and classification, with an emphasis on diabetic foot infection. Emergency Radiology 25, 175–188.

Miller, W.B., Murphy, W.A., Gilula, L.A., 1979. Brodie abscess: Reappraisal. Radiology 132, 15–23.

Moodie, R.L., 1923. Paleopathology: An introduction to the study of ancient evidences of disease. Urbana.

Nowakowski, D., Stefaniak, K., 2015. Pathological changes of the cranium of a young female cave bear (*Ursus spelaeus* R.) – A case study (the Sudety Mts, Poland). International Journal of Osteoarchaeology 25, 119–125.

Ortner, D., 2012. Differential diagnosis and issues in disease classification. In: Grauer, A.L. (Ed.), A companion to paleopathology. Chichester, 250–267.

Palombo, M.R., Zedda, M., 2016. Surviving in a predator-free environment: Hints from a bone remodelling process in a dwarf Pleistocene deer from Crete. Comptes Rendus Palevol 15, 255–264.

Pascalev, M., 1999. Radiological picture of post-traumatic osteomyelitis in dogs. Veterinarski Arhiv 69, 149–159.

Piddo, C., Reed, M.H., Black, G.B., 2000. Premature epiphyseal fusion and degenerative arthritis in chronic recurrent multifocal osteomyelitis. Skeletal Radiology 29, 94–96.

Resnick, D., Niwayama, G., 1995a. Osteomyelitis, septic arthritis, and soft tissue infection: Mechanisms and situations. In: Resnick, D. (Ed.), Diagnosis of bone and joint disorders. Philadelphia, 2325–2418. (3rd ed.)

1995b. Enostosis, hyperostosis, and periostitis. In: Resnick, D. (Ed.), Diagnosis of bone and joint disorders. Philadelphia, 4396–4466. (3rd ed.)

Resnick, D., Goergen, T.G., Niwayama, G., 1995a. Physical injury: Concepts and terminology. In: Resnick, D. (Ed.), Diagnosis of bone and joint disorders. Philadelphia, 2561–2692. (3rd ed.)

Resnick, D., Kyriakos, M., Greenway, G.D., 1995b. Tumors and tumor-like lesions of bone: imaging and pathology of specific lesions. In: Resnick, D. (Ed.), Diagnosis of bone and joint disorders. Philadelphia, 3628–3938. (3rd ed.)

Riise, Ø.R., Kirkhus, E., Handeland, K.S., Flatø, B., Reiseter, T., Cvancarova, M., Nakstad, B., Wathne, K.O., 2008. Childhood osteomyelitis – incidence and differentiation from other acute onset musculoskeletal features in a population-based study. BMC Pediatrics 8, 45.

Roberts, C.A., 2019. Infectious disease: Introduction, periostosis, periostitis, osteomyelitis, and septic arthritis. In: Buikstra, J.E. (Ed.), Ortner's identification of pathological conditions in human skeletal remains. London, 285–319.

Roebroeks, W., Gaudzinski-Windheuser, S., Baales, M., Kahlke, R.-D., 2018. Uneven data quality and the earliest occupation of Europe – the case of Untermassfeld (Germany). Journal of Paleolithic Archaeology 1, 5–31.

Rothschild, B.M., Martin, L.D., 1993. Paleopathology. Disease in the fossil record. Boca Raton.

Sher, A.V., 1997. An Early Quaternary Bison population from Untermaßfeld: *Bison menneri* sp. nov. In: Kahlke, R.-D. (Ed.), Das Pleistozän von Untermaßfeld bei Meiningen (Thüringen). Teil 1. Monographien des Römisch-Germanischen Zentralmuseums 40, 1, 101–180, pls. 22–33.

Stephan, E., Uerpmann, H.-P., Cramer, B., 2001. Paläotemperaturbestimmungen nach δ^{18}O-Werten von Equiden-Resten aus dem Unterpleistozän von Untermaßfeld. In: Kahlke, R.-D. (Ed.), Das Pleistozän von Untermaßfeld bei Meiningen (Thüringen). Teil 3. Monographien des Römisch-Germanischen Zentralmuseums 40, 3, 917–930.

Tasnádi-Kubacska, A., 1962. Paläopathologie: Pathologie der vorzeitlichen Tiere. Jena.

Van Asperen, E.N., Kahlke, R.-D., 2017. Dietary traits of the late Early Pleistocene *Bison menneri* (Bovidae, Mammalia) from its type site Untermassfeld (Central Germany) and the problem of Pleistocene »wood bison«. Quaternary Science Reviews 177, 299–313.

Weaver, A.D., 1972. Chronic localized osteomyelitis of the bovine limb. British Veterinary Journal 128, 470–476

Woolley, M.R., Chinsamy, A., Govender, R., Bester, M.N., 2021. Microanatomy and histology of bone pathologies of extant and extinct phocid seals. Historical Biology 33, 1231–1246.

ALESSIO IANNUCCI

NEW RESULTS ON SUIDS FROM THE EARLY PLEISTOCENE SITE OF UNTERMASSFELD

Abstract

New and previously discussed suid remains recovered from Untermassfeld (Thuringia, Germany) are described. The late Early Pleistocene site of Untermassfeld yielded one of the most abundant samples of Epivillafranchian suids, and yet only a minimum number of 6 individuals has been identified. Untermassfeld suids have been considered the first representatives of the extant wild boar and ascribed to *Sus scrofa priscus*, then reassigned to *Sus* sp., and eventually to *Sus strozzii*. The latter attribution is favoured herein and some morphological differences from typical Villafranchian *S. strozzii* are noticed, which seem to be consistently present in other Epivillafranchian samples. At the moment it is not possible to evaluate whether these traits emerged abruptly in Epivillafranchian populations or were the result of gradual changes not recorded in the scanty post-Olduvai (c. 1.8 Ma) European suid record.

1. Introduction

Suids represent not but a fraction of the mammalian remains recovered from the late Early Pleistocene site of Untermassfeld (Thuringia, Germany), the minimum number of individuals (MNI) being only 6, including 3 juveniles and 3 adults (Kahlke 2006; in this volume), accounting for only 1.3 % of all large mammals. Yet they played an important role in shaping the palaeobiological thought on the dispersal of the extant wild boar, *Sus scrofa* Linnaeus, 1758, into Europe. Indeed, Untermassfeld suids were described as the earliest fossil evidence of *S. scrofa* and assigned to *S. scrofa priscus* Goldfuss, 1823 (Guérin and Faure 1997), culminating the hypothesis of a progressive trend towards smaller dimension and more complex molar morphology in the evolution of the species (Hünermann 1965; 1969; 1975; Faure and Guérin 1983; Guérin and Faure 1997). Guérin and Faure (1997) suggested that the first representatives of *S. scrofa* coexisted with the last populations of *Sus strozzii* Forsyth Major, 1881, a typical suid of European Villafranchian faunas (Azzaroli 1954; Faure and Guérin 1984; Cherin et al. 2018; Iannucci et al. 2020a). The attribution of Epivillafranchian suids engendered controversy and debate, with different samples assigned to *S. scrofa*, *S. strozzii*, or *Sus* sp. (Iannucci et al. 2020a). Martínez-Navarro et al. (2015) argued that suids are basically absent from European localities dated between c. 1.8 and 1.2 Ma and interpreted the whole Epivillafranchian contingent as *S.* gr. *scrofa*, i.e., as the arrival of a species phylogenetically related to the extant wild boar. According to this hypothesis, the wild boar dispersal would be approximately concurrent with other faunal events chosen to

denote the beginning of the Epivillafranchian at c. 1.2 Ma (Kahlke 2006; 2007; 2009; Bellucci et al. 2015), and hence of great biochronological value. However, an almost complete mandible of *S. strozzii* was described from the post-Jaramillo (c. 0.99 Ma) locality of Frantoio (Bona and Sala 2016). Indeed, Van der Made et al. (2017) did not recognise any gap in the c. 1.8–1.2 Ma European suid fossil record, but classified the latest Early Pleistocene specimens as *Sus* sp., including the sample from Untermassfeld. Eventually, Cherin et al. (2020) argued that Epivillafranchian suids should be ascribed to *S. strozzii*, remarking that all lower male canines known from Epivillafranchian localities show a verrucosic section (i.e., with the two lateral sides almost equal in length and larger than the distal facet), different from the scrofic section of *S. scrofa* (i.e., with the lingual side shorter than the distal facet). In suids, canines are sexually dimorphic and in male individuals they project outwards and often upwards from the skull, with a variety of forms and shapes exhibited by extant and extinct species (e.g., Stehlin 1899–1900; Cooke and Ewer 1972; Hardjasasmita 1987; Orliac et al. 2010; Macdonald and Shaw 2018). The dichotomy between scrofic and verrucosic canines is of great relevance within *Sus*, although several authors recognised intermediate morphologies (e.g., Fujita et al. 2000; Dong 2008; Sun et al. 2021). The suid canines from Untermassfeld were described as scrofic (Guérin and Faure 1997), but later reconsidered as verrucosic (Van der Made et al. 2017; Cherin et al. 2020). However, Van der Made et al. (2017) did not provide further details, while Cherin et al. (2020) only rediscussed one previously published canine (Guérin and Faure 1997).
The suid material from Untermassfeld is redescribed herein and its taxonomic status is reviewed, also considering previously unpublished specimens.

2. Material and methods

The suid material described in this work is part of the Untermassfeld collection housed in the Senckenberg Research Station of Quaternary Palaeontology Weimar (IQW) in Germany. Most fossils were recovered from the south-eastern part of the excavation area, accordingly nicknamed »Schweineecke«, i.e., the »pig corner« (Ellenberg and Kahlke 1997).
Comparative material of *S. scrofa* and *S. strozzii* has been studied in the following institutions:
CMNH: Civic Museum of Natural History, Trieste (Italy); HNHM: Hungarian Natural History Museum, Budapest (Hungary); IGF: Natural History Museum of Florence, section of Geology and Paleontology (Italy); IsIPU: Italian Institute of Human Paleontology, Anagni (Italy); MACUS: Comparative Anatomy Museum »Battista Grassi«, Sapienza University of Rome (Italy); MANA: Civic Archaeological-Naturalistic Museum »Adolfo Klitsche de la Grange«, Allumiere (Italy); MNCN: National Museum of Natural Sciences, Madrid (Spain); MPM: Paleontological Museum, Accademia Valdarnese del Poggio, Montevarchi (Italy); MUST: University Museum of Earth Sciences (including the former MPUR: Museum of Paleontology), Department of Earth Sciences, Sapienza University of Rome (Italy); NMB: Natural History Museum, Basel (Switzerland); NMM: Natural History Museum, Mainz (Germany); NWHCM: Norwich Castle Museum (UK); PF: PaleoFactory Laboratory, Department of Earth Sciences, Sapienza University of Rome (Italy).
The expression »typical *S. strozzii*« is used throughout the text for referring to samples from localities customarily dated at c. 2.0–1.8 Ma, from which most fossils of the species have been recovered (Azzaroli 1954; Cherin et al. 2018; Iannucci et al. 2020a).
Measurements were taken with a digital calliper to the nearest 0.1 mm, mainly following Driesch (1976) and Van der Made (1996). The latter work is also the reference for the dental nomenclature used herein. When feasible, measurements were taken on both sides and the mean of measured values is used. When referring

to canines a superscript (for upper teeth) or subscript (for lower teeth) letter is used to denote males (m) or females (f) specimens (e. g., C_f = lower female canine). Measurements abbreviations are as follows:
LLD: labio-lingual diameter in incisors; MDD: mesio-distal diameter in incisors (maximum diameter of the crown in I_3); ID-MP: Infradentale-mental prominence (length of the mandibular symphysis); WMM_1: width of the mandible in front of m1; L: length; W: width (-m: mesial; -d: distal/second lobe; -t: third lobe); cLi: length of the lingual side in C_x; cLa: length of the labial side in C_x; cD: length of the distal side in C_x.

3. Description

3.1. Cranium and mandible

IQW 1985/20509 (Mei. 20028) is a juvenile cranium with both tooth series preserved, inclusive of DI^1-DI^3, P^1, DP^2-DP^4, M^1, and the germ of M^2 (**Fig. 1, Table 1**). Although fragmented, especially at the level of the maxilla, the preserved portion of the specimen is almost undeformed. In lateral view the vault of the cranium is dorsally convex and gently curved (**Fig. 1c**). The lacrimal bone has a sub-rectangular shape and preserves a lacrimal foramen. The orbit is wide, and its anterior margin ends above the distal margin of the germ of M^2. It preserves the opening of the supraorbital canal. In dorsal view the zygomatic arches are gradually departing from the maxilla, their anterior (maxillary) portion ends above DP^4, while the malar bone departs from the maxilla roughly above M^1. The parietal lines are well-separated, the minimum distance between them being 33.2 mm (**Fig. 1d**). Since the cranium belongs to a juvenile individual, most of these characters differs from the adult condition.
The juvenile mandible IQW 1980/16539 (Mei. 16060) belongs to the same individual of IQW 1985/20509 (Mei. 20028). It preserves part of the left ascending ramus, a good portion of both mandibular corpi, the symphysis, left and right DI_1-DI_2, germ of I_3, and DP_2-M_1 (**Fig. 2, Table 2**). Germs of the left M_2 and both P_4 are also present, although their morphology is not observable. The mandible is slender, and the symphysis is neither deep nor wide. In lateral view several small openings are evident, the most important being the main mental foramen below the P_1-DP_2 diastema (**Fig. 2c–d**).
The dentition of the individual provides information on its sex and age at death. In the well-preserved alveolus of the missing lower right permanent canine the lingual side is slightly larger than the labial facet and both have rounded margins, while the distal side is substantially smaller and has a small notch. This morphology suggests that it probably belonged to a female. Aging wild boars by dentition eruption stage is a consolidate practice in wildlife management (e. g., Matschke 1967; Clarke et al. 1992). P^1/P_1 and M^1/M_1 erupt approximately at the same time, followed by I^3/I_3 and canines, then by M^2/M_2, and so on. In IQW 1980/16539 (Mei. 16060) at least C_f is erupting (it should be considered that teeth classified as »erupting« in living animals are only those that protrude from the gum, while the alveolar opening in the bone is observable slightly earlier), while M_2 is still enclosed in the alveolus. In *S. scrofa* this would correspond to an age between 7 and 12 months, probably closer to 7 than 12, considering that I_3 is almost erupting but DI^3 is still present. Indeed, in *S. scrofa* upper teeth erupt slightly later than their lower counterparts (Kuşatman 1991).

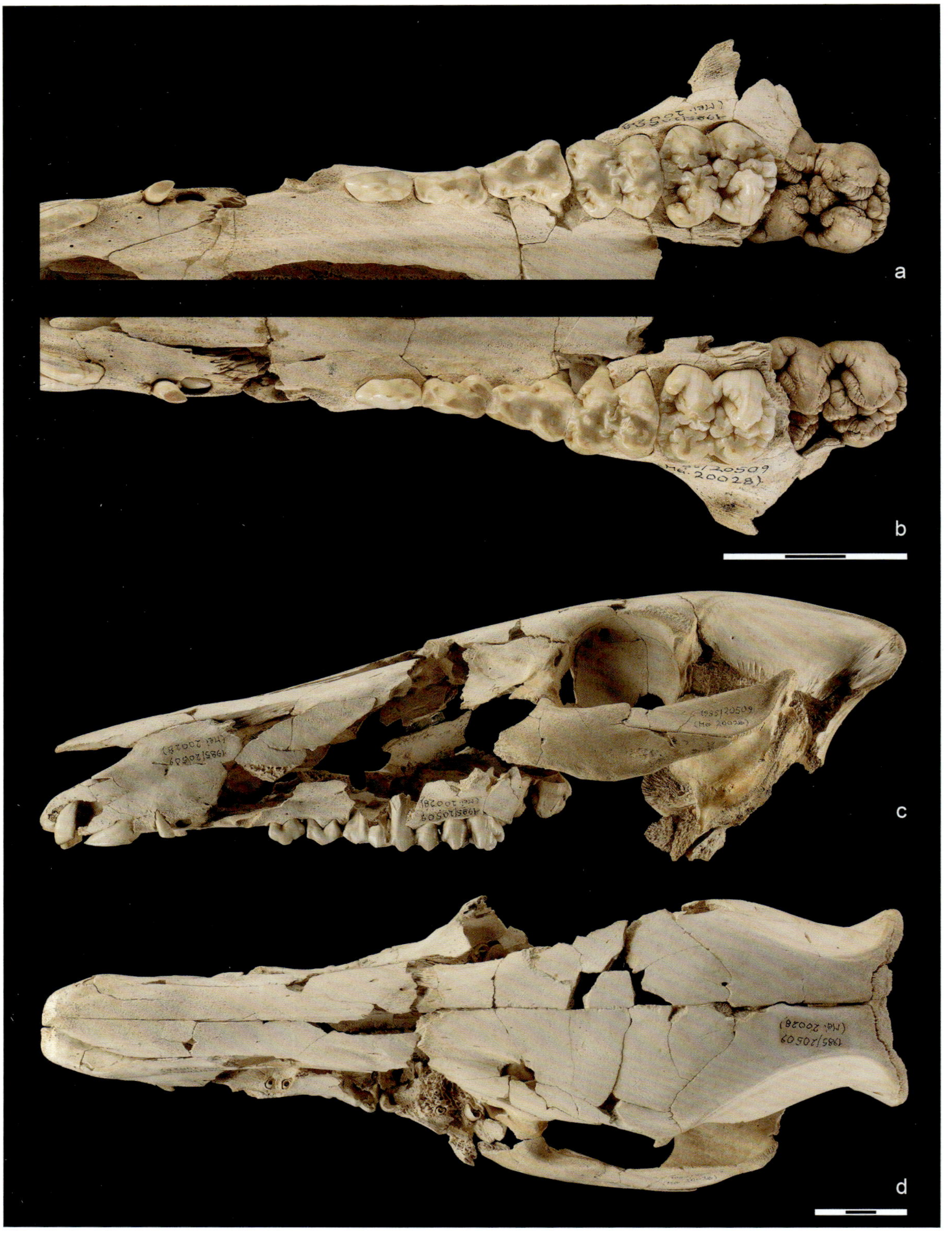

Fig. 1 *Sus strozzii* Forsyth Major, Untermassfeld. – **a–d** Juvenile cranium IQW 1985/20509 (Mei. 20028) with left and right DI^1-DI^3, P^1, DP^2-DP^4, M^1, germ of M^2, left occlusal, right occlusal, left lateral, dorsal views. – Scales: 30mm.

	IQW 1985/20 509 (Mei. 20 028)	IQW 1985/20 709 (Mei. 20 228)
DI¹ MDD	8.7	
DI¹ LLD	4.3	
DI² MDD	13.9	
DI² LLD	4.2	
DI³ L	4.5	
DI³ W	2.5	
DP² L	12.0	
DP² Wm	5.5	
DP² Wd	7.6	
DP³ L	14.4	
DP³ Wm	6.9	
DP³ Wd	10.9	
DP⁴ L	15.9	
DP⁴ Wm	13.9	
DP⁴ Wd	14.4	
P¹ L	11.5	
P¹ W	5.3	
P³ L		14.2
P⁴ L		13.9
P⁴ W		15.1
M¹ L	20.4	
M¹ Wm	17.7	
M¹ Wd	18.4	
M² L	27.1	
M² Wm	22.1	
M² Wd	20.0	
DP²-DP⁴	40.9	
DP⁴-M¹	31.6	

Table 1 *Sus strozzii* Forsyth Major, measurements (mm) of cranial remains and upper teeth, Untermassfeld. For abbreviations see section 2.

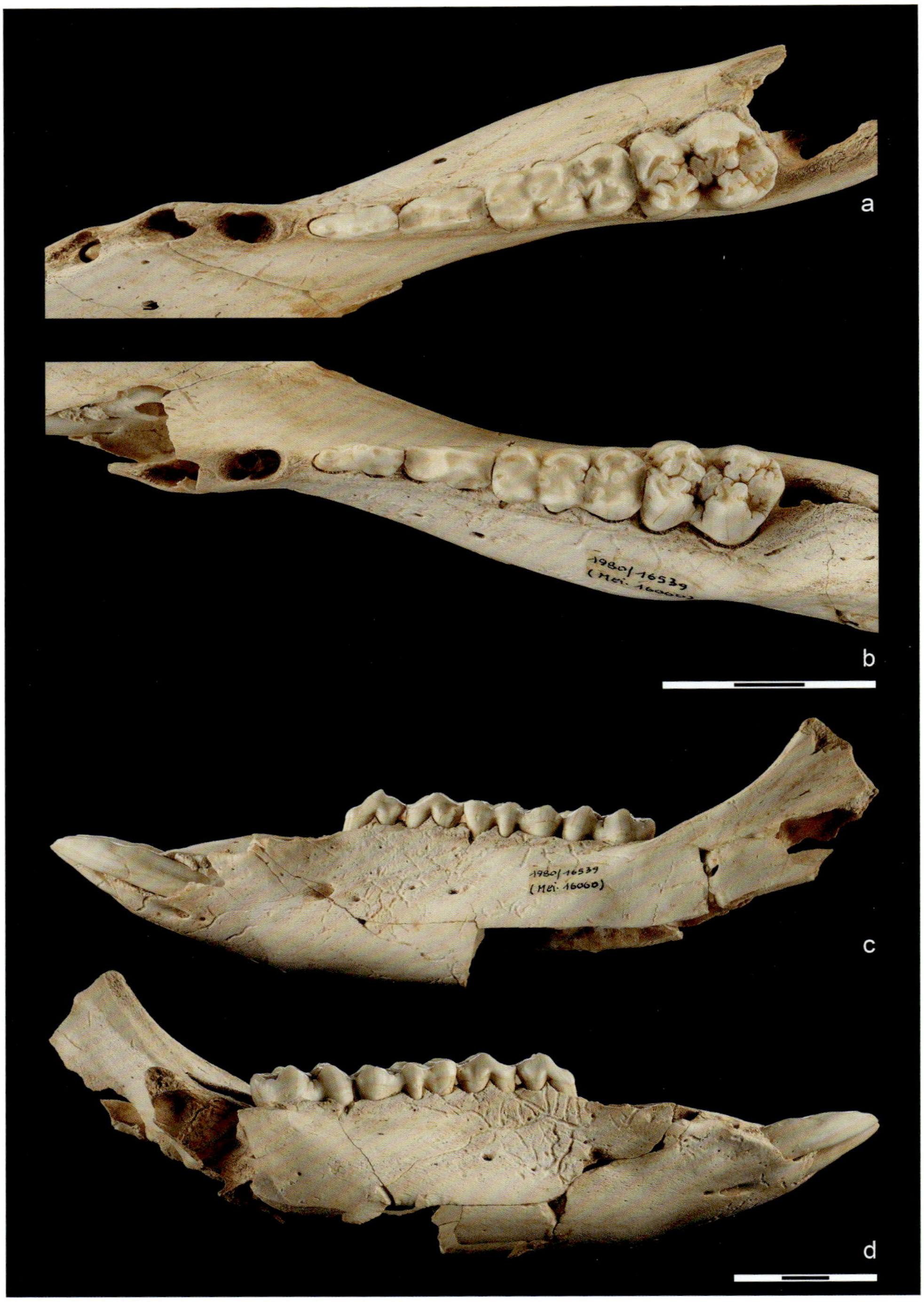

Fig. 2 *Sus strozzii* Forsyth Major, Untermassfeld. – **a–d** Juvenile mandible IQW 1980/16539 (Mei. 16060) with left and right DI_1-DI_2, germ of I_3, DP_2-M_1, right occlusal, left occlusal, left lateral, right lateral views. – Scales: 30 mm.

	IQW 1980/16539 (Mei. 16060)	IQW 1984/20309 (Mei. 19829) + IQW 1984/20310 (Mei. 19830)
DI_1 MDD	3.4	
DI_1 LLD	6.7	
DI_2 MDD	4.3	
DI_2 LLD	8.2	
DP_2 L	11.2	
DP_2 Wm	3.7	
DP_2 Wd	4.7	
DP_3 L	12.7	
DP_3 Wm	5.1	
DP_3 Wd	6.1	
DP_4 L	21.8	
DP_4 Wm	8.3	
DP_4 Wd	9.0	
DP_4 Wt	10.6	
P_2 L		14.0
P_2 Wm		5.7
P_2 Wd		6.3
P_3 Wd		9.7
P_4 L		15.3
P_4 Wm		10.3
P_4 Wd		12.4
M_1 L	20.2	17.6
M_1 Wm	13.0	12.7
M_1 Wd	14.5	14.6
DP_2-DP_4	45.5	
ID-MP	54.6	
WMM_1	22.3	

Table 2 *Sus strozzii* Forsyth Major, measurements (mm) of mandibles and lower teeth, Untermassfeld. For abbreviations see section 2.

3.2. Dentition

3.2.1. Deciduous dentition

Some deciduous teeth are preserved in the cranium IQW 1985/20509 (Mei. 20028) (**Fig. 1**) and in the mandible IQW 1980/16539 (Mei. 16060) (**Fig. 2**). DI^{1-2}/DI_{1-2} are similar to but smaller than their respective successional tooth. Unfortunately, they are severely worn, and hence details of crown morphology are not observable. DI^3 is a small peg with an oval occlusal outline. Its root is implanted in the premaxilla at an angle opposite to that of the other incisors, i.e., projecting distally rather than mesially.

DP^2 has a prominent central cusp, which separates two lower areas. The disto-lingual side of the tooth is enlarged and hosts a protocone. DP^3 is severely worn but preserves a characteristic piriform shape in occlusal view, with a separation between the mesial and distal sides of the tooth less marked than that of DP^2. DP^4 has a bilobed structure like that of a permanent molar, but its mesial side is oblique rather than perpendicular to the major axis of the tooth. DP_2 has a structure similar to that of P_2 but appears smaller and slenderer. DP_3 is severely worn. DP_4 is a trilobate tooth whose lobes are arranged in an opposite dimensional sorting than that of M_3, i.e., the mesial lobe is the narrowest and the distal lobe is the widest.

Data on deciduous dentition are in most cases too limited to provide useful comparative information, and Untermassfeld teeth are also severely worn. Nonetheless, in a study focused on DP_4 Iannucci et al. (2020a) recognised biometric differences between *S. scrofa* and *S. strozzii* comparable to those observable in the permanent dentition, i.e., with *S. strozzii* possessing proportionally wider teeth than those of *S. scrofa*. On that occasion Untermassfeld suids were treated as fossil *S. scrofa*, following Guérin and Faure (1997), but resulted outliers due to their large DP_4 Wm.

3.2.2. Upper permanent dentition

Isolated suid teeth from Untermassfeld are solely represented by permanent teeth and include several incisors and canines, and a few premolars and molars (**Fig. 3, Tables 3–5**). Upper incisors comprise a I^1, IQW 1986/21 130 (Mei. 20 749) (Guérin and Faure 1997, fig. 5, pl. 67) and a I^2, IQW 1984/20 307 (Mei. 19 827) (**Fig. 3a**). I^1 is covered by a cingulum-like development of the enamel along the border of its occlusal area, apart from the mesio-lingual portion. The cingulum is particularly pronounced on the disto-lingual portion, which is also its lowest point below the occlusal level. In I^2 the lingual cingulum develops in parallel to the mesio-distal axis and covers about two-thirds of the length of the tooth. The cingulum is clearly on a lower plane than the rest of the incisor. The labial side of I^2 hosts a series of three protruding structures and an elevated distal ridge. Being unworn, the tooth appears very elongated. A small opening in the premaxilla of IQW 1985/20 509 (Mei. 20 028), distal to DI^3, reveals the presence of a unerupted I^3, but its morphology is not observable (**Fig. 1a–c**). I^3 is present in *S. scrofa*, while in *S. strozzii* it has only been observed in a subadult from Senèze (NMB Se. 1775) but not in several adults of the species (Azzaroli 1954; Iannucci et al. 2020a).

P^1 is preserved on both sides of the cranium IQW 1985/20 509 (Mei. 20 028) and also represented by an isolated specimen in a slightly more advanced stage of wear, IQW 1984/19 978 (Mei. 19 498) (**Fig. 3b–b''**). P^1 has a main prominent central cusp, the paracone, and two lower areas hosting a paraprecrista (mesially) and a metacone (distally). Two well-separated roots develop below these areas. A small structure is present on the disto-lingual side (a potential protocone), more evident in IQW 1984/19 978 (Mei. 19 498) than in IQW 1985/20 509 (Mei. 20 028). In IQW 1985/20 509 (Mei. 20 028) there is no diastema between P^1 and DP^2. IQW 1985/20 709 (Mei. 20 228) comprises left and right germs of P^3 and P^4 of the same individual (**Fig. 3c–f**). P^3 are apparently not fully formed and not very informative. P^4 has a trapezoidal occlusal outline, with the mesial side oblique to the mesio-distal axis of the tooth. There are three prominent cusps, protocone (lingually), paracone (mesio-labially), and metacone (disto-labially). A protofossa separates the protocone from the metacone, but mesially the profossa is filled by accessory developments of protocone and paracone.

M^1 is preserved on both sides of the cranium IQW 1985/20 509 (Mei. 20 028). It is a bilobed tooth with four principal cusps and several crests that depart from them, resulting in a folded enamel pattern (**Fig. 1a–b**). Both germs of M^2 are also present in IQW 1985/20 509 (Mei. 20 028). Apart from their wrinkled (unworn) enamel and minor differences in the development of the cingula, they only differ in size from M^1.

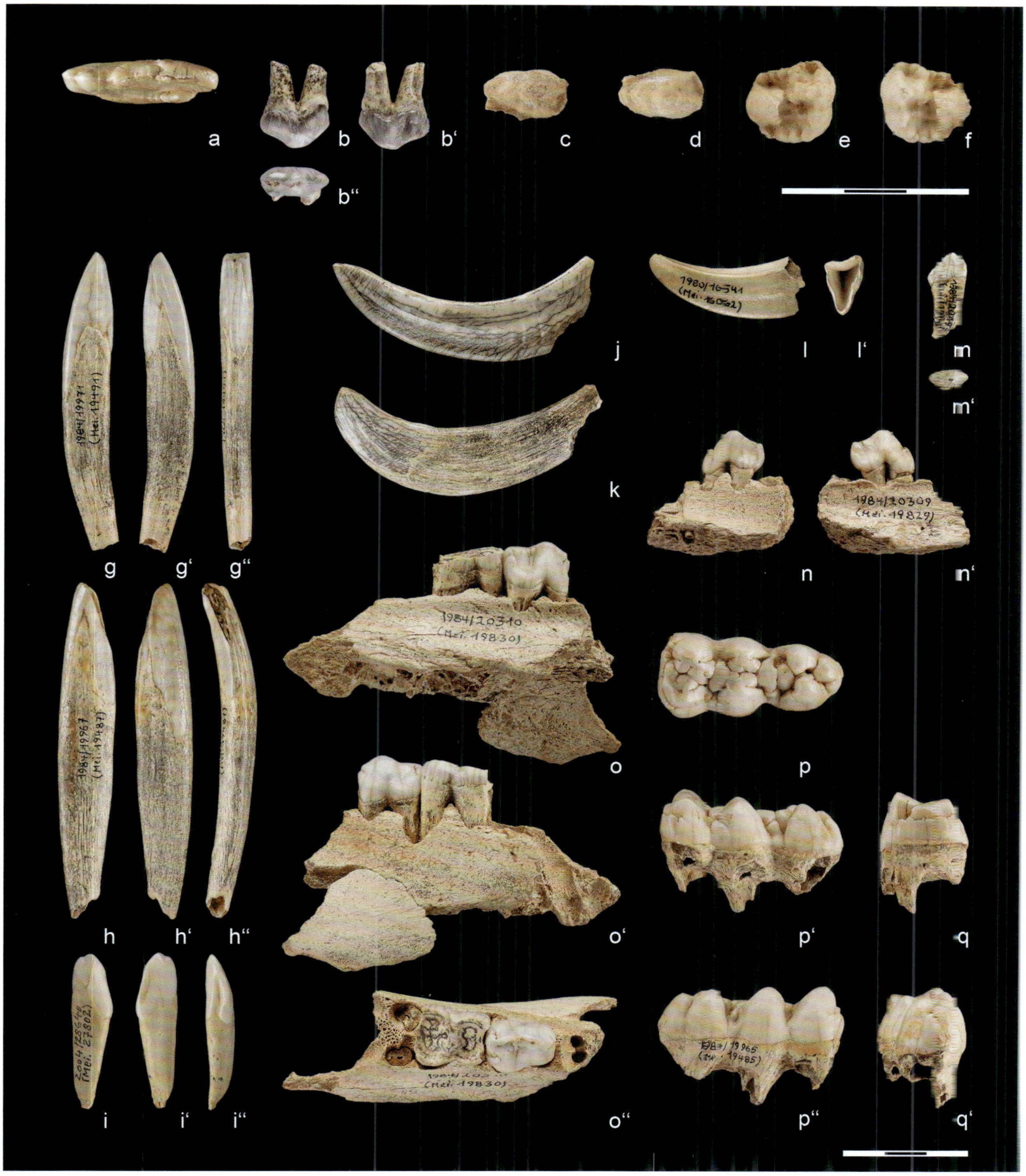

Fig. 3 *Sus strozzii* Forsyth Major, Untermassfeld. – **a** Left I^2 IQW 1984/20307 (Mei. 19827). – **b–b"** Right P^1 IQW 1984/19978 (Mei. 19498). – **c** Germ of left P^3 IQW 1985/20709 (Mei. 20228). – **d** Germ of right P^3 IQW 1985/20709 (Mei. 20228). – **e** Germ of left P^4 IQW 1985/20709 (Mei. 20228). – **f** Germ of right P^4 IQW 1985/20709 (Mei. 20228). – **g–g"** Left I_1 IQW 1984/19971 (Mei. 19491). – **h–h"** Right I_2 IQW 1984/19967 (Mei. 19487). – **i–i"** Right I_3 IQW 2004/28640 (Mei. 27802). – **j** Right C_m IQW 1984/20299 (Mei. 19819). – **k** Right C_f IQW 1984/19966 (Mei. 19486). – **l–l'** Left C_m IQW 1980/16541 (Mei. 16062). – **m–m'** Left P[illegible] IQW 1984/20197 (Mei. 19717). – **n–n'** Right P_2 IQW 1984/20309 (Mei. 19829). – **o–o"** Right hemimandible fragment with P[illegible]–M_1 IQW 1984/20310 (Mei. 19830). – **p–q'** Left M_3 IQW 1984/19965 (Mei. 19485). – **a, b", c, d, e, f, m', o", p** occlusal views; **b, l, n', o, p'** labial views; **b', g", h", i", j, k, m, n, o', p"** lingual views; **g, h', i', q'** distal views; **g', h, i, q** mesial views; **l'** cross-section view. – Scales: 30 mm

3.2.3. Lower permanent dentition

Lower incisors are relatively abundant (**Fig. 3g–i'', Table 3**). In lingual view I_1 is an elongated tooth with a central ridge (endocristid) separating two lower areas (**Fig. 3g–g''**). The cementoenamel junction is closer to the tip of the crown on the mesial side than on the distal side. I_2 has a similar basic structure, but it is markedly asymmetric in lingual view, with an internal (mesial) concavity (**Fig. 3h–h''**). I_3 is a much smaller tooth with a low and asymmetric crown (**Fig. 3i–i''**). The endocristid is relatively prominent and shifted distally, but clearly separated from the distal side by a groove. In *S. scrofa* the tooth is even more asymmetric and usually lacks prominent ridges. In typical *S. strozzii* I_3 is a stouter tooth with marked ridges, and other lower incisors are larger.

In addition to IQW 1984/19622 (Mei. 19492) + IQW 1984/19970 (Mei. 19490), already described by Guérin and Faure (1997, figs. 1–2, pl. 66) and rediscussed by Cherin et al. (2020, fig. 11), there are several C_m (**Fig. 3j, l–l', Table 4**). Some specimens are small because they belong to subadults or juveniles, but even in the latter case the section is clearly verrucosic, excluding an attribution to *S. scrofa* (**Fig. 3l'**). There is also a C_f, IQW 1984/19966 (Mei. 19486), in which the lateral facets are almost equal in length while the distal side is substantially smaller (**Fig. 3k**). Unlike C_m, C_f has a properly developed root, and the distal side of the canine is covered by enamel and engraved by a groove.

In the preserved P_1 IQW 1984/20197 (Mei. 19717) the main cuspid is shifted mesially, giving an asymmetric appearance to the tooth (**Fig. 3m–m'**). In this specimen there is only one root, while the P_1 alveolus present in IQW 1980/16539 (Mei. 16060) has a small occlusal notch and below it only one opening in the bone, suggesting that two partly fused roots were present (**Fig. 2a–b**). The absence of partly separated roots in IQW 1984/20197 (Mei. 19717) seems related to a reduction in size in comparison to P_1 of typical *S. strozzii*, observable in other Epivillafranchian samples as well (Bona and Sala 2016; Cherin et al. 2020). The remaining permanent premolars are observable in IQW 1984/20309 (Mei. 19829) and IQW 1984/20310 (Mei. 19830), which are part of an adult mandible preserved in several detached fragments that were recovered in close proximity from each other (from grid square Q 337, 1.07 m below site 0-level; Kahlke 2020; Kahlke et al. 2020). In lateral views P_2 is a triangular tooth with a high central cuspid and a mesial side lower than the distal side (**Fig. 3n–n'**). This tooth appears more elongated and slenderer than that of typical *S. strozzii*, although not at the level observed in extant *S. scrofa*. An elongated P_2 is also present in other Epivillafranchian samples (Bona and Sala 2016; Cherin et al. 2020). Only a not very informative distal fragment of P_3 is present. P_4 is preserved on both sides. Its crown has a trenchant morphology, with the main cuspids approximately aligned along the same mesio-distal axis, and high mesial and distal sides (**Fig. 3o–o''**). The tooth is wide, especially the distal lobe.

M_1 has a squared occlusal outline but it is severely worn, to the extent that its mesial side is concave (**Fig. 3o''**). In the mandible IQW 1980/16539 (Mei. 16060) M_1 is well-preserved (**Fig. 2a–b**). Its structure is similar to M^1 but differs in several details. M_1 is more elongated and symmetric along the mesio-distal axis, the notch between the two lobes is more pronounced on the labial side, the distal lobe is clearly wider than the mesial, and in each lobe the lingual cuspid is the highest above the occlusal level. M_3 is represented by a fragment of the distal lobe, IQW 1984/19793 (Mei. 19313), and a complete specimen, IQW 1984/19965 (Mei. 19485) (**Fig. 3p–q'**). M_3 is a trilobate tooth in which the mesial lobe is the widest. Cuspids that develop distally to the second lobe are rounded, as the enamel is not folded. The third lobe, separated from the second by a pronounced constriction, hosts five cuspids of relatively large size and two smaller cusplets. The first two large cuspids (mesially) are aligned along the same sagittal axis. This couple is pragmatically considered here a subdivided pentapreconulid because it would have been likely impossible to recognise two separate forms, had the tooth been in a more advanced stage of wear. The other cuspids are named topographically according to Van der Made (1996), although the homology of these structures is not clear.

	Tooth	MDD	LLD
IQW 1986/21130 (Mei. 20749)	I^1		c. 8.0
IQW 1984/20307 (Mei. 19827)	I^2	24.7	7.2
IQW 1984/19964 (Mei. 19484)	I_1	6.4	9.7
IQW 1984/19971 (Mei. 19491)	I_1	6.9	11.8
IQW 1984/19972 (Mei. 19492)	I_1	6.5	11.1
IQW 1984/19974 (Mei. 19494)	I_1	6.3	11.0
IQW 1984/19976 (Mei. 19496)	I_1	6.0	10.4
IQW 1984/19967 (Mei. 19487)	I_2	7.3	13.1
IQW 1984/19968 (Mei. 19488)	I_2	7.5	12.9
IQW 1984/19973 (Mei. 19493)	I_2	8.2	14.2
IQW 1984/19975 (Mei. 19495)	I_2	8.9	13.9
IQW 1984/20194 (Mei. 19714)	I_3	17.2	6.0
IQW 1984/20308 (Mei. 19828)	I_3	17.4	6.2
IQW 2004/28640 (Mei. 27802)	I_3	17.2	6.5

Table 3 *Sus strozzii* Forsyth Major, measurements (mm) of incisors, Untermassfeld. For abbreviations see section 2.

	Tooth	cLi	cLa	cD
IQW 1980/16541 (Mei. 16062)	C_m	14.8	13.1	8.3
IQW 1984/19622 (Mei. 19492) + IQW 1984/19970 (Mei. 19490)	C_m	21.6	19.8	13.9
IQW 1984/20192 (Mei. 19712)	C_m	23.9	22.2	15.7
IQW 1984/20193 (Mei. 19713)	C_m	16.4	14.3	11.6
IQW 1984/20299 (Mei. 19819)	C_m	16.6	15.4	12.2
IQW 1984/19966 (Mei. 19486)	C_f	15.1	15.1	6.4

Table 4 *Sus strozzii* Forsyth Major, measurements (mm) of lower canines, Untermassfeld. For abbreviations see section 2.

	Tooth	L	Wm	Wd	Wt
IQW 1984/19978 (Mei. 19498)	P^1	10.9	5.1		
IQW 1984/20197 (Mei. 19717)	P_1	9.8	4.3		
IQW 1980/16527 (Mei. 16048)	P_4		11.1		
IQW 1984/19793 (Mei. 19313)	M_3				> 12.3
IQW 1984/19965 (Mei. 19485)	M_3	44.8	20.9	19.4	16.4

Table 5 *Sus strozzii* Forsyth Major, measurements (mm) of premolars and molars, Untermassfeld. For abbreviations see section 2.

There is a small pentaectoconulid on the labial side of the pentapreconulid, pentaconid and hexaconid are oriented transversally to the main axis of the tooth, and heptaconid and octaconid are also present. The pronounced notch of the third lobe is also present on M_3 from Frantoio (Bona and Sala 2016) and Le Vallonnet (Cherin et al. 2020), but not in typical *S. strozzii* (housed in IGF, MANA, MPM, NMB; pers. obs., N = 28). In general, Epivillafranchian samples have relatively slenderer M_3 than that of typical *S. strozzii*. In *S. scrofa* M_3 is extremely variable in size and morphology of the distal complex (e. g., Iannucci et al. 2020b; Cherin et al. 2020), and hence resemblances or differences should be considered with caution.

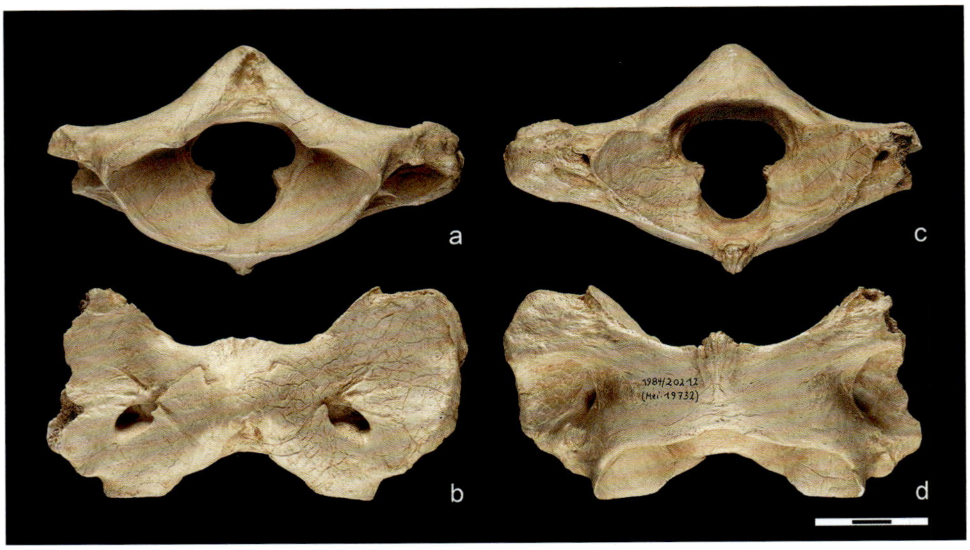

Fig. 4 *Sus strozzii* Forsyth Major, Untermassfeld. – **a–d** Atlas IQW 1984/20212 (Mei. 19732), cranial, dorsal, caudal, ventral views. – Scale: 30 mm.

3.3. Postcranial material

IQW 1984/20212 (Mei. 19732) is an almost complete atlas, only the right wing is slightly damaged (**Fig. 4**). A narrowing partly divides the vertebral foramen into two halves. The ventral half is approximately circular, while the dorsal half is elongated laterally. In dorsal view, the tuberculum is placed in the middle of the bone and the alar foramina are elliptic and diagonally elongated. In ventral view, the atlantal fossae are deep and their medial margins develop slightly beyond the lateral edges of the cranial and caudal articulations. In general, the morphology is similar to that of *S. scrofa* apart from the alar foramina, which in the wild boar tend to be more circular and are often partly subdivided by a small notch. In the latter feature IQW 1984/20212 (Mei. 19732) more closely resembles *S. strozzii*, but other traits differ, including the position of the dorsal tuberculum and the shape of the vertebral foramen (Azzaroli 1954). The atlas of *S. scrofa* is polymorphic in its proportions, in the relative positioning of the dorsal tuberculum, and in the depth of the atlantal fossae (Azzaroli 1954; pers. obs.).
IQW 1984/19927 (Mei. 19447) is a right fragmentary tibia lacking the proximal epiphysis (**Fig. 5**). The stoutness and curvature of the diaphysis, as well as the outline of the distal epiphysis are characteristic of a suid.

4. Discussion and conclusions

Untermassfeld suids have long been regarded as the earliest fossil evidence of the wild boar (Guérin and Faure 1997), but the presence of several verrucosic C_m excludes an attribution to *S. scrofa* and best fits with an assignment to *S. strozzii* (Cherin et al. 2020). In general, suid teeth from Untermassfeld, especially premolars, appear proportionally wider than those of *S. scrofa*, also differing in related morphological features (e.g., greater development of the hypoconid in lower premolars). The presence of relatively wide teeth is characteristic of *S. strozzii* (Van der Made and Moyà-Solà 1989; Iannucci et al. 2021a), often useful for dis-

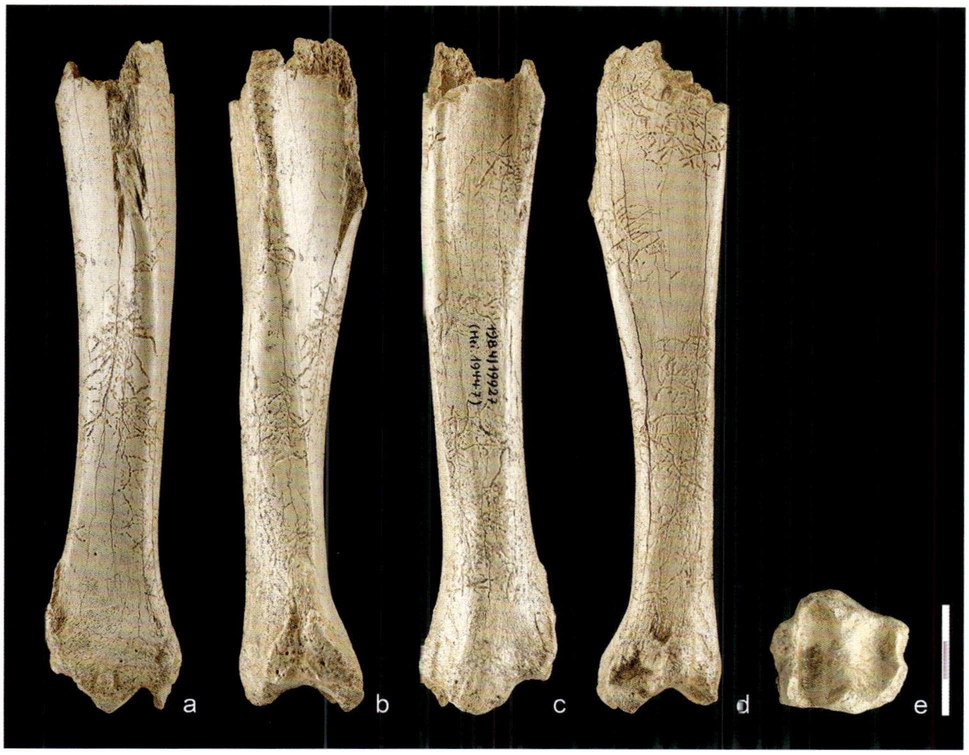

Fig. 5 *Sus strozzii* Forsyth Major, Untermassfeld. – **a–e** Right tibia, distal fragment IQW 1984/19927 (Mei. 19447), anterior, external, posterior, internal, distal views. – Scale: 30 mm.

criminating it from *S. scrofa* (Van der Made 1998; Iannucci et al. 2020a). On the other hand, Untermassfeld suids deviate from typical *S. strozzii* in some respects, with differences in the morphology of P_1, P_2, and M_3 which are consistently observed in other Epivillafranchian samples. Further research, discovery, and description of new material will likely identify other diverging features and provide more robust support for those recognised herein.

It is tempting to envision biochronological and palaeoecological implications for the observed morphological differences, but at the moment it is not possible to evaluate whether these traits emerged abruptly in Epivillafranchian populations or were the result of gradual changes not documented in the scanty post-Olduvai to pre-Jaramillo (c. 1.8–1.2 Ma) European suid record. In any case, while it is undeniable that suid fossils have not been recovered from several European localities dated between c. 1.8 and 1.2 Ma (Martínez-Navarro et al. 2015), there are some remains that could fill this apparent gap (Van der Made et al. 2017). Moreover, the unequal geographical distribution of Late Villafranchian palaeontological sites dated c. 1.8–1.2 Ma does not allow to rule out the possibility that suid populations survived in scarcely sampled areas (e.g., France and Germany).

It also remains to be answered what role played the dispersal of new suid populations from Asia, whose systematic placement is controverse. In fact, despite expressing divergent taxonomic opinions, several authors agreed in interpreting Epivillafranchian suids as the likely arrival of new populations (Martínez-Navarro et al. 2015; Van der Made et al. 2017; Cherin et al. 2020). A direct link was proposed by Moullé et al. (2006), who regarded Le Vallonnet suids as closely related to *Sus lydekkeri* Zdansky, 1928, defined from the Peking Man Site at Zhoukoudian. *Sus lydekkeri* has been considered a chronosubspecies of the extant wild boar, i.e.,

S. scrofa lydekkeri (Fujita et al. 2000), a geographical variant of *S. strozzii* (Dong 2008; Dong et al. 2013), or a valid and distinct suid species (Liu et al. 2017). These divergent opinions probably reflect the heterogeneous nature of the hypodigm of *S. lydekkeri* (Cherin et al. 2020). To clarify this issue an in-depth revision of *S. lydekkeri* and other Early Pleistocene Asian *Sus* species of comparable size (see, e.g., Fujita et al. 2000; Liu et al. 2017) would be needed, but it is conceivable that some remains ascribed to *S. lydekkeri* are indeed related to *S. strozzii* (see, e.g., Dong et al. 2013, fig. 4D), and hence the dispersal of Asian populations does not necessarily imply a species replacement.

In referring Untermassfeld suids to *S. scrofa priscus* Guérin and Faure (1997) crowned an evolutionary hypothesis according to which earlier representatives of the extant wild boar would have been characterised by low molar complexity (mainly in third molars) and large size (Hünermann 1965; 1969; 1975; Faure and Guérin 1983; Guérin and Faure 1997), but the existence of progressive trends in the evolution of these features is now challenged by several studies (Lister et al. 2010; Iannucci et al. 2020b). For instance, *S. scrofa* remains from the early Middle Pleistocene of Mosbach 2 are on average smaller than those recovered from the Late Pleistocene of Taubach (Küthe 1933; Hünermann 1977; pers. obs.). These considerations do not argue against the potential importance of biometric or morphological changes in Quaternary suids but should push to analyse large samples and adopt quantitative approaches to evaluate the biological significance of the observed differences.

Untermassfeld suids do not belong to *S. scrofa*, and the arrival of the wild boar into Europe does not characterise Epivillafranchian faunas since c. 1.2 Ma. The wild boar dispersal into Europe probably occurred close to the Early-Middle Pleistocene boundary (Van der Made et al. 2017; Cherin et al. 2020), in a period of marked environmental and faunal changes (Kahlke et al. 2011; Head and Gibbard 2015; Iannucci et al. 2021b). As for example evidenced by the impressive increase of wild boar populations in Europe over the last decades, *S. scrofa* is an extremely adaptable species, capable of coping with and taking advantage of environmental disturbances (Massei et al. 2015; Vetter et al. 2015). From this perspective, the *S. strozzii* - *S. scrofa* replacement was an important bioevent, representative of the faunal turnover occurred in Europe between the latest Early and early Middle Pleistocene.

Acknowledgements

I sincerely thank R.-D. Kahlke (Weimar) for the invitation to prepare this work, as well as for the warm hospitality and precious support during my stay in Weimar. I wish to further express my appreciation to G. Utschig (Weimar) and all the team of the Senckenberg Research Station of Quaternary Palaeontology in Weimar. S. Döring took the photos and E. Haase edited the figures (both Weimar). L. Pilot (Rome) kindly proofread the text for language editing. For access and support during the study of other collections I am also grateful to M. Aiglstorfer (Mainz), A. Aquiloni (Montevarchi), D. Arbulla (Trieste), L. Bellucci (Florence), M. Bentivoglio (Allumiere), L. Bruni (Anagni), R. Castiglia (Rome), E. Cioppi (Florence), L. Costeur (Basel), T. Engel (Mainz), S. Fraile Gracia (Madrid), M. Gasparik (Budapest), A. Medici (Allumiere), A. Oliver Pérez (Madrid), L. Riti (Rome), R. Sardella (Rome), and D. Waterhouse (Norwich). I also would like to thank R. Sardella (Rome) for his encouragement and advice. This research received support from the SYNTHESYS+ Project www.synthesys.info/ which is financed by the European Commission via the H2020 Research Infrastructure programme (grant DE-TAF-59 at the Senckenberg Research Station of Quaternary Palaeontology in Weimar; grant ES-TAF-2677 at National Museum of Natural Sciences in Madrid). Research support from the Italian Paleontological Society (Borsa di Studio SPI 2021) and Sapienza University of Rome – Avvio alla Ricerca (grant AR120172B7D44B9E) is also gratefully acknowledged.

References

Azzaroli, A., 1954. Filogenesi e biologia di *Sus strozzii* e *Sus minor*. Revisione della fauna dei terreni fluvio-lacustri del Valdarno Superiore V. Palaeontographia Italica 48 (1952–53), 41–76, pls. I–XII.

Bellucci, L., Sardella, R., Rook, L., 2015. Large mammal biochronology framework in Europe at Jaramillo: the Epivillafranchian as a formal biochron. Quaternary International 389, 84–89.

Bona, F., Sala, B., 2016. Villafranchian-Galerian mammal faunas transition in South-Western Europe. The case of the late Early Pleistocene mammal fauna of the Frantoio locality, Arda River (Castell'Arquato, Piacenza, Northern Italy). Geobios 49, 329–347.

Clarke, C.M.H., Dzieciolowski, R.M., Batcheler, D., Frampton, C.M., 1992. A Comparison of Tooth Eruption and Wear and Dental Cementum Techniques in Age Determination of New Zealand Feral Pigs. Wildlife Research 19, 769–777.

Cherin, M., Sorbelli, L., Crotti, M., Iurino, D.A., Sardella, R., Souron, A., 2018. New material of *Sus strozzii* (Suidae, Mammalia) from the Early Pleistocene of Italy and a phylogenetic analysis of suines. Quaternary Science Reviews 194, 94–115.

Cherin, M., Alba, D.M., Crotti, M., Menconero, S., Moullé, P.-É., Sorbelli, L., Madurell-Malapeira, J., 2020. The post-Jaramillo persistence of *Sus strozzii* (Suidae, Mammalia) in Europe: new evidence from the Vallparadís Section (NE Iberian Peninsula) and other coeval sites. Quaternary Science Reviews 233, 106234.

Cooke, H.B.S., Ewer, R.F., 1972. Fossil Suidae from Kanapoi and Lothagam, Northwestern Kenya. Bulletin of the Museum of Comparative Zoology 143, 149–237.

Driesch, A.v.d., 1976. A guide to the measurement of animal bones from archaeological sites. Peabody Museum Bulletin 1, III–IX, 1–137.

Dong, W., 2008. Early Pleistocene suid (mammal) from the Dajushan, Huainan, Anhui Province (China). Vertebrata PalAsiatica 46, 233–246.

Dong, W., Liu, J., Fang, Y., 2013. The large mammals from Tuozidong (eastern China) and the Early Pleistocene environmental availability for early human settlements. Quaternary International 295, 73–82.

Ellenberg, J., Kahlke, R.-D., 1997. Die quartärgeologische Entwicklung des mittleren Werratals und der Bau der unterpleistozänen Komplexfundstelle Untermaßfeld. In: Kahlke, R.-D. (Ed.), Das Pleistozän von Untermaßfeld bei Meiningen (Thüringen). Teil 1. Monographien des Römisch-Germanischen Zentralmuseums 40, 1, 29–62, pls. 1–18.

Faure, M., Guérin, C., 1983. Le *Sus scrofa* (Mammalia, Artiodactyla, Suidae) du gisement Pléistocène supérieur de Jaurens à Nespouls, Corrèze, France. Nouvelles Archives du Muséum d'Histoire Naturelle de Lyon 21, 45–63.

1984. *Sus strozzii* et *Sus scrofa*, deux mammiferes artiodactyles, marqueurs des paleoenvironnements. Palaeogeography, Palaeoclimatology, Palaeoecology 48, 215–228.

Forsyth Major, C.I., 1881. Studi sugli avanzi pliocenici del genere *Sus*. Atti della Società Toscana di Scienze Naturali, Process Verbali 2, 227.

Fujita, M., Kawamura, Y., Murase, N., 2000. Middle Pleistocene wild boar remains from NT Cave, Niimi, Okayama Prefecture, west Japan. Journal of Geosciences 43, 57–95.

Goldfuss, K.A., 1823. Ein Beitrag zur Kenntniss des Schweins der Vorzeit. (*Sus priscus*.) Nova Acta physico-medica Academiae Caesareae Leopoldino-Carolinae Naturae Curiosorum 11, 482–485.

Guérin, C., Faure, M., 1997. The Wild Boar (*Sus scrofa priscus*) from the Post-Villafranchian Lower Pleistocene of Untermaßfeld. In: Kahlke, R.-D. (Ed.), Das Pleistozän von Untermaßfeld bei Meiningen (Thüringen). Teil 1. Monographien des Römisch-Germanischen Zentralmuseums 40, 1, 375–383, pls. 63–67.

Hardjasasmita, H.S., 1987. Taxonomy and phylogeny of the Suidae (Mammalia) in Indonesia. Scripta Geologica 85, 1–68

Head, M.J., Gibbard, P.L., 2015. Early–Middle Pleistocene transitions: linking terrestrial and marine realms. Quaternary International 389, 7–46.

Hünermann, K.A., 1965. Die Suiden-Reste (Artiodactyla, Mammalia) des Altpleistozäns von Voigtstedt in Thüringen. Paläontologische Abhandlungen, Abteilung A Paläozoologie 2. Berlin, 427–432, pl. XXIII.

1969. *Sus scrofa priscus* Goldfuss im Pleistozän von Süßenborn bei Weimar. Paläontologische Abhandlungen, Abteilung A Paläozoologie 3. Berlin, 611–616.

1975. *Sus scrofa* Linné aus dem Pleistozän von Weimar-Ehringsdorf. Abhandlungen des Zentral Geologischen Instituts 23. Berlin, 251–263.

1977. *Sus scrofa* L. aus dem Jungpleistozän von Taubach bei Weimar in Thüringen. Quartärpaläontologie 2, 225–235, pls. XVIII–XX.

Iannucci, A., Gasparik, M., Sardella, R., 2020a. First report of *Sus strozzii* (Suidae, Mammalia) from the Early Pleistocene of Hungary (Dunaalmás) and species distinction based on deciduous teeth. The Science of Nature 107, 5.

Iannucci, A., Sardella, R., Strani, F., Mecozzi, B. 2020b. Size shifts in late Middle Pleistocene to Early Holocene *Sus scrofa* (Suidae, Mammalia) from Apulia (southern Italy): ecomorphological adaptations? Hystrix 31, 10–20.

Iannucci, A., Cherin, M., Sorbelli, L., Sardella, R., 2021a. Suidae Transition at the Miocene-Pliocene Boundary: a Reassessment of the Taxonomy and Chronology of *Propotamochoerus provincialis*. Journal of Mammalian Evolution 28, 323–335.

Iannucci, A., Mecozzi, B., Sardella, R., Iurino, D.A., 2021b. The extinction of the giant hyena *Pachycrocuta brevirostris* and a reappraisal of the Epivillafranchian and Galerian Hyaenidae in Europe: Faunal turnover during the Early-Middle Pleistocene Transition. Quaternary Science Reviews 272, 107240.

Kahlke, R.-D., 2006. Untermassfeld. A late Early Pleistocene (Epivillafranchian) fossil site near Meiningen (Thuringia, Germany) and its position in the development of the European mammal fauna. British Archaeological Reports, International Series 1578. Oxford.

2007. Late Early Pleistocene European large mammals and the concept of an Epivillafranchian biochron. In: Kahlke, R.-D., Maul, L.C., Mazza, P.P.A. (Eds.), Late Neogene and Quaternary biodiversity and evolution: Regional developments and interregional correlations. Proceedings of the 18th International Senckenberg

Conference (VI International Palaeontological Colloquium in Weimar). Volume II. Courier Forschungsinstitut Senckenberg 259, 265–278.

2009. Les communautés de grands mammifères du Pléistocène inférieur terminal et le concept d'un biochrone Épivillafranchien. Quaternaire 20, 415–427.

2020. Excavation progress and research at the Early Pleistocene site of Untermassfeld during the years 1997–2015. In: Kahlke, R.-D. (Ed.), The Pleistocene of Untermassfeld near Meiningen (Thüringen, Germany). Part 4. Monographien des Römisch-Germanischen Zentralmuseums 40, 4, 1031–1078.

in this volume. The Pleistocene of Untermassfeld – A synopsis of site origin, palaeobiodiversity, taphonomic characteristics, palaeoenvironment, chronostratigraphy, and significance in western Palaearctic faunal history. 1635–1734.

Kahlke, R.-D., Linnemann, U., Gärtner, A., 2020. New results on the origin and geological history of the Early Pleistocene site of Untermassfeld. In: Kahlke, R.-D. (Ed.), The Pleistocene of Untermassfeld near Meiningen (Thüringen, Germany). Part 4. Monographien des Römisch-Germanischen Zentralmuseums 40, 4, 1079–1104, Foldout I.

Kahlke, R.-D., García, N., Kostopoulos, D.S., Lacombat, F., Lister, A.M., Mazza, P.P.A., Spassov, N., Titov, V.V., 2011. Western Palaearctic palaeoenvironmental conditions during the Early and early Middle Pleistocene inferred from large mammal communities, and implications for hominin dispersal in Europe. Quaternary Science Reviews 30, 1368–1395.

Kuşatman B., 1991. The origins of pig domestication with particular reference to the Near East. PhD thesis, University College London, England.

Küthe, K.v., 1933. *Sus scrofa Mosbachensis*. Notizblatt des Vereins für Erdkunde und der Hessischen Geologischen Landesanstalt zu Darmstadt 14, 117–124.

Linnaeus, C., 1758. Systema Naturae Per Regna Tria Naturae, Secundum Classes, Ordines, Genera, Species, Cum Characteribus, Differentiis, Synonymis, Locis. Tomus I. Holmiae. (10th rev. ed.).

Lister, A.M., Parfitt, S.A., Owen, F.J., Collinge, S.E., Breda, M., 2010. Metric analysis of ungulate mammals in the early Middle Pleistocene of Britain, in relation to taxonomy and biostratigraphy: II: Cervidae, Equidae and Suidae. Quaternary International 228, 157–179.

Liu, W., Dong, W., Zhang, L., Zhao, W., Li, K., 2017. New material of Early Pleistocene *Sus* (Artiodactyla, Mammalia) from Yangshuizhan in Nihewan Basin, North China. Quaternary International 434, 32–47.

Macdonald, A.A., Shaw, D.J., 2018. Maxillary tooth growth in the adult male babirusa (genus *Babyrousa*). Comptes Rendus Biologies 341, 235–244.

Martínez-Navarro, B., Madurell-Malapeira, J., Ros-Montoya, S., Espigares, M.P., Medin, T., Hortola, P., Palmqvist, P., 2015. The Epivillafranchian and the arrival of pigs into Europe. Quaternary International 389, 131–138.

Massei, G., Kindberg, J., Licoppe, A., Gačić, D., Šprem, N., Kamler, J., Baubet, E., Hohmann, U., Monaco, A., Ozoliņš, J., Cellina, S., Podgórski, T., Fonseca, C., Markov, N., Pokorny, B., Rosell, C., Náhlik, A., 2015. Wild boar populations up, numbers of hunters down? A review of trends and implications for Europe. Pest Management Science 71, 492–500.

Matschke, G.H., 1967. Aging European Wild Hogs by Dentition. The Journal of Wildlife Management 31, 109–113.

Moullé, P.-E., Lacombat, F., Echassoux, A., 2006. Apport des grands mammifères de la grotte du Vallonnet (Roquebrune-Cap-Martin, Alpes-Maritimes, France) à la connaissance du cadre biochronologique de la seconde moitié du Pléistocène inférieur d'Europe. L'Anthropologie 110, 837–849.

Orliac, M.J., Antoine, P.-O., Ducrocq, S., 2010. Phylogenetic relationships of the Suidae (Mammalia, Cetartiodactyla): new insights on the relationships within Suoidea. Zoologica Scripta 39, 315–330.

Stehlin, H.G., 1899–1900. Ueber die Geschichte des Suiden-Gebisses. Abhandlungen der Schweizerischen Paläontologischen Gesellschaft 26–27, 1–527.

Sun, J.J., Zhang, B., Chen, X., Deng, L., Wen, J., Tong, H.W., 2021. New fossils of Late Pleistocene *Sus scrofa* from Yangjiawan Cave 2, Jiangxi, China. Vertebrata PalAsiatica 59, 64–80.

Van der Made, J., 1996. Listriodontinae (Suidae, Mammalia), their evolution, systematics and distribution in time and space. Mededelingen van de Werkgroep voor Tertiaire en Kwartaire Geologie 33, 3–254.

1998. Ungulates from Gran Dolina (Atapuerca, Burgos, Spain). Quaternaire 9, 267–281.

Van der Made, J., Moyà-Solà, S., 1989. European Suinae (Artiodactyla) from the Late Miocene onwards. Bollettino della Società Paleontologica Italiana 28, 329–339.

Van der Made, J., Rosell, J., Blasco, R., 2017. Faunas from Atapuerca at the Early-Middle Pleistocene limit: The ungulates from level TD8 in the context of climatic change. Quaternary International 433, 296–346.

Vetter, S.G., Ruf, T., Bieber, C., Arnold, W., 2015. What Is a Mild Winter? Regional Differences in Within-Species Responses to Climate Change. PLoS ONE 10, e0132178.

Zdansky, O., 1928. Die Säugetiere der Quartärfauna von Chou-K'ou-Tien. Palaeontologia Sinica, Series C, Volume V, Fascicle 4 1–146.

HANS VAN ESSEN

NEW RESULTS ON ELEPHANTIDS FROM THE EARLY PLEISTOCENE SITE OF UNTERMASSFELD – CRANIAL AND DENTAL REMAINS

Abstract

The small sample of identifiable elephantid dental remains so far recovered at Untermassfeld comprises a large majority (75 %) of milk premolars, most of which are unworn. Since Untermassfeld was already positioned as a site with a rich Epivillafranchian fauna, its elephantid remains could theoretically belong to either *M. meridionalis* (Nesti, 1825) or *M. trogontherii* (Pohlig, 1885), or to a third group, the so-called mosaic specimens. These started to appear around the same time as the immigrant *M. trogontherii* and show unexpected combinations of morphological characteristics of the two species mentioned, so that anagenetic evolution seems precluded as an explanation of their occurrence. They may be the result of crossbreeding instead. In order to avoid confusion at the taxonomical level, these forms are referred to *M. trogontherii* without subspecific distinction.
Given the above-mentioned three possibilities, an attempt was made to refer all dental remains from Untermassfeld to species on morphological grounds, since any expectations based on climatological data or the accompanying faunal and floral remains can never be decisive in this respect. Because evolutionary change in milk premolars is less clearly expressed than in true molars, referral to species is hampered by sometimes extensive range overlap. Nevertheless, the results of the analysis do confirm the presence of *M. meridionalis*, and suggest that the dentition of the juvenile cranium, in the exceptionally high HI value of the dP^3 that could be measured, shows a feature of mosaic character. Considered on their own, these results do neither underscore nor disprove the Epivillafranchian age attributed to the Untermassfeld fauna.

1. Introduction

Untermassfeld belongs to a series of European sites with Epivillafranchian faunas dated to a time bracket within which advanced mammoths of *Mammuthus trogontherii* morphology reached Europe, ultimately from eastern Asia, where they evolved from a regional *M. meridionalis* stock during the Early Pleistocene (Lister and Sher 2001; Wei et al. 2003; Lister et al. 2005; Kahlke 2007).

The dental elephantine remains from Untermassfeld are still limited in number. Initially they were all interpreted as belonging to *M. trogontherii trogontherii* (Pohlig, 1885) (Dubrovo 2001). This opinion was based on postcranial material and the dental element that had been recovered first: a complete isolated last milk premolar, initially regarded as a dP_4. On the basis of a more extensive morphological comparison this specimen was later reinterpreted as a dP^4 and referred to *M. meridionalis* (Nesti, 1825) (Van Essen 2011, 156ff.; **Tables 4–6** of this paper).

In addition to the dP^4, IQW 1984/20052 (Mei. 19572), the sample now comprises:

- A crushed juvenile cranium, IQW 1996/25646 (Mei. 25175), with both dP^2s and dP^3s in situ;
- Three unworn fragments of a left dP_3, IQW 1997/26303 (Mei. 25832), which were found close to the cranium and probably belong to its mandible; plates of the right dP_3 from the same mandible may be represented among a series of small fragments numbered IQW 1997/26304 (Mei. 25833);
- A number of apical fragments and small, unworn partial lamellae, likewise numbered IQW 1997/26304 (Mei. 25833), were also recovered from a spot near the cranium, and possibly belong to it as well;
- A well-worn posterior part of a right M_3 in situ in a mandible fragment, IQW 1999/26678 (Mei. 26207), the only tooth that belongs to an adult individual.

The MNI on the basis of dental remains alone is therefore now three. It is the objective of this paper to refer these specimens to species on the basis of their morphological characteristics.

In recent years it has become increasingly clear that elephantine material from European late Early to earliest Middle Pleistocene sites may comprise remains referable to both *M. meridionalis* and *M. trogontherii*. Apart from these samples, a small group of so-called mosaic molars has drawn attention. Some of their morphological traits, taken alone, would suggest referral to *M. meridionalis,* while others would indicate *M. trogontherii*, which may imply penetration of steppe mammoth genes into the European *M. meridionalis* populations (Lister et al. 2005; Van Essen 2011; Lister and Van Essen in prep.). Under these circumstances an a priori referral of all elephantine material from Untermassfeld to *M. meridionalis* on the basis of the dP^4 would not be convincing.

The earliest occurrences of *Palaeoloxodon antiquus* (Falconer et Cautley, 1847) within Europe and its periphery have so far been dated to the time bracket of 800–700 ka (Van der Made and Mazo 2003; Lister 2015). Although it may be too early to consider this range definitive, referral of the dental elements from Untermassfeld to the straight-tusked elephant is precluded by the absence of morphological details that point in this direction, and by the presence of details that indicate *Mammuthus*. The material was therefore compared to analogous specimens that belong to *M. meridionals* and *M. trogontherii*. Finally, a check on any anomalies that could lead to classification as mosaic elements appeared necessary. However, the interpretation of the dental remains added to the Untermassfeld sample after the recovery of the dP^4 presents some problems. Firstly, dP2 and dP3 do not possess the same potential for evolutionary change as ontogenetically later elements, especially M3, so that the existence of distinctive features that would allow referral of the two anteriormost milk premolars to a certain species is not self-evident. Secondly, the volume of the M_3 in the mandible fragment was reduced to roughly one third of the original by wear, and a fairly large area of its occlusal surface was crushed. Thirdly, with regard to the cranium, its original shape can hardly be studied in detail because of its present condition, and the samples available for a comparative analysis are small to virtually non-existent. In spite of this an attempt is made here to analyse the dental material by applying morphological criteria.

2. Material and methods

The elephantine dental remains from Untermassfeld conserved at the Senckenberg Research Station of Quaternary Palaeontology Weimar were measured and put through a comparative morphological analysis involving analogous specimens from various European sites. The sources of the comparative data presented in the text are public and private collections throughout Europe. Sample sizes differ per species and per element of the dentition, and specimens were selected for completeness as much as possible. Important and well-known fossiliferous regions and sites represented are, e.g., the Valdarno Superiore in Italy, various localities that pertain to the Cromer Forest-bed Formation in England (*M. meridionalis* and *M. trogontherii*), Süssenborn in Germany (*M. trogontherii*), and various localities within the Rhine-Meuse delta in the Netherlands, including parts of the bottom of the North Sea (*M. meridionalis*, *M. trogontherii*, *M. primigenius*, *P. antiquus*).

3. Descriptions and measurements

3.1. Abbreviations, symbols, and definitions

AEY	African Elephant Year(s), alternatively African Equivalent Year(s).
e	estimated original (max.) value on the basis of a measurement considered near this value, such as the length of a crown slightly shortened by wear, or the height of a scarcely worn or damaged lamella; an estimated value is intended to be closer to the original maximum than the actually measured value; the 'e' is also placed in front of ratio values that involve an estimated absolute value (see also 'ne' below)
ET	enamel thickness
H	height
HI	hypsodonty index
L	length
IGF	Geological Institute of Florence University
lam.	(true) lamella, i.e., the plate type that is the main component of elephantid molar crowns and constitutes the larger part of their volume; the dentine cores of true lamellae are continuous with the root dentine adjacent to the crown; the term 'lamella' excludes the platelets (see under p below) and the talon(id)s, the latter of which rise from the basal area of a lamella or of a larger talon(id), so that, in contrast to those of the other plates, their dentine cores are only indirectly continuous with the dentine of the roots
lat.	lateral, laterally
LF/LFi	lamellar frequency, LF being the number of true lamellae in 10 cm, LFi the analogous number inclusive of talon(id)s
LHI	length-height index
LWI	length-width index
max.	maximum
med.	median value, medially
min.	minimum
MNI	minimum number of individuals

mod.	modal value, modally
MsnpC	Muséum des sciences naturelles et de préhistoire, Chartres
NCM	Norwich Castle Museum
ne	number of estimated values
P (upper case)/Pi	plate number; with respect to dP2 there are two values for plate number: the traditional P is the number of true lamellae only, whereas Pi ('i' for 'inclusive') without differentiation comprises all plates present. The use of Pi could in principle be extended to all other elements of the dentition, albeit that, especially in M3, not all hindmost structures composed of dentine and enamel are technically plates, for instance if a so-called platelet ('p', next entry) consists of just a single digitation
p (lower case)	platelet; the posteriormost and small, usually platelike structure in M^3 and M_3, technically neither a true lamella nor a talon(id) (sensu Lister and van Essen 2003)
pers. obs.	personal observation(s)
plate	the general term for all parts of a molar crown that consist of a dentine core clad with a layer of enamel; three plate types are discerned here: 1) the (true) lamella (see under 'lam.' above), 2) the talons (in maxillary teeth) and talonids (in mandibular teeth) (see under 'lam.' above), and 3) the platelets that terminate M^3s and M_3s (see under 'p' above)
plate formula	shortest description of the composition of a molar crown in terms of plate types, in the order in which these are represented from front to back, e.g., x 6 x, $^{(x)}$x 9 x!$^{(x)}$, x 15 p
plate numbers ('P')	refer to the total number of true lamellae, which in the above plate formulae are 6, 9, and 15 respectively (see also 'P/Pi' above); the lamellae that remain in anteriorly worn or damaged crowns are counted in reverse order and indicated by Roman numbers; e.g., lam. III is the third true lamella in front of the posterior talon(id) (s) in such an incomplete crown
plate length	antero-posterior dimension of a plate; it comprises the dentine core and twice the enamel thickness, and is often synonymised with the thickness of a plate; the plate length increases in rootward direction
σ	standard deviation
V	coefficient of variation
Voi.	Voigtstedt
W	width
x	talon(id) in plate formulae
x!	talon(id) of large size
(x) or $^{(x)}$	talon(id) of comparatively small or very small size, the latter often borne by x!

Special abbreviations introduced to indicate morphological detail in the dental elements discussed and figured are explained in the respective figure captions.

3.2. Standard approach and terminology

60.9 etc.	Measurements are given in mm.

+ A plus sign is added to a measurement or a ratio if a plausible estimate of the original maximum cannot be given.

dP^2 etc. Elements from upper and lower jaws are indicated with their position numbers in superscript (dP^2) and subscript (dP_2) respectively; if this distinction is not made neither super- nor subscript is applied to the position number (dP2), so that both maxillary and mandibular specimens are implied.

▭O▭ etc. Wear figures are the figures produced at the occlusal surface of a molar crown as a result of its grinding function: in the course of their abrasion the plates reveal the structural details of their apical parts in successive near-horizontal cross sections through enamel walls and dentine cores. The resulting figures – together with other details of the crown – are often held to be indicative of certain elephantine genera, although counterexamples can sometimes be observed. Plates are normally tripartite, i.e., composed of a median pillar flanked by two lateral ones. Each pillar apically divides into short finger-like digitations. Within the apical region of a plate the pillars are separated by two clefts that reach a certain depth and therefore essentially cause three separate wear figures to appear at the occlusal surface until deeper abrasion of the crown produces a single elongate figure from side to side. Characterisations of wear figures name the shapes that represent the three pillars of the plate in wear. Initially the figures are annular (ring-shaped), as the occlusal surface cuts the tops of the digitations first. The wider pillars produce antero-posteriorly flat and medio-laterally elongate shapes, which are called lamellar. The combinations 'medially and laterally lamellar' (in short: med. lam. / lat. lam.), 'med. ann. / lat. lam.' (with a large finger-like median pillar), and 'med. lam. / lat. ann.' cover most of the wear figures that can be observed in molars of the genera *Mammuthus* and *Palaeoloxodon* (bipartite lamellae lead to two collateral lamellar wear figures). Representations in texts have so far been simple, viz.: - - -, - o -, and o - o respectively. In the present text these were substituted by ▭▭▭, ▭O▭, and O▭O respectively in order to make them stand out more clearly.

W values do not include the thickness of cover cement layers if these are present. The ratios HI and LWI will therefore turn out higher than would have been the case otherwise.
H values represent the maximal vertical extent of the enamel walls, i.e., from the apices of the plates to the parasagittal basal extremes of the lobe-shaped enamel joints between those plates (exemplified by the enamel joint under ile in **Fig. 3b**). The alternative is measured from the apices of the plates to the basal level where the lateral edges of the plates are linked to each other by their enamel cover, which level is usually referred to as the 'crown base' (c. 10 mm below the level indicated by ile in **Fig. 3b**). In most cases this measurement leads to lower HI and higher LHI values, very rarely the other way around.
Individual index values are multiplied by 100 and given to the nearest integer, and without decimal points; fractions are only retained in mean values.
With respect to dP2 there are two values for lamellar frequency: LF, the number of true lamellae per 10 cm, which is based on the P value, and LFi ('i' for 'inclusive'), which does not distinguish between true lamellae and talon(id)s, but is based on the total number of plates ('Pi').
For measurements ideally taken on the mandible the reader is referred to the scheme published by Beden (1983, 104).

3.3. Descriptions of specimens

3.3.1. Fragment of a juvenile cranium with left and right dP^2 and dP^3

The left lateral view of this fossil was figured by Kahlke (2001, 943, fig. 4; 2006, 30, fig. 18).

– 3.3.1.1. The cranium

What is left of the cranium IQW 1996/25 646 (Mei. 25 175) is fractured in many places and often reduced to tiny fragments by post-mortem tectonic pressure (**Fig. 1**). Although its overall shape remained recognisable, reliable measurements of its component parts cannot be given. Fractures locally coincide with sutures. As the animal died these were probably still lined with some cartilage, such as between the occipital and parietal bones. The preserved parts of the cranium comprise:

- The upper half of the premaxillary cover of the incisor alveoli, forced into a position which is slightly more upright than in analogous juvenile crania of *M. primigenius* (compare Maschenko 2002, 102 f.) and the living *E. maximus* (Deraniyagala 1955, pl. 21; pers. obs.);
- part of the nasal bones, which together with the upper parts of the premaxillary ones leave the shape of the nasal aperture more or less intact. Within the nasal aperture, the left of the two large and round foramina that lead to the nasal sinus is completely visible (compare Van der Merwe et al. 1995, 257, fig. 5A). Within the other half of the nasal aperture only part of the upper edge of the right foramen can be discerned. Between and below these foramina, the fragile and at this stage predominantly cartilaginous structures of the ethmoidal region did not survive. Maschenko (2002, 14) remarked that the nasal bones in woolly mammoth calves younger than three years are never preserved well, probably as a result of their still incomplete ossification, that most likely ended with the nasal process. Although its dental age has not reached three years by far, the cranium from Untermassfeld does include a bone fragment that represents the nasal process: from its original position on the median axis of the cranium it was both displaced and tilted to the right. It is recognisable by its median suture (**Fig. 1a**: np, st), that was formerly in line with the median suture observable along the left nasale, the medial edge of which is still in place (**Fig. 1a**: ns, st). All sutures in a juvenile cranium of *E. maximus* of practically the same dental age (Museon collection, The Hague; uncurated) remained unfused until the individual's early death, but the ossification of the bones in its nasal region, including the nasal process, shows no differences in comparison with the rest of the cranium;
- the frontal, parietal, and the upper part of the occipital region; the occipital condyles are missing;
- the left squamosal down to a level below the external auditory meatus;
- the bones of the left orbit, with the zygomatic process of the maxilla; the distorted right infraorbital foramen is visible in basal view;
- the larger part of the palate, which contains both left and right dP^2 and dP^3, in front of which four to five centimetres of maxillary bone remain; behind the dP^3s no trace remains of the borders of the V-shaped aperture of the combined choanae.

The tooth pairs are not in strictly symmetrical positions (**Fig. 1b**), the right pair being situated about 12.5 mm further to the front than the left one. This is taken to be the result of forces acting on the cranium as a whole, because the left dP^2 is slightly worn, and the right one scarcely at all. If asynchronous development and wear of the tooth rows were to have been the cause, one would expect the left row to be ahead of the right one, since a wear surface first developed on the left dP^2. With initial wear showing on the lamellae 1 and 2, and perhaps also 3, of the left dP^2, and on lam. 1 of the right specimen, the wear stage represents the older part of Laws' (1966) age group I. At this very same stage of development the lost alveoli behind

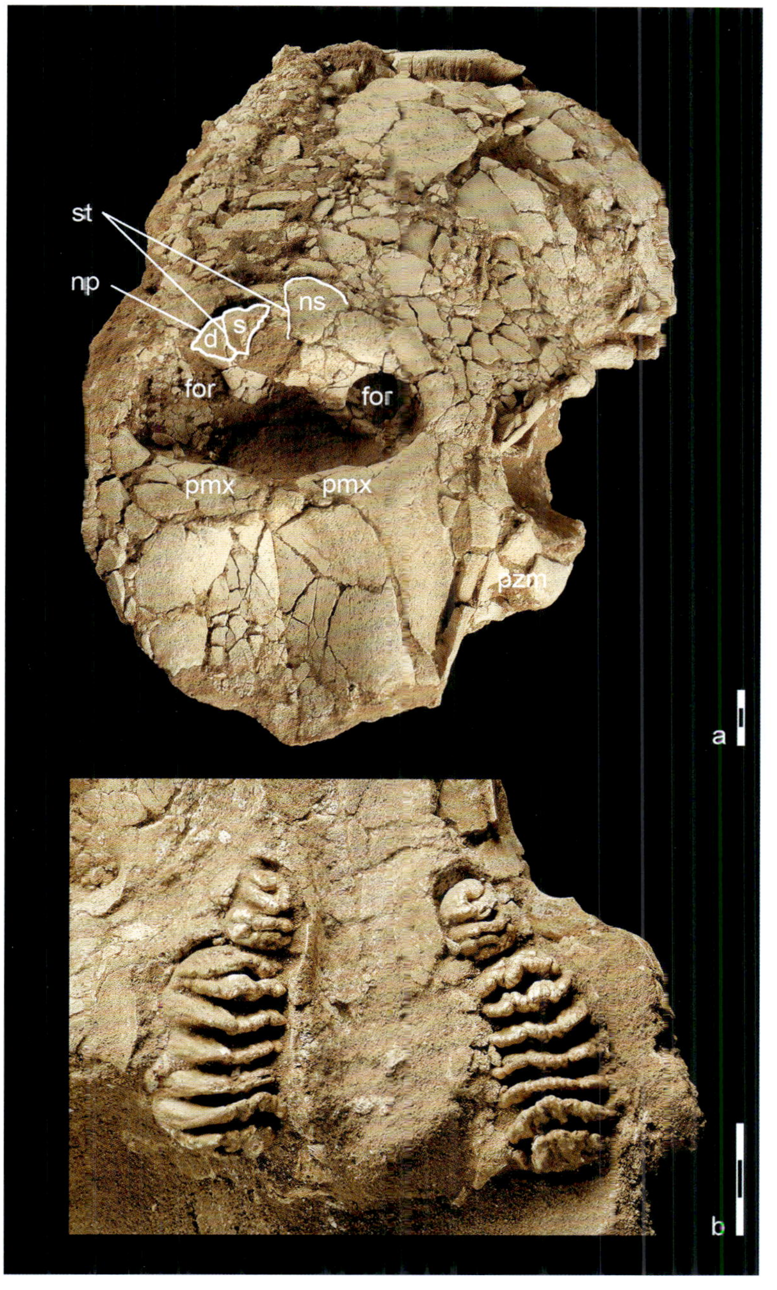

Fig. 1 *Mammuthus* cf. *meridionalis* (Nesti), Untermassfeld. – Fragmentary juvenile skull IQW 1996/25 646 (Mei. 25 175). – **a** Frontal view. – **b** Palatal view with left and right dP^2. – Scales: 30 mm. Abbreviations: for – foramen; np – processus nasalis with median suture indicated; ns – left os nasale with median suture indicated; pmx – praemaxillare; pzm – processus zygomaticus maxillae; st – suture.

the dP^3s probably contained a number of largely isolated plates of the dP^4 germs. Even the above-mentioned cranium of *E. maximus* figured by Deraniyagala (1955), which represents a foetal individual, shows the first growth stage of the dP_4 as a (post-mortem) disorderly row of still unconnected plates. In the specimen from Untermassfeld the dP^4 alveoli did not survive, but there are some isolated fragments of unworn lamellae that were recovered from the same block as the cranium and carry the collective number IQW 1997/26 304 (Mei. 25 833). Hence it seems possible that these fragments formed part of one or both of the still incomplete dP^4 germs or their mandibular counterparts, and – because the restored complete left dP_3 IQW 1997/26 303 (Mei. 25 832) described below is likely to belong to the scattered elements of the cranium as well – of the right dP_3.

Since the crown of the isolated left dP^4 IQW 1984/20052 (Mei. 19572) is complete, and cover cement and some interlamellar cement had already been deposited, it does not belong to the cranium, but to an individual of approximately one year (see section 3.3.4.).

At the stage of development indicated by the dP^2 and dP^3 specimens, the deciduous incisors (dI, the milk tusks or, colloquially, 'tushes') must have been present as well, and probably had already erupted beyond the anterior borders of their alveoli. The germs of the succeeding permanent incisors (I) may be expected to have been present at the same time, but in more receded positions within the alveoli, and closer to the median axis of the cranium, so that their tips would have been observable against the medial sides of the milk tusk bases. However, not a single milk tusk or permanent tusk has so far been recovered.

In the Untermassfeld cranium the permanent bony walls between the alveoli of the dP^2s and the dP^3s (Maschenko 2002, 56), clearly visible in the cranium of a newborn woolly mammoth calf from Sevsk (Brjansk, RUS) (l.c., 102, photo 3, fig. 2), cannot be detected at the level of the palate. Presumably they were superficially eroded to some depth, and are now hidden from view by sediment. On comparison, however, there is hardly any space left between the two milk premolars on either side of the Untermassfeld cranium, although the seemingly continuous alveolar space around the elements of the right maxilla does show a marked constriction between dP^2 and dP^3. Hence it seems possible that post-mortem tectonic pressure played a role here.

The alveolar edges along the tooth crowns to some extent cover the bases of the lamellae; they are in several places crumbled and somewhat sediment-encrusted. This suggests that the dental elements – or at least the not yet fully erupted dP^3s – do not merely show the stage of eruption that had been reached as the animal died, but again also the effects of pressure, which may have reduced the vertical distance between the alveolar borders and the apices of the crowns. The tilt of the elements towards the sagittal axis of the cranium, especially of those on the right, likewise suggests an effect of pressure.

Within the horizontal plane, both dP^2s have about retained their original positions relative to the dP^3s, because the lingual crown sides in both couples are about aligned. The apices of dP^2 and dP^3 on either side of the maxilla are about equidistant from the level of the palate. This may reflect a natural situation in which the death of the individual stopped the eruption of the dP^3s exactly at this point, or imply once again that pressure forced them back to this level. Comparison with the above-mentioned juvenile cranium of *E. maximus* (Museon collection, The Hague), with dP^2s and dP^3s at a wear stage hardly beyond that in the Untermassfeld cranium, speaks for the latter possibility: in the Asian elephant there is an appreciable difference between the apical levels of dP^2 and dP^3, with those of the latter further removed from the palate. The line drawing of the *M. primigenius* cranium from the Yamal peninsula, however, offers a fairly close parallel of the situation in the Untermassfeld specimen (Maschenko 2002, 15, fig. 6b).

– 3.3.1.2. The first and second milk premolars: general data and morphological approach

All four teeth have opaque, off-white enamel, which is largely covered by a reddish-brown film likely to have been derived from the sediment matrix.

The length of the preserved tooth rows is 82.5 mm on the left, and 84.0 mm on the right.

In terms of P values, the dP^2s as well as the dP^3s are complete. The posterior talon of the left dP^3 is largely missing.

The dP^2s have five plates in total. Scrutiny of relevant literature and personal observations have shown that this value is found in c. 75 % of the dP^2s of European fossil elephants on which data were available (29 out of 39 specimens, with 23 referable to *M. primigenius*). In the Untermassfeld specimens the number 5 is the equivalent of the traditional plate formula x 3 x. In view of the difficulties one can have in determining whether the anterior- and posteriormost plates in dP2s should be labelled talon(id)s or true lamellae, it is generally advisable to describe them as having four, five or six plates in total, and to refer to this alternative plate number with Pi ('i' for 'inclusive). This approach provides a better basis for comparisons and allows a uniform calculation of the lamellar frequency (LF) for all specimens published with any type of plate formula and a true max. length (L). As a consequence of the use of Pi, this alternative version of the LF is written as LFi. Because the often smaller first and last plates are counted in, LFi values are usually somewhat higher than LF values, which are based on true lamellae only.

The anterior- and posteriormost plates in the dP^{2}s from Untermassfeld are interpreted as talons, so that this term will continue to be used throughout the present text. The anterior talon of the left dP^2 carries a single minute digitation near the crown base. It is not considered to be a secondary talon, so that the plate formula does not appear as $^{(x)}$x 3 x or 6.

The plate formula of both unworn dP^{3}s is $^{(x)}$x! 5 x!$^{(x)}$ rather than $^{(x)}$ 7 $^{(x)}$: the bases of the anterior and posterior plates have become visible in the right specimen and show that the second plate merges with the third one at that level, and the hindmost larger plates of both show no gradual decrease in size with respect to the true lamellae, but leap to a still smaller format. The second and eighth plate in each specimen are therefore interpreted as large talons, which carry secondary talons of very small size. The fragments of the restored isolated dP$_3$ IQW 1997/26 303 (Mei. 25 332) – probably detached from the mandible that belonged to the cranium – and the isolated dP^4 IQW 1984/20 052 (Mei. 19 572) allow more detailed analyses of the analogous regions, which point in the same direction (see sections 3.3.2., 3.3.3., and 4.3.).

– 3.3.1.3. Complete left and right dP^2

The basic measurements are virtually the same for both specimens (**Fig. 1b**). At the author's request some more sediment was removed from the alveolus of the left one, so that its crown height and the various shape-determining ratios could be established. All dP^2 measurements therefore refer to the left specimen (**Table 1**).

On the lingual side the anterior talons are from top to base merged with lam. 1. As usual in such plate configurations, they remain separated buccally, so that a horseshoe-like figure is seen in occlusal view. As a consequence, the anterior talons have their highest point lingually, and their lowest buccally, so that they

		Mammuthus meridionalis dP^2 survey						*M. trogontherii* Voi.
		range	mean	σ	V	n	ne	
Pform	x 3 x							
Pi	5.00	5–6	5.17	0.37	0.07	6		5.00
L	22.60	20.8–25.8	22.88	1.97	0.09	6	2	e 18.00
W	17.60	17.6–20.4	18.73	1.06	0.06	6		14.40
H	e 17.00	19.0–21.0	20.00			2		e 16.00
HI	e 97.00	103.0–108.0	105.00			3	2	e 111.00
LWI	128.00	117.0–129.0	121.00	4.37	0.04	6	2	e 125.00
LHI	133.00	113.0–125.0	121.00			4	3	e 112.00
LFi	22.12	20.0–24.0	22.65	1.45	0.06	6	2	27.78
ET	c. 1.00	0.7–1.3	(min. 0.7–1.1; max. 1.0–1.3)					0.5–0.6

Table 1 The morphological data on the left dP^2 in the juvenile skull IQW 1996/25 646 (Mei. 25 175) from Untermassfeld compared to dP^2 specimens of *M. meridionalis* from various sites, and to the single dP^2 of *M. trogontherii* IQW 1965/4 019 (Voi. 2162) from Voigtstedt.

much resemble the anterior crescentoids known from many gomphotheriid molars. [Crescentoids in gomphotheriid upper molars rise from the posttrite (buccal) side of the anterior part of the crown base to merge with the protocone of the first pretrite (lingual) half-loph; see, e.g., Tobien 1973, 118, fig. 4].

At the developmental stage indicated by the milk premolars, neither interlamellar nor cover cement seems to have been deposited on the crowns of the dP^2s yet, although alternatively the first thin layer of cement – deposited on the anterior plates – may be hidden underneath the reddish encrustation layer or have been lost through erosion.

The tops of some digitations of the lamellae 1 and 2 of the left dP^2 are slightly worn. Whether the apical extreme of lam. 3 formed part of the occlusal surface is not certain. The right specimen was barely touched by wear at lam. 1 only. Buccally these lamellae are somewhat further worn than lingually, which is the normal situation in upper molars. It is also normal that either the left or the right specimen can be observed to have come into wear sooner than its lateral counterpart.

Because of the early wear stage, the enamel thickness (ET) is only measurable at a level where it is possibly slightly biased by apical thickening. The value listed in **Table 1** may therefore be considered an approximation.

The vertical furrows that characterise the lateral edge of enamel of most plates in dP3s and dP4s are hardly visible here, perhaps because of the film that covers the crowns, but the outer face of the enamel does show the familiar very fine horizontal wrinkles.

In both specimens the cement interval between the lamellae 1 and 2 is longer antero-posteriorly than those between the other plates, a peculiarity that was already observed in *M. meridionalis* by Friant (1959, 123), who called this presumably atavistic feature the »incisure profonde«.

As far as can be determined, the number of digitations per plate in the left specimen is five on lam. 2, and six on lam. 3; in the right one there are five on the lamellae 2 and 3, excluding an accessory digitation buccally in the interval between these.

The highest plate in each dP^2 is lam. 2. The basal length ('thickness') of the plates is 4.5 mm at most.

With the exception of the part bordering the crown base in the left dP^2, the roots of these specimens are invisible, so that the stage of their completion cannot be ascertained. By analogy with isolated specimens from other sites they are expected to have completed their growth.

– 3.3.1.4. Left and right dP^3

As in the dP^2s, deposition of interlamellar and cover cement is not observable (**Fig. 1b**). It may not have taken place or a thin initial layer remains hidden or was lost through erosion. At the lowermost levels of the crown the interlamellar spaces are filled with sediment, but since the dP^3s were in the course of eruption, it is assumed that their basal areas are completely consolidated, which is confirmed by at least the buccal crown base of the right specimen.

Both dP^3s comprise nine plates in total. As mentioned above, the plate formulae are here interpreted as $^{(x)}$x! 5 x!$^{(x)}$ (rather than $^{(x)}$ 7 $^{(x)}$). With regard to the status of the eighth plate in line there are arguments to interpret it as a large talon, in spite of the fact that the configuration of the pulp cavities cannot be inspected:

- The eighth plate is appreciably lower and narrower than those in front of it.
- An almost free-standing lateral digitation marks the lingual extreme of the eighth plate in dP^3 sin. As observable in the isolated dP_3 (see section 3.3.2.) – which has such a tall and separated digitation on either side of the analogous plate – and in the posterior talon(id)s of many other elephantid molars, these large lateral digitations show a tendency to be separated from the median pillar by a much deeper cleft than can be observed in the plates in front of them. This peculiarity is here interpreted as a confirmation that in the dP^3s of the juvenile cranium this eighth plate is a talon. Measurements are given in **Table 2**.

	Untermassfeld	*Mammuthus meridionalis* dP³ survey					
		range	mean	σ	V	n	ne
Pform	(x) x! 5 x! (x)						
P	5.0	5.0–6.0	5.6	0.49	0.09	10	
L	60.9	60.0–75.0	64.7	6.03	0.09	11	2
W	e 32.0	31.5–48.0	39.6	3.76	0.09	6	
H	42.5	34.0–44.0	38.6	2.84	0.07	8	4
HI	e 133.0	92–108	99.8	5.38	0.05	8	4
LWI	e 190.0	150–179	163.3	8.55	0.05	9	5
LHI	143.0	142–182	165.0	14.67	0.09	9	5
LF	11.9	8.8–11.4	9.9	0.64	0.06	11	1
ET		0.7–1.6	(min. 0.7–1.0; max. 1.1–1.6)				
		Mammuthus primigenius dP³ survey					
		range	mean	σ	V	n	ne
P		5.0–7.0	6.45	0.67	0.10	20	
L		46.0–71.0	56.50	4.98	0.09	31	10
W		30.0–43.0	35.50	3.25	0.09	33	3
H		33.5–47.0	39.40	3.72	0.09	18	5
HI		100–127	114.20	6.94	0.06	18	7
LWI		132–203	160.70	14.36	0.09	31	12
LHI		121–154	141.40	10.02	0.07	18	9
LF		11.7–15.5	13.70	0.89	0.06	21	
ET		0.4–1.2	(min. 0.4–0.7; max. 0.6–1.2)				

Table 2 The morphological data on the left dP³ in the juvenile skull from Untermassfeld IQW 1996/25 646 (Mei. 25 175) compared to dP³ specimens of *M. meridionalis* and *M. primigenius* from various European sites (from relevant literature and pers. obs.); partial overlap between *M. meridionalis* and *M. primigenius* suggests extensive overlap between the former and *M. trogontherii*.

In the left specimen the small posteriormost talon ((x)) was much damaged and is now represented by only a few parts against the basal posterior wall of the large preceding talon. In all probability the original size of the small talon was about equal to that of its counterpart in the right dP³, which remained undamaged and consists of a few minor digitations only.

In occlusal view most plates are not very straight medio-laterally. This is especially true of the anterior talon and lam. 1 of both specimens. Their tortuosity is a peculiarity very often seen in the anterior parts of other molars as well. The cause must be that the normal pressure between the various elements, which in bigger samples is demonstrated by the ubiquitous occurrence of anterior and posterior pressure scars, builds up before the anterior lamellae have become completely calcified.

The first true lamella in dP³ sin. and dex. appears to be bipartite, whereas the others are seemingly tripartite, the normal subdivision that produces the both medially and laterally lamellar wear figures predominant in *Mammuthus* (□□□). The cross sections of the buccal and lingual halves of the bipartite lamellae are anteriorly convex and posteriorly concave, as usual. The median end of the buccal half often incorporates a posterior accessory median digitation, which is the case here, too, although the digitations are not conspicuous. In general, bipartition of lamellae is a common phenomenon, especially in the milk premolars.

In *Mammuthus*, dP^3 and dP_3 may comprise up to five bipartite lamellae. These features are common in elephantine dP3s and dP4s, and again call to mind the gomphotheriid ancestors of the elephants, where analogous dental elements show bipartition throughout the crown by way of the so-called median sulcus.
Further accessory median digitations are found behind the lamellae 4 and 5 in dP^3 sin. and dex., and possibly also behind lam. 2 in dP^3 sin.
The number of apical digitations per plate is variable. The plates of dP^3 sin. from front to back show 12 on the large talon, and 10, 5, 6, 8, and 9 on the lamellae. In dP^3 dex. the counting is not so easy: > 7, > 6, > 7, > 9, -, > 9.
Another common feature in second and third milk premolars are the pronounced vertical furrows and ridges that characterise the enamel on the lateral edges of the plates. Because the specimens are unworn, it is not possible to measure their ET.
The tilted position of dP^3 dex. does not allow the measurement of its maximal width (W), so that it is assumed to be about equal to that in dP^3 sin.
The apical plate length in both specimens is about 3.0 mm. The basal plate length is about 8.0 mm.
No part of the (anterior) roots is visible in either specimen.

3.3.2. Three complementary unworn fragments of an isolated left dP_3
Three parts of an isolated dP_3, IQW 1997/26 303 (Mei. 25 832), were recovered. The two anterior ones, with the plate formulae $^{(x)}$x or $^{(x)}$1 and - 1 -, fit together, but the latter lost a large part on the lingual side, so that it cannot in any definitive way be attached to the third fragment, which has - 4 x!$^{(x)}$. In all probability these three fragments belong together because their alignment produces a plausible plate formula and the steady increase of crown width expected in a lower second milk premolar. The fragments also show the same range of colours. The length reconstruction of the tooth is based on the actual alignment of the constituent parts, as shown in **Fig. 2**.
The plate count then amounts to either $^{(x)}$x 5 x!$^{(x)}$, as in the dP^3s, or to $^{(x)}$ 6 x!$^{(x)}$, none of which is in wear. The plate formula x 7 x is rejected for the same reasons as given for the dP^3 described in section 3.3.1.4.:

- The eighth plate is markedly lower and narrower than those in front of it, so that it is interpreted as a large talonid; the difference in width is somewhat obscured by minor damage along the crown base, which in dorsal and ventral view (**Fig. 2a**, **c**) causes the false impression that especially the last true lamella is about as narrow as the large talonid.
- The large talonid is also much more deeply cleft than the true lamellae; the deep lateral clefts almost reach its base, so that the pulp cavities of the large lateral digitations were being filled with dentine individually (**Fig. 2c**: lpc and bpc).
- Because – unlike the in situ dP^3s – this dP_3 is isolated, it can also be observed that the enamel joint between the last true lamella and the large talonid is situated at a more apical level than the enamel joints between the true lamellae. This means that at the basal crown level a single pulp cavity is shared by two plates, which is the normal situation in combinations of a true lamella and a talon(id) (**Figs. 2b**, **c**). Whether the analogous situation is also to be assumed for the anterior extreme of the specimen is unclear, because a decision would depend upon the level at which the front ought to be attached to the rear of the crown. Because of the probability that the dP_3 is associated with the cranium, $^{(x)}$x 5 x!$^{(x)}$ is the more likely alternative.

Median accessory digitations of different sizes are present between the 4th and the 5th plate – by far the biggest one – and against the anterior walls of the 6th and 7th (**Figs. 2a–b**). The other two are mere hints of a central column.

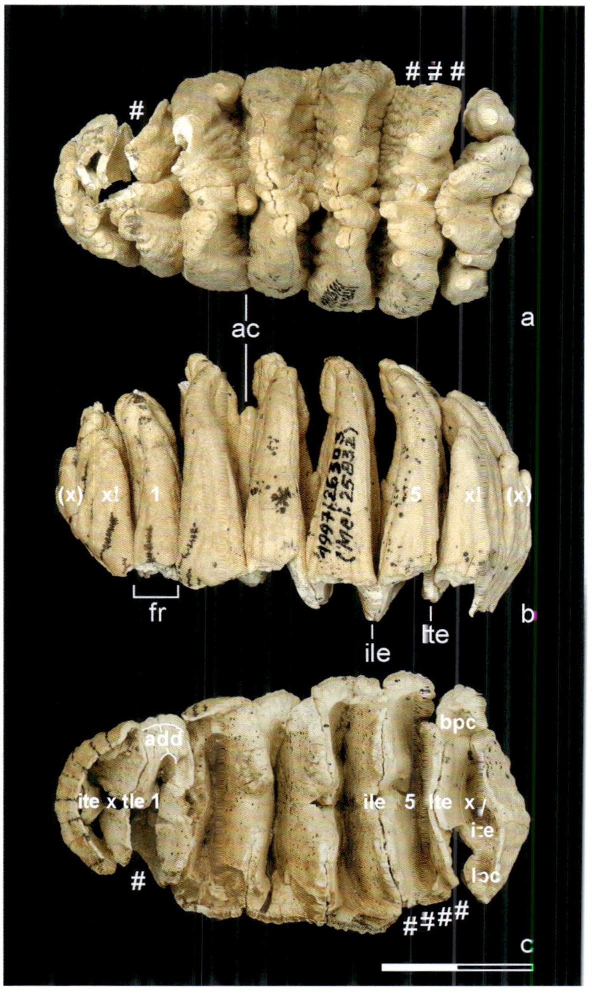

Fig. 2 *Mammuthus* cf. *meridionalis* (Nesti), Untermassfeld. – **a–c** Left dP_3 IQW 1997/26303 (Mei. 25832), occlusal, buccal, ventral views. – Scale: 20 mm. Abbreviations: ac – largest median accessory column; add – thickness of dentine deposition within anterior part of crown; bpc – separate buccal pulp cavity of large posterior talonid; fr – positions of fractures between the two anterior fragments; ile – interlamellar basal enamel joint; ite – intertalonid basal enamel joint; lpc – separate lingual pulp cavity of large posterior talonid; lte – basal enamel joint between posteriormost lamella and large talonid; tle – basal enamel joint between larger anterior talonid and first lamella; # – lingual zone of lateral damage.

The number of digitations per plate is about six in the 5th through the 7th plate, and nine in the larger talonid. Several good reasons can be given for assuming that this dP_3 belongs to the mandible that was associated with the cranium, but has so far not been found:

- The plate formula is plausibly identical with those of the dP^3s in the cranium.
- All plates are unworn, in the isolated dP_3 as well as in the in situ dP^3s, so that the dental ages suggested by these specimens virtually coincide; all plates had reached the stage at which their bases are connected by their enamel sheaths, and the deposition of interlamellar and cover cement was obviously at an incipient level.
- Both the anterior and the smaller posterior talonid of the dP_3 are very small, as are the minor talons in the dP^3s; they consist of a few short digitations only.
- There is no trace of the anterior root, which was presumably only represented by a thin, ring-shaped dentine wall that broke off. Since it is unworn, the specimen may still have been in the course of eruption, just as the in situ dP^3 specimens.
- The preservation of the fragments is akin to that of the in situ dP^3s.
- The fragments of the specimen were found close to the cranium (pers. comm. J.-A. Keiler); the probability that parts of a single individual became separated and were dispersed over a small area may be considered greater than the probability that any other individual of exactly the same age fell victim to the flash flood at the same time, after which its parts were deposited within the same small area. The same would

	Untermassfeld	*Mammuthus meridionalis* dP_3 survey					
		range	mean	σ	V	n	ne
Pform (x)	x 5 x!(x)						
P	5.0	5.0–7.0	6.0	0.45	0.07	20	3
L	61.5	59.0–85.0	69.0	7.80	0.11	17	6
W	34.5	33.0–44.0	37.7	3.16	0.08	21	1
H	42.5	30.0–43.0	36.4	3.44	0.09	15	12
HI	107.0	91.0–117.0	97.0	8.24	0.08	15	12
LWI	178.0	154.0–219.0	183.8	16.21	0.09	16	8
LHI	166.0	147.0–219.0	191.3	24.18	0.13	11	8
LF	11.0	8.5–11.4	9.9	0.80	0.08	19	
ET	e 1.0	0.4–1.7	(min. 0.4–1.4; max. 1.1–1.7)				
		Mammuthus trogontherii dP_3 survey					
		range	mean	σ	V	n	ne
P		5.0–6.0	5.7			3	1
L		52.2–68.5	59.9			3	2
W		31.3–39.2	35.5			3	
H		36.0				1	
HI		100.0				1	
LWI		164.0–175.0	168.7			3	2
LHI		164.0				1	
LF		10.3–12.2	11.5			3	
ET		0.7–1.7	(min. 0.7–1.0; max. 1.0–1.7)				

Table 3 The morphological data on the composite isolated left dP_3 IQW 1997/26303 (Mei. 25832) compared to dP_3 specimens of *M. meridionalis* and *M. trogontherii* from various European sites (from relevant literature and pers. obs.).

apply if the choice were between the remains of either a single already dead individual or of two such individuals of the same age, especially also because these could have died at different times and places.

Because upper teeth generally tend to be wider than lower ones, it might be considered an argument against the association of the dP^3s and the dP_3 that the latter is slightly wider than the former (34.5 vs. c. 32 mm respectively). Although the mean W values of upper and lower dP3s of *M. meridionalis* and *M. primigenius* comply with the rule, their ranges largely overlap and the mean upper and lower values differ by only 2 and 1 mm respectively, so that the occasional inverse relation seems possible.

The total absence of roots at any stage of development allows the observation that a certain quantity of dentine had already been deposited within the pulp cavities of the plates (**Fig. 2c**), but the pulp had not retreated from any plate core yet. Because the deposition of dentine proceeds from front to back of the crown, the thickness of the volume deposited decreases in posterior direction until all dentine cores of the plates have become solid. In this case, where measurable laterally, the thickness decreases from c. 3.5 mm within lam. 2 to c. 0.9 mm within the large posterior talonid.

The unworn state of the crown does not allow a full range of ET measurements. ET also varies according to the level within the crown, but a few fractured surfaces near the lingual bases of the 4th and 5th plates sug-

gest values around 1.0 mm. Vertical ridges and furrows down the lateral edges of the lamellae are present here as well as in the dP^3s. In addition, the enamel shows very fine horizontal wrinkles.
In spite of the fact that especially the large posterior talonid and some lamellae in front of it show subdivisions that can be described as 'medially lamellar and laterally annular' (O▢O), this is not interpreted as a detail that should lead to considering referral to *Palaeoloxodon*, because the observation applies to the apical level only: the lamellae have the medio-laterally broadly arched shape of *Mammuthus* molars, while the axes of the large clefts are perpendicular with respect to the crown base. If they had come into wear, these plates would therefore soon have shown lamellar figures laterally. This is confirmed by the cross sections of the lamellae at the basalmost level, which lack the gradual increase in length ('thickness') towards the median axis that characterises those of *P. antiquus* molars, and suggest instead that the tripartition of the lamellae follows the pattern indicative of *Mammuthus*, with both median and lateral pillars lamellar in cross section (▢▢▢; **Fig. 2c**, lam. 4). A dP_3 of *P. antiquus* would also be longer, to the effect that its LWI would not be 178, but possibly as high as 220–230.

3.3.3. Two isolated cf. dP4 fragments
Two unworn apical fragments of either upper or lower tripartite lamellae [IQW 1997/26 304 (Mei. 25 833)] have remaining heights of c. 30 and c. 35 mm, and widths of c. 35 and 40 mm respectively. The tripartition is of the type that would have led to lamellar wear figures of (sub-) equal width (▢▢▢). The above size values of the fragments lie somewhat below the maxima of the height and width ranges based on dP^3 and dP_3 specimens referred to *M. meridionalis* and *M. trogontherii*, which in these respects are apparently indistinguishable (**Tables 2–3**). Yet there are no indications that the fragments were about to narrow again in rootward direction after having reached their maximal width at the levels where they broke off. Instead they seem to have been wider beyond the fractures, so that they probably belonged to the next bigger category, dP4, or possibly M1. Since the dP^4 described in section 3.3.4. shows a cleft depth of c. 10 mm, attribution of the fragments to a dP4 is not contradicted by their cleft depths, which range from c. 12 to c. 15 mm. The fragments were found near the cranium, which must have contained unconsolidated dP^4 germs when still complete. In spite of the probability that the fragments belong to either the cranium or the associated mandible, the determination can go no further than cf. dP4.

3.3.4. Isolated complete left dP^4
The identification of this specimen IQW 1884/20 052 (Mei. 19 572) as a dP^4 of *M. meridionalis* (Van Essen 2011, 84, 156 ff.; **Table 4**) is at variance with the interpretation as a dP_4 of *M. trogontherii trogontherii* by Dubrovo (2001, 590). The main arguments against the interpretation of the tooth as a lower specimen are:

- The development of the first root on the buccal side of the crown instead of on the median axis;
- The low divergence index (DI) of the lamellae (i. e., the ratio between the mean apical and the mean basal LF value, which is 109 in this specimen and indicates near-parallel lamellae);
- The low length-width index (LWI) of 174. The LWI data on dP_4 of *M. trogontherii* lie well above this value because lower specimens are on average narrower than uppers (range: 209–259; mean: c. 237; n = 11). The value found in the Untermassfeld specimen also lies below the lower range extreme found for dP_4 of *M. meridionalis* (188–249; mean: 219; n = 21). and with respect to dP^4 it is the new lower range extreme of this species (174–224; mean: 194.1; n = 20). The former minimum was close, however: c. 177 for an upper specimen from the Cromer Forest-bed Formation (Norfolk, GB) (Lister and Van Essen, unpublished data).
- It may be added that the low hypsodonty index (HI) of 114.6 (**Table 6**) clearly excludes a dP4 cf *M. trogontherii* (range: 132–156, mean: 142.6, n=9, e 5), but not one of *M. meridionalis* (range: 100–131, mean: 116.2, n=14, e 9).

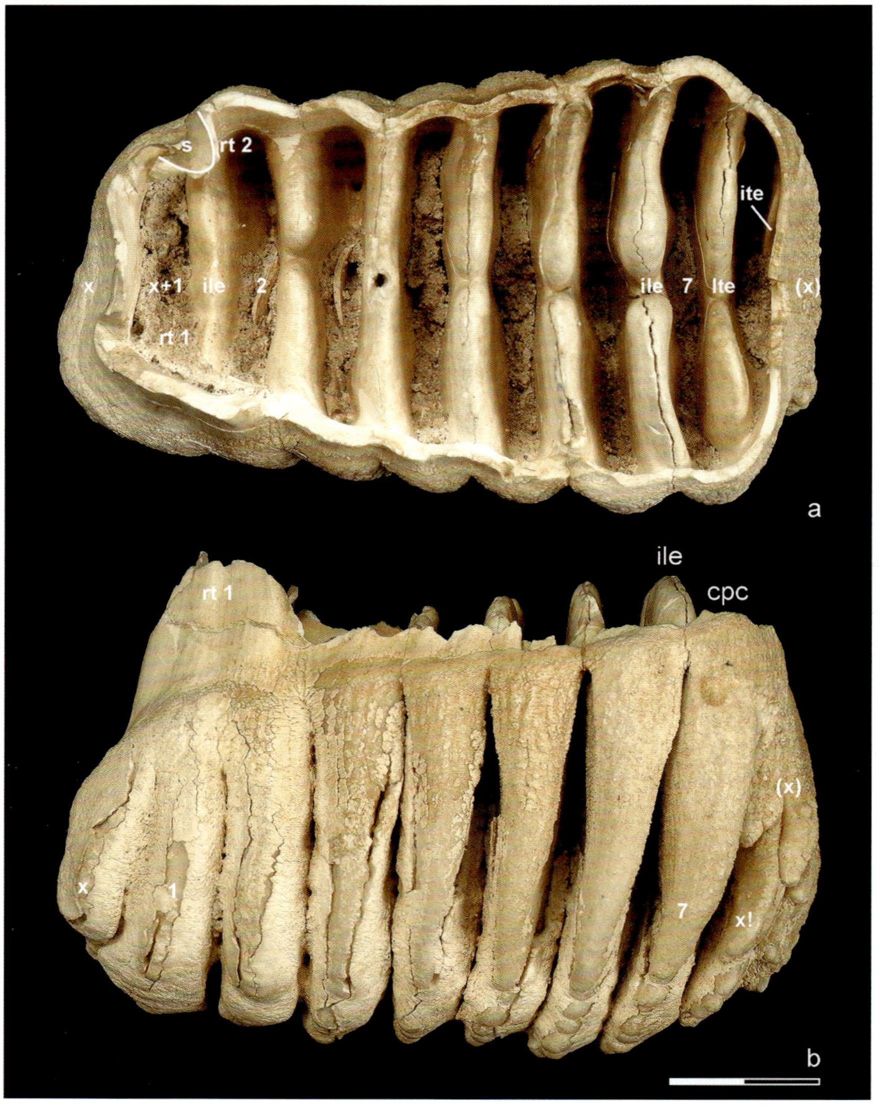

Fig. 3 *Mammuthus meridionalis* (Nesti), Untermassfeld. – **a–b** Complete left dP[4] IQW 1984/20052 (Mei. 19572), dorsal, buccal views. – Scale: 20mm. Abbreviations: cpc – combined pulp cavity of lam. 7, x! and (x); ile – interlamellar basal enamel joint; ite – intertalon basal enamel joint; lte – basal enamel joint between lam. 7 and large talon; rt – root; s – lingual sinus in outer root dentine.

The plate formula was re-examined after an earlier assessment (see below). Especially in buccal view (**Fig. 3b**) it becomes apparent that there are seven main extensions of the pulp cavity into the crown, which implies seven true lamellae. Hence the small first, the ninth, and the tenth plate are necessarily talons. This is confirmed by the unmistakable size differences with respect to the rest of the plates, which are true lamellae.
The pulp cavities of the talons branch from those of the first and seventh true lamellae, and from the larger posterior talon. The anteriormost pulp cavity is filled with sediment, but the talon in front of lam. 1 is clearly visible from the outside. It should be noted that the enamel joint between lam. 7 and the larger posterior talon (**Fig. 3a**, lte) is invisible in lateral view (**Fig. 3b**, cpc). It is, as it were, retracted to a level further away from the roots, so that the pulp cavities of lam. 7 and the larger posterior talon coincide above the level of the enamel joint between them (lte). The larger talon carries a smaller one, and in dorsal view the enamel joint between them is visible as a tiny, dentine-covered ridge within the pulp cavity of the larger talon

(**Fig. 3a**, ite). The pulp cavity of the secondary talon had nearly been filled with dentine, but is still visible. At the level of the enamel joint marked lte in **Fig. 3a**, the pulp cavity is therefore shared by three plates: lam. 7, the primary, and the secondary posterior talon. The rootward continuation of the posterior ending of the tooth beyond the fracture zone shows that the shared pulp cavity at the said level is just as long as those of the true lamellae (**Fig. 3b**). Hence the plate formula of this specimen is considered to be x 7 x [x)] rather than the 9 x initially advocated by Dubrovo (2001), and the x 8 (x) the present author then proposed in a reaction to that interpretation (full quote in Van Essen 2011, 84; **Table 5**). Further discrepancies between measurements published in 2001 and later are either insignificant or due to differences in approach. The index values given vary accordingly.

The tooth is complete and – but for the cover cement – virtually undamaged. The degree of completion and the fact that it is still unworn suggest a stage of development just prior to first wear (age class III according to Laws 1966). A fully functional, approximately half-worn dP^3 was therefore in front of it in life, so that the dP^4 represents an individual of about one AEY (G. Craig and G. Haynes in Haynes 1993, 338). This precludes association with the juvenile cranium and the isolated fragments of the unworn dP_3.

Because the dP^4 is unworn, its ET, although listed as 1.5 mm by Dubrovo (2001), cannot be measured accurately. The deposition of the cover cement had proceeded to the bases of the anterior talon and the lamellae 1 through 3, and after these becomes increasingly limited to the more apical regions of the crown.

Pform	(x)	7 x	(x)	L	108.5	LWI	174	LF	8.54
P		7		W	62.3	HI	114.6*	ET	not measurable
				H	71.4*	LHI	152	DI	109

Table 4 Plate count, basic measurements and ratios of the isolated left dP^4 IQW 1984/20052 (Mei. 19572). – * The alternative H value, measured from apex to lateral basal enamel extreme, is 56.4 mm, which leads to a HI of 91 and a LHI of 192.

Untermassfeld	P value	7				
	range	mean	n	ne	σ	V
M. meridionalis	7–8	7.20	20	7	0.40	0.06
M. trogontherii	8–10	9.00	10	1	0.63	0.07
M. primigenius	9–11	10.20	24	1	0.72	0.07
P. antiquus	8–10	9.25	4	2	0.83	0.09

Table 5 Comparison of the P value of the isolated left dP^4 IQW 1984/20052 (Mei. 19572) from Untermassfeld with analogous data based on European dP^4 specimens referable to *M. meridionalis*, *M. trogontherii*, *M. primigenius*, and *P. antiquus*.

Untermassfeld	crown value	114.6	(from H – 71.4* and W – 62.3)			
	range	mean	n	ne	σ	V
M. meridionalis	94–124	111.4	18	9	6.53	0.06
M. trogontherii	137–174	149.0	9	4	18.79	0.09
M. primigenius	130–174	152.5	19	3	12.55	0.08
P. antiquus	164–211	182.0	4	3	18.51	0.10

Table 6 Comparison of the HI value of the isolated left dP^4 IQW 1984/20052 (Mei. 19572) from Untermassfeld with analogous data based on European dP^4 specimens referable to *M. meridionalis*, *M. trogontherii*, *M. primigenius*, and *P. antiquus*. – * The alternative H value, measured from apex to the maximal lateral basal enamel extreme, is only 56.4 mm; this leads to a HI of 91.

The presence of vertical ridges and furrows in the basal lateral enamel lends a certain rugosity to the sides of the lamellae, but to a lesser extent than in the dP^3s of the juvenile cranium.

Apically the lamellae show a tripartite division (med. lam. / lat. lam.), with clefts of approximately 10 mm deep as far as they are visible. The total number of digitations per lamella is 6 on lam. 4, 7 on lam. 5, 11 on lam. 6, and 10 on lam. 7 and the large posterior talon.

The plates that correspond to the longest (buccal) side of the first root are the anterior talon, lam. 1, and lam. 2.

Root development is still at an initial stage, outlined in a damaged dentine wall of about 2 to 3 mm thick. Lingually the maximal longitudinal extent of the first root is marked by a sinus in the outer dentine wall, next to the basal joint of the lamellae 1 and 2 (**Fig. 3a**, s). Behind this sinus lies the space that was to be taken by the second root, presumably of triangular cross section. Buccally the first root extends about twice as far in antero-posterior direction as lingually, so that its cross section was likewise triangular in layout, and centered on the buccal half of the crown. This is the normal situation in the large majority of upper molars. Exceptions to this 'rule' are only, but commonly, found in dP^3 specimens of *M. primigenius*, where the first root is often as wide as the anterior crown base and much flattened in antero-posterior direction. For lack of well-preserved dP^3s referable to *M. trogontherii* and *M. meridionalis* it is not clear when this feature emerged within the *Mammuthus* lineage.

3.3.5. Complete remains of right M_3 in mandible fragment

The buccal side of the mandible fragment IQW 1999/26678 (Mei. 26207) in which the specimen is situated shows many cracks, but largely retains its original shape (**Fig. 4**). At the lingual side most of its volume has

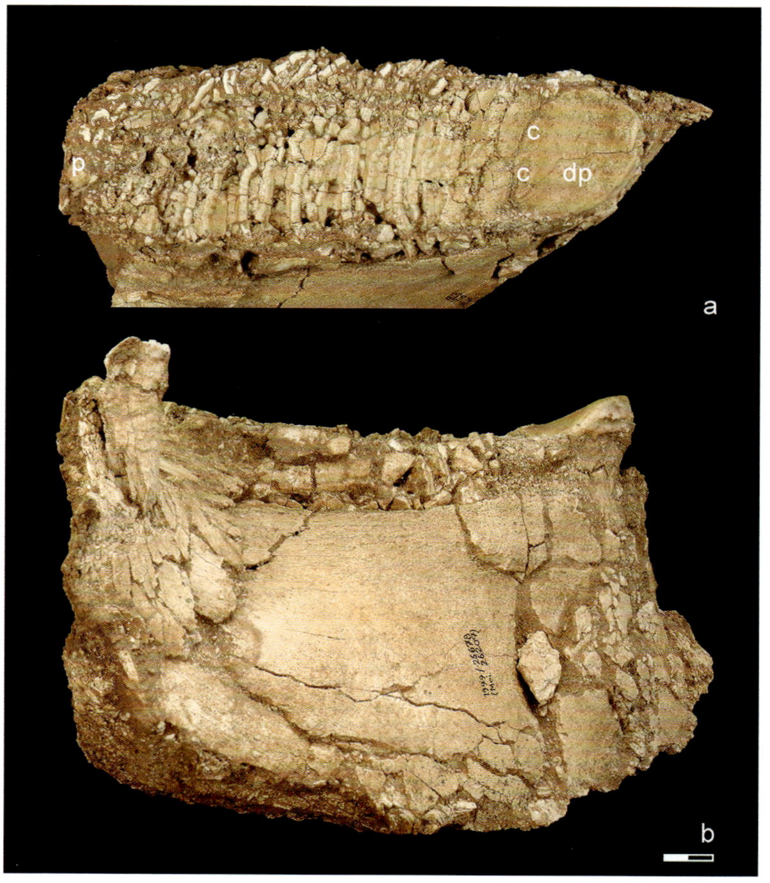

Fig. 4 *Mammuthus* cf. *meridionalis* (Nesti), Untermassfeld. – **a–b** Complete remains of right M_3 in mandible fragment IQW 1999/26678 (Mei. 26207), occlusal, buccal views. – Scale: 20 mm. Abbreviations: c – basal lobe of cement valley, encased in the enamel joint between the lam. IX and X; dp – dentine platform; p – posterior platelet.

crumbled away, so that the likewise crumbled lingual face of the M_3 lies bare. The ventral side of the horizontal ramus is missing, so that the measurements E (height of horizontal ramus at anterior alveolar border) and F (height of horizontal ramus at base of ascending ramus) according to Beden (1983, 104) cannot be taken, and therefore cannot be related to the horizontal distance between E and F either. The posterior extreme of the jaw fragment is formed by the base of the anterior border of the ascending ramus. As far as it is visible, the front edge of this basal part is very slightly concave. The symphysis is missing, and the area where the foramina mentalia are normally to be found is severely fractured at the base and largely shattered at higher levels.
The M_3 (**Table 7**) is represented by its medio-posterior part, which has a length of 224 mm. This is nearly as long as the mandible fragment. At the anterior end of the occlusal surface there is a dentine platform, behind which two cement islets are visible: these are the buccal and lingual basal lobes of the cement valley between the lam. IX and X. All further plates are in wear, so that the occlusal surface covers the entire remaining length of the tooth. Within c. 50 % of the occlusal area, however, the plates are completely shattered.
The rhythm of the lamellar parts still in place indicates that the length of the dentine platform is equivalent to the combined length of two lamellae and the cement intervals adjacent to them (i. e., ½ interv. + 1 lam. + 1 interv. + 1 lam. + ½ interv.). In the same manner, the space available within the fragmented posterior area suffices for one lamella and a robust posterior platelet. The plate formula is therefore reconstructable as $^{2}\infty$ 9 p, so that eleven lamellae and a posterior platelet are taken to be represented (**Fig. 4a**).
The wear figures of the lamellae are medially annular and laterally lamellar (▭O▭). Because the depths of the two main clefts of the tripartite lamellae decreased during the evolution of the mammoth lineage, the values observed may indicate whether a molar is relatively primitive or advanced. Even though the clefts in the worn M_3 from Untermassfeld were arguably at least 30 mm deep, this does not suffice to refer the specimen to *M. meridionalis*, because this is a mandibular tooth. Similar values are uncommon in M_3s of the advanced species, but may yet sometimes occur there (pers. obs.).

Pform	e	$^{2}\infty$	9 p	L		224	+	LF	5.32 +
P	e		11 +	W	e	80	(+: lam. IV)	ET	min. 2.9; mod. 3.2; max. 3.5
				H	e	60–75	(+: posterior lamellae)		

Table 7 Plate count and basic measurements of the complete remains of right M_3 IQW 1999/26678 (Mei. 26207).

	min.	mod.	med.	max.
M. meridionalis				
range	1.9–3.5	2.2–4.1	2.2–4.1	2.5–5.0
mean	2.51	2.81	2.93	3.34
n	60	37	59	61
σ/V	0.39/0.15	0.40/0.14	0.42/0.14	0.56/0.17
Untermassfeld	2.9	3.2	3.2	3.5
M. trogontherii				
range	1.5–2.5	1.8–3.0	1.8–2.9	2.0–3.4
mean	2.04	2.37	2.33	2.65
n	18	12	17	18
σ/V	0.25/0.12	0.33/0.14	0.27/0.12	0.34/0.13

Table 8 ET ranges based on M_3 specimens referable to *M. meridionalis* and *M. trogontherii*, compared to the individual range of the Untermassfeld specimen IQW 1999/26678 (Mei. 26207).

Enamel thickness (ET) is well measurable at the occlusal surface. The measurements therefore comprise values at higher as well as at lower levels (**Table 8**), but a marked decrease in thickness at the lowest levels is not visible. As far as can be detected, there are minor median dilatations in the posterior enamel bands of the lam. II, IV, V, and VI, and in the anterior one of lam. V. Frequency and amplitude of the enamel folds (after Beden 1983, 105) are 2–3 and 0–2 respectively. Folding is therefore moderate.
The lamellar frequency (LF) was calculated for the median axis of the occlusal surface. Because the lamellae in posterior parts of M_3s usually diverge towards the crown base, the LF value there is mostly lower than that of the tooth as a whole, which in this specimen was probably not higher than 6.0 (see section 4.4. for a reasoned approximation of this value).
The lingually bared root portion is superficially damaged, but its remains suggest a coalesced, roughly triangular and medio-laterally flattened posterior complex, preceded by a series of half-incorporated or completely individual roots. This configuration is often observable in much-worn M_3 specimens.

4. Discussion, comparisons, and referral to species

4.1. Fragment of juvenile cranium with left and right dP^2 and dP^3

The wear stage of the anterior milk premolars indicates that the cranium IQW 1996/25646 (Mei. 25175) should be placed in Laws' age group I, which corresponds to an individual age below two and a half months (G. Craig and G. Haynes in Haynes 1993, 338). The largely absent interlamellar and cover cement in dP^2 as well as in dP^3 probably imply an age well below the said limit.
Fairly close analogies to the Untermassfeld cranium are found in crania of newborn calves of *M. primigenius*: one from Sevsk (Paleontological Institute of the Russian Academy of Sciences, Moscow, no. 4353-2614), with an estimated individual age of two weeks, and another from the Yamal peninsula (Geological Institute of the Russian Academy of Sciences, Moscow, no. 77), aged about four weeks (Maschenko 2002, 15, fig. 6, 102, photo 3, 103, photo 4). The dental age of the cranium from the Yamal peninsula is scarcely greater than that of the Untermassfeld specimen because it has both dP^2s in early wear, and its unworn dP^3s still lack significant quantities of interlamellar and cover cement. Its individual age was estimated at about one month, so that the Untermassfeld specimen would even be slightly younger if an exactly parallel dental development could be counted on. However, basing himself on tooth formation stages in relation to the beginning of wear, Maschenko (2002, 29, 56) supposed that the milk dentition of woolly mammoths developed quicker than that of *E. maximus* and *M. meridionalis* and had a shorter functional life. If so, it implies that it took a late Early Pleistocene mammoth longer to reach the same wear stage, so that by comparison the age of the Untermassfeld juvenile would probably have to lie between one and two and a half postnatal months. The uncertainty about the exact stage of eruption of the somewhat dislocated dP^3s also makes this a rough estimate.

4.1.1. Complete left and right dP^2

Fossil elephantine dP^2s have become known from many places throughout Eurasia. Not surprisingly, the large majority of the finds are referable to *M. primigenius*. Well-known localities are, e.g., Předmostí (Moravia, CZ; Maška 1885; Musil 1968), Niederweningen (Zürich, CH; Lang 1892; Furrer 2005), caves in Britain and on the European mainland (Leith Adams 1879; Lister 2001; Münzel 2001), Sevsk, and the Russian Plain in general (Pohlig 1891, 318f., pl. A, 5-5'; Maschenko 2002; Maschenko et al. 2006). There are but a few dP^2s on record where the earlier mammoths are concerned. Available for comparison were measurements of specimens referable to *M. meridionalis* from Khapry (Rostov on Don, RUS), the Valdarno Superiore (Tuscany, I), an East Anglian crag deposit (possibly near Bramerton, Norfolk, GB), and Maasvlakte I near Rotterdam (S. Holland, NL). The number of specimens referable to *M. trogontherii* regrettably does not suffice to suggest any cluster outline based on morphological details or their combination. Pohlig (1891, 443, footnote; 444, fig. 145) and Dubrovo (1975, 128, tab. 1) mentioned two mandibular specimens from the Mosbach Sands (Hessen, D) and Tiraspol (Moldova) respectively. The former used to be in the Senckenberg collection in Frankfurt, but could not be retrieved (pers. comm. O. Kullmer, 2010), and the whereabouts of the latter are unknown (pers. comm. A. V. Sher, 2007). Just a single maxillary specimen then remains: the one from Voigtstedt (Thuringia, D; age estimated at about 0.7 Ma), which Dietrich (1965) referred to his new subspecies, *M. meridionalis voigtstedtensis* (Dietrich, 1958). It is in the collection of the Senckenberg Research Station of Quarternary Palaeontology Weimar – IQW 1965/4019 (Voi. 2162) – and, because it plots within the *M. primigenius* dP^2 clusters, was considered to be morphologically advanced with respect to *M. meridionalis* (Van Essen 2011, 202 ff.). Especially its individual ET range of 0.5–0.6 mm (median value 0.55 mm) does not match the overall range based on four specimens referable to *M. meridionalis*: 0.7–1.3 mm (median value 1.00/0.98 mm, from respectively overall range and four individual values). Hence, and because only *M. trogontherii* (including mosaic specimens) is positively identifiable within the rest of the small molar sample from Voigtstedt, the dP^2 was attributed to *M. trogontherii*. Its absolute length and the high LFi value that results from the greater compaction of its five short plates mark it as a comparatively small specimen. However, *M. trogontherii* is known for its on average large individuals, so it is likely that its so far unknown dP^2 size range overlaps with that of *M. meridionalis*. The above-mentioned lower specimen described by Pohlig – with x 4 x within 23 mm crown length – confirms this with respect to mandibular specimens. Whereas ratios such as HI, LWI and LHI will often separate ontogenetically later teeth at the species level, they do not have this capacity where dP2s are concerned (see, e. g., **Table 1**). A bivariate graph with HI vs. LWI does separate dP^2s from dP_2s very well (Van Essen 2011, 203, fig. 32), but not according to species.

The only observation regarding the Untermassfeld dP^2s that points in the direction of *M. meridionalis* is perhaps the comparatively moderate crown height, which leads to a low HI value and a high LHI value, but referral of the dP^2s to species cannot be separated from the verdict on the associated dP^3s, for which see section 4.1.2.

4.1.2. Complete left and right dP^3

Comparison of the dP^3s with analogous specimens that belong to *M. trogontherii* is only in part possible because no really complete dP^3s of this species were available: the Süssenborn sample – also in the collection of the Senckenberg Research Station of Quaternary Palaeontology Weimar – comprises three more than half-worn specimens, nos. IQW 1965/2151 (Süß. 6348), IQW 1964/2152 (S. 4482), and IQW 1965/2775 (S. 9733). The first two of these have the reconstructed P value 6, which is directly observable in the third. The first yielded a reliable W value (38 mm), a LF of 11.0, and an ET range of 1.0–1.2 mm; the second a LF value of 10.0 and a modal ET of 1.0. The L value of the last is an estimated 63 mm, to which a LF value of 12.0 and an ET range of 0.81–1.26 can be added. Most of these values fall within the respective ranges found for *M. meridionalis*, so that a fairly extensive overlap, mainly caused by similar individual sizes, is to be expected. With somewhat less confidence the HI value of the second Süssenborn specimen, via an estimated H value of c. 41 mm, lands at the higher extreme of the *M. meridionalis* range: 108, which still suggests overlap, especially as n = 4 only for the latter.

It is the considerable H value against the moderate L value that causes the LHI of the specimen from Untermassfeld to coincide with the *M. meridionalis* minimum. However, it is in HI that it markedly differs from

all other *M. meridionalis* specimens: while L and W suggest a comparatively small specimen if it should be referable to *M. meridionalis*, the H value lies near the maximum of the *M. meridionalis* range. This raises the HI value well above the maximum of *M. meridionalis*, and even slightly above that of *M. primigenius* (**Table 2**). Similarly, the low W value leads to a LWI of 190, which lies above the *M. meridionalis* maximum of 179 and nearly halfway to the *M. primigenius* maximum of 203.
The dP^3 in the cranium from Untermassfeld can therefore be characterised as particularly high-crowned. When plotted with *M. meridionalis* dP^3s it is an outlier in this respect, and in comparison with the patchy record on *M. trogontherii* it turns out to be closer to IQW 1965/2 775 (Süß. 9733) than to any specimen referable to *M. meridionalis*: the plate numbers are the same, the length values are very close to one another, and the LF values – that slightly surpass the *M. meridionalis* maximum – just about coincide. It should be noted, however, that IQW 1965/2 775 (Süß. 9733) does not yield a HI value, whereas that of IQW 1965/2 151 (Süß. 6348) may coincide with the *M. meridionalis* maximum (108).
It is the question now whether referral of the Untermassfeld dP^3 to *M. trogontherii* is feasible on the basis of the advanced characteristics of this specimen. It would mean that the shape of its crown and the degree of compaction of the lamellae within it are accepted as evidence in spite of the absence of a sufficiently large analogous *M. trogontherii* sample from Süssenborn and elsewhere. Moreover, the full range of variation may not be represented by the eight available *M. meridionalis* specimens either, but as it stands, a t-test classifies the HI value of the Untermassfeld specimen (133) as significantly different from the analogous values that compose the *M. meridionalis* range (mean 99.75, max. 108), even though the underlying H and W values of the Untermassfeld specimen individually fall within the respective *M. meridionalis* ranges, in the sense that H is near the upper limit, and W near the lower. Some heightening and narrowing of the crown is also observable in a few individuals that belong to so-called post-typical populations, i.e., populations younger, but not necessarily more advanced morphologically because of this, than those from the Valdarno Superiore in Italy. Among these is the one from Saint-Prest (F). Its age (c. 1.0 Ma) is similar to that of Untermassfeld. In M^3s of this sample, minor shifts of H and W values in opposite directions led to individual HI values that occasionally coincide with the higher extreme of the typical Valdarno range or surpass it by a few points, whereas the rest remained well within the typical range. Since this marginal increase in HI did not lead to a significant difference between the M^3 means of these two groups (Lister and Sher 2001, fig. 3b; Van Essen 2011, 141), it is questionable whether an analogous increase in the HI value of an individual dP^3 would at all be visible, let alone that it would surpass *M. primigenius* in this respect. Referral to an anagenetically advanced form of *M. meridionalis* would therefore seem improbable, also because the few specimens so far recovered at Untermassfeld might not characterise the entire population.
Another alternative could be that – both without and in combination with the dP^2s in front of them – the dP^3s are argued to signal the morphological inhomogeneity called 'mosaic'. Mosaic molars, known from a number of late Early and early Middle Pleistocene occurrences such as those from Sinyaya Balka (Taman peninsula, NE Black Sea shore, RUS; Lister and Sher 2001), Dorst (N Brabant, NL; Van Essen 2011, 161 ff.) and Stuttgart-Rosenstein (Baden-Württemberg, D; l.c., 170 ff.) essentially combine 'contradictory' P and HI values, in the sense that one of these is meridionaloid and the other trogontherioid, which possibly indicates (the aftermath of) secondary contact between the species. In the present case, a detail that could be argued to be characteristic of *M. meridionalis* – in particular the P value, which shows overlap between *M. meridionalis* and *M. trogontherii* (**Table 2**) – cannot be pinpointed.
In view of the fact that on the side of *M. trogontherii* there is hardly any material suitable for a necessarily more extensive morphological comparison, referral to species remains, as it were, in mid air.

4.2. Three complementary unworn fragments of an isolated left dP_3

Again, there are preciously few data on specimens referable to *M. trogontherii*, and the dP_3 from Untermassfeld meets almost any of the criteria set by the strongly overlapping ranges of *M. meridionalis* and *M. trogontherii* (**Table 3**). Within the available samples, dP_3 size in the steppe mammoth is somewhat smaller, but then there are only three specimens to go by. The lack of really high L values for *M. trogontherii* leads to narrow LWI and LHI ranges, which the Untermassfeld specimen surpasses by a few points. The *M. meridionalis* LWI and LHI ranges are much wider and suggest that in this case the true size range is covered by the much greater number of specimens represented.

The basic similarity between the Untermassfeld specimen and dP_3s of both *M. meridionalis* and *M. trogontherii*, including the almost total overlap of ET ranges, would very probably translate into overlapping clusters in diagrams if there were enough relevant data on *M. trogontherii*, and to the Untermassfeld specimen falling within the overlap zone. Unambiguous referral to species is therefore impossible in this case, as it proved to be in that of the dP^3s.

4.3. *M. meridionalis*, isolated unworn complete left dP^4

As regards referral to species, 7 is the modal P value in dP^4s of *M. meridionalis* and has so far not been observed in dP^4 specimens of the later mammoth species and *P. antiquus*. Had the value been 8, there would have been some overlap with the *M. trogontherii* and *P. antiquus* ranges, but then the low HI would still only have allowed referral to *M. meridionalis*: even in dP^4 there is a perceptible gap between the low-crowned *M. meridionalis* and its more high-crowned descendants, not to mention *P. antiquus* (**Tables 5–6**).

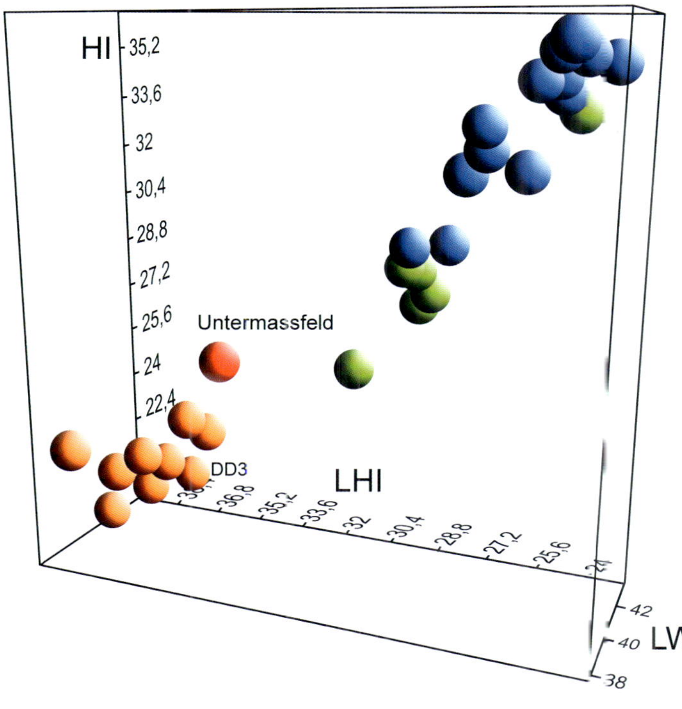

Fig. 5 The relationships between the ratios LWI, LHI, and HI in dP^4 specimens referable to representatives of the mammoth lineage in Europe. The ratios are represented as percentages of their sum; bright red/Untermassfeld – IQW 1984/20052 (Mei. 19572); coral/DD3 – Dorn-Dürkheim 3; coral – *M. meridionalis*; olive – *M. trogontherii*; blue – *M. primigenius*.

Compared to the mean size values of the rest of the *M. meridionalis* sample, the specimen from Untermassfeld turns out to be relatively short (95.5 % of mean L), wide (108 % of mean W), and high (113 % of mean H). Its H value (71.4 mm) is even the new maximum of the *M. meridionalis* dP^4 range (52.0–71.4 mm; mean 63.4 mm; n = 18), so that it appears to be relatively advanced. This is also suggested by its position in a 3D diagram that shows LWI, HI, and LHI as percentages of their sum (**Fig. 5**): because of its moderate length and considerable height, the specimen's LHI (152) is the lowest of those that represent *M. meridionalis* in the diagram, while its LWI (174) is the lowest of all analogous *Mammuthus* values represented. These shape characteristics boost the relative importance of the HI and thus pull the plot away from the central area of the *M. meridionalis* cluster, which includes the probably younger specimen from Dorn-Dürkheim 3. Still there is an unmistakable gap between the crown shapes of *M. meridionalis* – including the specimen from Untermassfeld – and its higher-crowned descendant species, which is consistent with indications that the morphological transition between them did not take place in Europe (Lister and Sher 2001; Lister et al. 2005).

Whether the observed changes in shape should be interpreted as a manifestation of mere intraspecific variability or as something akin to the mosaic contrasts between P and HI in some molars of late Early to early Middle Pleistocene mammoths remains to be determined. There are no compelling arguments, but a few remarks in favour of plain variability can be made:

- The Untermassfeld specimen may have widened some morphological ranges (H, LWI, LHI), but in its absolute measurements it strongly resembles some typical specimens from the Valdarno Superiore and elsewhere, e.g., IGF 140/12793 and NCM 76.94 (2017), respectively from the Valdarno Superiore and the Cromer Forest-bed Formation or the North Sea. No stratigraphic data exist on the latter, so that no date other than 'Early Pleistocene' can be attached to it. These specimens are here contrasted with the specimen from Untermassfeld, in the same order: L – 110; e 110; 108.5; W – 59.; 60.5; 62.3; H – e 67; 68; 71.4. The differences vary between 1.4 and 6.6 % (on average 3.8 %) with respect to the data on the Untermassfeld dP^4, while the HI values virtually coincide at a level scarcely above the *M. meridionalis* mean (e 114; 112; 115; sample mean: 111.4).
- The L value of a crown is in fact the sum of the variable lengths of plates and cement intervals, and also depends on the number of these. Small individual plate and interval lengths in combination with a fairly high and wide crown suffice to explain the very low LHI and LWI values. If any characteristic were to be considered advanced, it would have to be the on average reduced lengths of plates and cement intervals.
- In the 3D diagram (**Fig. 5**) the *M. trogontherii* and *M. primigenius* clusters are both strongly elongate because of the percentage ranges for the spread of HI and LHI values. The *M. meridionalis* cluster is somewhat less elongate, the cause of which need not be discussed here. Since it is impossible to refer the *M. trogontherii* specimen plotted in the upper right hand corner to *M. primigenius* because it belongs to the Süssenborn sample, it should be no problem at all to interpret the *M. meridionalis* cluster as morphologically coherent as well, and to refrain from considering taxonomical consequences for the specimen from Untermassfeld.

The morphological characteristics of dP^4 of *M. meridionalis* so far do not allow a clear distinction between stratigraphically older and younger specimens (see section 5.). Those from Untermassfeld and the presumably later Dorn-Dürkheim 3 suggest that the features of the overall shape of their dP^4 crowns are random and gradational with respect to the other dP^4s in the sample.

In general the specimens that may have resulted from secondary contact between *M. meridionalis* and *M. trogontherii* have so far been characterised by P and HI values that represent both species in mosaic, i.e., contrasting combination. In the dP^4 from Untermassfeld, both P and HI are eminently characteristic of *M. meridionalis*.

4.4. *M.* cf. *meridionalis*, complete remains of right M_3 in mandible fragment

The shape and ontogenetic growth stages of mandibles – especially of those that do not comply with the rules of thumb generally applied (as in Lister and Sher 2015) – have not been fully explored, and it is obvious that there is much individual, in part sexual, variability. The fragmentary state of the specimen from Untermassfeld (**Table 7**) does not allow a well-founded opinion on its primitive or advanced characteristics: neither the symphysis nor the shape of the ascending ramus can be compared to those of a series of well-dated mandibles, and an assessment of the degree to which the horizontal ramus was possibly shortened with respect to values found for mandibles of *M. meridionalis* is not possible because its height cannot be measured.

Especially because the medially annular and laterally lamellar wear figures (▭O▭) apparently characterised all lamellae of the M_3, and because this feature is combined with those of a low crown, a considerable posterior crown width, and moderately folded enamel without any detail that suggests *P. antiquus*, *Mammuthus* is clearly indicated.

Advanced wear deletes most of the detail that normally allows referral to species, so that the measurements still possible must serve as the basis for a reconstruction of the original maxima.

The ET range of the M_3 in each of its subdivisions (min., mod., med., and max.) lies above the analogous subranges found for *M. trogontherii*, and falls within those of *M. meridionalis*, slightly above the respective means (**Table 8**). The reconstruction of this molar therefore draws from data on *M. meridionalis* only. A determination of the wear stage of the M_3, and hence an approximation of its age in AEY, requires a plausible estimate of its original plate formula and length. Beyond the 11 lamellae and the posterior platelet accounted for, the next higher values of P in *M. meridionalis* (12 and 13) would lead to lengths that indicate rather short specimens. Although this is not impossible, it is rendered improbable for two reasons: firstly, the posterior W value suggests a crown maximum of c. 95 mm, very near the species' mean (c. 93 mm; n = 68). Secondly, the anterior ends of complete remains such as this one – characterised by advanced wear (with the posteriormost plates in early wear) and by a keel-shaped posterior root complex with a single pulp cavity – normally do not represent a basal point directly behind the front of the M_3, but suggest a length reduction of c. 25 %. This observation is based on comparisons of complete remains of M_3s of *M. primigenius* with truly complete crowns characterised by similar width and lamellar frequency values. Because plate numbers vary, the mean P of the M_3 sample employed was used as a reference value. The said percentage is therefore a rough estimate. Applied to the M_3 from Untermassfeld, it suggests an original L value around 299 mm and a LWI of c. 315.

The given max. P value 15, corresponding to a plate formula x 15 p, provides another possibility to approach the original situation, because the actual specimen would need a complementary part with the plate formula x 4 to reach that maximum. The LF given then determines how much longer the molar maximally was: 84 mm. The L value this leads to is 308 mm (224+84 mm), but it is too high because it is based on the LF value of the incomplete molar, which is lowered with respect to the original overall value because of some basal divergence of the posterior lamellae. The lamellae within the anterior part of the crown may be assumed to have been more compacted, so that a modest length reduction to 295 mm is plausible. This then takes the LF from 5.32 to 5.59. In its calculation, the anterior talonid was considered the equivalent of half a lamellar length, and the posterior platelet was given the value of a lamella because it presumably is of about the same length; x 15 p then equals 16.5 plates, and 1650 : 295 leads to the overall LF value 5.59. This is within range, and 0.60 above the mean frequency for M_3 of *M. meridionalis*, which is 4.99 (n = 64; $\sigma = 0.65$, V = 0.13).

The LWI of the reconstructed molar is 311, virtually the mean value (308) of a sample of truly complete M3 specimens referable to *M. meridionalis* ($n = 15$; $\sigma = 29.13$, $V = 0.09$).
Since the specimen IQW 1999/26678 (Mei. 26207) is in late wear, it should be noted that its upward rotation during the wear process now causes its length to be measured in a direction quite different from that formerly required for the complete molar: the length of a complete crown is measured between its anterior and posterior extreme, at a level about halfway up from its baseline. The measurement of the rather worn Untermassfeld specimen, however, follows a line that runs obliquely from the upper region of the roots at the front (i. e., the dentine platform below the level of the enamel joints of the lamellae) to a level near the tops of the hindmost plates. The longitudinal axis of the restored specimen is again that of a complete M3, not simply an extrapolation of the oblique measurement of the complete remains. Both lines meet at the rear end of the molar, at an angle of c. 30°. The original crown lost about a quarter of its length, and roughly two thirds of its volume. The wear stage observed, based on the reconstructed P and L values, is approximately XXVII according to Laws' schematic drawings of wear stages during mandibular tooth replacement in African elephants (1966). This implies an individual age of 52–54 AEY (G. Craig and G. Haynes in Haynes 1993, 338).
The specimen is referred to *M.* cf. *meridionalis* because theoretically it could still be mosaic. Trogontherioid characteristics might for instance be represented in the shape of the mandible or in a 16th true lamella, which would take the molar beyond the P range of *M. meridionalis*.

5. Synthesis and conclusions

Referral to species of the small sample of dental remains of mostly juvenile mammoths from Untermassfeld has in part proven problematic. This is mainly caused by the fact that there is much overlap in absolute measurements of the two anteriormost milk premolars of *M. meridionalis*, *M. trogontherii*, and even *M. primigenius*.
On the basis of purely morphological criteria, a certain inhomogeneity has turned out to be part of the population's dental characteristics. This may relate to secondary contact between *M. meridionalis* and *M. trogontherii* at some time and place, not necessarily Untermassfeld itself.
In more detail the following can be said about the specimens:

- The juvenile cranium with both dP^2s and dP^3s [IQW 1996/25646 (Mei. 25175)] cannot be compared to analogous specimens that belong to *M. meridionalis* and *M. trogontherii*. It was crushed and to some extent deformed. Hence there is no craniometric assessment of the parts that remain.
- The dP^2 specimens are of intermediate size and relatively low-crowned. They can only be compared to a single specimen referable to *M. trogontherii*, the smaller specimen from Voigtstedt. When considered separately, the morphological detail of the Untermassfeld specimens would lead to a referral to *Mammuthus* sp., or perhaps, if more data on *M. trogontherii* were available, to *M.* cf. *meridionalis*. In the end, however, they must follow the interpretation of the dP^3s.
- The dP^3 specimens would have made a similar impression if the remarkable HI value of c. 133 had not interfered. Declaring this irregularity an example of normal intraspecific variability, i. e., a chance meeting of a near-maximal H and a near-minimal W value, both within range of the known *M. meridionalis* dP^3s, would not be satisfactory because the known HI range maximum for *M. meridionalis* dP^3s lies so far below the value of the Untermassfeld specimen that a t-test confirms the significance of the difference. Theoretically possible options are either 1. that the specimen, and hence the cranium, is referred to *M. trogontherii* because of the significantly different HI value of the dP^3 in comparison with *M. meri-*

dionalis, even though referral to *M. trogontherii* of the Süssenborn morphotype cannot be substantiated because the specimens available for comparison are too incomplete and show range overlap, or 2. that it is for the same reason considered to belong to a population that as a whole could be called mosaic. Both options encompass the dP²s as well, and neither of them can claim any preference on purely morphological grounds. Remains of more or less contemporaneous European populations exclusively referable to *M. trogontherii* have not yet been recovered, whereas there are indications that the influx of trogontherioid elements into *M. meridionalis* populations had already begun around that time. Because of this and the undoubted presence of *M. meridionalis* at Untermassfeld, option 2 is preferred. Although it is difficult to render the complexity of the said biological variation in terms of Linnean taxonomy, the solution *M.* cf. *meridionalis* is chosen here and in the figure captions. Behind this choice is the consideration that essentially a single trogontherioid feature can be argued to have appeared within a population that does comprise specimens referable to *M. meridionalis* and *M.* cf. *meridionalis*, but has so far lacked indications of the presence of full *M. trogontherii* morphology. Future finds will hopefully throw some more light on this matter.

- The near-complete isolated left dP_3 IQW 1997/26303 (Mei. 25832) is likely to have formed part of the mandible that belonged to the cranium, but it does not show a conspicuously high HI. The difficulty to refer it to either *M. meridionalis* or *M. trogontherii* is caused by no other morphological circumstance than range overlap. Again, as motivated for the dP^3 in the skull, the solution *M.* cf. *meridionalis* is chosen.
- The same may apply to the minor lamellar fragments IQW 1997/26304 (Mei. 25833) that possibly represent one or two of the four dP4 germs that must have formed part of the dental elements within cranium and mandible.
- The complete isolated left dP^4 IQW 1984/20052 (Mei. 19572) does not show any traits that would justify referral to any other taxon than *M. meridionalis*. The species having been established, the question remains whether there was any evolutionary change in dental morphology between c. 2.0 and 1.0 Ma that could support the late Early Pleistocene age of the Untermassfeld assemblage, especially with regard to this isolated dP^4, which in every detail can be compared to a series of equally complete and isolated analogous specimens. The general picture is that none of the populations that succeeded the typical form of *M. meridionalis* from the Italian Upper Valdarno sites varied beyond the morphological range of that form in any character, but that there was a slight upward shift in average P and HI values. This applies to M^3 specimens of larger samples first and foremost, because it is in these that morphological changes become most clearly observable. Notwithstanding the stable range widths of the various morphological characters, a number of supposedly advanced *M. meridionalis* subspecies were created over the past century, but from the present point of view the criteria that should lead to their identification have proved insufficient because of small sample sizes, misinterpretations, methodological flaws and newer insights (Van Essen 2011, 132ff., 216, 255). If appropriate, it seems better therefore to speak of an advanced, evolved or derived form rather than of an advanced subspecies.

Size does not carry much weight in the interpretation of single specimens, but when it comes to morphological detail, the P values of *M. meridionalis* dP^4 specimens found at 'post-typical' sites, as opposed to those from the Valdarno Superiore and chronological equivalents, confirm that little or no evolutionary change in P should be expected to become apparent within this group: five specimens that belong to the younger sample do not differ at all from fifteen specimens from the older one in mean P (7), standard deviation (0.4) and coefficient of variation (0.06) (**Table 5**).

For a comparison of HI values the younger group is too small (n = 2). Nonetheless it is remarkable that even one of the latest European samples of *M. meridionalis*, that of Dorn-Dürkheim 3 (c. 0.8 Ma; Franzen et al. 2000), comprises a dP^4 with a HI of approximately 114 (**Fig. 5**, DD3). Even early Middle Pleistocene *M. meridionalis* molars from S. Giminiano and Soave (I) do not show any derived features (Ferretti 1999, 508 f.; Palombo and Ferretti 2005, 114), so that late *M. meridionalis* populations are not necessarily morphologically advanced.

- The complete remains of a right M_3 in the mandible fragment IQW 1999/26678 (Mei. 26207) are referable to *M. meridionalis* because of the ET range observed. This is modified to *M.* cf. *meridionalis* because some mosaic trait might be hidden because of incompleteness. Although not quite certain, the most plausible reconstruction of this specimen's plate formula is x 15 p within a length of 295–300 mm. Together with the reconstructed max. W value of c. 95 mm these characteristics classify the specimen as one of about average size.

Although the P value 15 has remained the maximum observed in M_3s of *M. meridionalis*, and late populations of this species do not have to be morphologically advanced, the frequency of the occurrence of the higher P values in M3s does seem to have somewhat increased in post-typical populations. Hence it could be argued that the P value reconstructed for the M_3 from Untermassfeld (15) – if correct – will sooner be indicative of a relatively late population than of a typical one. However, since the relevant M_3 samples are small, and a specimen with the P value 12 forms part of the more or less contemporaneous Saint-Prest population (MsnpC 28.SPR.78; pers. obs.), a greater number of complete specimens could possibly undermine the said argument.

More observations that lie beyond the morphology of the Untermassfeld sample but may still contribute to its interpretation are the following: Remains of more or less contemporaneous European populations comprising undoubted representatives of both *M. meridionalis* and *M. trogontherii* from a demonstrably single horizon have not yet been described, but such assemblages probably do exist. The first candidate that should be named is the afore-mentioned site of Sinyaya Balka, Taman peninsula, where a bimodally distributed sample with preservation characteristics unrelated with morphology was recovered, albeit from a reworked deposit. The site was dated to c. 1.0 Ma and yielded M3 specimens with 14 to 19 lamellae, and HI values from 130 to 180 (Lister and Sher 2001, fig. 3; Lister et al. 2005, 110, 113). The second candidate is Dorn-Dürkheim 3 in the Upper Rhine valley (D; c. 0.8 Ma), where a similar species combination occurred, but likewise in a reworked stratigraphical setting (Franzen et al. 2000; Lister et al. 2005). Instead of the ideal undisturbed single-level assemblages, however, there are numerous indications that isolated advanced traits, especially high HI and P values, may occur in fossil assemblages that also comprise dental remains referable to *M. meridionalis*, such as that from Dorst (NL) at c. 0.95 Ma (Van Essen 2011, 161 f.). Because of this, the above-mentioned second interpretation of the two alternatives possible with regard to the high HI of the dP^3 is preferred: its occurrence as a mosaic trait emerging within a population at least in part or predominantly characterised by meridionaloid features.

With *M. meridionalis*, *M.* cf. *meridionalis* and a presumably mosaic anomaly indicated by the dental components of its mammoth sample, Untermassfeld takes its place within the group of late Early Pleistocene and early Middle Pleistocene sites that signify the presence of individuals with morphological characteristics that in one or more aspects distinguish them from the resident *M. meridionalis* populations. Most important in this connection is the sudden rise in HI, which is unlikely to have been the result of anagenetic evolution within Europe. Should it become necessary in future to acknowledge the presence of *M. trogontherii* as well, this would not influence the outcome of the present analysis.

In order to avoid the nomenclatorial confusion that would arise if the now available subspecific taxon *voigtstedtensis* were to be transferred from *M. meridionalis* to *M. trogontherii* with the express intent to cover the mosaics of supposedly hybrid origin, the present author chose to refer these to the steppe mammoth and proposed to add an explanatory formula in brackets to point out their special status: *M. trogontherii* [*M. trogontherii* (Pohlig, 1885) x *M. meridionalis* (Nesti, 1825)] (Van Essen 2011, 216f.). It is not implied here that this referral is applicable to the cranium or even its dP^3, because the lack of a contrasting feature otherwise exclusively found in *M. meridionalis* dP^3s precludes this application. Furthermore, the individual that died in the flash flood was not necessarily the direct result of crossing between parents that did not belong to the same species. Inherited genes may have caused the salient feature of the high HI value via one of them.

On the basis of finds from Italy, the persistence of the species *M. meridionalis*, or at least largely meridionaloid mammoths in southern Europe, can be traced into the part of the Galerian biochron which is dated to the early Middle Pleistocene (Ferretti 1999; Palombo and Ferretti 2005, 114; Lister et al. 2005). Lister and Stuart (2010, 201ff.) discussed the complete remains of an M_2 fragment from the West Runton Freshwater Bed (GB) and considered it very likely that it belongs to *M. meridionalis*, which would likewise imply an early Middle Pleistocene age (Cromerian Complex Stage, c. 0.7 Ma). The time around 1.0 Ma, at which *M. trogontherii* first appeared in Europe, was therefore presumably followed by an episode of c. 300 ka before *M. meridionalis* finally became extinct and the exchange of genes between representatives of this species and of *M. trogontherii* – which may explain the existence of mosaic character combinations in mammoths – stopped. This implies that mammoth remains that show mosaic peculiarities of any kind may in principle not only be expected to date back to the latter half of the Epivillafranchian sensu Kahlke (2007, 273, fig. 8), but also to a part of the Galerian biochron beyond the Brunhes-Matuyama boundary. The Untermassfeld mammoth population as such therefore neither corroborates nor disproves the Jaramillo date based on other evidence.

Acknowledgements

I would like to thank the following persons for their much appreciated assistance in matters relevant to the content of this paper: A. Azzaroli † (Florence); P. Boudier (Chartres); A. Currant (London); V. Dijkstra (Arnhem); M. Ferretti (Florence); T. Henskens (Oss); R.-D. Kahlke; J.-A. Keiler, T. Korn, G. Utschig (all Weimar); O. Kullmer (Frankfurt/M.); A. Lebreton (Chartres); A. Lister (London); T. Stuart and colleagues (Norwich); T. Sutcliffe † (London); and L. Talstra (Beetsterzwaag). A. Lister is especially thanked for reviewing this paper and for suggesting various improvements. NERC grant no. GR3/8248 (1992) is acknowledged with respect to data on material from the Cromer Forest-bed Formation.

References

Azzaroli, A., 1977. Evolutionary patterns of Villafranchian elephants in Central Italy. Atti della Accademia Nazionale dei Lincei, Classe di Scienze Fisiche, Matematiche e Naturali, Serie VIII, Memorie 14, 149–168.

Beden, M., 1983. Family Elephantidae. In: Harris, J.M. (Ed.), Koobi Fora Research Project. Vol. 2. The fossil Ungulates: Proboscidea, Perissodactyla, and Suidae. Oxford, 40–129.

Haynes, G., 1993. Mammoths, Mastodonts, & Elephants. Biology, Behavior, and the Fossil Record. Cambridge. (2rd ed.)

Deraniyagala, P.E.P., 1955. Some Extinct Elephants, Their Relatives and the Two Living Species. Colombo.

Dietrich, W.O., 1965. Fossile Elephantenzähne von Voigtstedt in Thüringen. Paläontologische Abhandlungen, Abteilung A Paläozoologie 2. Berlin, 521–536, pls. XXXII–XXXV.

Dubrovo, I.A., 1975. *Mammuthus trogontherii trogontherii* (Pohlig) aus dem Pleistozän von Tiraspol. Quartärpaläontologie 1, 125–143, pls. I–VI.

2001. Remains of Elephantidae from the Lower Pleistocene site of Untermassfeld. In: Kahlke, R.-D. (Ed.), Das Pleistozän von Untermaßfeld bei Meiningen (Thüringen). Teil 2. Monographien des Römisch-Germanischen Zentralmuseums 40, 2, 589–605, pls. 95–101.

Ferretti, M.P., 1999. *Mammuthus meridionalis* (Mammalia, Proboscidea, Elephantidae) from the »Sabbie Gialle« of Oriolo (Cava la Salita, Faenza, Northern Italy) and other European late populations of southern mammoth. Eclogae Geologicae Helvetiae 92, 503–515.

Franzen, J.L., Gliozzi, E., Jellinek, T., Scholger, R., Weidenfeller, M., 2000. Die spätaltpleistozäne Fossillagerstätte Dorn-Dürkheim 3 und ihre Bedeutung für die Rekonstruktion der Entwicklung des rheinischen Flußsystems. Senckenbergiana lethaea 80, 305–353.

Friant, M., 1959. A propos du développement des molaires de l'*Elephas meridionalis* Nesti. Annales de la Societé Géologique du Nord 79, 113–128.

Furrer, H., 2005. Niederweningen, die bedeutendste Mammutfundstelle der Schweiz. Neufunde und eigenes Mammutmuseum. Bulletin für angewandte Geologie 10, 61–69.

Kahlke, R.-D., 2001. Die unterpleistozäne Komplexfundstelle Untermassfeld – Zusammenfassung des Kenntnisstandes sowie synthetische Betrachtungen zu Genesemodell, Paläoökologie und Stratigraphie. In: Kahlke, R.-D. (Ed.), Das Pleistozän von Untermaßfeld bei Meiningen (Thüringen). Teil 3. Monographien des Römisch-Germanischen Zentralmuseums 40, 3, 931–1030, suppl. I–XV.

2006. Untermassfeld. A late Early Pleistocene (Epivillafranchian) fossil site near Meiningen (Thuringia, Germany) and its position in the development of the European mammal fauna. British Archaeological Reports, International Series 1578. Oxford.

2007. Late Early Pleistocene European large mammals and the concept of an Epivillafranchian biochron. In: Kahlke, R.-D., Maul, L.C., Mazza, P.P.A. (Eds.), Late Neogene and Quaternary biodiversity and evolution: Regional developments and interregional correlations. Proceedings of the 18th International Senckenberg Conference (VI International Palaeontological Colloquium in Weimar), Volume II. Courier Forschungsinstitut Senckenberg 259, 265–278.

Lang, A., 1892. Geschichte der Mammutfunde. Ein Stück Geschichte der Paläontologie, nebst einem Bericht über den schweizerischen Mammutfund in Niederweningen 1890/1891. Neujahrsblatt der Naturforschenden Gesellschaft in Zürich 94, 3–35, figs. 1–12.

Leith Adams, A., 1879. Monograph on the British Fossil Elephants. Part II. Dentition and Osteology of *Elephas primigenius* (Blumenbach). Palaeontographical Society Monographs 33, 69–146, pls. VI–XV.

Lister, A.M., 2001. Age Profile of Mammoths in a Late Pleistocene Hyaena Den at Kent's Cavern, Devon, England. Publications in Anthropology 22, 35–43.

2015. Dating the arrival of straight-tusked elephant (*Palaeoloxodon* ssp.) in Eurasia. Bulletin du Musée d'Anthropologie Préhistorique de Monaco, supplément 6, 13–18.

Lister, A.M., Sher, A.V., 2001. The Origin and Evolution of the Woolly Mammoth. Science 294, 1094–1097.

2015. Evolution and dispersal of mammoths across the Northern Hemisphere. Supplementary materials. Science 350, 805–809, 3–58, 1–11.

Lister, A.M., Stuart, A.J., 2010. The West Runton mammoth (*Mammuthus trogontherii*) and its evolutionary significance. Quaternary International 228, 180–209.

Lister, A.M., van Essen, H., 2003. *Mammuthus rumanus* (Ştefănescu), the earliest mammoth in Europe. In: Petculescu, A., Stiucă, E. (Eds.), Advances in Vertebrate Paleontology 'Hen to Panta'. Bucharest, 47–52.

Lister, A.M., Sher, A.V., van Essen, H., Wei, G., 2005. The pattern and process of mammoth evolution in Eurasia. Quaternary International 126–128, 49–64.

Maschenko, E.N., 2002. Individual development, biology and evolution of the woolly mammoth. Cranium 19, 3–120.

Maschenko, E.N., Gablina, S.S., Tesakov, A.S., Simakova, A.N., 2006. The Sevsk woolly mammoth (*Mammuthus primigenius*) site in Russia. Taphonomic, biological and behavioral patterns. Quaternary International 142–143, 147–165.

Maška, K.J., 1885. Mammuthmilchzähne von Předmost in Mähren. Sitzungsberichte der niederrheinischen Gesellschaft für Natur- und Heilkunde in Bonn, 339–343.

Münzel, S.C., 2001. Seasonal hunting of mammoth in the Ach-Valley of the Swabian Jura. In: Cavarretta, G., Gioia, P., Mussi, M., Palombo, M.R. (Eds.), The World of Elephants. Proceedings of the 1st International Congress. Rome, 318–322.

Musil, R., 1968. Die Mammutmolaren von Předmostí (ČSSR). Paläontologische Abhandlungen, Abteilung A Paläozoologie 3. Berlin, 3–192, pls. 1–44.

Palombo, M.R., Ferretti, M.P., 2005. Elephant fossil record from Italy: knowledge, problems, and perspectives. Quaternary International 126–128, 107–136.

Pohlig, H., 1888, 1891. Dentition und Kranologie des *Elephas antiquus* Falc., mit Beiträgen über *Elephas primigenius* Blum. und *Elephas meridionalis* Nesti (1888). Nachträge zur Dentition des *Elephas antiquus* mit Beiträgen über *Elephas meridionalis* und *Elephas primigenius* (1891). Nova Acta der Kaiserlich Leopoldinisch-Carolinischen Deutschen Akademie der Naturforscher 57, 1–282 (1888), 286–466 (1891).

Tobien, H., 1973. The Structure of the Mastodont Molar (Proboscidea, Mammalia). Part 1. The Bunodont Pattern. Mainzer geowissenschaftliche Mitteilungen 2, 115–147.

Van der Made, J., Mazo, A.V., 2003. Proboscidean dispersals from Africa towards Western Europe. Deinsea 9, 437–452.

Van der Merwe, N.J., Bezuidenhout A.J., Seegers, C.D., 1995. The skull and mandible of the African elephant (*Loxodonta africana*). Onderstepoort Journal of Veterinary Research 62, 245–260.

Van Essen, H., 2011. Tracing Transitions. An overview of the evolution and migrations of the genus *Mammuthus* Brookes, 1828 (Mammalia, Proboscidea). PhD thesis, Leiden University.

Wei, G., Taruno, H., Jin, C., Xie, F., 2003. The earliest specimens of steppe mammoth, *Mammuthus trogontherii*, from the Early Pleistocene Nihewan Formation, North China. Earth Science (Chikyū-kagaku) 57, 289–298.

NURIA GARCÍA · PAUL P. A. MAZZA

NEW RESULTS ON URSIDS FROM THE EARLY PLEISTOCENE SITE OF UNTERMASSFELD

Abstract

The present study is a revision of the currently available fossil bear remains from the Epivillafranchian locality Untermassfeld (Germany). The authors have made a detailed description and short reappraisal of the taxonomic status of the specimens by comparison with bear material from other localities. The Untermassfeld bears share a number of common features with both modern brown bears, as well as with the Early Pleistocene *Ursus etruscus*. They differ from *U. etruscus* in exhibiting reduced number of premolars, as is also the case with *U. arctos*, with the *U. deningeri-spelaeus* lineage and with *U. dolinensis* from Atapuerca (Spain). Only *U. etruscus* retains the primitive condition with a full set of premolars. The Untermassfeld ursid is similar to the Atapuerca Gran Dolina *U. dolinensis* in many respects. Both share a number of primitive arctoid-like features, while others link them to the cave bear lineage. For this the Untermassfeld bear is here referred to as *Ursus* ex gr. *U. dolinensis*. Untermassfeld provides us with a window into European bears during the Epivillafranchian. They constituted a metapopulation of interconnected subpopulations characterised by a mosaic of primitive and derived traits. These features were differentially inherited in the descendant lineages that finally separated from one another starting sometime from the latest Early Pleistocene through the early Middle Pleistocene, leading to typical brown and cave bears.

1. Introduction

Excavations conducted from 1979 to 1988 at Untermassfeld (Thüringen, Germany) have yielded abundant fossil bear remains (Kahlke 1997; 2001a; 2001b). The bone-bed was dated to about one million years. Lithologically, palaeomagnetically and biostratigraphically, the Untermassfeld fossiliferous sands are referred to the Marine Isotope Stage (MIS) 31, a quite warm stage that overlaps the base of the Jaramillo polarity subchron (Kahlke 2006; Maul et al. 2007).

The Untermassfeld ursid material was first studied by Musil (2001). The collection included one skull, five mandibular remains with teeth and postcranial material of some 5 to 8 individuals. The fossils were described as a new species, *Ursus rodei*. The diagnosis of *U. rodei* Musil, 2001, characterises the Untermassfeld bear as medium to large size species, with a flat glabella like that of *U. arctos*, and arctoid dentition. Musil's

(2001, 656) description and differential diagnosis emphasises the total absence of spelaeoid features, while stressing typical arctoid ones indicative of phylogenetic proximity to *U. etruscus*.
The present revision includes not only the specimens originally studied by Musil (2001), but also new material recovered from 1989 onwards. Musil's (2001) new bear discoveries from Untermassfeld added to other ones of similar age (late Early Pleistocene) from Sierra de Atapuerca (García and Arsuaga 2001; García García 2003), Le Vallonnet (France, Moullé 1992), Vallparadís (Spain, Madurell-Malapeira et al. 2009), Deutsch-Altenburg (Austria, Rabeder et al. 2009) and called for a thorough re-examination of the traditional phylogeny of European bears (Freudenberg 1914; Erdbrink 1953; Kurtén 1968). Wagner and Čermák (2012) revised an important collection of early Middle Pleistocene dental material, including bears from both spelaeoid and arctoid lineages. The authors proposed new taxonomic attributions and new explanations for the evolutionary history and possible co-occurrence of spelaeoid and arctoid lineages. The bear sample analysed by Wagner and Čermák (2012) included non-fossil remains of *U. arctos* and *U. thibetanus* and early Middle Pleistocene ones of *U. deningeri*.
The brown bear is one of the most variable of large carnivores. It first appeared in the Middle Pleistocene (Davison et al. 2011). However, several variants with brown bear affinities are reported from various Epivillafranchian localities (Atapuerca Gran Dolina TD4/5, c. 0.8 Ma, García and Arsuaga 2001; Untermassfeld, c. 1.0 Ma, Musil 2001; Deutsch-Altenburg, c. 1.4 Ma, Rabeder et al. 2010). Rabeder et al. (2010) placed the bears from Deutsch-Altenburg, Ceyssaguet (France), Süssenborn (Germany), Untermassfeld and Atapuerca Gran Dolina in the subspecies *U. arctos suessenbornensis* Soergel, 1926, based on metapodial similarities with brown bears. Rabeder et al.'s (2010) classification considers only a limited number of features and does not present a comprehensive morphological comparison with the other Epivillafranchian bears mentioned above. Madurell-Malapeira et al. (2009) follow Moullé (1992) and relate the bears from Le Vallonnet to ancient representatives of *U. deningeri*. They also attribute the bears from CGRD6 Cal Guardiola deposits (c. 0.8 Ma, MIS 21), in the Cuenca del Vallès-Penedès (Spain), to an archaic *U. deningeri* on account of their similarities with the bears from Le Vallonnet. Baryshnikov and Boeskorov (2004) report the occurrence of two chrono-subspecies, the early Middle Pleistocene *U. arctos kamiensis* and the late Middle and Late Pleistocene *U. arctos priscus* (syn. *U. arctos karmalkiensis*). The diagnostic characters, patterns of variation and phylogenetic significance of these bears, as well as the reliable autapomorphic characters that help discriminate between the arctoid and spelaeoid lineages, still have to be clarified.

2. Material and methods

The sample analysed for the present work includes all the dentognathic bear remains recovered from Untermassfeld since 1979. A selection of both extant and extinct bear specimens was used for comparisons. These include remains of modern brown bears from several Iberian regions (including the Pyrenees) as well as from North America (Wyoming, Arizona, California, Alaska). Fossils of brown bears come from a range of Late Pleistocene to Holocene European sites. Most of the material of *Ursus deningeri* originates from the about 400 ka year-old levels of Sima de los Huesos (Atapuerca, Spain, Bischoff et al. 2003) as well as from early Middle Pleistocene deposits preserved in several institutions: Aristotle University of Thessaloniki, Greece; Hungarian Natural History Museum, Budapest, Hungary; Museum of Vertebrate Zoology, Berkeley, USA; Natural History Museum, London, Great Britain; Natural History Museum, Madrid, Spain; Senckenberg Research Station of Quaternary Palaeontology, Weimar, Germany (IQW); Smithsonian Institution, Washington, USA. The present study also includes comparisons with García and Arsuaga's (2001) *U. dolinensis*, from the Early Pleistocene levels of Atapuerca Gran Dolina (Spain). Upper and lower teeth are indicated by upper-case and lower-case letters, respectively. Measurements followed García García (2003) and Santos et al. (2014). Cuspids on the lingual side of P_4 are called 'metaconid'; 'entoconid- and metaconid-complex' is used for M_1 and 'buccal constriction and distal convexity' for M_3 following Wagner and Čermák (2012). Measurements were taken with engineering digital caliper with 0.1 mm accuracy. Specimens previously published by Musil (2001) are re-analysed, with particular remarks when required, while new finds are described for the first time. All measurements, unless otherwise indicated, were taken by the writers.

2.1. Untermassfeld cranial remains and upper dentition

Material in Musil (2001): partial senile skull IQW 1983/19 544 (Mei. 19 064), holotype of *Ursus rodei* (**Fig. 1**) with left I^3, C, alveoli of P^1 and P^3, P^4-M^2 and right C, P^3 alveolus, P^4-M^2.
New material: skull fragment IQW 1994/24 599 (Mei. 24 128) with left I^1-M^2 and right I^1-I^3 (**Fig. 2**);
left maxillary fragment IQW 2002/27 455 (Mei. 26 960) with complete tooth row (I^1-M^2) (**Fig. 3**);

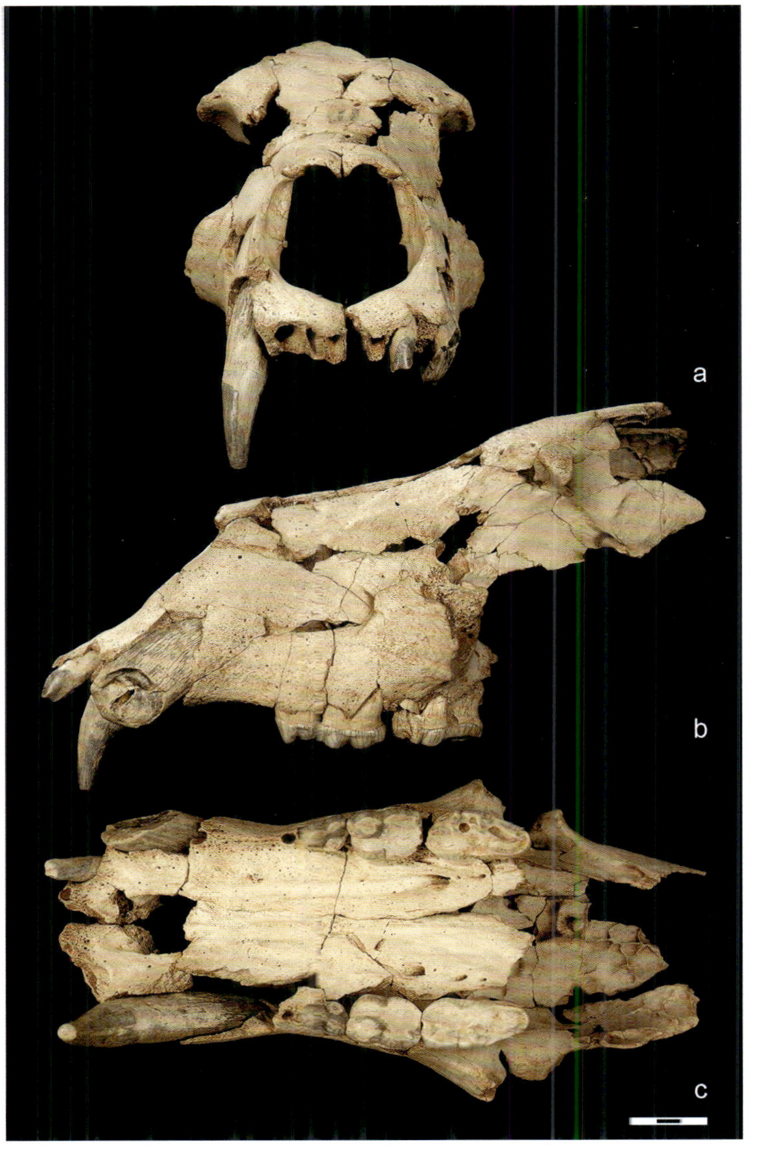

Fig. 1 *Ursus* ex gr. *U. dolinensis*, Untermassfeld. – **a–c** Partial skull IQW 1983/19 544 (Mei. 19 064) with left I^3, C fragment, alveoli of P^1 and P^3, P^4-M^2 and right C, alveolus of P^3, P^4-M^2, frontal, left lateral, palatal views. – Scale: 30 mm.

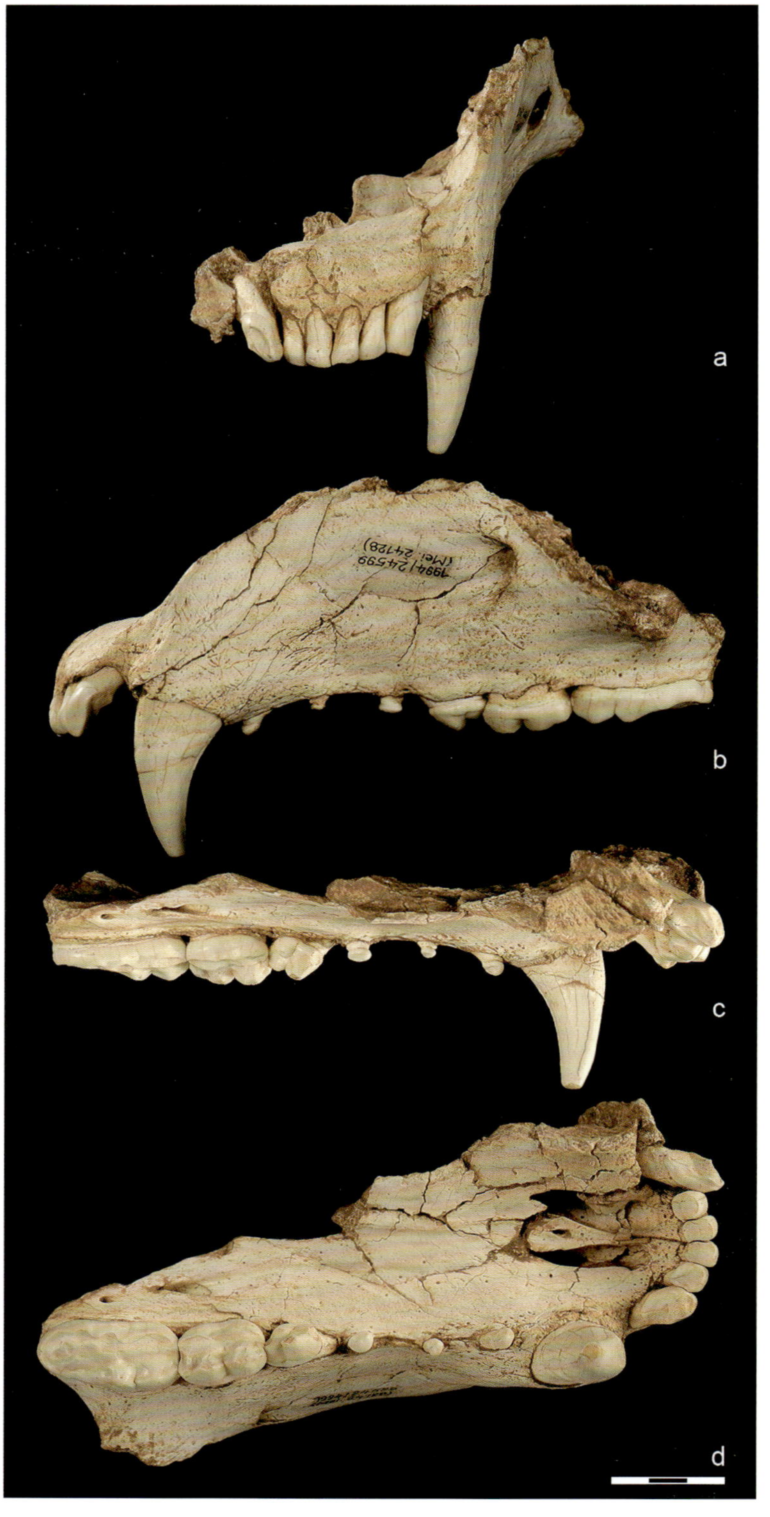

Fig. 2 *Ursus* ex gr. *U. dolinensis*, Untermassfeld. – **a–d** Skull fragment IQW 1994/24 599 (Mei. 24 128) with left I^1-M^2 and right I^1-I^3, frontal, left lateral, lingual, palatal views. – Scale: 30 mm.

Fig. 3 *Ursus* ex gr. *U. dolinensis*, Untermassfeld. – **a–d** Skull fragment IQW 2002/27455 (Mei. 26960) with left I^1-M^2, frontal, left lateral, lingual, palatal views. – Scale: 30mm.

maxillary fragment IQW 1989/23351 (Mei. 22870) with right P^4 and M^2 (**Fig. 4**);

fragmentary left maxilla IQW 1989/23353 (Mei. 22872) with M^1 – same individual as IQW 1989/23351 (Mei. 22870) – and left M^2 IQW 1989/23352 (Mei. 22871) (**Fig. 5**);

skull IQW 1995/24844 (Mei. 24373) with deformed brain case and articulated mandible (Kahlke 2001c, Fig. 6; 2006, Fig. 31); comparisons of palatal thickness at level of M^2 in **Fig. 6**.

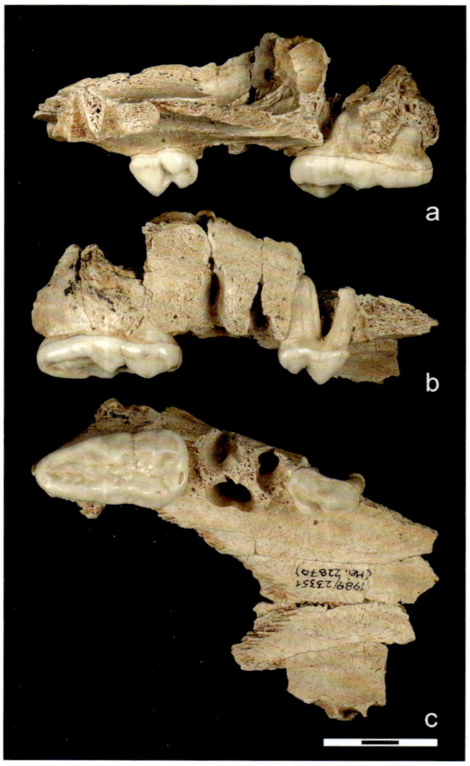

Fig. 4 *Ursus* ex gr. *U. dolinensis*, Untermassfeld. – **a–c** Skull fragment IQW 1989/23351 (Mei. 22870) with right P^4 and M^2, lingual, right lateral, palatal views. – Scale: 30 mm.

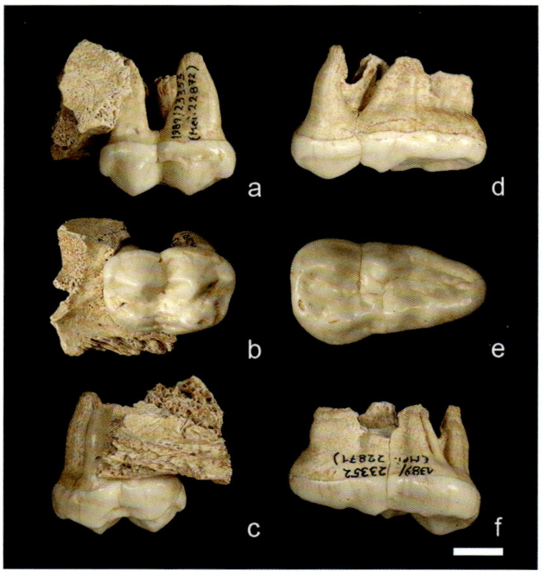

Fig. 5 *Ursus* ex gr. *U. dolinensis*, Untermassfeld. – **a–c** Maxilla fragment IQW 1989/23353 (Mei. 22872) with left M^1, buccal, occlusal, lingual views. – **d–f** Left M^2 IQW 1989/23352 (Mei. 22871), buccal, occ usal, lingual views. – Scale: 10 mm.

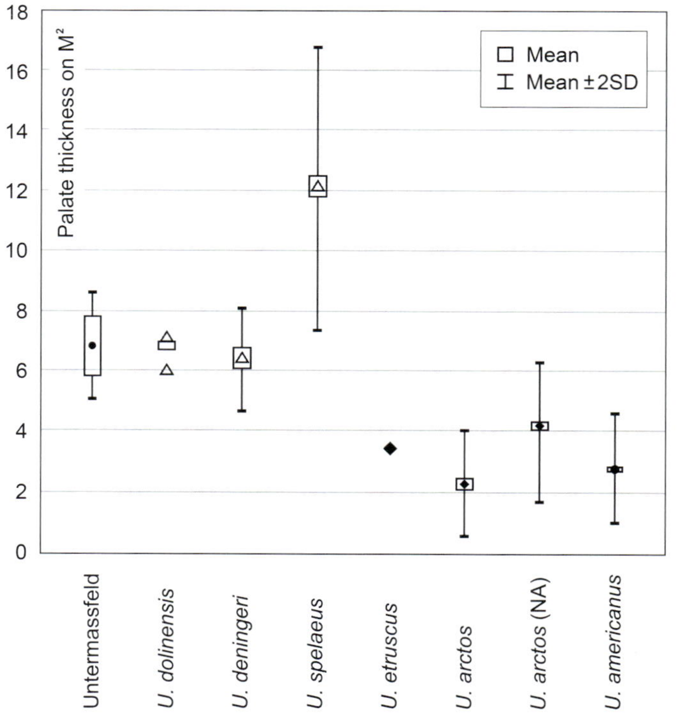

Fig. 6 Box and whiskers plot of the palatal thickness (mm) at level of M^2, in the Untermassfeld bears, in *Ursus dolinensis* and *U. deningeri* from Sierra de Atapuerca, in *U. spelaeus* and *U. etruscus*, in modern Iberian and North American *U. arctos*, and in *U. americanus* (comparison data partially after Santos et al. 2014).

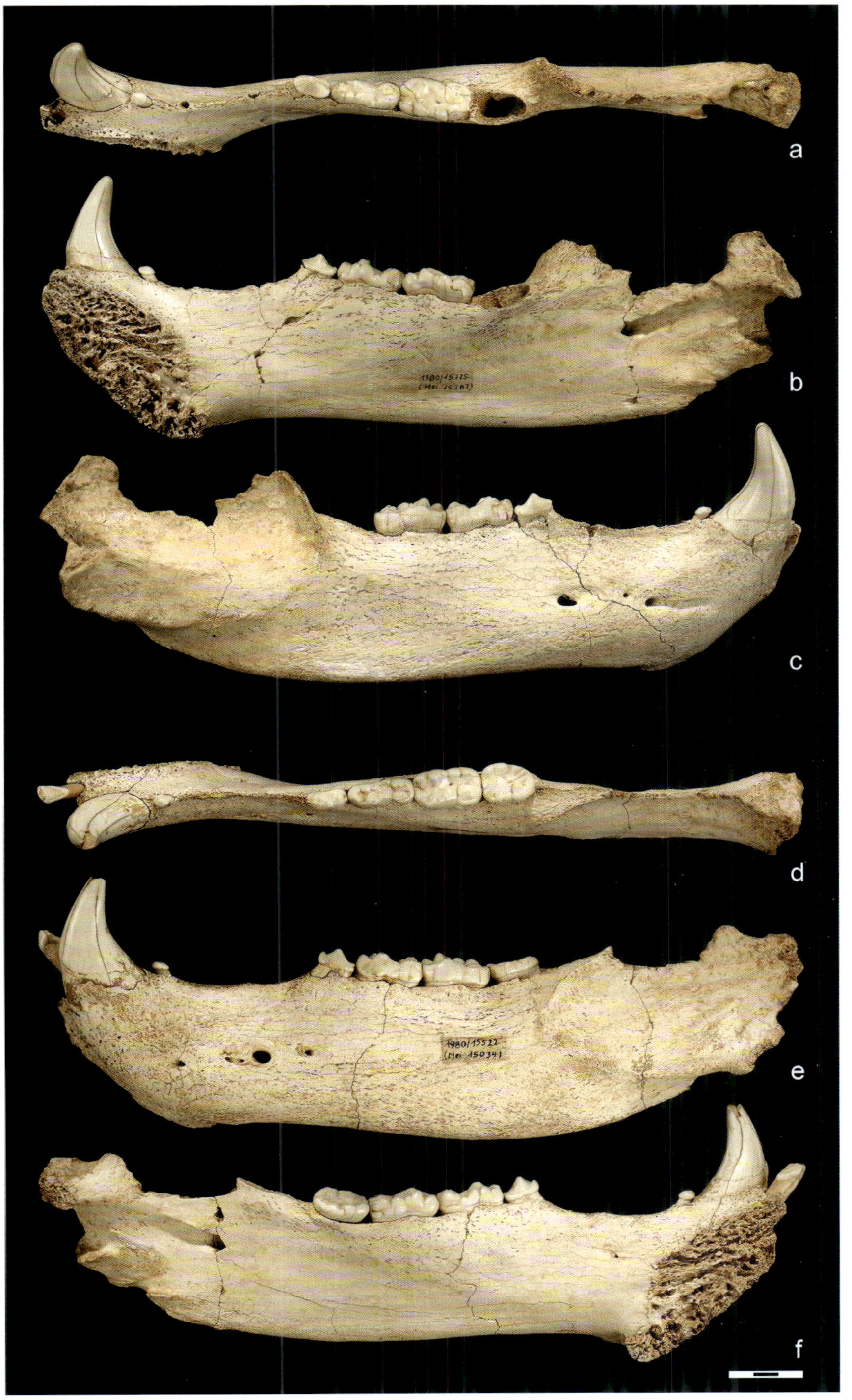

Fig. 7 *Ursus* ex gr. *U. dolinensis*, Untermassfeld. – **a–c** Right hemimandible IQW 1980/15 775 (Mei. 15 287) with alveoli of I_3, C, P_1, alveoli of P_2 and P_3, P_4-M_2, alveolus of M_3, occlusal, lingual, buccal views. – **d–f** Left hemimandible IQW 1980/15 522 (Mei. 15 034) with I_3, C, P_1, P_4-M_3, occlusal, buccal, lingual views. – Scale: 30 mm.

2.2. Untermassfeld mandibular remains and lower dentition

Material in Musil (2001): left hemimandible IQW 1980/15 522 (Mei. 15 034) with I_3, C, P_1, P_4-M_3 (**Fig. 7d–f**); right hemimandible IQW1980/15 775 (Mei. 15 287) (**Fig. 7a–c**).

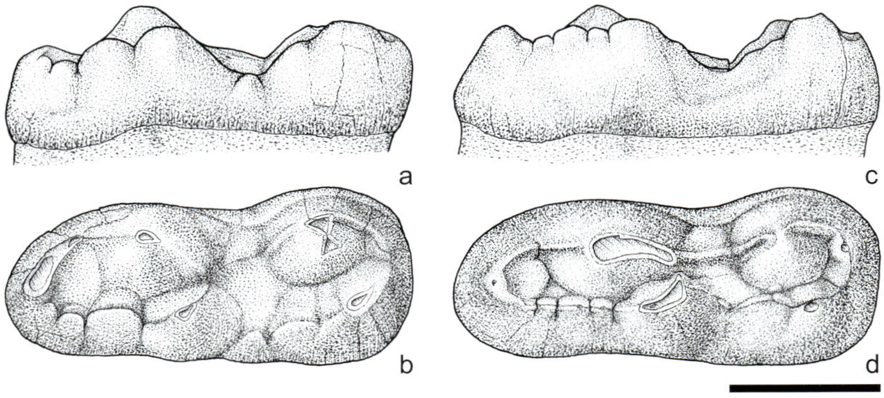

Fig. 8 *Ursus* ex gr. *U. dolinensis*, Untermassfeld. – **a–b** Right M_1 IQW 1989/15 775 (Mei. 15 287), lingual, occlusal views. – **c–d** Right M_1 IQW 1995/25 029 (Mei. 24 558), lingual, occlusal views. – Scale: 10 mm.

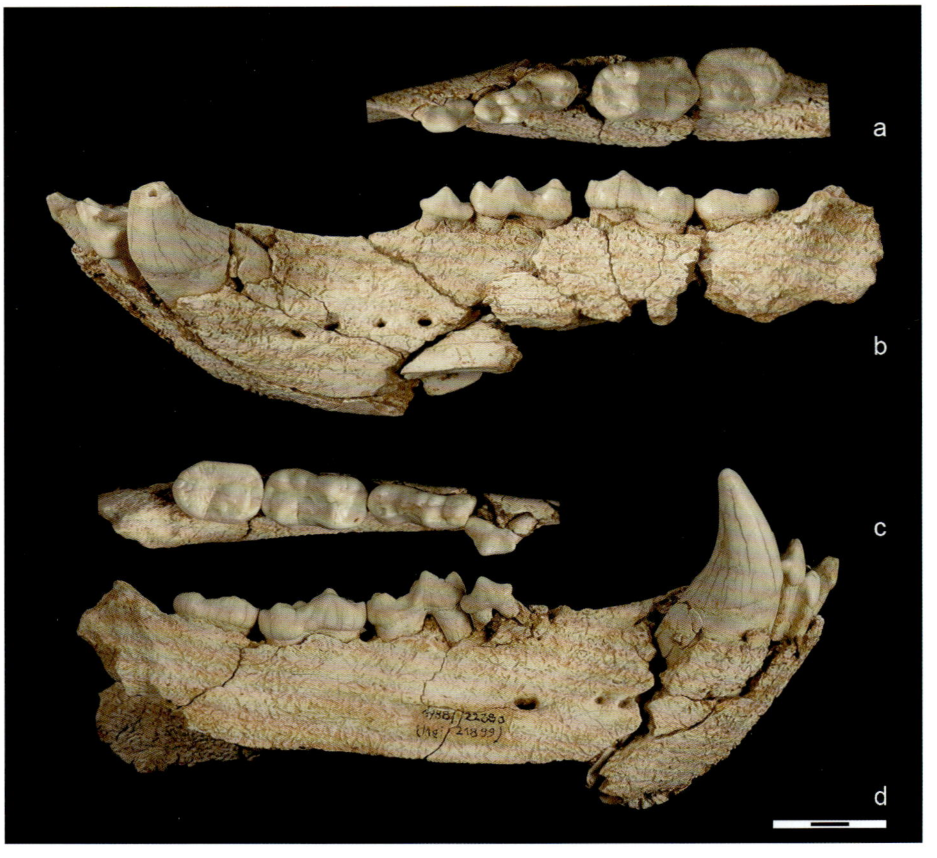

Fig. 9 *Ursus* ex gr. *U. dolinensis*, Untermassfeld. – **a–b** Left hemimandible fragment IQW 1988/22 380 (Mei. 21 899) with I_1-C, P_4-M_3, occlusal, buccal views – **c–d** Right hemimandible fragment IQW 1988/22 380 (Mei. 21 899) with I_1-C, P_4-M_3, occlusal buccal views. – Scale: 30 mm.

New material: right M_1 IQW 1989/15 775 (Mei. 15 287) (**Fig. 8a–b**); right M_1 IQW 1995/25 029 (Mei. 24 558) (**Fig. 8c–d**);
left hemimandible fragment IQW 1988/22 380 (Mei. 21 899) with I^1-C, P^4-M^3 (**Fig. 9a–b**); right hemimandible fragment IQW 1988/22 380 (Mei. 21 899) with I_1-C, P_4-M_3 (**Fig. 9c–d**).

3. Description of the remains

3.1. Morphology of the cranial remains and of the upper dentition

Calvarium IQW 1983/19544 (Mei. 19064) published by Musil (2001):
The specimen shows large canines, the left of which broken, the alveolus of left P^1 – the right region is not preserved – and the alveoli of both P^3s. The alveolus of P^1 is adjacent to canine alveolus, without any diastema in between.
Left and right P^4: all cusps are highly worn but with well-defined features. Straight mesio-distally, with triangular outline and without accessory cusps. In latero-medial view, the paracone is the highest cusp, with straight anterior border, followed in height by the protocone. The deuterocone is the lowest cusp; it is placed posteriorly and connected with the lingual cingulum. The premolars have a very thin and low cingulum all around their bases.
Left and right M^1: both teeth are very worn. Their parastyles and metastyles are similar-sized, robust, and straight dorso-ventrally.
Left and right M^2: both teeth are very worn. Their paracones and metacones are flattened by wear, but a clear notch can still be distinguished between them. The protocones and hypocones are totally worn. The teeth have a cingulum all round the crown, without any visible groove. The talons are well-developed, almost half as long as the molars, they are acute distally, straight and without torsions.
The rest of cranial material from Untermassfeld is more informative than IQW 1983/19544 (Mei 19064): upper dentitions IQW 1989/23351 (Mei. 22870), IQW 1989/23353 (Mei. 22872), IQW 2002/27455 (Mei. 26960), and IQW 1994/24599 (Mei. 24128).
P^4: mesio-distally straight, with triangular outline and without accessory cusps. The paracone is the highest cusp, sharp, with straight anterior border; the metacone is second in height. The paracone and metacone are aligned to form a labial blade. Although being the lowest cusp, the protocone (= deuterocone) is as sharp and well-defined as the other cusps.
M^1: mesio-distally straight, with nearly rectangular outline and parallel sides. The metastyle and parastyle are small but distinct and have straight dorso-ventral margins. The metastyle is larger than the parastyle. The mesocone is placed between the hypocone and the entocone.
M^2: paracone and metacone are often well-distinguished, divided by groove. Both cusps converge occlusally towards the sagittal line of tooth, and so do also the protocone and the hypocone. A cingulum runs all round and shows no grooves. The talon is well-developed, almost half as long as the tooth, it is acute distally and straight, without torsions.

3.2. Morphology of the mandibular remains and of the lower dentition

The mandibles are large with anterior premolars separated by long diastemas. The ascending rami of most specimens are not preserved; for this reason, the shapes of the mandibular condyles are unknown, as are also the mandibular heights. The anterior premolars (P_1-P_2-P_3) vary in number. There are cases where they are totally absent [IQW 1988/22380 (Mei. 21899), IQW 1983/18862 (Mei. 18382)], others in which there are small, widely spaced alveoli of P_1 and P_3 with no trace of P_2 in between [IQW 1983/18901 (Mei. 18421), IQW 1983/18902 (Mei. 18422)] and still others that exhibit only a tiny alveolus of P_1. Only one specimen with both hemi-mandibles [IQW1980/15522 (Mei. 15034) + IQW1980/15775 (Mei. 15287)] displays three small widely spaced alveoli on the right ramus, with P_1 still in place, and only the alveolus of P_1 on the left ramus

P_4: tooth with elliptic outline. The protoconid has accessory cuspids and a linguo-distal crest, there are no mesial cuspids. Three specimens, IQW 1980/15 775 (Mei. 15 287), IQW 1983/ 18 901 (Mei. 18 421) and IQW 2007/29 133 (Mei. 28 295), have a linguo-distal cuspid and a weak distal cuspid issuing from the linguo-distal crest.

M_1: the paraconid is simple and somewhat inclined distally. The protoconid is simple, and is the highest cuspid. The metaconid is variable: single-cusped in IQW 1980/15 522 (Mei. 15 034) and IQW 1988/22 380 (Mei. 21 899), unequally bicuspid, with the distal cuspid larger than the mesial cuspid, in IQW 1983/18 862 (Mei. 18 382) and IQW 2007/29 133 (Mei. 28 295), and tricuspid, with different possible combinations, in IQW 1980/15 775 (Mei. 15 287) and IQW 2007/29 124 (Mei. 28 286). Though damaged, the metaconid of IQW 1992/23 989 (Mei. 23 518) shows a main cuspid associated with at least another distinct smaller one. IQW 1995/25 029 (Mei. 24 558) has a main distal cuspid and three tiny cuspulids, aligned to form a mesial crest (**Fig. 8c–d**). The entoconid is less variable: it usually has one dominant cuspid (End1) attached to the mesial arm. One specimen has a main entoconid cuspid (End1) associated with a smaller mesial cuspid (End2) and a mesial arm connected with the base of the metaconid. The hypoconid is simple and always higher than the entoconid.

M_2: generally with rectangular outline and with a slight central constriction. The protoconid is large, and slopes gently anteriorly into a mesial crest. IQW 2007/29 133 (Mei. 28 295) has a central protoconid, with two smaller cuspids, one on each side. The metaconid is complex and most variable, with a main cuspid, a smaller mesial cuspid that slopes into a crest or into a tiny cuspid, and a third, very small cuspid, either positioned distally [IQW 1980/15 522 (Mei. 15 034)] or mesially [IQW1980/15 775 (Mei. 15 287)] to the main cuspid. The metaconid is always higher than the protoconid. The entoconid is formed by two identical cuspids, while the hypoconid is single-cusped. The enamel surface at the centre of tooth is fairly smooth with no ridges nor wrinkles.

M_3: the tooth generally has a rounded to pentagonal-shaped outline, and has a flat occlusal surface without prominent cuspids nor torsions of the crown. In three specimens [IQW 1980/15 522 (Mei 15 034), IQW 1995/25 029 (Mei. 24 558), IQW 1980/15 775 (Mei. 15 287)] out of six slight lingual notches separate the trigonid from the talonid.

4. Metrical and morphological comparative analyses

The Untermassfeld bears exhibit a combination of primitive and advanced characters. Primitive ursid representatives, such as the extinct *Ursus minimus* and *U. etruscus*, but also the extant Asian black bear *U. thibetanus*, retain all three upper and lower anterior premolars P1-P3. The same occurs in *U. dolinensis* from TD4-Atapuerca. There is a general tendency to lose some to all the anterior premolars in more advanced ursid lineages. Starting from *U. deningeri*, in which at least upper and lower P2s are missing, anterior premolars are progressively lost during the transition to late and most derived cave bear representatives (Mazza et al. 1995; García et al. 1997). Upper and lower P2s are also missing in *U. arctos*, *U. maritimus* and *U. malayanus* (Mazza et al. 1995). The Untermassfeld representatives show a range of possibilities, from having the whole set of anterior premolars, to having none, or even three on one side and one on the other, see hemimandibles IQW 1980/15 522 (Mei 15 034) and IQW 1980/15 775 (Mei. 15 287).

Morphometrically, the P^4s of *U. deningeri*, brown bears, *U. dolinensis* (TD4-Atapuerca) and of the Untermassfeld representatives are very similar to one another, but in the Untermassfeld bears and in *U. deningeri* values peak. Metastylids tend to be larger in *U. deningeri* than in brown bears (Wagner and Čermák 2012) and in the Untermassfeld specimens metastylids are barely developed. The P^4s of the Untermassfeld bears

	Untermass-feld	*U. dolinensis* (TD4-Ata-puerca)	*U. deningeri*	*U. spelaeus*	*U. arctos* (Iberian)	*U. arctos* (North America)	*U. ameri-canus*
Palatal thickness at M^2	6.91 ± 0.86 (4) (5.81–7.90)	6.66 ± 0.56 (3) (6.03–7.09)	6.43 ± 0.84 (6) (5.57–7.34)	12.11 ± 2.37 (55) (7.39–18.00)	2.37 ± 0.86 (22) (1.44–4.55)	4.08 ± 1.24 (97) (1.10–7.93)	2.72 ± 0.82 (132) (1.20–5.30)

Table 1 Comparison of the palatal thickness (mm) at level of M^2 in several bear species. – (n), (min–max).

and of early *U. arctos* representatives have no buccal constriction, which by contrast is often observed on the P^4s of *U. deningeri*. The paracone is higher than the protocone on the P^4s of the Untermassfeld bears, as well as on those of *U. dolinensis* and *U. arctos*, while on the P^4s of *U. deningeri* paracone and protocone are approximately of equivalent height. In occlusal view, the P^4s of *U. dolinensis* (TD4-Atapuerca) and of the Untermassfeld bears are triangular-shaped with fairly straight margins, while *U. deningeri* and *U. arctos* have more rounded fourth upper premolars.

Parastyles and metastyles on the Untermassfeld M^1s are proportionally developed like in *U. dolinensis* and *U. arctos* and not as strong as in *U. deningeri* (García García 2003, 328–332; Wagner and Čermák 2012). The thickness of the palate is a diagnostic discriminant between brown and cave bears (Santos et al. 2014). In *U. arctos* the palate is thin and flat or slightly convex dorsally (García et al. 2006; 2007), while in the Sima de los Huesos (Spain) and Petralona (Greece) *U. deningeri* palate thickness is intermediate between the averages of brown and cave bears (Santos et al. 2014). The palatal thicknesses of the four Untermassfeld skulls and of three TD4-Atapuerca skulls of *U. dolinensis* are similar to those of *U. deningeri* and differ from those typical of *U. spelaeus* and brown bears (**Fig. 6**; **Table 1**).

Overall, the horizontal rami of the Untermassfeld mandibles are as high, dorso-ventrally, as those of *U. etruscus* and *U. dolinensis*, and more slender than those of cave bear members. The coronoid process inclines backwards like in *U. dolinensis*, somewhat more than in *U. etruscus* and quite more than in *U. arctos* and cave bears.

Regarding the lower dentition, in addition to what has been said above about the anterior premolars, the original, plesiomorphic pattern of P_4s is exhibited by early representatives of *U. etruscus*, where a distal and mesial crest issue from the main cuspid (protoconid) (García García 2003, 343–345; Wagner and Čermák 2012). In advanced specimens of *U. etruscus*, the distal crest is usually forked with well-developed 'linguo-distal crest' (sensu Wagner and Čermák 2012). In *U. arctos* P_4 generally exhibits a main cuspid (protoconid), the 'linguo-distal crest' and often also a well-developed linguo-distal cuspulid that issues from the 'linguo-distal crest'. In this respect, the Untermassfeld P_4s resemble those of *U. arctos* and differ from those commonly possessed by *U. deningeri*, where the 'antero-lingual cuspid' is relatively weak, while the paraconid (antero-lingual) first and then the metaconid (lingual to protoconid) grow progressively larger.

There is relatively lower statistical difference (p-value = 0.05) between the maximum widths and lengths of the P_4s from Untermassfeld and of extant and past brown bears (p-value < 0.001), and those of the P_4s of *U. deningeri*. The Untermassfeld P_4s are therefore proportionally closer to the P_4s of brown bears and *U. dolinensis* (TD4-Atapuerca) than to those of cave bears (**Table 2**).

Passing to the molars, the M_1s of *U. arctos* are largely characterised by having the entoconid dominated by a large cuspid (End1), associated with a smaller one (End2) from which issues a short mesial arm (Wagner and Čermák 2012). In the entoconids of the M_1s of *U. deningeri*, the two cuspids are arranged to form a sloping crest (End1 and End2) and the mesial arm is more extended; often even more complex patterns can

	Ursus deningeri Sima de los Huesos			*Ursus arctos* Iberian (Holocene and recent)			*Ursus dolinensis* TD4 (holotype)	*Ursus* ex gr. *U. dolinensis* Untermassfeld		
	mean	min	max	mean	min	max		mean	min	max
Length	258.9 ± 18.3 (n = 11)	225.0	286.4	233.7 ± 21.1 (n = 29)	197.0	276.0	255.0	276.0 ± 48.1 (n = 2)	242.0	310.0
Length diastema	42.9 ± 8.1 (n = 12)	29.1	54.6	36.1 ± 5.9 (n = 29)	25.6	45.7	49.5	57.2 ± 5.6 (n = 4)	50.1	63.0
Mandibular height P_4	56.1 ± 4.3 (n = 12)	49.5	64.0	44.9 ± 5.5 (n = 29)	36.6	55.8	54.5	56.4 ± 5.7 (n = 6)	50.0	63.5
Mandibular height M_1	56.9 ± 5.3 (n = 12)	48.3	65.0	42.9 ± 5.9 (n = 29)	30.4	54.8	56.6	58.6 ± 6.8 (n = 4)	52.2	65.5
Mandibular height M_3	57.8 ± 6.3 (n = 12)	48.6	67.0	47.9 ± 6.8 (n = 29)	36.6	59.2	54.5	60.9 ± 3.9 (n = 4)	56.0	65.0
Mandibular breadth P_4-M_1	19.8 ± 1.7 (n = 12)	17.8	24.0	14.7 ± 2.8 (n = 29)	11.2	21.0	14.8	17.1 ± 2.9 (n = 3)	15.1	19.6
Mandibular breadth M_2-M_3	25.4 ± 2.1 (n = 12)	22.5	28.4	19.4 ± 3.1 (n = 29)	13.2	25.9	19.6	21.8 ± 2.2 (n = 4)	19.7	24.0
Length C-M_3	137.8 ± 16.9 (n = 12)	118.9	175.0	129.2 ± 18.9 (n = 27)	78.7	164.0	36.5	162.0 ± 4.0 (n = 3)	158.0	166.0
Length P_4-M_3	91.3 ± 4.2 (n = 12)	83.8	98.5	78.4 ± 7.8 (n = 29)	56.0	88.3	80.6	89.7 ± 6.2 (n = 4)	83.0	98.0
Length M_1-M_3	75.9 ± 4.1 (n = 12)	70.6	83.1	66.7 ± 6.5 (n = 29)	44.6	74.3	69.7	73.5 ± 2.7 (n = 3)	70.4	75.2
Condylar breadth	53.4 ± 6.7 (n = 12)	45.5	72.0	51.1 ± 8.7 (n = 27)	37.7	68.5	43.6			
Condylar heigth	23.6 ± 4.5 (n = 12)	19.0	36.3	17.4 ± 2.3 (n = 29)	13.0	23.3	19.1	18.4 (n = 1)		
Mandibular heigth	121.6 ± 7.7 (n = 10)	110.9	132.6	105.5 ± 18.0 (n = 27)	73.6	142.1	104.2	100.0 (n = 1)		

	Ursus deningeri Sima de los Huesos			*Ursus arctos* Iberian (Holocene and recent)			*Ursus dolinensis* TD4 (holotype)	*Ursus* ex gr. *U. dolinensis* Untermassfeld		
	mean	min	max	mean	min	max		mean	min	max
Condylar height (coronion)	85.1 ± 8.6 (n = 8)	77.8	101.1	84.5 ± 12.1 (n = 12)	65.7	102.4	76.0	58.0 (n = 1)		
Length canine	20.4 ± 1.4 (n = 7)	17.7	22.3	22.5 ± 2.8 (n = 12)	16.2	26.0	20.8	24.8 ± 1.5 (n = 6)	22.3	27.0
Breadth canine	16.2 ± 1.4 (n = 9)	13.8	18.3	14.5 ± 2.1 (n = 25)	11.3	18.0	15.5	17.5 ± 1.9 (n = 6)	14.4	20.2
Length P_4	13.9 ± 1.2 (n = 11)	12.0	15.8	12.7 ± 1.0 (n = 26)	10.8	15.3	12.6	14.1 ± 0.7 (n = 6)	13.0	14.8
Breadth P_4	8.4 ± 0.6 (n = 8)	7.5	9.6	7.5 ± 0.9 (n = 11)	5.9	8.4	8.1	8.1 ± 0.4 (n = 6)	7.7	8.6
Length M_1	26.3 ± 1.1 (n = 10)	25.0	28.0	24.1 ± 1.7 (n = 10)	20.6	26.1	22.2	27.2 ± 1.7 (n = 6)	25.2	29.9
Length trigonid M_1	15.6 ± 1.1 (n = 7)	13.7	16.8	14. 7 ± 1.4 (n = 10)	12.3	17.2	12.8	17.5 ± 0.9 (n = 6)	16.0	18.6
Breadth M_1	12.8 ± 0.7 (n = 8)	11.7	13.6	11.6 ± 1.0 (n = 10)	9.7	13.0	10.0	11.7 ± 0.2 (n = 6)	11.4	11.9
Length M_2	26.5 ± 1.6 (n = 10)	23.7	29.2	24.7 ± 2.1 (n = 11)	20.1	27.9	22.8	27.4 ± 1.5 (n = 6)	25.6	29.2
Length trigonid M_2	14.9 ± 1.7 (n = 9)	13.1	17.9	14.1 ± 2.1 (n = 11)	10.7	17.6	14.2	17.1 ± 1.0 (n = 6)	15.9	18.2
Breadth M_2	16.3 ± 1.1 (n = 10)	14.6	17.8	14.8 ± 0.9 (n = 11)	12.9	15.9	14.9	15.9 ± 0.5 (n = 5)	15.3	16.4
Length M_3	22.6 ± 2.1 (n = 10)	19.5	26.2	19.2 ± 2.3 (n = 10)	16.0	23.5	22.7	22.6 ± 1.8 (n = 5)	20.5	25.0
Breadth M_3	17.7 ± 0.9 (n = 9)	16.4	18.9	14.9 ± 1.4 (n = 10)	12.0	16.8	14.0	17.1 ± 1.1 (n = 5)	16.0	19.0

Table 2 Metrical comparisons of mandible and lower dentition variables of *Ursus deningeri* from Sima de los Huesos, Iberian Holocene and recent brown bears, *U. dolinensis* (holotype from TD4-Atapuerca) and Untermassfeld specimens.

be observed. In *U. dolinensis* from TD4-Atapuerca, the entoconid is similar to those of the Untermassfeld specimens, with one dominant cuspid (End1) attached to a mesial arm. In the M_1s of *U. deningeri* the mesial metastylid generally bends lingually from the metaconid, while in *U. arctos*, the mesial metastylid is usually stretched mesially or even bent slightly labially from the metaconid. The metaconid complex in *U. dolinensis* from TD4 consists of a double cusp or a main cuspid and two small mesial ones.

Dimensionally, the Untermassfeld M_1s differ from those of modern and past brown bears, as well as from those of *U. dolinensis* from TD4-Atapuerca (**Table 2**). *U. deningeri* has larger M_1s than Iberian brown bears (statistical significance p-value = 0.05). This is confirmed using a larger dataset of isolated M_1s (> 30) of *U. deningeri* (García García 2003, 358) where mean values are even larger, ranging around 27 mm, which is in line with the observations of Wagner and Čermák (2012). The Untermassfeld M_1s are approximately as long as those of *U. deningeri* and are therefore longer than the M_1s of all the other species considered in this study.

In brown bears, M_2 metaconids are generally (1) double-cusped, with two cuspids of more or less similar size, or (2) exhibit a smaller, additional mesial cusp (García García 2003, 368). However, there are specimens with more complex patterns, similar to those shown by *U. deningeri* (Wagner and Čermák 2012, Fig. 5). The double metaconid can be observed in primitive bears, like *U. etruscus* and *U. minimus* (García García 2003, 367). Specimens from Le Vallonnet and *U. dolinensis* from TD4-Atapuerca, which date to around 1 Ma, have M_2s similar, morphologically, to those of the Untermassfeld bears, with single-cusped metaconids, but associated with smaller mesial and distal cuspids. Two specimens from Untermassfeld show the typical brown bear pattern, with double-cusped metaconid associated with a small accessory cuspulid.

The M_2s of *U. arctos* and *U. deningeri* are characterised by having a central constriction: in *U. deningeri* the constriction is very marked (and becomes even more so in derived cave bears, García García 2003, 367), while it is much weaker in *U. arctos*, whose M_2s appear more primitively rectangular-shaped occlusally. The Untermassfeld M_2s are only slightly constricted, in spite of being closer, dimensionally, to those of *U. deningeri* than to those of *U. arctos*, which in fact are slightly smaller and relatively broader (**Table 2**).

Rather than being a vestigial tooth, as suggested by Couturier (1954), M_3 can be particularly large, depending on the evolutive lineage that is considered. In black bears, it is simply a small, oval button-like tooth (Mazza and Rustioni 1994), while in *U. etruscus* it is slightly larger, and either shaped like that of black bears or elongated by a narrow talon, but it is never pentagonal-shaped. The M_3s of the Early Pleistocene bears from Le Vallonnet and TD4-Atapuerca are pentagonally-shaped, with simple outline and without torsions, as are the ones from Untermassfeld. The M_3s of *U. arctos* are usually squarish, often rounded more than pentagonal, and their outlines show no or only a slightly visible notch between trigonid and talonid. Patterns can be uniform in local populations; for example, the M_3s (n = 25) of the brown bears of the late Pleistocene site of Taubach (Germany) are all pentagonal-shaped. M_3s of *U. deningeri* vary from the simple pentagonal-shaped ones of early representatives to more complex patterns of more advanced Deninger bears, which may display a lingual notch combined with torsion. In the early Middle Pleistocene site of Voigtstedt (Germany), one specimen, IQW 1966/7 751 (Voi. 3 197), has pentagonal shape with straight borders, like in brown bears, and another specimen, IQW 1966/5 091 (Voi. 3 360), has pentagonal contour with slight lingual notch, which is the most common pattern observed in the representatives of *U. deningeri* from Sima de los Huesos.

The M_3s of brown and Deninger bears illustrated by Wagner and Čermák (2012, Fig. 6) show similar outlines, but *U. deningeri* tends to have more complex occlusal surfaces than *U. arctos.* The M_3s of *U. deningeri* are larger, on average, than those of *U. arctos* and, more specifically, *U. deningeri* has relatively broader third lower molars than *U. arctos*. Dimensionally, the Untermassfeld M_3s overlap the *U. deningeri* ones from Sima de los Huesos.

5. Discussion and conclusions

The bears from Untermassfeld have a number of common traits shared with brown bears, but some are different from *Ursus arctos*, particularly in the morphology of the metaconid and entoconid complex of M_1. There are both dimensional and morphological grounds for referring the bears from Untermassfeld to *U.* ex gr. *U. dolinensis* and to include it into the spelaeoid lineage. The reasoning behind these conclusions is multiple. The bears from Untermassfeld share such a high number of features with brown bears that Musil (2001), Palombo et al. (2008) or Rabeder et al. (2010) considered them early representatives of the arctoid lineage. However, those traits are primitive. At the same time, the Untermassfeld bears share with *U. deningeri* a number of advanced characters typical of the spelaeoid lineage. These include:

(1) Dimensions almost identical with those of early representatives of *U. deningeri* of early Middle Pleistocene age, P_4s follow a model of spelaeoid diversification;

(2) the sizes of the Untermassfeld cheek teeth overlap those of early Middle Pleistocene *U. deningeri* specimens;

(3) the general morphology and that of End1-c of M_1 are similar to those of early Middle Pleistocene *U. deningeri*;

(4) M_2s are intermediate in trigonid width/maximal length between the early and middle Middle Pleistocene *U. deningeri* and resemble an early spelaeoid form (García García 2003; Baryshnikov 2006; Argant 2009; Wagner and Čermák 2012).

Having said this, the Untermassfeld bears appear to represent a situation similar to that for which the Pan-African Evolution theory was proposed to explain the origin of *Homo sapiens* (Scerri et al. 2019; Bergström et al. 2021). The evolution of brown (and through them of white bears) and cave bears could be the result of dynamic changes in gene flow between early bears scattered across Eurasia; trying to find a single original location for brown and cave bears is perhaps hopeless. Molecular analyses reveal continuous gene flow between different populations of Late Pleistocene brown bears and other closely related species. Brown bears are shown to have dispersed after the last glaciation from interbreeding southern European populations and not from genetic types isolated in separate glacial refugia (Valdiosera et al. 2007). Mitochondrial analysis found that a large number of pre- and post-Last Glacial Maximum brown bears from Ireland grouped more closely with modern-day Clade 2 polar bears than with western and central European modern and ancient Clade 1 brown bears (Edwards et al. 2011).

During the late Early Pleistocene or so, bears probably formed the original pool of interbreeding populations, technically metapopulations, spread across Eurasia. Bears would have been moving around and mixing over long periods and wide geographic areas. Brown and spelaeoid bears finally began to appear sometime around 1.0–0.9 Ma. Did they both emerge in Europe, or one in Europe and the other in Asia? This is a reasonable question, but this geographic area cannot objectively be identified: when we consider the combination of characters exhibited by the fossils, and the Untermassfeld bear is a model of this, the single region of origin for either of them probably cannot be found. In place of a series of population splits branching off an ancestral tree, dynamic interbreeding between different populations over time seems a more reasonable possibility and may explain several diversity patterns not explained by current alternative models. It probably was a sort of melting pot where various bear populations alternatively separated and progressed genetically in reciprocal isolation or joined interbreeding under the drive of unsteady environmental/climatic circumstances.

Acknowledgements

This paper was made possible only with the invaluable assistance of many people. We would like to thank all those who participated in the excavations at Untermassfeld over the years, as well as to the numerous preparators for the tireless efforts cleaning and restoring the specimens. Our gratitude cannot be expressed enough to S. Döring and T. Korn for their highly professional photography and to E. Haase for the high-quality graphics work (all Weimar). For data compiling thanks go especially to G. Utschig (Weimar). We should further like to present our special thanks to H. Hemmer (Mainz) who reviewed the manuscript and provided valuable comments on the topic. N. G. is supported by a project (PGC2018-093925-B-C33) from the Ministerio de Economía y Competitividad (Spain) for research in bear evolution.

References

Argant, A., 2009. Biochronologie et grands mammifères au Pléistocène moyen et supérieur en Europe occidentale: l'apport des canidés, des ursidés et des carnivores en général. Quaternaire 20, 467–480.

Baryshnikov G., 2006. Morphometrical Variability of Cheek Teeth in Cave Bears. Scientific Annals of the School of Geology of Aristotle University of Thessaloniki, Special volume 98, 81–102.

Baryshnikov, G., Boeskorov, G., 2004. Skull of the Pleistocene brown Bear (*Ursus arctos*) from Yakutia, Russia. Russian Journal of Theriology 3, 71–75.

Bergström, A., Stringer, C., Hajdinjak, M., Scerri, E.M.L., Skoglund, P., 2021. Origins of modern human ancestry. Nature 590, 229–237.

Bischoff, J.L., Shamp, D.D., Aramburg, A., Arsuaga, J.L., Carbonell, E., Bermudez de Castro, J.M., 2003. The Sima de los Huesos Hominids Date to Beyond U/Th Equilibrium (>350 kyr) and Perhaps to 400–500 kyr: New Radiometric Dates. Journal of Archaeological Science 30, 275–280.

Couturier, M.A.J., 1954. L'ours brun: *Ursus arctos* L. Grenoble.

Davison, J., Ho, S.Y.W., Bray, S.C., Korsten, M., Tammeleht, E., Hindrikson, M., Austin, J., 2011. Late-Quaternary biogeographic scenarios for the brown bear (*Ursus arctos*), a wild mammal model species. Quaternary Science Reviews 30, 418–430.

Edwards, C.J., Suchard, M.A., Lemey, P., Welch, J.J., Barnes, I., Fulton, T.L., Barnett, R., O'Connell, T.C, Coxon, P., Monaghan, N., Valdiosera, C.E., Lorenzen, E.D., Willerslev, E., Baryshnikov, G.F., Rambaut, A., Thomas, M.G., Bradley, D.G., Shapiro, B., 2011. Ancient Hybridization and an Irish Origin for the Modern Polar Bear Matriline. Current Biology 21, 1251–1258.

Erdbrink, D.P., 1953. A Review of Fossil and Recent Bears of the Old World. 2 volumes and index volume. PhD thesis, Universiteit Utrecht, The Netherlands.

Freudenberg, W., 1914. Die Säugetiere des älteren Quartärs von Mitteleuropa, mit besonderer Berücksichtigung der Fauna von Hundsheim und Deutschaltenburg in Niederösterreich, nebst Bemerkungen über verwandte Formen anderer Fundorte. Geologische und palaeontologische Abhandlungen, Neue Folge 12, 1–129.

García, N., Arsuaga, J.L., 2001. *Ursus dolinensis*: a new species of Early Pleistocene ursid from Trinchera Dolina, Atapuerca (Spain) / *Ursus dolinensis*: une nouvelle espèce d'ursidé du Pléistocène inférieur de Trinchera Dolina, Atapuerca (Espagne). Comptes Rendus de l'Académie des Sciences, Series IIA, Earth and Planetary Science 332, 717–725.

García, N., Arsuaga. J.L., Torres, T., 1997. The carnivore remains from the Sima de los Huesos Middle Pleistocene site (Sierra de Atapuerca, Spain). Journal of Human Evolution 33, 155–174.

García, N., Santos, E., Arsuaga, J.L., Carretero, J.M., 2006. High-resolution X-ray Computed Tomography Applied to the Study of Some Endocranial Traits in Cave and Brown Bears. Scientific Annals of the School of Geology of Aristotle University of Thessaloniki, Special volume 98, 141–146.

2007. Endocranial morphology of the *Ursus deningeri* Von Reichenau 1904 from the Sima de los Huesos (Sierra de Atapuerca) Middle Pleistocene site. Journal of Vertebrate Paleontology 27, 1007–1017.

García García, N., 2003. Osos y otros carnívoros de la Sierra de Atapuerca. Oviedo.

Kahlke, R.-D. (Ed.), 1997. Das Pleistozän von Untermaßfeld bei Meiningen (Thüringen). Teil 1. Monographien des Römisch-Germanischen Zentralmuseums Mainz 40, 1, I–VI, 1–418, pls. 1–67.

(Ed.), 2001a. Das Pleistozän von Untermaßfeld bei Meiningen (Thüringen). Teil 2. Monographien des Römisch-Germanischen Zentralmuseums Mainz 40, 2, I–VIII, 419–698, pls. 68–131, suppl. I.

(Ed.), 2001b. Das Pleistozän von Untermaßfeld bei Meiningen (Thüringen). Teil 3. Monographien des Römisch-Germanischen Zentralmuseums Mainz 40, 3, I–VI, 699–1030, pls. 132–151, suppl. I–XV.

2001c. Die unterpleistozäne Komplexfundstelle Untermaßfeld – Zusammenfassung des Kenntnisstandes sowie synthetische Betrachtungen zu Genesemodell, Paläoökologie und Stratigraphie. In: Kahlke, R.-D. (Ed.), Das Pleistozän von Untermaßfeld bei Meiningen (Thüringen). Teil 3. Monographien des Römisch-Germanischen Zentralmuseums 40, 3, 931–1030, suppl. I–XV.

2006. Untermassfeld – A late Early Pleistocene (Epivillafranchian) fossil site near Meiningen (Thuringia, Germany) and its position in the development of the European mammal fauna. British Archaeological Reports, International Series 1578. Oxford.

Kurtén, B., 1968. Pleistocene Mammals of Europe. London.

Madurell-Malapeira, J., Aurell, J., Alba, D.M., Moyà-Solà, S., 2009. Los Niveles Fosilíferos Epivillafranchienses de Vallparadís (Terrassa, Barcelona, España). Paleolusitana 1, 237–243.

Maul, L.C., Heinrich, W.-D., Parfitt, S.A., Paunescu, A.-C., 2007. Comment on the correlation between magnetostratigraphy and the evolution of *Microtus* (Arvicolidae, Rodentia, Mammalia) during the Early and early Middle Pleistocene. In: Kahlke, R.-D., Maul, L.C., Mazza, P. (Eds.), Late Neogene and Quaternary biodiversity and evolution: Regional developments and interregional correlations. Proceedings of the 18th International Senckenberg Conference (VI International Palaeontological Colloquium in Weimar). Volume II. Courier Forschungsinstitut Senckenberg 259, 243–263.

Mazza, P., Rustioni, M., 1994. On the phylogeny of Eurasian bears. Palaeontographica Abteilung A, Palaeozoologie, Stratigraphie 230, 1–38.

Mazza, P., Rustioni, M., Boscagli, G., 1995. Evolution of ursid dentition, with inferences on the functional morphology of the masticatory apparatus in the genus *Ursus*. In: Moggi-Cecchi, J. (Ed.), Aspects of Dental Biology: Paleontology, Anthropology and Evolution. Leonardo series 7. Florence, 147–157.

Moullé, P.E., 1992. Les grands mammifères du Pléistocène inferieur de la grotte du Vallonnet (Roquebrune-Cap-Martin, Alpes-Maritimes). Étude paléontologique des Carnivores, Equidé, Suidé et Bovidés. PhD thesis, Muséum national d'Histoire naturelle, France.

Musil, R., 2001. Die Ursiden-Reste aus dem Unterpleistozän von Untermaßfeld. In: Kahlke, R.-D. (Ed.), Das Pleistozän von Untermaßfeld bei Meiningen (Thüringen), Teil 2. Monographien des Römisch-Germanischen Zentralmuseums Mainz 40, 2, 633–658, pls. 114–120.

Palombo, M.R., Sardella, R., Novelli, M., 2008. Carnivora dispersal in Western Mediterranean during the last 2.6 Ma. Quaternary International 179, 176–189.

Rabeder, G., Pacher, M., Withalm, G., 2010. Die altpleistozänen Bären von Deutsch-Altenburg (Niederösterreich) / Early Pleistocene Bear Remains from Deutsch-Altenburg (Lower Austria). Mitteilungen der Kommission für Quartärforschung der Österreichischen Akademie der Wissenschaften 17. Wien.

Santos, E., Garcia, N., Carretero, J.M., Arsuaga, J.L., Tsoukala, E., 2014. Endocranial traits of the Sima de los Huesos (Atapuerca, Spain) and Petralona (Chalkidiki, Greece) Middle Pleistocene ursids. Phylogenetic and biochronological implications / Traits endocrâniens des ursidés du Pléistocène Moyen de la Sima de los Huesos (Atapuerca, Espagne) et Petralona (Chalkidiki, Grèce). Phylogénétique et implications biochronologiques. Annales de Paléontologie 100, 297–309.

Scerri, E.M.L., Chikhi, L., Thomas, M.G., 2019. Beyond multiregional and simple out-of-Africa models of human evolution. Nature Ecology and Evolution 3, 1370–1372.

Valdiosera, C.E., García, N., Anderung, C., Dalén, L., Crégut-Bonnoure, E., Kahlke, R.-D., Stiller, M., Brandström, M., Thomas, M.G., Arsuaga, J.L., Götherström, A., Barnes, I., 2007. Staying out in the cold: Glacial refugia and mitochondrial DNA phylogeography in ancient European brown bears. Molecular Ecology 16, 5140–5148.

Wagner, J., Čermák, S., 2012. Revision of the early Middle Pleistocene bears (Ursidae, Mammalia) of Central Europe, with special respect to possible co-occurrence of spelaeoid and arctoid lineages. Bulletin of Geosciences 87, 461–496.

ALESSIO IANNUCCI · BENIAMINO MECOZZI
RALF-DIETRICH KAHLKE · RAFFAELE SARDELLA

NEW RESULTS ON HYAENIDS FROM THE EARLY PLEISTOCENE SITE OF UNTERMASSFELD

Abstract

Here we describe the remains of the giant hyena *Pachycrocuta brevirostris* (Gervais, 1850) recovered from the site of Untermassfeld (Thuringia, Germany), representing the most abundant European sample of this widespread and iconic carnivoran. The morphological description provides insights on the variability of the species in comparison to other extinct and extant bone-cracking hyenas. The biometric analysis reveals that a weak but significant increase in size occurred during the existence of the species in Europe. Being the largest in terms of body size among the compared samples, the population from the Epivillafranchian of Untermassfeld epitomises this pattern.

1. Introduction

Since its dispersal into Europe ~2.0–1.8 Ma BP the giant hyena *Pachycrocuta brevirostris* (Gervais, 1850) became a widely distributed species and a major taphonomic agent in the region (Turner and Antón 1996; Napoleone et al. 2001; 2003; Mazza et al. 2004; Sardella and Palombo 2007; Iannucci et al. 2021). The putative competition between *P. brevirostris* and Palaeolithic hominin populations for accessing and scavenging carcasses is an intriguing topic that captured the interest of several authors and fuelled the scientific debate (e. g., Turner 1992; Boaz et al. 2000; Lewis and Werdelin 2010; Martínez-Navarro 2010; Espigares et al. 2013; Madurell-Malapeira et al. 2017). The site of Untermassfeld (Thuringia, Germany) offers a unique window on European Epivillafranchian faunas (Kahlke 2006; 2007; 2009; Bellucci et al. 2015), and on the morphological variability, palaeobiology, and palaeoecology of *P. brevirostris*, as it yielded the most abundant European collection of the species.

Untermassfeld hyenas were first mentioned by H.-D. Kahlke (1980; 1982). Following these earlier publications, a detailed description and evaluation of the then 78 finds, consisting mainly of cranial and dental remains, was made by Turner (2001), as well as findings of around 80 coprolites by Keiler (2001). Further specimens were figured by Kahlke (2001; 2006), Kahlke and Gaudzinski (2005) as well as Keiler (2008) and Roebroeks et al. (2018). Since then, a total of 230 dental and skeletal elements of *P. brevirostris* and 435 corresponding coprolites have been recovered from Untermassfeld. Some of these were found in densely packed concentrations or in flat masses of compacted faeces (Kahlke in this volume, Figs. 24–25, **Foldout III**). A quantitative study of the hyaenid bone modifications on *Bison menneri* skeletal elements, recovered in large numbers from the site, has also been conducted (Arnold in this volume).

Remains of *Pachycrocuta* occur in all fossil-bearing sediments at Untermassfeld, i. e., in the Lower Fluviatile Sands (LFS), as well as in channel fill deposits with reworked sediments and the lower part of the Upper Fluviatile Sands [(UFS(l)] (Kahlke et al. 2020; in this volume). Sedimentological and fine-scale stratigraphical investigations show that the age difference of the taxonomically identical fossil assemblages recovered from both sediment units is not of biostratigraphic relevance (Kahlke et al. 2020, 1103). Thus, in the biostratigraphic sense, the *Pachycrocuta* finds of Untermassfeld originate from a single, synchronous palaeopopulation.

The minimum number of *Pachycrocuta* individuals (MNI) found at Untermassfeld is 24, including 18 juveniles and 6 adults (Kahlke 2006 and finds newly identified by G. Utschig). This means that 5.4 % of all large mammal individuals recorded at Untermassfeld belong to this hyena species. The unusually high proportion of juveniles at 75 % is indicative of a particular vulnerability of individuals of this age group to flood events (Kahlke 2006, 78 f.). It seems likely, therefore, that the hyena dens were at least partially located in the predominantly sandy-clayey areas of the Werra river floodplain. Turner (2001, 686 f.) identified five age stages according to the dental development of these juvenile individuals, all of which represent life spans of less than one year. Because the flooding events that led to the formation of the site occurred during the winter months (Kahlke 2001, 976 f.; 2006, 82; in this volume, 1714), a clear seasonal pattern in the breeding behaviour of the Untermassfeld hyenas has not been established.

In the following pages, we describe the most complete and significant new discoveries of *Pachycrocuta* dentition and skeletal elements and some already published finds, updating or expanding the discussion on diagnostic morphological features. A comparison with other European remains of *P. brevirostris* is carried out to evaluate the biometric variability of the species during the Pleistocene.

2. Notes on nomenclature and systematics

The name *Hyaena brevirostris* was first used by Auguste Aymard for referring to remains of a large hyena he discovered near Le Puy-en-Velay (France) in the 1840s, since then deposited at the Musèe Crozatier. However, as recently pointed out by Alba et al. (2015) this was not accompanied by a published description to serve as a valid designation. Indeed, the species was formally, although unintentionally, established by Paul Gervais in the section dedicated to carnivorans (issued in 1850) of his Zoologie et Paléontologie françaises (1848–1852). This brief description was only later expanded by Boule (1893). In the meantime, other fossils were recovered and some ascribed to different species, such as *Hyaena robusta* Weithofer, 1889. The name *Pachycrocuta* was coined by Kretzoi (1938) but it was not widely used until the 1970s–1980s (Ficcarelli and Torre 1970; Howell and Petter 1980). In the following years, with the publication of the monographic work by Werdelin and Solounias (1991) on the Hyaenidae, and the studies focused on *P. brevirostris* by Turner and Antón (1996) and Turner (2001), our knowledge on the evolution and systematics of the species has been much improved.

Currently, all European remains of the giant hyena are usually included in *P. brevirostris*, whereas different approaches exist in referring African and Asian occurrences to distinct species or subspecies, or within *P. brevirostris* itself (e. g., Qiu 1987; Turner 1990; Werdelin 1999; Iannucci et al. 2021; Liu et al. 2021).

3. Material and methods

The Hyaenidae sample of Untermassfeld is part of the Untermassfeld collection housed in the Senckenberg Research Station of Quaternary Palaeontology Weimar (IQW) in Germany. Measurements and morphological characters are taken from the published literature, cited throughout the text and in tables when appropriate, and/or based on the material studied by the authors in several institutions, i.e., CMNH: Civic Museum of Natural History, Trieste (Italy); HNHM: Hungarian Natural History Museum, Budapest (Hungary); IGF: Natural History Museum of Florence, section of Geology and Paleontology (Italy); MACUS: Comparative Anatomy Museum »Battista Grassi«, Sapienza University of Rome (Italy); MSSGG: Museo Scientifico Speleologico della Grotta Gigante, Trieste (Italy); NHMUK: Natural History Museum, London (UK); NMM Natural History Museum, Mainz (Germany); NWHCM: Norwich Castle Museum (UK); PF: PaleoFactory Laboratory, Sapienza University of Rome (Italy); RMCA: Royal Museum for Central Africa, Tervuren (Belgium).

In the description and morphological comparison, we considered *P. brevirostris*, *Pliocrocuta perrieri* (Croizet et Jobert, 1828), *Crocuta crocuta* (Erxleben, 1777), *Hyaena hyaena* (Linnaeus, 1758), and *Parahyaena brunnea* (Thunberg, 1820). For the ease of reference to genera with the same initials *Pachycrocuta* is abbreviated to *P.*, *Pliocrocuta* to *Pl.*, and *Parahyaena* to *Pa.*

In the biometric comparison we focused on the differences in P^4 and M^1 lengths between different samples of European *P. brevirostris*, but we provided length (L) and width (W) values of other teeth from Untermassfeld, as well as craniodental and postcranial measurements. These were taken with a digital calliper to the nearest 0.1 mm, mainly following Werdelin and Solounias (1991) and Driesch (1976). Details and abbreviations of the measurements are as follows:

P^4 Lps	length of parastyle of P^4,
P^4 Lpc	length of paracone of P^4,
P^4 Lm	length of metastyle of P^4,
P^4 Wa	width of P^4 at protocone (anterior width),
P^4 Wbl	width of P^4 between paracone and metastyle (blade width),
WP	width of palate between P^4 and P^3,
P^2-P^4	inclusive length of P^2-P^4,
Di	length of diastema C-P_2,
M_1 Lpa	length of paraconid of M_1,
M_1 Lpr	length of protoconid of M_1,
HbM_1	dorsoventral depth of mandible behind M_1,
P_2-M_1	inclusive length of P_2-M_1,
C_1-Cd	length of mandible from anterior end of canine to posterior extremity of condyle,
GL	greatest length,
Dp	greatest depth of the proximal end,
Bp	greatest breadth of the proximal end
DD	smallest depth of the diaphysis,
SD	smallest breadth of the diaphysis,
Dd	greatest depth of the distal end,
Bd	greatest breadth of the distal end,
DPC	ulna, greatest depth across the coronoid process,
HTr	femur, height from femur head to third trochanter.

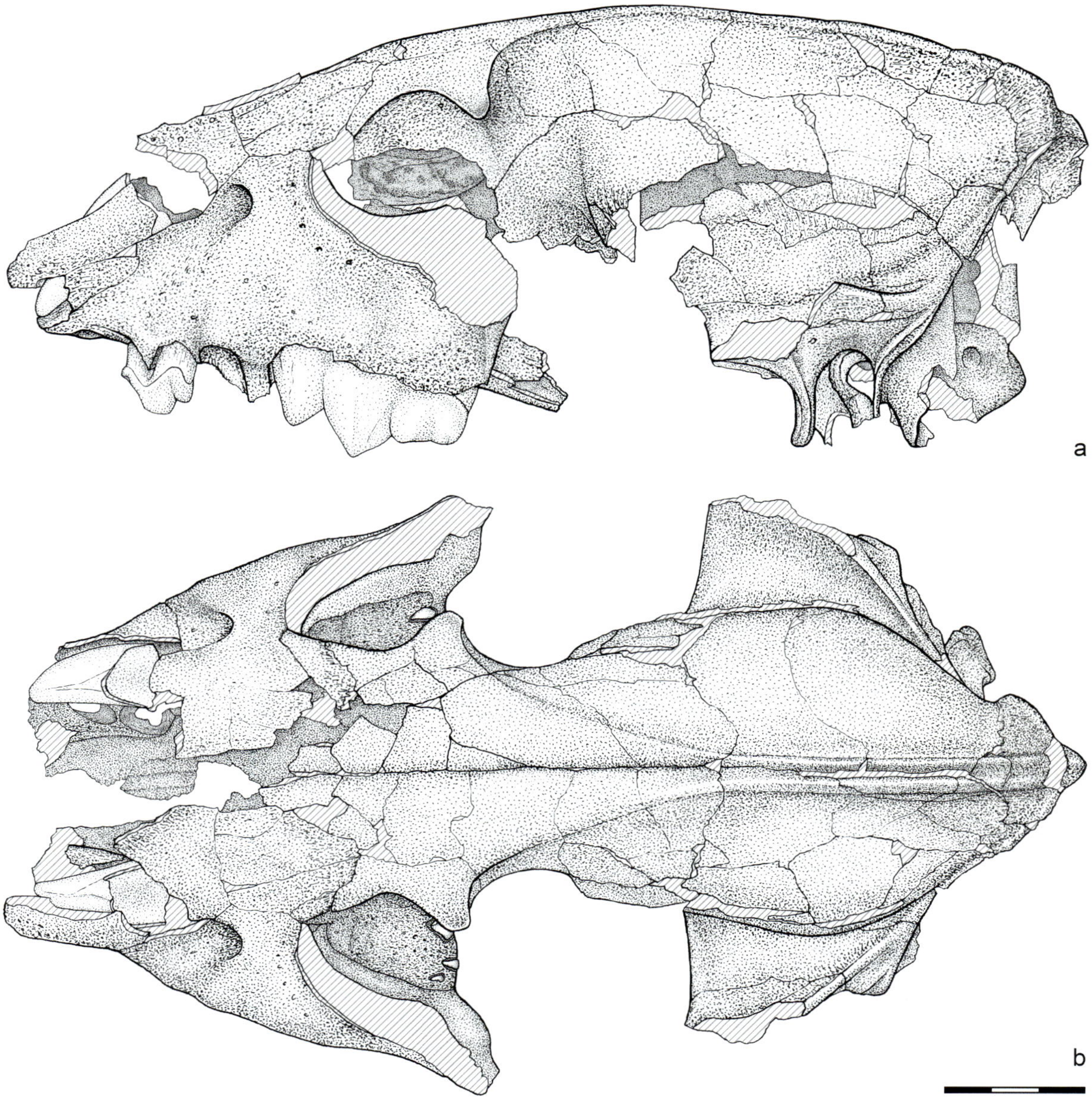

Fig. 1 *Pachycrocuta brevirostris* (Gervais), Untermassfeld. – **a–b** Juvenile cranium IQW 1996/25 647 (Mei. 25 176) with left erupting C^1, DP^2, P^4-M^1 and right erupting C^1, P^4-M^1, left lateral and dorsal views. – Scale: 30 mm.

4. Description

4.1. Crania

IQW 1996/25 647 (Mei. 25 176) is a juvenile partial cranium with erupting left C^1, DP^2, P^4-M^1 and erupting right C^1, P^4-M^1, finely restored from several detached fragments (**Fig. 1**; see also Roebroeks et al. 2018, fig. 14). It is the most complete cranial specimen from the site (**Fig. 2a–f**). The dorsal cranial profile is gently curved, not yet showing the concavity of the nasal region that characterises several adult specimens of *P. brevirostris* (Boule 1893; Ficcarelli and Torre 1970). The infraorbital foramen is placed above where P^3 would be erupted, and the anterior margin of the orbit ends roughly above the mesial margin of P^4. Despite

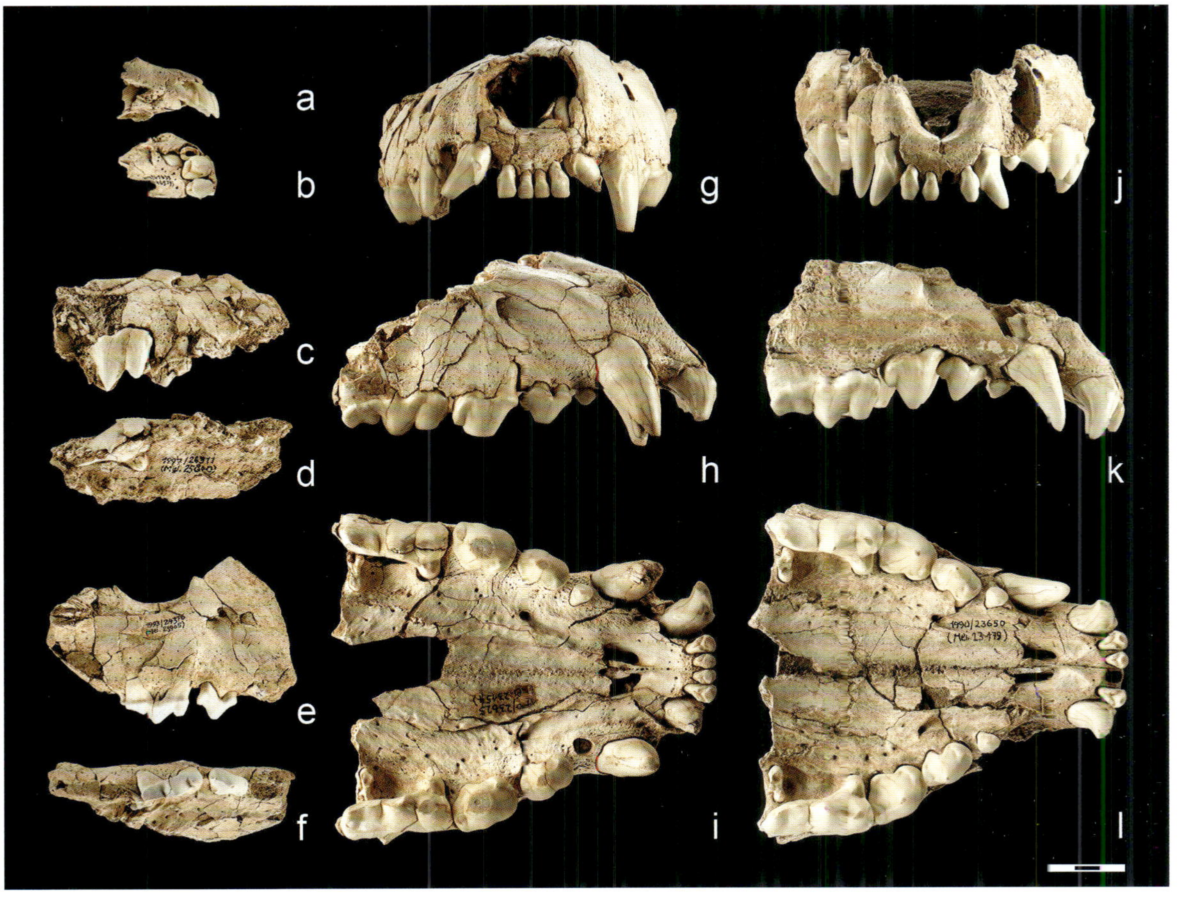

Fig. 2 *Pachycrocuta brevirostris* (Gervais), Untermassfeld. – **a–b** Right praemaxilla fragment IQW 2013/41639 (Mei. 46979) with I¹-I², root of DI³ and erupting I³. – **c–d** Right maxilla fragment IQW 1997/26311 (Mei. 25840) with germ of C¹, erupting P¹, DP²-DP³ fragments, P⁴ fragment. – **e–f** Right maxilla fragment IQW 1993/24376 (Mei. 23905) with DP²-DP³. – **g–i** Partial splanchnocranium IQW 1990/23625 (Mei. 23154) with left I¹-C¹, P²-M¹ and right I¹-M¹. – **j–l** Partial splanchnocranium IQW 1990/23650 (Mei. 23179) with left I²-I³, P¹-M¹ and right I¹-M¹ dex. – **a**, **c**, **e**, **h**, **k** right lateral views; **b**, **d**, **f**, **i**, **l** occlusal views; **g**, **j** frontal views. – Scale: 30 mm.

its fragmentary status and young age at death, some of the typical robustness characteristics of the species can be observed, including the wide area of attachment on the sagittal crest for the strong temporalis muscles, the marked curvature of the tooth rows, and the short and wide muzzle. The greatest palatal width is at the level of the palatine, but the relative widening of the anterior portion is also remarkable, and more similar to that of *C. crocuta* than to those of *H. hyaena* and *Pa. brunnea* (Werdelin and Solounias 1991). This trait is clearly observable on the adult specimens IQW 1990/23625 (Mei. 23154) and IQW 1990/23650 (Mei. 23179), which preserve most the ventral portion of the muzzle and associated dentition (**Fig. 2g–l**). In these specimens, the major palatine foramen opens at the level of P², but it is a bit variable in its exact position, even between the right and left sides of the same individual (**Fig. 2l**). The incisive fossae are wide and rounded, as those of extant *H. hyaena* and *Pa. brunnea* and unlike that of *C. crocuta*, although the fossa is narrower in IQW 1990/23650 (Mei. 23179) than in IQW 1990/23625 (Mei. 23154). IQW 1990/23625 (Mei. 23154) has a wide nasal cavity, which seems larger and more rounded than the preserved portion in IQW 1990/23650 (Mei. 23179) (**Fig. 2g, j**). These differences are comparable with the variability documented within samples of *Pl. perrieri* (Viret 1954; Kurtén and Crusafont Pairó 1977; Vinuesa et al. 2014) and *H. hyaena* (pers. obs.).

4.2. Mandibles

IQW 1993/24 369 (Mei. 23 898) and IQW 1994/24 584 (Mei. 24 113) are the left and right side, respectively, of a single almost complete mandible of an adult individual, only lacking the left I_1 (**Fig. 3**; Kahlke in this volume, Fig. 8). This is the most complete mandible, but there are several other remains, from juveniles (**Fig. 4**) to adults (**Fig. 5**). In *P. brevirostris* from Untermassfeld, the masseteric fossa is deep, almost reaching the talonid of M_1. In *H. hyaena* and *Pa. brunnea*, the fossa is usually comparably deep or less pronounced, whereas in *C. crocuta* is more variable and can be even stretched up to half the length of M_1 (pers. obs.). The height of the mandible is greater posteriorly than anteriorly but with few differences, resulting in a roughly straight mandibular corpus, as in *H. hyaena* and *Pa. brunnea* and unlike the curved and shallow corpus of *C. crocuta* (Palmqvist et al. 2011). The symphysis is massive, so dorso-ventrally deep that its ventral portion is observable as a bulge in buccal view. It is also antero-posteriorly broad, more than in extant bone-cracking hyenas.

The morphology of the symphysis is variable, from a sub-oval shape as in IQW 1999/26 624 (Mei. 26 153) (**Fig. 5c**) and IQW 1986/21 369 (Mei. 20 888) (Turner 2001, fig. 2, pl. 125), to forms relatively more antero-posteriorly elongated, with a posterior notch placed roughly at half the height of the mandible, e.g., IQW 1990/23 457 (Mei. 22 976) (**Fig. 5i**). The tooth rows have a pronounced lingual concavity along the antero-posterior axis of the mandible, with M_1 not aligned with the tooth row but slightly shifted lingually. These are adaptations to bone-cracking shared with *C. crocuta*, *H. hyaena*, and *Pa. brunnea* (Kurtén and Werdelin 1988; Werdelin and Solounias 1991). It is worth noting that the same also occurs in juvenile specimens, with DP_4 similarly displaced (**Fig. 4**). The deciduous cheek tooth row may be extremely crowded, to the extent

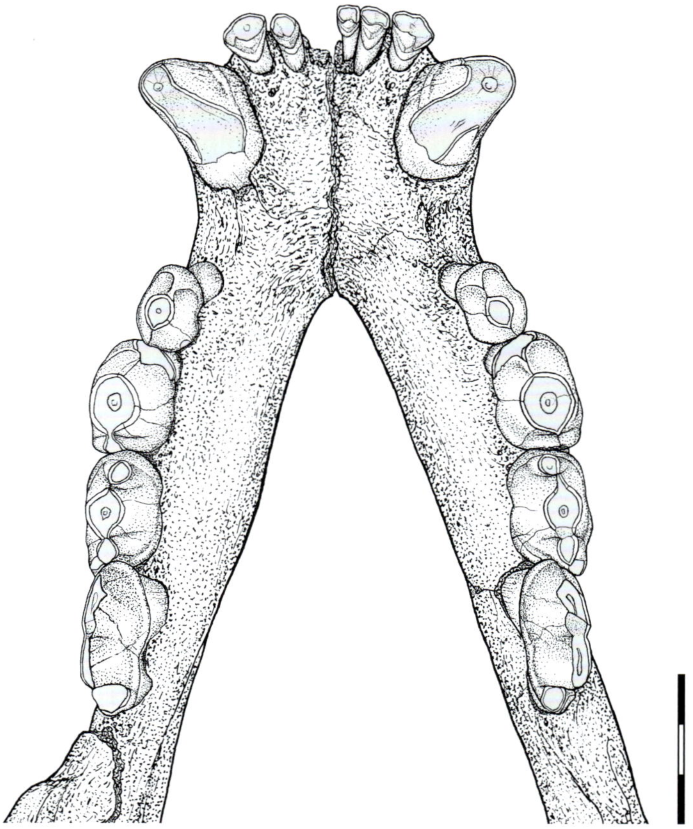

Fig. 3 *Pachycrocuta brevirostris* (Gervais), Untermassfeld. – Left hemimandible IQW 1993/24 369 (Mei. 23 898) with I_2-M_1 and right hemimandible IQW 1994/24 584 (Mei. 24 113) with I_1-M_1, occlusal view. – Scale: 30 mm.

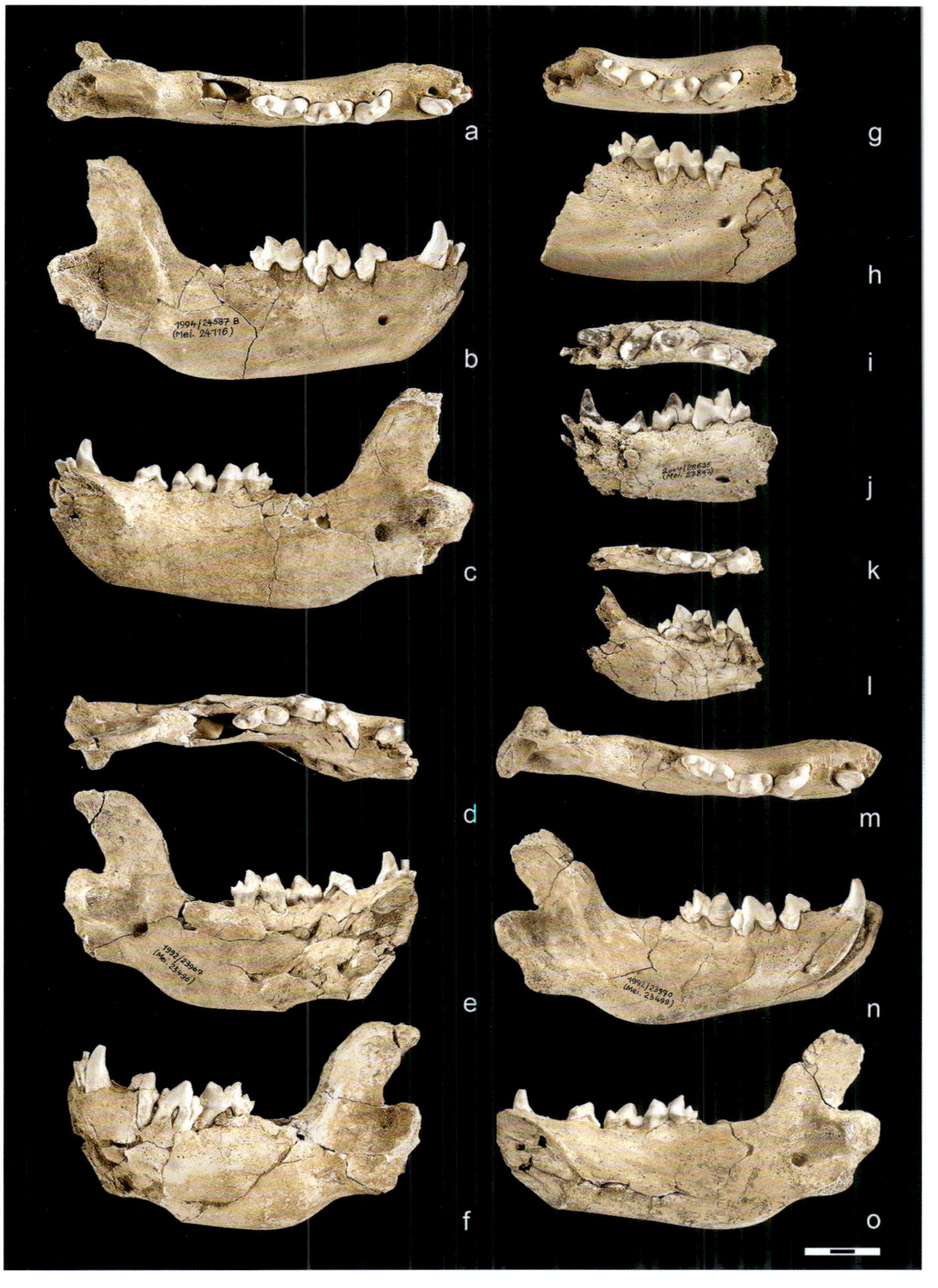

Fig. 4 *Pachycrocuta brevirostris* (Gervais), Untermassfeld. – **a–c** Right hemimandible IQW 1994/24587 B (Mei. 24116) with DI_2-DP_4 and erupting M_1. – **d–f** Left hemimandible IQW 1992/23967 (Mei. 23496) with DI_1-DP_4 and erupting M_1. – **g–h** Right hemimandible fragment IQW 2008/29841 (Mei. 29003) with DP_2-DP_4. – **i–j** Right hemimandible fragment IQW 2004/28685 (Mei. 27847) with DI_1-DI_2 and DC-DP_4. – **k–l** Right hemimandible fragment IQW 2010/31205 (Mei. 30367) with DP_3-DP_4. – **m–o** Right hemimandible IQW 1992/23970 (Mei. 23499) with DC-DP_4. – **a**, **d**, **g**, **i**, **k**, **m** occlusal views; **b**, **f**, **h**, **l**, **n** buccal views; **c**, **e**, **j**, **o** lingual views. – Scale: 30 mm.

that teeth appear imbricated in buccal view. This is part of the ontogenetic dental pattern of the species and indeed it seems more pronounced in younger individuals, e.g., as evident comparing IQW 1994/24587 B (Mei. 24116) with IQW 2004/28685 (Mei. 27847) (**Fig. 4a, i**).

Turner (2001) discussed this point, but it is important to stress that this arrangement differs from the condition observed in *C. crocuta*, whose juveniles have more spaced deciduous dentition (Bonifay 1971; Jimenez et al.

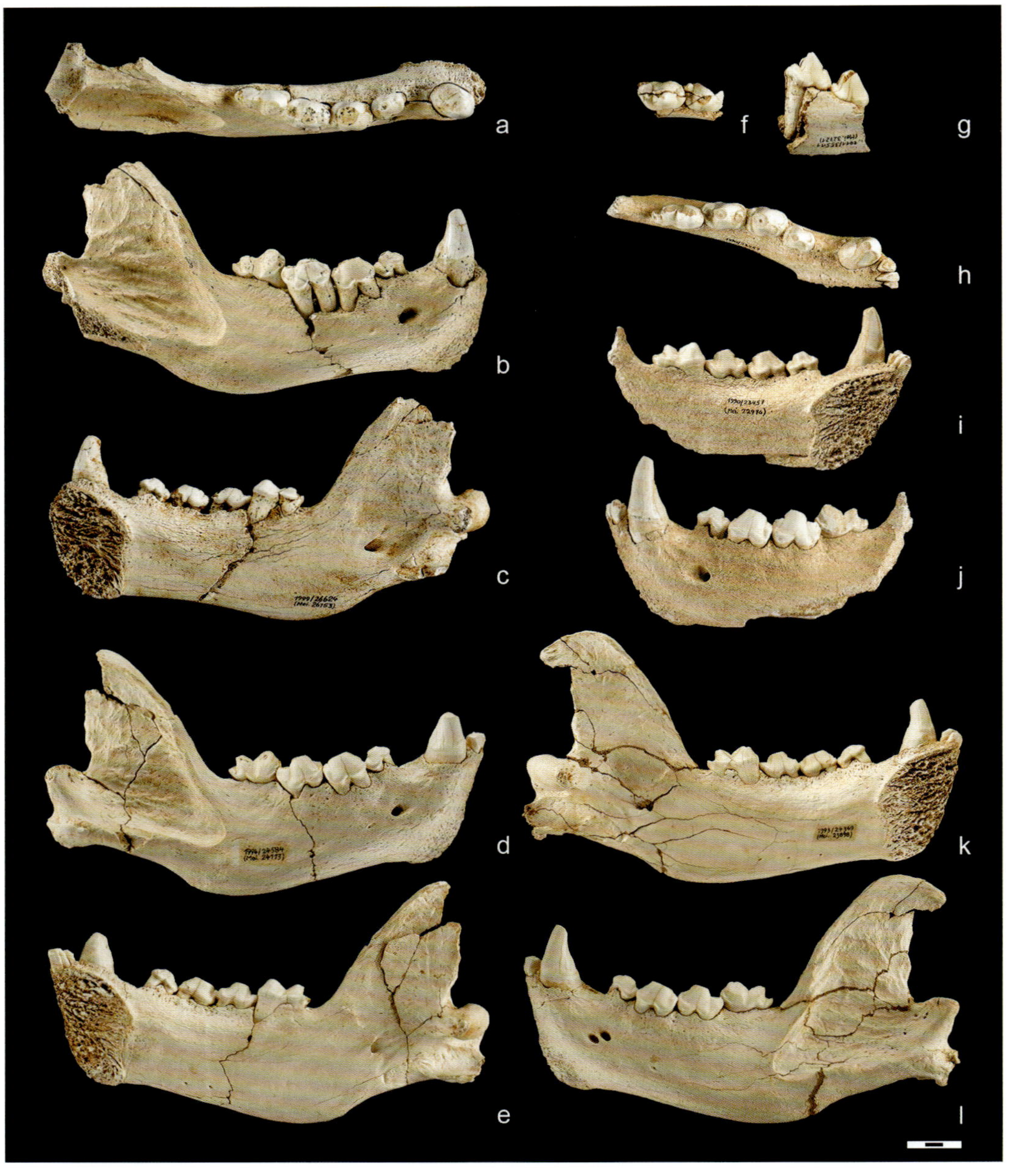

Fig. 5 *Pachycrocuta brevirostris* (Gervais), Untermassfeld. – **a–c** Right hemimandible IQW 1999/26 624 (Mei. 26 153) with C_1-M_1. – **d–e** Right hemimandible IQW 1994/24 584 (Mei. 24 113) with I_1-M_1. – **f–g** Left hemimandible fragment IQW 2011/35 511 (Mei. 32 721) with P_3-P_4. – **h–j** Left hemimandible fragment IQW 1990/23 457 (Mei. 22 976) with I_1-M_1. – **k–l** Left hemimandible IQW 1993/24 369 (Mei. 23 898) with I_2-M_1. – **a**, **f**, **h** occlusal views; **b**, **d**, **j**, **l** buccal views; **c**, **e**, **g**, **i**, **k** lingual views. – Scale: 30 mm.

2019). This is probably related to the size reduction of the deciduous dentition of the spotted hyena, a feature already noted by Kurtén (1956). The mental foramen is placed below P_2 (or DP_2 in juveniles), as in extant bone-cracking hyenas (**Figs. 4–5**). Only one foramen is usually present, but the left hemimandible IQW 1993/24 369 (Mei. 23 898) has a supernumerary aperture (**Fig. 5l**). It is worth noting that only one foramen is present on the right side of the same individual, IQW 1994/24 584 (Mei. 24 113), an interesting example of variability (**Fig. 5d**). Supernumerary mental foramina sometimes occur in several Hyaenidae, including the extant bone-cracking hyenas (Werdelin and Solounias 1991).

4.3. Dentition

Although several new hyaenid remains have been recovered from Untermassfeld since the previous publication by Turner (2001), he already provided a detailed description of the morphology of the teeth that would be redundant to repeat here. Nevertheless, some remarks are useful. Measurements taken on newly discovered and earlier published specimens (**Tables 1–2**) complement those provided by Turner (2001, tabs. 1–2). In most cases, size is the best distinction criterion from other related extinct and extant species, since the general morphology of the teeth is quite conservative. For instance, the mesio-lingual triangle formed by the enamel at the base of the crown of the upper canines, underlined by Turner (2001) as a salient feature of the tooth (**Fig. 2g**), is also present in *Pl. perrieri* and all extant bone-cracking hyenas (Viret 1954; pers. obs.). In all these species premolars – apart from the upper carnassial P^4, and P^1, which is a small one-rooted and monocuspid tooth – are basically conical structures, with an enlarged central cusp(id) and accessory mesial and distal cusplets. Nonetheless, in *C. crocuta* P^3 and P_3 are markedly enlarged and P^2 and P_2 relatively reduced (Kurtén 1956). Kurtén (1956) also highlighted the reduction in the distal accessory cusplet of P^3 of *C. crocuta*, and to a lesser extent in the distal accessory cusplet of P_3; Werdelin and Solounias (1991) considered the mesial accessory cusplet of P_4 large in *Pl. perrieri* and *P. brevirostris*, but small in extant bone-cracking hyenas. All these cusplets are accordingly well-developed in *P. brevirostris* from Untermassfeld (**Fig. 5**).

	IQW 1980/16085 (Mei. 15596)	**IQW 1990/23650 (Mei. 23179)**	**IQW 1990/23625 (Mei. 23154)**
I^3 L		17.4	17.7
I^3 W		12.9	13.0
C^1 L		25.3	20.5
C^1 W		c. 15.0	14.5
P^1 L		8.9	8.4
P^1 W		8.6	7.6
P^2 L	20.4	23.9	22.0
P^2 W	13.5	14.4	14.9
P^3 L	25.3	30.0	27.4
P^3 W	18.5	19.5	18.9
P^4 L	40.9	44.5	42.7
P^4 Wa	c. 26.0	22.2	22.6
P^4 Wbl	12.4	13.0	12.8
P^4 Lps	12.6	12.5	13.4
P^4 Lpc	13.4	15.0	13.3
P^4 Lm	15.3	16.2	15.1
M^1 L	4.6	7.0	5.9
M^1 W	14.4	18.6	17.3
P^2-P^4	92.8	98.7	105.0
WP		112.5	117.7

Table 1 *Pachycrocuta brevirostris* (Gervais), measurements (mm) of cranial remains and upper teeth, Untermassfeld (further measurements in Turner 2001, tab. 1). For abbreviations see section 3.

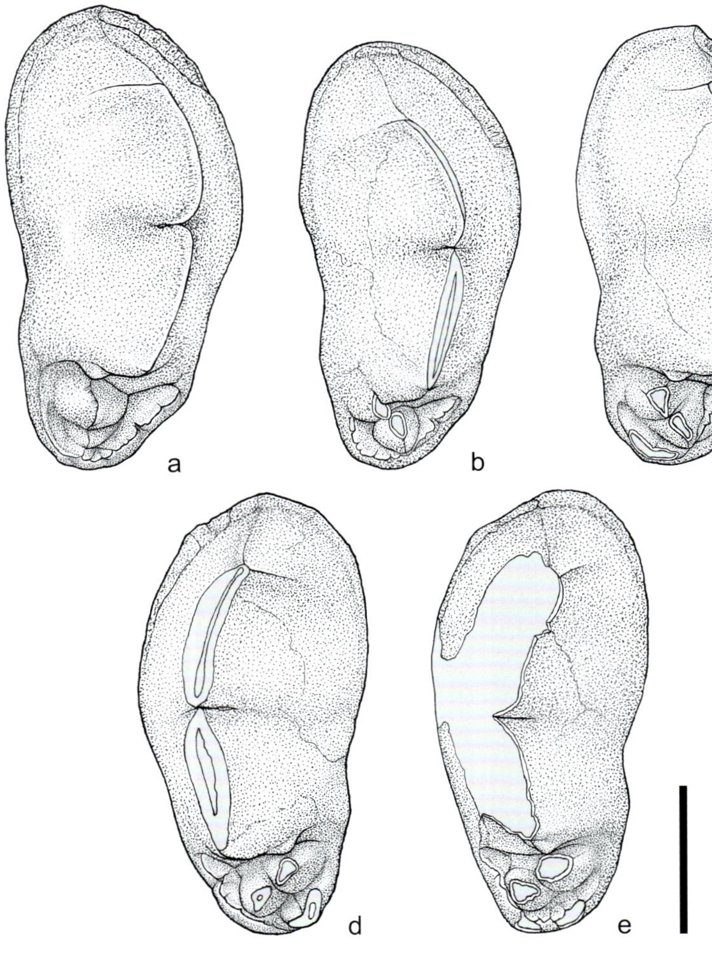

Fig. 6 *Pachycrocuta brevirostris* (Gervais), Untermassfeld. – **a** Right M_1 IQW 1988/22601 (Mei. 22120). – **b** Right M_1 IQW 1995/25336 (Mei. 24865). – **c–d** Right and left M_1 IQW 1986/21365 (Mei. 20888). – **e** Left M_1 IQW 1990/23457 (Mei. 22976), occlusal views. – Scale: 10mm

In *P. brevirostris,* M^1 is a functional tooth (**Fig. 2i, l**) as in *Pl. perrieri*, *H. hyaena*, and *Pa. brunnea*, and unlike in *C. crocuta*, where the tooth is seldom present and, if so, usually limited to a small vestigial peg (Viret 1954; Kurtén 1956; Werdelin and Solounias 1991). Contrary to what the often stressed »giant« size of the species might suggest, in P^4 and M^1 there is overlap between *P. brevirostris* and large individuals of *C. crocuta*, as the latter species has proportionally elongated carnassials (Kurtén 1956; Schütt 1971). However, the morphology of these teeth is different. In P^4 of *C. crocuta* the protocone is usually placed more mesially than in *P. brevirostris*, forming a wider angle with the paracone in occlusal view; the parastyle is reduced, similar in size to the protocone or even smaller; the paracone and metastyle are aligned along the same mesio-distal axis, whereas they converge lingually in *P. brevirostris*; and the metastyle is relatively longer (Kurtén 1956; Werdelin and Solounias 1991).
The identification of isolated lower carnassials is straightforward, since in M^1 of *C. crocuta* the talonid is extremely reduced. In brief, the elongation of the slicing component of the carnassials observed in *C. crocuta* reflects its more active-hunting behaviour in comparison to other extant bone-cracking hyenas, and presumably also to *P. brevirostris*. In the sample from Untermassfeld the metaconid is always absent, while two cuspids are usually present in the talonid (**Fig. 6**). Their relative size and development are variable, and

	Mandibula sin. IQW 1980/15 918 (Mei. 15 429)	Mandibula sin. IQW 1990/23 457 (Mei. 22 976)	Mandibula sin. + dex. IQW 1993/24 369 (Mei. 23 898) + IQW 1994/24 584 (Mei. 24 113)	Mandibula dex. IQW 1999/26 624 (Mei. 26 153)
C_1 L	22.0	21.8	22.2	21.7
C_1 W	16.6	16.5	16.4	17.5
Di	14.3	19.5	20.2	17.1
P_2 L	19.2	18.8	18.9	18.8
P_2 W	13.5	13.6	13.5	13.5
P_3 L	22.7	23.6	24.0	24.6
P_3 W	17.8	17.3	17.7	17.1
P_4 L	25.9	26.8	26.8	27.2
P_4 W	16.1	15.7	16.9	17.4
M_1 L	30.3	30.0	30.4	29.7
M_1 W	14.5	14.3	14.9	15.4
M_1 Lpa	16.2	15.4	16.1	16.0
M_1 Lpr	15.9	15.3	15.0	15.4
HbM_1	54.5		63.4	63.0
P_2-M_1	91.8	93.1	92.7	93.7
C_1-Cd	220.0		216.0	222.0

Table 2 *Pachycrocuta brevirostris* (Gervais), measurements (mm) of mandibles and lower teeth, Untermassfeld (further measurements in Turner 2001, tab. 1). For abbreviations see section 3.

indeed there is no widespread agreement on their level of homology (e. g., Kurtén 1956; 1972; Werdelin and Solounias 1991; Turner 2001; Liu et al. 2021). In IQW 1988/22 601 (Mei. 22 120) the mesial cuspid is larger, and the distal one is so small that it would have been impossible to detect it had the tooth not been completely unworn (**Fig. 6a**). In this specimen the two cuspids are aligned along an axis parallel to the major mesio-distal length of the tooth, similarly to IQW 1980/15 918 (Mei. 15 429) (Turner 2001, fig. 2, pl. 124) and IQW 1986/21 369 (Mei. 20 888) (Turner l.c., fig. 1, pl. 125), whereas in the other teeth this axis is tilted lingually (**Fig. 6b–e**). In IQW 1995/25 336 (Mei. 24 865) the distal cuspid is more developed than the mesial one (**Fig. 6b**). In the other specimens the two cuspids are roughly equally developed (**Fig. 6c–e**).

4.4. Postcranial material

IQW 1980/15 791 (Mei. 15 303) is a fragmentary left ulna, lacking the distal epiphysis and part of the proximal one, i. e., the olecranon and half of the trochlear notch (**Fig. 7**). The coronoid process and the lateral articular surface are preserved (**Table 3**). IQW 1980/15 791 (Mei. 15 303) is slightly smaller than remains from Pirro Nord (housed in PF; own measurements, N = 3, DPC = 56.7 mm).

IQW 1995/24 992 (Mei. 24 521), IQW 1995/24 993 (Mei. 24 522), IQW 1995/24 990 (Mei. 24 519), and IQW 1995/24 991 (Mei. 24 520) are the right metacarpals 2 to 5 of a single individual, all but Mc3 are almost

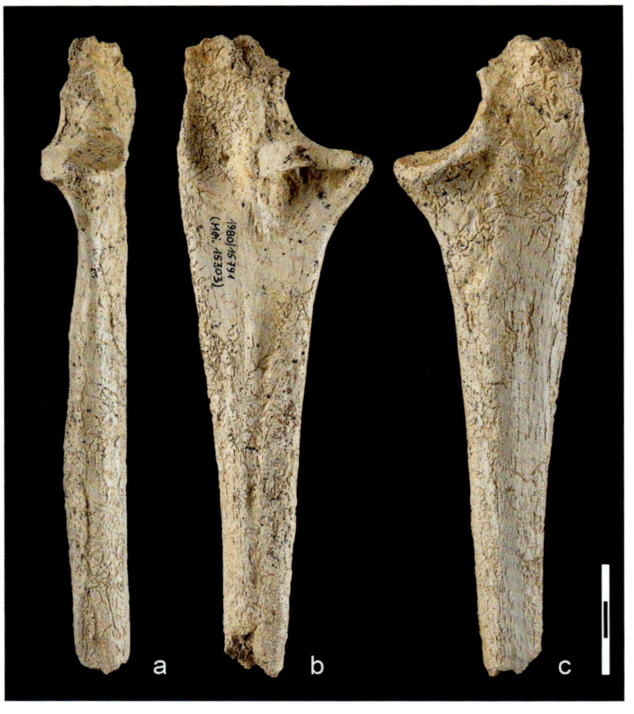

Fig. 7 *Pachycrocuta brevirostris* (Gervais), Untermassfeld. – **a–c** Left ulna fragment IQW 1980/15791 (Mei. 15303), anterior, external, internal views. – Scale: 30mm.

complete (**Fig. 8**). Their measurements are provided in **Table 4**. The morphology of the metacarpals is similar to that of the homologous elements of the extant bone-cracking hyenas, but they are more robust and stouter. For instance, the ratio Bp/GL for the Mc2 IQW 1995/24992 (Mei. 24521) is 0.21, whereas those calculated from the mean values of extant samples of *C. crocuta* (N = 8) and *H. hyaena* (N = 3) provided by Cardoso (1993) are both equal to 0.18. The same ratio for the Mc4 IQW 1995/24990 (Mei. 24519) is 0.18, while for *C. crocuta* is 0.15, and for *H. hyaena* only 0.13.

	Ulna sin. IQW 1980/15791 (Mei. 15303)	Femur sin. IQW 1992/24149 (Mei. 23678)	Femur dex. IQW 1990/23656 (Mei. 23185)
DPC	52.1		
HTr		71.0	70.6
DD	26.5	24.3	24.2
SD	16.8	25.1	26.3

Table 3 *Pachycrocuta brevirostris* (Gervais), measurements (mm) of ulna and femurs, Untermassfeld. For abbreviations see section 3.

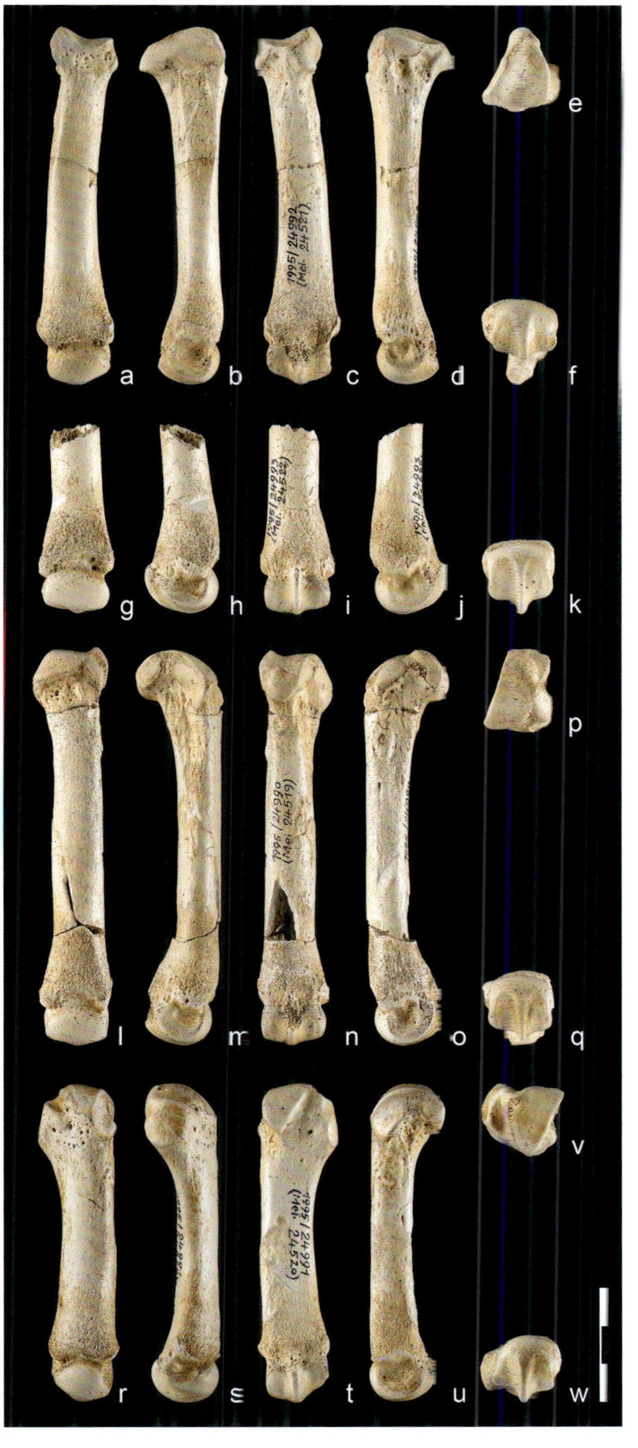

Fig. 8 *Pachycrocuta brevirostris* (Gervais), Untermassfeld. – **a–f** Right Mc2 IQW 1995/24992 (Mei. 24521). – **g–k** Right Mc3 distal fragment IQW 1995/24993 (Mei. 24522). – **l–q** Right Mc4 IQW 1995/24990 (Mei. 24519). – **r–w** Right Mc5 IQW 1995/24991 (Mei. 24520). – **a**, **g**, **l**, **r** dorsal views; **b**, **h**, **m**, **s** external views; **c**, **i**, **n**, **t** palmar views; **d**, **j**, **o**, **u** internal views; **e**, **p**, **v** proximal views; **f**, **k**, **q**, **w** distal views. – Scale: 30 mm

	Mc2 dex. IQW 1995/24 992 (Mei. 24 521)	Mc3 dex. IQW 1995/24 993 (Mei. 24 522)	Mc4 dex. IQW 1995/24 990 (Mei. 24 519)	Mc5 dex. IQW 1995/24 991 (Mei. 24 520)
GL	97.3		107.8	87.3
Dp	22.6		22.6	19.7
Bp	19.0		18.2	21.2
DD	11.0		11.3	11.5
SD	13.6		13.6	14.7
Dd	16.4	19.8	c. 18.6	18.3
Bd	20.6	20.2	19.0	19.5

Table 4 *Pachycrocuta brevirostris* (Gervais), measurements (mm) of a series of metacarpals of the same individual, Untermassfeld. For abbreviations see section 3.

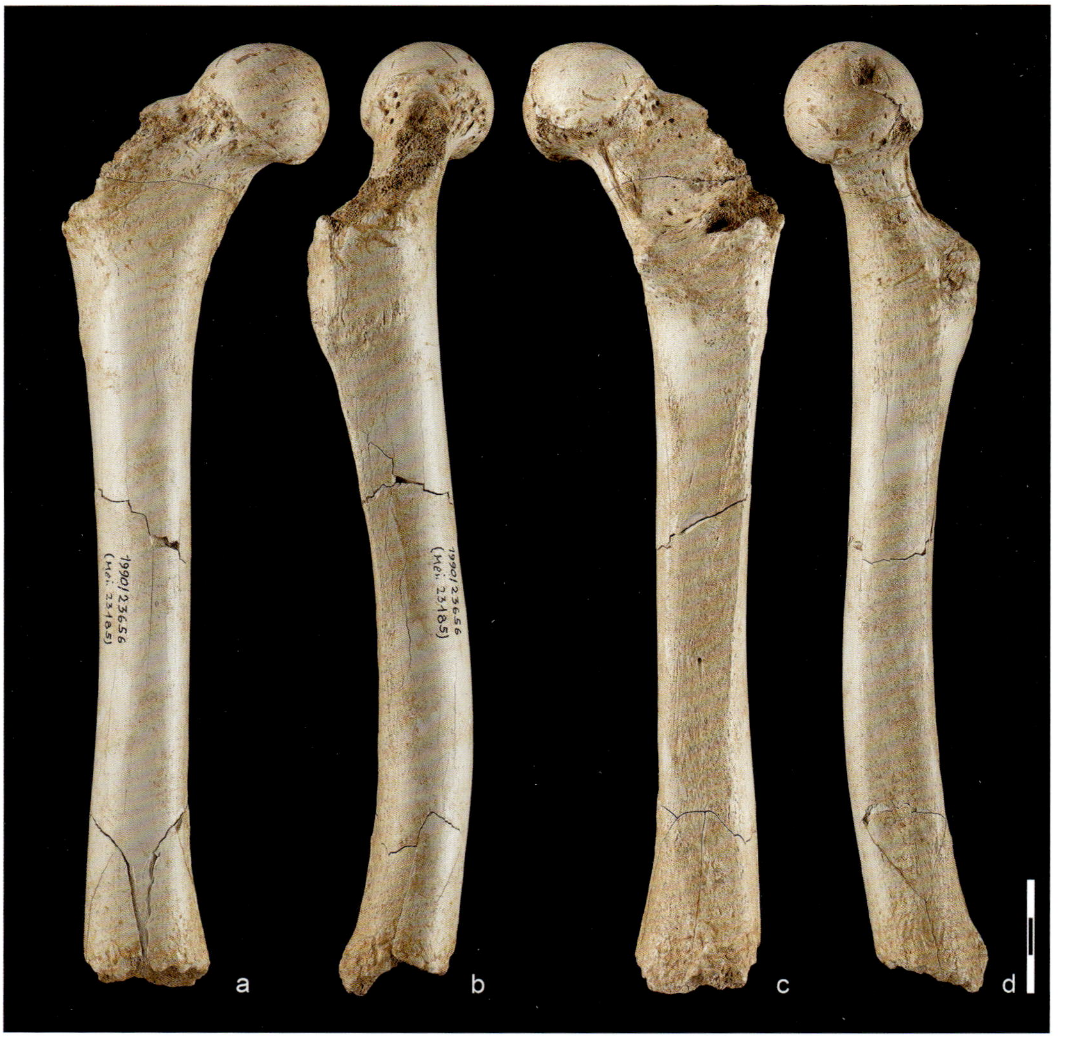

Fig. 9 *Pachycrocuta brevirostris* (Gervais), Untermassfeld. – **a–d** Right femur fragment IQW 1990/23 656 (Mei. 23 185), anterior, external, posterior, internal views. – Scale: 30 mm.

5. Comparative biometric discussion

The giant hyena *P. brevirostris* was a relatively common faunal element during the Early Pleistocene of Europe, but a few sites yielded abundant remains (Kurtén and Garevski 1989; Turner and Antón 1996). Although some localities have only tentative chronological constraints and/or may include time-averaged ensembles, it is reasonably possible to subdivide them into a Late Villafranchian and an Epivillafranchian group (**Table 5**). Comparing P^4 and M_1 lengths of the two reveals that the sample from Untermassfeld includes the largest *P. brevirostris*, and that an increase in size occurred during the late Early Pleistocene (**Fig. 10a, c**). However, as the overlap between the two groups is not negligible, plotting each locality separately depicts a more complex picture (**Fig. 10b, d**). Data for P^4 are quite scanty (N = 26), with most occurrences being represented by isolated specimens. The sample from the Upper Valdarno is relatively small and that from Untermassfeld large, but the remains from Venta Micena encompass all the observed variation (**Fig. 10b**).

Country	Locality	Abbreviation	Biochronology	References	P^4	M_1
Czech Republic	Stránská Skála	SS	Epivillafranchian (?)	Kurtén 1972	1	4
France	Grosse Marguerite	GM	Epivillafranchian	Fourvel and Lateur 2015		1
	Le Vallonnet	Va	Epivillafranchian	de Lumley et al. 1988		2
	Saint-Prest	SP	Epivillafranchian	Guérin et al. 2003	1	
	Sainzelles	Sa	Epivillafranchian	Boule 1893; Turner and Antón 1996	2	1
	Trois Pigeons	TP	Epivillafranchian	Fourvel and Lateur 2015		1
Germany	Untermassfeld	Un	Epivillafranchian	Turner 2001; this paper	5	6
	Würzburg-Schalksberg	WS	Epivillafranchian	Schütt 1974	1	
Greece	Kalamotó	Ka	Epivillafranchian	Tsoukala and Chatzopoulou 2005		1
Italy	Slivia	Sl	Epivillafranchian	this paper		1
Macedonia	Manastirec	Ma	Epivillafranchian (?)	Kurtén and Garevski 1989		2
Montenegro	Trlica	Tr	Epivillafranchian (?)	Vislobokova and Agadjanian 2015	1	1
Ukraine	Nogaisk	No	Epivillafranchian	Marciszak et al. 2021	1	
Russia (European)	Akhtanizovskaya	Ak	Epivillafranchian	Sotnikova and Titov 2009		[illegible]
Slovakia	Gombasek	Go	Epivillafranchian (?)	Turner and Antón 1996	1	4
Spain	Cal Guardiola	CG	Epivillafranchian	Madurell-Malapeira et al. 2009	1	
	Cueva Victoria	CV	Epivillafranchian	Madurell-Malapeira et al. 2015	1	3
	Incarcal	In	Epivillafranchian (?)	Julia and de Villalta 1984; Galobart et al. 2003		2
Greece	Gerakarou	Ge	Late Villafranchian	Koufos 1992		[illegible]
Italy	Pirro Nord	PN	Late Villafranchian	Petrucci et al. 2013; this paper		4
	Upper Valdarno	UV	Late Villafranchian	Turner and Antón 1996; Liu et al. 2021; this paper	7	8
Russia (European)	Liventsovka	Li	Late Villafranchian	Sotnikova et al. 2002		[illegible]
Spain	Venta Micena	VM	Late Villafranchian	Martínez-Navarro 1991; Medin et al. 2017	4	12

Table 5 *Pachycrocuta brevirostris* (Gervais), European localities with remains of carnassial teeth. P^4, M^1 – number of specimens.

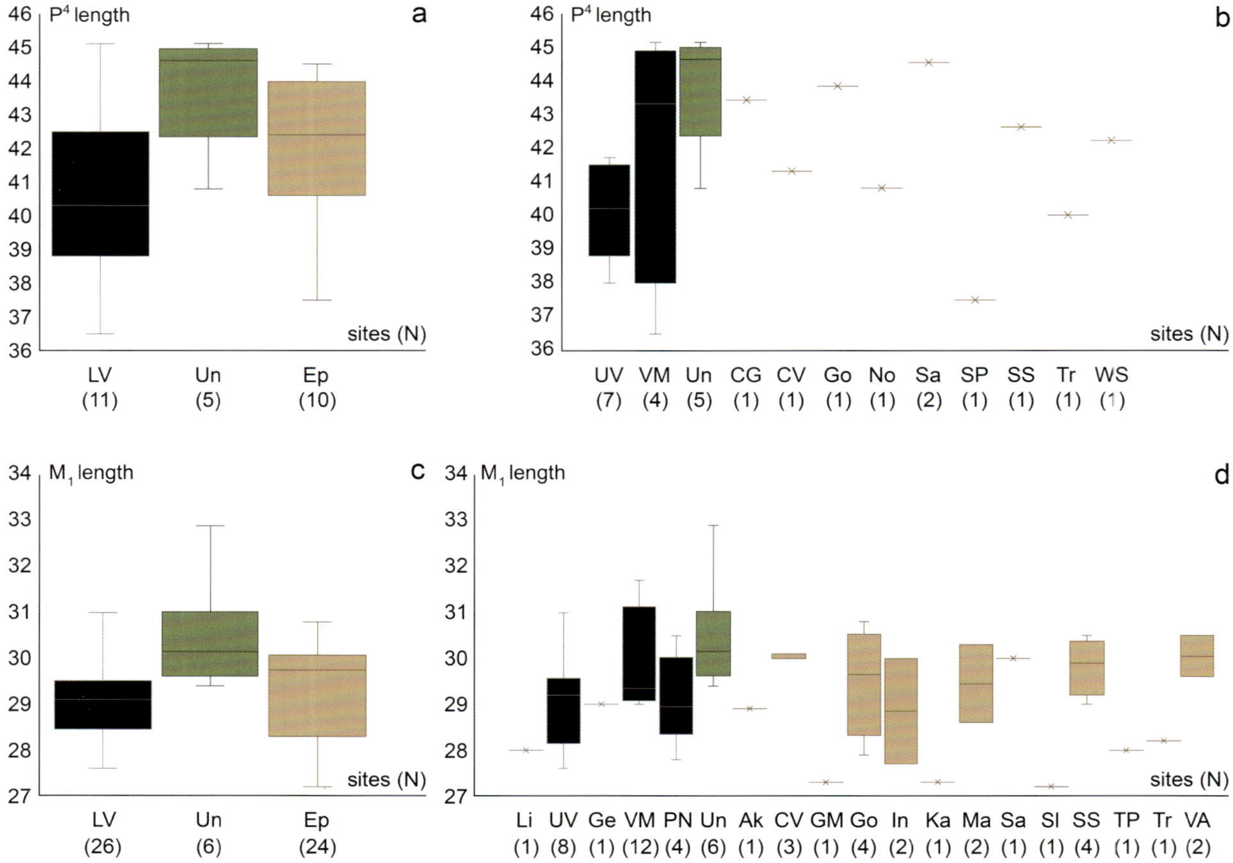

Fig. 10 *Pachycrocuta brevirostris* (Gervais), Untermassfeld and other European localities. – **a–b** Boxplots of P^4 L. – **c–d** Boxplots of M_1 L. – LV: Late Villafranchian. – Ep: Epivillafranchian, for other abbreviations and references: see **Table 5**. – N: Number of specimens for each locality. – Measurements: mm.

M_1 is more abundant (N = 56) and hence it allows a better evaluation of the pattern. The sample from Untermassfeld is characterised by the largest size, but the other values are highly heterogeneous, although a slight increase in the Epivillafranchian can be noted (**Fig. 10c**).

In a previous study Kurtén and Garevski (1989) found no substantial chronological differences in morphology during the existence of *P. brevirostris* in the Pleistocene of Europe. Now that more data are available, they provide support for the identification of a weak but significant trend towards an increase in size, exemplified by the palaeopopulation of Untermassfeld. Clearly, not all localities adhere to this tendency, and the presence of large or (relatively) small individuals of *P. brevirostris* at a given site cannot be used for biochronological correlations. Arguably, other factors, such as climate fluctuations, availability of trophic resources, and/or competition with other species might have played a role, and a few sites have adequate constraints for evaluating these variables. Moreover, the possible existence of sex-related differences should also be taken into account, as the debated »monomorphism« of *C. crocuta* is hardly generalisable to other species (Hamilton et al. 1986).

6. Conclusions

The attribution of the Hyaenidae sample from Untermassfeld to the giant hyena *P. brevirostris* is beyond doubt, considering the evident large size of the fossil remains and their morphology. Adaptations to bone-cracking common to extinct and extant hyenas belonging to this ecomorphotype (sensu Werdelin and Solounias 1996) are of course present in *P. brevirostris* from Untermassfeld, and in some cases extremely developed, as for the curvature of the tooth rows and the depth of the mandible (especially at the symphysis). Moreover, the species was also characterised by its unique combination of morphological features, which although somehow overshadowed by its superior dimension can be important in the identification of isolated remains.

Untermassfeld yielded not only the most abundant European sample of *P. brevirostris* but also the one characterised by the largest size. This is paradigmatic for the slight but seemingly significant dimensional increase occurred during the late Early Pleistocene, perhaps continuing a trend started since the separation of *P. brevirostris* from *Pl. perrieri*. This corroborates with the view of a progressive specialisation in the giant hyena, which could have made it a perfectly adapted component of the Late Villafranchian and Epivillafranchian mammal communities of Europe, but unable to cope with intense climate shifts of the Middle Pleistocene, eventually favouring its extinction (Iannucci et al. 2021).

Acknowledgements

We thank D. Rössler, M. Stache and R. Michalik (all Weimar) for the preparation of the new hyena finds discussed here. The photos were kindly taken by S. Döring and the drawings were created by E. Haase (both Weimar). G. Utschig (Weimar) compiled collection data. We are grateful for the support, assistance, and hospitality received during the study of museum collections, and we would like to thank M. Aiglstorfer (Mainz), D. Arbulla (Trieste), R. Castiglia (Rome), E. Cioppi (Florence), A. Currant (London), T. Engel (Mainz), M. Gasparik (Budapest), E. Gilissen (Tervuren), and D. Waterhouse (Norwich). We are thankful to L. Bellucci (Florence), J. Conti (Rome), D.A. Iurino (Rome), and F. Strani (Rome) for support and discussion. We are grateful to M. Cherin (Perugia) for comments and suggestions. This research received support from the SYNTHESYS project http://www.synthesys.info/ which was financed by European Community Research Infrastructure Action under the FP7 »Capacities« Program (HU-TAF-707 and BE-TAF-708 to RS). Parts of the manuscript were linguistically edited by C. Nielsen-Marsh (Leipzig).

References

Alba, D.M., Vinuesa, V., Madurell-Malapeira, J., 2015. On the original author and year of description of the extinct hyaenid *Pachycrocuta brevirostris*. Acta Palaeontologica Polonica 60, 573–576.

Arnold, J., in this volume. Bovid bone modifications by *Pachycrocuta* from the Early Pleistocene site of Untermassfeld. 1423–1464.

Bellucci, L., Sardella, R., Rook, L., 2015. Large mammal biochronology framework in Europe at Jaramillo: The Epivillafranchian as a formal biochron. Quaternary International 389, 84–89.

Boaz, N., Ciochon, R.L., Quinqi, X., Jinye, L., 2000. Large Mammalian Carnivores as a Taphonomic Factor in the Bone Accumulation at Zhoukoudian. Acta Anthropologica Sinica 19 (Supplement), 224–234.

Bonifay, M.-F., 1971. Carnivores quaternaires du Sud-Est de la France. Mémoires du Muséum National d'Histoire Naturelle, Nouvelle Série, Série C, 21, 43–377.

Boule, M., 1893. Description de l'*Hyaena brevirostris* du Pliocène de Sainzelles près le Puy (Haute-Loire). Annales des Sciences Naturelles, Zoologie et Paléontologie, 7. Série, 15, 85–97.

Cardoso, J.L., 1993. Contribuição para o conhecimento dos grandes mamíferos do Plistocénico Superior de Portugal. Oeiras.

Croizet, J.-B., Jobert A.-C.-G., 1828. Recherches Sur Les Ossemens Fossiles Du Departement Du Puy-De-Dome. Paris.

de Lumley, H., Kahlke, H.-D., Moigne, A.-M., Moullé, P.-E., 1988. Les faunes des grands Mammifères de la Grotte du Vallonnet Roquebrune-Cap-Martin, Alpes-Maritimes. L'Anthropologie 92, 465–496.

Driesch, A.v.d., 1976. A guide to the measurement of animal bones from archaeological sites. Peabody Museum Bulletin 1, III–IX, 1–137.

Erxleben, I.C.P., 1777. Systema Regni Animalis Per Classes, Ordines, Genera, Species, Varietates Cum Synonymia Et Historia Animalium, Classis 1 Mammalia. Lipsia.

Espigares, M.P., Martínez-Navarro, B., Palmqvist, P., Ros-Montoya, S., Toro, I., Agustí, J., Sala, R., 2013. *Homo* vs. *Pachycrocuta*: Earliest evidence of competition for an elephant carcass between scavengers at Fuente Nueva-3 (Orce, Spain). Quaternary International 295, 113–125.

Ficcarelli, G., Torre, D., 1970. Remarks on the taxonomy of hyaenids. Palaeontographia Italica 66, Nuova Serie 36, 13–33; pls. IV–XXII.

Fourvel, J.B., Lateur, N., 2015. Note sur deux mandibules de *Pachycrocuta brevirostris* (Aymard, 1846) du Pléistocène ancien du sud-est de la France: la grotte de la Grosse Marguerite (Aiguèze) et l'aven des Trois Pigeons (Nîmes). Paleo 26, 187–200.

Galobart, A., Antón, M., Maroto, J., 2003. Hiénidos y canidos de los yacimientos de Incarcal (Girona, NE de la Peninsula Ibérica). Una aproximacion a la paleobiologia del hiénodo del Pleistoceno inferior. Paleontologia i Evolució 34, 79–98.

Gervais, P., 1848–1852. Zoologie Et Paléontologie Françaises (Animaux Vertébrès). Nouvelles Recherches Sur Les Animaux Vivants Et Fossiles De La France. Tome I. Contenant L'Énumerération Méthodique Et Descriptive Des Espèces Ainsi Que Les Principes De Leur Distribution Géographique Et Paléontologique. Paris.

Guérin, C., Dewolf, Y., Lautridou, J.-P., 2003. Révision d'un site paléontologique célèbre: Saint-Prest (Chartres, France). Geobios 36, 55–82.

Hamilton, W.J., Ronald, L.T., Frank, L.G., 1986. Sexual Monomorphism in Spotted Hyenas, *Crocuta crocuta*. Ethology 71, 63–73.

Howell, F.C., Petter, G., 1980. The *Pachycrocuta* and *Hyaena* lineages (Plio-Pleistocene and extant species of the Hyaenidae). Their relationships with Miocene ictitheres: *Palhyaena* and *Hyaenictitherium*. Geobios 13, 579–623.

Iannucci, A., Mecozzi, B., Sardella, R., Iurino, D.A., 2021. The extinction of the giant hyena *Pachycrocuta brevirostris* and a reappraisal of the Epivillafranchian and Galerian Hyaenidae in Europe: Faunal turnover during the Early-Middle Pleistocene Transition. Quaternary Science Reviews 272, 107240.

Jimenez, I.J., Sanz, M., Daura, J., De Gaspar, I., García, N., 2019. Ontogenetic dental patterns in Pleistocene hyenas (*Crocuta crocuta* Erxleben, 1777) and their palaeobiological implications. International Journal of Osteoarchaeology 29, 808–821.

Julia, R., de Villalta J.F., 1984. El yacimiento de vertebrados del Pleistoceno inferior de Crespià (Girona, NE de la Península Ibérica). Acta Geològica Hispànica 19, 129–138.

Kahlke, H.-D., 1980. Flußpferde an der Werra. Urania-Universum 26. Leipzig, Jena, Berlin, 495–500.

1982. *Hippopotamus antiquus* Desmarest, 1822 aus dem Pleistozän von Meiningen in Südthüringen (Bezirk Suhl). Zeitschrift für geologische Wissenschaften 10, 943–949.

Kahlke, R.-D., 2001. Die unterpleistozäne Komplexfundstelle Untermaßfeld – Zusammenfassung des Kenntnisstandes sowie synthetische Betrachtungen zu Genesemodell, Paläoökologie und Stratigraphie. In: Kahlke, R.-D. (Ed.), Das Pleistozän von Untermaßfeld bei Meiningen (Thüringen). Teil 3. Monographien des Römisch-Germanischen Zentralmuseums 40, 3, 931–1030, suppl. I–XV.

2006. Untermassfeld. A late Early Pleistocene (Epivillafranchian) fossil site near Meiningen (Thuringia, Germany) and its position in the development of the European mammal fauna. British Archaeological Reports, International Series 1578. Oxford.

2007. Late Early Pleistocene European large mammals and the concept of an Epivillafranchian biochron. In: Kahlke, R.-D., Maul, L.C., Mazza, P.P.A. (Eds.), Late Neogene and Quaternary biodiversity and evolution: Regional developments and interregional correlations. Proceedings of the 18th International Senckenberg Conference (VI International Palaeontological Colloquium in Weimar). Volume II. Courier Forschungsinstitut Senckenberg 259, 265–278.

2009. Les communautés de grands mammifères du Pléistocène inférieur terminal et le concept d'un biochrone Épivillafranchien. Quaternaire 20, 415–427.

in this volume. The Pleistocene of Untermassfeld – A synopsis of site origin, palaeobiodiversity, taphonomic characteristics, palaeoenvironment, chronostratigraphy, and significance in western Palaearctic faunal history. 1635–1734.

Kahlke, R.-D., Gaudzinski, S., 2005. The blessing of a great flood: differentiation of mortality patterns in the large mammal record of the Lower Pleistocene fluvial site of Untermassfeld (Germany) and its relevance for the interpretation of faunal assemblages from archaeological sites. Journal of Archaeological Science 32, 1202–1222.

Kahlke, R.-D., Linnemann, U., Gärtner, A., 2020. New results on the origin and geological history of the Early Pleistocene site of Untermassfeld. In: Kahlke, R.-D. (Ed.), The Pleistocene of Untermassfeld near Meiningen (Thüringen, Germany). Part 4. Monographien des Römisch-Germanischen Zentralmuseums 40, 4, 1079–1104, Foldout I.

Keiler, J.-A., 2001. Die Koprolithen aus dem Unterpleistozän von Untermaßfeld. In: Kahlke, R.-D., (Ed.), Das Pleistozän von Untermaßfeld bei Meinigen (Thüringen), Teil 2. Monographien des Römisch-Germanischen Zentralmuseums 40, 2, 691–697, pls. 128–131.

2008. Bergung, Präparation und Konservierung fossiler Säugetierschädel aus dem Unterpleistozän von Untermaßfeld/Thüringen. Eine methodologische Betrachtung. Der Präparator 54, 62–83.

Koufos, G.D., 1992. The Pleistocene carnivores of the Mygdonia Basin (Macedonia, Greece). Annales de Paléontologie 78, 205–257.

Kretzoi, M., 1938. Die Raubtiere von Gombaszög nebst einer Übersicht der Gesamtfauna. Annales Musei Nationalis Hungaric 31, 89–157.

Kurtén, B., 1956. The Status and Affinities of *Hyaena sinensis* Owen and *Hyaena ultima* Matsumoto. American Museum Novitates 1764, 1–48.

1972. Fossil Hyaenidae from the excavations at Stránská Skála. Anthropos 20, nova seria 12, 113–120.

Kurtén, B., Crusafont Pairó, M., 1977. Villafranchian Carnivores (Mammalia) from la Puebla de Valverde (Teruel, Spain). Commentationes Biologicae 85, 1–39.

Kurtén, B., Garevski R., 1989. Giant hyena, *Hyaena brevirostris* Aymard (Mammalia, Carnivora), from the Middle Pleistocene of Manastirec, Yugoslavia. Annales Zoologici Fennici 26, 145–147.

Kurtén, B., Werdelin, L., 1988. A review of the genus *Chasmaporthetes* Hay, 1921 (Carnivora, Hyaenidae). Journal of Vertebrate Paleontology 8, 46–66.

Lewis, M.E., Werdelin, L., 2010. Carnivoran Dispersal Out of Africa During the Early Pleistocene: Relevance for Hominins? In: Fleagle, J.G., Shea, J.J., Grine, F.E., Baden, A.L., Leakey, R.E. (Eds.), Out of Africa I: The First Hominin Colonization of Eurasia. Vertebrate Paleobiology and Paleoanthropology. Dordrecht, 13–26.

Linnaeus, C., 1758. Systema Naturae Per Regna Tria Naturae, Secundum Classes, Ordines, Genera, Species Cum Characteribus, Differentiis, Synonymis, Locis. Tomus I. Holmiae. (10th rev. ed.)

Liu, J., Liu, J., Zhang, H., Wagner, J., Jiangzuo, Q., Song, Y., Liu, S., Wang, Y. Jin, C., 2021. The giant short-faced hyena *Pachycrocuta brevirostris* (Mammalia, Carnivora, Hyaenidae) from Northeast Asia: A reinterpretation of subspecies differentiation and intercontinental dispersal. Quaternary International 577, 29–51.

Madurell-Malapeira, J., Alba, D.M., Moyà-Solà, S., 2009. Carnivora from the late Early Pleistocene of Cal Guardiola (Terrassa, Vallès-Penedès Basin, Catalonia, Spain). Journal of Paleontology 83, 969–974.

Madurell-Malapeira, J., Morales, J., Vinuesa, V., Boscaini, A., 2015. Úrsidos, hiénidos y félidos del Pleistoceno inferior de Cueva Victoria (Cartagena, Murcia). Mastia (2012–2014) 11–13, 401–432.

Madurell-Malapeira, J., Alba, D.M., Espigares, M.P., Vinuesa, V., Palmqvist, P., Martínez-Navarro, B., Moyà-Solà, S., 2017. Were large carnivorans and great climatic shifts limiting factors for hominin dispersals? Evidence of the activity of *Pachycrocuta brevirostris* during the Mid-Pleistocene Revolution in the Vallparadís Section (Vallès-Penedès Basin, Iberian Peninsula). Quaternary International 431, 42–52.

Marciszak, A., Semenov, Y., Portnicki, P., Derkach, T., 2021. First record of *Pachycrocuta brevirostris* (Gervais, 1850) from Ukraine on the background of the European occurrence of the species. Journal of Iberian Geology 47, 535–549.

Martínez-Navarro, B., 1991. Revisión sistemática y estudio cuantitativo de la fauna de macromamíferos del yacimiento de Venta Micena (Orce, Granada). PhD thesis, Universidad Autónoma de Barcelona, Spain.

2010. Early Pleistocene Faunas and Hominin Dispersals. In: Fleagle, J.G., Shea, J.J., Grine, F.E., Baden, A.L., Leakey, R.E. (Eds.), Out of Africa I: The First Hominin Colonization of Eurasia. Vertebrate Paleobiology and Paleoanthropology. Dordrecht, 207–224.

Mazza, P.P.A., Bertini, A., Magi, M., 2004. The late Pliocene site of Poggio Rosso (central Italy): taphonomy and paleoenvironment. Palaios 19, 227–248.

Medin, T., Martínez-Navarro, B., Rivals, F., Madurell-Malapeira, J., Ros-Montoya, S., Espigares, M.P., Figueirido, B., Rook, L., Palmqvist, P., 2017. Late Villafranchian *Ursus etruscus* and other large carnivorans from the Orce sites (Guadix-Baza basin, Andalusia, southern Spain): taxonomy, biochronology, paleobiology, and ecogeographical context. Quaternary International 431, 20–41.

Napoleone, G., Albanielli, A., Azzaroli, A., Mazzini, M., 2001. The Poggio Rosso locality calibrated to the end-Pliocene and its significance for dating the late Villafranchian mammal faunas of the Upper Valdarno, Central Italy. Rivista Italiana di Paleontologia e Stratigrafia 107, 287–296.

Napoleone, G., Albanielli, A., Azzaroli, A., Bertini, A., Magi, M., Mazzini, M., 2003. Calibration of the Upper Valdarno Basin to the Plio-Pleistocene for correlating the Apennine continental sequences. Il Quaternario 16, 131–166.

Palmqvist, P., Martínez-Navarro, B., Pérez-Claros, J.A., Torregrosa, V., Figueirido, B., Jiménez-Arenas, J.M., Patrocinio Espigares, M., Ros-Montoya, S., De Renzi, M., 2011. The giant hyena *Pachycrocuta brevirostris*: Modelling the bone-cracking behavior of an extinct carnivore. Quaternary International 243, 61–79.

Petrucci, M., Cipullo, A., Martínez-Navarro, B., Rook, L., Sardella, R., 2013. The Late Villafranchian (Early Pleistocene) carnivores (Carnivora, Mammalia) from Pirro Nord (Italy). Palaeontographica, Abteilung A 298, 113–145.

Qiu, Z., 1987. Die Hyaeniden aus dem Ruscinium und Villafranchium Chinas. Münchner Geowissenschaftliche Abhandlungen Reihe A, Geologie und Paläontologie 9, 1–108.

Roebroeks, W., Gaudzinski-Windheuser, S., Baales, M., Kahlke, R.-D., 2018. Uneven Data Quality and the Earliest Occupation of Europe – the Case of Untermassfeld (Germany). Journal of Paleolithic Archaeology 1, 5–31.

Sardella, R., Palombo, M.R., 2007. The Pliocene-Pleistocene boundary: Which significance for the so called »Wolf Event«? Evidences from Western Europe. Quaternaire 18, 65–71.

Schütt, G., 1971. Die Hyänen der Mosbacher Sande (Altpleistozän, Wiesbaden/ Hessen) mit einem Beitrag zur Stammesgeschichte der Gattung *Crocuta*. Mainzer Naturwissenschaftliches Archiv 10, 29–76.

1974. Die Carnivoren von Würzburg-Schalksberg. Mit einem Beitrag zur biostratigraphischen und zoogeographischen Stellung der altpleistozänen Wirbeltierfaunen vom Mittelmain (Unterfranken). Neues Jahrbuch für Geologie und Paläontologie, Abhandlungen 147, 61–90.

Sotnikova, M., Titov, V., 2009. Carnivora of the Tamanian faunal unit (the Azov Sea area). Quaternary International 201, 43–52.

Sotnikova, M.V., Baigusheva, V.S., Titov, V.V., 2002. Carnivores of the Khapry Faunal Assemblage and Their Stratigraphic Implications. Stratigraphy and Geological Correlation 10, 375–390.

Thunberg, C.P., 1820. Beskrifning och Teckning på ett nytt species, *Hyaena* brunnea. Kungliga Svenska Vetenskapsakademiens Handlingar, Seria 3, Tomus 8, 59–65.

Tsoukala, E., Chatzopoulou, K., 2005. A New Early Pleistocene (Latest Villafranchian) Site with Mammals in Kalamotó (Mygdonia Basin, Macedonia, Greece) – Preliminary Report. Mitteilungen der Kommission für Quartärforschung der Österreichischen Akademie der Wissenschaften 14, 213–233.

Turner, A., 1990. The evolution of the guild of larger terrestrial carnivores during the Plio-Pleistocene in Africa. Geobios 23, 349–368.

1992. Large carnivores and earliest European hominids: changing determinants of resource availability during the Lower and Middle Pleistocene. Journal of Human Evolution 22, 109–122.

2001. Remains of *Pachycrocuta brevirostris* (Carnivora, Hyaenidae) from the Lower Pleistocene site of Untermaßfeld. In: Kahlke, R.-D., (Ed.), Das Pleistozän von Untermassfeld bei Meinigen (Thüringen), Teil 2. Monographien des Römisch-Germanischen Zentralmuseums 40, 2, 673–690, pls. 121–127.

Turner, A., Antón, M., 1996. The giant hyaena *Pachycrocuta brevirostris* (Mammalia, Carnivora, Hyaenidae). Geobios 29, 455–468.

Vinuesa, V., Madurell-Malapeira, J., Ansón, M., Alba, D.M., 2014. New cranial remains of *Pliocrocuta perrieri* (Carnivora, Hyaenidae) from the Villafranchian of the Iberian Peninsula. Bollettino della Società Paleontologica Italiana 53, 39–47.

Viret, J., 1954. Le Loess A Bancs Durcis De Saint-Vallier (Drome) Et Sa Faune De Mammifères Villafranchiens. Nouvelles Archives du Muséum d'Histoire naturelle de Lyon 4, 1–200.

Vislobokova, I.A., Agadjanian, A.K., 2015. New data on large mammals of the Pleistocene Trilca fauna, Montenegro, the Central Balkans. Paleontological Journal 49, 651–667

Weithofer, K.A., 1889. Die fossilen Hyänen des Arnothales in Toskana. Denkschriften der Kaiserlichen Akademie der Wissenschaften, Mathematisch-Naturwissenschaftliche Classe 55, 337–360, pls. I–IV.

Werdelin, L., 1999. *Pachycrocuta* (hyaenids) from the Pliocene of east Africa. Paläontologische Zeitschrift 73, 157–165.

Werdelin, L., Solounias, N., 1991. The Hyaenidae: taxonomy, systematics and evolution. Fossils and Strata 30, 1–104.

1996. The Evolutionary history of hyaenas in the Miocene of Europe and Western Asia. In: Bernor, R.L., Fahlbusch, V, Mittmann, H.-W. (Eds.), The Evolution of Western Eurasian Neogene Mammal Faunas. New York, 290–306.

JESSICA ARNOLD

BOVID BONE MODIFICATIONS BY *PACHYCROCUTA* FROM THE EARLY PLEISTOCENE SITE OF UNTERMASSFELD

Abstract

In this study, 2,169 *Bison menneri* bone samples from the fossil site of Untermassfeld were analysed for bite marks produced by the hyena species *Pachycrocuta brevirostris*. Of prime importance were quantitative data collections and as much detail concerning the localisation and categorisation of the bite marks as possible. The study investigates the extent to which characteristics and regularities can be established in the feeding strategies of Hyaenidae in general and *P. brevirostris* of Untermassfeld in particular. The analyses make it clear that the hyenas of Untermassfeld favour nutrient-rich skeletal elements for which the minimum amount of mechanical effort is needed. At the same time, the *B. menneri* material of Untermassfeld demonstrates a clear preference of certain feeding trends of the hyenas regarding different long bones. The gnaw and bite marks on the *B. menneri* material were also assigned to groups using the bone modification categories described by Maguire et al. (1980) and their frequency compared with recent bone-breaking hyena species and *P. brevirostris* from other fossil sites. The investigations should support future transregional comparisons with other Hyaenidae sites.

1. Introduction

Within the context of the Early Pleistocene fossil site of Untermassfeld and the studies on their formation, bone modifications play an important role since they are part of the taphonomic processes in all fossil deposits. To a considerable extent, the cause of these modifications is the Pleistocene hyena species *Pachycrocuta brevirostris* (Gervais, 1850), which is represented by numerous skeletal elements and coprolites in the Untermassfeld fossil record (Keiler 2001; Turner 2001; Iannucci et al. in this volume; Kahlke in this volume).

The following study focused on quantitative data collection including the most accurate possible localisation of the bone modifications. The feeding behaviour of *P. brevirostris* was also investigated in order to draw conclusions about possible characteristics and regularities. An integral part of the study was also the classification of the bite marks into seven different categories of bone modification. On the one hand, this provides an overview of the different bite marks and the manner of their formation and, on the other, it enables a direct comparison to be made with existing findings on the feeding behaviour of the various Hyaenidae.

2. Material und method

Results on bone modification by *P. brevirostris* on remains of *Hippopotamus antiquus* from Untermassfeld have previously been published (Kahlke 1997b, 282 f.). Thus, the influence of this hyena species on the site is already known (Kahlke 2001c, 956 f.; 2006, 81). In the current study, which focuses on *B. menneri,* a new part of the Untermassfeld collection has been investigated with regard to bone modifications caused by *P. brevirostris* and thus a further step towards the detailed analysis of the site has been achieved. The bones of *B. menneri* are particularly suitable for this investigation because of their representative number and massive size, which increases the likelihood of useable remains surviving following the processing of the bones by hyenas (Zapfe 1939).

In addition to the monographs on the Untermassfeld site by Kahlke et al. (1997a; 2001a; 2001b) the theoretical foundation for this investigation is notably based on the following studies: the work of Zapfe (1939), which deals with recent and Pleistocene spotted hyenas; Sutcliffe (1970), who described bone modifications using recent spotted hyenas in East Africa; Maguire et al. (1980), in which the authors deal with studies of modifications by recent bone-breaking Hyaenidae; Haynes (1980a; 1980b; 1983), whose study focuses on the traces of various recent and Pleistocene carnivores, especially in North America; Hill (1989), who studied the dens of recent spotted hyenas in Amboseli National Park, Kenya; Saunders and Dawson (1998), whose analyses were based on a Plio-Pleistocene hyena den (Haro River Quarry, Pakistan); Palmqvist et al. (2011) and Madurell-Malapeira et al. (2015), which deal with the activities of *P. brevirostris* in Venta Micena (Spain) and the Vallparadís Section (Iberian Peninsula); Fourvel et al. (2012), who focus on the dens of Pleistocene cave hyenas in France; and Fosse et al. (2012), whose study describes the taphonomic signature of *Canis lupus* in Poland (Bialowieza, Bieszczady).

The categorisation of the surface modifications caused by *P. brevirostris* was carried out according to the method developed by Maguire et al. (1980) and later adopted by Saunders and Dawson (1998). The nine categories of key bone modifications are described as follows.

Category 1: Ragged-edged chewing
These are typical spiral fractures recorded in long bones and irregular, serrated bone edges that are the result of the chewing motion of incisors and canines.

Category 2: Shallow-pitting
According to the aforementioned authors, this type of surface modification is presumably caused by the premolars of juvenile hyenas, which chew on the bone during an early stage of rearing. This activity forms many small round or triangular outlines.

Category 3: Punctate depressions or perforations
Described here are typical round depressions or perforations caused by the bone-breaking premolars, i. e. the P^3 and $P_{3–4}$ of adult hyenas.

Category 4: Lunate or crescent-shaped fracture scars
These lunate or crescent-shaped scars are remnants of the third category described above. They arise when the line of fracture from the hyena's bite runs along a perforation. In this case, so-called back flakes or pointed flakes, i. e., sickle-shaped fragments, are left behind.

Category 5: Striations or gouge marks
These are usually short, straight tooth tracks that are perpendicular to the longitudinal axis of the bone and therefore run approximately parallel to each other. From observations, these are attributed to juvenile hyenas.

Category 6: Contiguous or close, irregular and randomly-oriented grooves
These include irregular and randomly oriented grooves that are positioned close to each other. They often occur at the ends of bones and are considered the result of persistent chewing.

Category 7: Scooping or hollowing out
This is the hollowing out of the cancellous bone by incisors, canines and the tongue of hyenas.

Category 8: Acid-etching and erosion of bone
This type of modification is seen in regurgitated fragments and is caused by the hyena's strong gastric acidity. After digestion, the splinters have a polished appearance and exhibit fine fissures, which are a result of the etching (Zapfe 1939, 124 f.).

Category 9: Splintering or shatter-cracking (comminuted fractures)
These are usually jagged transverse fractures in shafts of limb bones, they are also called comminuted fractures.

In the present study, categories 1 to 7 were taken into particular consideration. Although category 8, bone splinters etched by gastric acid, were also present in the Untermassfeld material (Kahlke 2001c, 956; in this volume, Fig. 25e), they could not be examined under the aspect of a single species, in this case *B. menneri*, owing to their limited size. In the case of category 9, the assignment of these modifications to the activities of *P. brevirostris* is difficult, as such comminuted fractures and breaks can also be caused by other taphonomic processes, e. g., trampling, and geological processes, such as sediment pressure (Kahlke 2001c, 950; Fernández-Jalvo and Andrews 2016, 4).

A complete and detailed overview of the investigated material is available in the Senckenberg Research Station for Quaternary Palaeontology library (Arnold 2018).

The MNI values were determined using the metacarpalia III + IV dex. for the adult individuals and on the basis of the mandibulae or, more precisely, the dP_4 dex. for the juvenile animals. For juveniles, an MNI of 18 is recorded and for adults an MNI of 51. The MNI total for *B. menneri* in Untermassfeld is 69 (as of January 2018).

The following table (**Table 1**) gives a summary overview of the examined bones and teeth. With regard to the table values, it should be noted that there are sample groups which comprise several bones, e. g., vertebrae or parts of limbs, but which are entered under a common catalogue number. These groups are counted as a single find in the NISP-total. The table entry of the ossa sesamoidea includes both the forelimb and the hindlimb samples, as an accurate determination in this regard was limited. The same applies to the phalanges, which is why the phalanges of the front and hind legs as well as the phalanges I to III were summarised as one item in the statistics. In addition, metacarpalia III + IV and metacarpalia V as well as metatarsalia III + IV and metatarsalia II have been considered together, since there are not many mc V and mt II and as

		NISP			body site	
	total	out of this juvenil	fragments	sin.	dex.	indet.
Entire material	2,169	148	806	778	943	450
Cranium	18	0	16	1	0	17
Maxilla	28	8	26	10	18	1
Mandibula	73	33	56	35	36	3
Dens (isolated)	193	55	29	67	66	60
Vertebra	76	0	29			76
Costa	19	0	19	0	0	19
Scapula	13	0	6	4	4	5
Humerus	102	4	91	35	52	15
Radius	69	5	57	21	46	2
Ulna	36	2	29	8	27	1
Radius/Ulna (fused)	20	0	7	6	14	0
Carpalia (total)	274	0	12	115	146	13
Os carpale sec. + tert.	50	0	0	22	28	0
Os carpale quartum	44	0	0	17	27	0
Os carpi accessorium	28	0	0	11	17	0
Os carpi intermedium	47	0	0	25	22	0
Os carpi radiale	45	0	0	19	26	0
Ossa sesamoidea Mc	4	0	0	0	4	0
Os carpi ulnare	43	0	0	21	22	0
Metacarpalia III + IV	145	13	74	54	64	27
Metacarpale V	13	0	0	1	1	11
Pelvis	44	1	42			44
Femur	46	1	38	19	24	3
Patella	2	0	1	2	0	0
Tibia	89	6	72	37	47	5
Tarsalia (total)	534	2	63	209	231	94
Astragalus	105	1	12	51	51	3
Calcaneus	105	1	37	43	57	5
Os tarsi centrotarsale	2	0	0	2	0	0
Os tarsi malleolare	73	0	0	34	39	0
Os tarsi scaphocuboideum	84	0	12	38	47	0
Ossa sesamoidea (total)	79	0	0	0	0	79
Os tarsale primum	6	0	0	2	4	0
Os tarsale sec. + tert.	74	0	0	40	33	1
Metatarsale II	4	0	0	0	1	3
Metatarsalia III + IV	115	9	81	49	44	22
Phalanx I	123	5	33	53	60	10
Phalanx II	75	4	13	30	34	11
Phalanx III	57	0	12	23	28	6

Table 1 Synopsis of the entire material of *Bison menneri* from Untermassfeld.

they were also unaffected by bite marks, they have no influence on the present study. In contrast, astragali and calcanei were considered separately from the other tarsal bones, since of the tarsal bones only these were affected by bite marks and therefore a separate investigation makes sense.

Fig. 1 shows the distribution of the investigated material of *B. menneri* from Untermassfeld with respect to the various skeletal elements. Noticeable here is a dominance of the carpal and tarsal bones, which make up a total of 37 % of the finds, as well as other compact elements such as teeth (9 %) and phalanges (12 %). Less common, but represented with more than 100 finds are humeri (5 %) and metacarpalia (7 %) as well as metatarsalia (5 %). In addition, with more than 50 recovered items, radii (3 %), tibiae (4 %), mandibulae (4 %) and vertebrae (3 %) show a high presence. Rarely represented are ulnae and femora as well as pelvic fragments, each of which accounts for about 2 % of the bones. The rarest are cranial fragments, which include the processus cornualis, and maxillae (1 % each), as well as in the postcranial skeleton finds of scapulae, ribs, radius and ulna in fusion, and especially the patellae, with a share of less than 1 %.

The side of the body which the skeletal elements belonged to is, as far as could be determined, indicated. In total, 1,721 finds could be assigned to a body side, corresponding to 79 % of all finds. With 448 finds (21 %) a side could not be determined. The number of bones assigned to the right side of the body is 943 and accounts for 43 % of the total *B. menneri* findings. 780 finds, i.e., 36 % of the skeletal elements, are assigned to the left side of the body (**Fig. 2**).

The percentage of finds found to be skeletal elements of juvenile individuals amounts to 7 % and that of senile individuals to less than 1 %. The remaining finds from adult and subadult individuals account for 93 % of the total material. The term »various« comprises both adult and subadult material, as well as those bones in which the individual age could not be fully determined, due to strong fragmentation and the ambiguous connection in Bovidae between the closing of the epiphyseal plates and the growth of the individual skeletal elements (Koch 1935). This should ensure that the most accurate MNI possible can be ex-

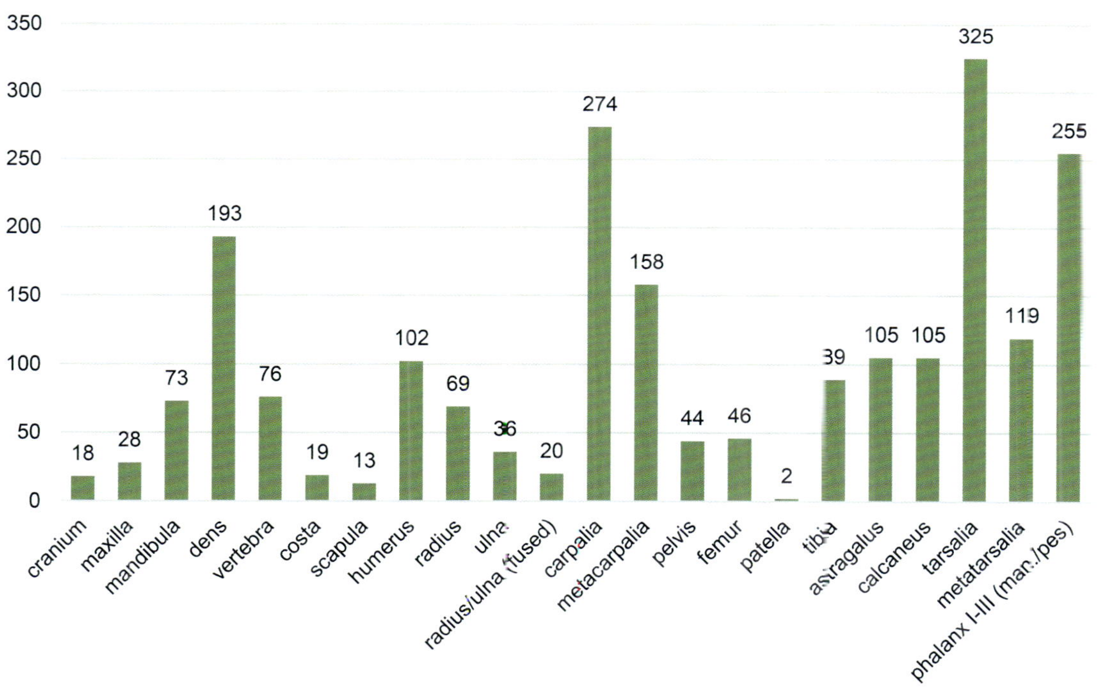

Fig. 1 Distribution according to skeletal elements of the material of *Bison menneri* from Untermassfeld examined in the present study.

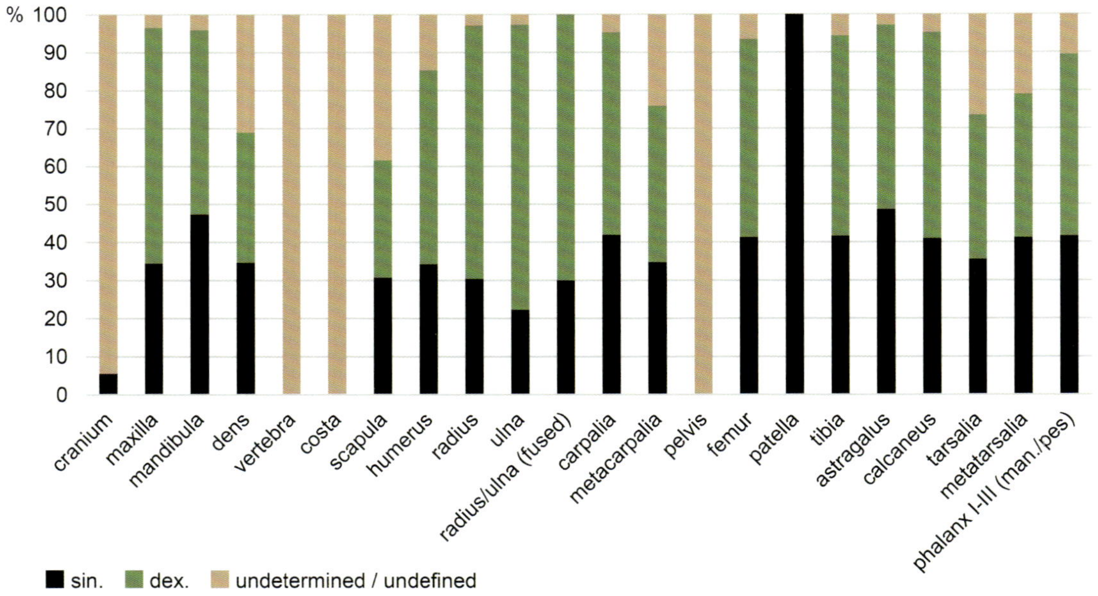

Fig. 2 Percentage distribution of the material of *Bison menneri* from Untermassfeld by body side.

pected. Overall, the proportion of finds identified as juvenile or senile differs significantly within the various skeletal elements (**Fig. 3**). 29 % of maxillae findings are from juveniles and 4 % from senile animals. In the case of the mandibulae, 45 % are assigned to juvenile bones and 5 % to senile bones. Within the isolated teeth, about 28 % are deciduous teeth. In the postcranial area, significantly fewer pieces from juvenile individuals were identified, with the metacarpalia and metatarsalia at 8 % representing the largest number. This is followed by radii (7 %), ulnae (6 %) and tibiae (7 %). Humeri (4 %), pelves and femora (2 % each), phalanges (4 %), astragali and calcanei (1 % each) have the lowest share of bones assigned to juvenile animals. The remaining skeletal elements, i. e. vertebrae, ribs, scapulae, radius and ulna in fusion, patellae and carpal resp. tarsal bones, with the exception of astragali and calcanei, could not be assigned to any juvenile finds. Bones of senile individuals were not identified in the postcranial skeleton.

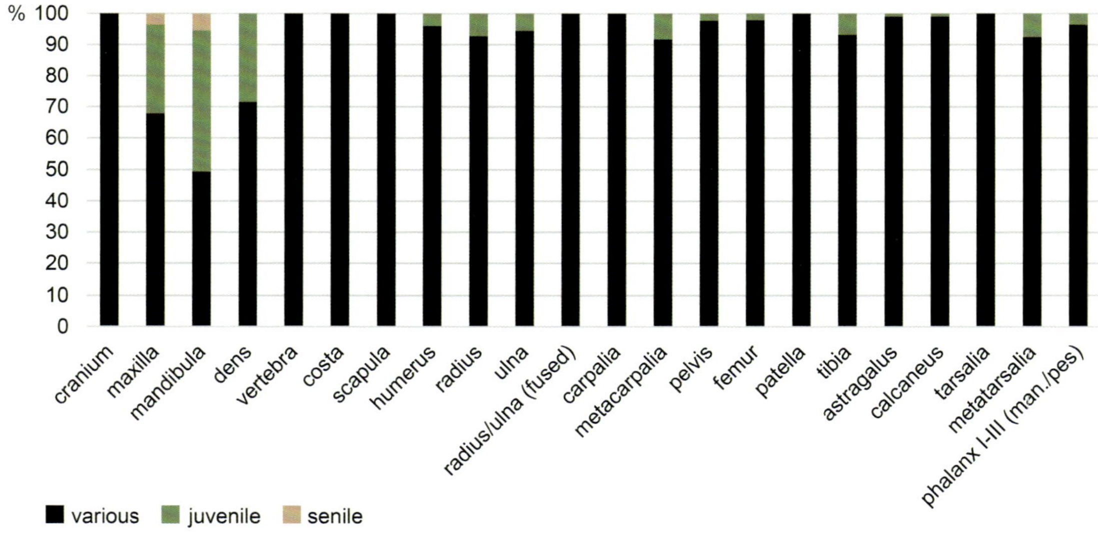

Fig. 3 Age distribution of the material of *Bison menneri* from Untermassfeld.

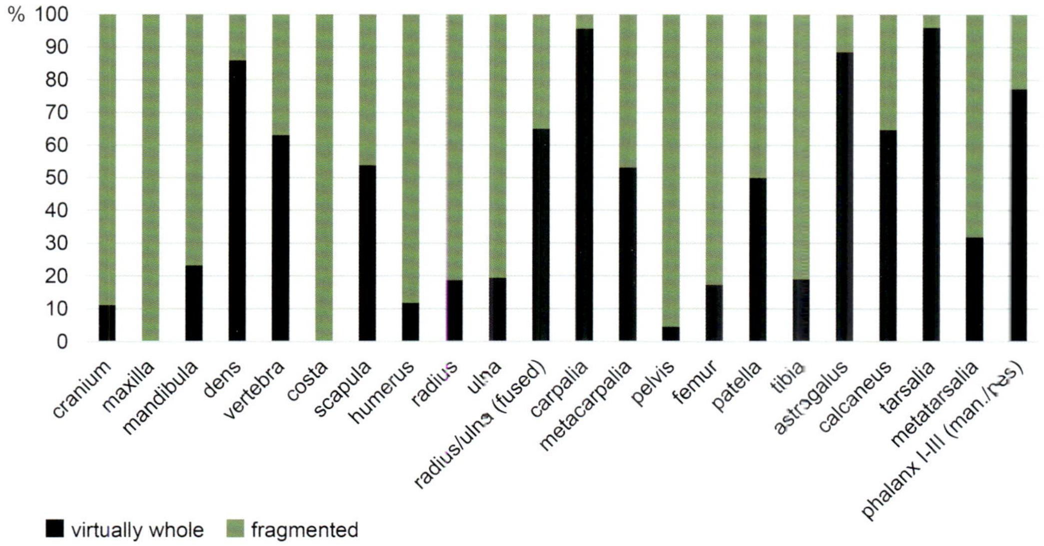

Fig. 4 Percentage share of the fragmented material of *Bison menneri* from Untermassfeld.

The proportion of fragmented bones comprises about 37 % of the total material. Depending on the skeletal element, the degree of fragmentation of the finds differs (**Fig. 4**). Maxillae (100 %) and mandibulae (77 %) are most frequently fragmented. This also applies to ribs (100 %), pelves (95 %) and the long bones, such as humeri, which are 88 % fragmented, followed by femora at 83 %. In the case of radii, ulnae and tibiae, 81 % of the finds are fragments. Even in metatarsal bones (68 %) and patellae (50 %) at least half or more of the finds are in a fragmented state. In vertebrae (37 %), scapulae (46 %), radius and ulna in fusion (35 %), metacarpalia (47 %), calcanei (35 %) and the phalanges (23 %), less than 50 % of bones are fragmented. Teeth (14 %) and the carpal and tarsal bones (4 % each), including the astragali (11 %), show the least fragmentation, at less than 15 %.

Depending on the type of long bone, various parts of the bone dominate the existing fragments. Approximately 54 % of long bone fragments are distal epiphyses. The proportion of proximal epiphyses is 40 %.

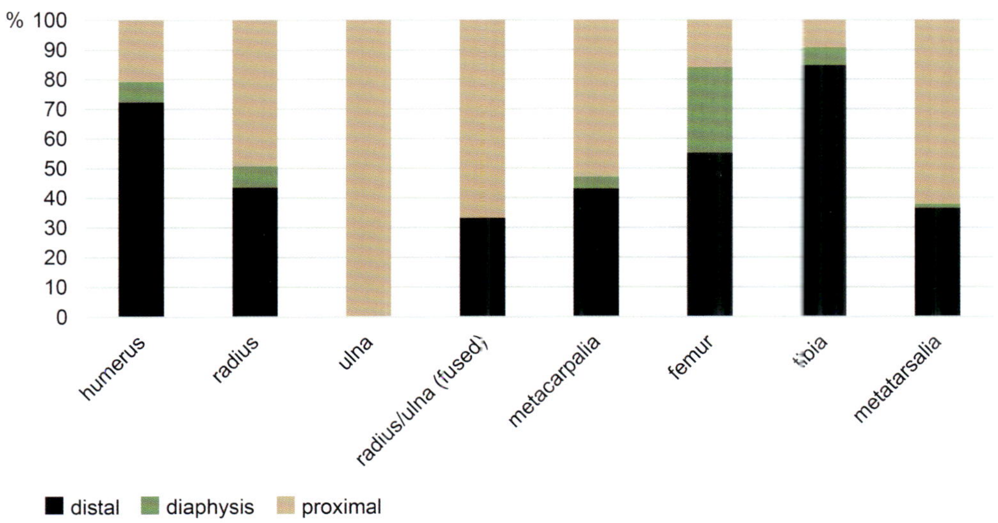

Fig. 5 Percentage distribution of the present bone fragments within the long bones of *Bison menneri* from Untermassfeld.

Finds with remaining diaphysis and missing epiphyses are especially rare, at 6 %. **Fig. 5** shows the percentage amounts of the bone sections of the different long bones.

The largest number of individual diaphyses occurs in the femora. Here, 29 % of the fragments are a diaphysis without epiphyses. More commonly, proximal or distal epiphyses are found, sometimes with sections of the diaphysis. Nevertheless, the distribution also differs here. In the case of radius and ulna in the fused state (63 %) and especially in the ulna (93 %), the proximal part of the bone is predominantly present as a fragment. This also applies to the metapodia in which 54 % of the fragments are proximal epiphyses. The other long bone fragments are mainly distal epiphyses, especially tibiae (78 %) and humeri (70 %), but also femora (55 %).

3. Results

In total, of the 2,169 bones and tooth remains examined, 273 finds clearly show demonstrable modifications due to the feeding activity of hyenas. This corresponds to 12.6 % of the total *Bison menneri* material from the Untermassfeld site. Radius and ulna in fusion have the highest number, with 45 % of the bones displaying bite marks (**Fig. 6**). For the patellae, the proportion is above 50 %, but it should be noted that this is based on only two finds. This is followed by tibiae (40 %) and humeri (36 %) as well as the isolated radii (22 %) and ulnae (39 %). Such modifications are also common to the mandibulae (33 %), femora (28 %) and pelves (30 %). Metacarpalia (19 %) and metatarsalia (21 %) as well as calcanei (16 %), vertebrae (12 %) and ribs (11 %) show fewer traces of feeding. The maxillae (7 %), astragali (3 %), carpal bones (2 %) and phalanges (7 %), meanwhile, are hardly affected by such traces. The few existing cranial bones, including the horn cores and the isolated teeth, as well as the tarsalia, with the exception of calcanei and astragali, are completely devoid of any visible bone modifications by predators.
Of the total finds of mandibulae that display bone modifications caused by hyena activity, 50 % are juvenile and 13 % are senile. In addition, humeri (5 %), radii (7 %), metapodia (mc 13 %, mt 12 %) and calcanei (6 %) all possess a low percentage of bones from juvenile individuals with bite marks. Within the other skeletal elements, no juvenile remains were recorded with bite marks.

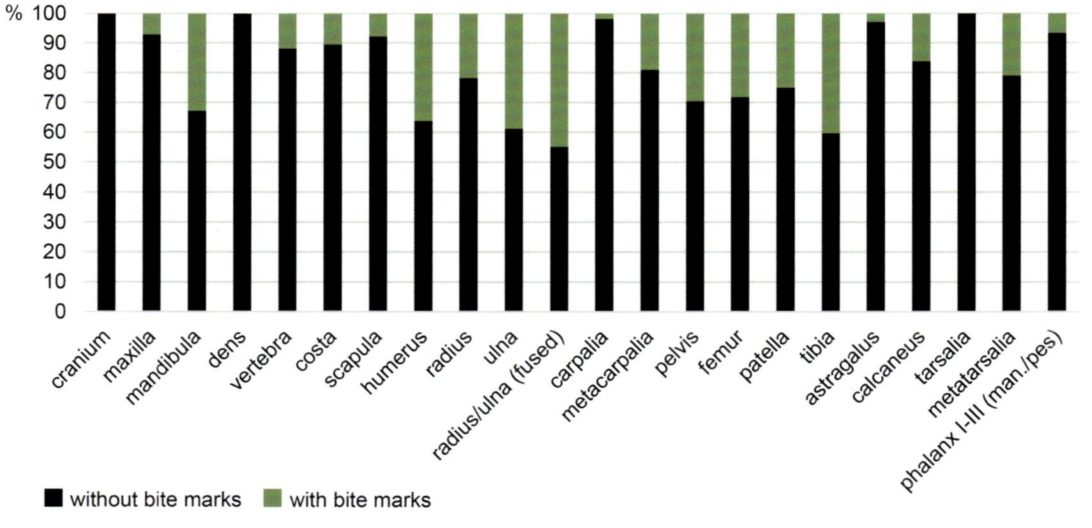

Fig. 6 Percentage share of bone modifications by *Pachycrocuta brevirostris* to skeletal elements of *Bison menneri* from Untermassfeld.

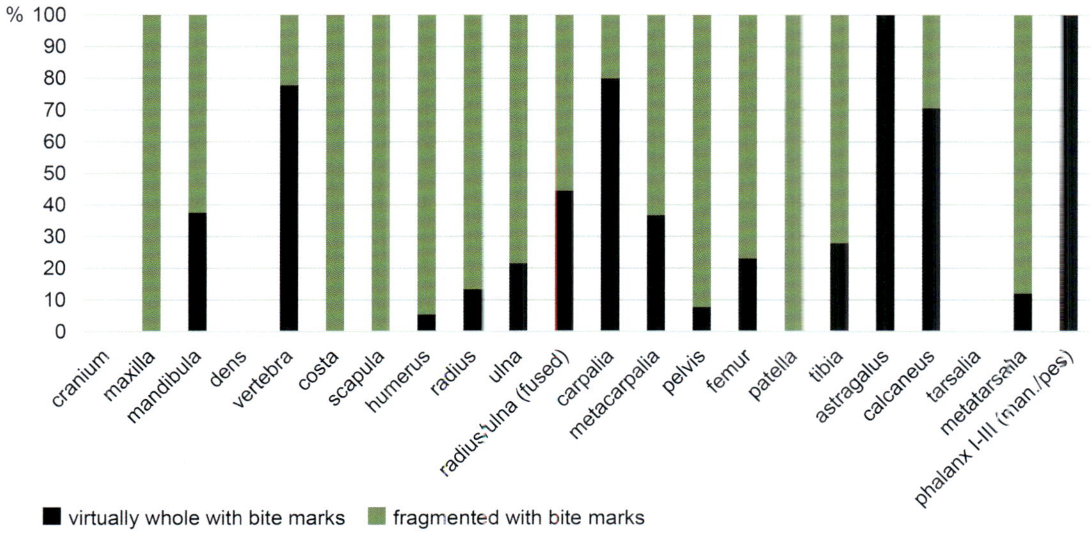

Fig. 7 Fragmentation state of the skeletal elements of *Bison menneri* from Untermassfeld modified by *Pachycrocuta brevirostris*.

The skeletal elements were fragmented to varying degrees by *P. brevirostris* (**Fig. 7**). Although astragali and phalanges show evidence of bite marks, the bones were left almost completely intact. This also occurs relatively often with vertebrae, calcanei and carpal bones. All other bones, especially the long bones, but also maxillae, mandibulae, and costal and pelvic bones are mostly fragmented through modification by *P. brevirostris*. Overall, 67 % of all *B. menneri* specimens found with evidence of bite marks by *P. brevirostris* are fragmented. 33 % are in a predominantly complete state.
The distribution of the long bones remaining after modification by *P. brevirostris* shows that overall more distal (47 %) and proximal epiphyses (40 %) remain following feeding. Less common are diaphyses with broken epiphyses (12 %). Different distributions can be seen, depending on the skeletal element (**Fig. 8**). Distal epiphyseal remains constitute the main share of the fragments of the humeri (83 %) and tibiae (92 %). Proximal epiphyses are mainly present in radii (62 %) and ulnae (100 %) as well as metapodia (mc 79 %, mt 84 %). Diaphyseal remains occur mainly in femora (50 %).

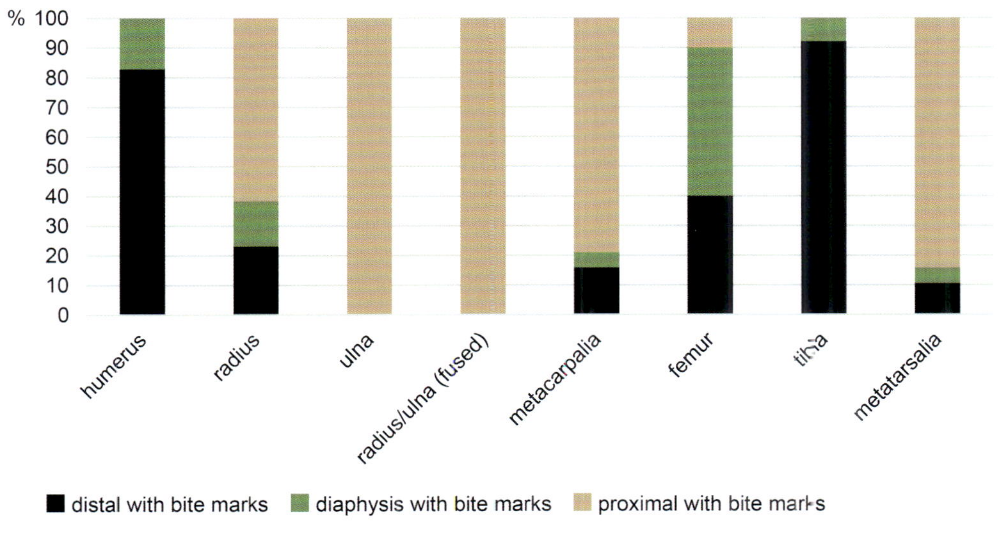

Fig. 8 Percentage distribution of bone fragments within the long bones of *Bison menneri* from Untermassfeld modified by *Pachycrocuta brevirostris*.

3.1. Localisation of bone modifications

A detailed localisation analysis of all bone modifications caused by *P. brevirostris* on the *B. menneri* material reveals the areas of the skeleton and skeletal elements that are predominantly affected by feeding activities (**Fig. 9**).

3.1.1. Cranial elements

Bite marks on the maxillae and remaining cranial bones are rare or absent. In contrast, the mandibulae of the bison collection from Untermassfeld show clear bite marks of hyenas, which are concentrated on certain areas of the lower jaw. Particularly noticeable are the tooth marks near the jaw angle (**Fig. 9**), the edge region of which is almost always missing. Furthermore, there are often modifications by *P. brevirostris* below the teeth of the mandibula and these usually run parallel to the row of teeth. Traces of hyena feeding are also concentrated on the articular process and in the area of the symphysis, which is often bitten off.

3.1.2. Vertebrae and ribs

The vertebrae of Untermassfeld's bison collection are usually only slightly fragmented. Some of the vertebrae modified by hyenas are even found articulated, such as IQW 2003/28 190 (Mei. 27 352) and IQW 2004/28 696 (Mei. 27 858) (Kahlke in this volume, Fig. 14). The bite marks documented on vertebrae in the Untermassfeld material are mainly located on the processus spinosi that form the hump of the bison (**Fig. 9**). However, occasionally the vertebral bodies are also affected by hyena feeding.

Ribs are underrepresented at Untermassfeld in relation to the total material of *B. menneri*, and are always fragmented. Feeding traces on the only detectably affected rib IQW 2003/28 248 (Mei. 27 410) are close to the centre of the bone.

3.1.3. Forelimb long bones

Scapulae are underrepresented in the *B. menneri* samples from Untermassfeld. In the only scapula fragment from Untermassfeld that shows traces of biting [IQW 1984/19 946 (Mei. 19 466)], only the glenoid fossa is present. Tooth impressions are present in the proximal direction of the fracture edge of the glenoid fossa as well as on the joint (**Fig. 9**).

In the case of the humeri from *B. menneri*, it is the distal epiphyses and some diaphyses in particular which survive with modifications by *P. brevirostris* (**Fig. 8**). Proximal epiphyses with clear bite marks are not documented. On the humeri, the toothmarks are located especially on the broken edges of the diaphyses and in the caudal area of the distal epiphyses (**Fig. 9**).

In the radii, only small amounts of distal epiphyses and diaphyses have bite marks. In the case of ulnae, as well as the radius and ulna in the fused state, bite marks were only identified on the proximal epiphyses (**Fig. 8**). Looking at the bite marks in **Fig. 9**, it becomes clear that the olecranon of the ulnae was preferentially chewed. It is also noticeable that the radii are affected by biting in the distal part and in the area of the diaphysis, but rarely at the proximal epiphysis.

The carpal and tarsal bones as well as the metapodia and the phalanges have been summarised and are dealt with in the hand and foot bones section.

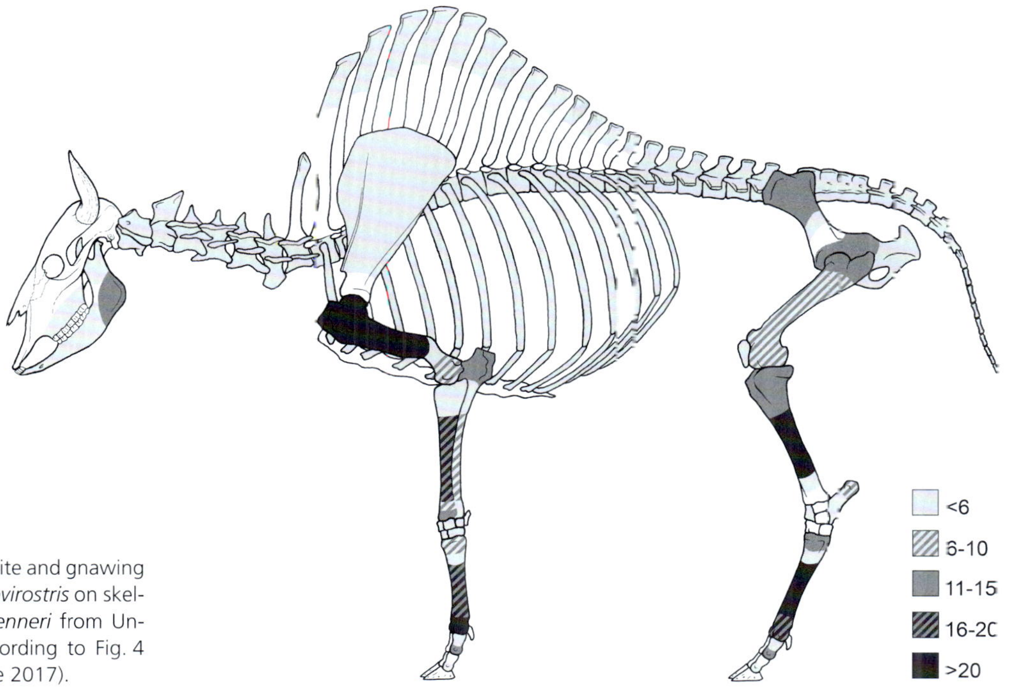

Fig. 9 Accumulation of bite and gnawing marks by *Pachycrocuta brevirostris* on skeletal elements of *Bison menneri* from Untermassfeld (modified according to Fig. 4 in Van Asperen and Kahlke 2017).

3.1.4. Hindlimb long bones

On the pelves, the ilium and the os ischii were often bitten off by the Untermassfeld hyena (**Fig. 9**). The os pubis also shows clear signs of biting. Only the parts around the acetabulum and at the transition from os ilium to the acetabulum normally remain. However, on these fragments, near the acetabulum, there are many smaller traces of bite marks by *P. brevirostris* (**Fig. 9**).

The Untermassfeld *B. menneri* femora finds possess a relatively even distribution of the distal epiphysis and the diaphysis remains. The proximal epiphysis is underrepresented (**Fig. 8**). In some existing proximal epiphyses as well as the proximal areas of the diaphysis, the bites are mainly on the trochanter major, caput and collum respectively, or directly below these when the proximal epiphysis has been chewed or broken off (**Fig. 9**). Tooth traces of *P. brevirostris* are also found in the area of the diaphysis, especially at the edges of the fracture that mark the end of the bitten piece. The distal side of the femora also shows clear modifications, especially in the area of the medial condyle and lateral condyle.

The remaining tibiae fragments are, for the most part, distal epiphyses with pieces of the diaphysis, and individual diaphyses (**Fig. 8**). Proximal epiphyses with bite marks of *P. brevirostris* are not recorded. The remaining distal ends with parts of the diaphysis have a distinctly tapered shape (**Fig. 10**). The tibial diaphysis is nearly completely covered with bite marks from *P. brevirostris* (**Fig. 9**). Often, the tooth impressions at the fracture point characterise the area of the bone where the proximal section was broken off. However, numerous bone modifications are also observed on the diaphysis of almost complete pieces. The epiphyses of the tibiae have no bite marks even when they are present (**Fig. 9**).

Neither complete finds of entire fibulae nor their fragments are documented in the fossil record of Untermassfeld.

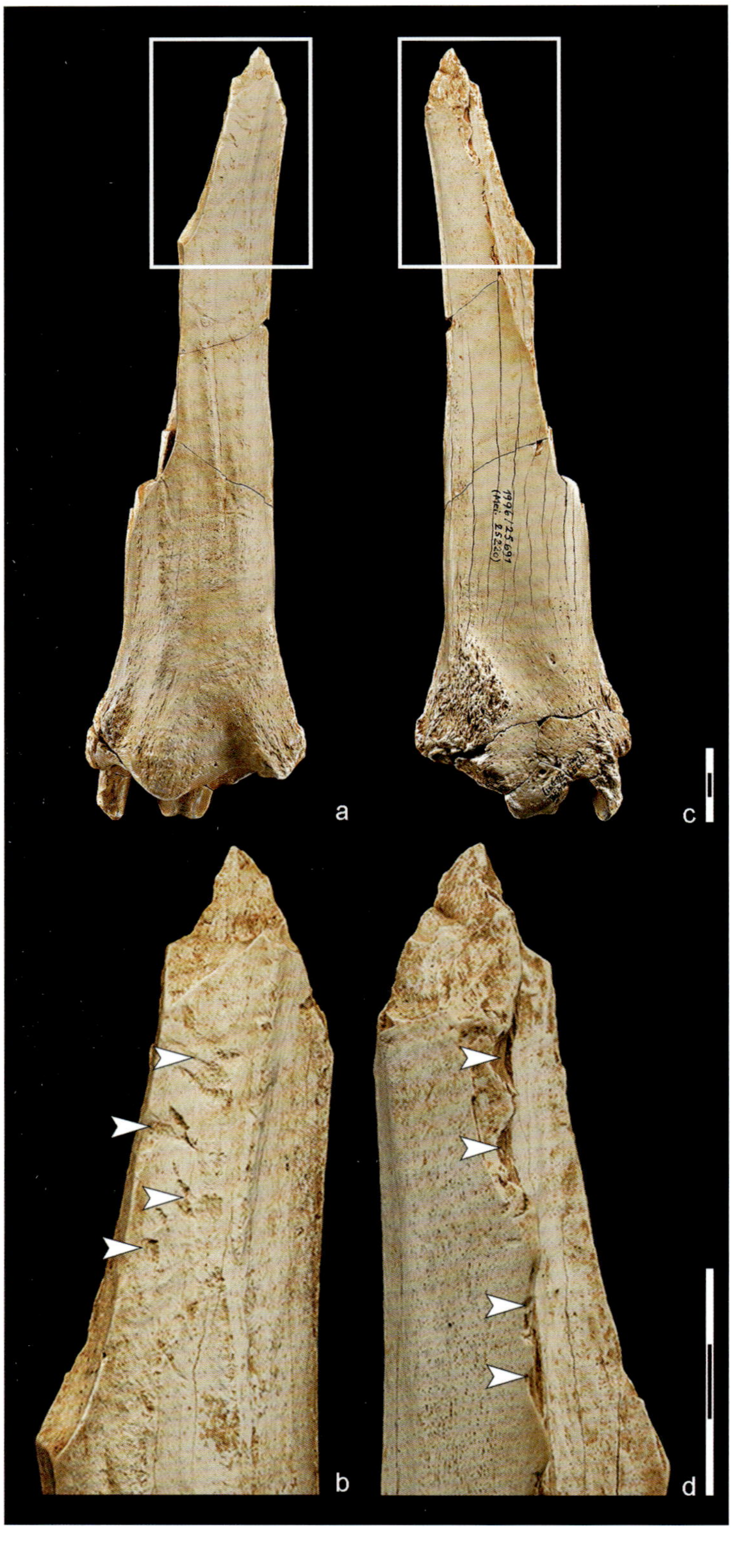

Fig. 10 *Bison menneri*, Untermassfeld. – **a–d** Distal tibia fragment dex. IQW 1996/25 691 (Mei. 25 220) with bite marks by *Pachycrocuta brevirostris* and a distinctive tapering fracture edge, caudal, cranial views. – Scales: 30 mm.

3.1.5. Hand and foot bones

In comparison to their large number, the carpalia and tarsalia are generally less affected by fragmentation and modification in the Untermassfeld fossil record. An exception to this is observed in the astragali and calcanei. Of all the bones of the carpalia and tarsalia, these possess the highest proportion of finds with bite marks from *P. brevirostris*. The calcanei recovered from Untermassfeld display predominantly broken off tubera calcanei and marks which are confined to the area below the partly missing proximal segment of the bone (**Fig. 9**).

Both metacarpalia and metatarsalia clearly outnumber the proximal ends and few distal epiphyses are preserved. Occasionally diaphyseal remains without epiphyses were documented (**Fig. 8**). In addition, the data shows that in metacarpals, a greater percentage of bones that possess feeding marks are mostly complete. In the case of the metatarsals, on the other hand, the proportion of fragmented and modified bones is greater than in the case of the radii, ulnae, femora and tibiae (**Fig. 7**). Only the humeri have more fragmented pieces with bite marks.

The metapodia reveal bone modifications in almost all parts of the bone. The diaphysis is a most equally affected in metacarpal and metatarsal bones. However, in metacarpals, the bite marks are clearly focused on the distal epiphysis (**Fig. 9**). In contrast, in the case of the metatarsals feeding marks often appear on the proximal epiphysis.

The bite marks on phalanges I and II are concentrated on the distal region (**Fig. 9**). Traces of hyena teeth are also very occasionally found in the middle section and at the proximal end. Despite the presence of activity by *P. brevirostris*, there was little fragmentation of either phalanges I and II. The phalanges III are not affected by bone modifications caused by *P. brevirostris*.

3.2. Categories of bone modifications

Following the studies of Maguire et al. (1980) and Saunders and Dawson (1998), the different types of gnaw marks were classified into seven different categories. The percentage of the frequency of the categories refers to the entirety of the documented skeletal elements of *B. menneri* with recorded bite marks (**Fig. 11**). Each find may have modifications belonging to one or more categories.
When looking at all Untermassfeld finds of *B. menneri* that display evidence of bite marks (**Fig. 11**), spiral fractures and ragged-edged chewing predominate (category 1 according to Maguire et al. 1980) (**Fig. 12.1**) with a share of 45 %. This is followed by contiguous or close, irregular and randomly-oriented grooves (category 6 according to Maguire et al. 1980) (**Fig. 12.6**) with approximately 22 %; next are lunate or crescent-shaped fracture scars (category 4 according to Maguire et al. 1980) (**Fig. 12.4**) and striations or gouge marks perpendicular to the longitudinal axis of the bones (category 5 according to Maguire et al. 1980) (**Fig. 12.5**) with approx. 10 % each. Scarcely represented are the other categories, which all show values below 10 %. Punctate depressions or perforations (category 3 according to Maguire et al. 1980) (**Fig. 12.3**) with 4 %, and hollowing out of the cancellous bone (category 7 according to Maguire et al. 1980) (**Fig. 12.7**) with 7 % are recorded relatively frequently, while shallow-pitting (category 2 according to Maguire et al. 1980) (**Fig. 12.2**) is extremely rare at about 1 %.
The frequency of the individual categories differs within the various skeletal elements of *B. menneri*. **Fig. 13** shows a percentage distribution of this occurrence in the Untermassfeld material. In the schematic depiction, the skeletal elements that have enough bite marks to identify concentrations in particular areas have been especially considered. The local distributions and the frequencies of modifications of the various categories on the Untermassfeld bison material are shown for the various skeletal elements in **Figs. 14, 17, 19, 20, 22, 23, 25, 29,** and **30**. For all skeletal elements except the pelves, the bones of the left side of the

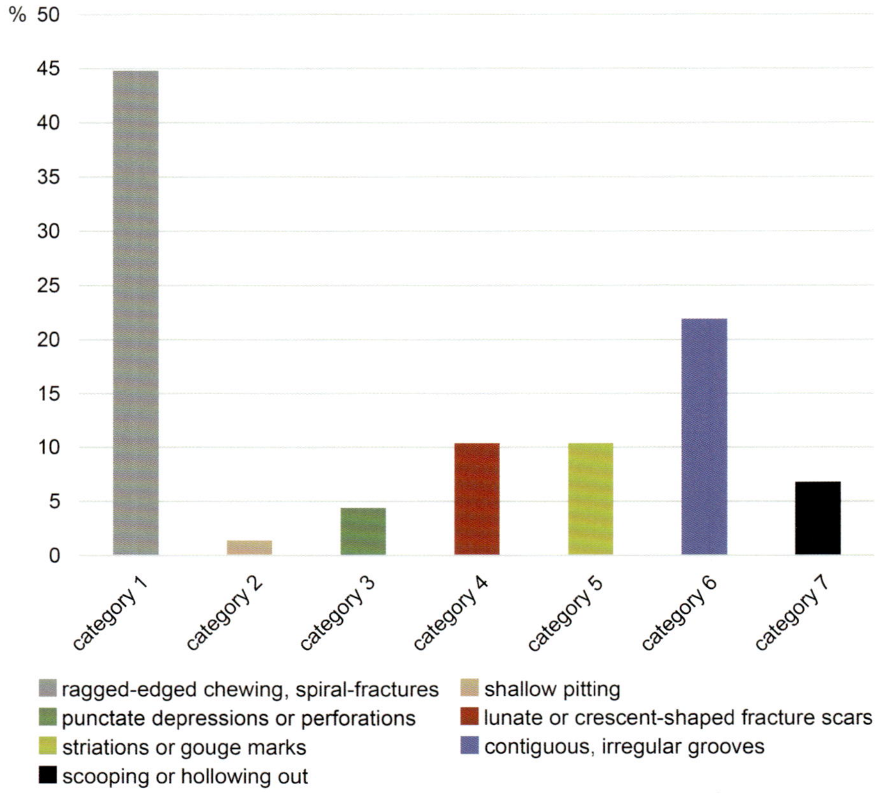

Fig. 11 Percentage share of the categories of bone modifications (according to Maguire et al. 1980) within the total modified material of *Bison menneri* from Untermassfeld.

body are shown in the diagrams. In cases where the positions of the bite marks were on the right side of the body these have been projected onto the left side of the body for the respective skeletal element. Not every skeletal element has the same number of bite marks and the same categories of bone modifications. Since the legend for each figure is the same, not all numerical ranges (n) and types of modifications (category) shown in the legend appear in each of the figures.

3.2.1. Cranial elements

For the maxillae, there appears to be a clear tendency for two categories. The most common category at 67 % is category 1 (ragged-edged chewing), caused by chewing movements (**Fig. 13**). The remaining 33 % of the marks are caused by category 6 (close, irregular grooves).

In the case of the mandibulae, a broader range of categories can be found, although category 1 still dominates at 46 %. This is followed by category 4 (lunate or crescent-shaped fracture scars) at 32 %, then category 6 at 15 % and category 2 (shallow-pitting) at 7 %. Marks assigned to categories 1 and 4 can be found mainly at the angle of the jaw (**Fig. 14**), which was to some extent often broken off by the feeding activities of *P. brevirostris*. In addition, these marks are normally concentrated below, and running parallel to, the row of teeth (**Fig. 15**). On the articular process, in contrast, multiple small triangular depressions are recorded, corresponding to the description of category 2 according to Maguire et al. (1980) (**Fig. 16**). In the area of the symphysis, more close, irregular and randomly-oriented grooves of category 6 are recorded, as well as below the row of teeth and scattered on the base of the temporalis muscle.

Fig. 12 Overview of the categories of bone modifications (according to Maguire et al. 1980) by means of examples of the *Bison menneri* material from Untermassfeld. – **1** category 1, spiral-fracture. – **2** category 2, shallow pitting. – **3** category 3, punctuate depression. – **4** category 4, crescent-shaped fracture scars. – **5** category 5, short gouge marks, perpendicular to the long axis of the bone. – **6** category 6, irregular grooves. – **7** category 7, hollowing out of the cancellous bone. – Scales: 10 mm.

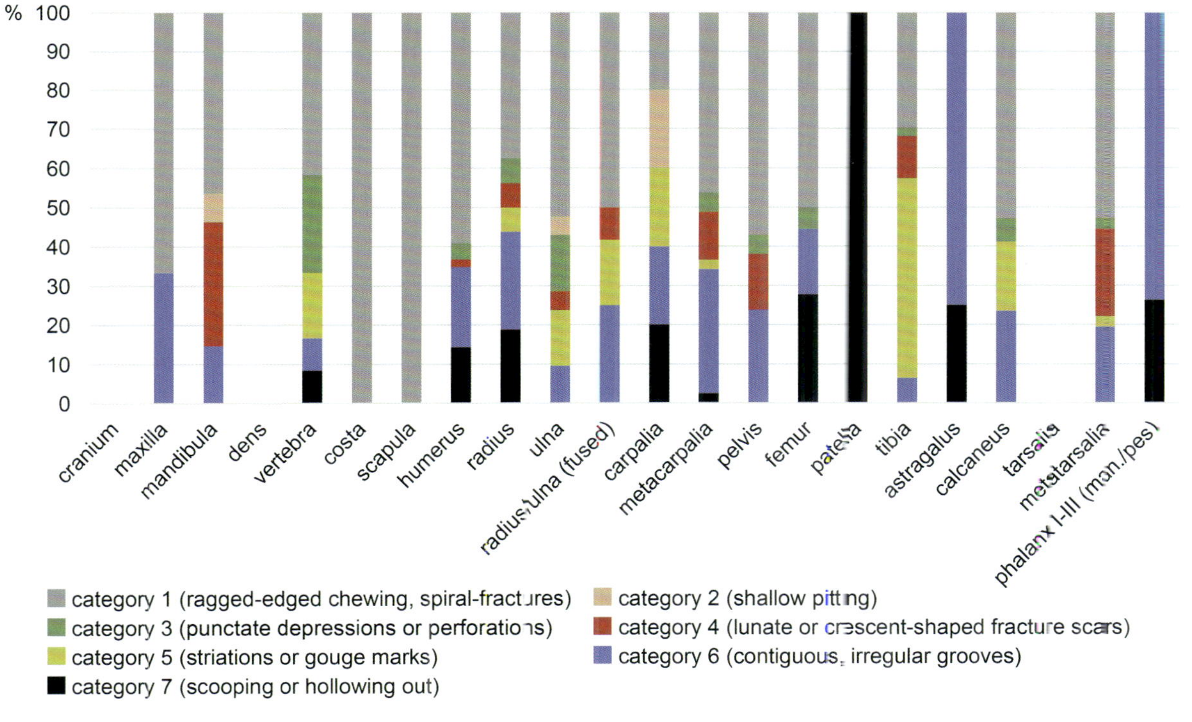

Fig. 13 Percentage distribution of the seven categories of bone modifications (according to Maguire et al. 1980) based on the skeletal elements of *Bison menneri* from Untermassfeld.

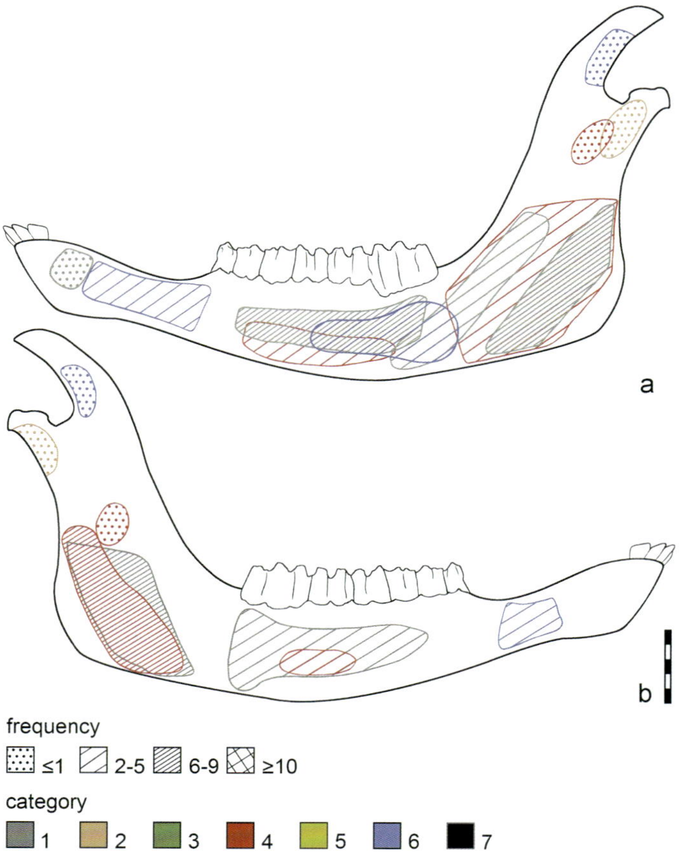

Fig. 14 *Bison menneri*, Untermassfeld. – **a–b** Mandibula sin., schematic depiction of the positions and frequencies of bone modification types, lateral, medial views. The positions of the modifications to the bones of the right body side were projected onto the left side. – Scale: 50 mm.

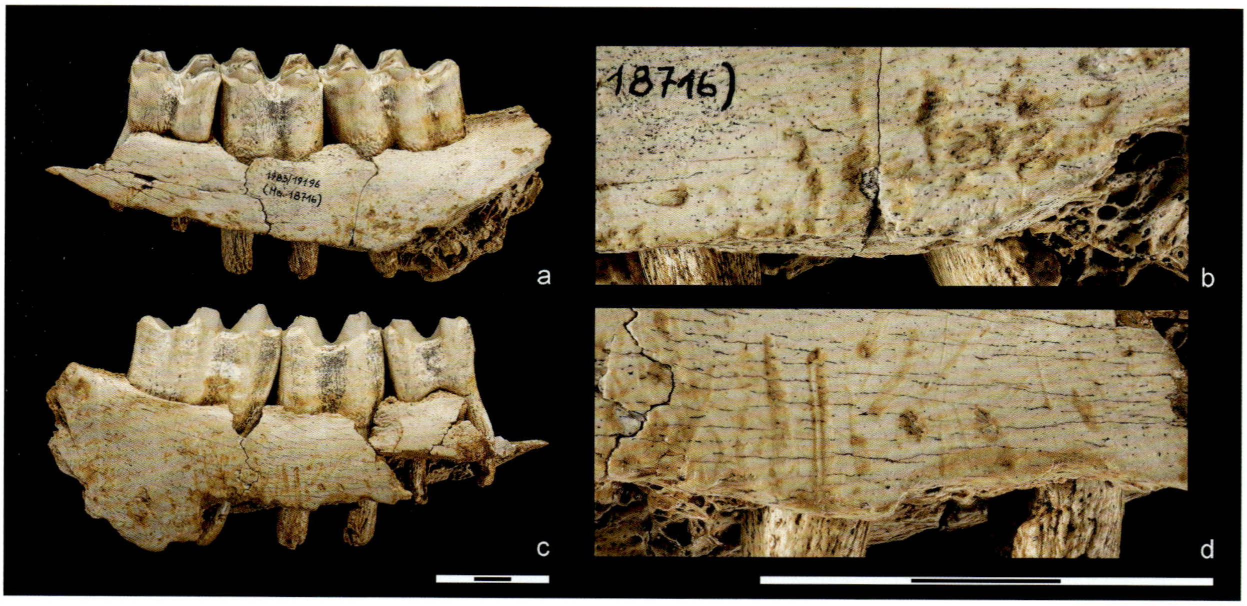

Fig. 15 *Bison menneri*, Untermassfeld. – **a–d** Mandibula fragment sin. IQW 1983/19 196 (Mei. 18 716) with bone modifications of the categories 1 (ragged-edged chewing), 4 (crescent-shaped fracture scars) and 6 (irregular grooves), lateral, medial views. – Scales: 30 mm.

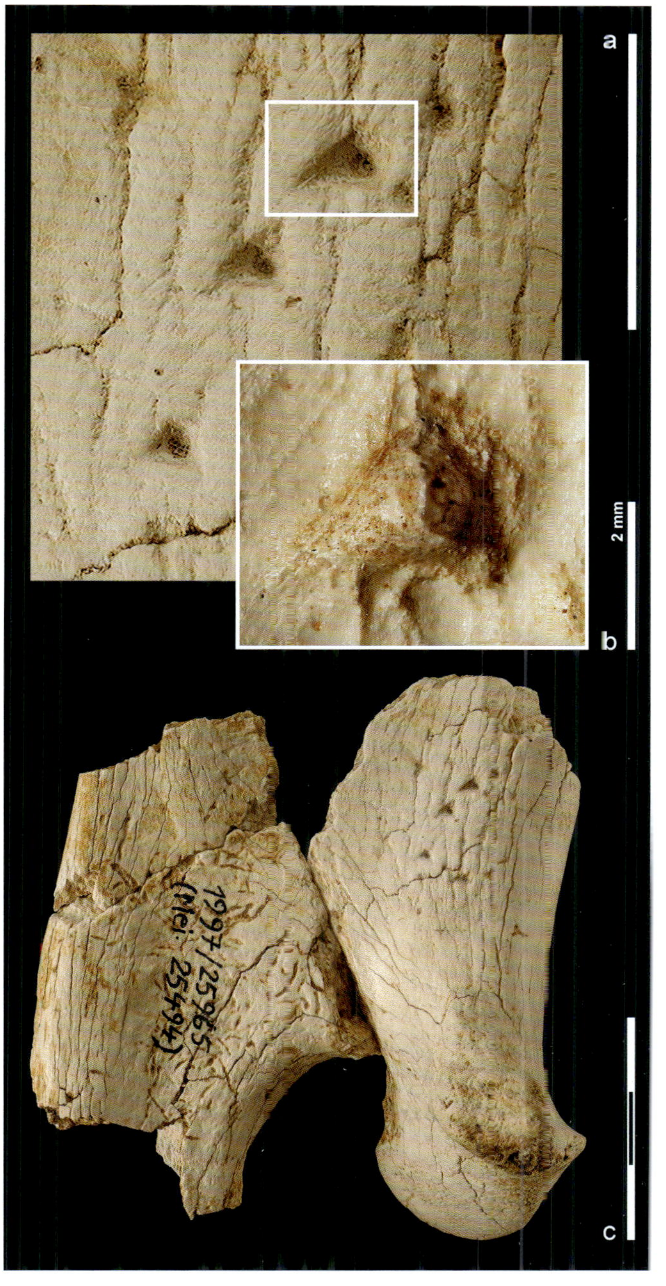

Fig. 16 *Bison menneri*, Untermassfeld. – **a–c** Mandibula (condyle) fragment sin. IQW 1997/25965 (Mei. 25494) with bone modifications of category 2 (shallow pitting), lateral view. – Scales: **a** 10 mm – **b** 2 mm – **c** 30 mm.

3.2.2. Vertebrae and ribs

In the case of the vertebrae, category 1 dominates at around 42 % (**Fig. 13**). Punctate depressions or perforations of category 3 are also present (25 %) along with category 5 striations or gouge marks (17 %). Less common are modifications assigned to categories 6 (close, irregular grooves) and 7 (hollowing out of the cancellous bone), which each account for 8 %. Ragged-edged chewing, which suggests damage belonging to category 1, occurs mainly on the processes of the vertebrae. Category 3 and 7 marks predominate on the vertebral bodies.

On the single affected rib, only category 1 bite marks are recorded, and these are located at various points along the bone, but especially in the middle section.

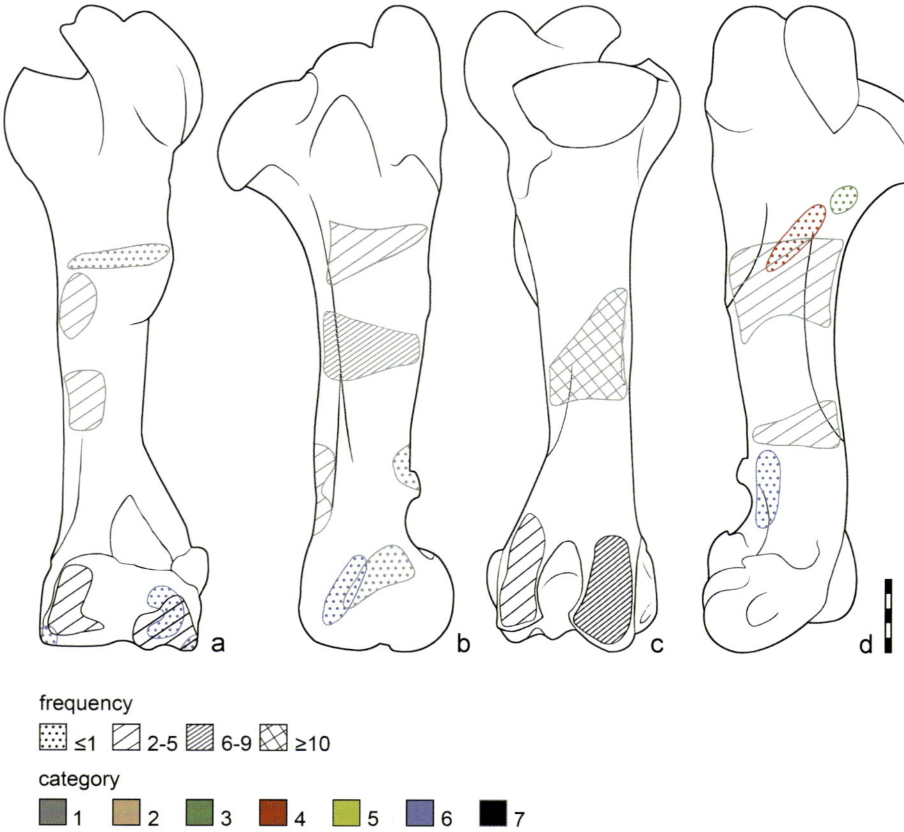

Fig. 17 *Bison menneri*, Untermassfeld. – **a–d** Humerus sin., schematic depiction of the positions and frequencies of bone modification categories, cranial, medial, caudal, lateral views. The positions of the modifications to the bones of the right body side were projected onto the left side. – Scale: 50 mm.

3.2.3. Forelimb elements except phalanges

Scapulae of *B. menneri* are relatively rare in the Untermassfeld fossil record. The single find with identified modifications has bite marks belonging to category 1 only, and these occur mainly at the fracture edges proximal to the glenoid fossa.

Ragged-edged chewing and spiral fractures of category 1 dominate on the *B. menneri* humeri and are attributed to 59 % of the bite marks on this skeletal element (**Fig. 13**). Category 6 (20 %) and 7 (14 %) are also relatively frequently represented. Less common are punctate depressions or perforations at 4 %, and lunate or crescent-shaped fracture scars at 2 %.

Marks belonging to category 1 are to be found in a particularly large number on the corpus humeri (**Figs. 17, 18a, c–d**), as well as in the area of the crista tuberculi and also occasionally on the distal portion of the bone (**Fig. 17**). Randomly-oriented grooves of category 6 occur mainly in the distal part of the bone (**Fig. 18b**). The distinctive marks of category 7 (hollowing out of the cancellous bone) also accumulate clearly in the distal portion, especially in the supracondylar ridge and the condylus humeri (**Fig. 18c**). In the less commonly recorded categories 3 and 4 no accumulations could be detected.

In isolated radii and ulnae, category 1 dominates with a share of 38 % and 52 %, respectively, and has a share of 50 % in the fused bones (**Fig. 13**). For category 2 only a 5 % occurrence is recorded in ulnae and category 3 was localised exclusively to isolated radii (6 %) and ulnae (14 %). Radii and ulnae possess a 6 % and 5 % share of observed bite marks for category 4, respectively, with fused bones at 8 %. Modifications

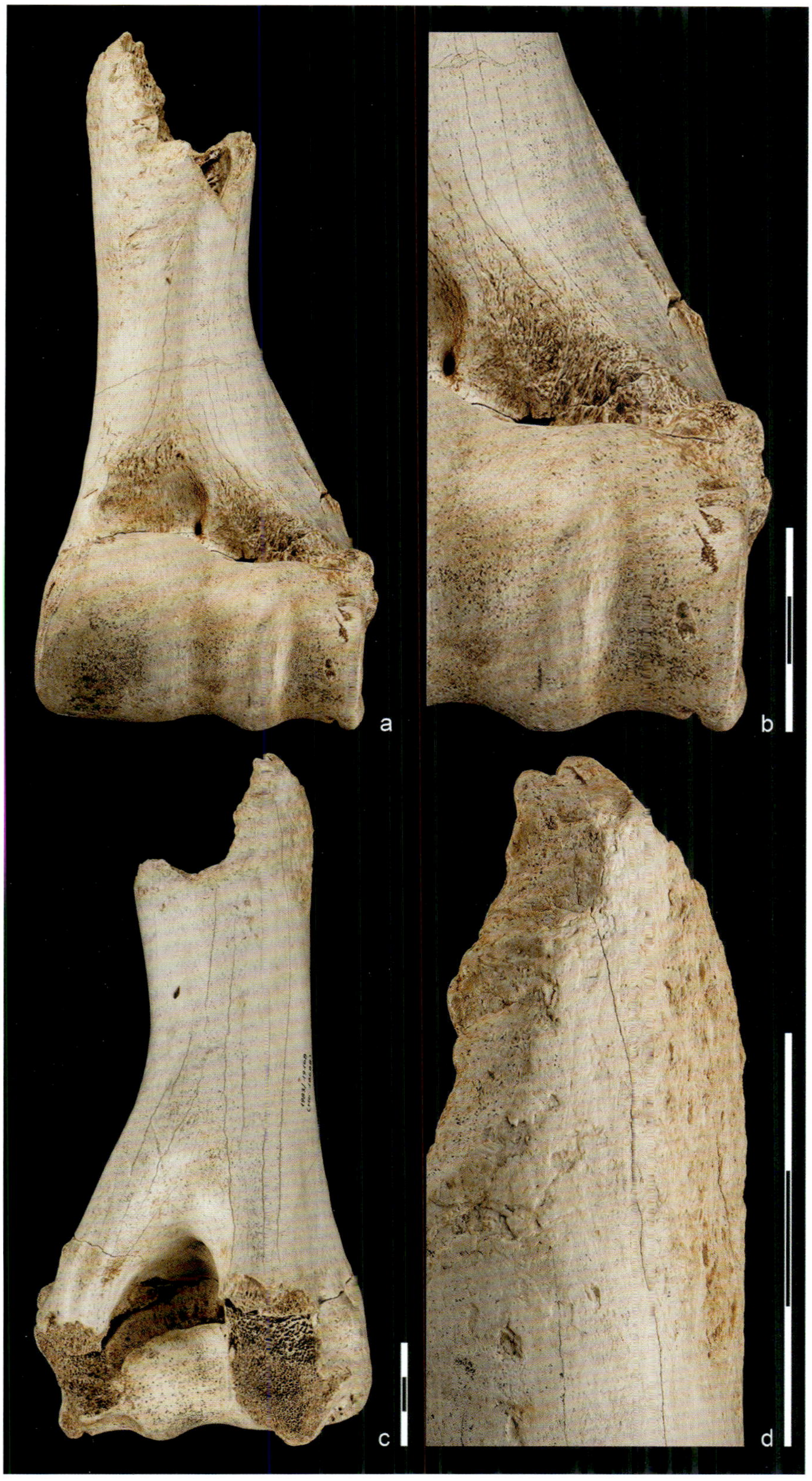

Fig. 18 *Bison menneri*, Untermassfeld. – **a–d** Distal humerus fragment sin. IQW 1983/19168 (Mei. 18688) with various bone modifications of categories 1 (spiral-fracture), 6 (irregular grooves) and 7 (hollowing out of the cancellous bone), cranial, caudal views. – Scales: 30mm.

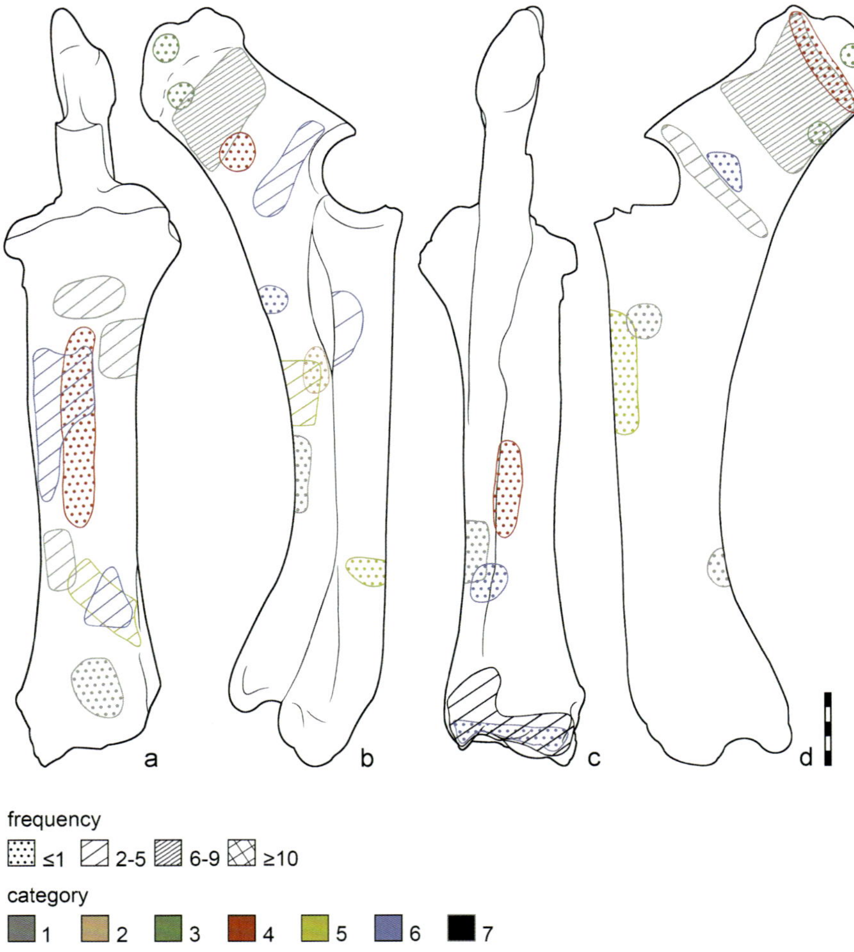

Fig. 19 *Bison menneri*, Untermassfeld. – **a–d** Radius and ulna sin., schematic depiction of the positions and frequencies of bone modification categories, cranial, medial, caudal, lateral views. The positions of the modifications to the bones of the right body side were projected onto the left side. – Scale: 50mm.

assigned to category 5 are present at a share of 6 % for radii, 14 % for ulnae, and 17 % for radius and ulnae in fusion. Category 6 is relatively seldom recorded in the case of the ulnae with a share of only about 10 %, compared to the isolated radii and the radius and ulna in a fused state, which have a share of 25 % each (**Fig. 13**). Category 7 could only be detected in isolated radii at a share of 19 %.

Accumulations of bite marks by *P. brevirostris* in certain categories can also be seen on some parts of the bones of the radii and ulnae of *B. menneri* (**Fig. 19**). Category 1 (ragged-edged chewing) was recorded on the ulnae, along with categories 3 (puncture depressions or perforations), 4 (lunate or crescent-shaped fracture scars) and 6 (close, irregular and randomly-oriented grooves), in particular in the area of the olecranon. In addition, category 1 marks occur on the corpus radii and ulnae as well as on the proximal and distal epiphysis of the radius. The small tooth indentations (shallow-pitting) of category 2 are rare and only localised in the area of the corpus ulnae. Category 4 lunate or crescent-shaped fracture scars are also found in the area of the corpus radii and corpus ulnae, as well as category 5 striations or gouge marks. Category 6 marks are distributed relatively irregularly and are located at the diaphyses and also occasionally in areas of the epiphyses. Modifications attributed to category 7 (hollowing out of the cancellous bone) are found only in the distal joint of the radius and ulna.

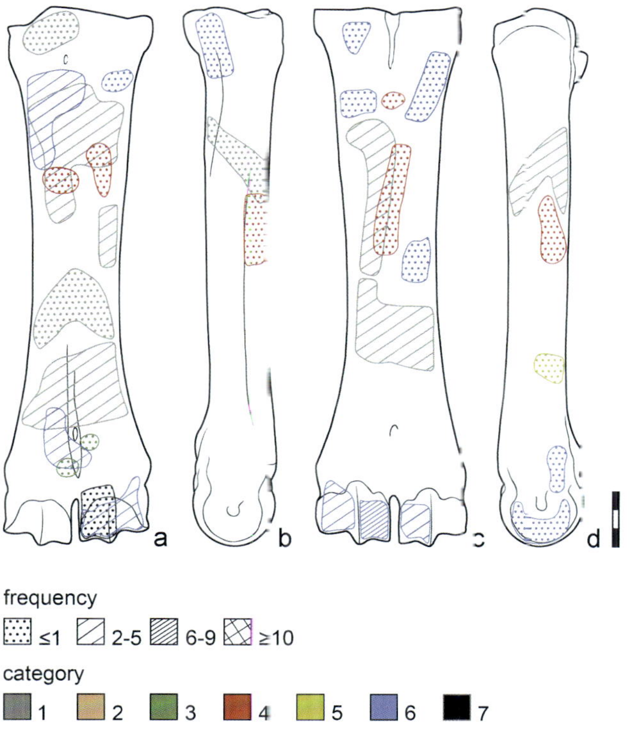

Fig. 20 *Bison menneri*, Untermassfeld. – **a–d** Metacarpal III + IV sin., schematic depiction of the positions and frequencies of bone modification categories, cranial, medial, caudal, lateral views. The positions of the modifications to the bones of the right body side were projected onto the left side. – Scale: 30 mm.

A very low number of bite marks by *P. brevirostris* are observed on the carpal bones. However, in contrast, they display a rather large variance in categories of bone modifications. All existing categories occur at a frequency of 20 % (**Fig. 13**) and these are found on various carpal bones. The categories recorded here are 1 and 2 as well as 5 to 7.

Bite marks attributable to category 1 comprise 46 % of the modifications located on the metacarpal bones (**Fig. 13**). The next largest share is category 6 (32 %). Rarer are category 4 (12 %), category 3 (5 %), category 5 (2 %), and category 7 (2 %).

In the case of the metacarpals, category 1 (ragged-edged chewing and spiral fractures) is distributed over the diaphysis area (**Figs. 20–21**). Epiphyses are not affected by category 1 modifications except in a single case. Premolar perforations attributed to category 3 are less common and focus on the contact area between the diaphysis and the distal epiphysis (**Fig. 20a**). The crescent-shaped fracture scars of category 4 occur mainly in the middle and proximal parts of the diaphysis (**Fig. 20**). Category 5 (striations or gouge marks perpendicular to the longitudinal axis of the bone) occurs only once in the area of the diaphysis (**Fig. 20d**). The irregular and randomly-oriented grooves belonging to category 6 are common and found almost everywhere scattered along the entire length of the bone. Only at the distal end is an accumulation recognisable (**Fig. 20**), and this is also where evidence of hollowing out of the cancellous bone by hyena teeth is also observed (category 7).

The phalanges of the anterior and posterior extremities are summarised in the hindlimb elements section.

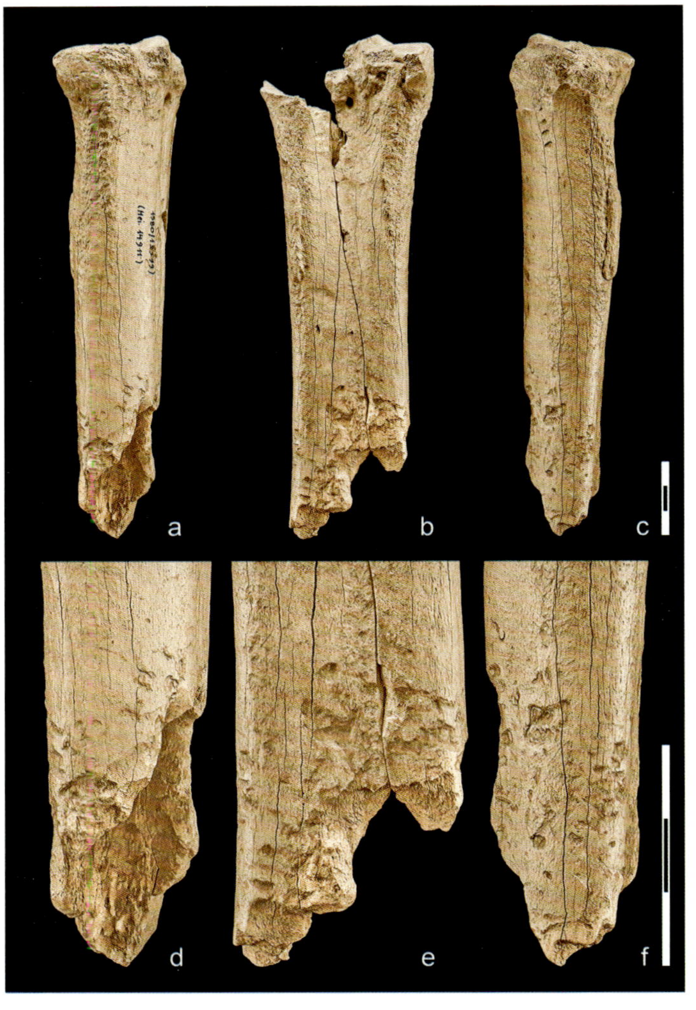

Fig. 21 *Bison menneri*, Untermassfeld. – **a–f** Proximal metacarpus fragment dex. IQW 1980/15 399 (Mei. 14 911) with bone modifications of category 1 (spiral-fracture and ragged-edged chewing), lateral, caudal, medial views. – Scales: 30 mm.

3.2.4. Hindlimb elements including phalanges

Bite marks on the pelvic bones are mainly assigned to categories 1 (57 %) and 6 (24 %), although category 4 (14 %) also occurs relatively frequently (**Fig. 13**). Category 3 (5 %) modifications are rare.

Ragged-edged chewing of category 1 can be identified particularly in the areas of the os ilium, os pubis and os ischii, but also on the acetabulum (**Fig. 22**). Furthermore, these modifications are occasionally localised near the symphysis. Category 4 (crescent-shaped fracture scars) also occurs in the ilium and acetabular area. The latter is also the location where the punctate depressions or perforations of category 3 can be found. The close, irregular grooves of category 6 are found distributed around different parts of the pelvis, such as the os ilium and the os pubis.

Category 1 modifications (50 %) can also be identified as the dominant modification on the femora. Although, noticeably, a relatively large share of the modifications (around 28 %) can be attributed to category 7 (hollowing out of the cancellous bone) (**Fig. 13**). The remaining 23 % is divided into categories 3 (6 %) and 6 (17 %). Ragged-edged chewing and spiral fractures assigned to category 1 are distributed almost over the entire surface of the femora. Nevertheless, in some places there are clear accumulations, especially in the area between trochanter major and t. minor and in the areas of the caput ossis femoris through collum, and fossa trochanterica. The section of the bone at the level of the linea pectinea also shows such modifications. Bite marks assigned to category 7 focus exclusively on the two epiphyses (**Fig. 23**). In particular, the caput ossis femoris, the collum ossis femoris, and the trochanter major and t. minor are proximally affected. In the distal area, the modifications are located in the area of the medial condyle and lateral condyle (**Fig. 24**). The isolated perforations of category 3 are located on the caput ossis femoris and above the medial condyle (**Fig. 23b**). In addition, the seldom recorded modifications of category 6 are found on the collum, the caput ossis femoris, and the distal epiphysis (**Fig. 23c**).

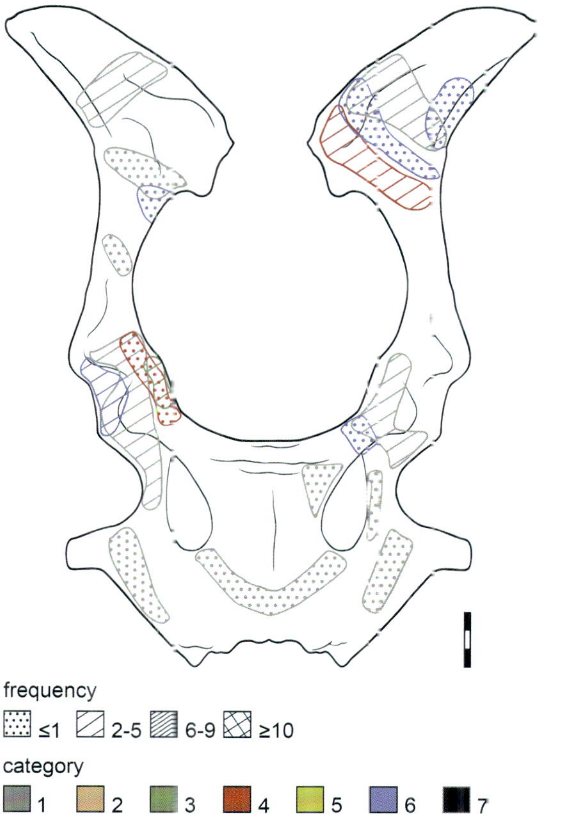

Fig. 22 *Bison menneri*, Untermassfeld. – Pelvis, schematic depiction of the positions and frequencies of bone modification categories, ventral view. – Scale: 30 mm.

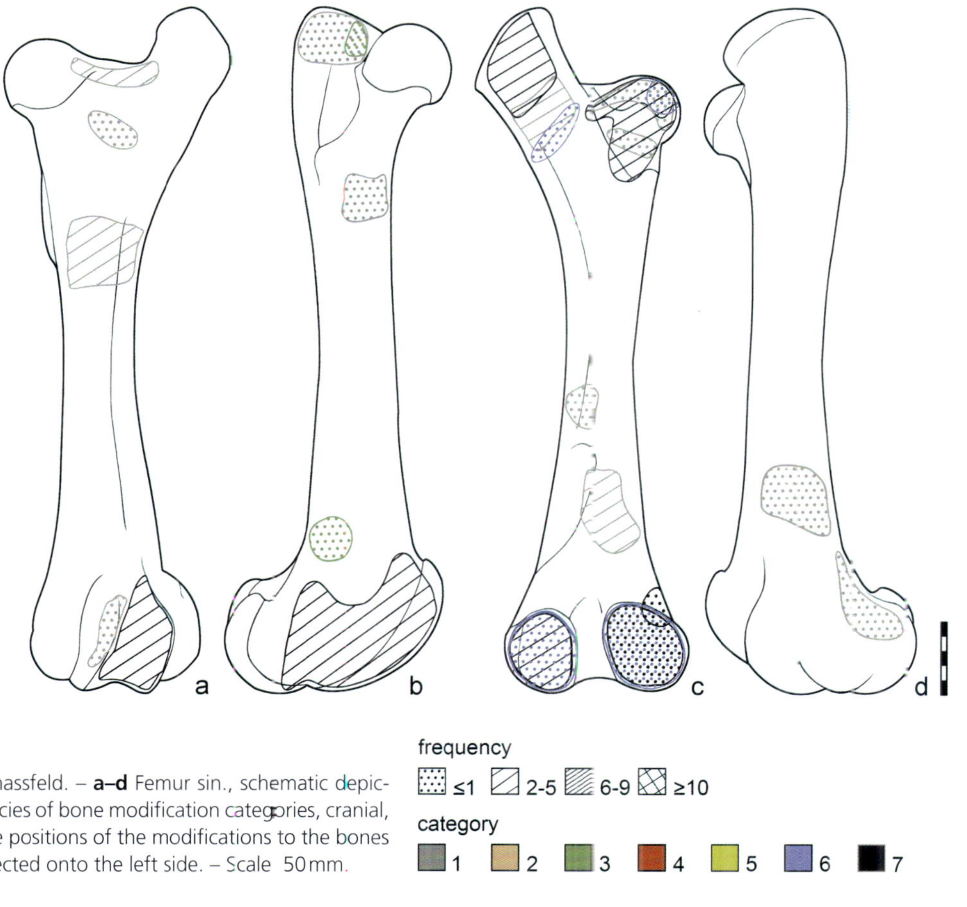

Fig. 23 *Bison menneri*, Untermassfeld. – **a–d** Femur sin., schematic depiction of the positions and frequencies of bone modification categories, cranial, medial, caudal, lateral views. The positions of the modifications to the bones of the right body side were projected onto the left side. – Scale 50 mm.

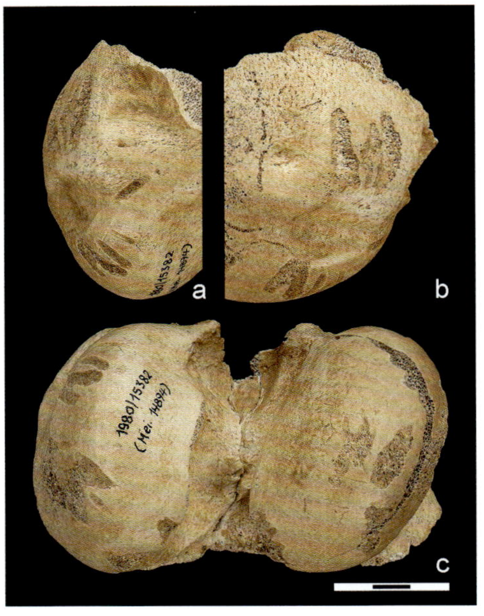

Fig. 24 *Bison menneri*, Untermassfeld. – **a–c** Distal femur fragment dex. IQW 1980/15382 (Mei. 14894) with bone modifications of the categories 6 (irregular grooves) and 7 (hollowing out of the cancellous bone), cranial, lateral, distal views. – Scale: 30 mm.

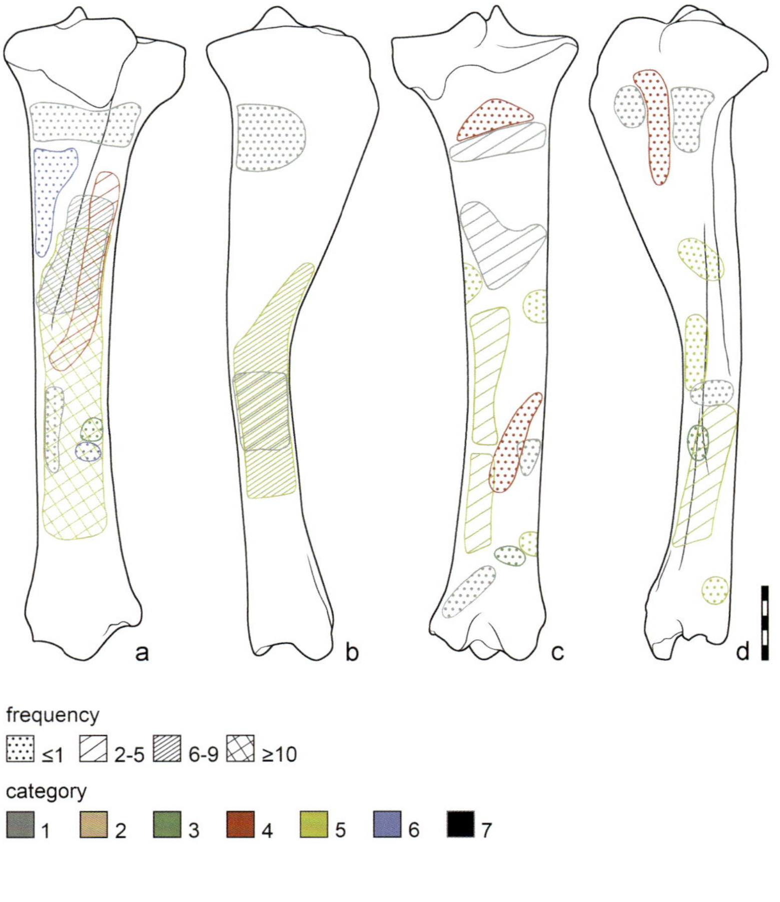

Fig. 25 *Bison menneri*, Untermassfeld. – **a–d** Tibia sin., schematic depiction of the positions and frequencies of bone modification categories, cranial, medial, caudal, lateral views. The positions of the modifications to the bones of the right body side were projected onto the left side. – Scale: 50 mm.

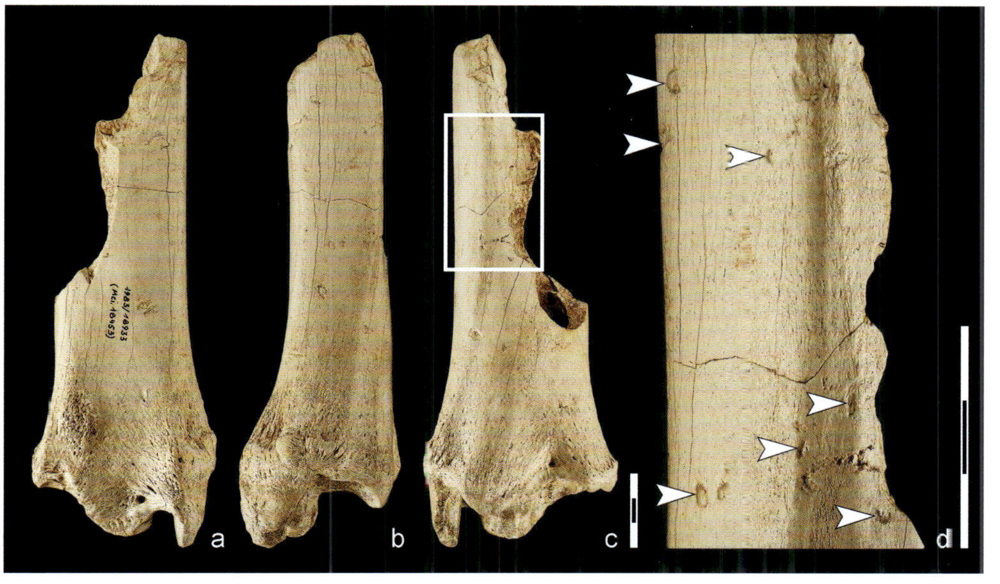

Fig. 26 *Bison menneri*, Untermassfeld. – **a–d** Distal tibia fragment dex. IQW 1983/18933 (Mei. 18453) with bone modifications of categories 1 (spiral-fracture, ragged-edged chewing) and 4 (crescent-shaped fracture scars), cranial, medial, caudal views. – Scales: 30 mm.

The patellae are strongly underrepresented with only two finds, and of these two finds, only one has bite marks. These are assigned to category 7 and are distributed over the entire bone.

The most commonly documented modifications to the tibiae of *B. menneri* are category 5 (51 %). In second place, with a share of 30 %, is category 1 (**Fig. 13**). Rarer is category 4, which has a share of 11 %, followed by bite marks assigned to category 6 at 6 %. The remaining 2 % can be assigned to category 3.

Bite marks localised on the tibia are largely concentrated in the areas between the two epiphyses (**Fig. 25**). Category 1 bite marks are often found in the upper and lower part of the tibial corpus and below the proximal epiphysis. Category 4 modifications are also localised in these areas (**Figs. 26c–d**). The individual tooth marks assigned to category 3 are located above the distal epiphysis and in the lower part of the corpus tibiae. Of all the bone modifications, category 5 (striations or gouge marks) occurs most frequently in the tibiae and could only be detected in the area of the corpus tibiae (**Fig. 27**), but were strongly concentrated there in almost all areas (**Figs. 25a–b**). The articular surfaces of the epiphyses were not affected.

In contrast to all other tarsal bones of *B. menneri*, bone modifications caused by *P. brevirostris* were identified on the astragali and calcanei. In the astragali such modifications are rather rare and are limited to categories 6 (75 %) and 7 (25 %) (**Fig. 13**). They are mainly located at the articular surfaces of the lateral malleolus and at the trochlea to the tibia. The articular surfaces towards the os naviculare and calcaneus are also affected.

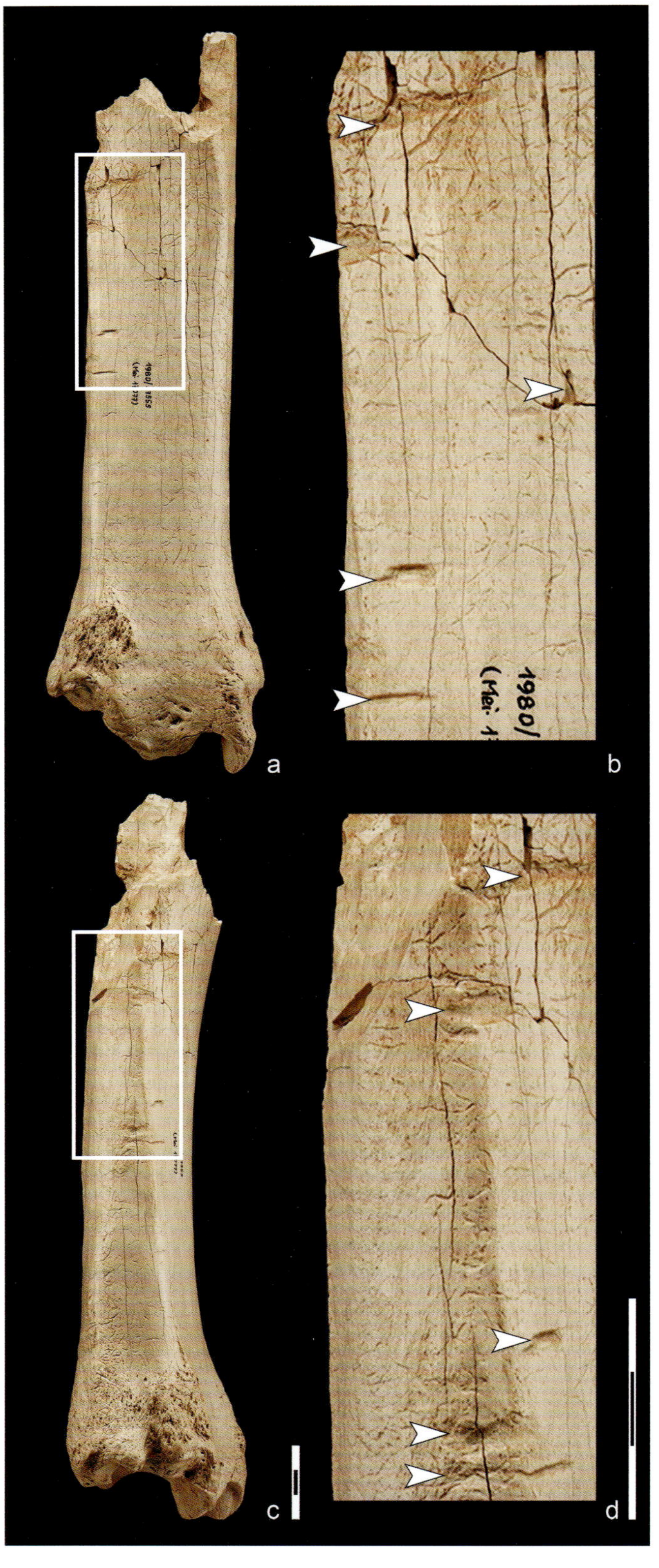

Fig. 27 *Bison menneri*, Untermassfeld. – **a–d** Distal tibia fragment dex. IQW 1980/17555 (Mei. 17077) with bone modifications of category 5 (gouge marks, perpendicular to the long axis of the bone), cranial, lateral views. – Scales: 30mm.

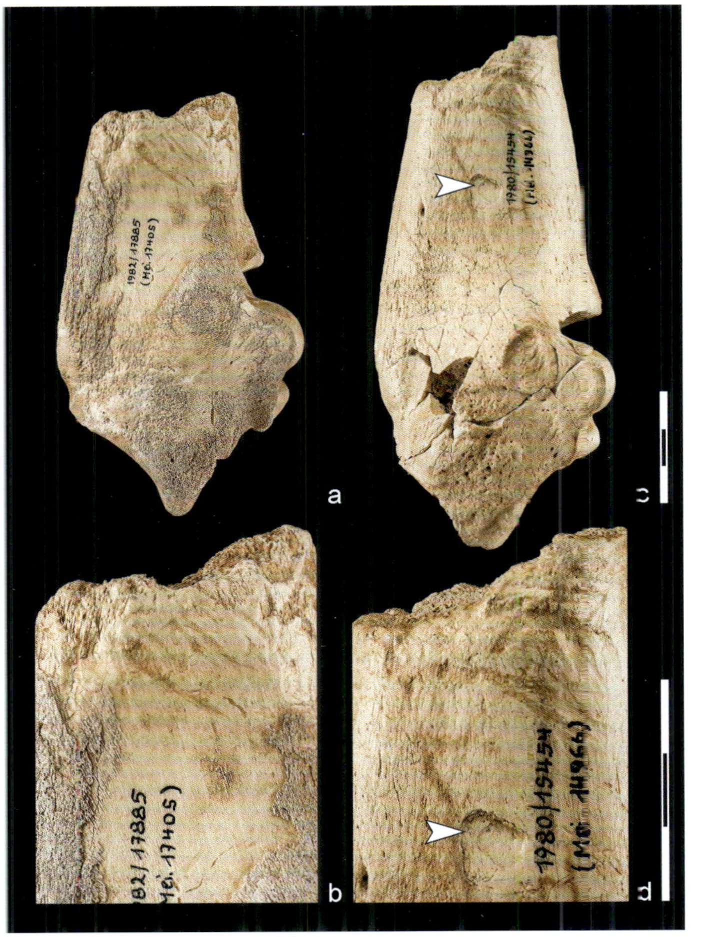

Fig. 28 *Bison menneri*, Untermassfeld. – **a–b** Calcaneus dex. IQW 1982/17885 (Mei. 17405) with bone modifications of category 1 (ragged-edged chewing), lateral views. – **c–d** Calcaneus dex. IQW 1980/15454 (Mei. 14966) with bone modifications of categories 1 (ragged-edged chewing) and 3 (punctate depression), lateral views. – Scales: 30 mm.

The calcanei are frequently affected by bite marks. Here, ragged-edged chewing (category 1) dominates at 53 % (**Fig. 13**). In second place, with just under 24 %, are the irregular grooves of category 6. This is followed by category 5 (striations or gouge marks) with 18 %, and category 3 (punctate depressions or perforations) with 6 % (**Figs. 28c–d**). In this skeletal element, the various categories of bite marks are located in particular in the vicinity of the calcaneal tuberosity (**Fig. 28**). The area of the bone that is articulated with other skeletal parts possesses almost no bite marks.

For the metatarsal bones, category 1 dominates at 53 % (**Fig. 13**). Although crescent-shaped fracture scars of category 4 (22 %) and randomly-oriented grooves of category 6 (19 %) are also well represented. For the remaining 6 % categories 3 and 5 have an equal share at 3 % each. Modifications of category 1 are found more frequently at the proximal epiphysis and in the area of the diaphysis (**Fig. 29**) but can also be found in all sections. The individually occurring punctate depressions or perforations of category 3 were located at both the middle section and on the distal part of the diaphysis. Category 4 marks are distributed widely

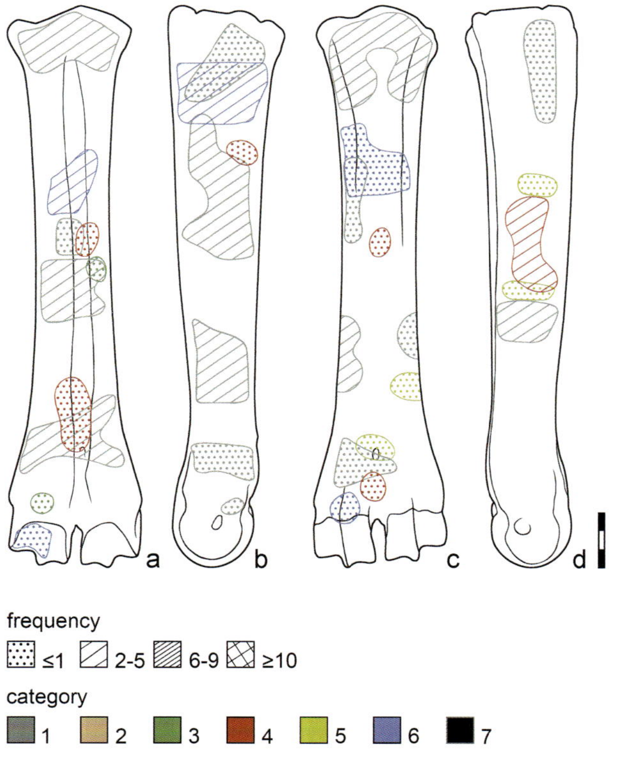

Fig. 29 *Bison menneri*, Untermassfeld. – **a–d** Metacarpal III + IV sin., schematic depiction of the positions and frequencies of bone modification categories, cranial, medial, caudal, lateral views. The positions of the modifications to the bones of the right body side were projected onto the left side. – Scale: 30 mm.

in the area of the diaphysis, especially in the middle section on the lateral side (**Fig. 29**). The same applies to the striations or gouge marks (category 5). The frequent irregular and randomly-oriented grooves of category 6 are detected in all areas of the bone, from the proximal to the distal. They are often documented in the upper part of the diaphysis, and, in addition, also occur directly on the articular surfaces of the distal epiphysis (**Fig. 29a**).

In the phalanges I, modifications of both categories are mainly at the distal epiphysis of the bones, but are occasionally also found at the proximal epiphysis and in the middle section (**Fig. 30a–c**). In phalanges II, only bite marks belonging to category 6 were found, and these were only on the distal part of the bone (**Fig. 30d**). No bite marks were identified on the phalanges III.

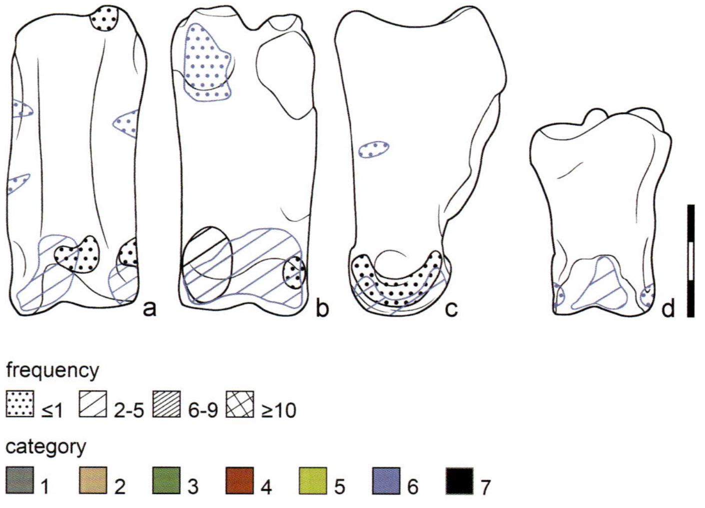

Fig. 30 *Bison menneri*, Untermassfeld. – **a–c** Phalanx I sin., schematic depiction of the positions and frequencies of bone modification categories, cranial, caudal, lateral views. – **d** Phalanx II sin., schematic depiction of the positions and frequencies of bone modification types, cranial view. The positions of the modifications to the bones of the right body side were projected onto the left side. – Scale: 30 mm.

4. Discussion

A striking feature of modifications induced by bone-breaking hyenas is the high degree of fragmentation in long bones compared to other carnivores (Fosse et al. 2012), as well as the fact that these Hyaenidae usually also completely break up the other skeletal elements with their teeth (Haynes 1983). Approximately two-thirds of the bison material from Untermassfeld showing bite marks is at least partially fragmented and the examined long bones often exhibit spiral fractures (**Fig. 13**). Moreover, hyenas leave behind stronger as well as more varied bite marks than other carnivore species and show exceptional osteophage capacities in comparison to other vertebrate groups, as observed by Fosse et al. (2012) in their study of the taphonomic signature of *Canis lupus* in Poland (Bialowieza, Bieszczady). These attributes are made possible by a strong dentition (Schütt 1971) and additional strong masticatory muscles, as well as an enlarged temporalis muscle (Palmqvist et al. 2011). Furthermore, hyena teeth often cause furrows and indentations in the trabecular structures within the bone shaft, which are generally conical to round in shape, and which in adult individuals are 3 to 5 mm wide and 3 to 5 mm deep (Haynes 1983). This data corresponds to the dental impressions made by *P. brevirostris* on the Untermassfeld bison material (see, e.g., **Fig. 12.3**). In addition, the compact bone of specimens modified by hyaenids almost always exhibits tooth marks (Haynes 1983) and in the case of hyena activities, high-frequency bite marks are present on the entire bone material (Fosse et al. 2012).

By comparison, Canidae only very rarely bite through larger and more massive bones, such as those belonging to Bovidae (Peters 1993). This behaviour should also apply in the case of *Canis mosbachensis*, which is common at Untermassfeld (Sotnikova 2001; Keiler 2001, 694; Kahlke in this volume, 1668 f.). Noteworthy similarities are demonstrated in Hyaenidae and Canidae with respect to the skeletal elements and bone sections affected by feeding marks. In the Untermassfeld bison bones the angle of the jaw, the dorsal processes of the vertebrae, the proximal tuberositas of the humerus, the olecranon, the cranial edge of the bone of the os ilium, the os pubis, and the caudal edge of the os ischii, trochanter major and tuber calcanei among others, are particularly often affected by biting. According to Haynes (1980a; 1980b; 1982), the same applies to the carcasses dissected by *Canis lupus*. These similar approaches can be explained by the instabilities of certain joints and ligaments and are therefore similar to the sequence of natural disarticulation (Toots 1965; Hill 1979a; 1979b; Münzel 1987).

Compared to hyenas, the teeth of the documented Untermassfeld Felidae (Hemmer 2001) are almost blade-like, less massive and not cone-shaped (Ewer 1973). Accordingly, it is extremely rare to find dental traces of felids on compact bone (Haynes 1983). In addition, dental traces of felids are described as approximately v-shaped (Haynes 1983) or punctiform and occur, at most, on the limb bones of a carcass (Fourvel et al. 2012).

Bite marks from Ursidae, which are also represented in the fossil record of Untermassfeld (García and Mazza in this volume), were not documented. Ursidae mainly leave traces of gnawing on the limb bones (Fourvel et al. 2012), though, as with the Felidae, they rarely make modifications to compact bones. They also grind the bones instead of biting them like Hyaenidae (Haynes 1983). The teeth impressions of ursids have an approximately square shape (l.c.).

It is probable that a part of the washed-up necromass at Untermassfeld comes from the dens of *P. brevirostris* and their surroundings near the Pleistocene Werra (Kahlke 2006, 78 f.). That these dens, which among other things serve for the rearing of the juveniles, were also flooded, can be deduced from the high proportion of juvenile individuals of *P. brevirostris* present at Untermassfeld (Kahlke 2006, 78). In contrast, the dens of the species *Canis mosbachensis* must have lain outside the flood area, since a significantly lower proportion of juvenile individuals were detected (Kahlke 2006, 79).

A further indication of nearby activities of *P. brevirostris* is provided by studies of coprolites, which are documented at the site of Untermassfeld and are also attributed to the hyena species *P. brevirostris*. For this purpose, comparisons were made with recent animals and investigations support this assumption (Keiler 2001, 697). The faeces were probably deposited by the hyenas in the immediate vicinity of the site (Keiler 2001, 696; Kahlke in this volume, 1694 ff.), which in turn suggests that *P. brevirostris* lived and ate nearby and that, to a large extent, the bite marks examined in this study originate from these feeding activities. The reason for the assumption is that the coprolites were found in almost intact condition (l.c.). Investigations confirm that coprolites only survive for short stretches in water (Larkin et al. 2000) and are almost impossible to find without damage following prolonged transport (Korth 1979). These observations support the morphological analysis concerning the origin of the site and confirm the conclusions regarding the activities of *P. brevirostris* (Kahlke in this volume).

On account of these facts and the characteristics present that are typical of hyenas with regard to the degree of fragmentation of the modified bones, the frequency of the modifications and the shape and size of the tooth impressions, it can be concluded that the feeding marks detected on the Untermassfeld *B. menneri* bones were caused by *P. brevirostris*. In no case could a feeding mark be asigned with certainty to any other species than the mentioned hyena.

4.1. Regularities of bone modifications by *Pachycrocuta brevirostris*

Examinations of the bone material of the species *B. menneri* from Untermassfeld revealed that 12.6 % of bones and teeth show bite marks belonging to *P. brevirostris*. Earlier analyses of the Untermassfeld material on bone modifications caused by *P. brevirostris* showed a similar value of 11.6 % for the bones of *Hippopotamus antiquus* (Kahlke 2001c, 956). The total number of finds of *H. antiquus* is significantly lower than that of *B. menneri* (Kahlke and Gaudzinski 2005; Kahlke in this volume).

The amount of bite marks on the Untermassfeld finds of *B. menneri* differs markedly depending on the skeletal element (**Fig. 6**) and bone section (**Fig. 8**). Long bones in particular are affected by modifications caused by *P. brevirostris*, but so too are mandibulae, pelves and calcanei. In contrast, maxillae, carpal and tarsal bones rarely possess bite marks and are often almost present in their entirety.

A determining factor for the selection of certain skeletal elements by *P. brevirostris* is their hunting and feeding behaviour. Among others, the anatomical position of the bones plays an important role for this, because of its relation to the order of disarticulation (Palmqvist and Arribas 2001).

Another cause for the preference of some skeletal elements by extinct as well as recent hyenas is to be found both in the muscle meat portion of the different body regions (Madurell-Malapeira et al. 2015) and in the total marrow yield of the various bones (e. g., Zapfe 1939; Haynes 1980a; 1980b; 1982; Palmqvist et al. 2011; Fosse et al. 2012). Looking at the percentages of the various bone fragments of *B. menneri* from Untermassfeld, it becomes clear that the humeri, femora and tibiae, as well as radii and ulnae are more frequently affected by *P. brevirostris* bone modifications than metapodia (**Fig. 6**). This clearly demonstrates a preference by *P. brevirostris* for the long bones with an average higher marrow yield (details in Brink 1997). Both extinct and recent hyena species also select according to the bone mineral density of the various skeletal elements (Palmqvist et al. 2011) as the strength of the bones depends on this factor (Madurell-Malapeira et al. 2015), which in turn leads to the recorded variance in the fragmentation degree of the different bones (Binford and Bertram 1977). The value of bone density depends on the size of the animal species, individual age, proportion of cancellous bone, the season, and the related nutritional status of the individual (examples in Münzel 1987).

Finds that clearly originated from juvenile individuals of *B. menneri* are rare in the material of Untermassfeld, especially in the postcranial area (**Fig. 3**). The reason for this is not only their possible removal by carnivores (Kahlke 1997c), but also, to some extent, the high degree of fragmentation, which makes an exact classification difficult.

In the mandibulae, which could most reliably be assigned to the young by their deciduous dentition, 50 % of the bones and bone fragments of juvenile individuals possessed bite marks from *P. brevirostris*. Since Sher (1997) showed an age and gender distribution for the remains of *B. menneri* that correspond to a natural population, it can be assumed that the bison groups had a herd-like structure when they died (Kahlke 2000; 2006, 77; Kahlke and Gaudzinski 2005). The proportion of adult animals in a natural population, and thus also within the Untermassfeld material, is higher than that of juveniles (Kahlke 2000; 2006, 77; Kahlke and Gaudzinski 2005). Due to the increased proportion of juvenile findings with bite marks, it may therefore be assumed that juveniles were the preferred prey of *P. brevirostris*.

If one considers the distribution of bone modifications by *P. brevirostris* within the various skeletal elements of the species *B. menneri*, the Untermassfeld material exhibits striking regularities that indicate recurrent patterns in the feeding behaviour of the hyenas.

4.1.1. Cranial elements

In the case of the mandibula, aside from the areas near the symphysis and below the row of teeth (**Fig. 15**), the angle of the jaw is most predominantly affected by tooth marks (**Fig. 14**). The latter was to a large extent consumed by the hyenas. This is consistent with data from Haynes (1980a; 1980b; 1982), according to which the mandibular angle is one of the first places to be destroyed by carnivores.

4.1.2. Vertebrae and ribs

The results of the present study show great differences in the degree of fragmentation for the vertebrae and ribs (**Fig. 7**). While the ribs in the Untermassfeld fossil record are all fragmented, more than 75 % of the vertebrae, and especially the vertebral bodies, are almost complete, despite the presence of tooth marks. This can be explained by the compact shape of the vertebral bodies and the accompanying difficulty in biting these bones. Only the narrower processes of the vertebrae were, to a large degree, frequently broken off, especially the spinal processes, as these form the nutrient-rich hump of *B. menneri*. Similar results are shown by Zapfe (1939), who carried out observations on recent *Crocuta crocuta* (Schönbrunn Zoo, Austria) and investigations of Pleistocene bones (Teufelslucke near Roggendorf, Austria), as well as researches conducted by Haynes (1980a; 1980b), who published them on bones from both Pleistocene as well as recent North American mammals. It is also striking that, despite the presence of bite marks, a vertebral column segment was found in an articulated state at Untermassfeld (Kahlke in this volume, Fig. 14). This is due to the pronounced articulation of the vertebrae, which means that the spine is usually one of the last parts to decay (Toots 1965).

Complete ribs of *B. menneri* are strongly underrepresented in the fossil record of Untermassfeld (**Fig. 1**), especially when comparing their number (n = 19) with the theoretically expected amount based on the total MNI of 69 (Kahlke 2001c, 959). This may be due, on one hand, to the natural disarticulation, during which ribs are one of the first skeletal elements to become disassociated from the rest of the remains (Toots 1965; Kahlke 2001c, 959), and on the other, because carnivores often open carcasses up in the abdominal area (Blumenschine 1986; Kahlke 2001c, 959). This may have led to *P. brevirostris* fragmenting the very narrow ribs in relation to other bones to such an extent that these fragments can no longer be identified as *B. menneri*. In the only costal fragment of the Untermassfeld collection of *B. menneri* found with bite marks, they are located in the middle section of the bone. Since carnivores preferentially attack the ribs of their prey on the cartilage and the ventral ends (Haynes 1980a; 1980b; 1982), it is precisely these middle sections that remain.

4.1.3. Forelimb long bones

Similar to the ribs, *B. menneri* scapulae are rare at Untermassfeld (**Fig. 1**). This can be explained by the fact that the proximal part of a scapula is easier to bite into pieces in comparison to other limb bones and, in addition, there is a high proportion of muscle meat in this area of the body. This is supported by the fact that in the only find of a scapula, IQW 1984/19946 (Mei. 19466), with modifications by *P. brevirostris*, only the fossa glenoidalis is present and the tooth marks are found at the fracture edge to the missing proximal section. Recent *C. crocuta* seems to show a similar type of behaviour, as in the study by Zapfe (1939) the glenoid fossa was the remaining piece of the only fed scapula.

In the *B. menneri* humeri, a clear pattern can be seen in the feeding behaviour of *P. brevirostris*, starting at the proximal and moving to the distal end of the bones (**Fig. 8**). The remaining distal epiphyses and the diaphyses of the *B. menneri* humeri often display tooth marks (**Fig. 18**). This coincides with the findings of Zapfe (1939), Espigares (2010), Palmqvist et al. (2011) and Madurell-Malapeira et al. (2015), who in their studies describe the frequent survival of the distal epiphyses with pieces of the diaphysis in the case of the humeri. According to Zapfe (1939), the regularity with which the hyenas first open the proximal epiphyses of their prey can be explained by the fact that fatty and cancellous tissue is the most easily accessible there. The distal ends are less attractive to all hyenas as they have a higher mineral density and lower fat content (Arribas and Palmqvist 1998; Palmqvist et al. 2011). The fact that bone-breaking hyenids only partially fragment some skeletal elements is due to the fact that the long bones of *B. menneri,* and bovids of similar size (see Zapfe 1939), are massive bones. The effort of fragmenting larger bones, in comparison to those from smaller species, causes hyenas to tire (Zapfe 1939, 113). The degree of fragmentation associated with bone size (Madurell-Malapeira et al. 2015) and the ratio of effort required to obtain expected nutrients (Fosse et al. 2012) varies with each species and the skeletal element involved.

Of the radii and ulnae in the bison material of Untermassfeld, the proximal areas of the bone dominate (**Fig. 8**). An exception is the area of the olecranon ulnae, which is absent in many finds (**Fig. 9**). Due to the tooth marks in the area of the fractured edge, it can be assumed that this section was often broken off by *P. brevirostris*. Additional bone modifications by *P. brevirostris* are found in particular in the area of the diaphyses and the distal epiphyses of the radii, and occasionally in the distal areas of the ulnae. The order in which *P. brevirostris* broke up the radii and ulnae at Untermassfeld is similar to descriptions given by Zapfe

(1939), who described how the olecranon ulnae was first broken off, followed by the radius subsequently being bitten from the distal end (Zapfe 1939, 113). Madurell-Malapeira et al. (2015) also describe frequent preservation of the proximal areas of the radii and ulnae as a result of feeding activities of *P. brevirostris*.

Carpal and tarsal bones, along with metapodia and phalanges, are summarised in the hand and foot bones section.

4.1.4. Hindlimb long bones

As the pelvis area of the prey also represents a suitable opening for carnivores (Kahlke 2001c, 959), this part of the skeleton in the *B. menneri* material from Untermassfeld also possesses a large number of bite marks (**Fig. 6**). The pelvic bones exist, for the most part, in a fragmented state (**Fig. 7**). In the areas of the os ilium, os ischii and os pubis parts of the bone were either broken off by *P. brevirostris* or modifications of the bone surface occurred through biting (**Fig. 9**). The latter also applies to the areas near the acetabulum. The places where *B. menneri* pelvic bones from Untermassfeld were bitten by *P. brevirostris* are comparable to observations of Zapfe (1939) on the feeding behaviour of the recent hyena *C. crocuta* and the studies by Haynes (1980a; 1980b; 1982).

In the case of the *B. menneri* femora at Untermassfeld, the proximal epiphyses are underrepresented (**Fig. 8**). It can be concluded, therefore, that, as with the humeri, *P. brevirostris* preferred to begin at the proximal end of the bone and continued feeding in a distal direction. The feeding sequence often begins with the trochanter major and the caput (**Fig. 9**), and this is followed by the modification and fragmentation of the diaphysis. However, noticeably the femora contain a high proportion of diaphyses, of which both epiphyses are broken off (**Fig. 8**). It is to be assumed that the aim of this sequence of breaking up the femora for *P. brevirostris* was reaching the nutrient-rich cancellous bone in the epiphyses. Previous studies on the feeding strategies of recent and extinct hyena species show different results with respect to the femora. In his work of 1939, Zapfe describes how recent *C. crocuta* begin to bite at the proximal end of the bone, leaving distal epiphyses with fragments of the diaphysis as remains. The leftover distal epiphysis was usually not modified by the hyenas (Zapfe 1939, 113). In contrast, authors such as Palmqvist et al. (2011) and Madurell-Malapeira et al. (2015) show that the femora are bitten in a varying direction and thus, in addition to the distal epiphysis, the diaphysis often remains.

Compared to the total number of bones and the MNI, the patellae are clearly underrepresented (**Fig. 1**). The knee joint and the popliteal fossa belong to the skeletal elements that rank first in the carnivore feeding order (Haynes 1980a; 1980b; 1982; Münzel 1987) and therefore it is possible that the patellae of the carcasses, which were washed up during the formation of the Untermassfeld fossil site, had previously been partially bitten by *P. brevirostris*. Another reason for the low number of patellae could be due to the fact that they are roundish in shape, and easily transported skeletal elements were carried away from the area where the site of Untermassfeld originated (Kahlke 2006, 81).

Among the tibiae modified by *P. brevirostris* from Untermassfeld, only distal epiphyses with more or less large sections of the diaphysis exist in the *B. menneri* find material (**Fig. 8**). A particular peculiarity is that the fractured edge of the diaphysis of the existing distal bone portion of the tibia is often obliquely broken, so that a partly sharp, proximally directed tip is formed (**Fig. 10**). This distinctive type of break is described not only in the bones of Untermassfeld modified by *P. brevirostris*, but also in the feeding activities of recent *C. crocuta* and cave hyenas (Zapfe 1939, 115, 122 f.) and illustrates a characteristic form of hyena species bone modification. In addition, bite marks of *P. brevirostris* on *B. menneri* are found mainly in the area of the diaphysis of the tibiae (**Fig. 9**). The fact that the distal epiphyses of the tibia have no bite marks is due to that hyenids preferentially bite this skeletal element from the proximal to the distal end, leaving the distal epiphyses remaining, as confirmed by further studies (e.g., Zapfe 1939; Palmqvist et al. 2011; Madurell-Malapeira et al. 2015). Haynes (1983) also reports that the distal end of the tibia often remained without any identifiable carnivore bite marks, making them difficult to identify as carnivore bitten bones.

The reason why complete bones or bone fragments of the *B. menneri* fibula are not found at Untermassfeld is probably due to the fact that they are relatively narrow and thus easier to break than other bones (Zapfe 1939, 112; Kahlke 2001c, 959). It is therefore to be assumed that this skeletal element was so severely fragmented by carnivores, or else by other taphonomic processes, that the fragments could not be assigned to either a particular skeletal element or to any particular species.

4.1.5. Hand and foot bones

The carpal as well as tarsal bones of *B. menneri* in Untermassfeld are very often more or less complete (**Fig. 4**) and rarely have any evidence of tooth marks on the bone surface (**Fig. 6**). The reason for this is almost certainly due to the unprofitability of these skeletal elements in terms of nutrients and their relatively compact shape, which increases the biting effort needed. Together, these characteristics mean that *P. brevirostris* often ignored these bones. This is not only witnessed for the Untermassfeld hyenas, but also at other sites (Madurell-Malapeira et al. 2015) and for other hyena species (Zapfe 1939). An exception to this rule is given by the calcanei, where 16 % of the recovered *B. menneri* bones possess bite marks (**Fig. 6**). By comparison, only 2 % or 3 % of the remaining carpal resp. tarsal bones have modifications by *P. brevirostris*. In preference, the tubera calcanei is broken off (**Fig. 28**), and this nutrient-obtaining strategy can be attributed to the fact that the process of the calcaneus protrudes from the tendon ligament surrounding the joint (Zapfe 1939, 115; Haynes 1980a; 1980b; 1982).

In the Untermassfeld *B. menneri* metapodia, 19 % of the metacarpals and 21 % of the metatarsals have bone modifications originating from feeding activities by *P. brevirostris* (**Fig. 6**). This is a small percentage in comparison to other long bones examined in this study. Considering all metapodia, it is evident that the remaining proximal fragments considerably predominate (**Fig. 8**). It can therefore be assumed that *P. brevirostris* preferentially fractured the metapodia moving in the direction from the distal to the proximal end. From this observation, the feeding behaviour of the Untermassfeld hyenas differs from those of earlier studies on bone modifications. In the study by Palmqvist et al. (2011), based on the bones of Early Pleistocene hyenas from the fossil site of Venta Micena (Guadix-Baza Basin, southeastern Spain), the investigation found that in the case of metapodia there was no preferential direction to the bite marks, but that rather these elements were bitten in varying directions. Other studies, however, support the notion that hyenas predominantly utilise the proximal epiphysis (Hill 1989; Fosse 1994a; 1994b; Fosse et al. 2011; Fourvel et al. 2012). Overall, however, the remaining metapodia are most likely to be present as near-intact bone (Palmqvist et al. 2011; Madurell-Malapeira et al. 2015). In fact, in spite of *P. brevirostris* tooth marks, the metacarpal bones in the Untermassfeld bison material remained largely intact relative to most other long bones (**Fig. 7**). However, in contrast, nearly complete and bite-marked *B. menneri* metatarsal bones are less common than most other limb bones. Only the humeri of *B. menneri* show a higher percentage of fragmented and modified bones through the actions of *P. brevirostris* than is the case with the metatarsalia. The hyenas of Untermassfeld have therefore disassembled the metatarsal bones at a higher frequency than was the case, for example, for the material in the study of Palmqvist et al. (2011). This could be due to the fact that the metatarsal bones of the Untermassfeld *B. menneri* species are larger and therefore the proportion of marrow and nutrients is more abundant in the skeletal elements than is the case with smaller species. On the other hand, in the case of the species *B. bison*, metatarsal bones at 46 g have a higher marrow proportion than the metacarpal bones at 34 g (Brink 1997). It can be assumed that the same applies to *B. menneri*. This would result in the metatarsals being a more attractive skeletal element for *P. brevirostris* than the metacarpals, and therefore the hyenas more frequently sought to access the bone marrow of the metatarsal bones. Likewise, the selective transport of bones by hyenas, which is influenced by aspects such as meat content and bone mineral density (Palmqvist and Arribas 2001) or the available food supply, may also be a cause for preference (Fourvel and Mwebi 2011).

When considering the amount of the phalanges affected by bite marks, at only 7 % (**Fig. 6**) they are rarely modified in comparison to the other *B. menneri* limb bones. This, similar to the carpal and tarsal bones, can also be explained by factors such as a low nutrient content and relatively high bone density (Zapfe 1939, 115; Palmqvist and Arribas 2001; Madurell-Malapeira et al. 2015). Looking at the frequency of phalanges in the skeleton of *B. menneri*, these bones area also slightly under-represented at Untermassfeld. The cause of this could lie in the discharge of these relatively small bones from the site (Kahlke 1997c, 400). The modifications in both phalanges I and II are mostly concentrated at the distal end. The phalanges III examined in the present study show no evidence of modification by *P. brevirostris* (**Fig. 9**) as is to be expected, since they are situated within the hooves.

4.2. Categories of bone modifications by Untermassfeld

In conclusion, it may be stated that the bone modification categories 1 to 7 as described by Maguire et al. (1980) based on studies of recent hyenids, could be fully verified in the material of *B. menneri* from Untermassfeld. At the same time, it is clear that not all of the bite marks of the hyena species *P. brevirostris* fully comply with the descriptions given by Maguire et al. (1980). Modifications caused by juvenile hyenas were found to deviate significantly, since due to inexperience they show a greater variability in their bite marks than observed in that of the adult individuals (Binder and Van Valkenburgh 2000).

Category 1 modifications include spiral fractures (**Fig. 12.1**) and ragged-edged chewing (Maguire et al. 1980). In the entire Untermassfeld collection of *B. menneri* bone this category is the most frequent (44 %) (**Fig. 11**). Except for patellae, astragali and the phalanges, all skeletal elements have bite marks belonging to category 1 (**Fig. 13**). These are to be found in large part on the diaphyses of the long bones, which are spirally broken as according to this category, but also in the area of the jaw angle, mandibular, vertebrae, and ribs, as well as on the broken edge of the olecranon ulnae and the tuber calcanei. Category 1 bone modifications often restrict the sections at which *P. brevirostris* stopped biting on the long bones and are therefore predominantly in the diaphyseal area. *B. menneri* bones, where category 1 is particularly common, are also those that show a particularly high degree of fragmentation. This biting strategy demonstrates, therefore, firstly, where *P. brevirostris* sought to reach the marrow of the long bones and, secondly, the sites where less massive bones such as, e. g., mandibulae or ribs were crushed by chewing movements.

Category 2 shallow pitting is the rarest of the bone modification types caused by *P. brevirostris*, accounting for only 2 % of the bite marks recorded on the bison material of Untermassfeld (**Fig. 11**). The small, nearly conical teeth impressions (**Figs. 12.2**, **16**) reflect, in both size and spacing, the tips of the deciduous teeth of *P. brevirostris*. The premolars of adult individuals are clearly too large for such tracks, even if they are barely worn and pointed. This confirms the conclusion of Maguire et al. (1980) that these traces are attributable to juvenile hyenas. The distinctive bite marks of category 2 can be found, among others, on two bone fragments of the mandibulae. It is noticeable that in both cases the tooth impressions are close to each other and are located in the region of the articular process (**Fig. 14**). This area may be an appropriate starting point for hyenas in a juvenile stage of development (Maguire et al. 1980) if they occasionally chew on carcass remains. In addition, similar, but less obvious, category 2 modification marks were found on the ulna diaphysis and an os carpale quartum.

At a share of only 5 %, the punctate depressions or perforations (**Fig. 12.3**) caused by the premolars of the hyenas (Maguire et al. 1980) and which belong to category 3 are also rarely represented in the Untermassfeld bison material (**Fig. 11**). Nevertheless, such tooth impressions, albeit to only a small amount on the individual skeletal elements, can be found in almost all long bones as well as vertebrae and pelvic fragments. The premolar perforations of *P. brevirostris* were mainly localised in the epiphyses of the bones of *B. menneri* and also near to category 4 modifications. This is because category 4 lunate or crescent-shaped fracture scars arise from category 3 perforations. Thus, there is a possibility that category 3 is rare because the round perforations become crescent-shaped scars (category 4) due to the high degree of fragmentation of the bones, which is typical of the feeding behaviour of all hyenas (Fosse et al. 2012).

Looking at the distribution of the different categories in the complete *B. menneri* material from Untermassfeld (**Fig. 11**), the frequency of category 4 lies in the middle range (10 %) in relation to the other six modification types. The lunate or crescent-shaped fracture scars (**Fig. 12.4**) occur when the fracture line caused by the bite of a hyena falls along a punctate depression (category 3) (Maguire et al. 1980). Correspondingly, the crescent-shaped bite marks are always at fracture edges, both in long bones (**Figs. 10c–d, 26d**), in the areas of the jaw angle (**Fig. 14**), on the olecranon ulnae (**Figs. 19b, d**), and on the pelvis (**Fig. 22**). They are often located near to spiral-fractures and ragged-edged chewing modifications (category 1) and occur in highly fragmented skeletal elements. These crescent-shaped bite marks are not recorded in the carpal and tarsal bones, or the patellae and phalanges (**Fig. 13**). Among other reasons, this can be explained by the fact that these compact bones, with the exception of the calcanei, have no category 3 bite marks, which are a prerequisite for the development of category 4 modifications.

The striations or gouge marks of category 5, which run perpendicular to the longitudinal axis of the bones (**Fig. 12.5**), occur at a frequency of 10 % within the Untermassfeld bison material (**Fig. 11**), and are mainly localised on the diaphysis of some long bones. Significantly, they appear more frequently on the tibia (**Fig. 25**), usually in the area of the diaphysis (**Fig. 27**) and often in otherwise largely complete bones. Modifications of this category have not been documented in cranial bones, vertebrae, ribs, scapulae, humeri, pelvic fragments, femora, patellae or phalanges (**Fig. 13**). Due to the size and spacing between category 5 gouge marks, it can be assumed that their origin does not only lay in the chewing motions of juvenile hyenid individuals, as described by Maguire et al. (1980). It seems more likely that *P. brevirostris* produced larger modifications, which are similar to the descriptions by the referred authors but were caused by the slipping of the teeth of adult hyenas. This may happen during transport of the bones or in an attempt to bring the skeletal elements into the desired position for further processing or breaking.

Close, irregular and randomly-oriented grooves of category 6 (**Fig. 12.6**) are the second most commonly observed modification type (22 %) (**Fig. 11**). The grooves usually occur at the ends of the limb bones in the

Untermassfeld bison material. This is similar to results described by Sutcliffe (1970) based on studies of the recent *Crocuta crocuta* in East Africa. In the bones of *B. menneri* this mainly affects the metacarpals (**Fig. 20**) and phalanges (**Fig. 30**), but these randomly-oriented grooves also appear on other long bones and on the mandibulae (**Fig. 14**). According to Maguire et al. (1980), category 6 bone modifications are caused through prolonged chewing by Hyaenidae. The intensity of some of these marks recorded at Untermassfeld (e.g., **Figs. 12.6, 24**) suggests that adult *P. brevirostris* may have left these grooves as they attempted to reach the cancellous bone of the epiphyses of limb bones with their teeth.

Category 7 makes up about 7 % of the bone modification types recorded in the Untermassfeld *B. menneri* material (**Fig. 11**). The hollowing out of the cancellous bone (**Fig. 12.7**) by *P. brevirostris* using the incisors and canines as well as the tongue, took place on all large-sized long bones, and in particular on the epiphyses (**Figs. 17, 19, 23**), with the exception of the tibia. This can be attributed not only to the higher mineral density and lower fat content in the distal epiphysis of the tibia (Arribas and Palmqvist 1998; Palmqvist et al. 2011), but also possibly to the triangular profile of this skeletal element, which makes it difficult to bite. Within the Untermassfeld bison material, it is mainly the humeri (**Fig. 18**) and femora (**Fig. 24**) that are affected by hollowing out of the cancellous bone. As expected, this modification is recorded at sites with the highest percentage of nutrient-rich cancellous bone. There is also the possibility that *P. brevirostris*, especially in the case of the humeri, femora and other large skeletal elements, used this strategy to access the interior of the bones (Kahlke 1997b, 282).

4.3. Distribution of the bone modification categories by *Pachycrocuta brevirostris* in comparison with recent hyena species

Fig. 11 shows the percentage of bone modification categories in the whole *P. brevirostris*-modified Untermassfeld *B. menneri* material. The depiction of the different categories in this graph is comparable to similar statistics by Saunders and Dawson (1998) from *P. brevirostris* in the Haro River Quarry (Pakistan), a fossil hyena den within a sediment-filled intermontane sink from the Plio-Pleistocene (Saunders and Dawson 1998, Fig. 14D, E) and similar graphs from the study by Maguire et al. (1980), which refer to recent *Crocuta crocuta* (spotted hyena), *Hyaena hyaena* (striped hyena) and *Parahyaena brunnea* (brown hyena).

When comparing the frequencies of the categories from *P. brevirostris* at the fossil site of Untermassfeld with those of the same hyena species at Haro River Quarry (Saunders and Dawson 1998), in both cases category 1 is the most commonly documented. At Untermassfeld, ragged-edged chewing and spiral fractures account for 44 %, while at Haro River Quarry they account for slightly less than 30 % (Saunders and Dawson l.c.). Together, these two sites also display the frequently occurring irregular and randomly-oriented grooves of category 6, which make up 22 % of the bite marks at Untermassfeld and almost 20 % on average at Haro River Quarry. The average values of category 3, at about 8 %, and category 4, at about 10 % seen in the two graphs published by Saunders and Dawson (l.c., Fig. 14D–E) are also similar to the occurrence of these categories in the Untermassfeld *B. menneri* (**Fig. 11**). At Untermassfeld the percentage share of premolar perforations by adult individuals (category 3) lies at about 5 % and the crescent-shaped fracture scars (category 4) are around 10 %.

However, the recorded difference between category 1, as the most frequent category, and the proportionally following categories is greater at Untermassfeld than at Haro River Quarry. At Untermassfeld, category 6 as the second most common modification is only half as often recorded as category 1, at 22 %. In the Haro River Quarry, however, it is shallow pitting (category 2) with around 25 %, and thus a difference of only about 5 %, in comparison with the most frequently observed modification, category 1. Category 2 makes up only 2 % of the bone modifications at Untermassfeld. In addition, the gouge marks of category 5 (11 %) and the hollowing out of the cancellous bone of category 7 (7 %) are more abundant at Untermassfeld than at Haro River Quarry, where they account for less than 5 % each of the total modifications, as published by Saunders and Dawson (1998). However, a similarity in the distribution between the categories of bone modifications by *P. brevirostris* from Untermassfeld and Haro River Quarry (Pakistan) is still evident.

According to both Maguire et al. (1980) and Saunders and Dawson (1998), *Parahyaena brunnea* and *Hyaena hyaena* display a similar distribution pattern for the different categories. Both species record high values for category 1 (spiral fractures) and category 5 (striations or gouge marks) modifications, and a relatively even distribution exists for categories 2 to 4, at an average of 10 %. Very low values are presented for categories 6 and 7, at less than 5 % (Saunders and Dawson 1998, Fig. 14B–C). The reason for these clear similarities probably lies in the similar modes of life and behaviour of these two recent hyena species in terms of food acquisition and utilisation (Kruuk 1972; Mills 1990).

In contrast, the recent *Crocuta crocuta* differs markedly from *Parahyaena brunnea, Hyaena hyaena* (Maguire et al. 1980) and *P. brevirostris*, both at Haro River Quarry (Saunders and Dawson 1998) and Untermassfeld. The difference in the occurrence of category 1 is especially striking, as it accounts for less than 10 % of recent *C. crocuta* modifications and is therefore much less common than in all other described hyena species, where spiral fractures and ragged-edged chewing usually dominate within the bone modification categories. The associated low degree of fragmentation in long bones, despite bite marks, is also considered typical of cave hyenas (Fosse 1994a; 1994b). For recent *C. crocuta* the lowest share is recorded in category 3 modifications (less than 5 %) and the highest for those of category 5 (more than 35 %).

Untermassfeld *P. brevirostris* shows more similarity with *Parahyaena brunnea* and *Hyaena hyaena* than with recent *Crocuta crocuta*, especially in the case of categories 1, 3 and 5, with regard to bone modifications. It is therefore to be assumed that the lifestyle of *P. brevirostris* resembles more that of the recent *Parahyaena brunnea* or *Hyaena hyaena*, especially in terms of feeding behaviour. In the context of clarifying their way of life it is often the body measurements of *P. brevirostris*, which are the basis for a comparison with the recent *C. crocuta*, since this is the largest recent hyena species (Turner and Antón 1996). However, as *P. brevirostris* and the spotted hyena coexisted for a while (l.c.), there should have been ecological differences between the two. This is also demonstrated by the aforementioned differences observed between the two in terms of their preferred modification categories and thus their corresponding techniques in utilising bones. *P. brevirostris* should not, therefore, be universally compared to recent *C. crocuta* in terms of their lifestyle or feeding behaviour.

Looking at the all of the categories described, it is clear that category 1 (ragged-edged chewing and spiral-fractures) and category 6 (irregular and randomly-oriented grooves) are striking for *P. brevirostris*. This is shown by the results of the present Untermassfeld study and also by comparison with the data of Saunders and Dawson (1998). This consistency clearly differentiates *P. brevirostris* from the three recent hyena representatives. For bones of the Untermassfeld *B. menneri* bovine species, the greatest similarity is found in the distribution of the categories of bone modifications, except for the differences in the quantities of category 2 (shallow-pitting), 5 (striations or gouge marks) and 7 (hollowing out of the cancellous bone), to those of *P. brevirostris* from Haro River Quarry.

It is unclear why category 2 bite marks, which are believed to be caused by juvenile hyenas, are rarely found within the Untermassfeld bison material. A particularly high number of juvenile individuals of *P. brevirostris* were recorded at Untermassfeld (Kahlke 2006, 78; Iannucci et al. in this volume) and these could have been expected to have affected the bones. One possible reason for this discrepancy is the already mentioned massive form of bison bones compared to those from smaller species, making them more difficult for juvenile hyenas to handle.

The significant differences in the amount of category 5 striations or gouge marks recorded at both Untermassfeld and Haro River Quarry may be explained by the differences in their origins, as already mentioned. The dimensions and distance of some of the grooves on the Untermassfeld material suggest that not only juvenile hyenas but also adult individuals could have been the cause of these modifications. In particular, the comparison of several rows of teeth of *P. brevirostris* from the Untermassfeld material with some of the broad bite marks lying perpendicular to the longitudinal axis (**Fig. 27**), suggests that they are caused by the adult hyenas' teeth slipping on the bone. In the study by Haynes (1983) such teeth marks are also justified by the transport of the corresponding bones by carnivores.

The fact that hollowing out of the cancellous bone, i.e., category 7, is more frequently represented in the bison material investigated in the present study than at Haro River Quarry, is most likely a result of *B. menneri's* massive size. Hollowing out of cancellous bone seems to occur more frequently in large types of prey (Kahlke 1997b, 282), as *P. brevirostris* occasionally didn't completely bite through the bones. In contrast, no single large prey species was studied at Haro River Quarry (Saunders and Dawson 1998).

5. Conclusions

In the present study, 2,169 specimens of *Bison menneri* finds from Untermassfeld were examined for feeding marks caused by the hyena species *Pachycrocuta brevirostris*. These modifications were quantitatively recorded for the entire bison material as well as the separate skeletal elements and, with as much detail as possible recorded on their location. The study examines the influence of *P. brevirostris* on the bone material of *B. menneri* from the fossil site of Untermassfeld. The evaluation of the aforementioned material should provide information about possible consistencies or patterns in the feeding behaviour of this hyena species. Classification of the bone modifications was carried out according to Maguire et al. (1980) in order to identify possible characteristics in the feeding strategies of Hyaenidae in general and *P. brevirostris* from Untermassfeld in particular. The investigations should support future trans-regional comparisons with more Hyaenidae sites.

In the course of the investigation, it became clear that the Untermassfeld hyenas particularly favoured skeletal elements with a high meat and marrow content or a low bone mineral density. Therefore, the bone modifications caused by *P. brevirostris* also accumulate on the nutrient-rich limb bones while occurring extremely rarely on carpal as well as tarsal bones and phalanges, which have hardly any muscle meat or bone marrow. The Untermassfeld *B. menneri* material also demonstrates a clear preference in the hyenas for certain feeding directions for different long bones. In humeri and tibiae, the bones are predominantly bitten from the proximal to distal ends. In the case of radii and ulnae, in which the olecranon is often additionally absent, as well as in the metapodia, the opposite direction is favoured. Only the femora are inconsistent, i.e., either they resemble the humeri and tibiae and the biting direction is distal, or the hyenas break off both of the epiphyses and the diaphysis remains.

The *B. menneri* finds were assigned to the following categories according to Maguire et al. (1980) and the different frequency of their occurrence was documented. Category 1 (spiral fractures and ragged-edged chewing) was most commonly produced by *P. brevirostris*. This is followed in descending numbers by category 6 (irregular grooves), category 4 (crescent-shaped fracture scars), category 5 (gouge marks), and then category 7 (hollowing out of the cancellous bone). Category 3 (punctate depressions or perforations caused by the hyenas' premolars) is less frequently observed. Category 2 (shallow-pitting) is the rarest modification observed in the Untermassfeld bison material. A comparison of this data with that of Saunders and Dawson (1998) shows that the frequent occurrence of categories 1 and 6 together is typical for *P. brevirostris* both at Untermassfeld and Haro River Quarry (Pakistan), a fossil hyena den. Compared to recent Hyaenidae, the distribution of the categories made by *P. brevirostris* from Untermassfeld resembles that of *Parahyaena brunnea* and *Hyaena hyaena*. Whereas, a comparison between the modification patterns of the Untermassfeld *P. brevirostris* with recent *Crocuta crocuta* produces the largest differences.

With regard to the frequencies and the distribution of the categories of bone modifications on the Untermassfeld bison material, category 2 marks (shallow-pitting) are particularly conspicuous. In two cases they occur in a nearly identical position on the articular process of the mandibulae, but are otherwise only documented on ulnae and carpalia. Also noteworthy is the frequency of the occurrence of category 5 (gouge marks) on the diaphysis of the tibia.

For a closer look at the consistencies as well as the abnormalities in the feeding behaviour of *P. brevirostris,* further investigations on other possible prey species from Untermassfeld are desirable. Such investigations would allow conclusions to be drawn as to whether the causes of the characteristic behaviours of *P. brevirostris* were owing to the shape and size of the bones and the respective type of prey, or whether the reasons are actually to be found in the skeletal elements in question, their anatomical position and the hunting and feeding behaviour of this hyena species.

Acknowledgements

I would like to thank M.S. Fischer (Jena) for putting me in contact with the Senckenberg Research Station of Quaternary Palaeontology Weimar and as a result the opportunity to undertake a master's thesis, which formed the basis for this present study. My special thanks go to R.-D. Kahlke (Weimar) who facilitated and led the study. He supported the development of the work through his expertise, exhaustive knowledge of the literature, and his dedicated supervision. In addition, he was always ready and happy to assist with in-depth discussions on the subject and with manuscript reviews. I would also like to thank all the staff of the Senckenberg Research Station of Quaternary Palaeontology in Weimar for their suggestions and hints, which helped me in the preparation of this study. In addition, I thank S. Döring (Weimar), who was responsible for producing the photographs and the preparation of the diagrams, and E. Haase (Weimar), who was an indispensable help in the processing of photographs and graphics. Finally, I would also like to thank C. Nielsen-Marsh (Leipzig) for the translation of this manuscript into English.

References

Arnold, J., 2018. Modifikation frühpleistozäner Bisonreste von Untermassfeld (Südthüringen, Deutschland) durch *Pachycrocuta*. Master thesis, Friedrich-Schiller-Universität Jena, Germany.

Arribas, A., Palmqvist, P., 1998. Taphonomy and palaeoecology of an assemblage of large mammals: Hyaenid activity in the Lower Pleistocene site at Venta Micena (Orce, Guadix-Baza-Basin, Granada, Spain). Geobios 31, Supplement 3, 3–47.

Binder, W.J., Van Valkenburgh, B., 2000. Development of bite strength and feeding behaviour in juvenile spotted hyenas (*Crocuta crocuta*). Journal of Zoology 252, 273–283.

Binford, L.R., Bertram, J.B., 1977. Bone Frequencies and Attritional Processes. In: Binford, L.R. (Ed.), For Theory Building In Archaeology. Essays On Faunal Remains, Aquatic Resources, Spatial Analysis, and Systematic Modeling. Academic Press. New York, 77–156.

Blumenschine, R.J., 1986. Early Hominid Scavenging Opportunities: Implications of Carcass Availability in the Serengeti and Ngorongoro Ecosystems. British Archaeological Reports, International Series 283. Oxford.

Brink, J.W., 1997. Fat Content in Leg Bones of *Bison bison*, and Applications to Archaeology. Journal of Archaeological Science 24, 259–274.

Espigares, M.P., 2010. Análisis y modelización del contexto sedimentario y los atributos tafonómicos de los yacimientos pleistocénicos del borde nororiental de la cuenca de Guadix-Baza. PhD thesis, University of Granada, Spain.

Ewer, R.F., 1973. The carnivores. Ithaca, New York, Cornell University Press.

Fernández-Jalvo, Y., Andrews, P., 2016. Atlas of Taphonomic Identifications. 1001+ Images of Fossil and Recent Mammal Bone Modification. Heidelberg.

Fosse, P., 1994a. L'hyène des cavernes comme agent d'accumulation des ossements à Lunel-Viel 1 (Hérault, France): observations préliminaires. In: Patou-Mathis, M. (Ed.), 6e Table Ronde Taphonomie/Bone Modification, Paris, France, septembre 1991. Outillage peu élaboré en os et en bois de cervidés 4 / Artefacts 9. Treignes, 91–96.

1994b. Taphonomie paléolithique: les grands mammifères de Soleilhac (Haute-Loire) et de Lunel-Viel 1 (Hérault). Résumé de thèse par l'auteur. Préhistoire Anthropologie Méditerranéennes 3, 211–215.

Fosse, P., Avery, G., Selva, N., Smietana, W., Okarma, H., Wajrak, A., Fourvel, J.B., Madelaine, S., 2011. Taphonomie comparée des os longs d'ongulés dévorés par les grands prédateurs modernes d'Europe et d'Afrique (*C. lupus, P. brunnea*). In: Brugal, J.-P., Gardeisen, A., Zucker, A. (Eds.), Prédateur dans tous leurs états. Evolution, Biodiversité, Interactions, Mythes, Symboles. XXXIe rencontres internationals d'archéologie et d'histoire d'Antibes. Antibes, 127–156.

Fosse, P., Wajrak, A., Fourvel, J.B., Madeleine, S., Esteban-Nadal, M., Cáceres, I., Yravedra, J., Brugal, J.-P., Prucca, A., Haynes, G., 2012. Bone Modification by Modern Wolf (*Canis lupus*): A Taphonomy Study From their Natural Feeding Places. Journal of Taphonomy 10, 197–217.

Fourvel, J.B., Mwebi, O., 2011. Hyenas' level of dependence on livestock in pastoralist areas in the Republic of Djibouti and Kenya: relation between prey availability and bone consumption sequence. In: Brugal, J.-P., Gardeisen, A. Zucker, A. (Eds.), Prédateurs dans tous leurs états. Évolution, Biodiversité, Interactions, Mythes, Symboles. XXXIe rencontres internationales d'archéologie et d'histoire d'Antibes. Antibes, 351–370.

Fourvel, J.B., Fosse, P., Brugal, J.-P., Tournepiche, J.-F., Cregut-Bonnoure, E., 2012. Consumption of Ungulate Long Bones by Pleistocene Hyaenas: a Comparative Study. Journal of Taphonomy 10, 239–263.

García, N., Mazza, P.P.A., in this volume. New results on ursids from the Early Pleistocene site of Untermassfeld. 1385–1401.

Haynes, G., 1980a. Evidence of carnivore gnawing on Pleistocene and Recent mammalian bones. Paleobiology 6, 341–351.

1980b. Prey bones and predators: Potential ecologic information from analysis of bone sites. Ossa 7, 75–97.

1982. Utilization of Skeletal Distribution of North American Prey Carcasses. Arctic 35, 266–281.

1983. A guide for differentiating mammalian carnivore taxa responsible for gnaw damage to herbivore limb bones. Paleobiology 9, 164–172.

Hemmer, H., 2001. Die Feliden aus dem Epivillafranchium von Untermaßfeld. In: Kahlke, R.-D. (Ed.), Das Pleistozän von Untermaßfeld bei Meiningen (Thüringen). Teil 3. Monographien des Römisch-Germanischen Zentralmuseums 40, 3, 699–782, pls. 132–142.

Hill, A., 1979a. Butchery and natural disarticulation: An investigatory technique. American Antiquity 44, 739–744.

1979b. Disarticulation and scattering of mammal skeletons. Paleobiology 5, 261–274.

1989. Bone Modification by Modern Spotted Hyenas. In: Bonnichsen, R., Sorg, M.H., (Eds.), Bone modification. Orono, 169–178.

Iannucci, A., Mecozzi, B., Kahlke, R.-D., Sardella, R., in this volume. New results on hyaenids from the Early Pleistocene site of Untermassfeld. 1403–1422.

Kahlke, R.-D. (Ed.), 1997a. Das Pleistozän von Untermaßfeld bei Meiningen (Thüringen). Teil 1. Monographien des Römisch-Germanischen Zentralmuseums 40, 1, I–VI, 1–418, pls. 1–67.

1997b. Die *Hippopotamus*-Reste aus dem Unterpleistozän von Untermaßfeld. In: Kahlke, R.-D. (Ed.), Das Pleistozän von Untermaßfeld bei Meiningen (Thüringen). Teil 1. Monographien des Römisch-Germanischen Zentralmuseums 40, 1, 277–374, pls. 45–62.

1997c. Bisheriger Gesamtbefund zur Geologie, Paläozoologie, Taphonomie, Ökologie und Stratigraphie der unterpleistozänen Komplexfundstelle Untermaßfeld. In: Kahlke, R.-D. (Ed.), Das Pleistozän von Untermaßfeld bei Meiningen (Thüringen). Teil 1. Monographien des Römisch-Germanischen Zentralmuseums 40, 1, 385–418.

2000. The Early Pleistocene (Epivillafranchian) faunal site of Untermassfeld (Thuringia, Central Germany). Synthesis of new results. In: Lordkipanidze, D., Bar-Yosef, O., Otte, M. (Eds.), Early humans at the gates of Europe. Proceedings of the first international symposium. Dmanisi, Tbilisi (Georgia) Septembre 1998.

Etudes et Recherches Archéologiques de l'Université de Liège 92, 123–138.

(Ed.), 2001a. Das Pleistozän von Untermaßfeld bei Meiningen (Thüringen). Teil 2. Monographien des Römisch-Germanischen Zentralmuseums 40, 2, I–VIII, 419–698, pls. 68–131, suppl. I.

(Ed.), 2001b. Das Pleistozän von Untermaßfeld bei Meiningen (Thüringen). Teil 3. Monographien des Römisch-Germanischen Zentralmuseums 40, 3, I–VI, 699–1030, pls. 132–151, suppl. I–XV.

2001c. Die unterpleistozäne Komplexfundstelle Untermaßfeld – Zusammenfassung des Kenntnisstandes sowie synthetische Betrachtungen zu Genesemodell, Paläoökologie und Stratigraphie. In: Kahlke, R.-D. (Ed.), Das Pleistozän von Untermaßfeld bei Meiningen (Thüringen). Teil 3. Monographien des Römisch-Germanischen Zentralmuseums 40, 3, 931–1030, suppl. I–XV.

2006. Untermassfeld. A late Pleistocene (Epivillafranchian) fossil site near Meiningen (Thuringia, Germany) and its position in the development of the European mammal fauna. British Archaeological Reports, International Series 1578. Oxford.

in this volume. The Pleistocene of Untermassfeld – A synopsis on site origin, palaeobiodiversity, taphonomic characteristics, palaeoenvironment, chronostratigraphy, and significance in western Palaearctic faunal history. 1635–1734.

Kahlke, R.-D., Gaudzinski, S., 2005. The blessing of a great flood: differentiation of mortality patterns in the large mammal record of the Lower Pleistocene fluvial site of Untermassfeld (Germany) and its relevance for the interpretation of faunal assemblages from archaeological sites. Journal of Archaeological Science 32, 1202–1222.

Keiler, J.-A., 2001. Die Koprolithen aus dem Unterpleistozän von Untermaßfeld. In: Kahlke, R.-D. (Ed.), Das Pleistozän von Untermaßfeld bei Meiningen (Thüringen). Teil 2. Monographien des Römisch-Germanischen Zentralmuseums 40, 2, 691–698, pls. 128–131.

Koch, W., 1935. The age order of epiphyseal union in the skeleton of the European bison (*Bos bonasus* L.). The Anatomical Record 61, 371–376.

Korth, W.W., 1979. Taphonomy of microvertebrate fossil assamblages. Annals of the Carnegie Museum 48, 235–285.

Kruuk, H., 1972. The Spotted Hyena. A Study of Predation and Social Behavior. Chicago.

Larkin, N.R., Alexander, J., Lewis, M.D., 2000. Using Experimental Studied of Recent Faecal Material to Examine Hyena Coprolites from the West Runton Freshwater Bed, Norfolk, U.K. Journal of Archaeological Science 27, 19–31.

Madurell-Malapeira, J., Alba, D.M., Espigares, M.-P., Vinuesa, V., Palmqvist, P., Martínez-Navarro, B., Moyà-Solà, S., 2015. Were large carnivorans and great climatic shifts limiting factors for hominin dispersals? Evidence of the activity of *Pachycrocuta brevirostris* during the Mid-Pleistocene revolution in the Vallparadís section (Vallès-Penedès Basin, Iberian Peninsula). Quaternary International 431, 42–52.

Maguire, J.M., Pemberton, D., Collett, M.H., 1980. The Makapansgat Limeworks grey breccia: Hominids, hyaenas, hystricids or hillwash? Palaeontologia Africana 23, 75–98.

Mills, M.G.L., 1990. Kalahari Hyenas: Comparative Behavior and Ecology of Two Species. London.

Münzel, S. C., 1987. Umingmak, ein Moschusochsenjagdplatz auf Banks Island, N. W. T., Canada – Archäologische Auswertung des Areals I D. In: Müller-Beck, H., Hahn, J. (Eds.), Archäologische Forschungen auf Banks Island 1970 – 1975. Urgeschichtliche Materialhefte 5/2. Tübingen.

Palmqvist, P., Arribas, A., 2001. Taphonomic decoding of the paleobiological information locked in a lower Pleistocene assemblage of large mammals. Paleobiology 27, 512–530.

Palmqvist, P., Martínez-Navarro, B., Pérez-Claros, J.A., Torregrosa, V., Figueirido, B., Jiménez-Arenas, J.M., Espigares, M.P., Ros-Montoya, S., De Renzi, M., 2011. The giant hyena *Pachycrocuta brevirostris*: Modelling the bone-cracking behavior of an extinct carnivore. Quaternary International 243, 61–79.

Peters, G., 1993. *Canis lupus* Linnaeus, 1758 – Wolf. In: Stubbe, M., Krapp, F. (Eds.), Handbuch der Säugetiere Europas. Band 5/I Raubsäuger (Teil I). Wiesbaden, 47–106.

Saunders, J.J., Dawson, B.K., 1998. Bone damage patterns produced by extinct hyena, *Pachycrocuta brevirostris* (Mammalia: Carnivora), at the Haro River Quarry, Northwestern Pakistan. In: Tomida, Y., Flynn, L.J., Jacobs, L.L. (Eds.), Advances in Vertebrate Palaeontology and Geochronology. National Science Museum Monographs 14. Tokyo, 215–242.

Schütt, G., 1971. Die Hyänen der Mosbacher Sande (Altpleistozän, Wiesbaden/Hessen) mit einem Beitrag zur Stammesgeschichte der Gattung *Crocuta*. Mainzer Naturwissenschaftliches Archiv 10, 29–76.

Sher, A.V., 1997. An Early Quaternary bison population from Untermaßfeld: *Bison menneri* sp. nov. In: Kahlke, R.-D. (Ed.), Das Pleistozän von Untermaßfeld bei Meiningen (Thüringen). Teil 1. Monographien des Römisch-Germanischen Zentralmuseums 40, 1, 101–180, pls. 22–33.

Sotnikova, M.V., 2001. Remains of Canidae from the Lower Pleistocene site of Untermaßfeld. In: Kahlke, R.-D. (Ed.), Das Pleistozän von Untermaßfeld bei Meiningen (Thüringen). Teil 2. Monographien des Römisch-Germanischen Zentralmuseums 40, 2, 607–632, pls. 102–113.

Sutcliffe, A.J.,1970. Spotted Hyena: Crusher, Gnawer, Digester and Collector of Bones. Nature 227, 1110–1113.

Toots, H., 1965. Sequence of disarticulation in mammalian skeletons. Rocky Mountain Geology 4, 37–39.

Turner, A., 2001. Remains of *Pachycrocuta brevirostris* (Carnivora, Hyaenidae) from the Lower Pleistocene site of Untermaßfeld. In: Kahlke, R.-D. (Ed.), Das Pleistozän von Untermaßfeld bei Meiningen (Thüringen). Teil 2. Monographien des Römisch-Germanischen Zentralmuseums 40, 2, 673–690, pls. 121–127.

Turner, A., Antón, M., 1996. The giant hyaena, *Pachycrocuta brevirostris* (Mammalia, Carnivora, Hyaenidae). Geobios 29, 455–468.

Van Asperen, E.N., Kahlke, R.-D., 2017. Dietary traits of the late Early Pleistocene *Bison menneri* (Bovidae, Mammalia) from its type site Untermassfeld (Central Germany) and the problem of Pleistocene 'wood bison'. Quaternary Science Reviews 177, 299–313.

Zapfe, H., 1939. Die Lebensspuren der eiszeitlichen Höhlenhyäne. Die urgeschichtliche Bedeutung der Lebensspuren knochenfressender Raubtiere. Palaeobiologica 6, 111–146.

HELMUT HEMMER · RALF-DIETRICH KAHLKE

NEW RESULTS ON FELIDS FROM THE EARLY PLEISTOCENE SITE OF UNTERMASSFELD

Abstract

Numerous new findings since the first comprehensive publication on the felids from Untermassfeld (Hemmer 2001) give additional insights into pertinent problems regarding felid taxonomy, the phylogeography of the cat species recorded in the Untermassfeld fauna, and their predator and prey interactions.
Additional material of *Panthera onca gombaszoegensis* confirms the jaguar as the most frequently detected cat in fossil finds from Untermassfeld. The discovery of a mandible fragment from a cub, hardly older than three to four months, has broadened our understanding of the ecology of the Epivillafranchian jaguar. In solitary *Panthera* species, cubs of this age reside close to their lair, within the mother's home range. In the case of Untermassfeld, this must have included a section of the river valley within the range of the documented high flood situations. A nearly complete hemimandible represents the first finding of a lower P_2 reported for a Eurasian jaguar, which until now has only been very rarely found in lions and tigers. The size variability of the recorded parts of the Untermassfeld jaguar population does not correspond to that of a full population. As of yet, smaller individuals in particular are apparently missing from the site.
A second, largely complete but badly crushed, skull of *Acinonyx pardinensis pleistocaenicus* was also recovered, together with a substantial part of the individually associated forequarter skeleton comprising the nearly complete left forepaw and the vertebral column up to the fourth thoracic vertebra. This partial skeleton is preserved in its original find position in an in-situ block preparation. Individually related thoracic vertebrae X–XIII and lumbar vertebrae I–III of a large individual from Untermassfeld show the typical *Acinonyx* proportions, as do available skeletal elements of fore and hind legs. A small head, typical of cheetahs, completes the overall picture. Estimations for a mean-sized Epivillafranchian cheetah from Untermassfeld reveal a head and body length of c. 175 cm, a shoulder height of c. 110 cm, and a body mass of c. 125 kg. Although *A. pardinensis pleistocaenicus* was gigantic, it had anatomical prerequisites comparable to those of extant *A. jubatus*, suggesting that it too chased its prey at a sprint and then killed by suffocation.
A well-preserved partial puma skeleton forms the holotype of the subspecies *Puma pardoides brevipes* ssp. nov., differing from the Middle Villafranchian *P. pardoides pardoides*. The former was a morphologically generalised cat with no emphasised specialisation, and is neither comparable to the fossil cheetah-like pumas of the Americas, nor to the extant puma – a jumper with very long hind legs. The functional morphol-

ogy of *P. pardoides brevipes* suggests an adaptation to a life in wooded habitats. Its fibula is noteworthy due to the anteriorly directed expansion of its proximal end. This diagnostic synapomorphic feature of the puma evolutionary line confirms the generic affiliation. The dentition of the Untermassfeld puma has shed light on two previously unclear cases. The first is a mandibular fragment from the Early Pleistocene Pinjor zone of the Indian Siwaliks, which was believed to be a leopard, but is obviously a puma. The other is the »leopard« of the Late Pliocene Moroccan site of Ahl al Oughlam, whose puma nature is underlined by its considerable similarity to the Untermassfeld cat.
Well-preserved cervical and thoracic vertebrae of *Lynx* have the same relative vertebral length as that recorded in the extant *L. pardinus* and *L. lynx*, in contrast to the long-necked *L. issiodorensis*. This crucial difference in functional morphology supports an upgrade, together with the closely related Villafranchian South European lynxes, to that of a separate species, *L. valdarnensis*. As sister species of the Iberian *L. pardinus*, both evolved contemporaneously. The Epivillafranchian lynx from Untermassfeld has been allocated to a new chrono-subspecies, *L. valdarnensis spelaeoides* ssp. nov.
New material from the dirk-toothed cat *Megantereon* from Untermassfeld is scarce, but shows an extremely wide variability of size, indicating strong sexual dimorphism. The persistent issues on the identity of the evolutionary advanced *Megantereon* in Europa towards the end of the Villafranchian and during the Epivillafranchian are broadly discussed. The classification of corresponding forms as a separate species, *M. adroveri*, is supported, whereas the hypothesis of a European spread of the African *M. whitei* is rejected.
A hemimandible of a juvenile homothere from Untermassfeld is one of the few available examples of scimitar cubs in Europe and was in the latter part of the second half-year of its life. Furthermore, the new *Homotherium* material from Untermassfeld includes a second, well-preserved mandible of an adult individual, which shows great similarity to a hemimandible from the late Early Pleistocene of Semibalki in the Azov Sea region. In view of the still controversial taxonomy of Pleistocene *Homotherium*, a two species concept is supported here, which keeps the Middle to early Late Villafranchian populations of Eurasia as *H. crenatidens*, the robustly built earlier species. Because of clear differences in the respective functional morphology, it is separated from the more gracile *H. latidens*. This geochronologically younger species is subdivided into the latest Early to Middle Pleistocene *H. latidens ultimum*, the Middle to Late Pleistocene European to East Beringian *H. latidens latidens*, and the spatially adjacent subspecies *H. latidens serum* of southern North America. The Untermassfeld scimitar cat is to be assigned to the first-named taxon, which roamed across Europe and into East Asia, surviving there up to the Middle Pleistocene.

1. Introduction

Twenty years have now passed since the first comprehensive publication on the felids from the Epivillafranchian of Untermassfeld appeared (Hemmer 2001). By then, the recovered fossil material of this extraordinary site included the remains of all of the recognised six cat species with body masses of more than 10 kg from the European late Early Pleistocene. Systematic excavations continued (Kahlke 2020) and again produced an overwhelming number of fossils, some of them spectacular. Following completion of the field work in 2019 (Kahlke in this volume), the aim of the current study is to present the knowledge gained from the site on the palaeobiology of the jaguar (*Panthera onca*), the giant cheetah (*Acinonyx pardinensis*), the Old World puma (*Puma pardoides*), the lynx (*Lynx valdarnensis*), the dirk-toothed cat (*Megantereon adroveri*), and the scimitar cat (*Homotherium latidens*) during the late Early Pleistocene in Central Europe.

The first paper on the Untermassfeld felids (Hemmer l.c.) initiated the taxonomic attribution of the Eurasian jaguar (then *P. gombaszoegensis*) to the American species *P. onca*, and proved the existence of the Eurasian puma, whose fossil remains were mistakenly attributed to several other genera and species. These investigations are continued here, based on the unique discovery of a partial puma skeleton that was excavated during the last field seasons at Untermassfeld.

New finds and corresponding research results have also been published from several other fossil sites across Europe, which have contributed to a better understanding of the Untermassfeld felid finds. The early Middle Pleistocene of Château (France) supplied a rich series of jaguar bones, including a nearly complete skeleton (Argant et al. 2007; Argant and Argant 2011). The Italian Late Villafranchian locality of Pantalla produced two nearly complete cheetah skulls (Cherin et al. 2014); another partial skull, virtually extracted from its host matrix by synchrotron radiation, was found at Monte Argentario (Cherin et al. 2018). The Spanish sites of Cueva Victoria and Vallparadís contributed important Epivillafranchian puma and lynx materials (Madurell-Malapeira et al. 2010; Boscaini et al. 2016). A re-evaluation of the well-known complete dirk-toothed cat skeleton from the Middle Villafranchian of Senèze (France) has provided extensive knowledge on the anatomy of *Megantereon cultridens* (Christiansen and Adolfssen 2007). The Spanish Late Villafranchian site of Incarcal became famous for its magnificent collection of scimitar cat remains, which show an enormous morphometric variability (Antón and Galobart 1999; Galobart et al. 2003; Antón et al. 2005; 2014). And, finally, Russian homotheres (Sotnikova and Titov 2009) proved to be key for the classification of the Untermassfeld species, as will be shown in the following pages.

During the past two decades, tremendous molecular genetic progress has been made in understanding hybridisation and gene flow as important elements in general phylogenetic development. This research has provided insight into a truly complex pattern, not only of divergent but also convergent evolutionary processes. For example, multiple post-speciation admixture has contributed genetic material between nearly each of the extant *Panthera* species (Figueiró et al. 2017). Molecular evidence of natural hybridisation in areas of range overlap has also been proved for other carnivores (e.g., Johnson et al. 1999; Edwards et al. 2011; Rutledge et al. 2015; VonHoldt et al. 2016). Corresponding advances in genetics have been incorporated into the discussions below.

2. New fossil material and its significance

The structure of the presented study follows that of the first paper published on Untermassfeld felids (Hemmer 2001). Each species section is introduced with preliminary notes, which provide short summaries on the current state of knowledge and identify current unresolved issues that may be solved based on the data obtained from the new finds. The felid remains that have been recently excavated at Untermassfeld or identified in the existing collection are listed here. The total material of the individual species of the Untermassfeld fauna is therefore the sum of the material lists given by Hemmer (l.c.) and those in the present article. The description of fossils relevant to the knowledge of corresponding species is supplemented by illustrations and tables of measurements. If the measured distances are not self-explanatory, they follow Schmid (1940) for dental elements and Driesch (1976) for the postcranial skeleton, especially for vertebrae. Concluding discussions try to fit the results into the current hypotheses on the evolution of the individual felid species, at least during the Early and early Middle Pleistocene, and draw conclusions for their respective taxonomic classification. Body masses are estimated by prediction equations and have been used to reconstruct the presumable prey size spectrum and matching species within the Untermassfeld herbivore fauna. Notes on habitat use conclude each species presentation.

2.1. *Panthera onca gombaszoegensis* (Kretzoi, 1938)

2.1.1. Introduction

Late Villafranchian 'panthers' of the Apennine peninsula, placed between tigers and jaguars (Ficcarelli and Torre 1968) and classified as *Panthera toscana* (Schaub, 1949), were considered half a century ago as conspecific with the early Middle Pleistocene European 'lion' *Panthera gombaszoegensis* (Kretzoi, 1938) (Hemmer and Schütt 1969). Subsequently, this taxon was recognised as a European jaguar (Hemmer 1971). Differences in the dental morphology of early and later palaeopopulations led to a division on a chrono-subspecific level, i.e., *P. gombaszoegensis toscana* and *P. gombaszoegensis gombaszoegensis* (Hemmer 1972). Growing evidence of only minor differences between Pleistocene Eurasian and North American jaguars was discussed on the occasion of the first description of corresponding finds from Untermassfeld (Hemmer 2001). This initiated the taxonomic decision to refer all Eurasian jaguars to the American species *P. onca*, i.e., *P. onca toscana* for Late Villafranchian and *P. onca gombaszoegensis* for the Epivillafranchian to Middle Pleistocene forms (Hemmer 2001; Hemmer et al. 2010). Some later authors followed this suggestion (e.g., Baryshnikov 2011; Mol et al. 2011; Vislobokova 2018), others did not (e.g., Argant and Argant 2011; Marciszak 2014; Madurell-Malapeira et al. 2015). Another approach prefers a three species concept, i.e., a subdivision into *P. toscana*, *P. gombaszoegensis*, and *P. onca* (Argant et al. 2007).

The decision of how many species a genus contains is certainly not a question of right or wrong, but a question of concepts used by different authors to define the former. Nevertheless, in palaeontological studies classifications at a species level should be conclusive in order to trace and characterise phylogenetic lines (Hemmer 2003). Rejecting the name *P. onca* for fossil jaguars because it's the name of the present-day jaguar, called »biological species«, in contrast to the »palaeontological species« *P. gombaszoegensis*, whilst retaining *P. onca* for both Pleistocene and extant American jaguars (Argant and Argant 2011), is an inconsistent approach. The use of the name *P. gombaszoegensis* for all Eurasian jaguars of Early to Middle Pleistocene ages, parallel to the simultaneous use of *P. toscana* for the earlier and *P. gombaszoegensis* for the later populations (Marciszak 2014), is, at the very least, extremely confusing.

There is agreement, however, on the jaguar's phylogeny (Hemmer et al. 2010; Argant and Argant 2011; Hemmer and Heidtke 2013). The earliest jaguars, obviously descending from an African *Panthera* ancestor population, evidently dispersed across Europe during the Olduvai polarity subchron (1.95–1.77 Ma) and reached Southwest Asia by the end of that period (Hemmer et al. 2010). Genetic assessment estimates the date of divergence of the jaguar line from the lion and leopard ancestor population at around 2.06 Ma (Johnson et al. 2006; see also corrective notes by Davis et al. 2010). These results coincide for a 2 Ma history of *P. onca*. Just one million years after its first arrival in Eurasia, the jaguar reached the intercontinental transit route to North America within a late Early Pleistocene glacial period (Hemmer et al. 2010). Finally, the species only dispersed to South America during the Middle Pleistocene (Eizirik et al. 2001; Arroyo-Cabrales 2002; Hemmer and Heidtke 2013). The jaguar didn't just survive in Eurasia until the end of the early Middle Pleistocene, as previously assumed (Hemmer et al. 2003), but persisted into the Holstein interglacial period in the Caucasus (Kudaro Cave, Baryshnikov 2011) and possibly even longer in Poland (Biśnik Cave, Marciszak 2014).

Whereas, at several other European sites more or less complete cranial and postcranial skeletal elements of *P. onca gombaszoegensis* have occasionally been found (e.g., Artenac, Tournepiche 1996; Château, Argant et al. 2007; Argant and Argant 2011), the Epivillafranchian series from Untermassfeld consists of mainly fragmented elements. Nevertheless, the Untermassfeld material is unique because of its relatively high minimum number of represented individuals (MNI, see section 2.1.3.), whose remains were all deposited during a very limited span of time (see section 2.1.2.). A largely complete *P. onca gombaszoegensis* skeleton recovered from the Château site (Argant and Argant 2011), and used as a size reference, enables the calculation

of the mean percent deviation for each of the Untermassfeld skeletal elements, as far as the corresponding measurements are available. This procedure leads to a relatively exact idea of the relative size distribution and variability within an Epivillafranchian jaguar population.
Such knowledge could even be of some taxonomic interest, as body size once played a central role in the discussion as to whether Pleistocene North American jaguars (*P. onca augusta*) should merit a subspecies or species status (Seymour 1993).
Extant jaguars prefer riparian landscapes in both forested and open grassy areas (Nowell and Jackson 1996). Such a habitat was also assumed for *P. onca gombaszoegensis* (Hemmer 2001; 2004; Hemmer et al. 2003), but this was later questioned (Argant and Argant 2011). A mandibular fragment from an Untermassfeld jaguar cub provides the perfect opportunity to re-examine these issues in an unexpected way, supplemented by new insights into the jaguar's use of its habitat.

2.1.2. Material

C sup. sin. IQW 1994/24 528 (Mei. 24 057),
C sup. sin. fragment IQW 1986/21 797 (Mei. 21 316),
P^3 dex. fragment IQW 1992/24 253 (Mei. 23 782),
Mandibula dex. (juv.) IQW 2009/29 962 (Mei. 29 124),
Mandibula sin. IQW 2010/30 721 (Mei. 29 883),
P_3 sin. IQW 1986/21 787 (Mei. 21 306) + IQW 1986/21 788 (Mei. 21 307),
P_3 dex. IQW 1984/19 684 (Mei. 19 204),
P_4 dex. IQW 1986/21 779 (Mei. 21 298),
Atlas IQW 2002/27 338 (Mei. 26 843),
Vertebra cervicalis V IQW 2011/35 493 (Mei. 32 703),
Humerus dex. diaphyseal fragment IQW 2018/45 338 (Mei. 50 647),
Humerus dex. distal fragment IQW 1994/24 434 (Mei. 23 963),
Pelvis dex. IQW 1983/18 981 (Mei. 18 501),
Pelvis dex. IQW 2009/30 453 (Mei. 29 615),
Pelvis dex. IQW 2011/35 491 (Mei. 32 701),
Tibia sin. distal fragment IQW 2012/38 597 (Mei. 35 571),
Tibia dex. proximal fragment IQW 1986/21 749 (Mei. 21 268),
Calcaneus dex. IQW 2007/29 116 (Mei. 28 278),
Naviculare (Tc) dex. IQW 1982/18 048 (Mei. 17 568),
Metatarsale (Mt) 4 dex. IQW 1994/24 520 (Mei. 24 049),
Metatarsale (Mt) 4 dex. proximal fragment IQW 2010/31 032 (Mei. 30 194)

The concentration of fossil material at the site was induced by the accumulating effect of a clastic mudflow fan. Almost the entire jaguar finds originate from the sediments of an erosion channel (see section 2.1.4.) that was more likely to have filled over a timespan of days or weeks rather than months or years (Kahlke et al. 2020). A minor exception is the fragmentary right mandible IQW 1995/25 343 (Mei. 24 872) from square Q 686 (see Hemmer 2001, 701 ff., pl. 134, 2–3) originating from the Lower Fluviatile Sands (LFS). This single specimen is, therefore, slightly older than the mass of the other finds. A few specimens from the years 1979 (catalogued in 1980 and 1982) cannot be assigned with any certainty to either of the two sand bodies (Kahlke 1997a, 8). However, the corresponding age difference within the entire material, of presumably not more than a few decades, is not of biostratigraphic relevance (Kahlke et al. 2020). Overall, the jaguar material from Untermassfeld represents one single biological population.

2.1.3. Description and comparison

The new jaguar material from Untermassfeld includes a fragmentary right hemimandible from a cub, IQW 2009/29 962 (Mei. 29 124) (**Fig. 1**). Both of its inner deciduous incisors are in place. These teeth are the first to shed during tooth replacement. Consequently, this cub must have been younger than the onset of the loss of the deciduous dentition. According to the allometric equation for extant felids [onset of tooth replacement (months) = 0.37 × body mass (g)$^{0.27}$] (Hemmer 1979), tooth replacement should have started in a species in the size range of the Untermassfeld jaguar (body mass estimations see below) at an age of 8 to 10 months. In the Untermassfeld jaguar cub the DP_3, the last tooth in the eruption sequence of the felid deciduous dentition, is present (**Fig. 1**). As it erupts in a pantherine cat of that size at around the age of two months (compilation in Hemmer 1966), this animal must have been older than that. The lack of any macrowear on the DP_3 points to the earliest stage within the period between the completion of the deciduous dentition and the start of tooth replacement with DI_1 and DI_2 loss, corresponding with stage III as defined for the eruption sequence of deciduous and permanent teeth in *Homotherium serum* (Rawn-Schatzinger 1983). An age of three to four months has been suggested for this stage in the scimitar cat (Rawn-Schatzinger l.c.), i.e., in a cat somewhat larger than the European jaguar, setting the maximum presumed life span of the Untermassfeld jaguar cub at three to four months.

The size of the mandibular bone, as indicated by the length of the diastema and the corresponding height of the mandible at that section (**Table 1**), is considerably smaller than in an early Middle Pleistocene European *P. onca gombaszoegensis* cub from l'Escale (Bonifay 1971). The description of this specimen points to the very beginning of tooth replacement (corresponding to the onset of Rawn-Schatzinger's stage V). A Late Pleistocene American jaguar cub (*P. onca augusta*) from Haile VII in Florida (Kurtén 1965) seems to represent the end of stage IV. This cat is also much larger than the Untermassfeld animal. Thus, the small size of the latter confirms its younger age at death compared to that of the l'Escale and Haile VII individuals. Moreover, the DP_3 of the Untermassfeld mandibular fragment is larger than in those cubs, indicating that it was not simply a very small cat. Overall, the evidence points to an individual age for the Untermassfeld cub of certainly less than half a year and presumably not older than three to four months.

A further newly discovered find is the complete left hemimandible IQW 2010/30 721 (Mei. 29 883) (**Fig. 2**), which supplements the previously described series of mandibular fragments and isolated teeth from Untermassfeld jaguars (Hemmer 2001, 701 ff.). Within the local series, as well as within the total range of known Eurasian jaguars, this hemimandible represents the upper region of the size variability (**Table 2**). Wear marks are limited to the molar, indicating an early mature age for the individual. The dentition shows the typical features of *P. onca gombaszoegensis*. However, this find is the first mandible from a Eurasian jaguar that originally had a P_2 as an additional premolar. The tooth itself is missing, but the open and sharp-edged alveola indicates its existence during the lifetime of the individual.

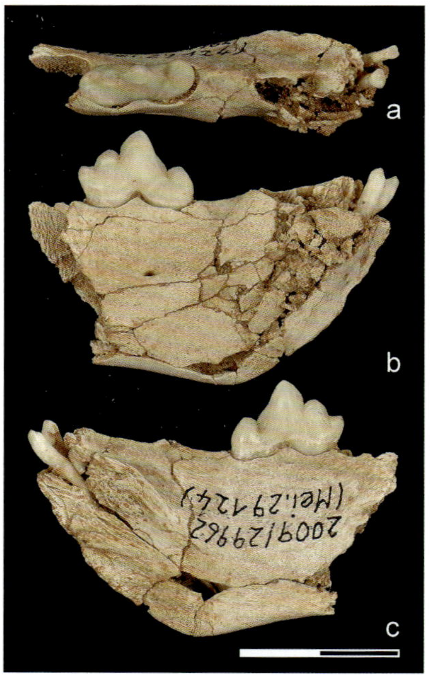

Fig. 1 *Panthera onca gombaszoegensis* (Kretzoi), Untermassfeld. – **a–c** Right hemimandible fragment IQW 2009/29 962 (Mei. 29 124) with DI_1, DI_2 and DP_3, occlusal, buccal, lingual views. – Scale: 20 mm.

	Panthera onca gombaszoegensis Untermassfeld, IQW 2009/29 962 (Mei. 29 124)	*Panthera onca gombaszoegensis* L'Escale CD–5 (Bonifay 1971)	*Panthera onca augusta* Reddick (Florida, USA), UF 2565 (Kurtén 1965)	*Panthera onca augusta* Haile VII (Florida, USA), UF 8455 (Kurtén 1965)
Ontogenetic stage as after Rawn-Schatzinger (1983)	III	V	III or IV	IV
Mandible height in the diastema	22.3			30.1
Diastema length	6.7	10.2		9.5
DI_1 transversal diameter	1.8			
anterior-posterior diameter	2.2			
DI_2 transversal diameter	2.5			
anterior-posterior diameter	2.2			
DP_3 length	14.2	13.2	12 7	13.3
maximal width	5.3	6.2	5 3	5.5
anterior width	4.5			
protoconid length	5.9		5 7	6.1
protoconid height	7.7			

Table 1 Pleistocene European and North American jaguars, comparative measurements (mm) of deciduous dentition.

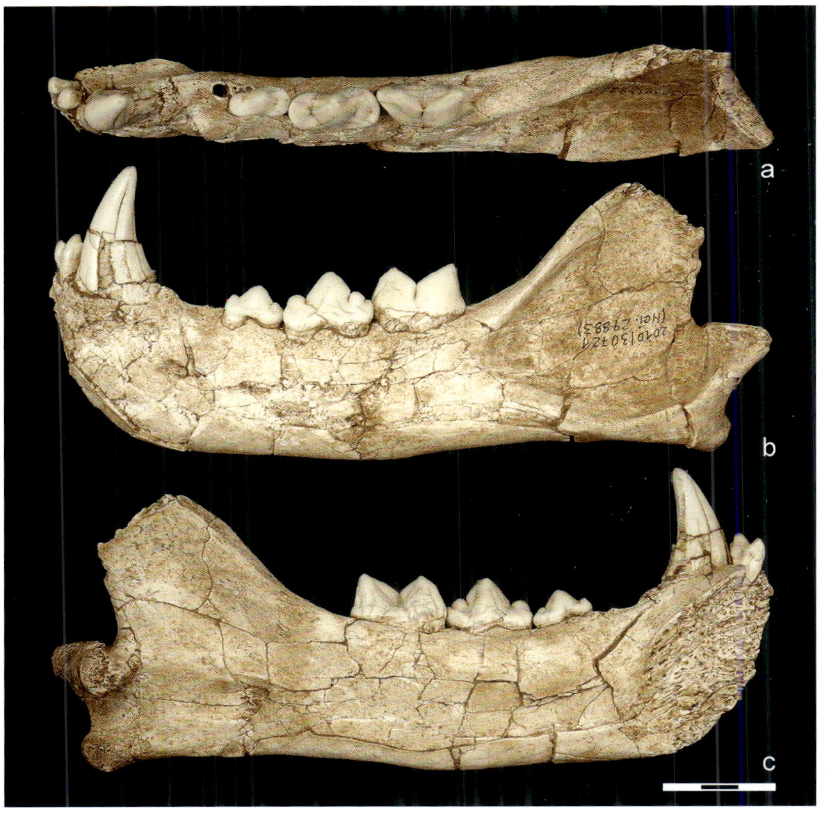

Fig. 2 *Panthera onca gombaszoegensis* (Kretzoi), Untermassfeld. – **a–c** Left hemi-mandible IQW 2010/30 721 (Mei. 29 883) with I_2-M_1, occlusal, buccal, lingual views. – Scale: 30 mm.

	IQW 2010/30721 (Mei. 29883)	Earlier published Untermassfeld jaguars: min. – max. (n) (Hemmer 2001)	IQW 1986/21787 (Mei. 21306) + IQW 1986/21788 (Mei. 21307)	IQW 1984/19684 (Mei. 19204)	IQW 1986/21779 (Mei. 21298)
Mandible length Processus condylaris-Infradentale	c. 190.0				
Depth behind M_1	38.5	37.5 (1)			
Depth in the diastema	39.5	33.5–33.7 (2)			
Thickness below P_4/M_1	17.4				
Alveolar length P_3-M_1	64.6	58.6–59.5 (2)			
Alveolar length P_2-M_1	68.4				
Distance C-P_3	16.5	16.0–22.0 (3)			
Diastema length C-P_2	14.1				
I_2 transversal width	4.3				
antero-posterior	6.3				
height	7.5	6.0–7.7 (2)			
I_3 transversal width	5.6	5.9 (1)			
antero-posterior	6.5	6.5 (1)			
height	12.8	9.2 (1)			
C length	c. 20.0	17.3–18.5 (3)			
width		13.0–13.8 (3)			
height	c. 35.5	28.5–38.1 (3)			
P_2 alveolar length	4.3				
alveolar width	3.2				
P_3 length	16.1	14.8–18.3 (4)	17.9	14.9	
maximal width	8.3	7.2–9.1 (4)	9.0	7.5	
anterior width	6.5	6.3–7.5 (4)	7.9	6.5	
protoconid length	8.5	8.2–10.7 (4)	10.3	8.2	
protoconid height	9.4	8.4–9.6 (4)	9.4	8.5	
P_4 length	24.0	21.8–24.5+ (4)			
maximal width	11.6+	10.7–12.2 (4)			12.2
anterior width	10.3	8.8–10.7 (4)			
protoconid length	11.7	11.3–11.8 (4)			
protoconid height	14.1	13.4–15.2+ (4)			
M_1 length	25.2	22.3–27.7 (4)			
width	12.5	11.2–13.4 (4)			
protoconid length	15.2	13.3–16.0 (4)			
paraconid length	13.8	12.1–15.0 (4)			
protoconid height	14.4	12+–15.0 (4)			
paraconid height		12.5+–14.2+ (4)			
depth at notch		6.3–7.6 (4)			

Table 2 *Panthera onca gombaszoegensis* (Kretzoi), measurements (mm) of the hemimandible IQW 2010/30721 (Mei. 29883), compared to the previously published sample of Untermassfeld jaguars (Hemmer 2001) and newly found teeth

	IQW 1980/17482 (Mei. 17004)	IQW 1983/18981 (Mei. 18501)	IQW 2009/30453 (Mei. 29615)	IQW 2011/35491 (Mei. 32701)
Acetabulum length on rim	36.4	37.2	35.5	40.2
Ilium min. height on shaft	39.2	39.3	40.9	40.1

Table 3 *Panthera onca gombaszoegensis* (Kretzoi), pelvis measurements (mm), Untermassfeld.

The MNI represented in the Untermassfeld jaguar series, as verified by mandible fragments and teeth, was previously determined to be 5 (Hemmer 2001). Based on the new material this number has been increased to 7.

Two cervical vertebrae are assigned to jaguars. An atlas fragment, IQW 2002/27338 (Mei. 26843), has the following measurements (mm): facies articularis cranialis width 53, facies articularis caudalis width 60, dorsal anterior-posterior diameter of neural arch 39.7, neural arch height 44. There is another, somewhat smaller atlas, IQW 1980/15470 (Mei. 14982), in the Untermassfeld jaguar collection, which was erroneously assigned to *Homotherium crenatidens* (Hemmer 2001, 752). A cervical vertebra V, IQW 2011/35493 (Mei. 32703), measures (mm): centrum length c. 30, length anterior to posterior zygapophyses 54.5, width across the anterior zygapophyses 70.5, width across the posterior zygapophyses 64.6, facies terminalis cranialis width 32.8, facies terminalis cranialis height c. 23, facies terminalis caudalis width 34.4, facies terminalis caudalis height c. 24.

The foreleg is represented in the new material by only two humerus fragments. A shaft, IQW 2013/45338 (Mei. 50647), has a minimal transverse diameter of 26.4 mm. The trochlear width of distal fragment IQW 1994/24434 (Mei. 23963) measures around 50 mm.

Three right-side pelvis fragments belong to the same size category as one previously documented (Hemmer 2001, 701) (**Table 3**). Compared to the skeleton of a recent large male South American jaguar, the Untermassfeld pelves are clearly larger.

A proximal tibia fragment, IQW 1986/21749 (Mei. 21268), is considerably larger than a tibia found at the Château site (**Fig. 3**, **Table 4**). The latter was compared with the tibia of an extant lioness of the same

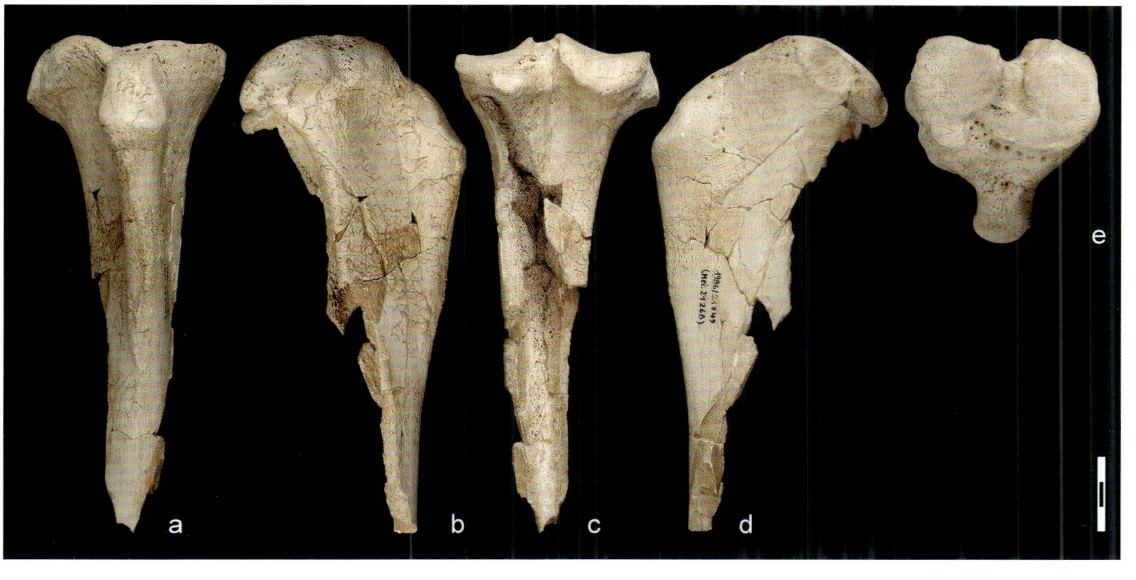

Fig. 3 *Panthera onca gombaszoegensis* (Kretzoi), Untermassfeld. **a–e** Right tibia fragment IQW 1986/21749 (Mei. 21268), anterior, external, posterior, internal, proximal views. – Scale: 30 mm.

	Untermassfeld IQW 1986/21749 (Mei. 21268)	Château CHA.1-00-F.8-40r (Argant and Argant 2011)	*P. onca augusta* North America min. – max. (n) (Kurtén 1973a)
Maximal proximal TD	79.0	75.3	71.0–76.C (3)
Maximal proximal APD	81.8	78.1	

Table 4 *Panthera onca gombaszoegensis* (Kretzoi), proximal tibia measurements (mm), Untermassfeld and Château, compared to *P. onca augusta*. TD – transverse diameter; APD – anterior-posterior diameter.

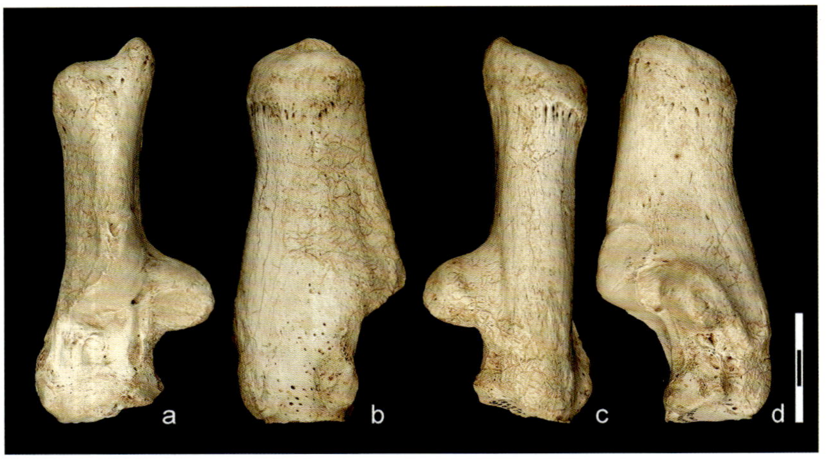

Fig. 4 *Panthera onca gombaszoegensis* (Kretzoi), Untermassfeld. – **a–d** Right calcaneus IQW 2007/29116 (Mei. 28278), dorsal, external, plantar, internal views. – Scale: 30mm.

	Untermassfeld IQW 2007/29116 (Mei. 28278)	Château min. – max. (n) (Argant and Argant 2011)	Kudaro cave min. – max. (n) (Baryshnikov 2011)	*P. onca augusta* North America min. – max. (n) (Kurtén 1965; 1973a)
Length	105.5	96.3–99.3 (2)	73.5–93.6 (2)	c. 74.0–104.0 (12)
Maximal proximal TD	50.0	44.6–46.8 (4)	37.1 (1)	31.4–45.0 (13)

Table 5 *Panthera onca gombaszoegensis* (Kretzoi), calcaneus measurements (mm), Untermassfeld and Château, compared to *P. onca augusta*. TD – transverse diameter.

length and found to be much stouter (Argant and Argant 2011). The Untermassfeld tibia also surpasses those of North American Pleistocene jaguars (**Table 4**).

A huge calcaneus of an Untermassfeld jaguar, IQW 2007/29116 (Mei. 28278), is also much bigger than the largest calcaneus found at Château (**Fig. 4**, **Table 5**) and just exceeds the upper limit of the recognised variability in Eurasian jaguars (Marciszak 2014, Fig. 6). The same is true for the comparison with *P. onca augusta* (**Table 5**). A naviculare, IQW 1982/18048 (Mei. 17568), is less conspicuous. This applies also for a slender metatarsal 4, IQW 1994/24520 (Mei. 24049) (**Fig. 5**, **Table 6**) and a fragmentary Mt4 IQW 2010/31032 (Mei. 30194).

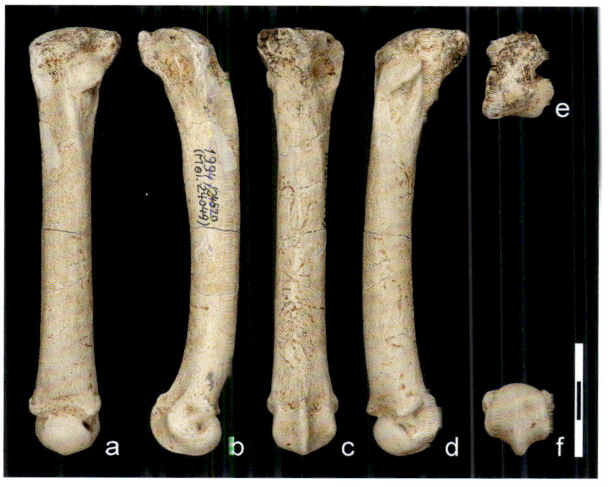

Fig. 5 *Panthera onca gombaszoegensis* (Kretzoi), Untermassfeld. – **a–f** Right metatarsal 4 IQW 1994/24 520 (Mei. 24 049), dorsal, external, plantar, internal, proximal, distal views. – Scale: 30 mm.

	Untermassfeld Mt4 dex. IQW 1994/24 520 (Mei. 24 049)	Untermassfeld Mt4 dex. IQW 2010/31 032 (Mei. 30 194)	Château min. – max. (n) (Argant and Argant 2011)	*P. onca augusta* North America min. – max. (n) (Kurtén 1965; 1973a)
Maximal length	113.1		112.4–119.9 (3)	102.0–118.0 (6)
Maximal proximal TD	19.5	19.0	19.9–22.2 (5)	
Maximal proximal APD	22.9		22.1–25.5 (4)	21.8–22.3 (2)
Minimal diaphyseal TD	12.8		13.4–15.7 (4)	13.1–15.0 (8)
Maximal distal TD	18.9		20.2–21.0 (4)	18.5–20.9 (6)
Maximal distal APD	18.8		18.5–19.8 (4)	

Table 6 *Panthera onca gombaszoegensis* (Kretzoi), metatarsal 4 measurements (mm), Untermassfeld and Château, compared to *P. onca augusta*. TD – transverse diameter; APD – anterior-posterior diameter.

The use of a largely complete skeleton of *P. onca gombaszoegensis* from the early Middle Pleistocene of Château (Argant and Argant 2011) as a size reference allows both the single dental and skeletal remains to be combined in a joint size scale that covers the Untermassfeld jaguar population from the smallest to the largest represented individuals. If available, the dimension of each Château element to be compared with an Untermassfeld equivalent is set at 100 %. From the divergence arises a positive or negative percent deviation. However, due to occasional differences in measuring techniques and the existence of allometries, i.e., size-related shifting of proportions, as given e.g., for the relative size and proportions of the P_3 in the lower dentition (Hemmer et al. 2003), this method might be somewhat inaccurate. Nevertheless, it gives a good idea of the relative size variability within a fossil population, such as the one from Untermassfeld (**Table 7**).

2.1.4. Discussion

The span between the smallest and the largest Château reference related size values of Untermassfeld jaguar remains, given as percent points, is 19 (from -7 % to +12 %, **Table 7**). Although this value indicates considerable individual size variability within the Untermassfeld population that has so far been identified, it is much less than expected for a complete *Panthera* population. Representative values were obtained for extant East African lion (*P. leo*) and leopard (*P. pardus*) populations from corresponding skull series stored

	Compared skeletal element and dimensions	Château reference related size (% difference)	Presumable sex and size category
IQW 1984/20268 (Mei. 19788)	Maxilla: L P^4	0	large female / small male
IQW 1983/19169 (Mei. 18689)	Mandible sin.: L tooth row and P_3, P_4, M_1	-6	female
IQW 1992/24137 (Mei. 23666)	Mandible sin.: L P_3, P_4	+1	large female / small male
IQW 2010/30721 (Mei. 29883)	Mandible sin.: L tooth row and P_3, P_4, M_1	+4	male
IQW 1995/25343 (Mei. 24872)	Mandible dex.: L tooth row and P_3, P_4, M_1	-4	female
IQW 1986/ 21782 Mei. 21301) IQW 1986/21789 (Mei. 21308) IQW 1986/ 21780 (Mei. 21299)	L P_3 dex., P_4 sin., M_1 dex.	+12	large male
IQW 1986/21787 (Mei. 21306) + IQW 1986/21788 (Mei. 21307)	L P_3 sin.	+11	large male
IQW 1984/19684 (Mei. 19204)	L P_3 dex.	-7	female
IQW 1980/17157 (Mei. 16678)	L M_1 dex.	+4	male
IQW 1980/16247 (Mei. 15758)	Humerus: TD distal, trochlea (sub-adult)	0	(body mass: young male)
IQW 1980/15513 (Mei. 15025)	Humerus: TD distal, trochlea	-2	large female / small male (body mass: small male)
IQW 1994/24434 (Mei. 23963)	Humerus: TD trochlea	-4	female
IQW 1980/16625 (Mei. 16146) IQW 1980/16621 (Mei. 16142) IQW 1980/16622 (Mei. 16143) IQW 1980/16624 (Mei. 16145) IQW 1980/16623 (Mei. 16144)	Metacarpal 1, 2, 3, 4, 5: TD proximal, diaphyseal (sub-adult)	-2	(body mass: young male)
IQW 1986/21749 (Mei. 21268)	Tibia: TD proximal	+5	male
IQW 2007/29116 (Mei. 28278)	Calcaneus: L	+8	large male
IQW 1982/18048 (Mei. 17568)	Naviculare: L	-3	female
IQW 1994/24520 (Mei. 24049)	Metatarsal 4: TD diaphyseal, distal	-4	female

Table 7 *Panthera onca gombaszoegensis* (Kretzoi), Untermassfeld, Château reference related size (basic data by Hemmer 2001; this paper). L – length; TD – transverse diameter; size dimensions: rounded means.

in the Bavarian State Collection of Zoology Munich. The percent point span (minimum to maximum value, each related to the mean) for the basal length of 31 lion skulls is 29, and that of 35 leopard skulls 34. If a similar span is set for *P. onca gombaszoegensis* from Untermassfeld, a substantial part of the former variability is not represented in the available fossil record. This is likely to affect the upper range of the size variability to a lesser degree, as a number of individual teeth (Hemmer 2001, Tab. 5) and postcranial elements (tibia, calcaneus) from Untermassfeld are among the largest finds of Eurasian Pleistocene jaguars. It is the smaller individuals that are most likely missing in the Untermassfeld series. The lower end of the variability could be indicated by a jaguar mandible of, probably, also Epivillafranchian age from Neuleiningen (Germany, Hemmer and Heidtke 2013). Its Château reference related size is -16. A combination of this value with that of the Untermassfeld series results in a percent point span of 28 for a total Epivillafranchian population, fully comparable to the lion and leopard results given above.
The diagnosis of sex is usually easy in both the lower and upper range of a pantherine cat's variability. This task is facilitated by the fact that, as a rule, the sexes only slightly overlap in body mass estimations using prediction equations. This is not the case when using linear measurements. Body mass calculations are based on the smallest diameters of the humerus and femur diaphyses (Hemmer 2001) and of metacarpal shafts (Hemmer et al. 2011), as the cross-sections of these bones are adapted to the load exerted on them by the body mass, both while resting as well as during locomotion. The long bones reflect the individual body mass more directly than the M_1 length-based predictor (Van Valkenburgh 1990) used in the first mass estimations of *P. onca gombaszoegensis* (Hemmer 2001; 2003; 2004; Argant and Argant 2011).
The largest published humerus of a European jaguar from the Epivillafranchian of Le Vallonnet (de Lumley et al. 1988) points to a raw body mass of 206 kg, which should be given as 200 kg to avoid any illusion of accuracy being out of reach of such a statistical calculation. It is surely reasonable to round body mass estimates up to the nearest 10 kg in the range of 50 to 200 kg and to the nearest 50 kg for all higher figures (Hemmer 2004). The humerus from Le Vallonnet has a Château reference related size (see above) of +8, i.e., the same as the enormous calcaneus from Untermassfeld. Consequently, 200 kg seems to be a reliable body mass for the largest male European jaguars. Three humerus shafts of the Untermassfeld series allow estimations to be made of 100, 160 and 170 kg respectively. The latter calculation results from a sub-adult individual as the mean of 140 kg based on earlier published metacarpals (Hemmer 2001, Tab. 3). The early Middle Pleistocene site of L'Escale (Bonifay 1971) supplied shafts of two humeri and four femora providing body mass values of 90, 100, 110, 130, 140 and 160 kg respectively. The first two figures correspond to Château reference related sizes of -12 and -5. The accumulation around 100 kg most likely indicates the centre of the female masses, whereas all values from 140 kg upwards presumably belong to males. Parallels in the extant pantherines are found with *P. leo persica* – females 110–120 kg, males 160–190 kg, and *P. tigris corbetti* – females 100–130 kg, males 150–195 kg (compilation by Nowell and Jackson 1996). These data support the body mass diagnosis of sex in *P. onca gombaszoegensis*. In accordance with these findings, the Untermassfeld jaguar remains are dominated by male individuals. When the corresponding MNI of 7 is projected onto the list of the Château reference related size (**Table 7**), the most probable scenario concluded from the Untermassfeld fossil record is the existence of at least one very large old male, one mean-sized adult male, one relatively small adult male, one sub-adult male and two adult females, as well as a small cub.
The seven recorded individuals of the former Untermassfeld jaguar population demonstrate a conspicuously high variability of premolar and molar structures, thus indicating a broad genetic diversity (Hemmer 2001; this paper). This fact should exclude a genetic bottleneck cause for the P_2 incidence in the mandible described above (see section 2.3.1., **Fig. 2**). On the other hand, the existence of a mandibular P_2 is extremely rare in pantherine cats. Two male Cape lion (*P. leo melanochaita*) skulls in the collections of the Natural His-

tory Museum, London (Mazak 1975) are the only examples within the subgenus *Panthera* (*P. leo*, *P. pardus*, *P. onca*) known to the authors. As the geographically limited former Cape lion population, »separated from other populations of lion by a complex of various physico-geographical factors« (Mazak l.c.), embodies more features that are unusual for *P. leo*, strong genetic bottlenecking might be a plausible explanation. Another comparable case of an additional P_2 within the genus *Panthera* was found in a Burmese tiger (*P. tigris*) (Lydekker 1878).

Based on the determined MNI, *P. onca gombaszoegensis* is the most common felid of the Untermassfeld fossil record. The relatively low number and mostly fragmentary condition of its skeletal elements are remarkable. The bulk of the jaguar finds originate from the filling of an erosion channel (see section 2.1.2.). The fossils were found scattered over a large number of grid squares (Q 27, 29, 31, 33, 38, 39, 46, 50, 51, 154, 268, 502, 524, 596, 627, 651, 660, 684, 692, 709, and 751) and distributed over a fairly extended depth interval, from 2.31 m below to 0.20 m above site 0-level (for field documentation see Kahlke 2020; in this volume, Fig. 13, Foldout II).

The ratio of recorded dental and skeletal remains in proportion to the corresponding MNI results in c. 7 for the Untermassfeld jaguar, in contrast to c. 24 for the puma and even c. 45 for the cheetah. Whereas the latter two species are represented by partial skeletons in different disarticulation stages (see sections 2.2.3., 2.3.3.), jaguar remains were mostly flushed into the area of the site (see section 2.1.3.) as single bones. The quotient of total number of skeletal elements of a species to its recorded MNI can be considered as flood vulnerability ratio (FVR). Within certain limits, this ratio can be used to analyse the resistance of large vertebrates to flood events. Some felid species that are not familiar with the destructive effect of exceptional water flows, as reconstructed for the site of Untermassfeld (Kahlke 2001b, 966 ff.; in this volume), may be severely affected and more often fall victim to the flood than cats that live along, and even in, riverine systems as part of their central habitats. These animals are experienced with flood events and are able to escape in time. Obviously, the European jaguar was such a cat, as is the extant jaguar.

The discovery of a mandibular fragment of a young cub less than half a year and presumably not more than three to four months old, provides an etho-ecological insight into the Untermassfeld jaguars' way of life. In solitary species of the genus *Panthera* (*P. tigris*, *P. pardus*), cubs of this age remain in close vicinity of their lair within the mother's home range, even when the latter is hunting. In the case of Middle Asian tigers, such lairs can be found on small, flood-protected riverine islands (Heptner and Sludskij 1980). Telemetric studies in the Brazilian Pantanal (Quigley and Crawshaw 2002) show that *P. onca* cubs up to half a year old also stay close together. When transferred to the Untermassfeld jaguar cub, such behaviour suggests that its lair must have been situated within the flood plain of the river valley, where the individual fell victim to the flooding. Accordingly, the home range of its mother included part of the valley along the river. This in turn indicates that riparian landscapes were the central habitat of *P. onca gombaszoegensis*, just as in extant jaguars.

The herbivore fauna of Untermassfeld offered a broad spectrum of potential prey. Cats are opportunistic hunters. However, if given a choice, they focus on different prey sizes based on their own body mass. The prey focus mass assessment can be calculated using a prediction equation based on estimated cat body mass [log {prey focus mass (kg)} = 1.917 log {cat body mass (kg)} - 1.710] and presents the results as prey mass classes in logarithmic scaling. The regularly hunted prey of a cat species ranges from three prey size classes below the focus class to one size class above (Hemmer 2004; 2007a). Thus, females of the Untermassfeld jaguar population would have focused on prey from around 10 to 200 kg, whereas males would have preferred to hunt prey of around 50 up to 1,000 kg. The three most frequently recorded Untermassfeld large herbivore species, *Eucladoceros guilii*, *Pseudodama vallonnetensis*, and *Bison menneri* (MNI 91, 86 and 69 respectively; Kahlke in this volume, Table 8) match this size range (see section 3.3.). Whereas the fallow deer-like cervid is more in the prey size range of female jaguars, the bison is more likely to fit into that of

male individuals. This indicates that the earlier published diagram showing the *P. onca gombaszoegensis* potential prey spectrum in the Untermassfeld fauna (Hemmer 2001, Fig. 3) remains valid.
The mean daily quantity a jaguar eats from a carcass often comprises only a small part of the total prey. The intake may also be estimated with a body mass-based prediction equation [log {daily food intake (kg)} = 0.695 log {cat body mass (kg)} - 0.732] (Hemmer 2004; 2007a). So Untermassfeld female jaguars could have consumed around 4.5 kg of meat per day and average-sized males around 6.5 kg, whereas the largest individuals could have consumed up to 7.5 kg. Even when calculating for multiple amounts of these quantities, taken immediately from a fresh kill and during repeated returns over subsequent days, a considerable amount of a large prey species would have remained to be used by other carnivores.

2.2. *Acinonyx pardinensis pleistocaenicus* (Zdansky, 1925)

2.2.1. Introduction

The first description of the giant Untermassfeld cheetah (Hemmer 2001) was based taxonomically on a monospecific concept for all Eurasian Late Pliocene and Early Pleistocene cheetahs, which were assigned to *Acinonyx pardinensis* (Croizet et Jobert, 1828) (Viret 1954; Schütt 1970a; Ficcarelli 1984).
The mandible and lower dentition of the first Untermassfeld specimen demonstrate considerable correspondence with middle to late Early Pleistocene North Chinese finds from Yuanchuhsien (Shanxi) and Nihewan (Hebei), the former being the holotype of *Cynailurus pleistocaenicus* Zdansky, 1925. The clear morphological differentiation of this eastern Eurasian cheetah from the western Villafranchian form was taxonomically expressed as two different geographic subspecies, *A. pardinensis pleistocaenicus* and *A. pardinensis pardinensis*; the Untermassfeld cheetah being attributed to the former of the two (Hemmer 2001).
The description of a series of early Early Pleistocene cheetahs from Longdan (Gansu, North China), dated to 2.55–2.15 Ma, older than *A. pardinensis pleistocaenicus*, reverted to a multispecies concept and shifted *A. pardinensis* out of the genus *Acinonyx* to the genus *Sivapanthera* Kretzoi, 1929 (Qiu et al. 2004). In addition to the new Longdan *Sivapanthera linxiaensis* Qiu, Deng et Wang, 2004, the later Early Pleistocene Chinese cheetah was upgraded to the species *Sivapanthera pleistocaenica*. If this classification is accepted, the Untermassfeld cheetah should also be integrated into the latter species. The taxonomic framework that includes the Longdan cheetah, as well as the latest Pliocene North African *A. aicha* from Ahl al Oughlam (Morocco, Geraads 1997), could have supported a multispecific as well as a monospecific concept, based on the limited knowledge of the time. Therefore, a macrospecies concept, i. e., *A. pardinensis* in the broadest sense, was suggested until new finds could provide better insight into Early Pleistocene cheetah evolution and thus facilitate a decision between the contrasting ideas (Hemmer et al. 2008; 2011). This concept was accepted by later work (Cherin et al. 2014). On the other hand, species ranking of the Untermassfeld cheetah as *A. pleistocaenicus* has been considered plausible, and a generic separation of *Sivapanthera linxiaensis* as a member of the *Acinonyx* stem-group has not been excluded (Spassov 2011).
Nevertheless, it should be kept in mind that a monospecies concept, which even included the extant species *A. jubatus,* was proposed (Turner and Antón 1997). However, some functionally important anatomical peculiarities of the modern cheetah's skull look more like a late acquisition in cheetah evolution, and are not present in *A. pardinensis* (Cherin et al. 2014; Geraads 2014; Grohé et al. 2018), justifying the rank of a separate species. In contrast, the distinctive characteristics of different Early Pleistocene forms »are perhaps insufficient to warrant species distinction from *A. pardinensis*« (Geraads 2014). Given all these arguments, a revision of cheetah taxonomy seems highly desirable. For the European scene, this revision will be based on the availability of new comparative material from different intervals of the Early Pleistocene: a second skull

from Saint-Vallier (France, Geraads 2014; c. 2.4 Ma), a partial mandible from Fonelas (Spain, Garrido 2008; c. 1.8 Ma), two skulls and a hemimandible from Pantalla (Italy, Cherin et al. 2014; c. 1.8–1.7 Ma), and a partial skull from Monte Argentario (Italy, Cherin et al. 2018; c. 1.5 Ma). The approximately 1.07–1.06 Ma old site of Untermassfeld has contributed a second cheetah skull and a corresponding hemimandible associated with large parts of the postcranial skeleton, as well as the predominantly postcranial remains of a third individual. This new material is dealt with here.

2.2.2. Material

Calvarium IQW 2004/28 616 (Mei. 27 778),
I^3 dex. IQW 2017/45 017 (Mei. 50 326),
C sup. sin. IQW 2015/42 019 (Mei. 47 328),
C sup. dex. IQW 2016/43 801 (Mei. 49 110),
C inf. sin. IQW 2017/44 983 (Mei. 50 292),
Mandibula sin. with P_3 IQW 2015/42 011 (Mei. 47 320),
P_4 sin. IQW 2016/43 806 (Mei. 49 115),
M_1 sin. IQW 2017/45 018 (Mei. 50 327),
Atlas fragment IQW 2004/28 618 (Mei. 27 780),
Vertebra thoracica IV IQW 2015/42 129 (Mei. 47 438),
Vertebra thoracica VIII IQW 2010/30 708 (Mei. 29 870),
Vertebra thoracica IX IQW 1980/16 626 (Mei. 16 147),
Vertebra thoracica X IQW 1980/16 629 (Mei. 16 150),
Vertebra thoracica X IQW 2015/41 935 (Mei. 47 244),
Vertebra thoracica XI IQW 2015/41 936 (Mei. 47 245),
Vertebra thoracica XII IQW 2015/41 940 (Mei. 47 249),
Vertebra thoracica XIII IQW 2015/41 939 (Mei. 47 248),
Vertebra lumbalis I IQW 2015/41 934 (Mei. 47 243),
Vertebra lumbalis II IQW 2016/43 844 (Mei. 49 153),
Vertebra lumbalis III IQW 2017/44 920 (Mei. 50 229),
Vertebra caudalis (proximal) IQW 1980/16 126 (Mei. 15 637),
Vertebra caudalis (transitional) IQW 1982/17 923 (Mei. 17 443),
Vertebra caudalis (transitional) IQW 1980/15 381 (Mei. 14 893),
Scapula sin. IQW 2016/43 953 (Mei. 49 262),
Scapula dex. IQW 2016/43 953 (Mei. 49 262),
Humerus dex. IQW 2016/43 819 (Mei. 49 128),
Humerus dex. distal fragment IQW 2017/45 019 (Mei. 50 328),
Ulna dex. IQW 2004/28 674 (Mei. 27 836),
Radius sin. diaphyseal fragment IQW 2010/30 678 (Mei. 29 840),
Trapezoid (C2) dex. IQW 1980/15 442 (Mei. 14 954),
Capitatum (C3) dex. IQW 2012/38 641 (Mei. 35 615),
Metacarpale (Mc) 2 dex. IQW 2015/42 082 (Mei. 47 391),
Metacarpale (Mc) 3 dex. IQW 2015/42 081 (Mei. 47 390),
Metacarpale (Mc) 4 sin. IQW 2012/38 493 (Mei. 35 467),
Femur dex. diaphyseal fragment IQW 1980/17 023 (Mei. 16 544),
Tibia sin. diaphyseal fragment IQW 2002/27 323 (Mei. 26 828),
Calcaneus dex. IQW 2016/43 793 (Mei. 49 102),

Cuneiforme 1 (T2) dex. IQW 1980/16258 (Mei. 15769),
Metatarsale (Mt) 2 dex. IQW 2017/44982 (Mei. 50291),
Metatarsale (Mt) 3 dex. IQW 2009/30526 (Mei. 29688),
Metatarsale (Mt) 3 dex. IQW 2016/43818 (Mei. 49127),
Metatarsale (Mt) 4 dex. IQW 2009/30527 (Mei. 29689),
Metatarsale (Mt) 4 dex. IQW 2017/45020 (Mei. 50329),
Phalanx I IQW 2016/43822 (Mei. 49131).

Individual elements within the preparation IQW 2014/41900 (Mei. 47209) (for the polyspecific bone concentration in situ and when lifting by crane see Kahlke 2020, Fig. 5a–b, for completed block preparation see frontispiece):
Vertebra cervicalis II (epistropheus),
Vertebra cervicalis III,
Vertebra cervicalis IV,
Vertebra cervicalis V,
Vertebra cervicalis VI,
Vertebra cervicalis VII,
Vertebra thoracica I,
Vertebra thoracica II,
Vertebra thoracica III,
Vertebra thoracica IV,
6 costae and proximal fragments of anterior ribs,
Scapula dex.,
Ulna sin.,
Radius sin.,
Scapholunare (Cr + Ci) sin.,
Triquetrum (Cu) sin.,
Pisiforme (Ca) sin,
Trapezium (C1) sin.,
Trapezoid (C2) sin.,
Capitatum (C3) sin.,
Hamatum (C4) sin.,
Metacarpale (Mc) 2 sin.,
Metacarpale (Mc) 3 sin.,
Metacarpale (Mc) 4 sin.,
Metacarpale (Mc) 5 sin.,
8 Ossa sesamoidea,
Phalanx I dig. 1 sin. man.,
Phalanx I dig. 2 sin. man.,
Phalanx I dig. 3 sin. man.,
Phalanx I dig. 4 sin. man.,
Phalanx I dig. 5 sin. man.,
Phalanx II dig. 2 sin. man.,
Phalanx II dig. 5 sin. man.,
Phalanx III dig. 2 sin. man.,
Phalanx III dig. 3 sin. man.,

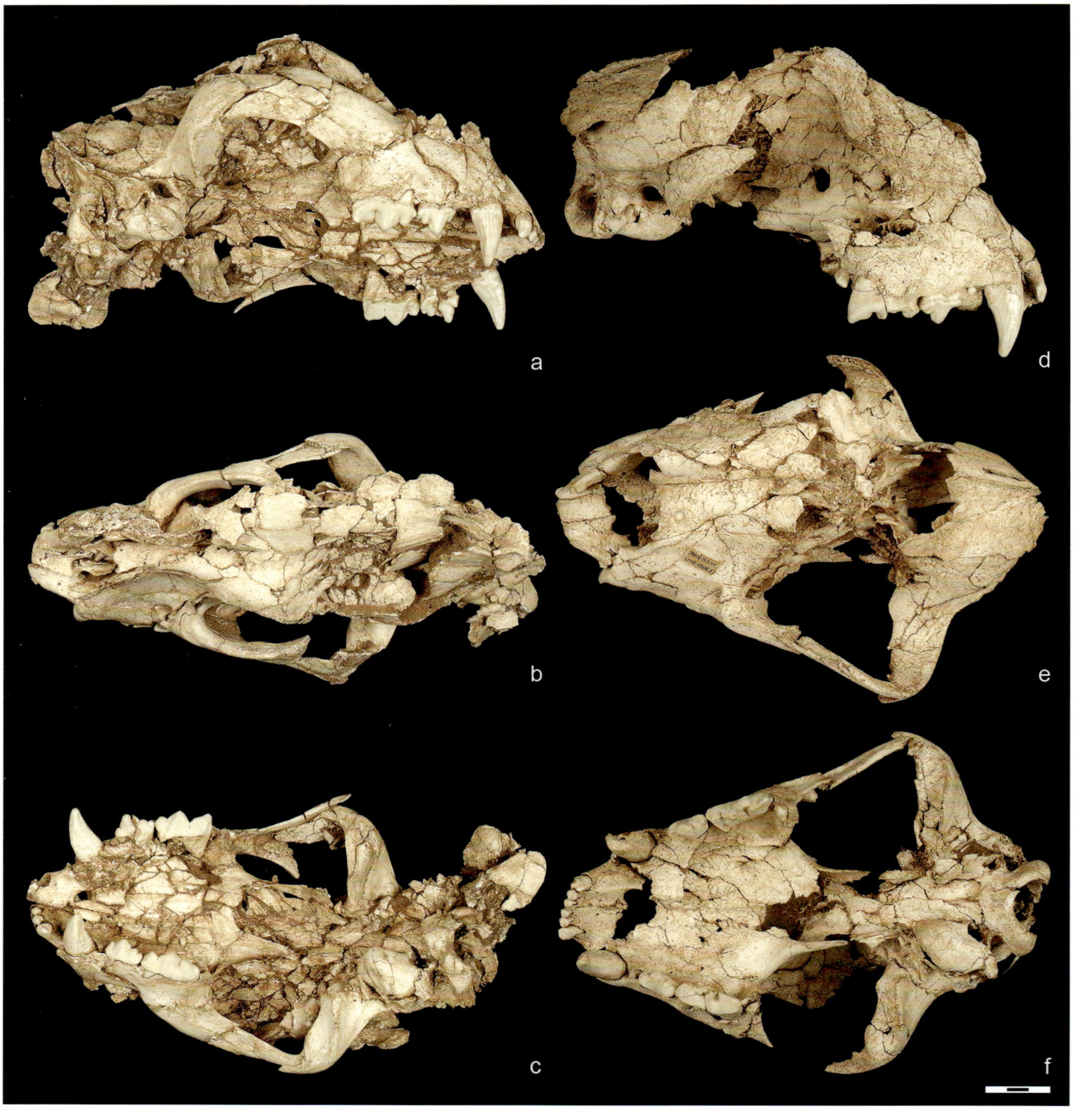

Fig. 6 *Acinonyx pardinensis pleistocaenicus* (Zdansky), Untermassfeld. – **a–c** Calvarium IQW 2004/28616 (Mei. 27778), probably female. – **d–f** Calvarium IQW 1980/16350 (Mei. 15861), probably male. – **a**, **d** right lateral views; **b**, **e** dorsal views; **c**, **f** ventral views. – Scale: 30 mm.

Phalanx III dig. 5 sin. man.,
Metacarpale (Mc) 1 dex.,
Metacarpale (Mc) 2 dex.,
Metacarpale (Mc) 3 dex.,
Metacarpale (Mc) 4 dex.,
Metacarpale (Mc) 5 dex.,
Phalanx I dig.1 dex. man.,
Phalanx I dig. 2 dex. man.,
Phalanx III dig. 1 dex. man.

2.2.3. Description and comparison

The second giant cheetah skull discovered at Untermassfeld (**Fig. 6a–c**) is even more badly crushed and distorted than the previously published specimen (Hemmer 2001, 712 ff., Tab. 7, pl. 135). Therefore, several measurements could only be taken or estimated with considerable caution, or were not possible at all. Because of this, a detailed comparison with the limited number of other available Early Pleistocene cheetah skulls does not make much sense. The new skull is only a little shorter than the first (c. 1–4 %), but is considerably narrower (**Table 8**). The absolute size of the Untermassfeld skulls matches that of the North Chinese early Early Pleistocene Longdan skulls and also fits the size category of later Early Pleistocene Chinese cheetahs. Late Pliocene European and North African fossils also show comparably enormous sizes, as do Early Pleistocene cheetahs from the South Asian Pinjor zone. In contrast, the Early Pleistocene European cheetahs older than the Late to final Villafranchian Farneta to Pirro faunal units in Italy are smaller (Monte Argentario, Cherin et al. 2018) (**Table 9**). Skull proportions are essentially the same in the previously published Untermassfeld specimen as they are in the somewhat smaller individuals of the latter European group (measurements Saint-Vallier: Viret 1954, but see Geraads 2014; Pantalla: Cherin et al. 2014). The teeth of the new Untermassfeld skull are only a little worn and thus indicate a young adult age.
The upper dentition of the newly discovered skull IQW 2004/28616 (Mei. 27778) is smaller than that of the earlier described specimen (**Fig. 7**, **Table 10**). The length divergence between both individuals spans from about 3 % in the P³ and 6–7 % in the P⁴, up to a considerable 14 % in the upper canine. The latter shows no longitudinal grooves, but a quite sharp rear edge. The P³ is remarkable by its anterior cusp (protocone) which is nearly as high as the posterior and stands in a lingual position (**Fig. 7a**), but close to the median

	IQW 2004/28616 (Mei. 27778)	IQW 1980/16350 (Mei. 15861) (Hemmer 2001)
Condylobasal length	c. 208.0	c. 216.0
Basal length	197.5	c. 200.0
Median palatal length (Prosthion-Staphylion)	c. 95.0	
Basifacial axis (Prosthion-Synsphenion)		c. 142.0
Basicranial axis (Synsphenion-Basion)		c. 72.5
Zygomatic width	≥ c. 150.0	c. 173.0
Facial width, C-C over alveoli		c. 70.0
Palatal width, P⁴-P⁴ over alveoli	c. 82.0	c. 102.0
Neurocranial width		c. 76.0 / 77.0
Mastoid width		c. 79.5
C-P⁴ alveolar length	65.5 / 66.0	74.5 / –
Bulla maximal diameter	33.5	40.0
Bulla minimal diameter	c. 26.0	26.0
Width over the occipital condyles		47.0
Foramen magnum width	c. 24.0	25.0
Foramen magnum height		21.0

Table 8 *Acinonyx pardinensis pleistocaenicus* (Zdansky), skull measurements (mm), Untermassfeld.

	Basal length	P^4 length	M_1 length
Early Pleistocene, Epivillafranchian: Europe			
Untermassfeld (Hemmer 2001, this paper)	197.5–c. 200.0 (2)	27.7–29.8 (4)	21.9–23.9 (3)
Early Pleistocene, Late Villafranchian: Europe			
Fonelas (Garrido 2008)			19.7 (1)
Pantalla (Cherin et al. 2014)	175.5 (1)	28.3–29.9 (4)	21.3 (1)
Olivola (Kurtén and Crusafont-Pairó 1977)		27.5 (1)	
Upper Valdarno (Kurtén and Crusafont-Pairó 1977)		27.6 (1)	
Monte Argentario (Cherin et al. 2018)		30.0 (1)	22.6 (1)
Early Pleistocene, Late Villafranchian: Asia			
Yuanqu (Zdansky 1925)			23.8 (1)
Gongwangling (Hu and Qi 1978)			24.5 (1)
Early Pleistocene, Middle Villafranchian: Europe			
Saint-Vallier (Viret 1954; Argant 2004)	165.0 (1)	25.9–28.8 (6)	20.1–21.0 (4)
Puebla de Valverde (Kurtén and Crusafont-Pairó 1977)		25.9 (1)	
Senèze (Kurtén and Crusafont-Pairó 1977)		25.8 (1)	
Livenzovka (Sotnikova et al. 2002)		30.1 (1)	
Early Pleistocene, Middle Villafranchian: Asia			
Longdan (Qiu et al. 2004)	200.0–202.0 (3)	26.9–31.0 (4)	19.4–23.2 (3)
Tuozidong (Nanjing Museum and Institute of Archaeology 2007)		c. 28.2 (1)	20.0–23.5 (3)
Kuruksai (Sharapov 1986)			18.0–19.0 (2)
Early Pleistocene, Pinjor: Asia			
Siwalik Hills (Lydekker 1884; measurements H.H.)			21.7–23.4 (3)
Late Pliocene, Early Villafranchian: Europe			
Villafranca d'Asti (measurements H.H.)		26.0 (1)	20.4 (1)
Les Étouaires (Schaub 1949)			24.5 (1)
Ardé (Schaub 1949)			20.3 (1)
Villarroya (Villalta 1952)			22.9–23.4 (2)
Montopoli (Ficcarelli 1984; Cherin et al. 2014)		26.7–27.5 (2)	19.6 / 20.0 (1)
Late Pliocene, Early Villafranchian: Africa			
Ahl al Oughlam (Geraads 1997)		28.3–32.1 (4)	20.8–22.7 (2)

Table 9 *Acinonyx* sp., comparative basic measurements (mm) of skull and dentition (n), Pliocene and Early Pleistocene of Eurasia and North Africa.

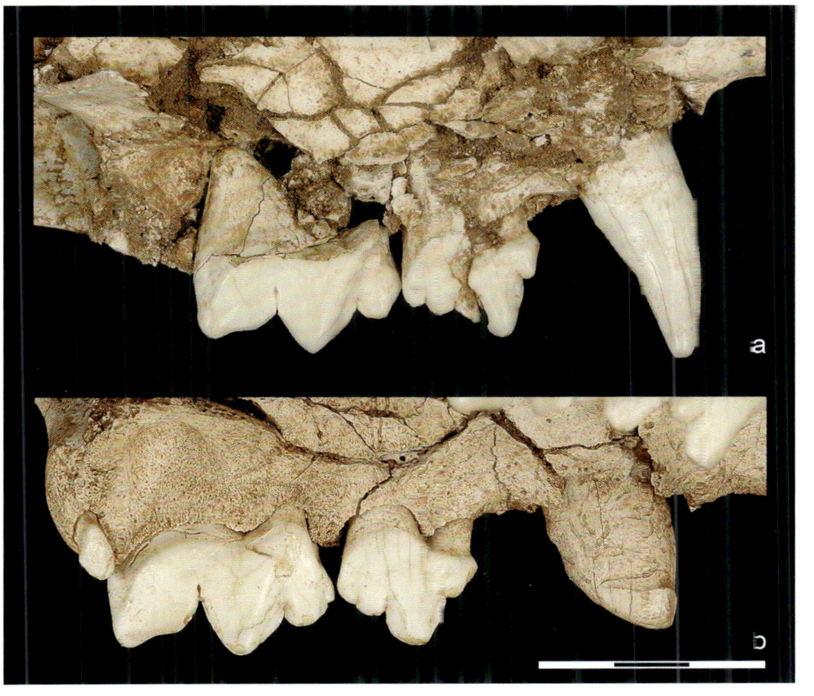

Fig. 7 *Acinonyx pardinensis pleistocaenicus* (Zdansky), Untermassfeld. – **a** Left C-P⁴ IQW 2004/28616 (Mei. 27778), probably female, lingual view. – **b** Left C-M¹ IQW 1980/16350 (Mei. 15861), probably male, lingual view. – Scale: 30mm.

	IQW 2004/28616 (Mei. 27778)		IQW 1980/16350 (Mei. 15861) (Hemmer 2001)		IQW 2015/42019 (Mei. 47328), IQW 2016/43301 (Mei. 49110)	
	sin.	dex.	sin.	dex	sin.	dex.
C sup. length	13.4	13.3		15.4	15.0	15.2
width	c. 10.5	10.3		12.1	11.7	11.6
crown height		26.1		27.0+		26.9
total height						61.9
P² length				3.7		
width				4.2		
P³ length		17.6	18.0	18.1		
maximal width	8.9	8.8	9.5	9.8		
anterior width	7.7	7.9	8.5	8.6		
paracone length		8.4	8.0	8.0		
paracone height		13.4	13.0+	13.0+		
P⁴ length	27.8	27.7	29.7	29.8		
length at protocone			28.6	28.5		
maximal width	c. 11.2	11.2	12.5			
width behind protocone	9.5	9.4	9.7			
maximal posterior width	9.1	9.1	10.0			
paracone length	10.9	11.2	11.7	c. 11.7		
metacone length	11.2	10.9	12.6	12.3		
paracone + metacone length	20.5	20.5	22.0	c. 23.2		
paracone height		15.9	15.0+			
M¹ length			4.5	4.2		
width			8.8	8.1		

Table 10 *Acinonyx pardinensis pleistocaenicus* (Zdansky), measurements (mm) of the upper dentition, Untermassfeld.

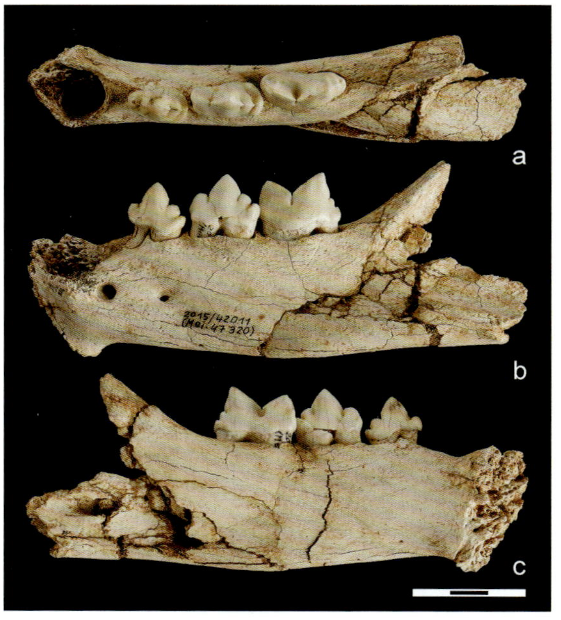

Fig. 8 *Acinonyx pardinensis pleistocaenicus* (Zdansky), Untermassfeld. – **a–c** Left hemimandible IQW 2015/42 011 (Mei. 47 320) with P_3, P_4 IQW 2016/43 806 (Mei. 49 115) and M_1 IQW 2017/45 018 (Mei. 50 327), probably female, occlusal, buccal, lingual views. – Scale: 30 mm.

line of the tooth in the occlusal view. This morphology was also similarly noted and discussed for the first Untermassfeld skull (Hemmer 2001, 720). In contrast, the P^3 protocone of the Pantalla cheetah is very small (Cherin et al. 2014). The protocone morphology of the Monte Argentario P^3 (Cherin et al. 2018) is similar to the Untermassfeld tooth. The P^4 of the new Untermassfeld skull is characterised by an extreme reduction of the protocone, a strong ectoparastyle – larger than in the first skull, and a wide buccal pit between the paracone and metacone. An isolated right upper canine IQW 2016/43 801 (Mei. 49 110) of a third Untermassfeld cheetah also has no grooves, but presents front and rear edges at its inner wall.

The newly excavated hemimandible IQW 2015/42 011 (Mei. 47 320) (**Fig. 8**) fits the second skull IQW 2004/28 616 (Mei. 27 778) (**Fig. 6a–c**) by size, matching bite, and tooth wear, indicating that both the calvarium and mandible belong to the same individual. This jaw was found with only the P_3 in place, the P_4 and the M_1 were excavated as isolated teeth and later combined due to their exact fit. The tooth length divergence between both Untermassfeld mandibles spans from around 3 % in the P_3, 6 % in the P_4, and 7 % in the M_1. The same sequence is found from the third premolar to the carnassial in both of the skulls (see above). Given the same situation for the lower canine in the new mandible, its estimated length should be roughly around 12 (+) mm. An isolated canine IQW 2017/44 983 (Mei. 50 292) does not belong to this jaw and must come from a larger individual (**Table 11**).

The most noticeable feature is the steep symphysis protruding downwards from the mandibular corpus (**Fig. 8b–c**). The anterior mental foramen is noticeably larger than the posterior (**Fig. 8b**). Both characteristics are also visible in the first Untermassfeld mandible IQW 1980/15 503 (Mei. 15 015) and IQW 1980/15 504 (Mei. 15 016) (Hemmer 2001, Figs. 4a, 5a, pl. 136). In European late Early Pleistocene *Acinonyx*, a nearly vertical anterior border is also present in the Monte Argentario specimen (Cherin et al. 2018) and a much older mandibular fragment from Fonelas (Spain, Garrido 2008) shows some similarity. Steep symphysis and large anterior mental foramen are also typical attributes of the Chinese late Early Pleistocene cheetahs (Qiu et al. 2004) and the same morphognostic features are presented by Siwalik Pinjor zone fossils (Lydekker 1884, pl. 44).

The very limited wear recorded in the dentition of the hemimandible IQW 2015/42 011 (Mei. 47 320) provides a perfect overview of the lower premolar and molar morphology. Both premolars are high and slender. In the P_3 the anterior cusp is a little smaller than the posterior, whereas the reverse proportions can be seen

	IQW 2015/42011 (Mei. 47320) sin.	IQW 2017/44983 (Mei. 50292) sin.	IQW 1980/15504 (Mei. 15016) sin.	IQW 1980/15503 (Mei. 15015) dex.
Mandible length Processus condylaris-Infradentale			177.0	181.5
Mandible height Processus angularis-Coronion			90.0+	91.5
Depth behind M_1	37.2		32.3	33.0
Depth in the diastema	28.6		30.5	31.5
Thickness below P_4/M_1	16.6		18.0	17.0
Alveolar length P_3-M_1	c. 55.0		58.2	59.8
Diastema length	c. 12.0		c. 10.5	c. 9.5
C inf. length		14.2	14.0+	13.0+
width		11.1	11.0	11.0
height			24.5+	
P_3 length	15.5		16.0	16.0
maximal width	9.7		9.7	9.5
anterior width	6.7		6.7	6.3
protoconid length	7.5		8.0	7.9
protoconid height	12.0		12.0+	11.5+
P_4 length	18.7		19.8	19.8
maximal width	9.4		9.8	9.7
anterior width	8.4		8.0	8.3
protoconid length	8.5		8.8	9.0
protoconid height	13.6		12.5+	13.0+
M_1 length	21.9		23.4	23.9
maximal width	10.5		10.0	10.3
protoconid length	14.4		14.4	14.6
paraconid length	10.9		11.5	11.4
protoconid height	14.0		15.9–	15.6
paraconid height	12.0		13.0	13.9

Table 11 *Acinonyx pardinensis pleistocaenicus* (Zdansky), measurements (mm) of mandible and lower dentition, Untermassfeld, hemi-mandible IQW 2015/42011 (Mei. 47320) completed with P_4 IQW 2016/43806 (Mei. 49115), and M_1 IQW 2017/45018 (Mei. 50327).

in the P_4, where the rear cingulum gives the impression of an additional cusp. The lower carnassial is conspicuous by the shape of its back profile. A small talonid is followed in a stairs-like manner by a small, but well individualised metaconid, separated from the protoconid by a tiny, but clearly visible incision. The same characteristic is present in the earlier published Untermassfeld individual (Hemmer 2001, 722). This distinctive morphology might also exist in the Monte Argentario cheetah, but the published picture (Cherin et al. 2018, Fig. 5) does not allow for unequivocal differentiation on this point. The stairs-like talonid-metaconid complex typical for the Untermassfeld cheetahs is present in two Chinese fossils, i.e., the holotype of *Cynailurus pleistocaenicus* (Zdansky, 1925) and a mandibular fragment from Gongwangling (Shaanxi, Hu and Qi 1978). The trace of a metaconid on the lingual side of the M_1 is mentioned for a Saint-Vallier fossil (Viret

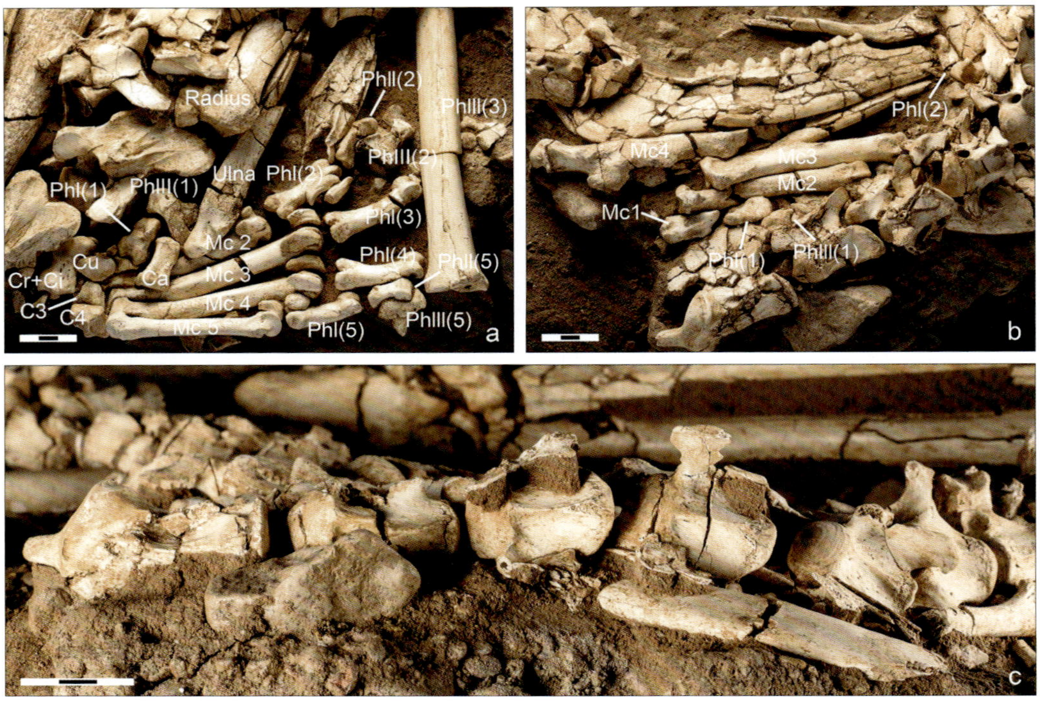

Fig. 9 *Acinonyx pardinensis pleistocaenicus* (Zdansky), in-situ block preparation IQW 2014/41 900 (Mei. 47 209), Untermassfeld. – **a** Left ulna, radius and manus. – **b** Elements of right manus. – **c** Cervical vertebra II to thoratic vertebra I. – Scales: 30 mm.

1954). The mandibular fragment from Fonelas presents a not clearly separated, weak talonid and metaconid without the stairs structure (Garrido 2008, Fig. 4), suggesting a precursor stage.

During preparation, part of the atlas remained connected to calvarium IQW 2004/28 616 (Mei. 27 778) described above. The other part of the atlas was found as an isolated fragment IQW 2004/28 618 (Mei. 27 780). Beginning with the epistropheus, a long section of the vertebral column encompassing C II to C VII and T I to T IV has been completely preserved in an in-situ block preparation IQW 2014/41 900 (Mei. 47 209) (**frontispiece**). This form of conservation (Kahlke 2020, 1049 ff.) was chosen in order to preserve a part of the unusually high concentration of bone finds at the base of the channel infill of the site [Upper Fluviatile Sands UFS(l) with reworked Younger Coarse Gravels YCG, detailed in Kahlke et al. 2020] and to protect numerous deformed and severely splintered skeletal elements from further disintegration.

The block also contains a large number of other postcranial elements from the corresponding cheetah individual (**Fig. 9**), including the complete left forepaw (see below). The length of the epistropheus including dens is 78 mm, 2.5 % less than the C II IQW 1980/16 353 (Mei. 15 864), which obviously belongs to the skull IQW 1980/16 350 (Mei. 15 861) (Hemmer 2001, 712 ff.). The newly discovered second cervical vertebra is much larger than its counterpart in the Late Pliocene cheetah skeleton of Les Étouaires (France), which measures 66 mm (Schaub 1939). Only a few reliable measurements could be taken along the vertebral column, due to partial embedding in the preparation's original sediment and multiple cracks and fissures. These are (in mm) for cervical vertebrae: centrum length C IV c. 30, C V c. 33; facies terminalis cranialis width C V c. 25, C VI 22.7; facies terminalis cranialis height C V c. 18, C VI c. 18; facies terminalis caudalis height C IV c. 20; for thoracic vertebrae: centrum length T I 28–29, T II 29, T III 29.

A series of thoracic and lumbar vertebrae (**Figs. 10–11, Tables 12–13**) comprises the whole front sector of the integrated region of the vertebral column between the diaphragmatic vertebra and the last lumbar (T X–L VII) (Randau da Costa Carvalho 2017), demonstrating the extraordinary length of the last thoracic

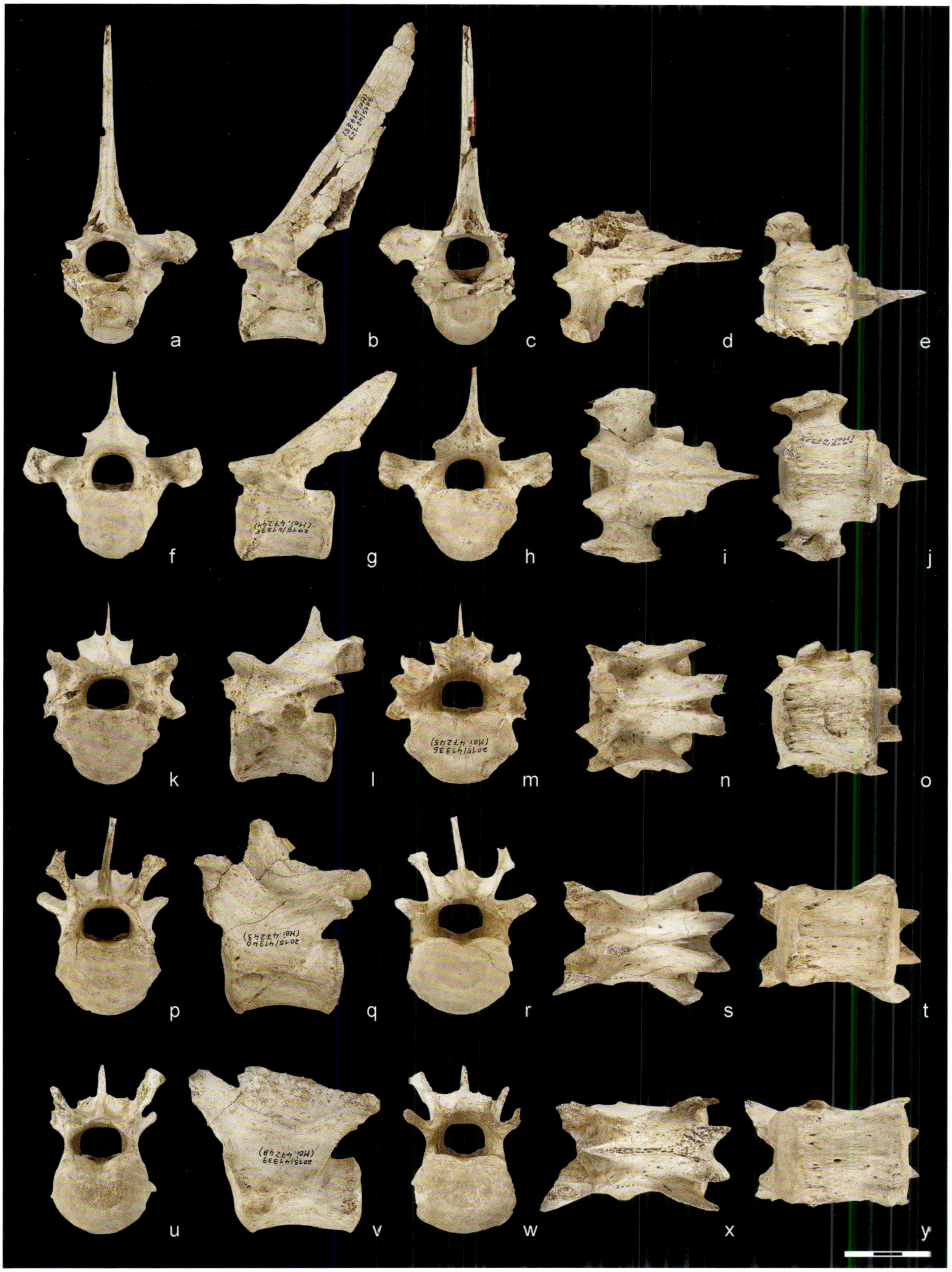

Fig. 10 *Acinonyx pardinensis pleistocaenicus* (Zdansky), Untermassfeld. – **a–e** Thoracic vertebra IV IQW 2015/42 129 (Mei. 47 438). – **f–j** Thoracic vertebra X IQW 2015/41 935 (Mei. 47 244). – **k–o** Thoracic vertebra XI IQW 2015/41 936 (Mei. 47 245). – **p–t** Thoracic vertebra XII IQW 2015/41 940 (Mei. 47 249). – **u–y** Thoracic vertebra XIII IQW 2015/41 939 (Mei. 47 248). – **a**, **f**, **k**, **p**, **u** anterior views; **b**, **g**, **l**, **q**, **v** left lateral views; **c**, **h**, **m**, **r**, **w** posterior views; **d**, **i**, **n**, **s**, **x** dorsal views; **e**, **j**, **o**, **t**, **y** ventral views. – Scale: 30 mm.

	T IV IQW 2015/42 129 (Mei. 47 438)	T VIII IQW 2010/30 708 (Mei. 29 870)	T IX IQW 1980/16 626 (Mei. 16 147)	T X IQW 1980/16 629 (Mei. 16 150)
Centrum length	29.3	30.2	33.0	36.0
Maximal height	115.0+			
Width at transverse processes				
Facies terminalis cranialis width			c. 31.0	
Facies terminalis cranialis height	23.0	c. 22.0	23.1	25.0
Facies terminalis caudalis width			c. 42.0	35.7
Facies terminalis caudalis height	22.6	23.3	23.5	
Neural canal cranial width				
Neural canal cranial height	c. 14.5	c. 11.1	13.6	
Neural canal caudal width				
Neural canal caudal height	c. 19.0			
	T X IQW 2015/41 935 (Mei. 47 244)	**T XI IQW 2015/41 936 (Mei. 47 245)**	**T XII IQW 2015/41 940 (Mei. 47 249)**	**T XIII IQW 2015/41 939 (Mei. 47 248)**
Centrum length	34.1	37.5	41.9	46.5
Maximal height	67.0+	63.0+		
Width at transverse processes	63.0	50.2		
Facies terminalis cranialis width	32.0	37.0	38.7	40.0
Facies terminalis cranialis height	24.5	27.8	29.0	31.0
Facies terminalis caudalis width	35.2	39.5	c. 38.6	
Facies terminalis caudalis height	27.0	27.5	29.0	30.0
Neural canal cranial width				
Neural canal cranial height	c. 13.2	12.3	15.2	17.4
Neural canal caudal width				
Neural canal caudal height	c. 12.7	16.9	16.5	14.5

Table 12 *Acinonyx pardinensis pleistocaenicus* (Zdansky), measurements (mm) of thoracic vertebrae, Untermassfeld.

and the lumbar vertebrae in this Untermassfeld giant cheetah. Its lumbar vertebrae are considerably longer than those of one of the world's largest extant cats, the Indian tiger (*Panthera tigris tigris*): L I 23.4 % above the maximum of five adult tigers of either sex, L II 22.8 %, L III 23.8 % (Tiwari et al. 2012). The exceptional relative length of the last thoracic and the lumbar vertebrae also defines the Late Pliocene *Acinonyx* vertebrae from Les Étouaires, as it is characteristic for the extant *A. jubatus* (Schaub 1939). Three nearly complete caudal vertebrae were recovered: a proximal, probably the mid-proximal vertebra (centrum length 39.6 mm), and two transitional vertebrae (centrum length 67.3 mm and 71.3 mm). One of these might represent the longest vertebra (terms after Russo 2015). Both of the transitional vertebrae are much larger than two caudal vertebrae from Les Étouaires (53.5 mm and 58.8 mm, described as being longer than the longest caudal vertebra of a lion skeleton, Schaub 1939).

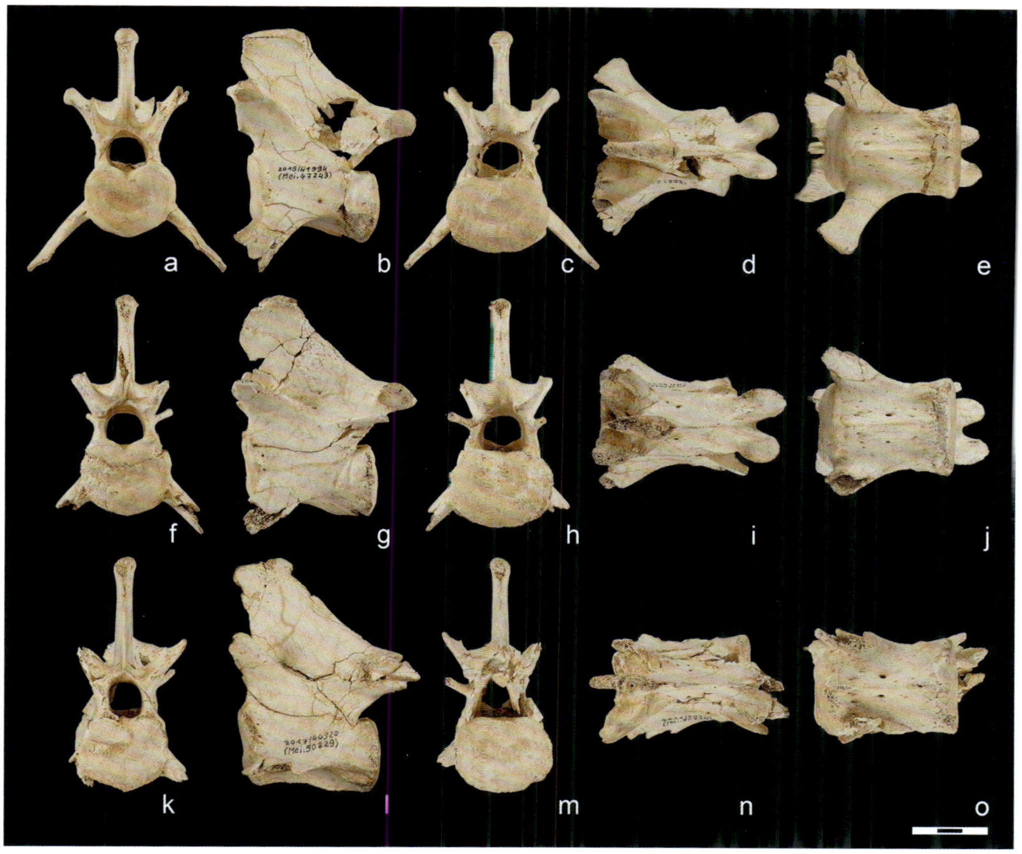

Fig. 11 *Acinonyx pardinensis pleistocaenicus* (Zdansky), Untermassfeld. – **a–e** Lumbar vertebra I IQW 2015/41 934 (Mei. 47 243). – **f–j** Lumbar vertebra II IQW 2016/43 844 (Mei. 49 153). – **k–o** Lumbar vertebra III IQW 2017/44 920 (Mei. 50 229). – **a**, **f**, **k** anterior views; **b**, **g**, **l** left lateral views; **c**, **h**, **m** posterior views; **d**, **i**, **n** dorsal views; **e**, **j**, **o** ventral views. – Scale: 30 mm.

	L I IQW 2015/41 934 (Mei. 47 243)	L II IQW 2016/43 844 (Mei. 49 153)	L III IQW 2017/44 920 (Mei. 50 229)
Centrum length	50.6	52.8	55.7
Maximal height	85.0	90.0	97.0
Width at transverse processes	81.0+		
Facies terminalis cranialis width	35.6	36.0	36.0
Facies terminalis cranialis height	32.0	33.0	30.0
Facies terminalis caudalis width	39.2	c. 40.0	40.0
Facies terminalis caudalis height	30.0	30.1	30.0
Neural canal cranial width	c. 17.0	c. 17.5	
Neural canal cranial height	c. 16.5	c. 17.0	c. 19.0
Neural canal caudal width			
Neural canal caudal height	c. 18.7	c. 18.0	

Table 13 *Acinonyx pardinensis pleistocaenicus* (Zdansky), measurements (mm) of lumbar vertebræ, Untermassfeld.

	IQW 2014/41 900 (Mei. 47 209)	IQW 2016/43 953 (Mei. 49 262) sin.	IQW 2016/43 953 (Mei. 49 262) dex.
Maximal length (along the spine)		c. 232.0	226.0+
Minimal length of the collum scapulae		44.9	44.1
Maximal length of the glenoid process	c. 50.0	53.3	53.3
Length of the glenoid cavity	c. 41.0	38.2	38.4
Width of the glenoid cavity	c. 30.0	32.0	

Table 14 *Acinonyx pardinensis pleistocaenicus* (Zdansky), scapula measurements (mm), Untermassfeld.

The in-situ block preparation IQW 2014/41 900 (Mei. 47 209) contains the right scapula of the smaller individual. Both scapulae of a second cheetah are embedded at the surface of another block preparation IQW 2016/43 953 (Mei. 49 262) (Kahlke et al. 2020, Fig. 4) and, except for their impressive size (**Table 14**), they completely resemble the scapula of extant *A. jubatus*.

The foreleg of *Acinonyx* is represented among the new discoveries by an ulna and a radius, as well as by the nearly complete left and the partial right paw of the smaller individual embedded in the block preparation (**Fig. 9a–b**). Isolated finds include a complete humerus, a distal humerus fragment, a complete ulna, the diaphysis of a radius, a trapezoid, a capitatum, and metacarpals 2, 3, and 4.

The humerus IQW 2016/43 819 (Mei. 49 128) is enormously slender in view of its extraordinary length (**Fig. 12a–f, Table 15**). This can be clearly illustrated by comparing it with the smallest of nine humeri from the Late Pleistocene North American *Panthera leo atrox*, the largest cat ever found. In the giant lion, this bone, although of similar length (328 vs. 320 mm), is stocky when compared to the giant cheetah: maximal proximal width 78.7 vs. 56.0 mm, diaphyseal transverse diameter 33 vs. 26.5 mm, maximal distal width 85.7 vs. 60.2 mm (*P. leo atrox*: Merriam and Stock 1932). The Untermassfeld humerus is however stockier than its much smaller (length from caput 216.2 mm) counterpart in the extant *A. jubatus*: minimal diaphyseal width in % length 8.5 vs. 6.7, maximal distal width in % length 19.4 vs. 17.3 (*A. jubatus* mean: Van Valk-

	Untermassfeld IQW 2016/43 819 (Mei. 49 128)	Untermassfeld IQW 2017/45 019 (Mei. 50 328)	Dmanisi (Hemmer et al. 2011)
Maximal length	320.0		281.5
Length from caput	310.0		273.0
Maximal proximal TD	56.0		58.5
Maximal proximal APD	79.6		73.5
Minimal diaphyseal TD	26.5	26.4	24.5
Maximal distal TD	60.2	61.4+	60.2
Maximal distal APD	44.7	44.7+	43.4
Trochlear TD	43.8	46.1+	45.5
Trochlear minimal thickness	23.6	24.1	22.8
Trochlear minimal height	19.8	21.8	20.1
Fossa olecrani max width	28.0	29.1	27.5

Table 15 *Acinonyx pardinensis pleistocaenicus* (Zdansky), humerus measurements (mm), Untermassfeld, Dmanisi cheetah humerus added for comparison. TD – transverse diameter; APD – anterior-posterior diameter.

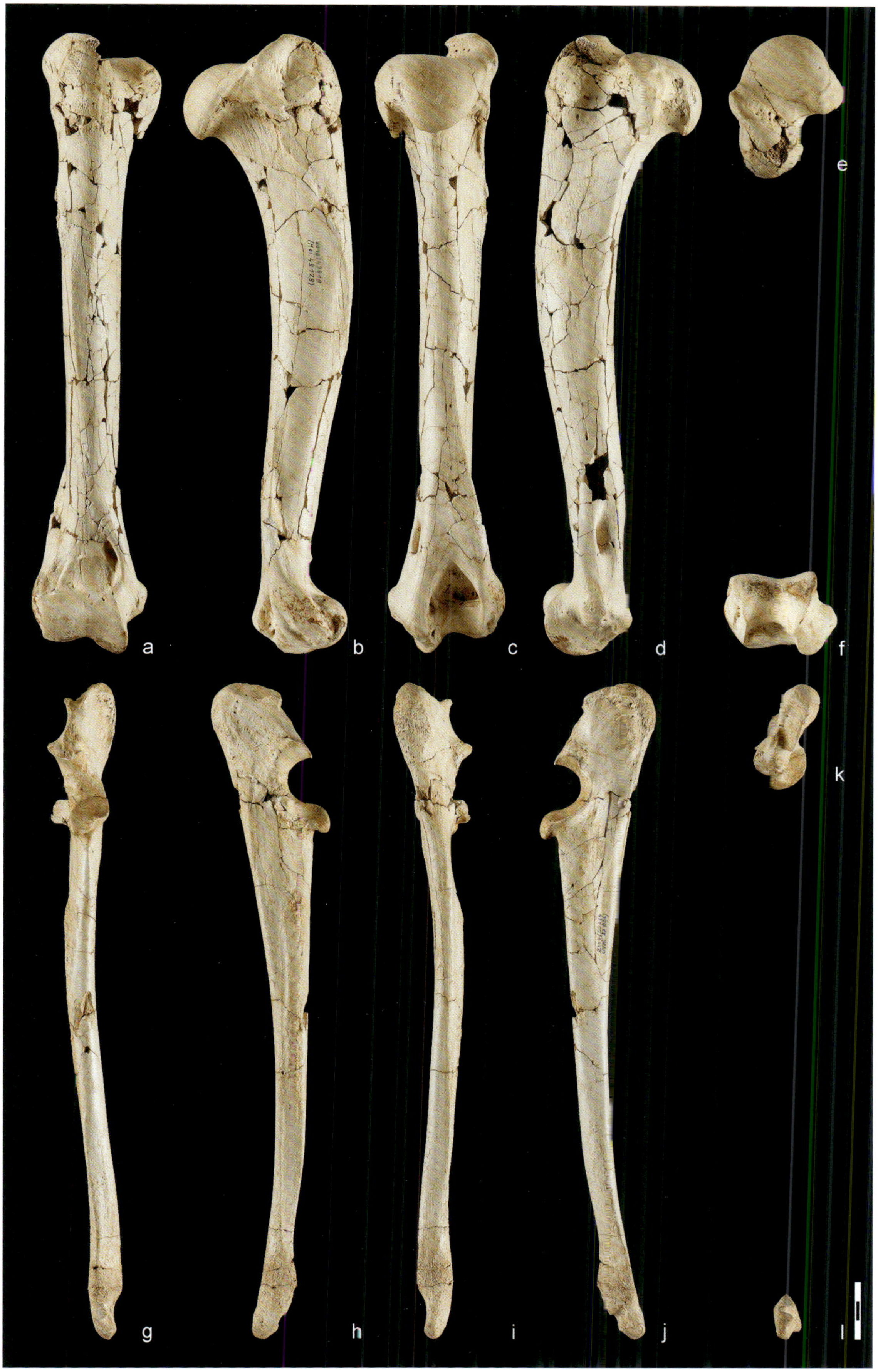

Fig. 12 *Acinonyx pardinensis pleistocaenicus* (Zdansky), Untermassfeld. – **a–f** Right humerus IQW 2016/43 819 (Mei. 49 128). – **g–l** Right ulna IQW 2004/28 674 (Mei. 27 836). – **a**, **g** anterior views; **b**, **h** external views; **c**, **i** posterior views; **d**, **j** internal views; **e**, **k** proximal views; **f**, **l** distal views. – Scale: 30 mm.

	Untermassfeld IQW 2004/28 674 (Mei. 27 836)	Dmanisi (Hemmer et al. 2011)
Maximal length	347.0	
Olecranon length	57.7	
Olecranon maximal TD	19.3	
Olecranon minimal APD	37.8	
APD over Processus anconaeus	43.9	
Height of trochlear notch	25.0	26.5

Table 16 *Acinonyx pardinensis pleistocaenicus* (Zdansky), ulna measurements (mm), Untermassfeld, Dmanisi cheetah ulna added for comparison. TD – transverse diameter; APD – anterior-posterior diameter.

enburgh et al. 1990). A somewhat shorter cheetah humerus from Dmanisi (Georgia, Hemmer et al. 2011) is even stockier than the Untermassfeld fossil: minimal diaphyseal width in % length 9.0, maximal distal width in % length 22.1. In *A. jubatus*, the ratio of humerus length / femur length scatters at around 0.86 (Van Valkenburgh et al. 1990). For the Untermassfeld cheetah this ratio is 0.84, implying similar proportions from foreleg to hind leg.

The ulna IQW 2004/28 674 (Mei. 27 836) is a long and slender bone (**Fig. 12g–l**, **Table 16**). The ratio for the lengths of ulna and humerus of an extant *A. jubatus* skeleton is 1.22, and that of a Late Pliocene skeleton from Les Étouaires c. 1.17 (Schaub 1939). The corresponding elements of the Untermassfeld cheetah result in a value of 1.08. This might indicate a somewhat shorter ulna than expected if both of the bones belong to the same individual. However, it is possible that the two bones belonged to individuals of different sexes (see section 2.2.4.).

The metapodials demonstrate different ratios of slenderness (**Figs. 13–15**, **Tables 17–18**), a very useful sexual characteristic (see section 2.2.4.) that can be immediately determined when comparable specimens are available for side by side analysis (**Figs. 14–15**). A comparative illustration is not possible for the Mc4 as a corresponding specimen is part of the previously mentioned block preparation; the figure of the available specimen (**Fig. 13**) might be useful for further work. The earlier described fifth metacarpal IQW 1980/16 555 (Mei. 16 076) from Untermassfeld (Hemmer 2001, 723) has a more evenly thick and less curved shaft than the corresponding element from the early Middle Pleistocene site of Hundsheim (Austria) and the comparatively small Mc5 of extant *A. jubatus* (Thenius 1954; Ficcarelli 1984). The Mc5 of the partial skeleton in the

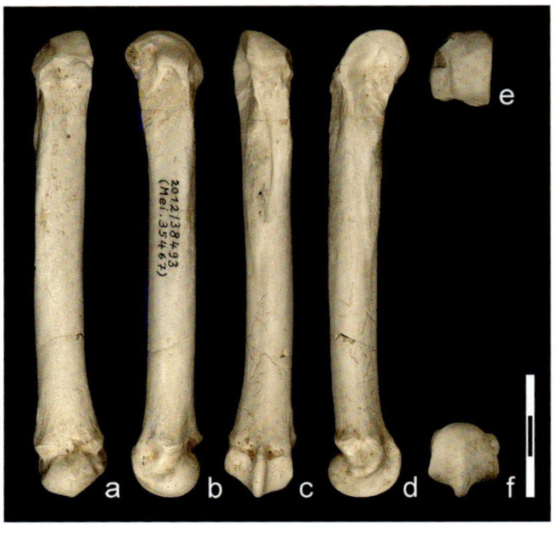

Fig. 13 *Acinonyx pardinensis pleistocaenicus* (Zdansky), Untermassfeld. – **a–f** Left metacarpal 4 IQW 2012/38 493 (Mei. 35 467), probably male, dorsal, internal, palmar, external, proximal, distal views. – Scale: 30 mm.

	Untermassfeld Mc2 dex. IQW 2015/42082 (Mei. 47391)	Untermassfeld Mc3 sin. IQW 1980/15656 (Mei. 15168)	Untermassfeld Mc3 dex. IQW 2015/42081 (Mei. 47390)	Untermassfeld Mc3 sin. IQW 2014/41900 (Mei. 47209)
Maximal length	104.3	122.5	123.1	c. 113.5
Maximal proximal TD	19.2	21.0	20.8	
Maximal proximal APD	25.0	21.0	21.1	
Minimal diaphyseal TD	12.5	15.0	14.1	c. 13.5
Maximal distal TD	18.7	20.0	20.1	
Maximal distal APD	17.4		19.3	

	Untermassfeld Mc3 dex. IQW 2014/41900 (Mei. 47209)	Untermassfeld Mc4 sin. IQW 2012/38493 (Mei. 35467)	Untermassfeld Mc4 sin. IQW 2014/41900 (Mei. 47209)	Untermassfeld Mc4 dex. IQW 2014/41900 (Mei. 47209)
Maximal length	c. 117.0	119.4	c. 111.0	c. 103.0
Maximal proximal TD		15.5		
Maximal proximal APD		18.9		
Minimal diaphyseal TD	c. 13.0	12.6	c. 11.5	c. 11.0
Maximal distal TD		17.9		
Maximal distal APD		18.9		

	Untermassfeld Mc5 sin. IQW 2014/41900 (Mei. 47209)	Untermassfeld Mc5 sin. IQW 1980/16555 (Mei. 16076)	Dmanisi Mc2 sin. D2002a (Hemmer et al. 2011)	Dmanisi Mc3 sin. D2002b (Hemmer et al. 2011)
Maximal length	c. 92.0	95.0	95.3	109.2
Maximal proximal TD		18.0	19.8	22.0
Maximal proximal APD		18.0	23.5	20.5
Minimal diaphyseal TD	c. 10.0	11.5	10.7	12.8
Maximal distal TD		17.5	16.7	17.8
Maximal distal APD			16.6	19.0

	Dmanisi Mc4 sin. D2002c (Hemmer et al. 2011)	Dmanisi Mc5 sin. D2002d (Hemmer et al. 2011)
Maximal length	106.9	86.3
Maximal proximal TD	13.7	15.5
Maximal proximal APD	20.4	18.2
Minimal diaphyseal TD	11.3	9.8
Maximal distal TD	16.8	16.5
Maximal distal APD	18.6	16.5

Table 17 *Acinonyx pardinensis pleistocaenicus* (Zdansky), metacarpal measurements (mm), Untermassfeld, Dmanisi cheetah metacarpals added for comparison. TD – transverse diameter; APD – anterior-posterior diameter.

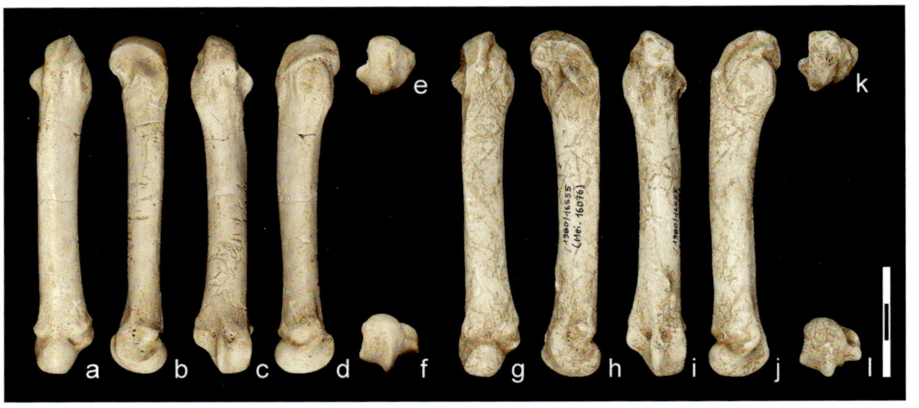

Fig. 14 *Acinonyx pardinensis pleistocaenicus* (Zdansky), Untermassfeld. – **a–f** Left metacarpal 5 from in-situ block preparation IQW 2014/41 900 (Mei. 47 209), probably female. – **g–l** Left metacarpal 5 IQW 1980/16 555 (Mei. 16 076), probably male. – **a**, **g** dorsal views; **b**, **h** internal views; **c**, **i** palmar views; **d**, **j** external views; **e**, **k** proximal views; **f**, **l** distal views. – Scale: 30 mm.

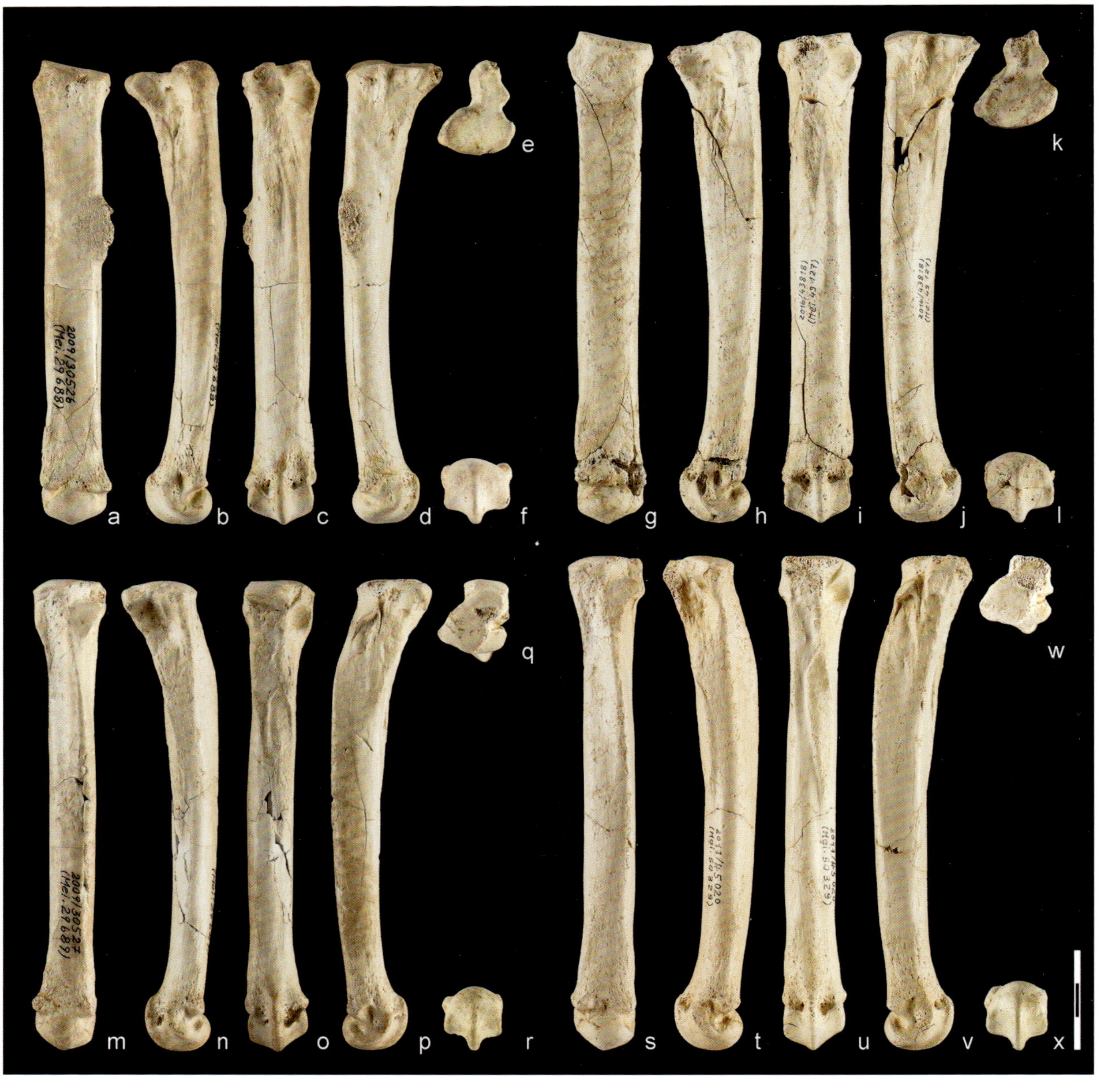

Fig. 15 *Acinonyx pardinensis pleistocaenicus* (Zdansky), Untermassfeld. – **a–f** Right metatarsal 3 IQW 2009/30 526 (Mei. 29 688), probably female. – **g–l** Right metatarsal 3 IQW 2016/43 818 (Mei. 49 127), probably male. – **m–r** Right metatarsal 4 IQW 2009/30 527 (Mei. 29 689), probably female. – **s–x** Right metatarsal 4 IQW 2017/45 020 (Mei. 50 329), probably male. – **a**, **g**, **m**, **s** dorsal views; **b**, **h**, **n**, **t** external views; **c**, **i**, **o**, **u** plantar views; **d**, **j**, **p**, **v** internal views; **e**, **k**, **q**, **w** proximal views; **f**, **l**, **r**, **x** distal views. – Scale: 30 mm.

	Mt2 dex. IQW 1980/15 797 (Mei. 15 308)	Mt2 dex. IQW 2017/44 982 (Mei. 50 291)	Mt3 dex. IQW 1980/15 796 (Mei. 15 307)	Mt3 dex. IQW 2003/30 526 (Mei. 29 688)
Maximal length	133.5	132.1		143.5
Maximal proximal TD	14.0	13.9	23.0	22.5
Maximal proximal APD	24.0	23.7+	31.0	29.1
Minimal diaphyseal TD	12.5	14.1	19.0	16.0
Maximal distal TD		20.1		21.9
Maximal distal APD		19.7		21.0
	Mt3 dex. IQW 2016/43 818 (Mei. 49 127)	**Mt4 dex. IQW 1980/15 655 (Mei. 15 167)**	**Mt4 dex IQW 2009/30 527 (Mei. 29 689)**	**Mt4 dex. IQW 2017/45 020 (Mei. 50 329)**
Maximal length	151.0	150.0	141.4	147.8+
Maximal proximal TD	25.0	19.0	20.6	20.8
Maximal proximal APD	31.2	24.0	23.6	23.6+
Minimal diaphyseal TD	17.9	15.5	12.9	14.6
Maximal distal TD	21.5	20.5	19.1	19.7
Maximal distal APD	22.0		20.0	20.7
	Mt5 dex. IQW 1980/15 657 (Mei. 15 169)			
Maximal length	129.0			
Maximal proximal TD	18.0			
Maximal proximal APD				
Minimal diaphyseal TD	12.0			
Maximal distal TD	19.0			
Maximal distal APD				

Table 18 *Acinonyx pardinensis pleistocaenicus* (Zdansky), metatarsal measurements (mm), Untermassfeld. TD – transverse diameter; APD – anterior-posterior diameter.

block preparation IQW 2014/41 900 (Mei. 47 209), slightly shorter and thinner than the first corresponding find from Untermassfeld (**Fig. 14**, **Table 17**), is indicative of the morphology of geochronologically later, smaller cheetahs. Despite the size difference, there are hardly any fundamental functional differences between the cheetahs from Untermassfeld and the younger forms.

The right Mc4 of the *Acinonyx* individual in block IQW 2014/41 900 (Mei. 47 209) shows an enormous lateral exostosis (**Fig. 9b**) that would have seriously handicapped this cat when running. Another, lengthwise extending exostosis on the internal side of the proximal part of a Mt3 (**Fig. 15a–d**) also looks like it would have been a severe handicap. The injuries sustained by the cheetah that led to these exostoses may have been due to a single accident, e.g., an encounter with a porcupine, assuming that the metacarpals and metatarsals in question belong to the same individual.

Skeletal elements representing the hind leg are rarer in the newly excavated material from Untermassfeld than those of the foreleg. Fragments of a femur and tibia, a calcaneus, a cuneiform, and five metatarsals (**Fig. 15**, **Table 18**) were found. The calcaneus IQW 2016/43 793 (Mei. 49 102) has the following measurements (mm): maximal length 105.5, maximal transverse diameter 41.0, maximal anterior-posterior diameter 42.0.

A phalanx I IQW 2016/43822 (Mei. 49121) has the following measurements (mm): maximal length 50.2, maximal proximal transverse diameter 18.4, maximal proximal anterior-posterior diameter 17.2, minimal diaphyseal transverse diameter 13.0, maximal distal transverse diameter 16.0., maximal distal anterior-posterior diameter 12.8. The discovery of the phalanges II and III of the fifth digit of the left forepaw (**Fig. 9a**) in the position of claw retraction supports the presumption (Hemmer 2001) that the claws of *A. pardinensis* were still retractile in adults, contrary to *A. jubatus*.

2.2.4. Discussion

The newly discovered *Acinonyx* remains from Untermassfeld confirm the morphology already established in the previously published material from this locality (Hemmer 2001). With the mandible as the only directly comparable skeletal element, the finds correspond to the characteristics of late Early Pleistocene Chinese cheetahs, first described as *Cynailurus pleistocaenicus* Zdansky, 1925. Consequently, there is no issue with classifying both the Epivillafranchian European and the Late Villafranchian East Asian palaeopopulations within the same taxon, ranked in the first description of the Untermassfeld felids as a subspecies of *A. pardinensis*, i.e., *A. pardinensis pleistocaenicus*, and differed from the nominate form *A. pardinensis pardinensis* (Hemmer 2001).

A set of easily recognisable features allow both of the aforementioned taxa to be distinguished. *A. pardinensis pleistocaenicus* is characterised by the conspicuous stairs-like metaconid-talonid complex of the M_1, the steep symphysis and the anterior mental foramen, which is noticeably larger than the posterior. This taxon was subsequently upgraded to species rank and shifted to the genus *Sivapanthera* as *S. pleistocaenicus* (Qiu et al. 2004).

The widely accepted revision of the systematics and taxonomy of the extant cat species (e.g., Chimento et al. 2014; Geraads 2014; but see Kitchener et al. 2017), based on genetic assessment (Johnson et al. 2006), combines all species of each of the, in total, eight extant felid evolutionary clades within one single genus, if they were found to have diverged from their respective sister taxa during the Pliocene and Pleistocene. Extending this conclusive concept to fossil species rejects the further use of any other generic name than *Acinonyx* for the evolution of the Plio-Pleistocene cheetah. Therefore, the Untermassfeld cheetah can undoubtedly be attributed to the genus *Acinonyx*. To make a conclusive decision on its taxonomic status at the species or subspecies level depends on what facts are available regarding its position in the Eurasian cheetah phylogeny.

The aforementioned Chinese palaeopopulation apparently existed at least between c. 1.9 and 1.3 Ma (Qiu 2006). This time span is immediately followed by the appearance of corresponding cheetahs in Europe, with the first known representative from Monte Argentario (Cherin et al. 2018). Their east-west dispersal over Eurasia seems to indicate a new advance into, until that time, an apparently *Acinonyx*-free Europe. The nominate taxon of *A. pardinensis* is best represented in Saint-Vallier (Viret 1954; Geraads 2014) and in Pantalla (Cherin et al. 2014), covering the time span from approximately 2.4 to 1.8–1.7 Ma. The Late Villafranchian Upper Valdarno localities supplied no really meaningful comparable mandibular material, so that only the absolute size may be tentatively used as a clue, although it is not an essential taxonomic feature. Nevertheless, by size, the Upper Valdarno cheetahs may also have been members of the earlier nominate form and could have been one of their youngest representatives. It seems quite possible that for a maximum of half a million years there was a geographical, allopatric coexistence of the western nominate and eastern Chinese forms of cheetah.

A comparable period of time for the development of phylogeographic intraspecific diversity is found in extant large cat species as indicated by molecular studies. For example, it was estimated that modern *Panthera onca* mtDNA lineages arose 0.28–0.51 Ma ago (Eizirik et al. 2001). Extant *Panthera pardus* lineages were estimated to have diverged about 0.47 Ma, when 2.00 Ma is used as a calibration point between lions and leopards (Uphyrkina et al. 2001). An application of these data to the *Acinonyx* evolution separates the contemporaneous western and eastern Middle to Late Villafranchian populations taxonomically at the subspecies level. The duration of the existence of the two mentioned cheetah populations is shorter than that of generally accepted felid species, which starting from the node between sister species can be specified to at least around one million years (deduced from data by Johnson et al. 2006). A prerequisite for the application of this time concept for cheetah evolution at the subspecific level is the rooting of both taxa in question within one and the same species, *A. pardinensis*.

The complete set of characteristic features that marks the mandible of *A. pardinensis pleistocaenicus* so well is obviously lacking in most of the Chinese cheetah finds, i. e., those from the Longdan site and including the finds from the Tuozi cave (Jiangsu, East China) of c. 2.0 Ma (Nanjing Museum and Institute of Archaeology 2007). Although, there is a partial exception within the Longdan series. The mandible IVPP V13537 found at Shitougou, a village near Longdan, has a metaconid-talonid complex in its carnassial in addition to a steeper symphysis when compared to the holotype mandible of *Sivapanthera linxiaensis* from the lower Longdan. Additionally, its coronoid process is strikingly different from the latter and shaped as in later *A. pardinensis* (Qiu et al. 2004, pls. 20, 21). If the two *Acinonyx* forms from the Longdan series actually lived in temporal proximity to each other, such a multiple morphognostic difference could well indicate the existence of two different species. Unfortunately, the exact chronological position of the Shitougou calvarium and mandible in question seem to be unknown. In Europe, a c. 1.8 Ma old mandible from Fonelas (Spain) also looks like a precursor stage of the later Chinese *A. pardinensis pleistocaenicus*. Lower jaws from the South Asian Pinjor zone of the Siwalik Series also show some of the Chinese/European late Early Pleistocene cheetah morphognostic features, but do not present all of them. Thus, the evolutionary time model for subspecies level acceptance seems to be applicable to the Eurasian Early Pleistocene *Acinonyx*. Consequently, the Untermassfeld cheetah is still classified as *A. pardinensis pleistocaenicus* and separated from *A. pardinensis pardinensis*. *A. pardinensis brachygnathus* (Lydekker, 1884) could be accepted as a third Eurasian Early Pleistocene subspecies from South Asia.

From the onset of the Middle up to the Late Villafranchian, Europe and western Asia seem to have been populated by large-sized cheetahs classified as *A. pardinensis pardinensis*. This nominate, western subspecies then disappeared before Europe was repopulated from East Asia by the gigantic *A. pardinensis pleistocaenicus*. It now appears acceptable to postulate the existence in the Pliocene of a separate *Acinonyx* ancestor species, the *Acinonyx* stem group (Spassov 2011), which gave rise to *A. pardinensis*. Such a consideration could replace the macrospecies concept which has been used until now (Hemmer et al. 2011). For this Pliocene ancestor, by priority the name *A. arvernensis* is available. Three taxonomic subdivisions may be established, namely *A. arvernensis arvernensis* (Croizet et Jobert, 1828) in Europe, *A. arvernensis aicha* Geraads, 1997 in North Africa, and *A. arvernensis linxiaensis* (Qiu, Deng et Wang, 2004) in China, the latter disappearing not earlier than c. 2.0 Ma. The evolution of cheetahs in Africa and the roots of *A. jubatus* are insufficiently documented, and so whether the extant cheetah is a sister species of *A. pardinensis* or an offshoot of the latter remains unanswered.

To get a good idea of the linear size of the *A. pardinensis pleistocaenicus* giant cheetahs, the mean mandibular length calculated from 10 modern African *A. jubatus* (122.1 mm) has been taken as a standard (Wayne et al. 1986). The first described Untermassfeld individual (Hemmer 2001, 711 ff.) is 1.45–1.49 times larger. A modern, clearly larger than average-sized and thus presumably male (means by Van Valkenburgh et al. 1990) skeleton (Schaub 1939) has been used to estimate the size relation between *A. jubatus* and the Untermassfeld cheetahs using the postcranial skeleton. The following ratios result from the Untermassfeld / *A. jubatus* bone lengths: humerus (see section 2.2.3., **Fig. 12a–f**, **Table 15**) 1.43, Ulna (see section 2.2.3.,

Fig. 12g–l, Table 16) 1.28, femur (Hemmer 2001, 716 ff.) 1.42, tibia (Hemmer l.c.) 1.41, presumably male Mt3 (see section 2.2.3., Fig. 15g–l, Table 17 and below) 1.39 and smaller, presumably female Mt3 (see section 2.2.3., Fig. 15a–f, Table 17) 1.32. As deduced from the ratio of the latter specimen, the available ulna IQW 2004/28674 (Mei. 27836) (Fig. 12g–l) could also be female.

An extrapolation based on the mean postcranial ratio of 1.41 for (presumably) male specimens and isometric calculation give a rough estimation of the linear body size of the studied *A. pardinensis pleistocaenicus*. The prerequisite for the head and body length extrapolation of the Untermassfeld cheetah is the proportioning of the spinal column sections corresponding to those in *A. jubatus*. The available thoracic and lumbar vertebrae from Untermassfeld (see section 2.2.2.) demonstrate the distinctive presence of these typical proportions, which have also been found in the late Pliocene Les Étouaires cheetah (Schaub 1939). Thus, there is no discernible impediment for the following calculation.

A total of 194 modern South and East African adult cheetah individuals of both sexes have a mean head and body length of 124.2 cm, with a minimum to maximum range of 105–152 cm (calculation based on Marker and Dickman 2003, Tabs. 2–3). When transferred to the Untermassfeld cheetah, this gives an estimate of around 175 cm for an average sized European Epivillafranchian *A. pardinensis pleistocaenicus*. Using the same process of extrapolation from modern *A. jubatus* long bones with a modern mean of c. 78 cm (calculation based on Marker and Dickman l.c.), an estimation of the shoulder height of the Untermassfeld cheetah results in around 110 cm. A head and body length in the range of 175 cm reveals a conspicuous feature of the cheetah's anatomy, i.e., its small head. A lion with a corresponding body length would have a cranial condylobasal length of about 306 mm. A tiger, which has a smaller head, still has a condylobasal length of around 262 mm (based on equations calculated by Hemmer 1974). In contrast, the corresponding lengths of the two described Untermassfeld cheetah skulls only measure around c. 208 and c. 216 mm. Thus, both fossil individuals show the typical *Acinonyx* features.

Physiologically, the most meaningful expression of body size is body mass. In the first paper on the Untermassfeld cats (Hemmer 2001), the M_1 length was used as a mass predictor (Van Valkenburgh 1990) as the results were comparable to those obtained from extant *A. jubatus* (Hemmer l.c.; Cherin et al. 2014). In a later publication, this measurement was still used as the only predictor for estimating the body mass of fossil cheetahs from various other sites (Cherin et al. 2014). When using dentition as an indicator of body mass, it should be remembered that measurements of teeth cannot give the actual load which the body mass exerted on a bone. Dentition measurements may allow to predict a genetically programmed frame of body size, but they can never give the real size an individual reached during its ontogeny (Hemmer 2007b). Thus, tooth dimensions should be applied extremely carefully, if at all, when trying to estimate the body mass of real individuals. For this reason, here we have used bone cross-sections, i.e., the minimal diaphyseal widths of the humerus, femur and metacarpals with the respective prediction equations (Hemmer 2001; 2004; Hemmer et al. 2011) to determine the body size of the Untermassfeld cheetahs.

The metacarpals of the partial skeleton IQW 2014/41900 (Mei. 47209) provide the following raw values: Mc3 sin. 110 kg, Mc4 sin. 107 kg, Mc5 sin. 79 kg, Mc3 dex. 101 kg, Mc4 dex. 96 kg, Mc5 dex. 99 kg. The right Mc4 value has been excluded in view of the extreme exostosis on this bone. The Mc5 results have also been discounted, as the body mass calculation based on this metacarpal led to an underestimation in *Acinonyx* (Hemmer et al. 2011). The remaining three values result in a mean of 106 kg, which should be rounded up to the next 10 kg to avoid false precision (Hemmer 2004). The body mass determination for the individual from the in-situ block preparation IQW 2014/41900 (Mei. 47209) thus leads to a value of c. 110 kg.

Larger metacarpals (see section 2.2.2., Table 17) resulted in values of 162 kg (Mc2), 137 kg and 120 kg (Mc3) and 132 kg (Mc4); 110 kg based on a large Mc5 has been discounted. These higher values for the body mass

are completed by 144 kg based on the femur (Hemmer 2001, Tab. 9a) in addition to 158 and 156 kg based on two humeri (**Table 15**). The resulting raw mean of 144 kg can be rounded down to 140 kg. In this way, values of 110 kg for the smaller and 140 kg for the larger cheetahs of Untermassfeld are obtained. Earlier estimations based on the molar length prediction (Hemmer 2001; Cherin et al. 2014) are clearly too low.
The range of reconstructed body weights from 110 kg to 140 kg corresponds to the expectation for females and males of the Untermassfeld palaeopopulation. The equation logBM♀ = 1.037 × logBM♂ - 0.14 (r = 0.969) is calculated using data from six extant cheetah populations of East, Southwest and South Africa (Marker and Dickman 2003). The extrapolation to a mean mass of 140 kg for male individuals indicates a mean mass for females at around 122 kg. Body mass dimorphism in the large reference series of Namibian cheetahs (Marker and Dickman l.c.) is 22.7 %. Based on a body mass of 140 kg for male individuals this would give a mean mass of around 114 kg for females. Both of these estimations are reasonably near to the 110 kg obtained for the smaller Untermassfeld individual.
Morphological differences between the relatively smaller and larger Untermassfeld cheetahs support their assignment to both sexes. The determined female-male mass dimorphism is much less than that seen in *Panthera onca gombaszoegensis* from the same site, with comparable female body masses (see section 2.1.4.). External morphometric data from the Namibian reference series give quite different percentages for sexual dimorphisms, with the greatest differences recorded in skull width and lower canine size, as well as in the legs for hind foot width (Marker and Dickman l.c.). This is paralleled by the considerably narrower skull, the smaller canines, and the slenderer metacarpals of the smaller Untermassfeld cheetah individual described here. As such, the cat from the in-situ block preparation IQW 2014/41 900 (Mei. 47 209) is most probably an average-sized female. All larger skeletal remains of the Untermassfeld material clearly belonged to two males. Based on the data, a mean body mass including both sexes would scatter at around 125 kg. This value is significantly less than the mass of a lion with a head and body length of 175 cm. This measurement corresponds to the maximum length of female African lions (140–175 cm) with body masses of 120–182 kg and is near the lower limit of male lions (170–250 cm) with body masses of 150–225 kg (Haas et al. 2005). The comparison with *Panthera leo* depicts *A. pardinensis pleistocaenicus* as a slender, lightweight cat, just like today's cheetah.
Modern-day *A. jubatus* is an extraordinarily specialised cat. Cheetahs are famous sprinters, facilitated by a light build with long limbs and a long, flexible vertebral column. Following a successful chase, the usual killing technique of suffocation by holding the throat is also due to a distinct specialisation. A sustained jaw clench is facilitated by a mandible with relatively short, blunt canines followed immediately by high-crowned premolars, which supply a steady row of sharp grasping points (Kingdon 1977; Hemmer 2007a). The Untermassfeld cheetah is much larger than the extant species, but has the same small head, a comparable light build, albeit stouter by size, functionally the same dentition suitable for specialised killing, a comparable vertebral column, and long legs for high-speed running. Consequently, the giant Epivillafranchian *A. pardinensis pleistocaenicus* was perfectly adapted for hunting using the same techniques as *A. jubatus*, i.e., a high-speed chase followed by suffocation killing – though not necessarily with the same famous speed and manoeuvrability, as supported by additional fine specialisations for gazelle hunting that may not have evolved before the Late Pleistocene (O'Regan 2002; Grohé et al. 2018). Neither the giant cheetah's hunting ability nor its killing equipment had anything in common with the functional adaptation of pantherine cats as earlier studies have tried to convey (Cherin et al. 2014). The potential prey of the giant Early Pleistocene cheetah cannot have been the same as that of its small Holocene relative, so different evolutionary fine-tuning of both predators' eco-physiological abilities would have taken place.
Body mass is a useful predictor in estimating the probable prey size of a fossil cat. The prey focus mass reconstruction method, which uses the cat's body mass as a predictor, provides, on a logarithmic scale,

not only an idea on the main preferred prey size (focus mass), but also outlines the range of prey regularly taken by the opportunistic hunter (Hemmer 2004; 2007a). Calculations of 110 kg and 140 kg as mean body masses for both sexes of the Untermassfeld cheetah results in a 20–500 kg class prey size range, with the focus on 200 kg animals. In view of the relatively small size of the cheetah's canines, a possible lower prey size class, i. e., 100 kg, should also be taken into consideration (Hemmer 2004; Hemmer et al. 2011). Be that as it may, it is the very abundant cervid *Pseudodama vallonnetensis* in the Untermassfeld fauna that best matches the predicted size range (see section 3.3.). This species, a mixed feeder (Breda et al. 2020, 1244), may have been the preferred target of the giant cheetahs. The larger *Eucladoceros guilii* most likely played a more marginal role as prey, as given that it was a browser it would have preferred more wooded habitats (Breda et al. 2020, 1244). Horses, however, would have also been possible prey in the cheetah's presumably more open hunting ground. Such a habitat preference is indicated, alongside the functional morphology of the felid species itself, by the flood vulnerability ratio (FVR, see section 2.1.4.). This ratio is 45 in the cheetah and supports the idea, when compared to the jaguar (ratio 7), that the rather dense vegetation of the river valley was not a substantial part of the Untermassfeld giant cheetah's home range. The previously published data of the presumed prey spectrum of *A. pardinensis pleistocaenicus* is therefore still valid (Hemmer 2001, Fig. 11).

Predictions regarding the daily meat consumption (Hemmer 2004; 2007a) of adult Untermassfeld cheetahs indicate an amount of 5 kg for females and 5.5 kg for males. A model calculation done for a probable male cheetah (*A. pardinensis pardinensis*) of about 100 kg body mass from the c. 1.8 Ma old Early Pleistocene site of Dmanisi (Georgia) suggests, with a mean mass of 50 kg per kill, a scenario of more than one ton of pure meat each year available to other consumers, in addition to around 4 ½ tons of other leftovers, such as bones, hide and intestines (Hemmer et al. 2011). A comparable situation is also conceivable for the Untermassfeld guild of competing carnivores. Like extant *A. jubatus*, the Epivillafranchian cheetah would have been a potential host for kleptoparasitism by species that dominated in the interspecific hierarchy. It can be assumed that despite its gigantic size, the cheetah would have tried to avoid any confrontations at its kill owing to its small head and the relatively small canines that were its main weapons. Nevertheless, its feeding situation was most probably satisfactory, given that even a severely handicapped individual (see section 2.2.3.) was able to survive.

2.3. *Puma pardoides brevipes* ssp. nov.

2.3.1. Introduction

Eurasian puma finds from Pliocene Mongolia were published in 1976 (Sotnikova 1976) and were also assumed to exist for Early Pleistocene Europe (Kurtén 1976). However, the real beginnings of European puma research started only 25 years later. Introductory to the description of the Untermassfeld puma, *Viretailurus schaubi* (Viret, 1954) from the Middle Villafranchian of Saint-Vallier (France) was synonymised with *Felis pardoides* Owen, 1846, and attributed to the genus *Puma* as *Puma pardoides* (Hemmer 2001). Eurasian puma finds were subsequently detected hidden amongst teeth from the Pliocene site of Kvabebi (Georgia), which had been formerly attributed to *Lynx*. Puma remains were also identified hidden amongst finds formerly assigned to *Panthera pardus* at Pliocene sites in East and South Africa (Hemmer et al. 2004). At the same time, the close affinities at the species level between the Saint-Vallier puma with the Untermassfeld cat were confirmed, as well as the assignment of these fossils to the genus *Puma* (Argant 2004). An additional *Puma* confirmation was presented based on Iberian material from the Middle Villafranchian of La Puebla de Valverde and from the Epivillafranchian of Cueva Victoria and Vallparadís. This material was also compared with *Panthera pardus* and *Panthera uncia*. It was noted that the *P. pardoides* dental remains from Cueva Victoria and Vallparadís more closely resemble extant pumas than those from Saint-Vallier and La Puebla de Valverde (Madurell-Malapeira et al. 2010). The reservation that the taxonomic attribution of these European cats to the genus *Puma* had never been supported by the identification of any synapomorphy with the extant puma (Cherin et al. 2013a) is now dispelled with the presentation of a mostly preserved skeleton from Untermassfeld. This find includes the synapomorphic anterior extension of the fibula head, confirming its generic puma affiliation. *P. pardoides* is actually classified in a *Puma* subgenus of its own, *Puma (Viretailurus) pardoides* (Spassov 2011; Chimento et al. 2014). Beside the nominate subgenus for *P. concolor*, other *Puma* subgenera have been established based on advancements made in felid taxonomy over the last two decades (Chimento et al. l.c.). Genetic assessment of the extant cat species supports the recognition of eight evolutionary clades in total, which presumably radiated in the Late Miocene. Species within such a clade, considered to have diverged during the Pliocene or Pleistocene, were joined taxonomically in a single genus. Thus, the small American jaguarundi was originally classified into its own genus, *Herpailurus*. Being clearly related to *P. concolor*, it was then attributed to the genus *Puma* as *P. yagouaroundi* (Johnson et al. 2006). The availability of a differentiating taxonomic unit between the rank of genus on the one hand and the rank of species on the other, i. e., the subgenus level, has been neglected in that process. Nevertheless, setting up separate subgenera seems to be a suitable instrument with which to express Pliocene divergence of extant felid species or species groups. This procedure helps to illustrate separate units in multi-species genera, as shown for the South American cats (Chimento et al. l.c.).

As a consequence, the jaguarundi was granted the subgeneric rank, *P.* (*Herpailurus*) *yagouaroundi* (Chimento et al. 2014). The earliest South American puma from the Late Pliocene of Argentina, *P.* (*H.*) *pumoides* (Castellanos, 1958), displays great similarities with the latter species, but was much larger, like a small *P. concolor*. The jaguarundi-like characters of this felid are combined with traits of the extant *P. concolor* and *Acinonyx jubatus*. Therefore, *P. pumoides* has been considered as representing the basic evolutionary level of the subgenus *Herpailurus*. It also displays similarities with the Late Pliocene and Pleistocene North and Central American cheetah-like pumas (Chimento et al. l.c.). These latter cats had been first assigned to pumas (Savage 1960; Kurtén 1976), then with cheetahs in the rank of the separate *Acinonyx* subgenus *Miracinonyx* (Adams 1979). Following this, they were raised to the rank of genus (Van Valkenburgh et al. 1990) and were later proved, using mitochondrial ancient DNA, to have diverged from the puma, but not the cheetah (Barnett et al. 2005). Consequently, they were classified as pumas, as originally assumed, belonging taxonomically to the genus *Puma* and given the rank of a further subgenus, *P.* (*Miracinonyx*) (Chimento et al. 2014).

The extant puma (subgenus *Puma*) is a large, slender cat characterised by very long hind legs and feet, a long tail, and a relatively small head: the build of a jumper. The jaguarundi (subgenus *Herpailurus*) is a small cat with some mustelid-type physique: an elongated head when compared to other small-sized cat species, relatively short legs, and a very long tail. These features are representative of an adaptation for moving easily through dense undergrowth. The subgenus *Miracinonyx* of the North American Plio-Pleistocene was characterised by a marked cheetah-like anatomy, pointing to a specialisation for running.

The enormous anatomical diversity between the three American subgenera focuses the question on the physique of the Plio-Pleistocene Old World pumas, where the phylogenetic origin of the genus and its divergence from the cheetah line may be presumed, and which may represent the origin of puma evolution better than any New World species. The build of the subgenus *Viretailurus* was, as yet, unknown. This gap is now closed by finds of nearly all diagnostically important elements of a puma skeleton mainly discovered during the Untermassfeld 2016 and 2017 field seasons (see section 2.3.2.). Some single finds from the years 2006–2008 could also be assigned to this individual (for excavation progress see Kahlke 2020; in this volume).

2.3.2. Material

Maxillare sin. fragment with P^2 IQW 2018/45 336 (Mei. 50 645) and associated C sup. IQW 2017/45 012 (Mei. 50 321),
C sup. dex. fragment IQW 2016/43 814 (Mei. 49 123),
I^3 dex. IQW 2018/45 414 (Mei. 50 723),
P^4 sin. distal fragment IQW 2008/29 823 (Mei. 28 985),
Mandibula sin. with P_3-M_1 IQW 2018/45 413 (Mei. 50 722),
M_1 dex. fragment IQW 1989/23 140 (Mei. 22 659),
Vertebra thoracica III IQW 2017/45 024 (Mei. 50 333),
Vertebra thoracica VI IQW 2017/45 329 (Mei. 50 638),
Vertebra thoracica IX IQW 2017/45 023 (Mei. 50 332),
Vertebra thoracica XII IQW 2017/44 984 (Mei. 50 293),
Sacrum IQW 2016/43 797 (Mei. 49 106),
Scapula dex. IQW 1995/25 289 (Mei. 24 818),
Humerus dex. IQW 2016/43 792 (Mei. 49 101),
Ulna dex. IQW 2017/45 013 (Mei. 50 322),
Radius sin. IQW 2016/43 817 (Mei. 49 126),
Radius dex. IQW 2017/45 022 (Mei. 50 331),
Scapholunare (Cr + Ci) sin. IQW 2017/44 941 (Mei. 50 250),
Pisiforme (Ca) sin. IQW 2016/43 803 (Mei. 49 112),
Trapezoid (C2) sin. IQW 2016/43 820 (Mei. 49 129),
Metacarpale (Mc) 1 sin. IQW 2016/43 807 (Mei. 49 116),
Metacarpale (Mc) 3 sin. IQW 2016/43 813 (Mei. 49 122),
Metacarpale (Mc) 3 dex. IQW 2016/43 811 (Mei. 49 120),
Metacarpale (Mc) 4 dex. IQW 2016/43 800 (Mei. 49 109),
Phalanx I dig. 1 sin. man. IQW 2016/43 808 (Mei. 49 117),
Phalanx I dig. 3 dex. man. IQW 2016/43 812 (Mei. 49 121),
Phalanx III dig. 1 dex. man. IQW 2016/43 810 (Mei. 49 119),
Phalanx III dig. 3 sin. man. IQW 2016/43 805 (Mei. 49 114),
Pelvis dex. IQW 1995/24 934 (Mei. 24 463),
Femur sin. IQW 2016/43 795 (Mei. 49 104),
Femur dex. IQW 2016/43 796 (Mei. 49 105),
Patella dex. IQW 2018/45 335 (Mei. 50 644),
Tibia sin. IQW 2016/43 794 (Mei. 49 103),
Fibula sin. proximal fragment, IQW 2016/43 887 (Mei. 49 196),
Fibula sin. distal fragment, IQW 2016/43 888 (Mei. 49 197),
Astragalus sin. IQW 1993/24 322 (Mei. 23 851),
Naviculare (Tc) sin. IQW 2007/29 104 (Mei. 28 266),
Phalanx I IQW 2018/45 411 (Mei. 50 720),
Phalanx I IQW 2009/30 464 (Mei. 29 626).

All *Puma* remains excavated during the 2016 and 2017 field seasons (catalogued 2016–2018, see above) were located in the five immediately adjacent grid squares Q 537, 547–549 and 556. The dental and skeletal elements were distributed over a depth interval ranging from 2.58 to 2.00 m below site 0-level (for field documentation see Kahlke 2020) and belong to a single skeleton (see section 2.3.3.). They originated from a sandy-gravelly amalgamation zone full of vertebrate remains, which was deposited in the deepest part of an erosion channel immediately following the highest point of a strong flood event (Kahlke 2001a; 2006; Kahlke et al. 2020). Additional puma remains that were sporadically found during the years 2006–2008 (catalogued 2007–2009, see above) also came from this channel filling. These remains were found about 6m upstream in the connected grid squares Q 25, 28 and 31, 1.93 to 1.23m below site 0-level (Kahlke in this volume, Fig. 13, Foldout III).

2.3.3. Description and comparison

The rich new collection of cranial and especially postcranial bones (see section 2.3.2.) obviously represents a single adult individual. A well-preserved upper left canine (height including the root 53.5+, crown height 23.2+, length 11.1, width 9.7; measurements in mm) and a P^2 (length 5.6, width 3.9) fit into a left maxillary fragment (**Fig. 16a–d**). This is the first known complete upper canine of a Eurasian puma. An immature canine (length 14.6mm, width 12.0mm) from the Middle Villafranchian locality of La Puebla de Valverde (Spain) has previously been described (Kurtén and Crusafont Pairó 1977).

The canine from Untermassfeld is a little worn and rounded at its tip. It is noticeably slender, in contrast to the usually plump canine of *P. concolor*. Longitudinal grooves are well marked. The extant puma demonstrates a wide variability in this character: in some cases, canine grooves are lacking and in others they are only weakly developed (Hemmer 1978). The Untermassfeld canine is significantly smaller than its counterpart in *P. concolor*. Its length is 3.0 s (standard deviation) less than a *P. concolor* mean (Van Valkenburgh et al. 1990, tab 1). Its right counterpart is incompletely preserved. An earlier excavated fragment of a left P^4 includes the inner wall of the rear part and the distal root.

The left hemimandible of the Untermassfeld puma skeleton is nearly completely preserved, lacking only the symphyseal part (**Fig. 16e–g, Table 19**). This lower jaw resembles mandibles of small extant Mexican pumas of the onza morph (see Greenwell 1986) on the one hand and Middle Villafranchian *P. pardoides* from Saint-Vallier on the other. The mandible is broken in the socket of the canine, so a nearly vertical position of the canine root is exposed, indicating a steep front profile of the symphysis. There are two mental foramina aligned horizontally, the larger one to the front is situated below the short diastema, the smaller one lies below the P_3. This configuration places the Untermassfeld puma outside the variability of extant *P. concolor*, but corresponds to the common pattern »for all fossil and living felids, other than pumas« (Savage 1960). There is a very small, *P. concolor*-like third mental foramen in the Saint-Vallier (Viret 1954, pl. 10) and the Cueva Victoria (Madurell-Malapeira et al. 2010) *P. pardoides* mandibles, whereas the size and position of the other two foramina are as those recorded in the Untermassfeld specimen. The shallow rostral end of the masseteric fossa reaches far under the molar, as is typical for *P. concolor*, in contrast to *Panthera pardus* (Hemmer 2007a, fig.1). This character is also found in the earlier published *P. pardoides* mandibular fragment from Untermassfeld (Hemmer 2001, 729ff., pl. 141, 1–3) and in the Saint-Vallier puma (Viret 1954). The front part of the deep masseteric pit is limited above by a sharp crest as is also described for a Saint-Vallier mandible (Viret l.c.) and also found in *P. concolor*. The ventral margin of the masseteric fossa forms a circular arc in its rear sector, which is typical for pumas (Hemmer 2007a, Fig. 1) but which is quite unlike the rather platter-like structure sloping out ventro-laterally to the angular process, as seen in leopards.

The lower dentition (**Table 20**) is about the size of a female extant puma (Kurtén 1973b). The lengths of the premolars and the molar are 0.4–1.6 s shorter than the respective overall mean values of *P. concolor* (Van Valkenburgh et al. 1990, Table 2). The premolars are very worn. The protoconid of both the P_3 and the P_4 is polished off and the wear slopes nearly to the groove in front of the rear accessory cusp. The latter is affected by considerable abrasion in both premolars. The same holds for the paraconid of the P_4. The molar is less worn, mainly at the tip of the paraconid. The wear facets of the M_1 are puma-like, parallel to the cutting

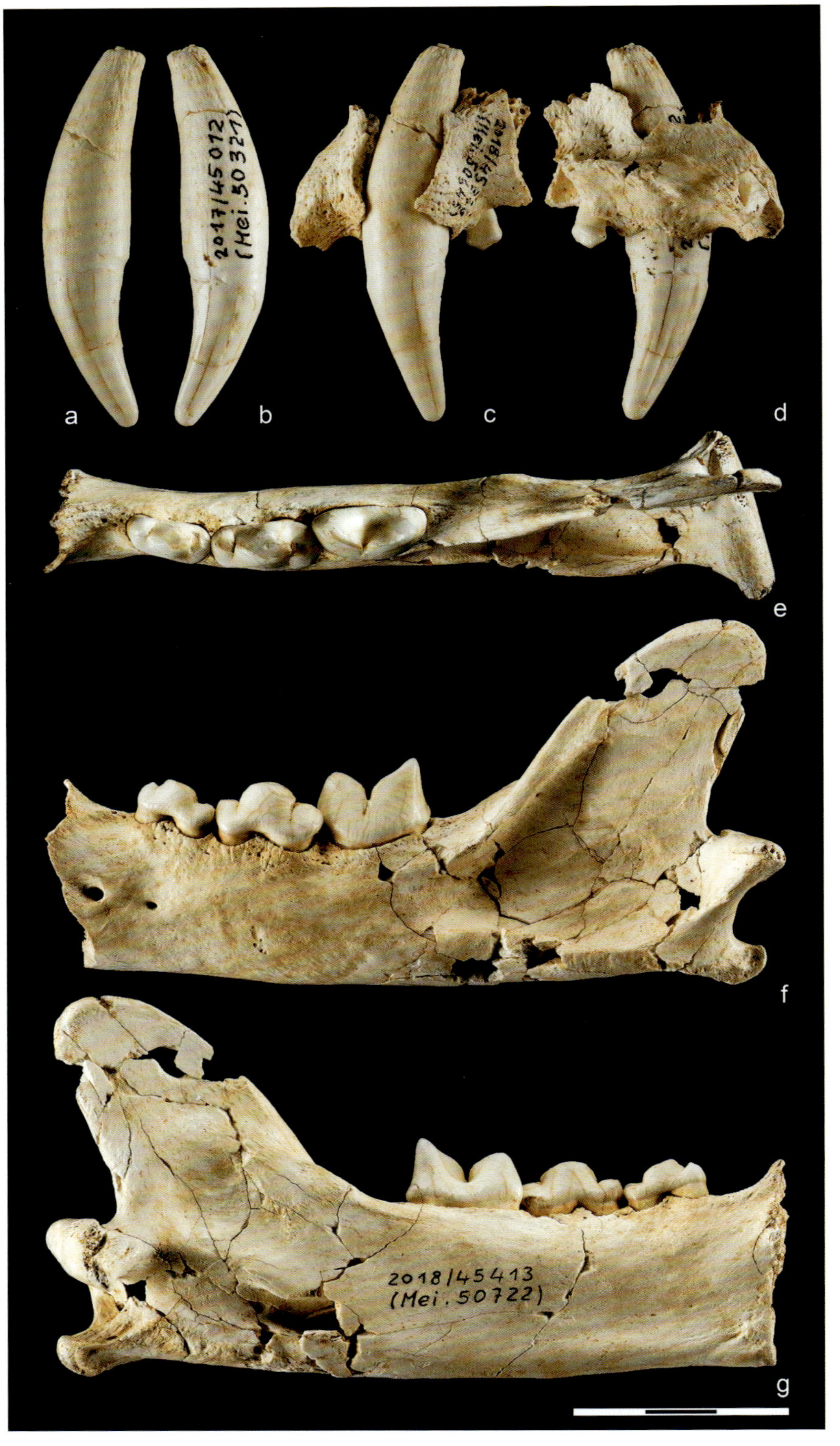

Fig. 16 *Puma pardoides brevipes* ssp. nov., Untermassfeld. – **a–d** Left upper canine IQW 2017/45012 (Mei. 50321), isolated and combined with maxilla fragment and left P^2 IQW 2018/45336 (Mei. 50645). – **e–g** Left mandible with P_3-M_1 IQW 2018/45413 (Mei. 50722). – **a**, **c**, **f** buccal views; **b**, **d**, **g** lingual views; **e** occlusal view. – Scale: 30 mm.

edges. This character, somewhat unusual for a felid, has already been mentioned in the original description of the Saint-Vallier mandibles (Viret 1954). The two premolars are well characterised by a prominent cingulum running around the base of the teeth, rising in front into the anterior accessory cusp. Identical morphology is found in the previously published Untermassfeld mandibular fragment (Hemmer 2001, 729 ff.). This feature is less marked in the Saint-Vallier and Kvabebi pumas, where the heel cingulum, especially of the P_4,

	IQW 2018/45413 (Mei. 50722)	IQW 1983/18556 (Mei. 18376)
Mandible height Processus angularis-Coronion	54.5	
Depth behind M_1	25.0	26.0
Depth in front of P_3	24.6	
Thickness below P_4/M_1	12.9	
Alveolar length P_3-M_1	43.1	
Diastema length	10.0	
P_3 length	11.8	
maximal width	5.9	
anterior width	4.7	
protoconid length	6.0	
protoconid height		
P_4 length	15.6	15.5
maximal width	7.6	7.7
anterior width	6.4	6.4
protoconid length	8.0	7.8
protoconid height		
M_1 length	16.6	16.9
maximal width	8.2	7.9
protoconid length	10.0	10.5
paraconid length	9.4	8.2
protoconid height	10.3+	
paraconid height	10.0+	

Table 19 *Puma pardoides brevipes* ssp. nov., measurements (mm) of the holotype left hemimandible, compared to an earlier described mandibular fragment (Hemmer 2001), Untermassfeld.

looks as if it has been folded. The M_1 of the cats of the latter two sites also have a bulging talonid in common, whereas this structure is smaller in both Untermassfeld specimens.

The vertebral column of the Untermassfeld skeleton is represented by four thoracic vertebrae evenly distributed over the total thoracic series and by the sacrum. The centra of the third, sixth and ninth thoracic are on average only 3 % longer than those found in the respective vertebrae of a very small extant puma used for comparison [IQW 2012/39542 (k. A. 36516)] but are 9 % greater in height. This suggests that the centra of these vertebrae in the Untermassfeld *P. pardoides* are relatively short and stocky. In contrast, the proportions of the twelfth thoracic of the latter are nearly the same as in *P. concolor* (**Fig. 17a–t, Table 21**). A short thoracic region relative to the lumbar region is characteristic for both the Early Pleistocene North American *P. inexpectatus* and the Late Pleistocene *P. trumani* when compared to *P. concolor* (Van Valkenburgh et al. 1990). When considering regionalisation of the vertebral column with a highly integrated region between the diaphragmatic vertebra and the last lumbar (thoracic vertebra X to lumbar vertebra VII) (Randau da Costa Carvalho 2017), the longer thoracic vertebra XII of *P. pardoides*, in contrast to thoracic vertebrae III, VI and IX, is also suggestive of longer lumbars.

The sacrum of the Untermassfeld puma skeleton (**Fig. 17u–y, Table 22**) greatly differs from that of *P. concolor*, being proportionally longer and narrower in its middle section. Its caudal articulation area (third vertebra) is much more robust, whereas its cranial articulation surface (first vertebra) is of the same size as that found in the small extant puma used for comparison. The third sacral vertebra is only moderately

	P_3 length	P_3 width	P_4 length	P_4 width	M_1 length	M_1 width
Early Pleistocene, Epivillafranchian						
Untermassfeld (Hemmer 2001 and new finds)						
IQW 2018/45413 (Mei. 50722)	11.8	5.9	15.6	7.6	16.6	8.2
IQW 1983/18556 (Mei. 18376)			15.5	7.7	16.9	7.9
Cueva Victoria and Vallparadís (Madurell-Malapeira et al. 2010)						
IPS 46144	12.6	6.3	16.7	8.7	18.6	8.9
IPS 46145	12.5	6.8				
EVT 04178			15.6	7.8		8.0
Early Pleistocene, Middle Villafranchian						
La Puebla de Valverde (Madurell-Malapeira et al. 2010)						
IPS 36127					19.8	8.9
Saint-Vallier (Viret 1954; measurements H.H.)						
QSV 137	10.7	5.6	14.1	7.0	16.3	7.8
QSV 138 dex.	13.0	6.8	16.2	8.0	18.9	8.2
QSV 138 sin. (cast NMB St.V.772)	13.2	7.3	c. 16.5	8.6	c. 18.6	9.0
Early Pleistocene, Pinjor						
Siwalik Hills (measurements H.H.)						
BM 16537a	13.0	6.4	17.0	8.0	c. 19.0	9.1
Late Pliocene, Early Villafranchian						
Ahl al Oughlam (Geraads 1997, assigned to *Panthera pardus*)						
AaO-13	11.3	5.6	15.6	7.5	17.2	8.0
AaO-1375			16.1	8.0		
AaO-1447			15.3	7.9		
AaO-2027			16.6	7.3		
AaO-2548			16.0	7.4		
AaO-2550			16.2	8.1		
AaO-3457	12.5	6.0	15.9	7.5	17.1	7.5
Kvabebi (Vekua et al. 2007; measurements H.H.)						
K 3035 dex.			16.1	8.1	17.8	7.8
K 3035 sin.			16.7	8.1		

Table 20 Comparative measurements (mm) of lower premolar and molar lengths and widths of Late Pliocene and Early Pleistocene *Puma pardoides* of Europe, Asia and North Africa.

	T III IQW 2017/45024 (Mei. 50333)	T VI IQW 2017/45329 (Mei. 50638)	T IX IQW 2017/45023 (Mei. 50332)	T XII IQW 2017/44984 (Mei. 50293)
Centrum length	18.0 [17.0]	18.3 [17.6]	19.0 [19.2]	24.0 [22.0]
Facies terminalis cranialis width	18.5 [15.7]	16.3 [13.5]	16.0 [16.1]	– [20.3]
Facies terminalis cranialis height	c. 13.0 [10.7]	c. 12.0 [11.6]	13.8 [12.8]	13.5 [13.0]
Facies terminalis caudalis width	21.0 [21.7]	– [23.2]	c. 22.5 [25.3]	c. 22.0 [21.4]
Facies terminalis caudalis height	13.5 [11.8]	c. 13.0 [12.2]	13.5 [13.0]	16.0 [c. 15.0]
Maximal height	– [61.0+]	– [57.0]	– [50.0+]	40.0 [36.6]
Neural canal cranial width	14.0 [c. 13.0]	c. 12.0–13.0 [c. 10.0–11.0]	13.0 [c. 10.0–11.0]	13.0 [12.0]
Neural canal cranial height	9.0 [8.8]	– [9.0]	9.0 [9.0]	9.0 [7.5]
Neural canal caudal width	– [c. 13.0–14.0]	– [c. 11.0–12.0]	c. 12.0–13.0 [c. 10.0–11.0]	c. 12.0–13.0 [c. 12.0]
Neural canal caudal height	– [c. 10.0]	– [c. 10.6]	11.3 [10.0]	10.7 [8.0]

Table 21 *Puma pardoides brevipes* ssp. nov., measurements (mm) of thoracic vertebrae of the Untermassfeld puma skeleton, compared [in brackets] with a very small extant *Puma concolor* IQW 2012/39542 (k. A. 36516).

reduced with regard to that of the first. This is in contrast to short-tailed cats (*Lynx*, also *Homotherium*: Rawn-Schatzinger 1992) and indicates the existence of a tail at least as long as that found in *P. concolor*. The Untermassfeld puma sacrum is singular due to an unusual feature, as its third vertebra is only partially fused with the second on only the left side. A comparable case of incomplete fusion has been described for *Smilodon populator* (Rawn-Schatzinger l.c.).

Part of a right scapula does not belong to the new Untermassfeld puma skeleton that was found in the erosion channel (see section 2.3.2.), but may be attributed to the previously described individual (Hemmer 2001). Skeletal elements of the latter puma were all discovered in Lower Fluviatile Sands (LFS) that were deposited on flooded surfaces outside of the channel (Kahlke 2001; 2006; Kahlke et al. 2020). The scapula lay in grid square Q 921, 0.70 m above site 0-level (Kahlke in this volume, Fig. 13, Foldout III). Nevertheless, it is not preserved well enough to supply important comparative information. The maximal width of the articulation surface measures c. 20.5 mm.

The right humerus of the Untermassfeld puma skeleton (**Fig. 18a–f**) is 1.4 s shorter than the *P. concolor* mean value (Van Valkenburgh et al. 1990, tab 3), but nevertheless ranges within the variability of the latter species. The shape of the distal part of the humerus is fully comparable with that of the earlier described specimen IQW 1996/25710 (Mei. 25239), but the new fossil is about 6 % smaller (**Table 23**). There is conspicuous morphological similarity with the Pliocene South American *P.* (*Herpailurus*) *pumoides* distal humerus, but not with that of *P. concolor* (Chimento et al. 2014, Fig. 8). On the other hand, the distal epiphysis of the humerus of *P. pumoides* differs from the Untermassfeld puma and the other known *P. pardoides* specimens by looking somewhat less compressed antero-posteriorly (quotient antero-posterior diameter/transverse diameter ≤ 0.60 in *P. pardoides*, ≥ 0.60 in *P. concolor*, 0.68 in *P. pumoides*, Chimento et al. 2014). The size of the Untermassfeld humeri is close to that of the Middle Villafranchian specimen from La Puebla

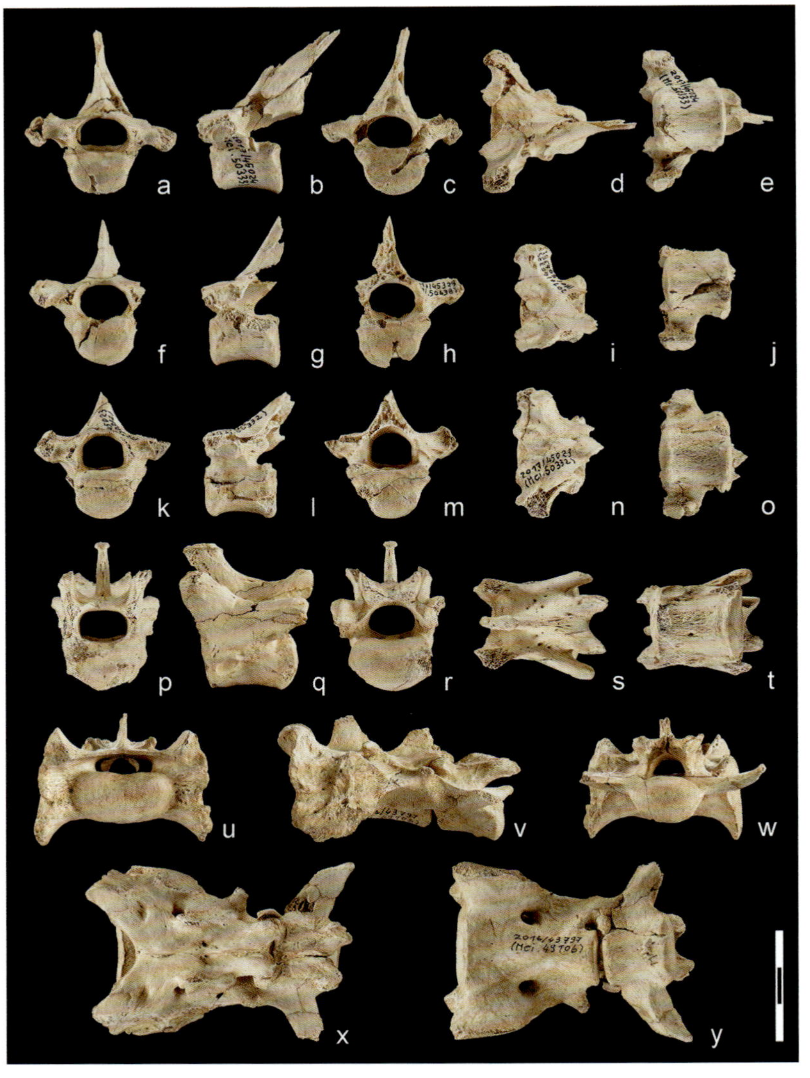

Fig. 17 *Puma pardoides brevipes* ssp. nov., Untermassfeld. – **a–e** Thoracic vertebra III IQW 2017/45 024 (Mei. 50 333). – **f–j** Thoracic vertebra VI IQW 2017/45 329 (Mei. 50 638). – **k–o** Thoracic vertebra IX IQW 2017/45 023 (Mei. 50 332). – **p–t** Thoracis vertebra XII IQW 2017/44 984 (Mei. 50 293). – **u–y** Sacrum IQW 2016/43 797 (Mei. 49 106). – **a**, **f**, **k**, **p**, **u** anterior views; **b**, **g**, **l**, **q**, **v** left lateral views; **c**, **h**, **m**, **r**, **w** posterior views; **d**, **i**, **n**, **s**, **x** dorsal views; **e**, **j**, **o**, **t**, **y** ventral views. – Scale: 30 mm.

	S 1–S 3 centrum IQW 2016/43 797 (Mei. 49 106)
Maximal length	71.0 [–]
Total centrum length	59.9 [49.0]
Width across 1st transverse process	47.6 [45.9]
1st centrum Facies terminalis cranialis width	26.1 [c. 26.0]
1st centrum Facies terminalis cranialis height	14.1 [13.9]
3rd centrum Facies terminalis caudalis width	18.9 [15.9]
3rd centrum Facies terminalis caudalis height	12.5 [10.6]
Neural canal cranial width	15.0 [15.5]
Neural canal cranial height	5.1 [4.9]
Neural canal caudal width	8.0 [7.8]
Neural canal caudal height	6.2 [4.9]

Table 22 *Puma pardoides brevipes* ssp. nov., measurements (mm) of the sacrum of the Untermassfeld puma skeleton, compared [in brackets] with a very small extant *Puma concolor* IQW 2012/39 542 (k. A. 36 516).

Epivillafranchian *P. pardoides brevipes* ssp. nov.	Untermassfeld IQW 2016/43792 (Mei. 49101)	Untermassfeld IQW 1996/25710 (Mei. 25239)	
Maximal length	189.5		
Length from caput	187.0		
Maximal proximal TD	37.6		
Maximal proximal APD	44.3		
Minimal diaphyseal TD	16.0		
Maximal distal TD	44.6	47.0	
Maximal distal APD	26.8	28.0*	
Trochlear TD	30.0	32.0	
Trochlear minimal thickness	13.8	15.0	
Middle Villafranchian *P. pardoides pardoides*	La Puebla de Valverde IPS 27272 (Madurell-Malapeira et al. 2010)	Saint-Vallier SV.916.490 (Argant 2004)**	Saint-Vallier MHNL 161.830 (QSV 120) (Argant 2004)**
Maximal length		230.7	253.0
Maximal proximal TD		46.0	50.1
Maximal proximal APD		60.0	64.3
Minimal diaphyseal TD	15.9	19.2	21.1
Maximal distal TD	41.4	56.2	58.5
Maximal distal APD	24.8	31.1	34.3

Table 23 *Puma pardoides* (Owen), humerus measurements (mm). TD – transverse diameter; APD – anterior-posterior diameter. * – Erroneously given as 23.5 in Hemmer (2001). ** – Erroneously listed as *Acinonyx pardinensis* in Argant (2004), see Hemmer et al. (2011).

Ulna	IQW 2017/45013 (Mei. 50322)	IQW 1995/25335 (Mei. 24864) (Hemmer 2001)
Maximal length	198.0	
Olecranon length	34.6	
Olecranon maximal TD	14.9	
Olecranon minimal APD	22.3	c. 26.0
APD over processus anconaeus	28.5	30.0
Height of trochlear notch	18.5	
Width of trochlear notch	22.7	
Radius	IQW 2016/43817 (Mei. 49126)	IQW 2017/45022 (Mei. 50331)
Maximal length	162.5	
Maximal proximal TD	20.9	20.2
Maximal proximal APD	14.9	
Minimal diaphyseal TD	15.0	
Maximal distal TD	30.8	
Maximal distal APD	19.9	

Table 24 *Puma pardoides brevipes* ssp. nov., ulna and radius measurements (mm), Untermassfeld. TD – transverse diameter; APD – anterior-posterior diameter.

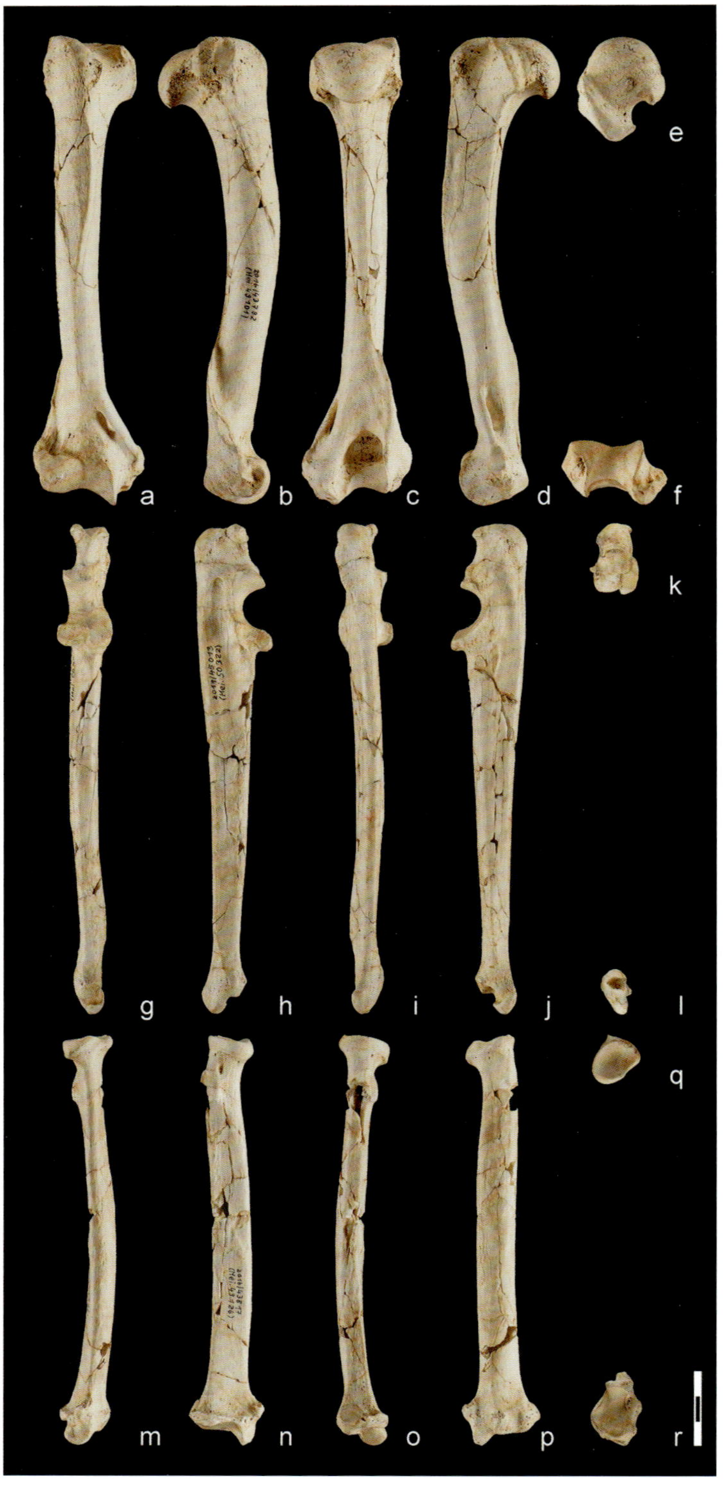

Fig. 18 *Puma pardoides brevipes* ssp. nov., Untermassfeld. – **a–f** Right humerus IQW 2016/43792 (Mei. 49101). – **g–l** Right ulna IQW 2017/45013 (Mei. 50322). – **m–r** Left radius IQW 2016/43817 (Mei. 49126). – **a**, **g**, **m** anterior views; **b**, **h**, **p** external views; **c**, **i**, **o** posterior views; **d**, **j**, **n** internal views; **e**, **k**, **q** proximal views; **f**, **l**, **r** distal views. – Scale: 30 mm.

de Valverde (about 6–12 % smaller), but they are substantially smaller than two fossils from Saint-Vallier (**Table 23**), which surpass the latter by around 30–39 %.

The shape of the right ulna (**Fig. 18g–l**, **Table 24**) best corresponds to the ulna of *P. concolor*, when compared to other American Pliocene and Pleistocene puma species (Chimento et al. 2014, Fig. 9, proximal part; Van Valkenburgh et al. 1990, Fig. 8, distal part). The new piece is a little smaller than that previously recorded from Untermassfeld (Hemmer 2001, 729).

The left radius of the Untermassfeld skeleton is completely preserved (**Fig. 18m–r**), whereas the right is missing the distal part (**Table 24**). It is 1.2 s shorter than the *P. concolor* mean value (Van Valkenburgh et al. 1990, tab 3), therefore falling within the variability of the latter species. Its proximal articular surface shows particular similarity to that of *P. pumoides* (Chimento et al. 2014, Fig. 10). The humeroradial index [radius (maximal length)/humerus (physiological length = length from caput) × 100] of the right humerus is 85.9, which is smaller than one of the published means of *P. concolor* (89.5), but still falls within its extremely large standard deviation (Gonyea 1976). It is larger, however, than another published *P. concolor* mean (84.7, calculated from tables 6 and 7 by Spoor 1985), demonstrating the large variability in *P. concolor*. The *P. pardoides* index is considerably smaller than the humeroradial index of *P. inexpectatus* (91.1; lengths by Van Valkenburgh et al. 1990) and falls within both the overlapping variabilities of *Panthera pardus* and *Panthera onca*. It is significantly smaller than the respective indices of *Panthera leo* and *Panthera uncia*, and it tends to be larger than the index of *Neofelis nebulosa* (Gonyea 1976; Spoor 1985).

The carpals are represented by a pisiforme, a scapholunare and a trapezoid (**Fig. 19a–f'**). The pisiform bone is longer (20.5 mm) than in *P. concolor* (data by Morales-Mejía et al. 2010), but is not as long as in *P. inexpectatus* or *P. trumani* (data by Van Valkenburgh et al. 1990), when compared to the scapholunare's width (23+ mm). The scapholunare itself seems to be most like its counterpart from *P. inexpectatus* (Van Valkenburgh et al. l.c., Fig. 9).

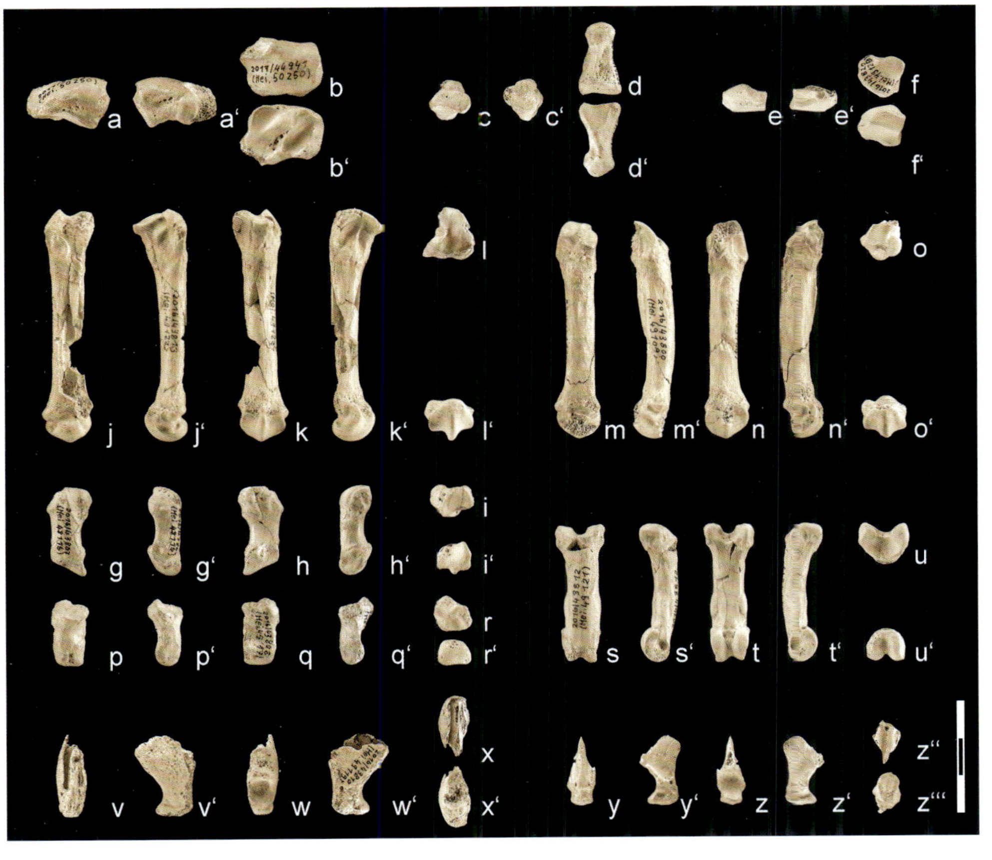

Fig. 19 *Puma pardoides brevipes* ssp. nov., Untermassfeld. – **a–b'** Left scapholunare IQW 2017/44941 (Mei. 50250). – **c–d'** Left pisiforme IQW 2016/43803 (Mei. 49112). – **e–f'** Left trapezoid IQW 2016/43820 (Mei. 49129). – **g–i'** Left metacarpal 1 IQW 2016/43807 (Mei. 49116). – **j–l'** Left metacarpal 3 IQW 2016/43813 (Mei. 49122). – **m–o'** Right metacarpal 4 IQW 2016/43800 (Mei. 49109). – **p–r'** Phalange I (1) sin. (man.) IQW 2016/43808 (Mei. 49117). – **s–u'** Phalange I (3) dex. (man.) IQW 2016/43812 (Mei. 49121). – **v–x'** Phalange III (1) dex. (man.) IQW 2016/43810 (Mei. 49119). – **y–z'''** Phalange III (3) sin. (man.) IQW 2016/43805 (Mei. 49114). – **a, c, e, g, j, m, p, s, v, y** dorsal views; **g', j', n', p', t', w', y'** internal views; **a', c', e', h, k, n, q, t, w, z** palmar views; **h', k', m', q', s', v', z'** external views; **b, d, f, i, l, o, r, u, x, z''** proximal views; **b', d', f', i', l', o', r', u', x', z'''** distal views. – Scale: 30 mm.

Epivillafranchian *P. pardoides brevipes* ssp. nov.	**Untermassfeld** Mc1 sin. IQW 2016/43807 (Mei. 49116)	Untermassfeld Mc3 sin. IQW 2016/43813 (Mei. 49122)	Untermassfeld Mc3 dex. IQW 2016/43811 (Mei. 49120)	Untermassfeld Mc4 dex. IQW 2016/43800) (Mei. 49109)
Maximal length	24.2	64.0	59.5	61.0+
Maximal proximal TD	12.1	13.1	14.4	
Maximal proximal APD	8.8	14.4	14.6	
Minimal diaphyseal TD	7.6	c. 8.4	9.9	7.6
Maximal distal TD	10.0	13.6	13.8	13.2
Maximal distal APD	8.7	11.5	11.2	
Middle Villafranchian *P. pardoides pardoides*	La Puebla de Valverde Mc4 dex. IPS 27285a (Madurell-Malapeira et al. 2010)			
Maximal length	71.4			
Maximal proximal TD	12.6			
Maximal proximal APD	14.5			
Minimal diaphyseal TD	8.3			
Maximal distal TD	14.6			
Maximal distal APD	13.4			

Table 25 *Puma pardoides* (Owen), metacarpal measurements (mm). TD – transverse diameter; APD – anterior-posterior diameter.

The metacarpals (**Fig. 19g–o'**) are short, much shorter than in extant pumas. Significantly, the Mc3 length is 3.8 s less than the *P. concolor* mean value (Van Valkenburgh et al. 1990, tab 5) and thus outside the variability of the extant species. In view of their absolute thickness, the metacarpals are stout, not slender. Their minimal transverse diaphyseal diameters fall within the *P. concolor* variability (Van Valkenburgh et al., l.c.). A Mc4 from the Spanish Middle Villafranchian site of La Puebla de Valverde is about 10 % larger than its Untermassfeld counterpart (**Table 25**).

The humero-metacarpal index [Mc3 (maximal length)/humerus (physiological length = length from caput) × 100] is 34.2 in the Untermassfeld cat. This is less than in *P. concolor* (mean values 36.7 and 38.8, respectively; data by Spoor 1985 and Van Valkenburgh et al. l.c.) and in *P. inexpectatus* (39.1, based on data by Van Valkenburgh et al. l.c.). It is higher than in *Neofelis nebulosa* (30.5), *Panthera onca* (31.1) and *Panthera pardus* (32.0) (calculated with data given by Spoor l.c.). The Mc3/Mc1 length index (Mc1/Mc3 × 100) is 38 in the Untermassfeld *P. pardoides*, 32 in *P. inexpectatus* (Van Valkenburgh et al. l.c.) and also 32 in *P. concolor* (Morales-Mejía et al. 2010). This indicates a considerably long Mc1, when compared to both American species. In contrast, the relative length of the Untermassfeld puma Mc1 matches the extant jaguar (Mc3/Mc1 length index 38; data by Morales-Mejía et al. l.c.).

A few phalanges from the forepaw were recovered, including the proximal and the distal phalanx of digit 1 and digit 3 (**Fig. 19p–z'''**). The dew claw is considerably larger than the latter, which is laterally compressed and sharp (**Table 26**). The proximal phalanx of digit 1 is shorter in *P. pardoides* than in *P. concolor*, when comparing the length of Mc1 (index 75.6 vs. 86.7; basic data for the extant puma by Morales-Mejía et al. 2010).

Phalanx I	dig. 1 sin. man. IQW 2016/43 808 (Mei. 49 117)	dig. 3 dex. man. IQW 2016/43 812 (Mei. 49 121)
Maximal length	18.3	38.2
Maximal proximal TD	11.0	13.6
Maximal proximal APD	8.2	10.3
Minimal diaphysis TD	8.7	8.4
Maximal distal TD	8.9	10.7
Maximal distal APD	6.8	8.1
Terminal (claw) phalanx (III)	dig. 1 dex. man. IQW 2016/43 810 (Mei. 49 119)	dig. 3 sin. man. IQW 2016/43 805 (Mei. 49 114)
Maximal length	c. 23.0	
Maximal TD	8.4	7.5

Table 26 *Puma pardoides brevipes* ssp. nov., forepaw phalanx measurements (mm), Untermassfeld. TD – transverse diameter; APD – anterior-posterior diameter.

No part of the pelvis was found with the new Untermassfeld puma skeleton from the channel filling. However, the right counterpart of the previously published left pelvic half (Hemmer 2001, 729 ff.) has recently been discovered in the collection of not yet described Untermassfeld finds. Both of these pelvic parts were found in Lower Fluviatile Sands (LFS) outside the channel (Kahlke 2001; 2006; Kahlke et al. 2020) in grid squares Q 921 (sin.) and 924 (dex.), 0.4–1.0 m above site 0-level (Kahlke in this volume, Fig. 13, Foldout III; see scapula description above). The newly discovered right half (inner length of the acetabulum 25.3 mm, smallest dorso-ventral diameter of the ilium 27.1 mm) is somewhat less well preserved than the former (**Fig. 20**). As stated earlier (Hemmer l.c.), the pelvis of *P. pardoides* is not as long as in *P. concolor*. This is true

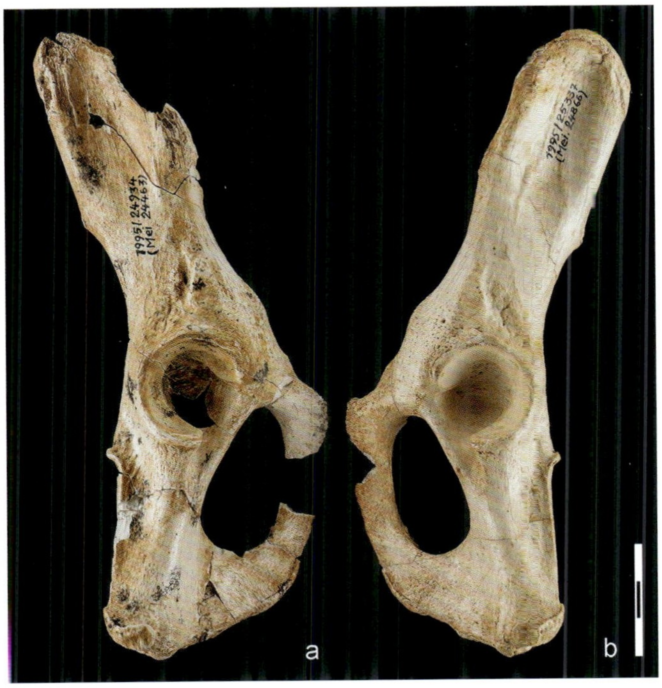

Fig. 20 *Puma pardoides brevipes* ssp. nov., Untermassfeld. – **a** Right side of pelvis IQW 1995/24 934 (Mei. 24 463), external view. – **b** Left side of pelvis IQW 1995/25 337 (Mei. 24 866), external view. – Scale: 30 mm.

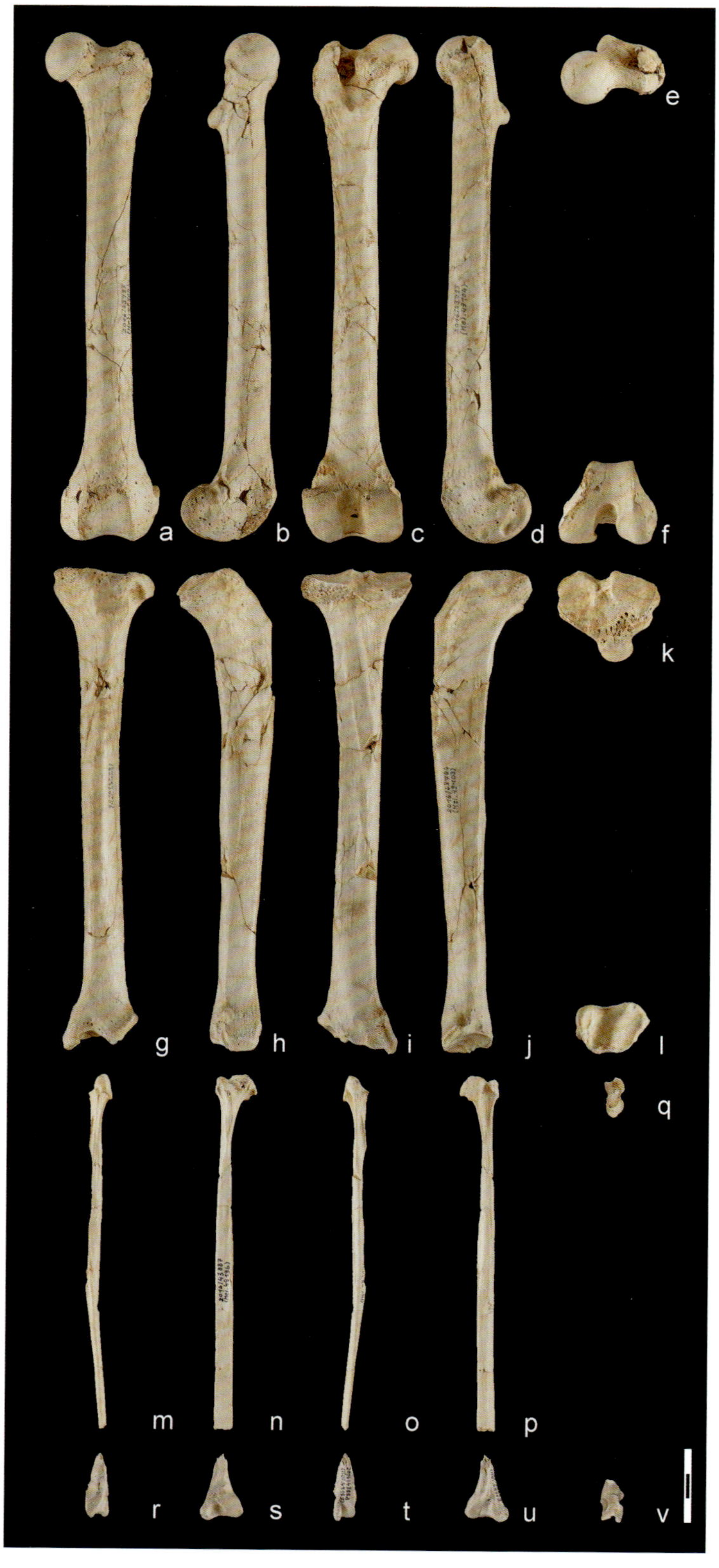

Fig. 21 *Puma pardoides brevipes* ssp. nov., Untermassfeld. – **a–f** Left femur IQW 2016/43 795 (Mei. 49 104). – **g–l** Left tibia IQW 2016/43 794 (Mei. 49 103). – **m–q** Left fibula proximal fragment IQW 2016/43 887 (Mei. 49 196). – **r–v** Left fibula distal fragment IQW 2016/43 888 (Mei. 49 197). – **a**, **g**, **m**, **r** anterior views; **b**, **h**, **n**, **s** internal views; **c**, **i**, **o**, **t** posterior views; **d**, **j**, **p**, **u** external views; **e**, **k**, **q** proximal views; **f**, **l**, **v** distal views. – Scale: 30 mm.

for the ilium (about three times as long as the acetabulum in *P. pardoides*, about four times in *P. concolor*) as well as the ischium (almost two times as long as the acetabulum vs. about three times). Thus, the pelvis proportions are very unlike those of *P. concolor*. In contrast, they are nearly the same as those in *Neofelis*

Epivillafranchian *P. pardoides brevipes* ssp. nov.	Untermassfeld IQW 2016/43795 (Mei. 49104)	Untermassfeld IQW 2016/43796 (Mei. 49105)	Untermassfeld IQW 1996/25711 (Mei. 25240) (Hemmer 2001)	Untermassfeld IQW 1996/25618 (Mei. 25147) (Hemmer 2001)
Maximal length (from caput)	210.9	211.6		
Length from trochanter	207.4	209.0		
Maximal proximal TD	43.9	43.5		
Caput APD	22.5	22.4		
Minimal diaphyseal TD	17.6	17.2	18.5	18.5
Maximal distal TD	41.5	41.0		
Maximal distal APD	36.0	36.0		
Middle Villafranchian *P. pardoides pardoides*	La Puebla de Valverde IPS 27343 (Madurell-Malapeira et al. 2010)	Saint-Vallier SV.98.759 (Argant 2004)	Saint-Vallier SV.161.851 (Viret 1954; Argant 2004)	Les Étouaires, Prr 192 (Kurtén and Crusafont Pairó 1977)
Maximal length	271.0		209.6	275.0
Maximal proximal TD			42.6	
Caput APD		19.8		
Minimal diaphyseal TD	23.9		16.4	21.9
Maximal distal TD	48.5	38.1	39.7	49.5
Maximal distal APD	47.1		38.2	

Table 27 *Puma pardoides* (Owen), femur measurements (mm). TD – transverse diameter; APD – anterior-posterior diameter.

nebulosa (compare **Fig. 20** with Tujek 2008, Fig. 145 for extant puma; Carlon and Hubbard 2012, Fig. 3 for clouded leopard).

Both femora of the Untermassfeld *P. pardoides* skeleton are well preserved (**Fig. 21a–f**). Their proximal morphology is more like that of *P.* (*Herpailurus*) *pumoides* than *P. concolor* (Chimento et al. 2014, fig. 12). The distal aspect looks more similar to that of the extant puma. The mentioned Untermassfeld femora (**Table 27**) are significantly shorter than the femora of *P. concolor*, with a 3.7 s difference when compared to a mean value of the latter species (data by Van Valkenburgh et al. 1990). The previously published Untermassfeld puma femora (Hemmer 2001, 729 ff.) are only a little larger. Earlier described and much older femora from the Middle Villafranchian of Spain (La Puebla de Valverde, Madurell-Malapeira et al. 2010) and France (Saint-Vallier, Argant 2004) prove the existence of two different size classes within *P. pardoides*. A specimen from La Puebla de Valverde is about 28 % longer than the femora of Untermassfeld; two femora from Saint-Vallier have nearly the same length as the bones described here. In relation to the humerus, the femur of *P. pardoides* is very short. The corresponding index (humerus/femur × 100) is 113.2. This meets the value found for *Panthera uncia* (113.1), is near to *Neofelis nebulosa* (113.8), and somewhat smaller than in *Panthera pardus* and *Panthera onca* (115.2, 115.3, respectively; comparative values calculated with data by Spoor 1985). In contrast, *P. concolor* is characterised by a long femur (118.3 based on Spoor l.c., 125.1 based on Van Valkenburgh et al. 1990); the same is true for *P. inexpectatus* (124.2 using data from Van Valkenburgh et al. l.c.).

The patella of the Untermassfeld puma looks more like that of *P. concolor* than that of *P. inexpectatus* or *P. trumani* (compared with Van Valkenburgh et al. 1990, Fig. 12).

Tibia **Epivillafranchian** *Puma pardoides brevipes* ssp. nov.	Untermassfeld IQW 2016/43794 (Mei. 49103)	Untermassfeld IQW 1996/25713 (Mei. 49103) (Hemmer 2001)	
Maximal length	201.1	215.0	
Maximal proximal TD	42.1		
Maximal proximal APD	c. 40.5		
Minimal diaphyseal TD	14.8	17.5	
Maximal distal TD	30.0	34.5	
Maximal distal APD	20.9	21.5	
Tibia **Middle Villafranchian** *Puma pardoides pardoides*	La Puebla de Valverde IPS 27345 (Madurell-Malapeira et al. 2010)	Saint-Vallier SV.97.238 (Argant 2004)	Saint-Vallier SV.99.550 (Argant 2004)
Maximal length		216.6	
Maximal proximal TD		40.9	41.2
Maximal proximal APD		38.8	45.6
Minimal diaphyseal TD	22.7	20.3	18.8
Maximal distal TD	37.3	30.8	
Maximal distal APD	24.3	18.4	
Fibula	Untermassfeld IQW 2016/43887 (Mei. 49196)	Untermassfeld IQW 2016/43888 (Mei. 49197)	Saint-Vallier SV.97.625 (Argant 2004)
Proximal APD	16.9+		
Proximal TD	8.5		
Distal TD		17.2	16.5
Distal APD		9.8	8.6

Table 28 *Puma pardoides* (Owen), tibia and fibula measurements (mm). TD – transverse diameter; APD – anterior-posterior diameter.

The left tibia of the Untermassfeld skeleton (**Fig. 21g–l**) is well preserved and is considerably shorter than the extant puma's tibia (**Table 28**), differing by 2.2 s from a *P. concolor* mean value (data by Van Valkenburgh et al. 1990, tab 3). The previously published tibia from the second individual from that site (Hemmer 2001, 729ff.) is 7 % larger. A Middle Villafranchian tibia from Saint-Vallier is of a comparable size to the latter, whereas a tibia of La Puebla de Valverde is about 28 % larger. Both the femur and the tibia of *P. pardoides* are similarly stout when compared to the long, slim *P. concolor* bones. The femoro-tibial index of the Untermassfeld puma (95.4) ranges within the variability of *P. concolor* (Gonyea 1976: 99.6 ± 3.83; data by Spoor 1985: 92.5). A comparable relationship between tibia and femur length is found in *P. inexpectatus* (based on Van Valkenburgh et al. 1990), *Neofelis nebulosa* (Gonyea l.c.: 99.7 ± 3.07; Spoor l.c.: 90.4), and in *Panthera pardus* (Gonyea l.c.: 94.8 ± 0.68; Spoor l.c.: 91.0); *Panthera onca* has a relatively shorter tibia, whereas the tibia of *Panthera uncia* is longer (Gonyea l.c.; Spoor l.c.).

The fibula (**Fig. 21m–v**, **Table 28**) is particularly striking due to an anteriorly directed expansion of its proximal end. The tip of this enlarged flange seems to be damaged. In its front view the presumed break shows the outline of a nearly rectangular, high and narrow structure. Such a prolongation of the fibula head, which structurally can be extremely variable, is found in the extant *P. (Puma) concolor* as well as in the small *P. (Her-*

pailurus) yagouaroundi and in the Pleistocene *P.* (*Miracinonyx*) *inexpectatus* and *P.* (*Miracinonyx*) *trumani*, representing an autapomorphic character of the puma lineage (Van Valkenburgh et al. 1990; Salles 1992). The fibula head of the Untermassfeld cat most resembles that found in the Early Pleistocene North American *P. inexpectatus* (Van Valkenburgh et al. 1990, Fig. 13). The synapomorphic anterior prolongation at the head of the fibula confirms the generic puma affiliation of the Untermassfeld cat.

A somewhat damaged left astragalus in articulation with the corresponding tibia (see above) and a navicular were also recovered. The latter has a transversal diameter of 17.2 mm and a dorsoplantar diameter of 20.5 mm, the size of a small *P. concolor*. The astragalus fits to the naviculare and is smaller than an astragalus from La Puebla de Valverde (Madurell-Malapeira et al. 2010).

2.3.4. Discussion

The comparison between the lower tooth rows of the Untermassfeld and the Saint-Vallier mandibles demonstrates a clear difference between the Epivillafranchian pumas on the one hand and the over a million years older Middle Villafranchian ones on the other. The heel cingulum of the premolars, especially the P_4, looks folded in the Saint-Vallier cat, whereas the prominent cingulum of the Untermassfeld puma is more rounded. The lower border of the tooth crown is marked by a moderate concave rounding over the gap of the roots in the latter, and by a rather acute-angled line in the former. Taken together, this gives the impression that the front and the rear part of the P_4 would have been anterior-posteriorly pressed together in the Middle Villafranchian form in contrast to the Epivillafranchian counterpart. The molar of the latter has a small talonid, which is larger and more bulging in the former. Differences in the dentition of the early and the late European pumas have also been mentioned for the Iberian record (Madurell-Malapeira et al. 2010). Consequently, these two populations should be taxonomically separated at the subspecies level. The stratigraphic position of the type molar of »*Felis*« *pardoides* Owen, 1846 from the English Red Crag locality of Newbourne (East Anglia) is not completely clear but is not later than MN 17 (Hemmer et al. 2004). Thus, the Middle Villafranchian European puma subspecies is the nominate one, *P. pardoides pardoides*. The Epivillafranchian form is described here as *P. pardoides brevipes* ssp. nov.

Puma pardoides brevipes ssp. nov.

Holotype:

Well preserved partial skeleton consisting of a set of the following components (*sensu* ICZN 1999, Art. 73.1.5.): Maxillare sin. fragment with P^2 IQW 2018/45336 (Mei. 50645) and associated C sup. IQW 2017/45012 (Mei. 50321), C sup. dex. fragment IQW 2016/43814 (Mei. 49123), I^3 dex. IQW 2018/45414 (Mei. 50723), P^4 sin. fragment IQW 2008/29823 (Mei. 28985), mandibula sin. with P_3-M_1 IQW 2018/45413 (Mei. 50722), vertebra thoracica III IQW 2017/45024 (Mei. 50333), vertebra thoracica VI IQW 2017/45329 (Mei. 50638), vertebra thoracica IX IQW 2017/45023 (Mei. 50332), vertebra thoracica XII IQW 2017/44984 (Mei. 50293), sacrum IQW 2016/ 43797 (Mei. 49106), humerus dex. IQW 2016/43792 (Mei. 49101), ulna dex. IQW 2017/45013 (Mei. 50322), radius sin. IQW 2016/43817 (Mei. 49126), radius dex. IQW 2017/45022 (Mei. 50331), scapholunare sin. IQW 2017/44941 (Mei. 50250), pisiforme sin. IQW 2016/43803 (Mei. 49112), trapezoid sin. IQW 2016/43820 (Mei. 49129), metacarpale 1 sin. IQW 2016/43807 (Mei. 49116), metacarpale 3 sin. IQW 2016/43813 (Mei. 49122), metacarpale 3 dex. IQW 2016/43811 (Mei. 49120), metacarpale 4 dex. IQW 2016/43800 (Mei. 49109), phalanx I dig. 1 sin. man. IQW 2016/43808 (Mei. 49117), phalanx I dig. 3 dex. man. IQW 2016/43812 (Mei. 49121), phalanx III dig. 1 dex. man. IQW 2016/43810 (Mei. 49119), phalanx III dig. 3 sin. man. IQW 2016/43805 (Mei. 49114), femur sin. IQW 2016/43795 (Mei. 49104), femur dex. IQW 2016/43796 (Mei. 49105), patella dex. IQW 2018/45335 (Mei. 50644), tibia sin. IQW 2016/43794

(Mei. 49 103), fibula sin. proximal fragment IQW 2016/43 887 (Mei. 49 196), fibula sin. distal fragment IQW 2016/43 888 (Mei. 49 197), astragalus sin. IQW 1993/24 322 (Mei. 23 851), naviculare sin. IQW 2007/29 104 (Mei. 28 266), phalanx I IQW 2018/45 411 (Mei. 50 720), phalanx I IQW 2009/30 464 (Mei. 29 626); stored in the type collection of the Senckenberg Research Station of Quaternary Palaeontology Weimar (**Figs. 16–19, 21**).

Etymology: *brevipes*, Latin: brevis – short, pes – foot
Type locality: Untermassfeld, Thuringia, Germany
Type layer: Upper Fluviatile Sands, lower part [UFS(l)], sandy-gravelly amalgamation zone at the base of the fossil bearing channel infill
Chrono- and biostratigraphic age: about 1.07–1.06 Ma, Epivillafranchian
Diagnosis: as in the diagnostic description above (see section 2.3.3.)

The individual unity of the listed components of the holotype is proved by the microstratigraphic documentation of the corresponding grid squares of the excavation, the embedding status and/or articulations of corresponding skeletal elements.

Due to its far-reaching skeletal completeness the Untermassfeld puma collection qualifies as a reference for Old World *P. pardoides* remains. Indeed, two enigmatic cases can be more easily understood using this material. The first is a mandibular fragment from the Indian Siwaliks Pinjor zone (Early Pleistocene) in the British Museum of Natural History, found in the 19th century (BMNH 16537a). This fossil had been previously associated with diverse taxa: allied or similar to *Panthera pardus* (Lydekker 1884; Matthew 1929), as a holotype of a new species *Panthera csomai* (Kretzoi 1929), and also as a paratype of *Sivafelis potens* (Pilgrim 1932). The holotype of the latter belongs to *Acinonyx pardinensis* (Schütt 1970a). Later, the mandibular fragment in question was thought to represent an archaic stage of leopard evolution, in view of its leopard size and a seemingly more jaguar-like long P_3 (Hemmer 1976). Following the latter, identification as a jaguar was argued as a more probable solution (Turner 1990). Finally, the fossil was tentatively attributed to *Panthera uncia* in view of its shortened diastema (Dennell et al. 2005). A comparison with the holotype mandible of *P. pardoides brevipes* (**Fig. 22**) demonstrates now the real nature of this piece, misidentified for more than a century. It does not originate in the genus *Panthera*, neither from a leopard, nor from a jaguar, nor from a snow-leopard. It is a puma, larger than the Untermassfeld individuals, but with fully comparable morphological features. The long P_3, the premolar cingulum, the short diastema, the rostral end of the masseteric fossa reaching far under the molar, all these are characters of *P. pardoides*. Even the pattern of the tooth wear is the same. The exact stratigraphic localisation of the Indian fossil within the Early Pleistocene is unknown. The premolar morphology puts it nearer to *P. pardoides brevipes* than to *P. pardoides pardoides*. Therefore *P. pardoides csomai* (Kretzoi, 1929) extends the dental characters of the late European form to South Asia.
The second open question pertains to the true nature of the so-called leopard of the North African Late Pliocene locality Ahl al Oughlam (Morocco). A series of dental remains and a few postcranial elements have been described under the name *Panthera pardus*, but placed in a group of leopard-sized cats with weak canines, such as »*Panthera schaubi*« from Saint-Vallier and »*Felis*, large species« (Barry 1987) from the East African Pliocene of Laetoli (Geraads 1997). *Panthera schaubi* is a synonym of *P. pardoides*. The above-mentioned African species was attributed to Old World pumas (Hemmer 2001; Hemmer et al. 2004). A comparative look at the lower tooth rows of the Untermassfeld *P. pardoides brevipes* [IQW 1983/18 556 (Mei. 18 376)] and the Ahl al Oughlam »leopard« (AaO-13) shows such a considerable similarity between the dentitions of these fossils it's as if they were twins (**Fig. 23**). The exact same applies for the upper canines found in these two localities [IQW 2017/45 012 (Mei. 50 321) vs. AaO-19]. Such an extent of correspondence is intriguing, as there is a chronological distance of about 1.5 Ma between the two sites. The functional

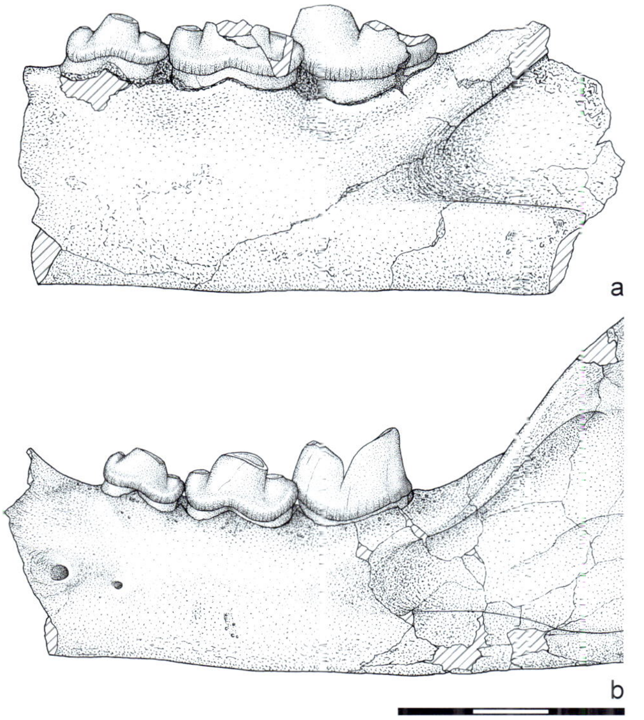

Fig. 22 *Puma pardoides csomai* (Kretzoi), Siwalik Hills (India). – **a** Left mandible fragment BMNH 16537a with P_3-M_1, buccal view. | *Puma pardoides brevipes* ssp. nov., Untermassfeld. – **b** Left mandible IQW 2018/45413 (Mei. 50722) with P_3-M_1, buccal view. – Scale: 30 mm.

morphology of *P. pardoides* should have been so successful that it overcame all ecological changes in the world of those epochs with very little morphological change. The specific identity of the Untermassfeld and the Ahl al Oughlam cats is beyond any doubt. Thus, there were no leopards in the Late Pliocene of Morocco, there were, however, pumas. Nevertheless, there is a difference between the Ahl al Oughlam felid and the Untermassfeld cat that originally supported the *Panthera pardus* diagnosis of the former (Geraads l.c.): the lower canine is not nearly as vertically inserted as it is in *P. pardoides brevipes* and, more obviously, *P. pardoides pardoides*. However, that character also applies for the Late Pliocene *P. pardoides* ssp. of Kvabebi (Georgia, Vekua et al. 2007). Thus, it does not affect the puma diagnosis.

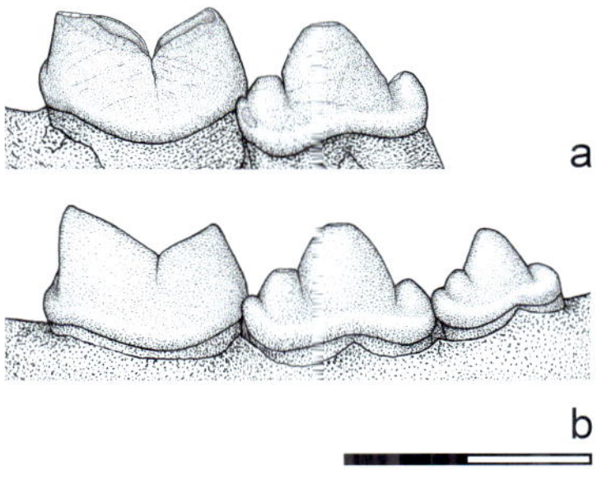

Fig. 23 *Puma pardoides brevipes* ssp. nov., Untermassfeld. – **a** Right P_4-M_1 IQW 1983/18556 (Mei. 18376), buccal view. | *Puma pardoides* ssp., Ahl al Oughlam (Morocco). – **b** Right P_3-M_1, buccal view. – Scale: 20 mm.

An important basis for understanding the functional morphology and ecological role of an animal is its body mass. For the new Untermassfeld puma the humerus, femur, and Mc3 and Mc4 minimal shaft widths are available as mass predictors. Their raw values are 36.7, 39.2, 36.3, 40.6 and 40.8 kg, giving a mean of 38.7 kg. Rounding up, this gives a body mass of about 40 kg. The previously published second individual was estimated as being a little heavier, about 45 kg (exclusively based on the femur shaft; Hemmer 2001: 40–45 kg, based on molar and femur prediction equations; but see Hemmer 2007a: dentition characters must be applied with care for body mass predictions). Estimated raw masses for *P. pardoides* from La Puebla de Valverde (based on

Tables 23, 25, 27) are 36.1, 50.0 and 109.0 kg, and from Saint-Vallier, 31.0, 62.2 and 81.6 kg. These data may be understood as the expression of an enormous sexual dimorphism: the female body masses varying between 30 and 50 kg, the male ones between 60 and 110 kg. This wide variability is completely covered by the body mass range of *P. concolor* (Nowell and Jackson 1996). No individuals of the upper size class have been identified at Untermassfeld, but large-sized lower dentitions at the Iberian Epivillafranchian localities also demonstrate the existence of a similar sexual dimorphism in the late Early Pleistocene Eurasian pumas. Consequently, the two identified puma individuals from Untermassfeld should be both females.

The discovery of a large number of skeletal elements of the Untermassfeld puma enables the build of *P. pardoides* to be reconstructed with a high degree of confidence. Its anatomy does not completely correspond to that of any extant species of large cats. The absolute lengths of the humerus and the radius, as well as the size of the mandible, are in accordance with the body mass estimated with respect to *P. concolor*. The length quotient of the radius and humerus is comparable with that of the extant puma, but also with that of the jaguar and the leopard, and, to some extent, even with that of the clouded leopard. In contrast, the metacarpals are substantially shorter in the European puma than in *P. concolor*. Nevertheless, the forefoot is not as short as in *Panthera pardus*, *Panthera onca* and *Neofelis nebulosa*. The length quotient of the humerus and the femur points to an extremely short femur, comparable to *Panthera uncia* and *Neofelis nebulosa* and similar to *Panthera pardus* and *Panthera onca*, which has a slightly longer femur. The short femur is supplemented by a short pelvis, as in the clouded leopard. The length quotient of the tibia and the femur corresponds to that observed in the extant puma, the leopard and the clouded leopard. In the jaguar, the tibia is comparatively somewhat shorter and is somewhat longer in the snow leopard (for comparative data see Gonyea 1976; Spoor 1985). Thus, the functional external morphology of the European puma has little in common with the extant puma and the snow leopard, both excellent jumpers. Some similarities exist with the clouded leopard, the best climber among the extant large cats (Hemmer 1968), and the sharp claws in *P. pardoides* could well support an ability for climbing. Using vertebral profiles of locomotor groups in felids (Randau da Costa Carvalho 2017), the base of the available thoracic vertebrae from Untermassfeld indicate a stronger affiliation to the terrestrial, than the arboreal or scansorial groups. Combining these results, the European puma may be considered as a morphologically generalised cat without a far-reaching specialisation, but well adapted to life in wooded habitats.

With an approximate body mass of 40–45 kg, the daily food requirement of the Untermassfeld pumas can be calculated as around 2.5 kg. Earlier published prediction equations enable the modal class of the body mass that constitutes the prey focus of a given cat species to be estimated with a high level of confidence. As cats are opportunistic hunters, which usually kill much more than just one prey species, regularly hunted prey can range from one size class above the calculated prey focus mass to three classes below (Hemmer 2004; 2007a). Occasionally, however, even prey animals outside this broad size range may have been taken. Therefore, the prey focus class of a 40–45 kg European puma is around 20 kg and its expected prey body mass would range between 2 and 50 kg, with occasional deviations towards smaller or larger prey. Another calculation based on the comparatively weak upper canine results in a somewhat lower raw value, but still indicates the 20 kg prey focus mass category. Compared to other large cat species, the small canine size would have undoubtedly influenced how the European puma brought down its prey. Functional compensation may have been provided in *P. pardoides* by the enlarged dewclaw, which would have been used for primary grasping the victim (see, e. g., discussion for *Homotherium* with extreme development of the dewclaw in Antón et al. 2005).

The transfer of the estimated prey body mass range to the actual Untermassfeld mammal population must take into account the presence of male European pumas in the former vicinity of the site, even if their remains were not found during the excavation. Consequently, the upper range of the supposed potential prey size variation should be increased by up to two additional classes of prey mass. The abundant deer species *Pseudodama vallonnetensis* and the rarer *Capreolus cusanoides* (MNI given by Kahlke in this volume, Table 8) should both be considered as a prime food source for *P. pardoides brevipes* (see section 3.3.). The complete range of prey, both regular and occasional, may have varied in size from large rodents to horses and larger-sized deer species (corresponding to Hemmer 2001, Fig. 14).

The Untermassfeld cat, whose skeleton provides so much information about *P. pardoides*, had a deformed right paw that must have, to some extent, impaired hunting. Missing toes or injured foot pads do not, however, hinder extant pumas from reaching old age (Young and Goldman 1946). A camera-trapping study has even shown that a female puma missing the left hind foot is able to survive in good condition (Huck and Fernandez-Duque 2012). Such a situation would also have been possible during the Early Pleistocene. The affected hunting behaviour of the puma in question could throw light on a noticeable observation concerning tooth wear. Although a killing weapon, the upper canine shows only slight wear at the tip, whereas the lower tooth row is considerably worn. Less killing, but sufficient eating, could point to more scavenging and/or large-scale preying on small mammals without the need for powerful canine action. Both strategies could have been successfully exercised in the Epivillafranchian Untermassfeld environment.

2.4. *Lynx valdarnensis spelaeoides* ssp. nov.

2.4.1. Introduction

When the Untermassfeld lynx was first described (Hemmer 2001), it was the only known representative of its genus in the European Epivillafranchian. It was named *Lynx issiodorensis* (Croizet et Jobert, 1828) ssp. ex aff. *spelaeus* (Boule, 1906). The classification as *L. issiodorensis* was based on its evolutionary archaic structure with stout mandible and somewhat stocky humerus. The addition »ex aff. *spelaeus*« is not part of the name as following the International Code of Zoological Nomenclature (ICZN), but was coined to express morphometric affinities to *L. spelaeus* of the Middle and the Late Pleistocene of Southern Europe. A mandible from the early Middle Pleistocene Marine Isotope Stage (MIS) 14 level of the Arago cave (French Pyrenees) has been determined as being closely associated with the Untermassfeld lynx based on archaic characters of the dentition and skeletal robustness. Nevertheless, this Arago lynx has been classified as belonging to an early group of *L. spelaeus* (Testu 2006). Later, the Untermassfeld lynx was taxonomically included in an unnamed chrono-subspecies of *L. issiodorensis* (*L. issiodorensis* ssp. 2, Cipullo 2010), which in turn was argued as representing part of a sequential evolutionary pattern leading eventually to *L. pardinus spelaeus*, which gave rise to the extant Iberian lynx *L. pardinus* (*L. issiodorensis* ssp. 1 of the Ruscinian – *L. issiodorensis issiodorensis* of the Early and Middle Villafranchian – *L. issiodorensis valdarnensis* of the early Late Villafranchian – and *L. issiodorensis* ssp. 2 of the latest Villafranchian and Epivillafranchian, Cipullo 2010; Cherin et al. 2013b; Petrucci et al. 2013).

Meanwhile, new finds from two sites in the eastern Iberian Peninsula (Vallparadís Section and Cueva Victoria) have been published that considerably widen our insight into the late Early Pleistocene lynx evolution in Southern and Central Europe. They indicate the existence of early representatives of *L. pardinus* in south-western Europe during the same period of time as the Untermassfeld population was living with features of the archaic *L. issiodorensis* in the centre of the continent (Boscaini et al. 2016). The evolutionary roots of *L. pardinus* can even be traced back to the Late Villafranchian based on a skull from the north-eastern Spanish Avenc Marvel cave (Boscaini et al. 2015). At the same time, the Italian lynx population, still classified as *L. issiodorensis valdarnensis* (Werdelin 1981), represents a taxonomic unit of its own with differing skull morphology (Cherin et al. 2013b), to which the Untermassfeld lynxes were also attributed (Garrido 2008). Finally, the use of the open nomenclature *Lynx* sp. instead of *L. issiodorensis* is suggested for European

lynxes dated to the interval 1.6–0.6 Ma that are not explicitly identified as *L. pardinus*. This also concerns the Untermassfeld remains, although the possibility of a confirmation of their *L. issiodorensis* nature by new finds is considered (Mecozzi et al. 2021).
New lynx remains from Untermassfeld now provide the opportunity to re-open the discussion on the position of these cats within the frame of European lynx evolution from the late Early to the early Middle Pleistocene.

2.4.2. Material

P_3 and P_4 dex. IQW 2012/38474 (Mei. 35448),
Vertebra cervicalis V IQW 1996/25868 (Mei. 25397),
Vertebra cervicalis VI IQW 1996/25867 (Mei. 25396),
Vertebra cervicalis VII IQW 1996/25866 (Mei. 25395),
Vertebra thoracica I IQW 1996/25865 (Mei. 25394),
Vertebra thoracica II IQW 1996/25905 (Mei. 25434),
Vertebra thoracica IV IQW 1996/25908 (Mei. 25437),
Vertebra thoracica V IQW 1996/25907 (Mei. 25436),
Vertebra thoracica VI IQW 1996/25903 (Mei. 25432),
Vertebra thoracica VII IQW 1996/25909 (Mei. 25438),
Vertebra thoracica VIII IQW 1996/25902 (Mei. 25431),
Vertebra thoracica IX IQW 1996/25904 (Mei. 25433),
Scapula dex. IQW 1996/25885 (Mei. 25414),
Humerus dex. diaphyseal fragment IQW 1984/20319 (Mei. 19839).

The lynx remains mainly originate from the LFS. Single finds, namely the above-listed right P_3 and P_4, as well as the humerus fragment, were recovered from the channel filling [UFS(l)].

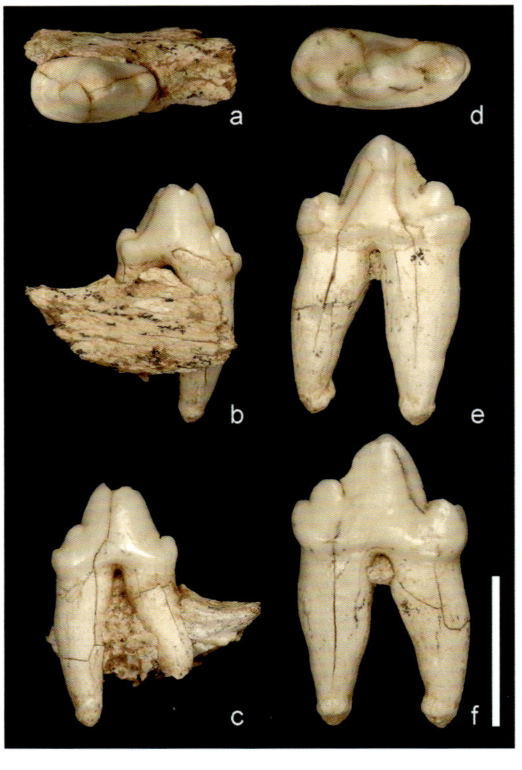

Fig. 24 *Lynx valdarnensis spelaeoides* ssp. nov., Untermassfeld. – **a–c** Right P_3 IQW 2012/38474 (Mei. 35448). – **d–f** Right P_4 IQW 2012/38474 (Mei. 35448). – **a, d** occlusal views; **b, e** lingual views; **c, f** buccal views. – Scale: 10mm.

2.4.3. Description and comparison

The two new lynx lower premolars represent a single, young adult individual. They show no signs of wear, but are a little damaged (**Fig. 24**). These teeth belong to the same category of size as the previously described dentitions (**Table 29**). Their unworn condition is demonstrated by the original structure of the cutting edges. In the P_4 a sharp crista-like thin blade, up to about 1mm wide, runs around the protoconid. The paraconid is similar in size to the distal accessory cusp, which forms in front of a fairly well-developed cingulum. The P_3 is dominated by a very high protoconid, the tip of which has broken away. The cutting edges are structured as described for the P_4. The proximal (paraconid) and the distal accessory cusps of the P_3 are small, the latter stretches upwards along the distal edge of the protoconid. The distal cingulum is small but clearly developed.

	IQW 2012/38474 (Mei. 35448) dex.	IQW 1994/24691 (Mei. 24220) sin.	IQW 1994/24691 (Mei. 24220) dex.	IQW 1980/16573 (Mei. 16094) dex.
Mandible length Processus condylaris-Infradentale		92.5	92.5	
Mandible height Processus angularis-Coronion		44.2		
Depth behind M_1		20.0	20.2	19.6
Depth in the diastema			17.4	
Alveolar length P_3-M_1		32.5	32.2	34.3
Diastema length		8.8	c. 8.0	> 6.0 / < 8.5
C height		14.4+	14.7	13.7+
length		7.6	7.7	8.0
width		6.3	6.3	6.3
P_3 length	8.2	8.0	7.8	9.3
maximal width	4.5	4.6	4.6	4.5
anterior width	3.7	3.3	3.3	3.5
protoconid length	5.1	4.7	4.8	4.3
protoconid height	6.3+		6.2	6.4
P_4 length	11.6	11.5	11.4	11.3
maximal width	5.3	5.2	5.4	5.5
anterior width	4.5	4.6	4.6	4.5
protoconid length	5.6	5.2	5.3	4.9
protoconid height	8.0		7.0+	
M_1 length		13.0	13.1	14.1+
maximal width		6.0	5.9	5.5
protoconid length		7.5	7.8	
paraconid length		6.5	6.4	6.9
protoconid height		7.6	7.5	
paraconid height		7.4		7.5+

Table 29 *Lynx valdarnensis spelaeoides* ssp. nov., measurements (mm) of premolars IQW 2012/38474 (Mei. 35448), of the holotype mandible IQW 1994/24691 (Mei. 24220), and of the hemimandible IQW 1980/16573 (Mei. 16094), Untermassfeld.

With the finding of these new teeth, the MNI of Untermassfeld lynxes has increased to three.

Besides the Untermassfeld material, Epivillafranchian lynx remains have been recovered from the north-east Iberian localities of Vallparadís and Cueva Victoria (Boscaini et al. 2016). The mandibular corpus of these Spanish lynxes is significantly less massive than in the Untermassfeld cats. The Iberian P_3 tends to be narrower, but both series overlap in this feature (**Table 30**). The length ratio of P_3 and M_1 shows different proportions in *L. issiodorensis* (Les Étouaires, Saint-Vallier and Upper Valdarno) compared to the extant species *L. pardinus* and *L. lynx* (Ficcarelli and Torre 1977, Fig. 2). The original measurements of the *L. pardinus* and Upper Valdarno material (Ficcarelli and Torre l.c., Table 1) are re-used here, not only to demonstrate the position of the Untermassfeld lynx in the context of earlier and later populations, but also to allow a direct comparison to be made with the diagram published by the mentioned authors (**Fig. 25**). In their size and proportions, the Untermassfeld teeth cover the span from the centre of *L. pardinus* variability to the range of the Iberian Epivillafranchian lynxes as well as that of the Italian Late Villafranchian series. The latest

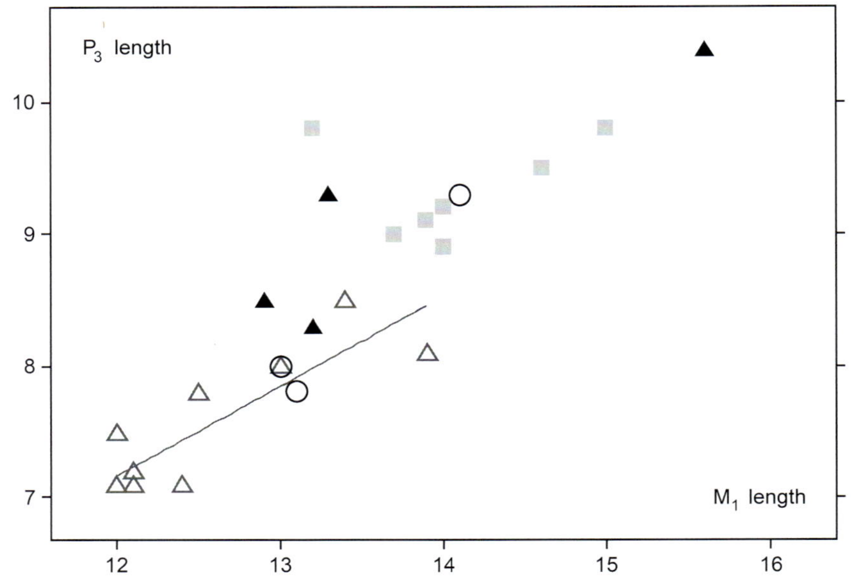

Fig. 25 Bivariate plot of the lengths (mm) of P_3 and M_1. – ○ Untermassfeld. – ▲ Cueva Victoria and Vallparadís. – ■ Upper Valdarno. – △ Extant *L. pardinus*. – Allometric line for *L. pardinus*: (P_3 length) = 0.4266 × (M_1 length)$^{1.135}$.

Early Pleistocene Spanish teeth and those of the late Early Pleistocene from Italy scatter above the axis of the Holocene Iberian lynxes, so that they fit into the archaic *L. issiodorensis* pattern (Ficcarelli and Torre l.c., Fig. 2). In size, the Untermassfeld dentitions fit the lower to central range of the Early Pleistocene Iberian and Italian series. The length and width profile of the Untermassfeld mandibular teeth matches the *L. issiodorensis* type (Kurtén and Werdelin 1984), as does the early Middle Pleistocene Arago C.M.I (Complexe Moyen I) mandible (Testu 2006).

The cervical and thoracic vertebrae of the Untermassfeld lynx (**Fig. 26**) resemble their extant counterparts in L. *pardinus*. There are exostoses on the right side of the centrum in the thoracic vertebrae V to VII and anterior-inferior in T VIII. These vertebrae are about 5 % larger (mean centrum lengths) than those found in a large male of the extant Iberian lynx (Gálvez Prada et al. 2013; for size comparison see Beltrán and Delibes

	Depth behind M_1 in % of P_3-M_1 length mean | min–max | n	P_3 protoconid height in % P_3 length mean | min–max | n	P_3 width in % length mean | min–max | n
Lynx valdarnensis spelaeoides ssp. nov., Epivillafranchian, Untermassfeld	60.4 | 57.1–62.7 | 3	> 75 | 68.8–79.5 | 3	55.2 | 49.5–59.0 | 4
Lynx pardinus, unnamed ssp., Epivillafranchian, Vallparadís and Cueva Victoria (Boscaini et al. 2016)	46.2 | 42.8–51.2 | 4		49.5 | 47.1–54.1 | 5
Lynx valdarnensis valdarnensis, Late Villafranchian (Fabrini 1896; Cherin et al. 2013b)	56.8 | – | 1	67.8 | 65–72 | 7	52.2 | 50.0–55.5 | 7

Table 30 Indices of mandible and dentition of the Untermassfeld lynx compared with other late Early Pleistocene European lynxes.

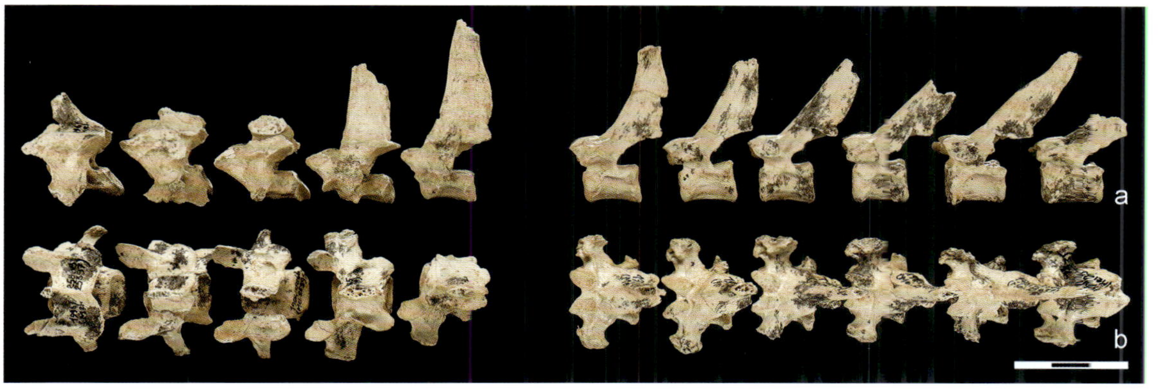

Fig. 26 *Lynx valdarnensis spelaeoides* ssp. nov., Untermassfeld. – **a–b** Cervical vertebra V IQW 1996/25868 (Mei. 25397), cervical vertebra VI IQW 1996/25867 (Mei. 25396), cervical vertebra VII IQW 1996/25866 (Mei. 25395), thoracic vertebra I IQW 1996/25865 (Mei. 25394), thoracic vertebra II IQW 1996/25905 (Mei. 25434), thoracic vertebra IV IQW 1996/25908 (Mei. 25437), thoracic vertebra V IQW 1996/25907 (Mei. 25436), thoracic vertebra VI IQW 1996/25903 (Mei. 25432), thoracic vertebra VII IQW 1996/25909 (Mei. 25438), thoracic vertebra VIII IQW 1996/25902 (Mei. 25431), thoracic vertebra IX IQW 1996/25904 (Mei. 25433), lateral, dorsal views. – Scale: 30 mm.

1993). The quotient of the centrum lengths of the cervical vertebrae V–VII / centrum lengths of the thoracic vertebrae I–II is 1.62 for the Untermassfeld cat (**Tables 31–32**), 1.60 for an Iberian lynx, and 1.55–1.63 for three individuals of *L. lynx* (Kurtén 1978, Table 5). These values contrast strongly with that of *L. issiodorensis* from the type site Les Étouaires, which is characterised by much longer neck vertebrae, giving a ratio of 1.77 (data by Kurtén l.c.).

The size of the right scapula fragment IQW 1996/25885 (Mei. 25414) is also larger, by the same order of magnitude (about 8 %), than in the above-mentioned large *L. pardinus* male. The measurements of the Untermassfeld scapula are (mm; in brackets for the *L. pardinus* male in question, after Gálvez Prada et al. 2013): length of the glenoid cavity, 21 [19.4]; width of the glenoid cavity, 16 [16.1]; maximal length of the glenoid process, 24 [22.6]; minimal length at the collum, 22 [18.7].

	C V IQW 1996/25868 (Mei. 25397)	C VI IQW 1996/25867 (Mei. 25396)	C VII IQW 1996/25866 (Mei. 25395)
Centrum length	17.5 [15.2]	15.5 [14.5]	14.8 [15.0]
Length anterior to posterior zygapophyses	22.0 [23.2]	21.5 [21.7]	21.9 [21.4]
Width across the anterior zygapophyses	– [27.5]	– [26.6]	26.5 [26.7]
Width across the posterior zygapophyses	26.9 [25.9]	– [26.1]	26.1 [23.6]
Facies terminalis cranialis width	c. 13.3* [12.7]	c. 12.1 [12.0]	c. 12.0 [10.5]
Facies terminalis cranialis height	11.0* [6.2]	– [6.2]	c. 7.5 [6.9]
Facies terminalis caudalis width	12.9 [12.4]	12.0 [11.5]	15.0* [11.7]
Facies terminalis caudalis height	8.1 [7.2]	8.5 [7.2]	9.4 [6.9]
Neural canal interior width	12.3 [–]	13.0 [–]	12.5 [–]
Neural canal interior height	9.0* [–]	11.0 [–]	9.7 [–]

Table 31 *Lynx valdarnensis spelaeoides* ssp. nov., measurements (mm) of cervical vertebrae of Untermassfeld, compared [in brackets] to a large male (for size comparison see Beltrán and Delibes 1993) of the extant *L. pardinus* (Gálvez Prada et al. 2013). * – exostosis.

	T I IQW 1996/25865 (Mei. 25394)	T II IQW 1996/25905 (Mei. 25434)	T IV IQW 1996/25908 (Mei. 25437)	T V IQW 1996/25907 (Mei. 25436)
Centrum length	14.8 [13.7]	14.7 [14.3]	14.8 [14.0]	14.5 [14.2]
Width at transverse processes	35.9 [34.3]	– [30.5]	30.9 [28.4]	31.2 [27.9]
Facies terminalis cranialis width	18.0 [18.0]	– [16.0]	c. 13.0 [12.0]	c. 12.0 [11.0]
Facies terminalis cranialis height	7.6 [7.0]	7.8 [7.3]	8.5 [7.8]	8.6 [8.2]
Facies terminalis caudalis width	16.4 [15.5]	18.0 [16.0]	18.0 [16.4]	18.8 [16.6]
Facies terminalis caudalis height	8.0 [8.0]	8.3 [7.6]	9.0 [8.4]	9.6 [8.3]
Neural canal interior width	12.0 [–]		10.0 [–]	9.7 [–]
Neural canal interior height	8.6 [–]	7.6 [–]	7.3 [–]	7.2 [–]
	T VI IQW 1996/25903 (Mei. 25432)	**T VII IQW 1996/25909 (Mei. 25438)**	**T VIII IQW 1996/25902 (Mei. 25431)**	**T IX IQW 1996/25904 (Mei. 25433)**
Centrum length	14.5 [13.8]	14.7 [14.2]	15.4 [14.5]	15.7 [15.3]
Width at transverse processes	– [27.7]	28.3 [27.6]	– [27.8]	28.7 [28.5]
Facies terminalis cranialis width	12.0 [11.5]	12.8 [11.9]	13.0 [12.6]	12.0–13.0 [12.3]
Facies terminalis cranialis height	9.2 [8.5]	9.3 [8.9]	9.6 [9.1]	9.8 [9.5]
Facies terminalis caudalis width	19.8 [17.1]	18.4 [–]	18.6 [17.2]	19.0 [17.5]
Facies terminalis caudalis height	9.2 [8.7]	9.7 [9.1]	9.8 [9.2]	9.8 [9.8]
Neural canal interior width	9.4 [–]	9.6 [–]	9.1 [–]	8.7 [–]
Neural canal interior height	7.2 [–]	7.0 [–]	7.0 [–]	6.9 [–]

Table 32 *Lynx valdarnensis spelaeoides* ssp. nov., measurements (mm) of thoracic vertebrae of Untermassfeld, compared [in brackets] to a large male (for size comparison see Beltrán and Delibes 1993) of the extant *L. pardinus* (Gálvez Prada et al. 2013), exostoses at the right side of the centrum in thoracic vertebrae V–VII, antero-inferior in thoracic vertebra VIII.

The previously described (Hemmer 2001, 735 ff.) humerus IQW 1996/25714 (Mei. 25243) (**Fig. 27a**), and the scapula and vertebrae discussed above, most probably belong to the same individual (size of the humerus c. 7 % larger than the *L. pardinus* humerus used for comparison: Gálvez Prada et al. l.c.). These skeletal elements were found during the 1993 field season (catalogued 1996) in the immediately adjacent grid squares Q 923 and 924, 0.30 m below to 0.30 m above site 0-level (Kahlke in this volume, Fig. 13, Foldout III), but were not fully available for the initial felid study. They originate from Lower Fluviatile Sands (LFS), which were deposited during surface flooding (Kahlke et al. 2020; Kahlke in this volume, 1649). The finds represent a cat a little bigger than the largest extant Iberian lynx male. With a minimal diaphyseal transverse diameter of 14.2 mm, the second fragmentary humerus IQW 1984/20 319 (Mei. 19 839) (**Fig. 27b**) increased current understanding concerning the size variability of the Untermassfeld lynxes to match that of the large Upper Valdarno specimens. It is comparable to a humerus from Il Tasso (n. 20, length 185 mm, see Fabrini 1896, 22, pl. II).

The relative humerus length of the Untermassfeld lynx is shorter than that found in *L. lynx*, is nearer in size to *L. pardinus* (humerus of the large skeleton presented by Gálvez Prada et al. 2013), corresponds to the minimum of *L. spelaeus*, but is much larger than that of *L. issiodorensis* from Les Étouaires (Hemmer 2001, 737, Tab. 17).

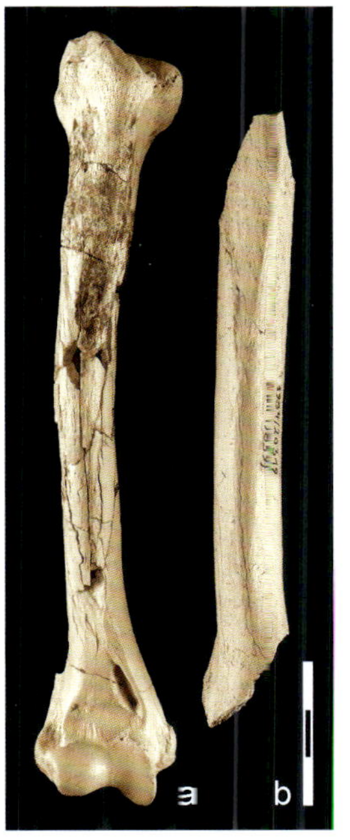

Fig. 27 *Lynx valdarnensis spelaeoides* ssp. nov., Untermassfeld. – **a** Right humerus IQW 1996/25 714 (Mei. 25 243), anterior view. – **b** Right humerus diaphyseal fragment IQW 1984/20 319 (Mei. 19 839), anterior view. – Scale: 30 mm.

2.4.4. Discussion

When considering their mandibular and femoral morphology, the Epivillafranchian lynxes of Untermassfeld are most similar to the Late Villafranchian lynx population of Italy. Both possess the archaic *L. issiodorensis* pattern in their dental proportions, a trait they also share with the Iberian Epivillafranchian lynxes. Nevertheless, they are otherwise clearly different (**Table 30**). A fictitious taxonomist of the time would presumably have recognised a single European lynx species with two geographic populations fitting the taxonomic category of separate sister subspecies: Iberian and Central European. In retrospect, however, an evolutionary trend linking the late Early Pleistocene Iberian population to the modern *L. pardinus* can be recognised. As a consequence, all Iberian lynxes from the Late Villafranchian to the extant population have been combined at the species level and attributed to *L. pardinus* (Boscaini et al. 2016). The assessment of nuclear and mitochondrial DNA assigns the phylogenetic node between the extant species *L. pardinus* and *L. lynx* to the late Early Pleistocene, within a 95 % confidence level at 1.98–0.70 Ma (Johnson et al. 2006). This fits perfectly with current palaeontological thinking; that Iberian lynxes evolved as a separate species during this time period.

The taxonomic decision to include the early Spanish lynxes in the species *L. pardinus* means that the combined Late Villafranchian Italian and the Epivillafranchian Central European populations must be upgraded to the rank of a species, i. e., a sister species of *L. pardinus*. This decision is strongly supported by the morphology of the axial skeleton, i. e., the proportion of the cervical and thoracic vertebrae, which cannot be assigned to *L. issiodorensis*. The taxon *L. issiodorensis valdarnensis* Werdelin, 1981 is therefore upgraded to *L. valdarnensis*. Two chrono-subspecies may be distinguished within this species. The nominate form *L. valdarnensis valdarnensis* include the Late Villafranchian lynxes of Italy. A lynx hemimandible from the Spanish Fonelas P-1 site (MNQ 18) corresponds to these Italian lynxes and is also attributable to *L. valdarnensis valdarnensis* (Garrido 2008). It indicates the Early Pleistocene phylogenetic rooting of *L. pardinus* within an Iberian population of the latter form. To classify the Fonelas lynx as *L. valdarnensis* rather than the basal form of *L. pardinus* gives later *L. valdarnensis* from outside the Iberian Peninsula the status of a surviving stem species. The Epivillafranchian lynx of Untermassfeld is described here as a new subspecies, *L. valdarnensis spelaeoides* ssp. nov.

Lynx valdarnensis spelaeoides ssp. nov.

Holotype:
Well preserved partial skeleton consisting of a set of the following components (*sensu* ICZN 1999, Art. 73.1.5.):
Mandibula with complete dentition IQW 1994/24 691 (Mei. 24 220), vertebra cervicalis V IQW 1996/25 868 (Mei. 25 397), vertebra cervicalis VI IQW 1996/25 867 (Mei. 25 396), vertebra cervicalis VII IQW 1996/25 866 (Mei. 25 395), vertebra thoracica I IQW 1996/25 865 (Mei. 25 394), vertebra thoracica II IQW 1996/25 905 (Mei. 25 434), vertebra thoracica IV IQW 1996/25 908 (Mei. 25 437), vertebra thoracica V IQW 1996/25 907 (Mei. 25 436), vertebra thoracica VI IQW 1996/25 903 (Mei. 25 432), vertebra thoracica VII IQW 1996/25 909 (Mei. 25 438), vertebra thoracica VIII IQW 1996/25 902 (Mei. 25 431), vertebra thoracica IX IQW 1996/25 904 (Mei. 25 433), scapula dex. IQW 1996/25 885 (Mei. 25 414), humerus dex. fragment IQW 1984/20 319 (Mei. 19 839); stored in the type collection of the Senckenberg Research Station of Quaternary Palaeontology, Weimar (Hemmer 2001, pl. 139, 1–2; **Figs. 26, 27b**).

Etymology: *spelaeoides*, Latin: spelaeum – cave, refers to the cave lynx *L. spelaeus*, spelaeoides – similar to the cave lynx
Type locality: Untermassfeld, Thuringia, Germany
Type layer: Lower Fluviatile Sands (LFS)
Chrono- and biostratigraphic age: about 1.07–1.06 Ma, Epivillafranchian
Diagnosis: medium sized lynx with stout mandible, high and broad P_3, *L. issiodorensis*-like length and width profile of the lower dentition, proportions of cervical and thoracic vertebrae as in modern lynx, humerus as in *L. pardinus*

In view of the close morphological affinity of the Untermassfeld *L. valdarnensis spelaeoides* and the lynx of Complex Moyen I (C.M.I) (MIS 14) of the southern French Arago cave (Testu 2006), the occurrence of the Epivillafranchian Central European lynx obviously lasted well into the early Middle Pleistocene. Lynx remains from the German sites of Mauer and Mosbach, originally attributed to *L. valdarnensis* (Werdelin 1981), may possibly also represent late forms of *L. valdarnensis spelaeoides*. Whether this new subspecies fully resembles *L. issiodorensis* ssp. 2 (Cipullo 2010) may be a matter for additional studies. A considerable morphological gap has been found between the Arago C.M.I lynx and later *L. spelaeus*, although this early Middle Pleistocene lynx was also attributed taxonomically to the latter species (Testu 2006). A detailed review of skull and postcranial features of *L. spelaeus* from different localities dated between the end of the early Middle and the Late Pleistocene does not show a clear evolutionary pattern over time and space, but stresses the general characterisation as a taxon of its own with a combination of *L. pardinus* and *L. lynx* features (Testu l.c.). Disclaiming this fact (Boscaini et al. 2016) fails to adequately consider the full set of characters of *L. spelaeus*. In contrast, the analysis of cytochrome b and control region fragments of ancient mt-DNA from Late Pleistocene to Holocene fossil lynx of Spain, southern France and Italy does not indicate the existence of independently evolved mt-DNA from *L. pardinus* and *L. lynx* haplotypes (Rodríguez-Varela et al. 2015). This fact may perfectly parallel results in extant North American *Canis* species. The Great Lakes region wolf (*Canis lycaon*) and the red wolf (*Canis rufus*) were found to represent admixed populations composed of various proportions of gray wolf (*Canis lupus*) and coyote (*Canis latrans*) ancestry (VonHoldt et al. 2016). Thus, *L. spelaeus* may represent hybrid populations composed of various proportions of *L. pardinus* and *L. valdarnensis* during the Middle Pleistocene and of *L. pardinus* and *L. lynx* during the Late Pleistocene. Hybridisation is not uncommon in extant lynxes. *L. canadensis* and *L. rufus* hybridise in Maine, Minnesota and New Brunswick where the natural distribution areas of both species overlap (Schwartz et al. 2004; Homyack et al. 2008). In principle, there is no reason not to expect the same for the European Pleistocene species in the border regions of their distribution areas.

The body mass of the Untermassfeld lynxes is best estimated using the prediction equation for the humerus minimal diaphyseal width (Hemmer 2001; 2004). The two available humeri, of very different individual sizes, have raw values of 17.2 and 26.0 kg. Rounding up to the nearest kg for cats weighing less than 20 kg and to the nearest 5 kg for the range between 20 and 50 kg (Hemmer 2004) avoids any illusion of accuracy beyond the possibilities of such a statistical method, resulting in values of 17 kg (15 kg as given by Hemmer 2001 was based on an earlier rounding method) and 25 kg. Even the lesser of the two body masses is considerably higher than that found in extant adult Iberian lynxes: 8.7–9.9 kg in females, 11.1–14.5 kg in males (Beltrán and Delibes 1993). The mandible length of the holotype of *L. valdarnensis spelaeoides* ranges within the upper part of the *L. pardinus* variability (92.5 vs. 82.0–96.9 mm), as does its M_1 length (13.0/13.1 vs. 12.2–13.4 mm, Beltrán and Delibes l.c.). This indicates somewhat heavier bodies for the Untermassfeld lynx than that of the more slender Iberian lynx. Linear measurements of the smaller individuals, i.e., of the females, of *L. valdarnensis spelaeoides* overlap with that of male extant *L. pardinus*.

Body mass is a useful predictor of life history and behavioural characteristics. Based on this, the modal class of the preferred size of a cat's prey and the wide range of regularly hunted prey can be estimated (for prey focus mass method see Hemmer 2004; 2007a). A 17 kg Untermassfeld lynx may have focused on prey species around 5 kg body mass and regularly preyed upon animals between 0.5 and 10 kg. For a 25 kg lynx the numbers are 10 kg and 1–20 kg, respectively. Even prey outside this wide mass variability may have fallen victim to these opportunistic hunters, at least occasionally. The prediction equation for the mean daily food intake (Hemmer 2004; 2007a) gives estimations of 1.3–1.8 kg. Projecting onto the real Untermassfeld fauna (Kahlke in this volume) suggests for this range of prey mass in the upper sector calves of *Pseudodama vallonnetensis*, may be occasionally even young calves of *Eucladoceros giulii*, and possibly piglets of *Sus scrofa priscus,* in the focus range *Capreolus cusanoides* and the lagomorphs *Hypolagus brachygnathus* and *Lepus* sp., and in the lowest sector, perhaps not so unimportant, the frequently recorded ground squirrel *Spermophilus*. An earlier published figure detailing the presumed prey spectrum of the Untermassfeld lynx (Hemmer 2001, Fig. 15) agrees well with the results of the predictor-based estimation of favoured prey size.

2.5. *Megantereon adroveri* Pons Moyà, 1987

2.5.1. Introduction

In the first description of the *Megantereon* remains from Untermassfeld (Hemmer 2001) they were attributed to the taxon *M. cultridens adroveri* Pons Moyà, 1987, originally described from the Late Villafranchian Spanish locality Venta Micena. However, a later study of the Venta Micena dirk-tooth cats tried to prove that they did not belong to the European *M. cultridens* but instead to the African *M. whitei* (Broom, 1937) (Martínez Navarro and Palmqvist 1995). This idea was based on absolute tooth size, which grouped the Venta Micena dentition with a small African collective series. Based on this assumption, the Venta Micena *Megantereon* was believed to represent a different taxon to the European species merely because of its smaller size. The original publication describing the Untermassfeld *Megantereon* rejected this methodology as not particularly conclusive (Hemmer 2001). Comparatively speaking, a much broader size variability exists in some extant large cat species than that recorded in the dirk-tooth cats. Instead, the favoured hypothesis was that the advanced form of *M. cultridens* of the late Early Pleistocene represented the last stage of a European evolutionary trend (Sardella 1998).

However, the African origin hypothesis for the last European dirk-tooth cats was repeated by other researchers (Martínez-Navarro 1996; Palmqvist 2002), and the Untermassfeld *Megantereon* was assigned to the species *M. whitei* (Palmqvist et al. 2007). Some authors generally adopted the *M. whitei* hypothesis while still using the original naming of the Untermassfeld taxon as *M. cultridens adroveri* (Sardella et al. 2008; Vislobokova 2018). A re-evaluation of the scenario whereby the African species colonised Europe and southwest Asia during the Early Pleistocene pointed again to the possibility that the latest European dirk-tooth cats represented nothing more than smaller-sized *M. cultridens* with derived features (*M. cultridens adroveri*) (Arribas and Garrido 2008). Finally, it was concluded that the advanced European *Megantereon* could be related to *M. cultridens* rather than to *M. whitei* and it was placed taxonomically in a separate species, *M. adroveri*, the original subspecies rank of this taxon having been upgraded (Lewis and Werdelin 2010).
Thus, there are three alternative classifications available for the Untermassfeld dirk-tooth cat: *M. cultridens adroveri*, *M. whitei* and *M. adroveri*. This situation requires a new evaluation of its position within European *Megantereon* evolution that is as conclusive as possible. New Untermassfeld fossils attributable to *Megantereon* are very scarce and limited to some postcranial elements. This re-evaluation will therefore be based on the previously published material (Hemmer 2001).

2.5.2. Material

Vertebra cervicalis II (epistropheus) IQW 2016/44 192 (Mei. 49 501),
Vertebra lumbalis II (juv.) IQW 1980/15 546 (Mei. 14 958),
Vertebra lumbalis VI (juv.) IQW 1980/16 572 (Mei. 16 093),
Vertebra lumbalis VII (juv.) IQW 1980/16 571 (Mei. 16 092),
Vertebra lumbalis VI? IQW 2008/29 665 (Mei. 28 827),
Sacrum (juv.) IQW 1980/16 570 (Mei. 16 091),
Scapula sin. IQW 1996/25 674 (Mei. 25 203),
Astragalus dex. IQW 2003/28 316 (Mei. 27 478).

2.5.3. Description and comparison

Fragments of an epistropheus and a lumbar vertebra of an adult cat and those of three vertebrae with the sacrum of a juvenile individual have been attributed to *Megantereon* (**Table 33**). Compared to the nearly complete Senèze skeleton (Se311) used as a reference (Christiansen and Adolfssen 2007), the Untermassfeld adult vertebra is around 10 % smaller. The following measurements are from the incomplete juvenile sacrum (mm): maximal width, 81.1, cranial articular width, 44.7, height, 24.
A left scapula, IQW 1996/25 674 (Mei. 25 203), is not complete but is, nevertheless, so well preserved that comparative measurements give a good impression of its extraordinary size (**Table 34**). Its length is 18 % larger than that of the Senèze skeleton and surpasses female lions (Christiansen and Adolfssen 2007). It is 46 % larger than that of a small dirk-tooth cat from Dmanisi.
An astragalus (**Fig. 28**) from Untermassfeld, IQW 2003/28 316 (Mei. 27 478), is of a similar size to one from Dmanisi (Vekua 1996) (mm, Dmanisi specimen in brackets): maximal length 37.6 [39.5], maximal depth 23.0 [22.5]. It is much smaller than the astragalus of the Senèze skeleton (Christiansen and Adolfssen 2007, Fig. 25).

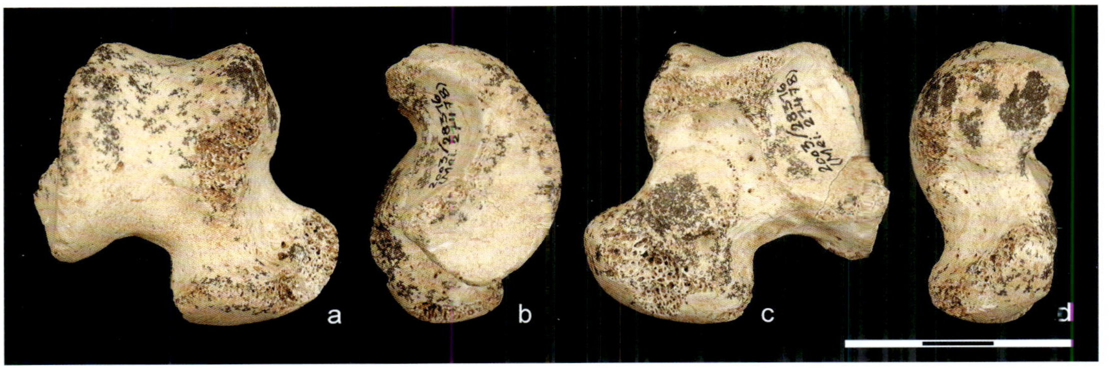

Fig. 28 *Megantereon adroveri* Pons Moyà, Untermassfeld. – **a–d** Right astragalus IQW 2003/28316 (Mei. 27478), dorsal, external, plantar, internal views. – Scale: 30 mm.

	L II (juv.) IQW 1980/15546 (Mei. 14958)	L VI (juv.) IQW 1980/16572 (Mei. 16093)	L VII (juv.) IQW 1980/16571 (Mei. 16092)	L VI? IQW 2008/29665 (Mei. 28827)
Maximal height				89.2
Centrum length		44.0+	43.3	48.5
Facies terminalis cranialis width	c. 35.0	c. 40.0		c. 40.0
Facies terminalis cranialis height			29.0	33.0
Facies terminalis caudalis width	c. 37.0	44.0	c. 43.0	43.3
Facies terminalis caudalis height		29.0	27.0	31.5
Neural canal cranial width				c. 20.0
Neural canal cranial height		c. 16.0		c. 14.9
Neural canal caudal width				
Neural canal caudal height		c. 13.5		c. 15.5

Table 33 *Megantereon adroveri* Pons Moyà, measurements (mm) of lumbar vertebrae, Untermassfeld.

	M. adroveri Untermassfeld IQW 1996/25674 (Mei. 25203)	*M. cultridens cultridens* Senèze, Se311 (Christiansen and Adolfssen 2007)	*M. falconeri* ssp. Dmanisi (Vekua 1996)
Maximal length (along the spine)	262.2	221.6	
Minimal length of the collum scapulae	52.8		
Maximal length of the glenoid process	59.4		
Length of the glenoid cavity	49.8		34.0
Width of the glenoid cavity	39.3		27.0

Table 34 *Megantereon*, scapula measurements (mm).

2.5.4. Discussion

Table 35 demonstrates that western European dirk-tooth cats ranging from the latest Plio- to earliest Early Pleistocene age (Early and early Middle Villafranchian), i.e., those that are to be classified chrono-stratigraphically between Villarroya and Les Étouaires on the one hand and Saint-Vallier on the other, form a joint group with relatively large upper and lower third premolars (P^3 length around half the P^4 length, P_3 length c. 60–70 % P_4 length), a large lower fourth premolar, and a relatively short lower carnassia ($P_4 > 90$ % M_1) respectively. Taxonomically, this group represents the chrono-subspecies *M. cultridens megantereon* (Bravard, 1828) (Hemmer 2001). Starting with the finds from Senèze, lower molars are found to be longer ($P_4 < 90$ % M_1). Multiple metric and morphognostic differences have been noted between the dirk-tooth cats from Saint-Vallier and those from Senèze. Based on these results, the idea of two independent European Middle Villafranchian *Megantereon* clades was derived (Liu 2005). The ages of the two French sites (Rook and Martínez-Navarro 2010; Kahlke et al. 2011; Nomade et al. 2014) indicate that a sufficient time span existed between them to generate corresponding differences brought about by evolutionary change. First illustrated with the Spanish Fonelas dirk-tooth cats at around 1.8 Ma, the P_3 became shorter (< 60 % P_4), with a strong reduction of its paraconid and the emergence of a very short diastema between the P_3 and P_4 (Arribas and Garrido 2008). This trend has been further supported by the dirk-tooth cats from Olivola and, subsequently, the sites of Upper Valdarno, where a very broad variability is recorded, spanning from archaic to progressive inter-tooth proportions. The P_3 is shortened by up to less than 50 % of that of the P_4, and the P^3 is down to 40 % of the P^4. As the lectotype of the chrono-subspecies *M. cultridens cultridens* (Cuvier, 1824) (Hemmer 2001) originates from the Late Villafranchian locality of Figline, Tasso Faunal Unit (FU) of the Upper Valdarno (Ficcarelli 1979), it represents the second evolutionary stage of European *Megantereon*. A separation of the Fonelas dirk-tooth cat as *M. cultridens roderici* Arribas and Garrido, 2008 may perhaps be justified for transitional forms between the earlier *M. cultridens megantereon* and the later *M. cultridens cultridens*.

In European dirk-toothed cats, the relative size of the upper canine did not change substantially over a period of one million years (**Table 35**). This functional structure, obviously successful over such a long span of time in succeeding ecosystems, became longer towards the end of the Late Villafranchian. A third evolutionary stage of *Megantereon* is therefore clearly recognisable, extending from the Italian Farneta FU to the Epivillafranchian, with the localities Venta Micena, Argentario, Pirro Nord, Apollonia and Untermassfeld (**Table 35**) (Sardella et al. 2008). In this stage, both the upper as well as the lower third premolars are even more reduced, there is a larger diastema between P_3 and P_4, and the upper canines are enlarged to an extreme size not found in stratigraphically earlier European dirk-toothed cats (canine length > 90 % P^4 length, > 260 % P^3 length). The evolutionary and taxonomic status of these cats is considered controversial to this day (see section 2.5.1.).

The set of traits typical for this advanced European *Megantereon*, i.e., the extremely long upper canines combined with a very short P^3, has already been observed as part of a wide variability in Chinese dirk-toothed cats from Longdan and Nihewan (**Table 35**), dated between 2.58 and 1.6 Ma (Qui 2006). This fact supports the idea of a gene flow from East Asia into late European populations. In African dirk-toothed cats, the size of the upper canines exceeds that of the corresponding teeth of the first two European evolutionary stages (i.e., *M. cultridens megantereon*, *M. cultridens cultridens*), but does not reach the extreme dimensions found in the most advanced European forms as well as in some Asian individuals. This observation alone should exclude an African ancestry for the advanced European dirk-toothed cats, thus confirming the previous rejection of the hypothesis of *M. whitei* invading Europe (Arribas and Garrido 2008; Lewis and Werdelin 2010). A very pronounced reduction of the upper and lower third premolars existed in both African and Asian populations before this trend also prevailed in Europe towards the end of the Late

	C in % P^4	C in % P^3	P^3 in % P^4	P_3 in % P_4	P_4 in % M_1
Europe					
Villaroya	60.6			59.9	93.8
Les Ètouaires	76.5	148.3	51.5	64.3	93.4
Perrier	77.5–78.1	149.7–158.3	49.3–51.8		
Pardines	82.1	158.3	51.9		
Saint-Vallier			50.3	69.7	95.7
Senèze	74.4	144.9–145.0	51.4	61.0	88.1
Fonelas				49.8–55.9	82.3–87.0
Bugiulesti	70.0	123.5	56.7		
Olivola			39.8		
Upper Valdarno	74.3	183.7	40.4–58.3	45.9–46.6	81.8–95.0
Urkút					88.0
Venta Micena					81.5
Argentario	93.2				83.8
Pirro Nord	106.3	260.9	40.7		
Apollonia				38.9	79.5
Untermassfeld	≤ 106.1	≤ 351.1	30.2		
Asia					
Kuruksay				55.9	92.3–97.4
Renzidong	81.2–82.6	159.5–161.4	50.9–51.2		
Lingyi	76.7	164.3	46.7		
Yushe	84.4	200.0	42.4		
Longdan	78.9–93.5	165.5–225.4	36.2–47.7	43.1–44.2	85.1–93.8
Nihewan	101.4	204.3	46.4	50.0	79.2
Dmanisi	79.3–80.7	165.7–208.9	38.6–47.9	37.5	83.3–84.2
Gongwangling				41.7	78.3
Zhoukoudian 1	71.4	156.3	45.7		
Siwaliks				38.1	80 8–96.4
Africa					
South Turkwel					84.8
East Turkana	86.7	238.9	36.3	35.9	82.4
Sterkfontein					68.9
Kromdraai A	84.4–90.0	212.5	39.7	37.1	85.9
Elandsfontein					72.8
Schurveburg				45.5	82.4

Table 35 Metric differentiation of Old World Pliocene and Pleistocene *Megantereon*. Calculations of inter-tooth proportions based on data and descriptions given by Villalta (1952), Viret (1954), Ewer (1955), Turner (1987), Martínez-Navarro and Palmqvist (1995), Sardella (1998), Liu (2005), and Palmqvist et al. (2007).

Villafranchian. Chinese dirk-toothed cats from Longdan and Nihewan have closely connected or fused P_3 roots (Qiu et al. 2004). This also applies in part to the cats from the Early Pleistocene South Asian Siwaliks (Lydekker 1884; Petter and Howell 1982). The P_3 of the Dmanisi *Megantereon* is also single-rooted (Vekua 1996). Furthermore, this character also appears in the Monte Argentario *Megantereon* which belongs to the advanced European type (Sardella et al. 2008). The quotients of the lower dentition of the European Apollonia dirk-toothed cats are almost the same as for the Chinese Gongwangling mandible.
To prove the status of the African dirk-toothed cat as a valid species *M. whitei*, separate from *M. cultridens*, it has been claimed »that the two *Megantereon* species have distinct tooth proportions, with *M. whitei* showing an upper canine which is more than 1.5 times longer mesiodistally than the fourth lower premolar« (Palmqvist et al. 2007, 167, Fig. 7). In fact, the only two African specimens where these measurements are possible, namely those from Kromdraai A and East Lake Turkana (Palmqvist et al. l.c., tab. 1), produced quotients of 1.47 and 1.48; just less than 1.5, but not more. A value just below 1.5 was also obtained from a skull from Longdan, China: 1.49. A second individual from the same site produced a ratio of 1.20. A quotient over 1.5 is only seen in the advanced European dirk-toothed cat of Argentario: 1.58.
A complete *Megantereon* skull from Dmanisi has a C/P_4 quotient of 1.39 (measurements by Vekua 1996), i.e., well below the 1.5 mark. The ratios of the upper canines to the P^4 and the P^3 of this and of another Dmanisi skull (**Table 35**) are also lower than those found in African and stratigraphically younger European individuals. So, the upper canines of the Dmanisi individuals are not as extremely long as those found in the European advanced *Megantereon*, and don't even reach the dimensions of the somewhat shorter canines of the African *M. whitei*. Consequently, as far as the characteristic feature of upper canine length supports the taxonomic attribution, the Dmanisi dirk-toothed cat would belong neither to the advanced European *Megantereon* nor to the African *M. whitei*. Oddly enough, it was exactly these Dmanisi skulls that played a crucial role in the discussion around the African *Megantereon* species in Europe and its presumed route out of Africa (Martínez-Navarro and Palmqvist 1995; 1996; Palmqvist 2002; Werdelin and Lewis 2002; Palmqvist et al. 2007; Lewis and Werdelin 2010). As there are more features that distinguish the Dmanisi *Megantereon* from the African *M. whitei* (Werdelin and Lewis 2002, Fig. 2; Lewis and Werdelin 2010, Fig. 2.2), the hypothesis of a dispersal of the African form via the Near East turntable to southwest Asia and on to Europe should be rejected.
Four alternative scenarios are possible in explaining the population change of the European dirk-toothed cats during the Late Villafranchian. Considerably accelerated regional evolution is one possible option which could explain the achievement of the same morphology that previously existed in the Asian population as part of a much broader variability. A second possible scenario is the extinction of the European *M. cultridens cultridens* population followed by a subsequent dirk-toothed cat re-occupation out of Asia. A corresponding extinction and replacement out of Africa is a third possible explanation that has previously been advocated at some length. The fourth possible scenario is a combination of the thinning out of the existing European population and infiltration of Asian or African cats with subsequent hybridisation and introgression. The third scenario mentioned, i.e., a replacement out of Africa, seems the most improbable (see above). The second and fourth scenarios, i.e., re-occupation by Asian *Megantereon* or hybridisation and introgression, look somewhat more likely than scenario one, namely independent regional evolution in Europe. Although, the current state of knowledge does not yet allow for any definitive clarification regarding the evolution of European dirk-toothed cats during the Late Villafranchian, a new level of European *Megantereon,* combining excessive upper canine size with very small upper and lower third premolars, did appear during this period.
The three Early Pleistocene *Megantereon* clades of the Old World, namely the European, the African and the Asian, may be distinguished taxonomically at the species level (*M. cultridens*, *M. whitei*, *M. falconeri*)

(Martínez-Navarro and Palmqvist 1995; Sardella 1998; Lewis and Werdelin 2010). This ranking would match the average longevity of extant cat species through time, as understood from genetic assessment (Johnson et al. 2006). A rational taxonomic classification of the advanced European *Megantereon* in a synopsis with the Asian *M. falconeri* or the African *M. whitei* excludes the subspecific rank of *M. cultridens*. In view of the minimum evolutionary age of a possible ancestral species at the time when the new *Megantereon* population appeared on the European stage, the corresponding rank as a species of its own (Lewis and Werdelin 2010) should be legitimate regardless of its phylogenetic ancestry. Consequently, the Untermassfeld dirk-toothed cat has been upgraded from its original classification as the subspecies *M. cultridens adroveri* (Hemmer 2001) to the species *M. adroveri*.

As with its taxonomic classification, the body mass of *Megantereon* has also been the subject of some controversy (Martínez-Navarro 1995; 1996; Hemmer 2001; Christiansen and Adolfssen 2007). A mass of 55 kg was estimated using a carnassial (M_1) prediction equation, whereas at least 100 kg was obtained by comparing the surface area of the diaphyseal cross-section of the humerus, and a mass approaching 200 kg was achieved by comparing the width of the distal epiphysis of the humerus with those of leopard and lion (Martínez-Navarro and Palmqvist 1996).

For the *M. cultridens cultridens* skeleton (Se311) from Senèze a body mass between 130 and 150 kg was calculated by using the least diaphyseal latero-medial diameters of humerus and femur (Hemmer 2001). The same individual was estimated to have been around 100–110 kg by a multivariate method based on 15 variables of the humerus, 3 of the ulna, 12 of the femur, and 6 of the tibia (Christiansen and Adolfssen 2007). Prediction equations based on the smallest diaphyseal diameters of the metacarpals (Hemmer et al. 2011; measurements of the Senèze individual by Sardella et al. 2008) result in masses of 131 to 83 kg, decreasing from Mc2 to Mc5, with a mean of 100 kg. If these calculations are considered in addition to those from the humerus and femur formulae, they result in a body mass of about 120 kg for the Senèze individual, i.e., a value similar to that of the earlier 100–110 kg estimation by Christiansen and Adolfssen (l.c.). However, the particularly powerful forequarters of *Megantereon* (Schaub 1925; Christiansen and Adolfssen l.c.) seem to produce a body mass over-estimation, not only when based on humerus measurements but also when considering the extraordinarily massive Mc2. The latter relationship is confirmed by a forepaw of the Pirro Nord *M. adroveri*, which gives an estimation of 134 kg based on the Mc2 vs. 85–99 kg when considering the Mc3, 4 and 5 (measurements by Sardella et al. 2008).

Based on a fragmentary humerus, the body mass of one of the Untermassfeld dirk-toothed cat individuals was calculated to be 160 kg (Hemmer 2001, 742 ff.). Taking into account the possible over-estimation for this skeletal element as mentioned above, the actual mass of the individual would be closer to 130–140 kg. Since the Senèze skeleton obviously belonged to a male (Christiansen and Adolfssen 2007), the Untermassfeld *Megantereon* must also, therefore, have been a large male individual. Its enormous size is also underlined by the scapula measurements (**Table 34**). Females would have ranged from about 60 kg upwards (Christiansen and Adolfssen l.c. for a skull from Perrier; Hemmer 2001 for a humerus from Dmanisi). This indicates a pronounced sexual dimorphism whose reliable assessment requires isochronous fossil material from one single site. Indeed, the enormous scapula IQW 1996/25674 (Mei. 25203) (see section 2.5.3.), in combination with the small astragalus IQW 2003/28316 (Mei. 27478) (**Fig. 28**), indicate the extreme size variability in the Untermassfeld *Megantereon*, which is even larger than that seen in extant lion and leopard populations (see section 2.1.4.).

A cat with a body mass of around 130 kg would have focused on prey in the 200 kg class and regularly hunted animals across the 20–500 kg classes (Hemmer 2004; 2007a). With its extremely long upper canines and its very powerful forequarters, a male dirk-toothed cat of the aforementioned body mass was functionally adapted to keep control over much larger prey than would be possible for any extant cat of comparable size. For the male individual from Untermassfeld this would mean a shift in the prey mass range by one size class, i. e., a range of 50–1,000 kg with the focus on 500 kg.

Modelling based on isotopic signatures suggests that more than 90 % of the biomass ingested by *M. adroveri* at Venta Micena was the result of preying on *Praemegaceros* cf. *verticornis*, *Stephanorhinus* sp. and *Equus altidens* (Palmqvist et al. 2008). All of these species fit centrally or at least marginally into the predicted prey body mass range.

However, in view of depleted $\delta^{13}C$ values (compilation by Palmqvist et al. l.c.) the horse result should be treated here with caution. Overall, it can be concluded that large to very large browsing ungulates seem to have been the preferred prey of dirk-toothed cats in closed, wooded habitats. Given their occurrence in large numbers, it can be deduced that in the Untermassfeld herbivore guild the large deer (*Eucladoceros giulii*) and the rhino (*Stephanorhinus hundsheimensis*) (Breda et al. 2020; Kotowski et al. 2020; Kahlke in this volume) would have been the main targets of male dirk-toothed cats. If female *Megantereon* were indeed substantially smaller than males, female and even juvenile rhinos are likely to have been out of their reach in the corresponding palaeoenvironment.

The mean daily meat consumption of a 130 kg dirk-toothed cat has been estimated at around 5.5 kg using the respective prediction equation (Hemmer 2004; 2007a). Even if the cat gorged a multiple of that number immediately following a kill, a considerable discrepancy in the prey focus mass remains. Around 40 % of the live mass of an extant fallow deer (*Dama dama*) is represented by meat, heart and liver (Hemmer et al. 2011, based on Reinken 1987). Assuming a roughly similar composition in a 250 kg *Eucladoceros* deer, this would have meant around 100 kg of potential food for a dirk-toothed cat. To compensate for the economic discrepancy between the amount of food available after a successful kill and the daily meat consumption, a predator has to return for several days to a carcass or to rest in its immediate vicinity for subsequent multiple meals. For the particularly heavily built *Megantereon,* maximum use of the prey must have been essential as its hunting success rate would have been limited in a wooded habitat. It is conceivable that its hunting success rate was even lower than that observed in the tiger (*Panthera tigris*), which probably means a ratio of at least twenty unsuccessful attempts to one successful killing (Schaller 1967). Staying for a long time near a carcass greatly enhances the chance of repeated challenges by potential kleptoparasitic scavengers. The defensive behaviour of a furious dirk-toothed cat with opened mouth and exposed white, shining super-canines would have been one of the most intimidating gestures displayed by carnivores of that time, and would have kept any potential competitors and scavengers at a distance in confrontational situations. With such imposing behaviour, *Megantereon* may have even been able to drive away other large carnivores from their kills and thus be successful without having to actually use their canines while hunting (Hemmer 2001).

Most authors agree that *Megantereon* preferred closed, wooded habitats (e. g., Marean 1989; Hemmer 2001; Palmqvist et al. 2008; Lewis and Werdelin 2010), and the limited Untermassfeld evidence does not contribute any essential new insight here. The flood vulnerability ratio (FVR, see section 2.1.4.) cannot be applied to the dirk-toothed cat, as the minimum number of individuals is too low. The total *Megantereon* material from Untermassfeld represents an MNI of 2. While the Lower Fluviatile Sands (LFS) contained elements of the larger male individual [e. g., the scapula IQW 1996/25 674 (Mei. 25 203), see above], the astragalus of the female [IQW 2003/28 316 (Mei. 27 478), see above] originated from the lower part of the Upper Fluviatile Sands [UFS(l)]. Nevertheless, the FVR of the Untermassfeld dirk-toothed cat (6) corresponds more closely to that of the jaguar, and not to the ratios of the cheetah or puma. Thus, some degree of water affinity and corresponding experience with the potential power of river water may be assumed for the Untermassfeld *Megantereon*.

2.6. *Homotherium latidens ultimum* (Teilhard de Chardin, 1936)

2.6.1. Introduction

The earlier published finds of *Homotherium* from Untermassfeld record the presence of at least three adult individuals at the site (Hemmer 2001, 750ff.). Despite such a limited number, the finds display a considerable range of morphometric variability, enabling them to be assigned to the species *H. nestianum* (Weithofer, 1889; not Fabrini, 1890 as commonly cited), *H. crenatidens* (Weithofer, 1889; not Fabrini, 1890 as commonly cited) and *H. latidens* (Owen, 1846) (Hemmer 2001, 756ff.). Consequently, this evidence reduces the three aforementioned taxa to nothing more than different morphotypes of one single species, since the extremely rapid genesis of the site (Kahlke et al. 2020, 1103) indicates the existence of a corresponding biological population of scimitar cats in the Untermassfeld area. In fact, the *nestianum* type seems to indicate males and the *crenatidens* type females, which up to now has been considered as a possible interpretation (de Bonis 1976). The first publication on the Untermassfeld scimitar cats follows the two species concept for Pleistocene *Homotherium* with *H. crenatidens* as the geochronologically older species and *H. latidens* the younger. The Untermassfeld finds were then attributed to *H. crenatidens*, while noting the presence of features also associated with the later *H. latidens*, such as the noticeable thin canine blade or the narrow P_4.

As with the small Untermassfeld series, the extraordinarily extensive material of homotheres from the Late Villafranchian Spanish site of Incarcal (Antón and Galobart 1999; Galobart et al. 2003; Antón et al. 2005; 2014) also displays an enormous morphometric variability, which has encouraged the grouping of Pliocene-Pleistocene scimitar cats into the single species *H. latidens*. The comparative description of a *Homotherium* mandible from Semibalki-3 in the Sea of Azov region (Rostov Oblast, Southern Russia), attributed to the Tamanian faunal unit with an age of 1.1–0.8 Ma (Sotnikova and Titov 2009), reveals close affinities in size and morphology with the first reported Untermassfeld mandible. These affinities are especially conspicuous in the strongly arched incisors-canine series. Both mandibles, together with some specimens recovered from Asia, are now attributed to *H. latidens* in view of similarities with an early Middle Pleistocene fragment from Hundsheim (Austria, Freudenberg 1914). Thus, the two species concept, i.e., the division into the Villafranchian *H. crenatidens* and the post-Villafranchian *H. latidens*, has been renewed (Sotnikova and Titov 2009). The study of scimitar cat teeth from the Epivillafranchian (early Galerian Faunal Unit) site of Monte Peglia (Italy) has recorded similarities with the extensive Late Villafranchian (Pirro Nord F. U.) homothere sample from Pirro Nord (Italy) and the Untermassfeld material. Using a monospecific concept for the latest Pliocene and Pleistocene European *Homotherium* as *H. latidens*, the scimitar cat group from Pirro Nord, Monte Peglia, and Untermassfeld has been classified as an advanced form of the *crenatidens* morphotype (Sardella and Iurino 2012). A review of Eurasian and American *Homotherium* with particular emphasis on the Incarcal series and Alaskan scimitar cats (Antón et al. 2014) finds no evidence of a species-level division in Eurasian Late Pliocene to Pleistocene homotheres. The review describes an easily recognisable pocketed anterior margin of the masseteric fossa of the mandible as a feature that characterises both Middle and Late Pleistocene European *H. latidens* and Late Pleistocene North American *H. serum* (Cope, 1893). However, neither the Untermassfeld homotheres (Hemmer 2001) nor the East European to West Asian post-Villafranchian group of scimitar cats (Sotnikova and Titov 2009) were included in this review.

The task of the *Homotherium* section presented here is not just to describe the new finds of the Untermassfeld scimitar cat, but to also determine its classification against the background of the aforementioned and sometimes quite controversial phylogenetic and taxonomic discussions.

2.6.2. Material

Mandibula dex. (juv.) IQW 2007/29 145 (Mei. 28 307),
Mandibula sin. IQW 2009/30 238 (Mei. 29 400),
Mandibula dex. IQW 2009/30 239 (Mei. 29 401),
Vertebra cervicalis II (epistropheus) IQW 1992/23 951 (Mei. 23 483),
Vertebra cervicalis III IQW 1980/15 204 (Mei. 14 686),
Vertebra cervicalis IV IQW 1980/16 674 (Mei. 16 195),
Scapula sin. IQW 1995/24 907 (Mei. 24 446),
Scapula dex. IQW 1992/23 872 (Mei. 23 401),
Scapula dex. IQW 1999/26 606 (Mei. 26 135),
Scapula dex. IQW 2011/32 263 (Mei. 31 426),
Humerus sin. diaphyseal fragment IQW 2008/29 685 (Mei. 28 847),
Humerus dex. diaphyseal fragment IQW 2004/28 628 (Mei. 27 790),
Ulna dex. IQW 1992/23 857 (Mei. 23 386),
Radius dex. IQW 1992/23 856 (Mei. 23 385),
Radius dex. diaphyseal fragment IQW 2009/30 166 (Mei. 29 328),
Pelvis sin. IQW 1993/24 288 (Mei. 23 817),
Tibia dex. proximal fragment IQW 2004/28 645 (Mei. 27 807),
Metatarsale (Mt) 4 dex. proximal fragment IQW 2012/38 435 (Mei. 35 409).

2.6.3. Description and comparison

The right hemimandible of a juvenile homothere IQW 2007/29 145 (Mei. 28 307) (**Fig. 29**, **Table 36**) is one of the few available examples of scimitar cubs from European sites (de Beaumont 1980; Arribas and Garrido 2008; Petrucci 2008; Petrucci et al. 2013). Its DP_4 is completely preserved, with the exception of some chipping at the labial side of the paraconid blade. Only the broken back root of the DP_3 remains. The tip of the M_1, which is visible in the mandibular corpus shows that this tooth was about to erupt. According to the stages for the eruption of deciduous and permanent teeth in *H. serum* (Rawn-Schatzinger 1983), this pattern corresponds to stage IV near the transition to stage V, i.e., to an individual in the latter part

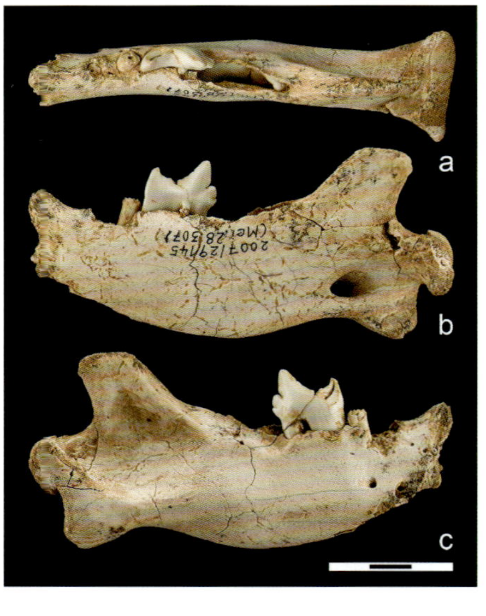

	IQW 2007/29 145 (Mei. 28 307)
Mandible length Processus condylaris-Infradentale	104.0+
Mandible height Processus angularis-Coronion	48.9+
DP_4 length	17.5
maximal width	6.4
protoconid height	c. 13.7

Table 36 *Homotherium latidens ultimum* (Teilhard de Chardin), measurements (mm) of a juvenile hemimandible, Untermassfeld.

Fig. 29 *Homotherium latidens ultimum* (Teilhard de Chardin). Untermassfeld. – **a–c** Right hemimandible IQW 2007/29 145 (Mei. 28 307) with erupted DP_4, occlusal, lingual, buccal views. – Scale: 30 mm.

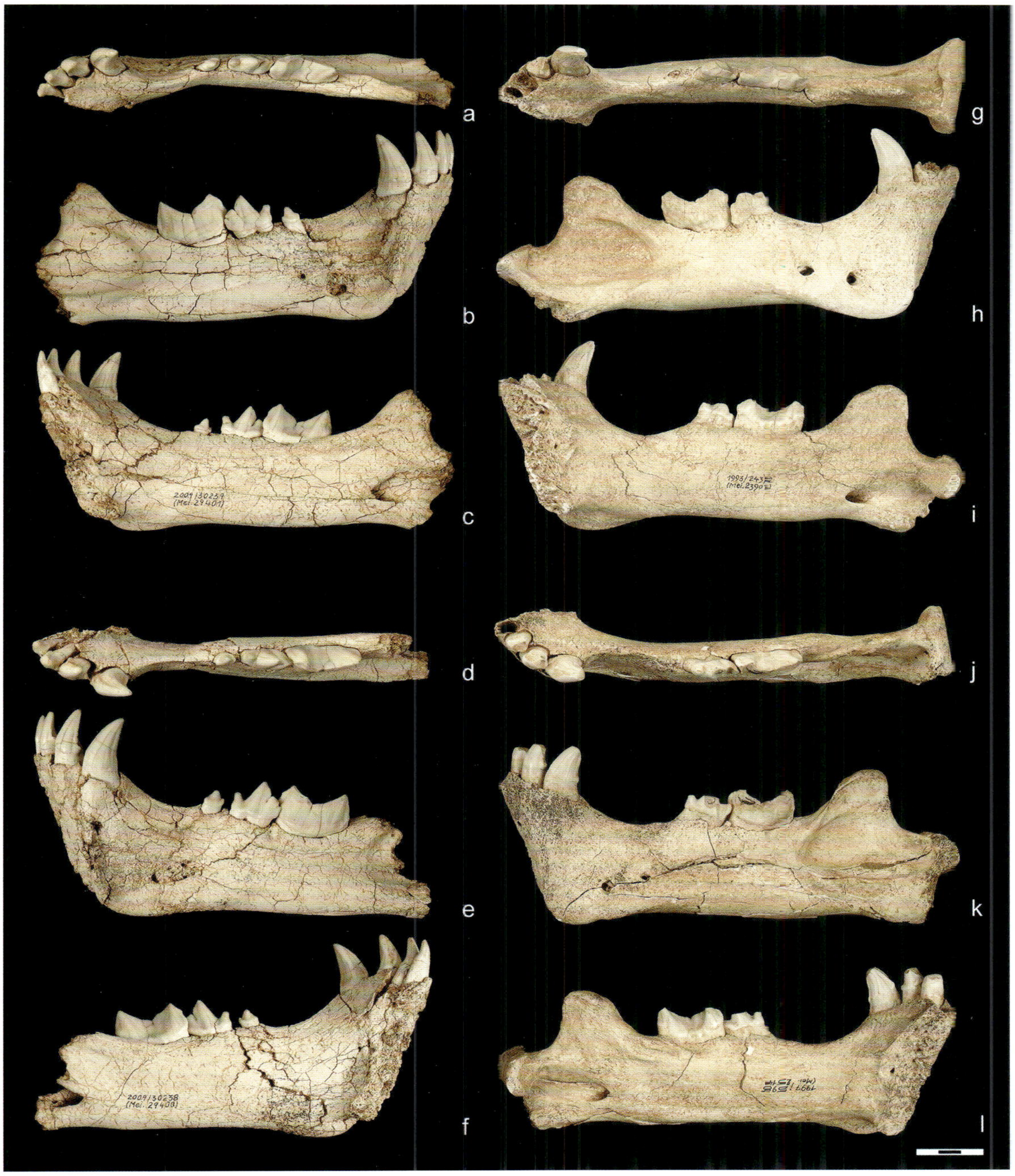

Fig. 30 *Homotherium latidens ultimum* (Teilhard de Chardin), Untermassfeld. – **a–c** Right hemimandible IQW 2009/30239 (Mei. 29401) with I_1-M_1. – **d–f** Left hemimandible IQW 2009/30238 (Mei. 29400) with I_1-M_1. – **g–i** Right hemimandible IQW 1993/24372 (Mei. 23901) with fragmentary I_3, C, P_4-M_1. – **j–l** Left hemimandible IQW 1997/25985 (Mei. 25514) with I_2-C, P_4-M_1. – **a**, **d**, **g**, **j** occlusal views; **b**, **e**, **h**, **k** buccal views; **c**, **f**, **i**, **l** lingual views. – Scale: 30 mm.

	IQW 2009/30238 (Mei. 29400)	IQW 2009/30239 (Mei. 29401)	IQW 1997/25985 (Mei. 25514) (Hemmer 2001)	IQW 1993/24372 (Mei. 23901) (Hemmer 2001)
Mandible length Processus condylaris-Infradentale			209.0	208.0
Mandible height Processus angularis-Coronion			c. 72.5	68.0
Depth behind M_1	43.9	44.8	48.0	42.5
Depth in front of P_3	46.7	45.0	41.5	38.0
Thickness below P_4/M_1	17.4	16.7		
Alveolar length P_3-M_1	66.0	67.0		62.0
Diastema (C-P_3) length	c. 40.0	33.4		
I_1 transversal width	6.2	5.9		
antero-posterior length	8.1	9.6		
height	14.3	11.3		
I_2 transversal width	8.6	8.5		
antero-posterior length	10.5	10.5		
height	17.1	17.5		
I_3 transversal width	12.3	11.9		
antero-posterior length	11.7	11.0		
height	21.0	22.5		
C inf. length	15.2	15.4	15.1	13.8+
width	10.6	10.9	10.0	10.1
height	26.7	25.9		20.0+
P_3 length	9.0	8.8		
maximal width	5.8	5.5		
protoconid length	6.3	5.6		
protoconid height	8.8	8.6		
P_4 length	21.0	21.3	21.0	19.4+
maximal width	9.3	9.2	8.6	8.5
anterior width	7.6	7.6	7.2	7.4
protoconid length	11.0	10.0	10.5	9.2
protoconid height	17.7	17.1		
M_1 length	32.9	32.7	30.5	30.4
maximal width	13.4	13.3	12.1	11.9
protoconid length	18.6	18.5	16.7	16.7
paraconid length	16.6	17.1	15.7	15.3
protoconid height	17.3+	17.4		
paraconid height	17.5+			

	IQW 1988/22597 (Mei. 22116)	IQW 1988/22595 (Mei. 22114)
M_1 length	c. 33.7	c. 34.0
maximal width	13.1	13.1
protoconid length	c. 17.5	c. 17.8
paraconid length	c. 16.7	c. 19.0

Table 37 *Homotherium latidens ultimum* (Teilhard de Chardin), measurements (mm) of the mandible IQW 2009/30238 (Mei. 29400) + IQW 2009/30239 (Mei. 29401), compared to mandible IQW 1997/25985 (Mei. 25514) + IQW 1993/24372 (Mei. 23901) (Hemmer 2001) and isolated molars, Untermassfeld.

of the second half-year of its life. The measurable blade of the erupting M_1 protoconid has a length of c. 13 mm, indicating that it would have been a relatively small individual had it reached maturity, presumably a female [the same measurement 16.5–17.3 mm in IQW 2009/30238 (Mei. 29400) + IQW 2009/30239 (Mei. 29401), see below].

The newly discovered *Homotherium* material from Untermassfeld includes a second, well preserved mandible of an adult individual (**Fig. 30a–f, Table 37**) that supplements the published finds (Hemmer 2001, 752ff.). It is somewhat larger than the mandible IQW 1993/24372 (Mei. 23301) + IQW 1997/25985 (Mei. 25514) described earlier (**Fig. 30g–l**). The new right hemimandible IQW 2009/30239 (Mei. 29401) is missing parts of the coronoid, condylar and angular processes. Its left counterpart IQW 2009/30238 (Mei. 29400) is broken behind the carnassial and both of the complete rows of teeth show only slight wear. The combined mandible looks very robust, even more than the previously described complete specimen, as the crest in the diastema section runs steadily upwards from the P_4 to the canine and does not recess in the diastema over the mental flange, as in the case of the aforementioned mandible. The flange itself is quite inconspicuous in lateral view, starting below the P_4 and protruding only a little below the symphysis. In the anterior section of the mandible its lower edge bends upwards at almost a right angle, so that the resulting structure supports the lower canine.

The straight front side of the symphysis runs at an inclined angle to the first incisor, ending far in front of the canine. In dorsal view, the alveoli of the incisors and canines are arranged in a slightly curved row. Thus, the complete incisors-canine series of the mandible IQW 2009/30238 (Mei. 29400) + IQW 2009/30239 (Mei. 29401) is strongly arched, more so than in most other homothere mandibles. In lateral and mesial views (**Fig. 30b–c, e–f**), each of the four teeth are therefore visible due to their evenly staggered arrangement. The masseteric fossae taper off into an open anterior margin below the protoconid of the carnassial. There are two small mental foramina, one below the P_3 and one below the diastema that opens into a round, shallow pit. On the whole, the shape of the second *Homotherium* mandible from Untermassfeld is most similar to that of a hemimandible from the late Early Pleistocene locality Semibalki-3 in the Azov Sea region (Sotnikova and Titov 2009, Figs. 3, 5).

The dentition of the mandible IQW 2009/30238 (Mei. 29400) + IQW 2009/30239 (Mei. 29401) only shows substantial wear at the M_1 paraconid, less wear can be seen at the protoconid of the carnassial, and minor wear at the rear edge of the P_4 protoconid. Faint signs of wear are also present on the tips of the incisors. The rest of the teeth are in excellent condition and show almost no signs of wear, a situation that is not encountered very often in *Homotherium* dental remains. P_4 and M_1 incline slightly backwards. Only the slender, high and pointed anterior cusp of the P_4 is vertically oriented, so that it splays out forwards away from the protoconid. The P_3 is a very small, single-rooted tooth with a high protoconid and a cingulum that appears to form anterior and posterior mini-cuspids. The incisors are caniniform and show fine crenelations at their mesial and distal ridges. The serration of the canine ridges is not as fine as in the incisors. Within the strongly arched incisors-canine row the tooth size increases from the I_1 to the C (**Fig. 30a–f, Table 37**).

The new scimitar cat finds from Untermassfeld include three cervical vertebrae (**Fig. 31, Table 38**). The best preserved of the three is an epistropheus that is about the same size as the one measured in the Middle Villafranchian Senèze skeleton. In contrast, the C III and C IV are smaller than those from Senèze (Ballesio 1963). When compared with two specimens from the Late Pleistocene North American Friesenhahn Cave (Texas, USA), the Untermassfeld epistropheus is smaller, whereas the C III and C IV fit in well with a small series of vertebrae from that site (Rawn-Schatzinger 1992).

Four scapulae, including fragments, representing at least three individuals, were also recovered. The two best preserved specimens, from which measurements could be taken (**Table 39**), could come from the left and right scapulae of the same individual. Their sizes range from a little below to just within the lowermost

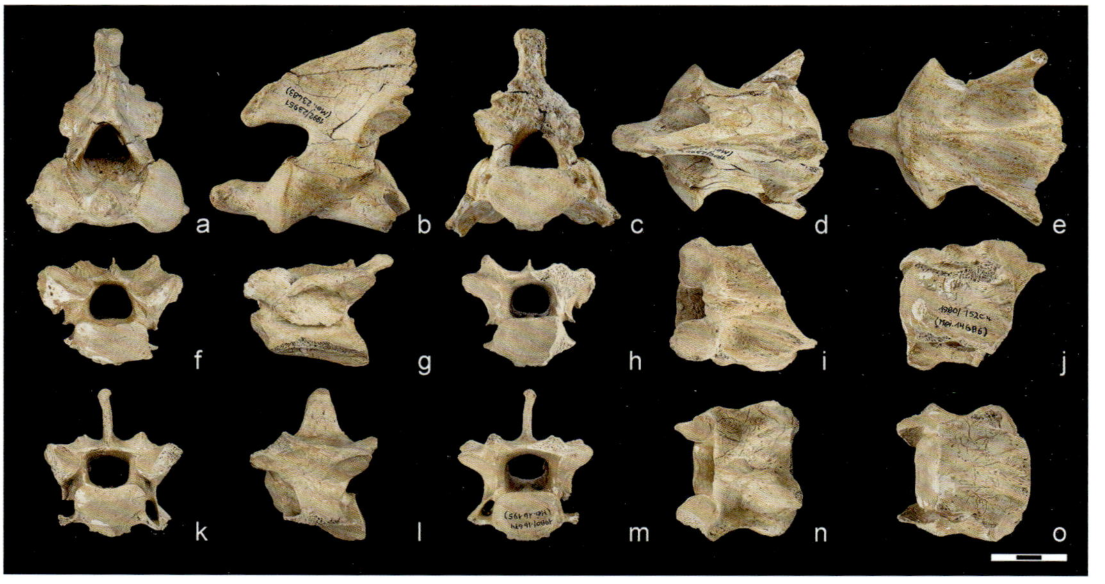

Fig. 31 *Homotherium latidens ultimum* (Teilhard de Chardin), Untermassfeld. – **a–e** Epistropheus IQW 1992/23951 (Mei. 23483). – **f–j** Cervical vertebra III IQW 1980/15204 (Mei. 14686). – **k–o** Cervical vertebra IV IQW 1980/16674 (Mei. 16195). – **a**, **f**, **k** anterior views; **b**, **g**, **l** left lateral views; **c**, **h**, **m** posterior views; **d**, **i**, **n** dorsal views; **e**, **j**, **o** ventral views. – Scale: 30 mm.

	C II (epistropheus) IQW 1992/23951 (Mei. 23483)	C III IQW 1980/15204 (Mei. 14686)	C IV IQW 1980/16674 (Mei. 16195)
Centrum length (C II including dens)	86.4	c. 37.0	c. 35.5
Arcus length	72.0		
Maximal height	c. 80.0	43.7	67.0
Length anterior to posterior zygapophyses		58.6	
Width across anterior zygapophyses		48.2	
Width across posterior zygapophyses			c. 56.0
Minimal width of vertebra	42.1		
Facies terminalis cranialis width	61.6	27.4	28.8
Facies terminalis cranialis height		19.0	19.5
Facies terminalis caudalis width	36.0		29.6
Facies terminalis caudalis height	26.6	20.4	20.6
Neural canal cranial width		c. 16.5	
Neural canal cranial height		c. 14.5	c. 14.6
Neural canal caudal height	c. 16.5	c. 15.0	c. 17.4

Table 38 *Homotherium latidens ultimum* (Teilhard de Chardin), measurements (mm) of cervical vertebrae II–IV, Untermassfeld. For size comparison: epistropheus centrum length including dens Friesenhahn Cave 79.9–83.8 (n = 2, Rawn-Schatzinger 1992); C III centrum length Incarcal 36.3 (n = 1, Galobart et al. 2003), Senèze 42 (n = 1, Ballesio 1963), Friesenhahn Cave 30.3–38.5 (r = 3, Rawn-Schatziger l. c.); cervical vertebra IV centrum length Senèze 39 (n = 1, Ballesio l. c.), Friesenhahn Cave 30.2–36.4 (n = 3, Rawn-Schatzinger l. c.).

	Untermassfeld IQW 1995/24907 (Mei. 24446)	Untermassfeld IQW 1992/23872 (Mei. 23401)	Friesenhahn Cave (Rawn-Schatzinger 1992)
Maximal length (along the spine)	295.0+		299.5–301.5 (n = 3)
Minimal length of the collum scapulae	59.4	59.2	
Maximal length of the glenoid process	75.6	74.4	
Length of the glenoid cavity	59.7	57.5	62.8–78.3 (n = 9)
Width of the glenoid cavity	37.5	40.0	37.2–43.8 (n = 8)

Table 39 *Homotherium latidens ultimum* (Teilhard de Chardin), scapula measurements (mm), Untermassfeld, compared to the Friesenhahn Cave series of *H. latidens serum* (Rawn-Schatzinger 1992).

region of the variability of scimitar cat scapulae from the North American Friesenhahn Cave (Rawn-Schatzinger 1992).

In addition to two newly excavated humerus shaft fragments and part of a radius, the foreleg of Untermassfeld homotheres is represented by an ulna fragment IQW 1992/23857 (Mei. 23386) (**Fig. 32a–d**) and a complete radius IQW 1992/23856 (Mei. 23385) (**Fig. 32e–j**), presumably from the same individual. The sizes of the latter specimens correspond to those of the Senèze skeleton and roughly fit the Incarcal and Friesenhahn Cave series (**Table 40**).

The hindleg is represented by a new proximal tibia fragment (**Fig. 33**), whose size matches the Senèze skeleton and fits the variability of the Friesenhahn Cave shin bones (**Table 41**).

The newly discovered postcranial elements from Untermassfeld document the existence of at least one *Homotherium* individual of the Senèze size category and another that is slightly smaller.

2.6.4. Discussion

There are no significant differences in size between Old World *Homotherium* lower carnassials throughout the whole Pleistocene (**Table 42**). Although the Middle Pleistocene homotheres were not especially small, given the very limited finds it is as yet not possible to determine whether early Early Pleistocene scimitar cats tended to be particularly tall. The total percent point span, i.e., the differences between mean and minimum values plus the difference between the maximal and mean values (in percent) of all Eurasian Late Villafranchian scimitar cat M_1 lengths is 33, as is the case for all Pleistocene representatives. This value almost corresponds to that (34) obtained for the basal length of a series of 35 East African leopard (*Panthera pardus*) skulls (see section 2.1.4.). Thus, the linear M_1 size variability of all *Homotherium* records of the European Pleistocene is not greater than that of one single population of extant leopards. Clearly, the carnassial size design of the Old World scimitar cats was so well optimised across all of their habitats that no morphometric changes occurred over two and a half million years. This is in contrast to the general structural diversity that emerged even within single populations during the evolution of the genus (e.g., Incarcal: Antón et al. 2014; Untermassfeld: Hemmer 2001; this paper) and makes the taxonomic separation of different entities difficult (e.g., Antón et al. l.c.). The situation in North America however is somewhat different. Here, the northern, i.e., Beringian (Alaska, Yukon Territory), homotheres fit within the Eurasian variability, whereas the southern form tends to have slightly smaller sized molars.

The taxonomically most distinguishable form are the homotheres of the European Middle to Late Pleistocene. Analysis of mandible finds from the early Middle Pleistocene Forest Bed (UK) and up to the Late Pleistocene of the North Sea (Backhouse and Lydekker 1885, Reumer et al. 2003), demonstrate that this

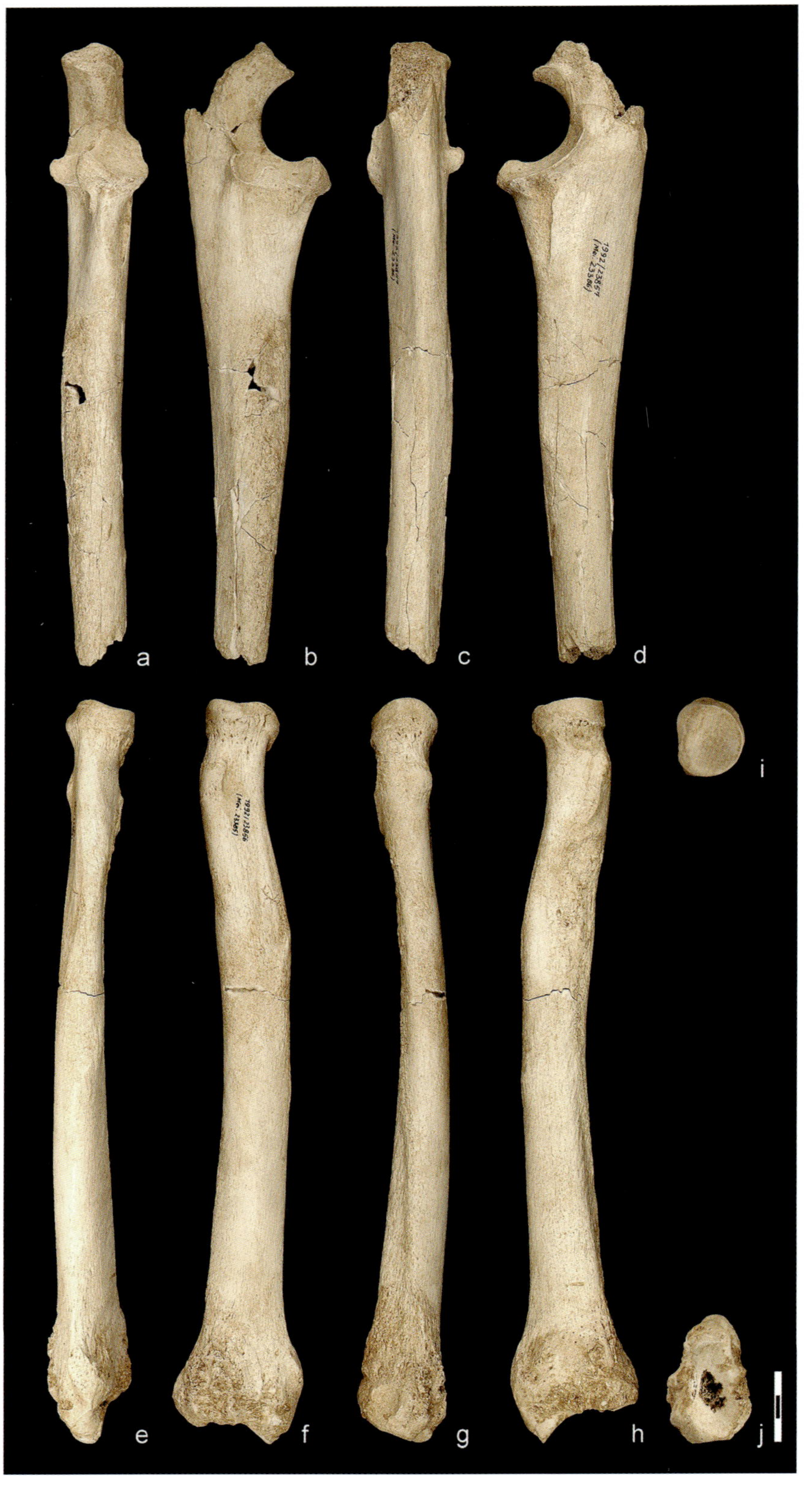

Fig. 32 *Homotherium latidens ultimum* (Teilhard de Chardin), Untermassfeld. – **a–d** Right ulna IQW 1992/23 857 (Mei. 23 386). – **e–j** Right radius IQW 1992/23 856 (Mei. 23 385). – **a**, **f** anterior views; **b**, **g** external views; **c**, **h** posterior views; **d**, **e** internal views; **i** proximal view; **j** distal view. – Scale: 30 mm.

Ulna	Untermassfeld IQW 1992/23857 (Mei. 23386)	Senèze (Ballesio 1963)	Friesenhahn Cave (Rawn-Schatzinger 1992)
APD over Processus anconaeus	59.5	59.0	60.1–67.3 (n = 4)
Height of trochlear notch	33.8	37.0	37.7–45.2 (n = 10)
Width of trochlear notch	41.0	35.0	34.3–45.1 (n = 8)

Radius	Untermassfeld IQW 1992/23856 (Mei. 23385)	Incarcal IN-V 83 (Galobart et al. 2003)	Incarcal IN-V 47, IN-V 158, IN-V 23 (Galobart et al. 2003)
Maximal length	316.0	286.0	
Maximal proximal TD	32.0	32.1	29.4–31.8
Maximal proximal APD	27.0	26.8	24.7–26.2
Maximal distal TD	54.5	50.7	
Maximal distal APD	34.5	32.4	

	Incarcal IN-I 545, IN-I 840, IN-I 29 (Galobart et al. 2003)	Senèze (Ballesio 1963)	Friesenhahn Cave (Rawn-Schatzinger 1992)
Maximal length		317.0	295.7–328 0 (n = 6)
Maximal proximal TD		34.0	37.0–57.4 (n = 8)
Maximal proximal APD		29.0	32.0–38.1 (n = 8)
Maximal distal TD	47.2–55.4	55.0	33.2–61.4 (n = 10)
Maximal distal APD	32.8–37.5	38.0 / 39.0	28.9–36.7 (n = 9)

Table 40 *Homotherium*, radius and ulna measurements (mm). TD – transverse diameter; APD – anterior-posterior diameter.

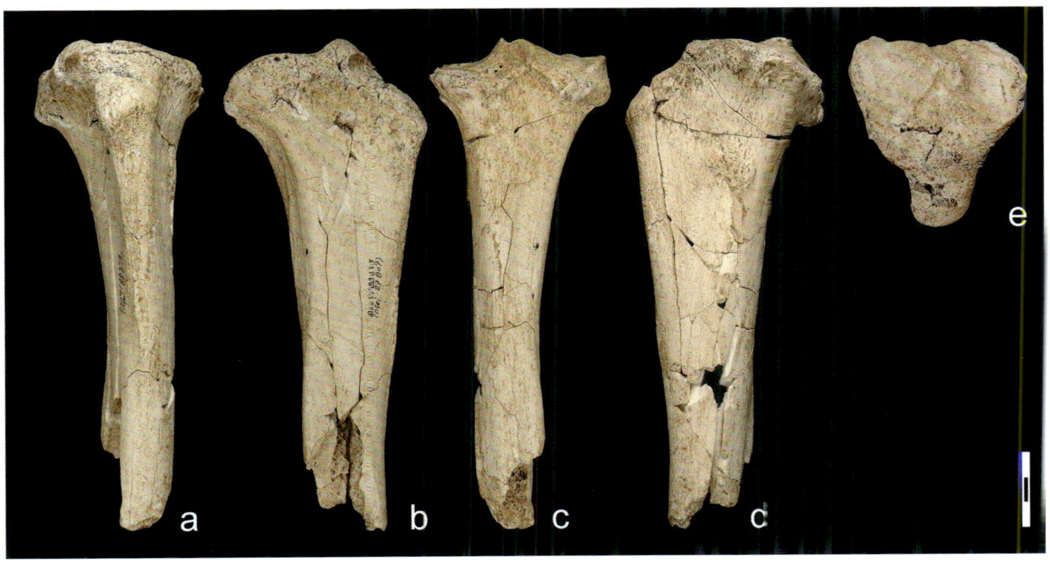

Fig. 33 *Homotherium latidens ultimum* (Teilhard de Chardin), Untermassfeld. – **a–e** Right tibia IQW 2004/28645 (Mei. 27807), anterior, external, posterior, internal, proximal views. – Scale: 30 mm.

	Untermassfeld IQW 2004/28645 (Mei. 27807)	Senèze (Ballesio 1963)	Friesenhahn Cave (Rawn-Schatzinger 1992)
Maximal proximal TD	71.4	69.0	67.4–76.0 (n = 6)
Maximal proximal APD	78.9	80.0	

Table 41 *Homotherium*, tibia measurements (mm). TD – transverse diameter; APD – anterior-posterior diameter.

group is characterised by a unique combination of several easily recognisable morphognostic features. The most eye-catching is the pocketed front margin of the masseteric fossa (Antón et al. 2014). The mandibles show a coronoid process that appears to be shifted forward in relation to the position of the condylar and angular processes and looks more compact and less hooked. The anterior mental foramen is very wide near the surface of the jaw. The mandibular incisors-canine series are gently arched and not all of the teeth are grouped in an evenly staggered arrangement (**Fig. 34**). Another noticeable and remarkable feature of the Middle Pleistocene European scimitar cats can be seen on the upper carnassial, which has a preparastyle (Sardella and Iurino 2012), although this varies in size and shape (Domegliara Selvavecchia, Italy, Sardella and Iurino l.c.; Stránská Skála, Czech Republic, Sardella and Iurino l.c.; Hundsheim, Austria, Freudenberg 1914; Artenac, France, Beden et al. 1984). For Middle to Late Pleistocene European homotheres, the post-cranial skeleton possesses a characteristic feature in the distal joint head of the third metacarpal, which is not spherical as is usual in felids, but cylindrical, recalling that of canids and indicating reduced lateral mobility of the toes, but increased cursorial ability (Schütt 1970b).

The taxonomic attribution of these European scimitar cats seems to be beyond dispute. The type series of *H. latidens* from Kents Cavern (UK) cannot be exactly dated because of the possible transportation of the fossil from outside the cave, but is likely to be related in some way to palaeolithic man (McFarlane and Lundberg 2013). It undoubtedly fits the spatiotemporal distribution framework of the European Middle to Late Pleistocene scimitar cats. These forms represent the nominate subspecies if the species *H. latidens* is taxonomically widened to include morphologically different geographic palaeopopulations. Morphologically largely identical mandibles from the North Sea (Reumer et al. 2003), a western homothere outpost, and from Alaska (Antón et al. 2014), an eastern outpost in the Holarctic tundra-steppe (Kahlke 2014), are the latest known representatives of *H. latidens latidens*. The previously mentioned morphological evidence is supported by partial mitogenome sequences from the North Sea mandible, dated to c. 28 ka BP, and two humeri from the Yukon Territory that proved to be beyond the limits of radiocarbon dating (Paijmans et al. 2017).

Inclusion of the eastern Beringian scimitar cat within the nominate subspecies of *H. latidens* raises the question as to the reasonable taxonomic classification of the North American *H. serum*. In the past, this cat was morphognostically grouped with the European *H. latidens* and the Chinese *H. ultimum*, which was described as being notably similar to *H. serum* at a time when the *Homotherium* multispecies concept was state of the art. These three species, European, East Asian and North American, were contrasted with the Early Pleistocene *H. crenatidens* (Churcher 1966).

The American taxon combines several traits typical to *H. latidens latidens*, such as the pocketed rostral margin of the masseteric fossa, the (variable) morphology of the coronoid process, the form of the mental flange, the varying preparastyle structure, and a large anterior mental foramen, but does differ in its broad frontals (Churcher 1966; Antón et al. 2014). The degree of arching of the lower incisors-canine rows is extremely variable. For example, the mandible of a skull, reconstructed with the aid of finds from Texas, Kansas and Oklahoma (Churcher l.c., Fig. 1), displays an evenly staggered configuration for the aforementioned

2.6–1.8 Ma Europe (Ballesio 1963; de Bonis 1976; Sotnikova et al. 2002)	1.8–1.2 Ma Europe (Galobart et al. 2003; Sardella and Iurino 2012; Antón et al. 2014)	1.2–0.9 Ma Untermassfeld (Hemmer 2001; this paper)	1.2–0.9 Ma Semibalki-3 (Sotnikova and Titov 2009)
32.0–35.4 (n = 3)	26.0–36.2 (n = 15)	30.4–c. 34.0 (n = 6)	29.9 (n = 1)
1.2–0.9 Ma Monte Peglia1 (Sardella and Iurino 2012)	1.2–0.9 Ma / 0.9–0.4 Ma Lakhuti-2 (Sharapov 1989)	0.9–0.4 Ma Europe (Freudenberg 1914; Bishop 1982; Sardella and Iurino 2012)	0.9–0.4 Ma Zhoukoudian (Teilhard de Chardin 1936; 1939)
34.0 (n = 1)	27.0–34.3 (n = 5)	29.0 (n = 1)	30.0–32.0 (n = 3)
< 0.4 Ma Alaska (Antón et al. 2014)	< 0.4 Ma North America (Meade 1961)		
30.0–32.0 (n = 3)	25.6–30.6 (n = 10)		

Table 42 Lower carnassial length variability (mm) of the Untermassfeld *Homotherium* compared to those of selected Pleistocene European, Asian and North American scimitar cats, with listed sites for the significant time span of 1.2–0.9 Ma (chronostratigraphic units sensu Kahlke et al. 2011).

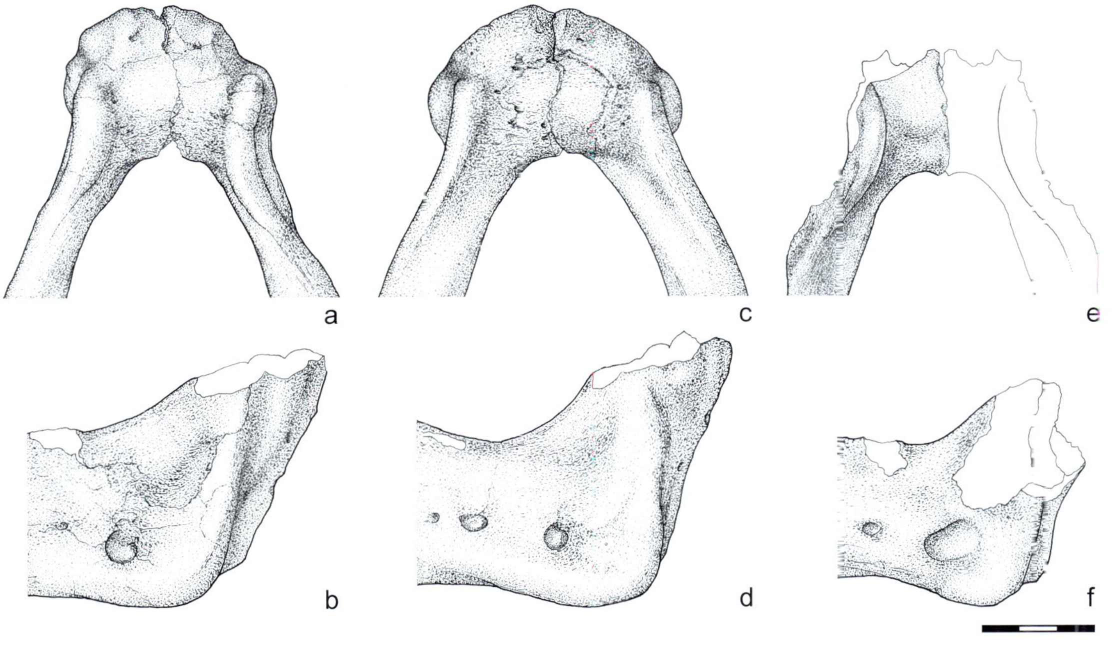

Fig. 34 *Homotherium latidens ultimum* (Teilhard de Chardin), Untermassfeld. – **a–b** Right hemimandible IQW 2009/30239 (Mei. 29401) and left hemimandible IQW 2009/30238 (Mei. 29400). – **c–d** Right hemimandible IQW 1993/24372 (Mei. 23901) and left hemimandible IQW 1997/25985 (Mei. 25514). | *H. latidens latidens* (Owen), North Sea. – **e–f** Right hemimandible (cast) IQW 2011/32128 (Nordsee 31290), graphically supplemented. – **a**, **c**, **e** ventral views; **b**, **d**, **f** right buccal views. – Scale: 30 mm.

teeth that corresponds to the observed arrangement in the Untermassfeld homotheres. In contrast, very little arching is present in a mandible from Friesenhahn Cave (Meade 1961, pl. 2), and certainly no more than that observed in the Middle Villafranchian mandible of the Senèze (France) skeleton (Ballesio 1963, Fig. 13). These observations are accompanied by conspicuous differences in the skull base (de Bonis 1976; Antón et al. 2014). Combined together, the two trait variations lead to a more pronounced morphological variability than is commonly found in a regional series of *Homotherium* remains. This suggests a possible influence of a more archaic North American homothere taxon on the evolutionary history of *H. serum*.

If *H. latidens*, as understood in the former narrow sense, is taxonomically downgraded to the subspecies *H. latidens latidens*, the same procedure seems reasonable for *H. serum*, which should then be ranked as another *H. latidens* subspecies, i.e., *H. latidens serum*. The first attempt to integrate *H. serum* in *H. latidens* on a genomic basis (Pajimans et al. 2017) was unfortunately not based on any fossil from the real, i.e., southern, population of *H. serum*, but used specimens from the northern Beringian population of the Canadian Yukon Territory. Nevertheless, the Beringian *H. latidens latidens* and the southern *H. latidens serum* might have experienced occasional genetic exchange due to individuals of the former taxon advancing south via an ice-free corridor. A DNA fragment extracted from a Yukon specimen corresponds to the DNA recovered from an individual found in the Tyson Spring Cave (Minnesota, USA) dated to c. 26.9 ka, a time when the Wisconsin ice margin was less than 60km away from the site (Barnett et al. 2005; Widga et al. 2012).

Downgrading the former species ranking of the Middle to Late Pleistocene European and North American scimitar cats to the corresponding subspecies of *H. latidens* would then require the same procedure to be applied to the former, similarly separated, species *H. ultimum* of the late Early and early Middle Pleistocene of Zhoukoudian (Beijing, China) Localities 9 and 13, which was considered to be closely related to *H. latidens serum* (Churcher 1966). Consequently, this Chinese form would have to be designated as *H. latidens ultimum*. An incomplete mandible (specimen C) from the probably Early Pleistocene Zhoukoudian Locality 13 (Shen et al. 2015), looks morphognostically very similar to the Untermassfeld and Semibalki *Homotherium* lower jaws (see sections 2.6.1., 2.6.3.). Just like these, it lacks the pocketed anterior margin of the masseteric fossa that is typical of *H. latidens latidens* (Teilhard de Chardin 1939, Fig. 4). The strongly fragmented condition of the holotype mandible from the Middle Pleistocene Zhoukoudian Locality 9 (Teilhard de Chardin 1936, Fig. 6, pl. 1, Fig. 2) unfortunately means that the absence of this essential structure remains unclear. The late Early Pleistocene Middle Asian site Lakhuti-2 (Tajikistan) produced a number of scimitar cat remains that were taxonomically assigned to the separate species *H. teilhardipiveteaui* Sharapov, 1989, and which were judged to be intermediaries between the chronostratigraphically older *H. crenatidens* and the younger European, East Asian and North American homotheres (Sharapov 1989). On the other hand, the Lakhuti scimitar cat has been described as being morphologically close to the Epivillafranchian finds from Semibalki and Untermassfeld (Sotnikova and Titov 2009). Its fragmentary skull is characterised by an evenly staggered arrangement of the teeth within the upper and lower incisors-canine arch, as found in the Untermassfeld *Homotherium* and related forms (Sotnikova and Titov l.c.), as well as in *H. latidens ultimum,* and also partly in *H. latidens serum*. The rostral mental foramen of the Lakhuti mandible opens out into a wide pit, recalling the North Sea find of *H. latidens latidens* and a *H. latidens serum* skull from the Friesenhahn Cave (Rawn-Schatzinger 1992, Fig. 1). The mental flange strongly resembles that of the smaller of the two Untermassfeld mandibles IQW 1993/24372 (Mei. 23901) + IQW 1997/25985 (Mei. 25514) (**Fig. 30g–l**) on the one hand and the corresponding structure of a *H. latidens latidens* mandibular fragment from Hundsheim (Freudenberg 1914) on the other. A thickening in front of the P^4 parastyle is interpreted as rudimentary preparastyle (Sharapov 1989). The lack of a preparastyle is characteristic of all Middle and Late Villafranchian *H. crenatidens*. Whether it is also generally lacking in the Epivillafranchian (i.e., the Untermassfeld) group (see section 2.6.1.), as demonstrated by the Monte Peglia dentition (Sardella and Iurino

2012), cannot be ascertained due to the sparsity of the material. In any case, it is also part of the variability observed in *H. latidens serum* (Churcher 1966). Unfortunately, the rostral part of the masseteric fossa is not preserved in the Lakhuti specimen.

The Middle Asian scimitar cat from Tajikistan discussed here bridges the gap between the Epivillafranchian records from Europe, considered as advanced forms of *Homotherium* (Sotnikova and Titov 2009; Sardella and Iurino 2012), and the later Middle to Late Pleistocene *H. latidens*. It cannot be overlooked, however, that this also applies to the Chinese *H. latidens ultimum*. In fact, the Lakhuti homothere appears as another *H. latidens ultimum*. Due to a largely matching combination of features, this late Early to early Middle Pleistocene form from Middle and East Asia can barely be distinguished from the Untermassfeld and Semibalki individuals. The Untermassfeld homothere is therefore assigned to *H. latidens ultimum*. In doing so, the spatio-temporal extension of this taxon, first described in China, is expanded considerably, from Europe to East Asia and from the Epivillafranchian to the Middle Pleistocene. The earliest European records, including those from Untermassfeld, seem to represent the stem type of *H. latidens ultimum*, while later Chinese fossils from the type locality correspond more closely to a more advanced type.

The attempt to expand the taxon *H. latidens,* as outlined here with three Holarctic subspecies, to earlier, i.e., Middle to Late Villafranchian scimitar cat palaeopopulations, would destroy its coherent taxonomic delimitation. Consequently, the name *H. crenatidens* is kept for the early homotheres, which means that a two-species concept for the genus is again advocated here. It must be kept in mind, however, that a corresponding systematic assignment cannot be made without subjective assessment in cases without biological proof of speciation. Given that isolating mechanisms are the prerequisite for speciation, the postcranial skeletal morphology could hint that a certain pre-mating isolation might have evolved in *H. latidens*.

An evolutionary change in the morphology of the distal metacarpal epiphyses indicates a modified function of these skeletal elements. The comparison of the generally more gracile metacarpals of *H. latidens latidens* from the early Middle Pleistocene sites of Mosbach, Mauer and Hundsheim, paralleled by similarly gracile metacarpals of *H. latidens serum* (Jefferson and Tejada-Flores 1993) with those of the more robust Middle Villafranchian *H. crenatidens* from Senèze, shows that the lateral mobility of the paws was restricted in the former. From this, a better adaptation to fast and persistent running can be deduced for *H. latidens latidens*. Regarding its locomotive characteristics, the morphology of the distal area of the metacarpals functionally places this scimitar cat between hyaenids and canids (Schütt 1970b). Given considerably higher rates of spontaneous locomotion in bears and wolves compared to modern large cats in captive environments (Hemmer 2001), it may be assumed that similar paw structures cause similar behaviour. This would mean that *H. latidens latidens* might have had a higher level of locomotion than *H. crenatidens*, which would have shaped the social behaviour of the two taxa in different ways.

The taxonomic classification given above suggests an outline for the phylogeographic background of the evolutionary history of *H. latidens*. Given the range of variation of Asian *H. crenatidens,* a stem population of scimitar cats with various advanced traits must have evolved during the late Early Pleistocene. This form, taxonomically listed as *H. latidens ultimum*, spread over a vast area ranging from Europe to East Asia during the Epivillafranchian. In terms of a surviving stem subspecies, this group of scimitar cats did not change significantly in the East Asian region up to the early Middle Pleistocene. In temporal proximity to the Early/ Middle Pleistocene boundary, a population of still quite generalised *H. latidens ultimum* changed into a significantly more advanced form with less variable morphological characteristics, taxonomically classified as *H. latidens latidens*. Towards the end of the late Middle Pleistocene an eastern Siberian branch of *H. latidens ultimum* migrated via Beringia to North America and once there dispersed into the south of the region, giving rise to *H. latidens serum*. Ultimately, *H. latidens latidens* successfully adapted to the conditions of the Eurasian mammoth steppe, so that the species crossed the Beringian corridor again and colonised ice-free areas north and west of the large American inland glaciers. This twofold expansion of a large cat species at different evolutionary stages finds a striking parallel in *Panthera leo* (for lion taxonomy see Hemmer 2011). Giant Middle Pleistocene *P. leo fossilis* crossed Beringia first and evolved into the American lion *P. leo atrox* in the south of North America. During the Late Pleistocene glacial period, smaller cave lions, *P. leo spelaea / P. leo vereshchagini*, also migrated to the Nearctic (Hemmer l.c.; Sotnikova and Foronova 2014).

Compared to modern large cats, *H. latidens* had high forequarters, digitigrade forepaws and sloping rear quarters with semiplantigrade hindpaws (described for *H. latidens serum* by Rawn-Schatzinger 1992). It was adapted to running in a way similar to that of hyenas (Rawn-Schatzinger l.c.; see also Antón et al. 2005). This anatomy clearly sets it apart from that of modern large cats and results in a differing array of mass distribution, as was seen when an attempt was made to calculate the body mass using the prediction equations for humerus and femur shaft least transversal diameters (Hemmer 2001). For example, the *H. crenatidens* skeleton from Senèze (Ballesio 1963) had body mass estimations of around 180 and 200 kg when the calculation was based on the humeri (and 200 kg as the mean of the values obtained by the Mc2-Mc5 equations), but 320 and 290 kg when based on the femora. This result, which gives a much lower mass calculation based on the forequarters than that based on the hindquarters was reproduced for *H. latidens serum*, with a humerus based mean of c. 230 kg (n = 5) and a femur-based one of c. 280 kg (n = 11) (measurements after Rawn-Schatzinger 1992). However, none of these values reflect the real body mass of the animals. If the mean of these results is used, it results in a mass of c. 250 kg for the Senèze skeleton, which could represent the mass of a female if the *crenatidens* morphotype denotes females and the *nestianum* type corresponds to males (see section 2.6.1.).

A similar value for the mean body mass calculated using both sexes is obtained for *H. latidens serum*. The prediction equations based on the least diaphyseal transverse diameter of the metacarpals (as calculated by Hemmer et al. 2011) result for the *H. latidens serum* series of measurements (by Rawn-Schatzinger l.c.) in values for the Mc2 that are clearly too high (mean 278 kg), and ones for the Mc5 that are too low (mean 205 kg), whereas the Mc3 and Mc4 both give medium values (means 235 kg respectively 250 kg) that match the humerus values (232 kg), so that these last three equations seem to be useful for mass calculations based on the animal's forequarters. The Mc2 and Mc5 equations for body mass calculations of felids should not be used for *Homotherium*, as demonstrated by a complete set of metacarpals from Pirro Nord (Petrucci et al. 2013). The corresponding calculations result in the following, greatly differing values: Mc2 – Mc3 – Mc4 – Mc5: 207 – 158 – 180 – 101 kg.

In order to obtain an impression of the sexual dimorphism of body mass, the distribution pattern of the mass values obtained using an extended series of measurements for the *H. latidens serum* humerus and corresponding Mc3 and Mc4 (total n = 23, Rawn-Schatzinger 1992) was evaluated. This process resulted in a female peak of around 210 kg and a male peak of around 290 kg, indicating a slightly lighter body than that of the *H. crenatidens* individual from Senèze. A Pirro Nord homothere had an even smaller mass, as determined from the Mc2 and Mc3 based prediction of the forequarter mass.

Taking into account the carnassial size, which reflects the linear body size and which remained unchanged over the whole Pleistocene (**Table 42**), a progressive slimming down is observed, beginning with the stout and robust *H. crenatidens* of the Middle and early Late Villafranchian. This evolutionary trend culminated in the still large, but more gracile and swifter *H. latidens latidens* of the Early Middle Pleistocene. The data provided by the humeri and Mc3 from Hundsheim, Mosbach and Mauer (Freudenberg 1914; Schütt 1970b) suggest a mass for the forequarters of around 140 (110–180) kg, which might include both sexes. Unfortunately, no useable measurements for such calculations are available from the Untermassfeld homotheres and other Epivillafranchian *H. latidens ultimum*.

It has been argued, that for body mass estimates in machairodontines it would be more appropriate to rely on comparative approaches, which, for *Homotherium,* would mean comparing skulls and associated skeletal elements with those of Pleistocene lions (Antón et al. 2005). Indeed, based on cranial and postcranial skeleton predictors, the body mass of Late Pleistocene cave lions has already been estimated to be between 140 and 400 kg, with a supposed female mean of around 200 kg and a supposed male mean of up to 350 kg (Hemmer 2011). These estimates are consistent with the *Homotherium* body mass calculations given here, proving the validity of using feline predictor-based methods on machairodontines.
Predicting the probable size range of prey species which might have been hunted by the Untermassfeld scimitar cats is problematic for two reasons. On the one hand, *Homotherium* was a predator that possessed a build and killing apparatus not paralleled in modern large cats, on the other hand we have insufficient knowledge about the mean body mass of male and female homotheres that once lived in the area around the site and in other localities during the Epivillafranchian. Thus, here, an attempt has been made that uses the little information gained from *H. latidens serum*. Here, the prey focus mass equation (Hemmer 2004) indicates an expected body mass range of between 50 and 1000 kg for the regularly hunted prey species of a 210 kg female scimitar cat, and between 100 and 2000 kg for that of a 290 kg male. These values include juvenile proboscideans, for which the homotheres of the North American *H. latidens serum* population were undoubtedly specialised (Rawn-Schatzinger 1992). At Untermassfeld, a comparable prey was available in the form of *Mammuthus meridionalis*. Rhinos (*Stephanorhinus hundsheimensis*) and hippos (*Hippopotamus antiquus*; but see section 3.3.) would have completed the potential main prey spectrum at this locality, as has already been depicted (Hemmer 2001, Fig. 18)
The specialisation of *H. latidens ultimum* in hunting pachyderms can be clearly seen in the optimisation of the strongly arched lower incisor-canine series. Combined with the upper incisor arch, the conical lower incisors and canines, which meshed perfectly with each other, formed a compact lattice with sharp, aligned points in front of the sabre canine. Like the tearing action performed mainly by pantherine cats (Leyhausen 1973), this apparatus was not only perfectly suited to stabbing and ripping large pieces of meat from compact, long-fibre muscle (Hemmer 2001), but also dragging heavy cadavers without the increased risk of breaking the sabres behind. The daily mean amount of meat consumed by these cats has been estimated using the appropriate prediction equation (Hemmer 2004) to about 7 ½ kg for average-sized females and 9 ½ kg for average-sized males.
Little to no immediate information is available concerning the habitat of the Untermassfeld scimitar cat. The Untermassfeld collection contains a MNI of 5 individuals of very different individual ages (Hemmer 2001; this paper). Based on the data, a flood vulnerability ratio (FVR) of 6 has been calculated, similar to the value obtained for *Panthera onca gombaszoegensis* (see section 2.1.4.). Thus, some familiarity with the possibly destructive effect of running water cannot be excluded for the Untermassfeld homotheres.
The existence of a *Homotherium* cub does not provide the same information regarding the lair within the river basin as has been discussed for the jaguar (see section 2.1.4.). This is because the juvenile scimitar cat was in the latter part of the second half-year of its life (see section 2.6.3.), making it conceivable that it was on a hunting trip with its mother at the time of its death (Rawn-Schatzinger 1983). Thus, riparian landscapes were not necessarily the central habitat of *H. latidens ultimum*. A presumably bear- to wolf-like gait, with a considerably higher spontaneous locomotion than that found in the *Panthera* species, would indicate relatively fast-paced patrolling through an extensive hunting ground rather than slow stalking (Hemmer 2001). Given such a pattern of behaviour, a river valley may have been a welcome continuous landscape component that was repeatedly crossed depending on the expected hunting success on either side of the river. This speculation, supported by the analysis of the metacarpal morphology, is in agreement with the discovery of evidence of positive selection in several genes involved in vision, cognitive function and energy

consumption from comparative genomic analyses of *H. latidens latidens*, which is thereby considered to have been a diurnal, cursorial hunter (Barnett et al. 2020). Following the new discoveries at Untermassfeld, nothing more can be added to the earlier discussion about the presumed type of hunting and killing of homotheres (Hemmer 2001).

3. Conclusions

A synopsis of the results and discussions obtained for the six felid species of the Epivillafranchian fauna from Untermassfeld has given additional insights into pertinent problems of felid taxonomy, the phylogeography of the Untermassfeld cats, and their predator and prey interactions.

3.1. Taxonomy

Species can be equated with metapopulations, i.e., sets of connected subpopulations. The limits of these metapopulations are set by the limits of interbreeding. One solution that has been proposed to eliminate conflicts among rival species theories adopts the general concept of species as separately evolving metapopulation lineages, and treats this aspect as the only necessary property of species (Queiroz 2005). Subspecies are understood as geographic (or chronologic) subpopulations within a metapopulation. These subspecies are distinguished from each other by one or more characters, not necessarily present in all individuals (75 % rule), and which have, or presumably had, limited genetic exchange with other allopatric populations. Extinct species are treated here in the same way as extant ones, i.e., as metapopulations. Unnecessary taxonomic inflation caused by splitting can be avoided if there is no conclusive evidence of reproductive isolation of geographically neighbouring or chronologically consecutive populations, as in the case of jaguars (see section 2.1.1.). On the other hand, a rigorous lumping procedure that ignores any fixed character differences, at least on the level of subpopulations within a metapopulation, does not make sense either. Different ideas concerning the necessary preconditions for the divergence of metapopulations on their way to speciation (see Queiroz 2005, Fig. 2) complicate the discussion. For the felids discussed here, this problem occurs with *Homotherium crenatidens* and *H. latidens* (see section 2.6.4.).
Another possible future controversy could arise for the evolution of the European lynx (see section 2.4.4.). Using morphology as a species criterion, all Iberian lynxes from the Late Villafranchian up to the extant population are to be assigned to *Lynx pardinus* (Boscaini et al. 2016), which correlates well with the splitting of a corresponding metapopulation as expected from genetic testing (Johnson et al. 2006). However, there is very limited evidence on when reduced fecundity (reproductive failure in male hybrids) became an effective post-mating isolation mechanism between sections of evolutionarily diverging *Lynx* metapopulations. The stratigraphic range of *L. valdarnensis spelaeoides* as described from Untermassfeld apparently lasted in the French Pyrenees until the early Middle Pleistocene of Marine Isotope Stage (MIS) 14 (Testu 2006). Not long after this, through contact of the Iberian *L. pardinus* with an eastern lynx subpopulation of a second metapopulation, a new phenotype, *L. spelaeus*, appeared in the aforementioned western Mediterranean region. We can consider the individuals of this form, characterised by various proportions of *L. pardinus* and *L. lynx* features, as a hybrid population of both. This may point to the development of a post-mating barrier between the parental lines, perhaps starting around one million years after their molecular node.
»Species or not a species?« has remained a central question in the taxonomy of fossil felids over the past decades. Another, although much less frequently, asked question was »Genus or not a genus?« The re-

vival of the genus *Sivapanthera* Kretzoi, 1929 for Early Pleistocene cheetahs (Qiu et al. 2004) (see sections 2.2.1., 2.2.4.) is an example of such a discussion. A multispecies concept due to splitting (see above) leads to a situation where some of the named species created by taxonomic inflation are transferred to another genus. A separate ancestral species, in this case *Acinonyx arvernensis* (vs. *A. pardinensis*) in the sense of a separate chrono-metapopulation, in no way justifies its assignment to a separate genus without any further argument. Generally, it seems reasonable to follow the taxonomic concept introduced by molecular assessment of extant felids (Johnson et al. 2006). This concept unites all species of each of the eight evolutionary clades into one single genus if their evolutionary nodes are of Plio-/Pleistocene age. In this sense, Miocene divergence justifies its own generic rank. Unfortunately, this concept neglects the undeniable existence of separate species groups within genera, e.g., the genus *Leopardus* with the morphologically as well as the genetically clearly distinguishable subgenera *Leopardus* and *Oncifelis* (Chimento et al. 2014). The classic taxonomic unit that differs from both the genus and the species level is the subgenus. Downgrading former felid genera to subgenus rank if divergence occurred during the Pliocene completes the taxonomical aids used to express the steps of felid evolution in a conclusive way. This procedure has been chosen here in order to indicate the special position of *Puma pardoides* (subgenus *Viretailurus*), with regard to the extremely different American puma lineages (subgenera Puma, *Miracinonyx* and *Herpailurus* (see section 2.3.1.).

3.2. Phylogeography

When jaguars recovered from Untermassfeld were first described there was no argument against the idea of an intra-European regional evolution from the initial *Panthera onca toscana* to the later *P. onca gombaszoegensis* (Hemmer 2001). This changed with the study of a jaguar mandible from Dmanisi (Georgia) (Hemmer et al. 2010), which supported a Central Asian localisation of the *P. onca gombaszoegensis* root population. Such jaguars could have dispersed into the west, reaching Central Europe during the Epivillafranchian at the latest. However, there is still no indication as to whether these immigrants then mixed there with the last members of *P. onca toscana*, or whether the early European jaguars had already disappeared prior to the arrival of *P. onca gombaszoegensis*.

An Asian origin for the Untermassfeld giant cheetah was discussed in the previous study (Hemmer 2001) and is now confirmed here. The previously existing Early Pleistocene European cheetah subspecies *Acinonyx pardinensis pardinensis* disappeared, and Europe was subsequently repopulated by *A. pardinensis pleistocaenicus*, which has been recovered from earlier sites in East Asia.

The evolutionary history of the Eurasian puma, however, still remains full of unanswered questions. What is clear, however, is that the European population of the early Early Pleistocene *Puma pardoides pardoides* disappeared over the course of the Early Pleistocene and was replaced by the Epivillafranchian *P. pardoides brevipes*. The South Asian *P. pardoides csomai* shows more similarities with the latter subspecies. The dentition of the Untermassfeld puma and that of the Late Pliocene North African Ahl al Oughlam »leopard«, which is older than the European *P. pardoides pardoides*, are morphologically very similar despite an age difference of 1.5 Ma. So, the Early Pleistocene puma repopulation of Europe may have made its way through the Near and Middle East, a region that often acted as a focal point for faunal exchange between Africa and southern Asia.

In contrast to the cheetah and puma, the Epivillafranchian Untermassfeld lynx is the result of regional evolution and not that of influx from outside the continent as its origins lie in the Late Villafranchian lynxes of Europe.

The evolution of the European dirk-toothed cats during the Late Villafranchian, including the origin of *Megantereon adroveri*, has not yet been sufficiently clarified. Replacement of earlier forms by Asian *M. falconeri* or hybridisation and introgression are conceivable scenarios. The European scimitar cat's population transformed towards the end of the Late Villafranchian from *Homotherium crenatidens* to *H. latidens ultimum*, the form recorded at Untermassfeld. This advanced homothere has clear eastern affinities. Its distribution extended over the Azov Sea region in Eastern Europe through Central to East Asia.

As this synopsis shows, the only species of the Untermassfeld felids with undisputed regional European evolutionary history is the smallest of the guild, i.e., the lynx. The three largest species, i.e., the scimitar cat, the jaguar, and the giant cheetah replaced earlier, morphologically different populations of their own (*Panthera*, *Acinonyx*) or of a closely related (*Homotherium*) supra-regional metapopulation. All have Asian affinities. The two medium-sized species, i.e., the puma and the dirk-toothed cat, also took over the European distribution range of predecessor relatives, but their phylogeographical origins remain unclear, although Asian roots are also probable in both cases. Body size seems to be the key to explaining the described situation.

Increasing body size is likely to have led to a decrease in the population density of the respective species. However, the body mass of felids influences about one third of the total variability of their highest population density, whereas habitat influence predominates (Hemmer 2004). Therefore, a body-mass based prediction equation can provide only approximate data on population density for extinct cat species. When the results of a general carnivore regression (Peters 1983, cited after Martin 1990) are additionally taken into account, the following possible density ranges (individuals / 100 km^2) result for the Untermassfeld species: jaguar 5–10, cheetah 6–10, puma 13–14, lynx 22–46, dirk-toothed cat 7–11, scimitar cat 2–7. The lower the population density, the more likely the extinction due to the long-term decline of effective sizes of relic populations caused by extreme deterioration of relevant climatic parameters. Herbivores usually have considerably higher population densities than carnivores. Their survival in refugial areas during periods of unfavourable ecological change is therefore more probable. Thus, the appearance or disappearance of large carnivores could indicate ecologically induced changes in biodiversity more sensitively than contemporaneous herbivores.

3.3. Behavioural ecology

Three herbivore species with very high minimum numbers of individuals (Kahlke in this volume) have been identified as the main food resource for the Untermassfeld carnivores, namely, the bovid *Bison* (*Poephagus*) *menneri* (Sher 1997, Bukhsianidze 2020) and the two cervids *Eucladoceros giulii* and *Pseudodama vallonnetensis* (H.-D. Kahlke 1997, Breda et al. 2020). The six differently sized and specialised cat species of the Untermassfeld fauna were outstanding hunters. Their presumable interspecific dominance hierarchy, as controlled by corresponding specific behavioural ecology and body size, has already been discussed in detail and depicted graphically (Hemmer 2001). The latest finds from Untermassfeld have not led to any substantial change in the relevant conclusions. However, advances have been made in the means of determining the probable prey size hunted by the individual species. The development of a prediction equation based on the cats' body mass now enables the estimation of the prey focus mass alongside a broader spectrum of less frequently utilised prey, made up of either significantly smaller or possibly even larger-sized animals. This, in turn, illustrates the widely diversified capacity of each felid to act as opportunistic hunters (Hemmer 2004, 2007a).

Body mass predictions for the three most common Untermassfeld artiodactyl species were determined using equations for the highly mass correlated proximal transverse diameters of metacarpals and metatarsals

(Scott 1990). *Bison menneri*, the archaic yak from Untermassfeld (Bukhsianidze 2020), is represented by a large number of sexed metapodials (Sher 1997). Using the bovid prediction equation, a body mass variability of c. 650 to < 950 kg, with a mean of c. 800 kg for adult females, and of > 900 to < 1,300 kg, with a mean of c. 1,100 kg for adult males was determined. The latter range is similar to that of adult males of the extant wild yak, *B.* (*Poephagus*) *mutus*, with > 800 to 1,200 kg (Leslie and Schaller 2009). Interpolation based on the birth and adult masses of female and male extant wisent, *B. bonasus* (Pucek 1986), and of three months old wild yak calves (Leslie and Schaller l.c.), leads to an estimate for the birth mass of *B. menneri* at around 45 kg. Yearlings would have reached the 200 kg class.
Body mass estimations based on the cervid equation (Scott 1990; basic data after H.-D. Kahlke 1997) provide no clearly separable female and male ranges for the fairly large *Eucladoceros giulii*. The most probable assignment of the resulting mass values indicates 360–430 kg for adult females and 470–580 kg for adult males. These are much higher mass estimations than those previously published (Kahlke and Gaudzinski 2005, Fig. 10) that were based on dentition, whereas the metapodial cross-sections used here better reflect the mass of the body (see sections 2.1.4., 2.2.4.). The completely documented data of the newly bred Neumühle-Riswicker domestic fallow deer (Hemmer 2015) were used as a reference in order to obtain clues about the individual development of the body mass of the two common Untermassfeld cervids. In this way, a birth mass of ≥ 25 kg and a body mass of 200 kg can be specified for *E. giulii* towards the end of the first year of life. Based on metapodial dimensions (H.-D. Kahlke 1997), the following estimations of body masses result for *Pseudodama vallonnetensis*: adult females 90–110 kg, adult males 120–160 kg, birth mass ≥ 7 kg, reaching 50 kg during the first year.
Table 43 compiles the prey size demands of both sexes of the Untermassfeld felid species with respect to the artiodactyls frequently detected at the site. The most extensive overlap of the reconstructed prey size spectra is observed in the 50 and 100 kg weight classes, followed by the 10, 20, and 200 kg classes. Top requirements in these prey categories meet top availabilities of the previously mentioned predominant large herbivore species. Only the 10 kg class, and to some extent, the 20 kg class are not sufficiently covered. Accordingly, female lynxes and, occasionally, the males need to hunt smaller prey. The correspondence between artiodactyl resources and felid food requirements does not mean, however, that all recorded cat species were only dependent on the three named prey species. As opportunistic hunters, the three big felines (*Panthera*, *Acinonyx*, *Puma*) would also have killed other types of prey, depending on availability. For machairodontines (*Megantereon*, *Homotherium*), which were highly specialised in different ways, other types of prey spectra would have been in focus.
The main feature that *Megantereon* and *Homotherium* have in common are their excessively high-crowned upper canines. They are easily distinguished from each other, however, owing to their different body shapes. The elongated canines would have borne a significantly increased risk of breaking compared to those of the conical-toothed felids (Van Valkenburgh and Ruff 1987). Two possible alternative strategies would have minimised this risk. The first involved bringing the prey to the ground and holding it tightly to prevent muscular movement while the canines were driven in. This method would have been safest when solitary prey was hunted, so that combative herd members, such as old male bovids, would not have attacked the cat. This strategy would have been utilised by the dirk-toothed cat with its particularly powerful forequarters, which becomes apparent when looking at the described body mass overestimation based on the humerus (see section 2.6.4.). An alternative killing method involved a rapid stabbing strike of the long, flat and sharp canine blade into the most vulnerable part of the prey, followed by its immediate withdrawal before a defensive movement of the victim could damage the tooth. Such a strategy was obviously applied by the scimitar cat, which with its more hyena-like appearance was not built for holding prey down for any length of time, as demonstrated by the body mass underestimation from the forequarter prediction equation (see

Prey size class (kg)	Assignment of the Untermassfeld felids to prey size classes								*Bison*	*Eucladoceros*	*Pseudodama*
0.5	L♀										
1	L♀	L♂									
2	Pu♀	L♀	L♂								
5	Pu♀	**L**♀	L♂	*M*♀							
10	Pa♀	A♀♂	Pu♀	*Pu*♂	L♀	**L**♂	*M*♀				newborn
20	Pa♀	A♀♂	**Pu**♀	*Pu*♂	L♂	*M*♀	M♂			newborn	< 1 year
50	Pa♀	Pa♂	A♀♂	Pu♀	*Pu*♂	***M***♀	M♂	H♀	newborn	< 1 year	≤ 1 year
100	**Pa**♀	Pa♂	**A**♀♂	***Pu***♂	*M*♀	M♂	H♀	H♂	< 1 year	< 1 year	adult ♀♂
200	Pa♀	Pa♂	**A**♀♂	*Pu*♂	**M**♂	H♀	H♂		1 year	≤ 1 year	adult ♂
500	**Pa**♂	A♀♂	M♂	**H**♀	H♂				adult ♀	adult ♀♂	
1000	Pa♂	H♀	**H**♂						adult ♀♂		
2000	H♂										

Table 43 Prey size demand of the Untermassfeld felid guild in relation to the most frequently represented (MNI) artiodactyls (prey size classes after Hemmer 2004). – L *Lynx valdarnensis spelaeoides*. – Pu *Puma pardoides brevipes*. – M *Megantereon adroveri*. – A *Acinonyx pardinensis pleistocaenicus*. – Pa *Panthera onca gombaszoegensis*. – H *Homotherium latidens ultimum*. – Bold: prey focus mass. – Italics: presumed prey size class assignment according to fossil assemblages from other sites, in cases where no corresponding felid individual was recovered at Untermassfeld.

section 2.6.4.). Lightning attacks of this kind might even have been possible when a potential prey individual, for example a juvenile elephant, moved just that little bit too far away from a protecting herd.

The Untermassfeld southern mammoths, *Mammuthus meridionalis*, originally attributed to *M. trogontherii trogontherii* (Dubrovo 2001; species attribution revised by Van Essen in this volume), were giants. A shoulder height of 350 cm was estimated for a female, and 440 cm for a male (Dubrovo 2001). A number of prediction equations (compiled by Roth 1990, Table 9.2) result in body mass estimations for the female of around 6.5 t. Corresponding calculations for the male animal, however, lead to possibly excessive values of around 12.5 t, in view of massive size extrapolation. Two-year-old calves of the African forest elephant, *Loxodonta africana cyclotis*, have shoulder heights of around 135 cm and body masses of some 500 kg (compilation in Haltenorth and Diller 1977). Based on the given shoulder height of the Untermassfeld female of 350 cm and the maximum height of adult female forest elephants of 260 cm (Haltenorth and Diller l.c.), for a two-year-old Untermassfeld mammoth a body mass of around 1,000 kg can be estimated. This is the exact prey focus mass predicted for male homotheres. Calves of extant African and Asian elephants begin to separate from the maternal herd at the age of two years, at which point they are at an increased risk of carnivore attacks. The determination of *Mammuthus* cf. *columbi* calves of this exact age as the main food source of *Homotherium latidens serum* in the Texas Friesenhahn cave underscores the close relationship between hunting behavior and predator prey focus mass for mammoths and scimitar cats (compilation by Rawn-Schatzinger 1992).

The Untermassfeld large mammal community included two more pachyderms, also presumably hunted by the machairodonts, i.e., the rhinoceros *Stephanorhinus hundsheimensis* (H.-D. Kahlke 2001, Kotowski et al. 2020) and *Hippopotamus antiquus* (R.-D. Kahlke 1997b, 2001c), both present in quite large numbers (Kahlke in this volume). The skull of an adult female Untermassfeld rhino has dimensions that mostly fall within the upper range of extant Indian (*Rhinoceros unicornis*) and black rhinos (*Diceros bicornis*) (Kotowski et al. 2020), which might indicate a body mass of 1.5–2 t for the former. The birth mass of the Untermassfeld rhinos would have fallen within the 50 kg body mass class (comparative data for the extant species in Haltenorth and Diller 1977; Laurie et al. 1983). Thus, the mass of the Untermassfeld rhinos fits the regularly hunted prey spectrum of *Homotherium* while the juveniles matched that of *Megantereon* (**Table 43**). The Untermassfeld hippo was larger and heavier than the extant *H. amphibius*. For adult males a body mass > 3 t was estimated (R.-D. Kahlke 1997b). The birth mass would have been in the 50 kg class (for *H. amphibius* see Haltenorth and Diller 1977). Extant hippos regularly leave the water's edge to graze nocturnally, ranging from relatively close by to much farther afield. If *H. antiquus* did the same, it would not have been a hunting target for the scimitar cat, which was presumably a diurnal hunter (Barnett et al. 2020). Adult hippos were too large for *Megantereon*, although the dirk-toothed cat might have occasionally tried to prey on the juveniles.

Acknowledgements

Work on the manuscript of this publication has been supported in numerous ways by the highly professional experts at the Senckenberg Research Station of Quaternary Palaeontology Weimar. Following further excavations carried out by the Weimar Institute, which took place over two decades and without any loss, the often time-consuming preparation and conservation work was then expertly performed by D. Rössler, M. Stache, and R. Michalik (all Weimar). Special thanks should also go to G. Utschig (Weimar) for his work identifying some essential finds in the, as yet, unanalysed part of the Untermassfeld collection, and for compiling find data of felid remains, carrying out a number of measurements on the material, and procuring comparative material as an aid to identification. T. Korn and S. Döring (both Weimar) are thanked for taking the numerous photographs necessary for this publication along with E. Haase (Weimar), who created the drawings, performed the digital processing of the photographs, and arranged the images for all of the illustrations. Special thanks go to D. Lordkipanidze and A. K. Vekua † (both Tbilisi), who graciously lent important comparative material from the Georgian excavation sites Kvabebi and Dmanisi. We also thank D. Geraads (Paris), who donated photographs and casts of felid dentitions from the Moroccan site of Ahl al Oughlam for comparative purposes, and L. Kirillova and M. Melnikova (both Moscow), who provided access to rare literature. B. Fiedler (Weimar) kindly procured numerous references and edited the final layout of the manuscript. And lastly, we are very grateful to C. Nielsen-Marsh (Leipzig) for language editing of the text. We would also like to thank D. J. Álvarez-Lao (Oviedo), who gave us valuable information as part of the reviewing.

References

Adams, D., 1979. The Cheetah: Native American. Science 205, 4411, 1155–1158.

Antón, M., Galobart, A., 1999. Neck function and predatory behavior in the scimitar toothed cat *Homotherium latidens* (Owen). Journal of Vertebrate Paleontology 19, 771–784.

Antón, M., Galobart, A., Turner, A., 2005. Co-existence of scimitar-toothed cats, lions and hominins in the European Pleistocene. Implications of the post-cranial anatomy of *Homotherium latidens* (Owen) for comparative palaeoecology. Quaternary Science Reviews 24, 1287–1301.

Antón, M., Salesa, M.J., Galobart, A., Tseng, Z.J., 2014. The Plio-Pleistocene scimitar-toothed felid genus *Homotherium* Fabrini, 1890 (Machairodontinae, Homotherini): diversity, palaeogeography and taxonomic implications. Quaternary Science Reviews 96, 259–268.

Argant, A., 2004. Les Carnivores du gisement Pliocène final de Saint-Vallier (Drôme, France). In: Faure, F., Guérin, C. (Eds.), Le gisement pliocène final de Saint-Vallier (Drôme, France). Geobios 37, Supplement 1, S133–S182.

Argant, A., Argant, J., 2011. The *Panthera gombaszogensis* story: the contribution of the Château Breccia (Saône-et-Loire, Burgundy, France). In: Argant, A., Argant, J. (coord.), Proceedings of the 16th International Cave Bear and Lion Symposium, Azé (Saône-et-Loire, Srancw), 22–26 septembre 2010. Quaternaire, Hors-série 4, 247–269.

Argant, A., Argant, J., Jeannet, M., Erbajeva, K., 2007. The big cats of the fossil site Château Breccia Northern Section (Saône-et-Loire, Burgundy, France): stratigraphy, palaeoenvironment, ethology and biochronological dating. In: Kahlke, R.-D., Maul, L.C., Mazza, P. (Eds.): Late Neogene and Quaternary biodiversity and evolution: Regional developments and interregional correlations. Proceedings of the 18th International Senckenberg Conference (VI International Palaeontological Colloquium in Weimar). Volume II. Courier Forschungsinstitut Senckenberg 259, 121–140.

Arribas, A., Garrido, G., 2008. Machairodontinae [*Megantereon cultridens roderici* n. ssp. and *Homotherium latidens* (Owen, 1846)] at the Fonelas P-1 site (Guadix Basin, Granada). In: Arribas, A. (Ed.): Vertebrados del Plioceno superior terminal en el suroeste de Europa: Fonelas P-1 y al Proyecto Fonelas. Cuadernos del Museo Geominero 10, 251–277.

Arroyo-Cabrales, J., 2002. Registro fósil del jaguar. In: Medellín, R.A., Equihua, C.A., Chetkiewicz, C., Crawshaw Jr., P.G., Rabinowitz, A., Redford, K.H., Robinson, J.G., Sanderson, E.W., Taber, A.B. (Eds.), El Jaguar en el Nuevo Milenio. México, 343–354.

Backhouse, J., Lydekker, R., 1886. On a Mandible of *Machaerodus* from the Forest-bed. Quarterly Journal of the Geological Society 42, 309–313.

Ballesio, R., 1963. Monographie d'un Machairodus du gisement villafranchien de Senèze: *Homotherium crenatidens* Fabrini. Travaux du Laboratoire de Géologie de la Faculté des Sciences de Lyon, Nouvelle Série 9, 1–129, pls. 1–4.

Barnett, R., Barnes, I., Phillips, M.J., Martin, L.D., Harington, C.R., Leonard, J.A., Cooper, A., 2005. Evolution of the extinct Sabretooths and the American cheetah-like cat. Current Biology 15, 589–590.

Barnett, R., Westbury, M.V., Sandoval-Velasco, M., Vieira, F.G., Jeon, S., Zazula, G., Martin, M.D., Ho, S.Y.W., Mather, N., Gopalakrishnan, S., Ramos-Madrigal, J., de Manuel, M., Zepeda-Mendoza, M.L., Antunes, A., Baez, A.C, de Cahsan, B. Larson, G., O'Brien, S.J., Eizirik, E., Johnson W.E., Koepfli, K.-P., Wilting, A., Fickel, J., Dalén, L., Lorenzen, E.D., Marques-Bonet, T., Hansen, A.J., Zhang, G., Bhak, J., Yamaguchi, N., Thomas, M., Gilbert, M.T.P., 2020. Genomic Adaptations and Evolutionary History of the Extinct Scimitar-Toothed Cat, *Homotherium latidens*. Current Biology 30, 1–8.

Barry, J.C., 1987. Large carnivores (Canidae, Hyaenidae, Felidae) from Laetoli. In: Leakey, M.D., Harris, J.M. (Eds.), Laetoli: A Pliocene site in Northern Tanzania. Oxford, 235–258.

Baryshnikov, G.F., 2011. Pleistocene Felidae (Mammalia, Carnivora) from the Kudaro Paleolithic cave sites in the Caucasus. Proceedings of the Zoological Institute of the Russian Academy of Sciences 315, 197–226.

Beden, M., de Bonis, L., Brunet, M., Tournepiche, J.-F., 1984. Première découverte d'un félin machairodonte dans le Pleistocène moyen des Charentes. Comptes Rendus de l'Académie des Sciences Paris 298, Séries II, 241–244.

Beltrán, J.F., Delibes, M., 1993. Physical Characteristics of Iberian Lynxes (*Lynx pardinus*) from Doñana, Southwestern Spain. Journal of Mammalogy 74, 852–862.

Bishop, M.J., 1982. The Mammal Fauna of the Early Middle Pleistocene Cavern infill Site of Westbury-Sub-Mendip, Somerset. Special Papers in Palaeontology 28, 1–108.

Bonifay, M.-F., 1971. Carnivores Quaternaires du Sud-Est de la France. Mémoires du Muséum National d'Histoire Naturelle, Nouvelle Série, Série C, 21, 43–377.

Boscaini, A., Madurell-Malapeira, J., Llenas, M., Martínez-Navarro, B., 2015. The origin of the critically endangered Iberian lynx: Speciation, diet and adaptive changes. Quaternary Science Reviews 123, 247–253.

Boscaini, A., Alba, D.M., Beltrán, J.F., Moyà-Solà, S., Madurell-Malapeira, J., 2016. Latest Early Pleistocene remains of *Lynx pardinus* (Carnivora, Felidae) from the Iberian Peninsula: Taxonomy and evolutionary implications. Quaternary Science Reviews 143, 96–106.

Breda, M., Kahlke, R.-D., Lister, A., 2020. New results on cervids from the Early Pleistocene site of Untermassfeld. In: Kahlke, R.-D. (Ed.), The Pleistocene of Untermassfeld near Meiningen (Thüringen, Germany). Part 4. Monographien des Römisch-Germanischen Zentralmuseums 40, 4, 1197–1249.

Bukhsianidze, M., 2020. New results on bovids from the Early Pleistocene site of Untermassfeld. In: Kahlke, R.-D. (Ed.), The Pleistocene of Untermassfeld near Meiningen (Thüringen, Germany). Part 4. Monographien des Römisch-Germanischen Zentralmuseums 40, 4, 1169–1195.

Carlon, B., Hubbard, C., 2012. Hip and Thigh Anatomy of the Clouded Leopard (*Neofelis nebulosa*) with Comparisons to the Domestic Cat (*Felis catus*). The Anatomical Record 295, 577–589.

Cherin, M., Iurino, D.A., Sardella, R., 2013a. Earliest occurrence of *Puma pardoides* (Owen, 1846) (Carnivora, Felidae) at the Plio/Pleistocene transition in western Europe: New evidence from the

Middle Villafranchian assemblage of Montopoli, Italy. Comptes Rendus Palevol 12, 165–171.

2013b. New well-preserved material of *Lynx issiodorensis valdarnensis* (Felidae, Mammalia) from the Early Pleistocene of Pantalla (central Italy). Bolletino della Società Paleontologica Italiana 52, 103–111.

Cherin, M., Iurino, D.A., Sardella, R., Rook, L., 2014. *Acinonyx pardinensis* (Carnivora, Felidae) from the Early Pleistocene of Pantalla (Italy): predatory behavior and ecological role of the giant Plio-Pleistocene cheetah. Quaternary Science Reviews 87, 82–97.

Cherin, M., Iurino, D.A., Zanatta, M., Fernandez, V., Paciaroni, A., Petrillo, C., Rettori, R., Sardella, R., 2018. Synchrotron radiation reveals the identity of the large felid from Monte Argentario (Early Pleistocene, Italy). Scientific Reports 8, 8338.

Chimento, N.R., Derguy, M.R., Hemmer, H., 2014. *Puma* (*Herpailurus*) *pumoides* (Castellanos, 1958) nov. comb. Comentarios sistemáticos y registro fossil. Estudios Geológicos, Serie Correlación Geológica 30, 92–134.

Christiansen, P., Adolfssen, J.S., 2007. Osteology and ecology of *Megantereon cultridens* SE311 (Mammalia; Felidae; Machairodontinae), a sabrecat from the Late Pliocene – Early Pleistocene of Senéze, France. Zoological Journal of the Linnean Society 151, 833–884.

Churcher, C.S., 1966. The affinities of *Dinobastis serus* Cope 1893. Quaternaria 8, 263–275.

Cipullo, A., 2010. L'evoluzione del genere *Lynx* dal Pliocene al Pleistocene Medio nell'area mediterranea. PhD thesis, Università degli Studi di Modena e Reggio Emilia, Italy.

Davis, B.W., Li, G., Murphy, W.J., 2010. Supermatrix and Species Tree Methods Resolve Phylogenetic Relationships Within the Big Cats, *Panthera* (Carnivora: Felidae). Molecular Phylogenetics and Evolution 56, 64–76.

de Beaumont, G., 1980. Säugerfaunen von der Grenze Pliozän/Pleistozän aus Rheinhessen 2. Les carnivores du Quaternaire ancien de Neuleiningen (Pfalz). Mainzer geowissenschaftliche Mitteilungen 8, 7–16.

de Bonis, L., 1976. Un félidé à longues canines de la colline de Perrier (Puy-de-Dôme): ses rapports avec les félinés machairodontes. Annales de Paléontologie, Vertébrés 62, 159–198.

de Lumley, H., Kahlke, H.-D., Moigne, A.-M., Moulle, P.-É., 1988. Les faunes de grands Mammifères de la Grotte du Vallonet Roquebrune-Cap-Martin, Alpes-Maritimes. L'Anthropologie 92, 465–496.

Dennell, R.W., Turner, A., Coard, R., Beech, M., Anwar, M., 2005. Two Upper Siwalik (Pinjor Stage) fossil accumulations from localities 73 and 362 in the Pabbi Hills, Pakistan. Journal of the Palaeontological Society of India 50, 101–111.

Driesch, A. v. d., 1976. A guide to the measurement of animal bones from archaeological sites. Peabody Museum Bulletin 1, III–IX, 1–137.

Dubrovo, I.A., 2001. Remains of Elephantidae from the Lower Pleistocene site of Untermassfeld. In: Kahlke, R.-D. (Ed.), Das Pleistozän von Untermaßfeld bei Meiningen (Thüringen). Teil 2. Monographien des Römisch-Germanischen Zentralmuseums 40, 2, 589–605, pls. 95–101.

Edwards, C.J., Suchard, M.A., Lemey, P., Welch, J.J., Barnes, I., Fulton, T.L., Barnett, R., O'Conell, T.C., Coxon, P., Monaghan, N., Valdiosera, C.E., Lorenzen, E.D., Willerslev, E., Baryshnikov G.F., Rambaut, A., Thomas, M.G., Bradley, D.G., Shapiro, B., 2011. Ancient Hybridization and an Irish Origin for the Modern Polar Bear Matriline. Current Biology 21, 1251–1258.

Eizirik, E., Kim, J.-H., Menotti-Raymond, M., Crawshaw, P.G., O'Brien, S.J., Johnson, W.E., 2001. Phylogeography, population history and conservation genetics of jaguars (*Panthera onca*, Mammalia, Felidae). Molecular Ecology 10, 65–79.

Ewer, R.F., 1955. The fossil carnivores of the Transvaal caves: Machairodontinae. Proceedings of the Zoological Society of London 125, 587–615.

Fabrini, E., 1896. La Lince del Plioceno italiano. Palaeontographia Italica 2, 1–24, pls. I–III.

Ficcarelli, G. 1979: The Villafranchian machairodonts of Tuscany. Palaeontographia Italica 71, Nova Seria 31, 17–26.

Ficcarelli, G., 1984. The Villafranchian cheetahs from Tuscany and remarks on the dispersal and evolution of the genus *Acinonyx*. Palaeontographia Italica 73, Nova Seria 41, 94–103.

Ficcarelli, G., Torre, D., 1968. Upper Villafranchian panthers of Tuscany. Palaeontographia Italica 64, Nova Seria 34, 137–184, pls. XIX–XXXIV.

1977. Phyletic relationships between *Lynx* group *issiodorensis* and *Lynx pardina*. Bollettino della Società Paleontologica Italiana 16, 197–202.

Figueiró, H.V., Li, G., Trindade, F.J., Assis, J., Pais, F., Fernandes, G., Santos, S.H.D., Hughes, G.M., Komissarov, A., Antunes, A., Trinca, C.S., Rodrigues, M.R., Linderoth, T., Bi, K., Silveira, L., Azevedo, F.C.C., Kantek, D., Ramalho, E., Brassaloti, R.A., Villela, P.M.S., Nunes, A.L.V., Teixeira, R.H.F., Morato, R.G., Loska, D., Saragüeta, P., Gabaldón, T., Teeling, E.C., O'Brien, S.J., Nielsen, R., Coutinho, L.L., Oliveira, G., Murphy, W.J., Eizirik, E., 2017. Genome-wide signatures of complex introgression and adaptive evolution in the big cats. Science Advances 3, e1700299.

Freudenberg, W., 1914. Die Säugetiere des älteren Quartärs von Mitteleuropa mit besonderer Berücksichtigung der Fauna von Hundsheim in Niederösterreich nebst Ausführungen über verwandte Formen von Mauer, Mosbach, Cromer und anderen Fundorten. Geologische und Palaeontologische Abhandlungen, Neue Folge 12, 453–671, pls. 1–20.

Galobart, A., Pons-Moyà, J., Antón, M., Maroto, J., 2003. Descripción del material de *Homotherium latidens* (Owen) de los yacimientos del Pleistoceno inferior de Incarcal (Girona, NE de la Península Ibérica). Paleontologia i Evolució 34, 99–141.

Gálvez Prada, F., Serrano, J.P., Sañudo Franquelo, B., Beltrán Gala, J.F., 2013. Atlas del esqueleto del lince ibérico (*Lynx pardinus*) [online]. Available: https://thevirtualmuseumoflife.com/lynx/lynx-es.html and http://thevirtualmuseumoflife.com/lynx/index-en.php. [Accessed 09 January 2019]

Garrido, G., 2008. Linces y guepardos (Felidae, Carnivora, Mammalia) en el Plioceno superior terminal del Fonelas P-1 (Cuenca de Guadix, Granada). In: A. Arribas (Ed.), Vertebrados del Plioceno superior terminal en el suroeste de Europa: Fonelas P-1 y el Proyecto Fonelas. Cuadernos del Museo Geominero 10, 231–249.

Geraads, D., 1997. Carnivores du Pliocène terminal de Ahl al Oughlam (Casablanca, Maroc). Geobios 30, 127–164.

2014. How old is the cheetah skull shape? The case of *Acinonyx pardinensis* (Mammalia, Felidae). Geobios 47, 39–44

Gonyea, W.J., 1976. Behavioral Implications of Saber-Toothed Felid Morphology. Paleobiology 2, 332–342.

Greenwell, R., 1986. Onza Specimen Obtained-Identity Being Studied. The ISC Newsletter 5, 1–6.

Grohé, C., Lee, B., Flynn, J.J., 2018. Recent inner ear specialization for high-speed hunting in cheetahs. Scientific Reports 8, 2301.

Haas, S.K., Hayssen, V., Krausman, P.R., 2005. *Panthera leo*. Mammalian Species 762, 1–11.

Haltenorth, T., Diller, H., 1977. Säugetiere Afrikas und Madagaskars. München, Bern, Wien.

Hemmer, H., 1966. Untersuchungen zur Stammesgeschichte der Pantherkatzen (Pantherinae). Teil I. Veröffentlichungen der Zoologischen Staatssammlung München 11, 1–121.

1968. Untersuchungen zur Stammesgeschichte der Pantherkatzen (Pantherinae). Teil II: Studien zur Ethologie des Nebelparders *Neofelis nebulosa* (Griffith 1821) und des Irbis *Uncia uncia* (Schreber 1775). Veröffentlichungen der Zoologischen Staatssammlung München 12, 155–247.

1971. Zur Charakterisierung und stratigraphischen Bedeutung von *Panthera gombaszoegensis* (Kretzoi, 1938). Neues Jahrbuch für Geologie und Paläontologie, Monatshefte 1971, 701–711.

1972. Zur systematischen Stellung von »*Jansofelis vaufreyi*« Bonifay, 1971, und »*Felis lunellensis*« Bonifay, 1971, aus dem Pleistozän Südfrankreichs (Carnivora, Felidae). Neues Jahrbuch für Geologie und Paläontologie, Monatshefte 1972, 215–223.

1974. Körperproportionsunterschiede bei Pantherkatzen (Pantherinae). Säugetierkundliche Mitteilungen 22, 233–236.

1976. Fossil history of living Felidae. In: Eaton; R.L. (Ed.), The World's Cats, Volume III, 2. Seattle, 1–14.

1978. The evolutionary systematics of living Felidae: present status and current problems. Carnivore 1, 71–79.

1979. Gestation period and postnatal development in felids. Carnivore 2, 90–100.

2001. Die Feliden aus dem Epivillafranchium von Untermaßfeld. In: Kahlke, R.-D. (Ed.), Das Pleistozän von Untermaßfeld bei Meiningen (Thüringen). Teil 3. Monographien des Römisch-Germanischen Zentralmuseums 40, 3, 699–782, pls. 132–143.

2003. Pleistozäne Katzen Europas – eine Übersicht. Cranium 20, 6–22.

2004. Notes on the ecological role of European cats (Mammalia: Felidae) of the last two million years. In: Baquedano, E., Rubio Jara, S. (Eds.), Miscelánea en homenaje a Emiliano Aguirre. Volume II. Paleontología. Zona Arqueológica 4, 2, 214–232.

2007a. Saber-tooth cats and cave lions – from fossils to felid performance and former living communities. In: Kahlke, R.-D., Maul, L.C., Mazza, P.P.A. (Eds.), Late Neogene and Quaternary biodiversity and evolution: Regional developments and interregional correlations. Proceedings of the 18th International Senckenberg Conference (VI International Palaeontological Colloquium in Weimar). Volume II. Courier Forschungsinstitut Senckenberg 259, 1–12.

2007b. Estimation of Basic Life History Data of Fossil Hominoids. In: Henke, W., Tattersall, I. (Eds.), Handbook of Paleoanthropology, Vol. 1: Principles, Methods, and Approaches. Heidelberg, 587–619.

2011. The story of the cave lion – *Panthera leo spelaea* (Goldfuss, 1810) – a review. Quaternaire, Hors-série 4, 201–208.

2015. Neumühle-Riswicker Damhirsche. Münster.

Hemmer, H., Heidtke, U.H.J., 2013. Die Pfalz – eine frühe Station auf dem langen Weg des Jaguars, *Panthera onca* (Linnaeus, 1758), nach Südamerika. Jahresberichte und Mitteilungen des oberrheinischen geologischen Vereins, Neue Folge 95, 373–390.

Hemmer, H., Schütt, G., 1969. Ein Unterkiefer von *Panthera gombaszoegensis* (Kretzoi, 1938) aus den Mosbacher Sanden. Mainzer Naturwissenschaftliches Archiv 8, 90–101.

Hemmer, H., Kahlke, R.-D., Keller, T., 2003. *Panthera onca gombaszoegensis* (Kretzoi, 1938) aus den frühmittelpleistozänen Mosbach-Sanden (Wiesbaden, Hessen, Deutschland) – Ein Beitrag zur Kenntnis der Variabilität und Verbreitungsgeschichte des Jaguars. Neues Jahrbuch für Geologie und Paläontologie, Abhandlungen 229, 31–60.

2008. Cheetahs in the Middle Pleistocene of Europe: *Acinonyx pardinensis* (sensu lato) *intermedius* (Thenius, 1954) from the Mosbach sands (Wiesbaden, Hesse, Germany). Neues Jahrbuch für Geologie und Paläontologie, Abhandlungen 249, 345–356.

Hemmer, H., Kahlke, R.-D., Vekua, A.K., 2004. The Old World puma – *Puma pardoides* (Owen, 1846) (Carnivora: Felidae) – in the Lower Villafranchian (Upper Pliocene) of Kvabebi (East Georgia, Transcaucasia) and its evolutionary and biogeographical significance. Neues Jahrbuch für Geologie und Paläontologie, Abhandlungen 233, 197–231.

2010. *Panthera onca georgica* ssp. nov. from the Early Pleistocene of Dmanisi (Republic of Georgia) and the phylogeography of jaguars (Mammalia, Carnivora, Felidae). Neues Jahrbuch für Geologie und Paläontologie, Abhandlungen 257, 115–127.

2011. The cheetah *Acinonyx pardinensis* (Croizet et Jobert, 1828) s.l. at the hominin site of Dmanisi (Georgia) – A potential prime meat supplier in Early Pleistocene ecosystems. Quaternary Science Reviews 30, 2703–2714.

Heptner, V.G., Sludskij, A.A., 1980. Die Säugetiere der Sowjetunion. Band III: Raubtiere (Feloidea). Jena.

Homyack, J.A., Vashon, J.H., Libby, C., Lindquist, E.L., Loch, S., McAlpine, D.F., Pilgrim, K.L., Schwartz, M.K., 2008. Canada Lynx-bobcat (*Lynx canadensis* x *L. rufus*) Hybrids at the Southern Periphery of Lynx range in Maine, Minnesota and New Brunswick. American Midland Naturalist 159, 504–508.

Hu, C., Qi, T., 1978. Gongwangling Pleistocene mammalian fauna of Lantian, Shaanxi. Palaeontologia Sinica 155, New Series C 21, 1–64, pls. 1–15.

Huck, M., Fernandez-Duque, E., 2012. Survival of a wild puma missing a foot. Cat News 57, 7–8.

Jefferson, G.T., Tejada-Flores, A.E., 1993. The Late Pleistocene Record of *Homotherium* (Felidae: Machairodontinae) in the Southwestern United States. PaleoBios 15, 37–46.

Johnson, W.E., Eizirik, E., Pecon-Slattery, J., Murphy, W.J., Antunes, A., Teeling, E., O'Brien, S.J., 2006. The Late Miocene Radiation of Modern Felidae: A Genetic Assessment. Science 311, 5757, 73–77.

Johnson, W.E., Pecon Slattery, J., Eizirik, E., Kim, J.-H., Menotti Raymond, M., Bonacic, C., Cambre, R., Crawshaw, P., Nunes, A., Seuánez, H.N., Martins Moreira, M.A., Seymour, K.L., Simon, F., Swanson, W., O'Brien, S.J., 1999. Disparate phylogeographic patterns of molecular genetic variation in four closely related South American small cat species. Molecular Ecology 8, S79–S94.

Kahlke, H.-D., 1997. Die Cerviden-Reste aus dem Unterpleistozän von Untermaßfeld. In: Kahlke, R.-D. (Ed.), Das Pleistozän von Untermaßfeld bei Meiningen (Thüringen). Teil 1. Monographien des Römisch-Germanischen Zentralmuseums 40, 1, 181–275, pls. 34–44.

2001. Die Rhinocerotiden-Reste aus dem Unterpleistozän von Untermaßfeld. In: Kahlke, R.-D. (Ed.), Das Pleistozän von Untermaßfeld bei Meiningen (Thüringen). Teil 2. Monographien des Römisch-Germanischen Zentralmuseums 40, 2, 501–555, pls. 79–91.

Kahlke, R.-D., 1997a. Zur Entdeckungs- und Erforschungsgeschichte der unterpleistozänen Komplexfundstelle Untermaßfeld. In: Kahlke, R.-D. (Ed.), Des Pleistozän von Untermaßfeld bei Meiningen (Thüringen). Teil 1. Monographien des Römisch-Germanischen Zentralmuseums 40, 1, 1–28.

1997b. Die *Hippopotamus*-Reste aus dem Unterpleistozän von Untermaßfeld. In: Kahlke, R.-D. (Ed.), Das Pleistozän von Untermaßfeld bei Meiningen (Thüringen). Teil 1. Monographien des Römisch-Germanischen Zentralmuseums 40, 1, 277–374, pls. 45–62.

2001a. Verbesserte geologische Standardprofile zur unterpleistozänen Komplexfundstelle Untermaßfeld. In: Kahlke, R.-D. (Ed.), Das Pleistozän von Untermaßfeld bei Meiningen (Thüringen). Teil 2. Monographien des Römisch-Germanischen Zentralmuseums 40, 2, 433–445, suppl. I.

2001b. Die unterpleistozäne Komplexfundstelle Untermaßfeld – Zusammenfassung des Kenntnisstandes sowie synthetische Betrachtungen zu Genesemodell, Paläoökologie und Stratigraphie. In: Kahlke, R.-D. (Ed.), Das Pleistozän von Untermaßfeld bei Meiningen (Thüringen). Teil 3. Monographien des Römisch-Germanischen Zentralmuseums 40, 3, 931–1030, suppl. I–XV.

2001c. Schädelreste von *Hippopotamus* aus dem Unterpleistozän von Untermaßfeld. In: Kahlke, R.-D. (Ed.), Das Pleistozän von Untermaßfeld bei Meiningen (Thüringen). Teil 2. Monographien des Römisch-Germanischen Zentralmuseums 40, 2, 483–500, pls. 77–78.

2006. Untermassfeld. A late Early Pleistocene (Epivillafranchian) fossil site near Meiningen (Thuringia, Germany) and its position in the development of the European mammal fauna. British Archaeological Reports, International Series 1578, Oxford.

2014. The origin of Eurasian Mammoth Faunas (*Mammuthus–Coelodonta* Faunal Complex). Quaternary Science Reviews 96, 32–49.

2020. Excavation progress and research at the Early Pleistocene site of Untermassfeld during the years 1997–2015. In: Kahlke, R.-D. (Ed.), The Pleistocene of Untermassfeld near Meiningen (Thüringen, Germany). Part 4. Monographien des Römisch-Germanischen Zentralmuseums 40, 4, 1031–1078.

in this volume. The Pleistocene of Untermassfeld – A synopsis of site origin, palaeobiodiversity, taphonomic characteristics, palaeoenvironment, chronostratigraphy, and significance in western Palaearctic faunal history. 1635–1734.

Kahlke, R.-D., Gaudzinski, S., 2005. The blessing of a great flood: differentiation of mortality patterns in the large mammal record of the Lower Pleistocene fluvial site of Untermassfeld (Germany) and its relevance for the interpretation of faunal assemblages from archaeological sites. Journal of Archaeological Science 32, 1202–1222.

Kahlke, R.-D., Linnemann, U., Gärtner, A., 2020. New results on the origin and geological history of the Early Pleistocene site of Untermassfeld. In: Kahlke, R.-D. (Ed.), The Pleistocene of Untermassfeld near Meiningen (Thüringen, Germany). Part 4. Monographien des Römisch-Germanischen Zentralmuseums 40, 4, 1079–1104, Foldout I.

Kahlke, R.-D., García, N., Kostopoulos, D.S., Lacombat, F., Lister, A.M., Mazza, P.P.A., Spassov N., Titov, V.V., 2011. Western Palaearctic palaeoenvironmental conditions during the Early and early Middle Pleistocene inferred from large mammal communities, and implications for hominin dispersal in Europe. Quaternary Science Reviews 30, 1368–1395.

Kingdon, J., 1977. East African Mammals. An Atlas of Evolution in Afrika. Volume IIIA. Carnivores. London.

Kitchener, A.C., Breitenmoser-Würsten, C., Eizirik, E., Gentry, A., Werdelin, L., Wilting, A., Yamaguchi, N., Abramov, A.V., Christiansen, P., Driscoll, C., Duckworth, J.W., Johnson, W., Luo, S.-J., Meijaard, E., O'Donoghue P., Sanderson, J., Seymour, K., Bruford, M., Groves, C., Hoffmann, M., Nowell, K., Timmons, Z., Tobe S., 2017. A revised taxonomy of the Felidae. The final report of the Cat Classification Task Force of the IUCN/SSC Cat Specialist Group. Cat News Special Issue 11, 1–80.

Kotowski, A., Stefaniak, K., Kahlke, R.-D., 2020. A rhinocerotid skull from the Early Pleistocene site of Untermassfeld. In: Kahlke, R.-D. (Ed.), The Pleistocene of Untermassfeld near Meiningen (Thüringen, Germany). Part 4. Monographien des Römisch-Germanischen Zentralmuseums 40, 4, 1273–1293.

Kretzoi, M., 1929. Feliden-Studien. Budapest.

Kurtén, B., 1965. The Pleistocene Felidae of Florida. Bulletin of the Florida State Museum 9, 215–273.

1973a. Pleistocene jaguars in North America. Commentationes Biologicae 62, 1–23.

1973b. Geographic variation in size in the puma (*Felis concolor*). Commentationes Biologicae 63, 1–8.

1976. Fossil puma (Mammalia: Felidae) in North America. Netherlands Journal of Zoology 26, 502–534.

1978. The Lynx from Etouaires, Lynx issiodorensis (Croizet & Jobert), late Pliocene. Annales Zoologici Fennici 15, 314–322.

Kurtén, B., Crusafont Pairò, M., 1977. Villafranchian carnivores (Mammalia) from La Puebla de Valverde (Teruel, Spain). Commentationes Biologicae 85, 1–39.

Kurtén, B., Werdelin, L., 1984. The relationships of *Lynx shansius* Teilhard. Annales Zoologici Fennici 21, 129–133.

Laurie, W.A., Lang, E.M., Groves, C.P., 1983. *Rhinoceros unicornis*. Mammalian Species 211, 1–6.

Leslie, D.M., Schaller, G.B., 2009. *Bos grunniens* and *Bos mutus* (Artiodactyla: Bovidae). Mammalian Species 836, 1–17.

Lewis, M.E., Werdelin, L., 2010. Carnivoran Dispersal Out of Africa During the Early Pleistocene: Relevance for Hominins? In: Fleagle, J.G., Shea, J.J., Grine, F.E., Baden, A.L., Leakey, R.E. (Eds.), Out of Africa I: The First Hominin Colonization of Eurasia. Dordrecht, 13–14.

Leyhausen, P., 1973. Verhaltensstudien bei Katzen. Berlin, Hamburg. (3rd ed.)

Liu, J.-Y., 2005. *Megantereon* fossil remains from Renzidong Cave, Fanchang County, Anhui Province, China. Vertebrata PalAsiatica 43, 122–134. (in Chinese, with English summary).

Lydekker, R., 1878. Aberrant Dentition of *Felis Tigris*. Journal of the Asiatic Society of Bengal 47, 2–3.

Lydekker, R., 1884. Siwalik and Narbada Carnivora. Palaeontologia Indica, Series 10 Indian Tertiary and post-Tertiary Vertebrata, Volume 2, Part 6, 178–355.

Madurell-Malapeira, J., Alba, D.M., Moyà-Solà, S., Aurell-Garrido, J., 2010. The Iberian record of the puma-like cat *Puma pardoides* (Owen, 1846) (Carnivora, Felidae). Comptes Rendus Palevol 9, 55–62.

Madurell-Malapeira, J., Morales, J., Vinuesa, V., Boscaini, A., 2015. Úrsidos, hiénidos y félidos del Pleistoceno inferior de Cueva Victoria (Cartagena, Murcia). Mastia 11–13, 401–432.

Marciszak, A., 2014. Presence of *Panthera gombaszoegensis* (Kretzoi, 1938) in the late Middle Pleistocene of Biśnik Cave, Poland, with an overview of Eurasian jaguar size variability. Quaternary International 326–327, 105–113.

Marean, C.W., 1989. Sabertooth cats and their relevance for early hominid diet and evolution. Journal of Human Evolution 18, 559–582.

Marker, L.L., Dickman, A.J., 2003. Morphology, Physical Condition, and Growth of the Cheetah (*Acinonyx jubatus jubatus*). Journal of Mammalogy 84, 840–850.

Martin, R.A., 1990. Estimating body mass and correlated variables in extinct mammals: travels in the fourth dimension. In: Damuth, J., MacFadden, B.J. (Eds.), Body Size in Mammalian Paleobiology: Estimation and Biological Implications. Cambridge, 49–68.

Martínez-Navarro, B., Palmqvist, P., 1995. Presence of the African Machairodont *Megantereon whitei* (Broom, 1937) (Felidae, Carnivora, Mammalia) in the Lower Pleistocene site of Venta Micena (Orce, Granada, Spain), with some Considerations on the Origin, Evolution and Dispersal of the Genus. Journal of Archaeological Science 22, 569–582.

1996. Presence of the African Saber-toothed Felid *Megantereon whitei* (Broom, 1937) (Mammalia, Carnivora, Machairodontinae) in Apollonia-1 (Mygdonia Basin, Macedonia, Greece). Journal of Archaeological Science 23, 869–872.

Matthew, W.D., 1929. Critical observations upon Siwalik mammals (Exclusive of Proboscidea). Bulletin of the American Museum of Natural History 56, 437–560.

Mazak, V., 1975. Notes on the black-maned lion of the Cape, *Panthera leo melanochaita* (Ch. H. Smith, 1842) and a revised list of the preserved specimens. Verhandelingen der Koninklijke Nederlandse Akademie van Wetenschappen, Afdeeling Natuurkunde, Reeks 2, 64, 1–44, pls. I–IX.

McFarlane, D.A., Lundberg, J., 2013. On the occurrence of the scimitar-toothed cat, *Homotherium latidens* (Carnivora; Felidae), at Kents Cavern, England. Journal of Archaeological Science 40, 1629–1635.

Meade, G.E., 1961. The Saber-toothed Cat, *Dinobastis serus*. Bulletin of the Texas Memorial Museum 2, 23–60.

Mecozzi, B., Sardella, R., Boscaini, A., Cherin, M., Costeur, L., Madurell-Malapeira, J., Pavia, M., Profico, A., Iurino, D.A., 2021. The tale of a short-tailed cat: New outstanding Late Pleistocene fossils of *Lynx pardinus* from southern Italy. Quaternary Science Reviews 262, 106840.

Merriam, J.C., Stock, C., 1932. The Felidae of Rancho La Brea. Carnegie Institution of Washington Publication 422, Washington.

Mol, D., Logchem, W.v., Vos, J. de, 2011. New record of the European jaguar, *Panthera onca gombaszoegensis* (Kretzoi, 1938), from the Plio-Pleistocene of Langenboom (The Netherlands). Cainozoic Research 8, 35–40.

Morales-Mejía, F.M., Arroyo-Cabrales, J., Polaco, O.J., 2010. Estudio comparativo de algunos elementos de las extremidades anteriores y posteriores y piezas dentales de puma (*Puma concolor*) y jaguar (*Panthera onca*). TIP Revista Especializada en Ciencias Químico-Biológicas 13, 73–90.

Nanjing Museum and Institute of Archaeology, 2007. The Early Pleistocene Mammalian Fauna at Tuozi Cave, Nanjing, China. Beijing. (in Chinese, with English summary)

Nomade, S., Pastre, J.F., Guillou, H., Faure, M., Guérin, C., Delson, E., Debard, E., Voinchet, P., Messager, E., 2014. $^{40}Ar/^{39}Ar$ constraints on some French landmark Late Pliocene to Early Pleistocene large mammalian paleofaunas: Paleoenvironmental and paleoecological implications. Quaternary Geochronology 21, 2–15.

Nowell, K., Jackson, P., 1996. Wild cats. Status survey and conservation action plan. Gland.

O'Regan, H., 2002. Defining Cheetahs, a multivariate analysis of skull shape in big cats. Mammal Review 32, 58–62.

Paijmans, J.L.A., Barnett, R., Gilbert, M.T.P., Zepeda-Mendoza, M.L., Reumer, J.W.F., de Vos, J., Zazula, G., Nagel, D., Baryshnikov, G.F., Leonard, J.A., Rohland, N., Westbury, M.V., Barlow, A., Hofreiter, M., 2017. Evolutionary History of Saber-Toothed Cats based on Ancient Mitogenomics. Current Biology 27, 3330–3336.e5.

Palmqvist, P., 2002. On the presence of *Megantereon whitei* at the South Turkwel Hominid Site, northern Kenia. Journal of Paleontology 76, 928–930.

Palmqvist, P., Torregrosa, V., Pérez-Claros, J.A., Martínez-Navarro, B., Turner, A., 2007. A re-evaluation of the diversity of *Megantereon* (Mammalia, Carnivora, Machairodontinae) and the problem of species identification in extinct carnivores. Journal of Vertebrate Paleontology 27, 160–175.

Palmqvist, P., Pérez-Claros, J.A., Janis, C.M., Figueirido, B., Torregrosa, V., Gröcke, D.R., 2008. Biogeochemical and Ecomorphological Inferences On Prey Selection and Resource Partitioning Among Mammalian Carnivores In An Early Pleistocene Community. Palaios 23, 724–737.

Petrucci, M., 2008. Analisi sistematica dei Macromammiferi di Pirro Nord (Apricena, Foggia, Puglia). PhD thesis, Università degli Studi di Torino, Italy.

Petrucci, M., Cipullo, A., Martinez-Navarro, B., Rook, L., Sardella, R., 2013. The Late Villafranchian (Early Pleistocene) carnivores (Carnivora, Mammalia) from Pirro Nord (Italy). Palaeontographica Abteilung A 298, 113–145.

Petter, G., Howell, F.C., 1982. Un Félidé machairodonte des formations plio-pleistocènes des Siwaliks: *Megantereon falconeri* Pomel/= *M. sivalensis* (F. et C.)/(Mammalia, Carnivora, Felidae). Comptes Rendus l'Academie des Sciences 295, Série II, 281–284.

Pilgrim, G.E., 1932. The fossil Carnivora of India. Palaeontologia Indica, New Series 18, i–iii, 1–232.

Pons-Moyà, J., 1987. Los carnívoros (Mammalia) de Venta Micena (Granada, España). Paleontologia i Evolució, Memòria Especial 1, 109–128.

Pucek, Z., 1986. *Bison bonasus* (Linnaeus, 1758) – Wisent. In: Niethammer, J., Krapp, F. (Eds.), Handbuch der Säugetiere Europas. Band 2/II, Paarhufer – Artiodactyla (Suidae, Cervidae, Bovidae). Wiesbaden, 278–315.

Qiu, Z., Deng, T., Wang, B., 2004. Early Pleistocene Mammalian Fauna from Longdan, Dongxiang, Gansu, China. Palaeontologia Sinica 191, New Series C 27, xi, 1–198, 34 pls.

Qiu, Z.-X., 2006. Quaternary environmental changes and evolution of large mammals in North China. Vertebrata PalAsiatica 44, 109–132.

Queiroz, K. de, 2005. Ernst Mayr and the modern concept of species. Proceedings of the National Academy of Sciences of the United States of America 102, supplement 1, 6600–6607.

Quigley, H.B., Crawshaw Jr., P.G., 2002. Reproducción, crecimiento y dispersión del jaguar en la region del Pantanal de Brasil. In Medellín, R.A., Equihua, C.A., Chetkiewicz, C., Crawshaw Jr., P.G., Rabinowitz, A., Redford, K.H., Robinson, J.G., Sanderson, E.W., Taber, A.B. (Eds.), El Jaguar en el Nuevo Milenio. México, 289–302.

Randau da Costa Carvalho, M., 2017. Drivers and constraints of shape evolution in the vertebral column of felids (Felidae, Carnivora). PhD thesis, University College of London, U.K.

Rawn-Schatzinger, V., 1983. Development and eruption sequence of deciduous and permanent teeth in the saber-tooth cat *Homotherium serum* Cope. Journal of Vertebrate Paleontology 3, 49–57.

1992. The Scimitar Cat *Homotherium serum* Cope: osteology, functional morphology, and predatory behaviour. Illinois State Museum Report of Investigations 47, I–VIII, 1–80.

Reinken, G., 1987. Damtierhaltung. Stuttgart. (2nd ed.)

Reumer, J.W.F., Rook, L., Borg, K. van der, Post, K., Mol, D., Vos, J. de, 2003. Late Pleistocene survival of the saber-toothed cat *Homotherium* in northwestern Europe. Journal of Vertebrate Paleontology 23, 260–262.

Rodríguez-Varela, R., Tagliacozzo, A., Ureña, I., García, N., Crégut-Bonnoure, E., Mannino, M.A., Arsuaga, J.L., Valdiosera, C., 2015. Ancient DNA evidence of Iberian lynx palaeoendemism. Quaternary Science Reviews 112, 172–180.

Rook, L., Martínez-Navarro, B., 2010. Villafranchian: The long story of a Plio-Pleistocene European large mammal biochronologic unit. Quaternary International 219, 134–144.

Roth, V.L., 1990. Insular dwarf elephants: a case study in body mass estimation and ecological inference. In: Damuth, J., MacFadden, B.J. (Eds.), Body Size in Mammalian Paleobiology: Estimation and Biological Implications. Cambridge, 151–179.

Russo, G.A., 2015. Postsacral Vertebral Morphology in Relation to Tail Length Among Primates and Other Mammals. The Anatomical Record 298, 354–375.

Rutledge, L.Y., Devillard, S., Boone, J.Q., Hohenlohe, P.A., White, B.N., 2015. RAD sequencing and genomic simulations resolve hybrid origins within North American *Canis*. Biology Letters 11, 20150303.

Salles, L.O., 1992. Felid phylogenetics: extant taxa and skull morphology (Felidae, Aeluroidea). American Museum Novitates 3047, 1–67.

Sardella, R., 1998. The Plio-Pleistocene Old World dirk-toothed cat *Megantereon* ex gr. *cultridens* (Mammalia, Felidae, Machairodontinae), with comments on taxonomy, origin and evolution. Neues Jahrbuch für Geologie und Paläontologie, Abhandlungen 207, 1–36.

Sardella, R., Iurino, D.A., 2012. The latest Early Pleistocene saber-toothed cat *Homotherium* (Felidae, Mammalia) from Monte Peglia (Umbria, central Italy). Bollettino della Società Paleontologica Italiana 51, 15–22.

Sardella, R., Petrucci, M., Rook, L., 2008. The African species *Megantereon whitei* from the Early Pleistocene of Monte Argentario (South Tuscany, Central Italy). Comptes Rendus Palevol 7, 601–606.

Savage, D.E., 1960. A survey of various late Cenozoic vertebrate faunas of the Panhandle of Texas. Part III. Felidae. University of California Publications in Geological Sciences 36, 317–343.

Schaller, G.B., 1967. The deer and the tiger. A study of wildlife in India. Chicago, London.

Schaub, S., 1925. Ueber die Osteologie von *Machairodus cultridens* Cuvier. Eclogae geologicae Helvetiae 18, 255–266.

1939. Was ist *Felis elata* Bravard? Eclogae geologicae Helvetiae 32, 187–193.

1949. Révision de quelques Carnassiers villafranchiens du Niveau des Etouaires (Montagne de Perrier, Puy-de-Dôme). Eclogae geologicae Helvetiae 42, 492–506.

Schmid, E., 1940. Variationsstatistische Untersuchungen am Gebiß pleistozäner und rezenter Leoparden und anderer Feliden. Zeitschrift für Säugetierkunde 15, 1–179.

Schütt, G., 1970a. Ein Gepardenfund aus den Mosbacher Sanden (Altpleistozän, Wiesbaden). Mainzer Naturwissenschaftliches Archiv 9, 118–131.

1970b. Nachweis der Säbelzahnkatze *Homotherium* in den altpleistozänen Mosbacher Sanden (Wiesbaden/Hessen). Neues Jahrbuch für Geologie und Paläontologie, Monatshefte 1970, 187–192.

Schwartz, M.K., Pilgrim, K.L., McKelvey, K.S., Lindquist, E.L., Claar, J.J., Loch, S., Ruggiero, L.F., 2004. Hybridization between Canada Lynx and Bobcats: Genetic Results and Management Implications. Conservation Genetics 5, 349–355.

Scott, K.M., 1990. Postcranial dimensions of ungulates as predictors of body mass. In: Damuth, J., MacFadden, B.J. (Eds.), Body Size in Mammalian Paleobiology: Estimation and Biological Implications. Cambridge, 301–335, 370–377.

Seymour, K.L., 1993. Size change in North American Quaternary jaguars. In: Martin, R.A., Barnosky, A.D. (Eds.), Morphological change in Quaternary mammals of North America. Cambridge, 343–372.

Sharapov, Sh., 1986. Kuruksajskij kompleks poznepliotsenovykh mlekopitayushchih Afgano-Tadzhikskoj depressii. Dushanbe. (in Russian)

1989. O novom vide sablezuboj koshki iz poznego Eoplejstotsena Afgano-Tadzhikskoj depressii i evolyuzii roda *Homotherium*. Paleontologicheskij zhurnal 1989, 73–84. (in Russian)

Shen, C., Zhang, X., Gao, X., 2015. Zhoukoudian in transition: Research history, lithic technologies, and transformation of Chinese Palaeolithic archaeology. Quaternary International 400, 4–13.

Sher, A.V., 1997. An Early Quaternary bison population from Untermassfeld: *Bison menneri* sp. nov. In: Kahlke, R.-D. (Ed.), Das Pleistozän von Untermaßfeld bei Meiningen (Thüringen). Teil 1. Monographien des Römisch-Germanischen Zentralmuseums 40, 1, 101–180, pls. 22–33.

Sotnikova, M.V., 1976. Verkhnepliotsenovye Carnivora Zentral'noj Azii. Izvestiya Akademii nauk SSSR. Seriya geologicheskaya 11, 133–137. (in Russian)

Sotnikova, M.V., Foronova, I.V., 2014. First Asian record of *Panthera* (*Leo*) *fossilis* (Mammalia, Carnivora, Felidae) in the Early Pleistocene of Western Siberia, Russia. Integrative Zoology 9, 517–530.

Sotnikova, M., Titov, V., 2009. Carnivora of the Tamanian faunal unit (the Azov Sea area). Quaternary International 201, 43–52.

Sotnikova, M.V., Baigusheva, V.S., Titov, V.V., 2002. Carnivores of the Khapry Faunal Assemblage and Their Stratigraphic Implications. Stratigraphy and Geological Correlation 10, 375–390.

Spassov, N., 2011. *Acinonyx pardinensis* (Croizet et Jobert) remains from the Middle Villafranchian locality of Varshets (Bulgaria) and the Plio-Pleistocene history of the cheetahs in Eurasia. Estudios Geológicos 67, 245–253.

Spoor, C.F., 1985. Body proportions in Hyaenidae. Anatomischer Anzeiger 160, 215–220.

Teilhard de Chardin, P., 1936. Fossil Mammals from Locality 9 of Choukoutien. Palaeontologia Sinica, Series C., Volume VII, Fascicle 4, 5–70.

1939. On two skulls of *Machairodus* from the Lower Pleistocene Beds of Choukoutien. Bulletin of the Geological Society of China 19, 235–256.

Testu, A., 2006. Étude paléontologique et biostratigraphique des Felidae et Hyaenidae pléistocènes de l'Europe méditerranéenne. Sites de la Caune de l'Arago, Orgnac 3, le Portel-Ouest, Bize-Tournal, l'Hortus, la Crouzade en France, la Covade de l'Arbreda en Espagne, Karaïn E en Turquie. Thèse de doctorat en Préhistoire, Université de Perpignan, France.

Thenius, E., 1954. Gepardreste aus dem Altquartär von Hundsheim in Niederösterreich. Neues Jahrbuch für Geologie und Paläontologie, Monatshefte 1953, 225–238.

Tiwari, Y., Taluja, J.S., Vaish, R., Pandey, A., Shrivastav, A.B., 2012. Biometry of Lumbar Vertebrae in Tiger (*Panthera tigris*). World Research Journal of Anatomy 2, 24–27.

Tournepiche, J.-F., 1996. Les grands mammifères Pleistocènes de Poitou-Charentes. Paléo 8, 109–141.

Tujek, A.G.I., 2008. Atlas radiographique et ostéologique du puma (*Puma concolor*). Thèse Docteur Vétérinaire Diplome d'État, Université Paul Sabatier Toulouse III, France.

Turner, A., 1987. *Megantereon cultridens* (Cuvier) (Mammalia, Felidae, Machairodontinae) from Plio-Pleistocene deposits in Africa and Eurasia, with comments on dispersal and possibility of a New World origin. Journal of Paleontology 61, 1256–1268.

1990. Late Neogene/Lower Pleistocene Felidae of Africa: Evolution and Dispersal. Quartärpaläontologie 8, 247–256.

Turner, A., Antón, M., 1997. The Big Cats and Their Fossil Relatives. An Illustrated Guide to Their Evolution and Natural History. New York.

Uphyrkina, O., Johnson, W.E., Quigley, H., Miquelle, D., Marker, L., Bush, M., O'Brien, S.J., 2001. Phylogenetics, genome diversity and origin of modern leopard, *Panthera pardus*. Molecular Ecology 10, 2617–2633.

Van Essen, in this volume. New results on elephantids from the Early Pleistocene site of Untermassfeld – Cranial and dental remains. 1355–1384.

Van Valkenburgh, B., 1990. Skeletal and dental predictors of body mass in carnivores. In: Damuth, J., MacFadden, B.J. (Eds.), Body Size in Mammalian Paleobiology. Estimation and Biological Implications. Cambridge, 181–203.

Van Valkenburgh, B., Ruff, C.B., 1987. Canine tooth strength and killing behaviour in large carnivores. Journal of Zoology 212, 379–397.

Van Valkenburgh, B., Grady, F., Kurtén, B., 1990. The Plio-Pleistocene cheetah-like cat *Miracinonyx inexpectatus* of North America. Journal of Vertebrate Paleontology 10, 434–454.

Vekua, A., 1996. Die Wirbeltierfauna des Villafranchium von Dmanisi (Georgien, Kaukasus) und ihre biostratigraphische Bedeutung. Jahrbuch des Römisch-Germanischen Zentralmuseums Mainz 42, 77–180, pls. 7–54.

Vekua, A., Lordkipanidze, D., Agusti, J., Rook, L., Oms, O., Kiladze, G., 2007. New paleontological discoveries in Kvabebi (East Georgia). Problems of Paleobiology 2, 146–164.

Villalta, J.F., 1952. Contribución al conocimiento de la fauna de mamíferos fósiles del Plioceno de Villarroya (Logroño). Boletín del Instituto Geológico y Minero de España 64, 3–204.

Viret, J., 1954. Le loess à bancs durcis de Saint-Vallier (Drôme), et sa faune de mammifères villafranchiens. Nouvelles archives du Muséum d'histoire naturelle de Lyon 4, 1–200, pls. 1–33.

Vislobokova, I.A., 2018. On a New Find of *Megantereon* (Carnivora, Felidae, Machairodontinae) from the Early Pleistocene of Trlica (Montenegro, the Central Balkans). Paleontological Journal 52, 1445–1449.

VonHoldt, B.M., Cahill, J.A., Fan, Z., Gronau, I., Robinson, J., Pollinger, J.P., Shapiro, B., Wall, J., Wayne, R.K., 2016. Whole-genome sequence analysis shows that two endemic species of North American wolf are admixtures of the coyote and gray wolf. Science Advances 2, e1501417.

Wayne, R.K., Modi, W.S., O'Brien, S.J., 1986. Morphological variability and asymmetry in the cheetah (*Acinonyx jubatus*), a genetically uniform species. Evolution 40, 78–85.

Werdelin, L., 1981. The evolution of lynxes. Annales Zoologici Fennici 18, 37–71.

Werdelin, L., Lewis, M.E., 2002. Species identification in *Megantereon*: A reply to Palmqvist. Journal of Paleontology 76, 931–933.

Widga, C., Fulton, T.L., Martin, L.D., Shapiro, B., 2012. *Homotherium serum* and *Cervalces* from the Great Lakes Region, USA: geochronology, morphology and ancient DNA. Boreas 41, 546–556.

Young, S.P., Goldman, E.A., 1946. The puma – mysterious American cat. New York.

Zdansky, O., 1925. Quartäre Carnivoren aus Nord-China. Palaeontologia Sinica, Series C, Volume II, Fascicle 2, 1–29.

LUTZ CHRISTIAN MAUL

NEW RESULTS ON MICROMAMMALS FROM THE EARLY PLEISTOCENE SITE OF UNTERMASSFELD

Abstract

Small mammal remains recovered from the important late Early Pleistocene fossil site of Untermassfeld in Central Germany were described in 2001. Here we discuss new finds excavated subsequently in the context of an overview on the entire small mammal material of this site. The most spectacular new remains are an articulated skeleton of the giant hamster *Cricetus runtonesis*, articulated remains of *Microtus thenii*, and several complete or partial skeletons of a ground squirrel, which are the basis for description of *Spermophilus kahlkei* n. sp. The beaver *Trogontherium* is now represented by a nearly complete skull and a mandible. *Desmana thermalis*, *Sorex minutus*, *Hypolagus brachygnathus* and *Sicista* sp. are additions to the updated faunal list of Untermassfeld. The stratigraphic position of the site at the base of the Jaramillo subchron is confirmed by new finds of arvicolids that enlarge our previous samples. The previous paleoenvironmental reconstruction – a mosaic pattern of open and wooded habitats – is further corroborated by the new data.

1. Introduction

The Pleistocene site of Untermassfeld yielded one of the most important fossil assemblages of the late Early Pleistocene of Europe. Based on palaeomagnetic and biostratigraphic evidence, this very productive site can be dated to c. 1.05 Ma. Large mammals display a faunal composition which led to the definition of a new stratigraphic unit – the Epivillafranchian (Kahlke 2007). Because of the great abundance of various fossil groups found there, and its stratigraphic and geographic position, Untermassfeld has served for years as a reference fauna for continental fossil sites.

Small mammals are rather well represented at Untermassfeld, and more than 7,000 remains (3,525 catalogue nos: 2,354 of excavations in 1978–2000, and 1,171 in 2001–2015) have been excavated. Preservation quality varies from exquisite to poor. Twenty years ago, results of the studies of the available small mammal assemblage were published (Maul 2001a) in the context of the monographic publication of the Untermassfeld material (Kahlke 1997; 2001a; 2001c; 2006). Since then, many additional remains have been unearthed, among them new small mammal material including partial and articulated skeletons.

The aim of the present paper is the analysis of the finds collected since 2001 after relocation to the southern area of the site (Kahlke 2020) in the context of a current overview of the entire small mammal fauna from Untermassfeld, with a complete list of all available identifiable specimens. Updated interpretations of taxonomy, biostratigraphy, palaeoecology and taphonomy will be discussed.

2. Material and methods

The site of Untermassfeld is situated c. 2 km south of the town of Meiningen in Southern Thuringia (Central Germany, Thuringia, 50°32'35.65" N, 10°25'9.49" E) on a high terrace of the river Werra. Basal coarse clastic gravels are overlain by the fine-grained so-called Lower Fluviatile Sands and the superimposed Upper Fluviatile Sands. All Pleistocene hitherto found small mammal remains originate from the Upper Fluviatile Sands (for details see Ellenberg and Kahlke 1997; Kahlke 2001b; Kahlke et al. 2020). A carbonate impregnation derived from lime-rich water that passed through Triassic sediments (Lower Muschelkalk) is responsible for the presevation of the fossils (Heinrich 1982; Ellenberg and Kahlke 1997).
The small mammal material was collected mainly by screen-washing of 1.0 × 1.0 × 0.25 m blocks (mesh size 2–0.71 mm, reduction of around 95 % of the volume). After 2001, a further 800 kg of sediment from Untermassfeld was sieved for small vertebrates using a mesh size of 0.8 mm (see Kahlke 2020). Skeletons were prepared from fossiliferous blocks in situ and in the lab. Microphotographs and measurements of small mammal specimens were taken with an OLYMPUS DP 71 camera with multiple focal planes. Measurement standards and anatomical nomenclature are the same as in Maul (2001a). Taxonomy of higher groups of small mammals follows mainly Wilson and Reeder (2005), that of Soricidae Reumer (1998), of Leporidae Lopez-Martínez (2008), of Sciuridae Helgen et al. (2009), and of Castoridae Korth (2001). The stratigraphical terms Biharian and Toringian are used in the sense of Fejfar et al. (1998). Abbreviations are explained in the text where they apply. Catalogue numbers are not given in the material lists in the taxonomy section. However, a complete list of specimens from the excavation seasons 1979–2016 with catalogue numbers can be found in the appendix. All small mammal material is housed in the collection of the Senckenberg Research Station of Quaternary Palaeontology in Weimar. Catalog numbers begin with IQW, followed by the year of cataloguing and unique catalog numbers. Comparative material of extant taxa has been used from the mammalogy collection of Senckenberg Frankfurt, and from the Phyletic Museum of the University of Jena.

3. Results

3.1. Taxonomy

Order Erinaceomorpha Gregory, 1910
Family Erinaceidae Fischer, 1814
Subfamily Erinaceinae Fischer, 1814
Genus *Erinaceus* Linnaeus, 1758
Erinaceus cf. *europaeus* Linnaeus, 1758
1 right P^3; 1 left and 1 right mandible fragment without teeth.

The premolar and the mandible fragments (presumably from one individual) exceed in size the small *Erinacaeus lechei* Kormos, 1930, *E. ostramosi* Jánossy, 1972, and *Hemiechinus auritus* (Gmelin, 1779), but are morphologically and metrically within the range of the extant *E. europaeus* Linnaeus, 1758, and *E. rouma-*

nicus Barrett-Hamilton, 1900, and the fossil *E. praeglacialis* Brunner, 1934, a possible synonym of *E. europaeus* (Maul 2001a). There are no new finds of hedgehogs from Untermassfeld. Because the specific assignment of our finds is still ambiguous we refer them to *E.* cf. *europaeus*.

Order Soricomorpha Gregory, 1910
Family Talpidae Fischer, 1814
Subfamily Talpinae Fischer, 1814
Tribus Talpini Fischer, 1814
Genus *Talpa* Linnaeus, 1758
Talpa cf. *europaea* Linnaeus, 1758
1 left and 2 right C^{sup}; 1 left mandible fragment with A_2-M_1; 1 left mandible fragment without teeth; 1 right mandible fragment with A_4-M_1; 3 A_{inf}; 1 right C_{inf}; 2 left M_3; 1 right M_{inf}; 6 left and 4 right humerus fragments; 1 right radius and ulna; 3 left and 2 right radius fragments; 6 distal phalanges; 6 phalanges of indeterminable position.

Talpa minor Freudenberg, 1914
2 left humerus fragments.

All humeri but two and all other postcranial elements from Untermassfeld fit the range of the medium-sized *T. europaea*, while the smaller humeri belong to *T. minor* according to data given by Koenigswald (1970). *T. praeglacialis* Kormos, 1930 and *T. fossilis* (Petényi, 1864) are morphologically similar to *T. europaea* and are commonly considered synonyms of it (Koenigswald 1970; Rabeder 1972; Rzebik-Kowalska 1972). However, Sansalone et al. (2015) showed with a geometric morphometric analysis that the pectoral ridge and teres tubercle are larger in humeri of *T. europaea* than in *T. fossilis*, implying the presence of two different species. In only two humeri from Untermassfeld could the teres tubercle be measured, and its length is within the range of the extant *T. europaea*. However, in none of the new specimens was there a complete, measurable pectoral ridge. The nine humeri have a diaphysis width between 4.19 mm and 4.74 mm (mean = 4.50 mm), and epiphysis width between 8.18 mm and 10.86 mm (mean = 9.29 mm), whereas in two other specimens (i. e., the ones referred to *T. minor*) the diaphysis width is 2.96 mm and 3.30 mm, and epiphysis width 6.39 mm and 6.56 mm. We have no new arguments to exluce the presence of other medium- and small-sized *Talpa* species in Untermassfeld (see Maul 2011), so we maintain the referral to *T.* cf. *europaea* and *T. minor*. Based on the number of left humeri, the minimum number of individuals (MNI) is 6 for *T.* cf. *europaea*, and 2 for *T. minor*.

Tribus Desmanini Thomas, 1912
Genus *Desmana* Güldenstädt, 1777
Desmana thermalis Kormos, 1930
1 right mandible fragment with isolated P_2, P_4, and M_1 of the same mandible.

Although very fragmented the mandible remain is typical for a desman in morphology and size. Two mental foramina are visible: one below the mesial margin of M_1 and on behind P_1. According to measurements given by several authors (Schreuder 1940; Thenius 1948; Heller 1954; Jánossy 1965; Rzebik-Kowalska 1971; Rümke 1985a; 1985b), the lengths of P_2 (1.99 mm), P_4 (2.22 mm), and M_1 (3.05 mm) are clearly smaller than in *D. moschata* (Linnaeus, 1758) but larger than in *Galemys semseyi* Kormos, 1913, and *G. kormosi* (Schreuder, 1940), and fit the ranges of *D. thermalis* and *D. nehringi* Kormos, 1913. However, the height of mandible behind P_4 on lingual side is 4.66 mm, which conforms to data of *D. thermalis* and excludes the smaller *D. nehringi*.

Family Soricidae Fischer, 1814
Subfamily Soricinae Fischer, 1814
Tribe Soricini Fischer, 1814
Genus *Sorex* Linnaeus, 1758
Subgenus *Sorex* Linnaeus, 1758
Sorex (*Sorex*) *minutus* Linnaeus, 1766
1 left humerus, distal fragment.

This fragment of a humerus has the typical shape of *Sorex* (Brunner, 1953) but is distinctively smaller than in *S. araneus*. It is therefore assigned to the small sized *S. minutus* Linnaeus, 1766.

Sorex (*Sorex*) *runtonensis* Hinton, 1910
1 left M^1; 1 left mandible with I_{inf}-M_3; 1 right mandible with I_{inf}-M_1; 1 left mandible with I_{inf}-P_4; 1 left and 1 right mandible with M_1-M_3; 2 left and 1 right mandibles with M_1-M_2; 1 left and 1 right mandible with M_2-M_3; 1 right mandible with M_2; 1 left mandible without teeth; 1 left and 2 right I_{inf}.
The remains of this red-toothed shrew can be identified as *Sorex* based on the tricuspulate incisors, an entoconid ridge on the molars, reduced M_3, and articular process with a morphology typical for this genus (Rzebik-Kowalska 1991). *S. araneus* Linnaeus, 1758, *S. coronatus* Millet, 1828, *S. isodon* Turov, 1924, *S. samniticus* Altobello, 1926, *S. kennardi hundsheimensis* Rabeder, 1972, *S. subaraneus* Heller, 1958 are larger than our finds, and *S. minutus* Linnaeus, 1766, *S. caecutiens* Laxmann, 1788, and *S. kennardi* Hinton, 1911 are smaller. The remains from Untermassfeld best fit the metrical range of *S. runtonensis* Hinton, 1911. Among the new material is one further mandible, but it is in the range of this species (height of processus coronoideus 3.83 mm and 3.99 mm, M_1-M_3 length 3.44 mm and 3.45 mm), closer in size to the Early Pleistocene than to the early Middle Pleistocene samples, and thus confirming the successive size increase of this taxon (Rzebik-Kowalska 2000; Popov 2003; Rofes et al. 2016). Thus we maintain our previous referral (Maul 2001a) to *S. runtonensis*. The number of left mandibles indicates the presence of at least 7 individuals.

Subgenus *Drepanosorex* Kretzoi, 1941
Sorex (*Drepanosorex*) *margaritodon* Kormos, 1930
1 left maxilla with P^4-M^2; 1 right I^{sup}; 1 right P^4; 1 left M^1; 2 left M^2; 1 left mandible with P_3-P_4; 2 left mandibles with P_4-M_3; 1 left and 2 right mandibles with M_1-M_3; 1 left and 1 right mandible with M_1-M_2; 1 femur fragment.

Fissident I^{sup}, light orange pigmentation of teeth, and size and shape of the articular process (exceeding even *S. araneus*) indicate a member of the subgenus *Drepanosorex* (Kretzoi 1965; Reumer 1984; Rzebik-Kowalska 1991). Rzebik-Kowalska (1991) considered *S. margaritodon* a synonym of *S. savini* because of metric overlap and the lack of diagnostic differences. However, already the size values of both species from their type localities (*S. margaritodon* – Betfia 2/Romania, județ Bihor; *S. savini* – West Runton/UK, Norfolk County) are strikingly different. According to Reumer (1985) four *Drepanosorex* species represent a common lineage characterised by a continuously increasing size of the coronoid process and the molar row length: Villányian *S. praearaneus* Kormos, 1934, Biharian *S. margaritodon* Kormos, 1930, and *S. savini* Hinton, 1911, Toringian *S. austriacus* (Kormos, 1937). Because the specimens from Untermassfeld overlap both the measurements of *S. margaritodon* and *S. savini*, we previously considered them a transitional form (Maul 2001a). However, new finds shifted the range closer to *S. margaritodon* (height of the coronoid process, minimum = 4.63 mm, mean = 5.03 mm, maximum = 5.30 mm, n = 4; M_1-M_3 length minimum = 3.60 mm, mean = 3.93 mm, maximum = 4.02 mm, n = 4) (cf. data in Jánossy 1972; Maul 2001a; Rofes and Cuenca-Bescós 2013). Therefore, we now refer our specimens to that species. The MNI is 5 calculated from the number of left mandibles.

Tribe Beremendiini Reumer, 1984
Genus *Beremendia* Kormos, 1934
Beremendia fissidens (Petényi, 1864)
2 left and 1 right I^{sup}; 1 right P^4; 1 right M^1; 1 left mandible with I_{inf}-M_3; 1 left and 1 right mandible with M_1-M_3; 1 right mandible with M_2-M_3 IQW; 1 left mandible with M_3; 1 left and 1 right I_{inf}; 1 femur fragment.

This large-sized soricid from Untermassfeld possesses the typical traits of *Beremendia fissidens* (Petényi, 1864), i.e., teeth with dark red- to brown-stained tips and robust mandibles. *Petenyia hungarica* Kormos, 1934 and *Beremendia minor* Rzebik-Kowalska, 1976 are smaller, *Macroneomys brachygnathus* Fejfar, 1966 and *Blarinoides mariae* Sulimski, 1959 differ morphologically, and *Beremendia ucrainica* (Pidoplichko, 1956) seems to be a synonym of *B. fissidens* (see Fejfar 1966; Rzebik-Kowalska 1976; Maul and Rzebik-Kowalska 1998; Maul 2001a). Since the size of the teeth and mandibles from Untermassfeld is within the range of *B. fissidens*, but the height of the coronoid process is slightly smaller, we previously referred the specimens to *B.* cf. *fissidens* (Maul 2001a). Subsequently the species has been recorded from many other sites throughout Europe (Suárez and Mein 2004; Reumer et al. 2005; Furió et al. 2010; Hellmund and Ziegler 2012; Rzebik-Kowalska 2013), which enlarged the total size range of this species. Now we have three specimens with a coronoid process height of 5.63 mm, 5.71 mm, and 5.83 mm, which is in the range of the finds from Atapuerca (minimum = 5.51 mm, mean = 6.01 mm, maximum = 6.64 mm, after Rofes and Cuenca-Bescós 2009). Thus we assign the remains from Untermassfeld to *Bermendia fissidens* without 'confer'. An MNI of 3 is concluded from the number of left mandibles.

Order Lagomorpha Brandt, 1855
Family Leporidae Fischer, 1817
Subfamily Archaeolaginae Dice, 1929
Genus *Hypolagus* Dice, 1917
Hypolagus brachygnathus Kormos, 1930
1 left P_3.

The only leporid specimen from Untermassfeld that can unambiguously be referred to *Hypolagus* is a lower P_3 with the diagnostic large hypoflexid invading half of the occlusal surface. Morphotype comparison with *Hypolagus* finds of various ages indicated that our find must be referred to *H. brachygnathus* (for details see Maul 2007), the Pleistocene species, which differs slightly from the more ancient *H. petenyii* Čermak et Fladerer, 2009 (Čermák 2009; Čermák et al. 2012).

Subfamily Leporinae Fischer, 1817
Genus *Lepus* Linnaeus, 1758
Lepus sp.
2 right P^2; 2 right humeri distal fragments.

The P^2s display a deep hypoflexus and a shallower mesoflexus, which is typical for *Lepus* and never occurs in *Hypolagus* (Fladerer 1987; Fladerer and Reiner 1996). The size of the distal fragments of humeri is within the variation of both genera from the locality Deutsch-Altenburg, but slightly better fits the somewhat larger *Lepus*. Therefore, we refer these Untermassfeld specimens to *Lepus* sp.

Hypolagus brachygnathus vel *Lepus* sp.
1 left P^3; 2 left P^{sup}; 1 right M^2; 2 left M^{sup}; 1 right dP_3; 2 femur fragments; 2 tibia fragments; 2 phalanges of indeterminable position; 1 vertebra; 4 costae.
These additional remains, typical in size and shape for leporids, cannot be referred with certainty either to *Hypolagus* or *Lepus*.

Order Rodentia Bowdich, 1821
Suborder Sciuromorpha Brandt, 1855
Family Sciuridae Fischer, 1817
Subfamily Sciurinae Fischer, 1817
Tribe Sciurini Fischer, 1817
Genus *Sciurus* Linnaeus, 1758
Subgenus *Sciurus* Linnaeus, 1758
Sciurus whitei Hinton, 1914
1 right I^{sup}; 1 left and 1 right $M^{1/2}$; 1 left $M_{1/2}$.

The molars show the typical pattern of *Sciurus*, with a moderately pronounced metaconid and a strong entoconid on M_{inf}. Both the extant *S. vulgaris* Linnaeus, 1758 and *S. warthae* Sulimski, 1964 from the Pliocene of Węże/Poland, województwo Łódź (Sulimski 1964) differ metrically from our finds. *S. whitei* Hinton, 1914 was described from West Runton on the basis of a single P^4, which, however, is clearly smaller than the same tooth in *S. vulgaris*. The P^4 and molars of *S. whitei hungaricus* from its type locality Tarkő/Hungary, megye Hevest are also smaller, as are those of *Sciurus* sp. from Kozi Grzbiet/Poland, województwo Kielce, Kővesvárad/Hungary, megye Borsod (Jánossy 1963) and Hohensülzen/Germany, Rheinland-Pfalz (Storch et al. 1973). The remains from Untermassfeld are even smaller than the teeth from Tarkő. Although the validity of *S. whitei* is not universally accepted (Reumer and Hoek Ostende 2003), we stress the observation that Biharian finds, to our knowledge, are always smaller than extant *S. vulgaris*, and consequently we refer our finds to *S. whitei*.

Subfamily Xerinae Osborn, 1910
Tribe Marmotini Pocock, 1923
Genus *Spermophilus* F. Cuvier, 1825
Spermophilus kahlkei n. sp.
38 skeletal units; 42 crania and cranial fragments; 3 maxilla fragments; 44 left and 47 right mandible fragments; 140 incisor fragments; 100 P^{sup}/M^{sup}; 105 P_{inf}/M_{inf}; 99 molar fragments; 25 scapula fragments; 80 humerus fragments; 48 ulna fragments; 43 radius fragments; 28 pelvis fragments; 51 femur fragments; 46 tibia fragments; 2 fibula fragments; 24 astragali; 24 calcanei; 327 metapodials/phalanges of indeterminable position; 3 sacrum fragments; 84 vertebra fragments; 26 costa fragments; 161 indeterminable postcranial fragments.

More than 1,000 specimens from Untermassfeld, including several articulated, more or less complete skeletons, belong to a ground squirrel (**Figs 1–3**). From the number of skulls in the southern part of the excavation, Kahlke (in this volume) implied an MNI of 45. Based on the counting of left P^4, a minimum of further 35 individuals has been recorded in the remaining areas. In Europe, these animals are represented by the genus *Spermophilus*. Compared to *Sciurus* their cheek teeth have higher tubercles, a more triangular outline of P^4, M^1 and M^2, para- and metastyle forming a 'V', and less pronounced entoconid on the lower molars. In the Early and early Middle Pleistocene of Europe four *Spermophilus* species have been described: *S. primigenius* (Kormos, 1934) (type locality Villany 3/Hungary, megye Baranya), *S. nogaici* (Topachevskij, 1957) (type locality Nogajsk/Ukraine, Zaporiz'ka oblast'), *S. polonicus* (Gromov, 1965) (type locality Kamyk/Poland, województwo Katowice), and *S. dietrichi* (Kretzoi, 1965) (type locality Voigtstedt/Germany, Thuringia). The morphological distinction between *S. primigenius* and *S. polonicus* seems to be less clear than described in the literature. According to Gromov et al. (1965) a vertical groove in the root of the P^3 is characteristic of *S. polonicus*, but it can be found also in Hungarian material of *S. primigenius*. Kormos (1934) claimed that well-developed mesostylics on lower molars are characteristic of *S. primigenius*, but they can be seen also in Polish material of *S. polonicus*. In the extensive collection from Untermassfeld, P^3 with both grooved and ungrooved roots occur, and mesostylids in lower molars are often present but lack in c. 10 % of the specimens. Gromov et al. (1965) referred all four species mentioned above to the subgenus *Urocitellus*, which they defined morphologically by (1) a distinct protocone and metaconule, (2) distal margin of protocone collinear with the posterior cingulum (in subgenus *Spermophilus* and *Colobotis* in a line with metastyle), (3) P_4 distally broader, and (4) ento- and hypoconid of M_3 less distinct than in other subgenera. According to these characters, also our material would belong to *Urocitellus*.

Helgen et al. (2009) recently revised the genus *Spermophilus*, stating that the extant species formerly assigned to *Spermophilus* have to be referred now to eight different genera, one of which is *Urocitellus*. At present we do not know whether or how these results can be applied to the Early Pleistocene species as well, so we defer a decision on whether the fossils from Untermassfeld are referable to *Urocitellus* in the new taxonomic sense.

However, a comparison of the size of cheek teeth of the aforementioned fossil species reveals a stratigraphic pattern, as intimated previously (Maul 2001a). Measurements of molars of *S. nogaici* and of *S. primigenius* are always smaller, and those of *S. dietrichi* larger, than those from Untermassfeld. Interestingly, the late Biharian sample of *S. polonicus* from Kozi Grzbiet clusters with the type of *S. dietrichi* from Voigtstedt and is distinctively larger than *S. polonicus* from the early Biharian site of Kamyk. For all tooth positions, the values of the Untermassfeld material are clearly intermediate between the measurements of these two Polish

Fig. 1 *Spermophilus kahlkei* n. sp., holotype, Untermassfeld. – Postcranial skeleton IQW 2012/38656 (Mei. 35630), left dorso-lateral view. – Scale: 30 mm. The head of the same individual is shown in **Fig. 2g–i** and **Fig. 3m–r**.

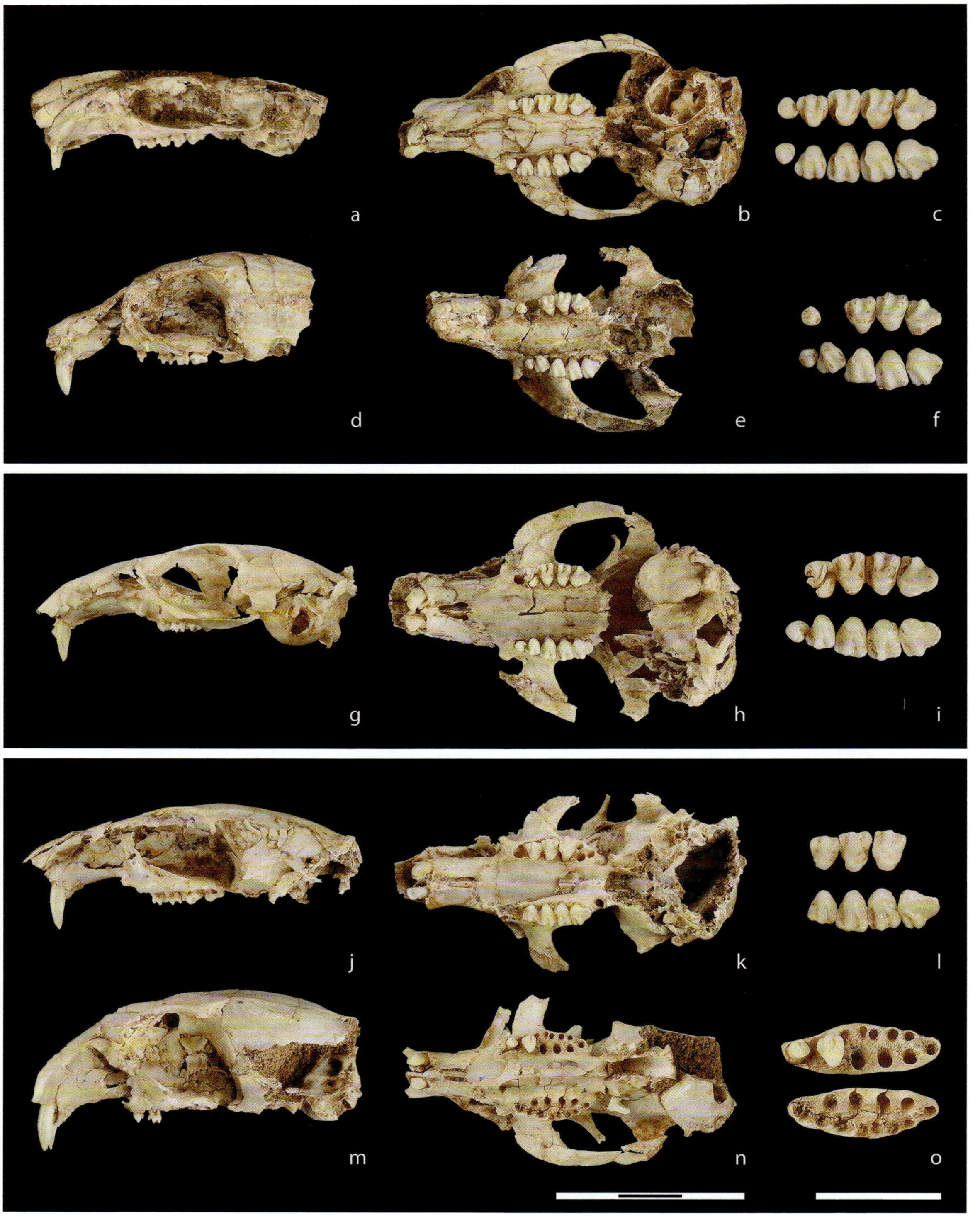

Fig. 2 *Spermophilus kahlkei* n. sp., type material, Untermassfeld. Skull fragments. – **a–c** IQW 2011/31351 (Mei. 30513) (paratype). – **d–f** IQW 2012/38419 (Mei. 35393) (paratype). – **g–i** IQW 2012/38657 (Mei. 35631) (holotype). – **j–l** IQW 2009/30627 (Mei. 29789) (paratype). – **m–o** IQW 2012/39320 (Mei. 36294) (paratype). – **a**, **d**, **g**, **j**, **m** left lateral views; **b**, **e**, **h**, **k**, **n** palatinal views; **c**, **f**, **i**, **l**, **o** occlusal views. – Scales: **left and middle column** 30 mm – **right column** 10 mm.

sites, best visible in P^4 and P_4 (**Figs 4–5**). This is consistent also with data on arvicolines from the same sites. The primitive arvicoline species *Microtus* (*Allophaiomys*) *deucalion* occurs in Kamyk, which is definitely less evolved and older than *Microtus thenii* from Untermassfeld (see section 3.4.), whereas the *Microtus* species

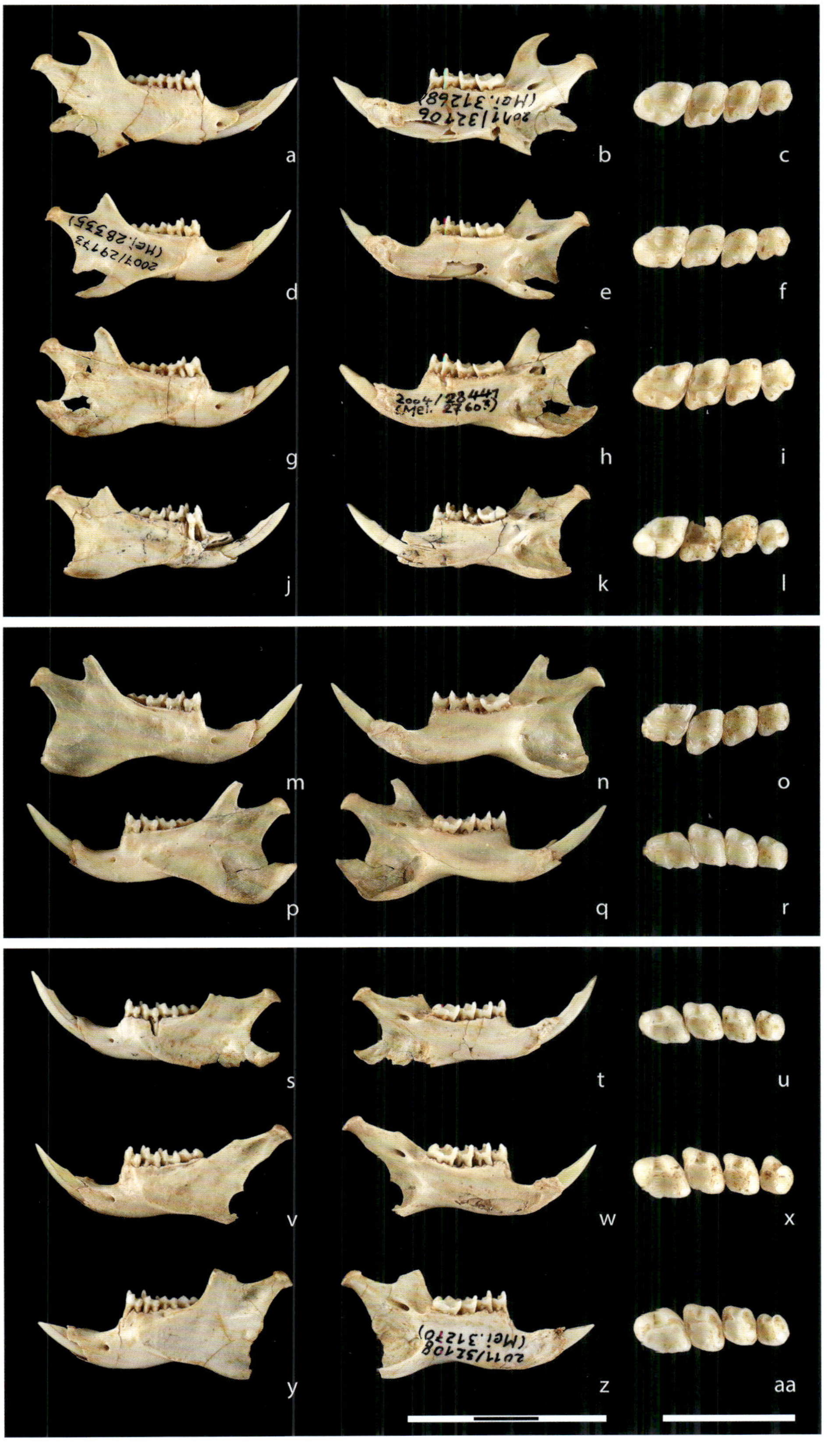

Fig. 3 *Spermophilus kahlkei* n. sp., type material, Untermassfeld. – **a–o** Right mandibles. – **p–aa** Left mandibles. – **a–c** IQW 2011/32 106 (Mei. 31 268) (paratype). – **d–f** IQW 2007/29 173 (Mei. 28 335) (paratype). – **g–i** IQW 2004/28 441 (Mei. 27 603) (paratype). – **j–l** IQW 2012/ 38 582 (Mei. 35 556) (paratype). – **m–r** IQW 2012/38 657 (Mei. 35 631) (holotype). – **s–u** IQW 2012/38 629 (Mei. 35 603) (paratype). – **v–x** IQW 2003/27 644 (Mei. 27 149) (paratype). – **y–aa** IQW 2011/32 108 (Mei. 31 270) (paratype). – **a**, **d**, **g**, **j**, **m**, **p**, **s**, **v**, **y** buccal views; **b**, **e**, **h**, **k**, **n**, **q**, **t**, **w**, **z** lingual views; **c**, **f**, **i**, **l**, **o**, **r**, **u**, **x**, **aa** occlusal views.– Scales: s. Fig. 2.

in Kozi Grzbiet and in Voigtstedt are more advanced than *M. thenii*. That the size increase in Early and early Middle Pleistocene *Spermophilus* follows the stratigraphic order, is supported by the type specimens of *S. primigenius* and *S. nogaici* which are both smaller and older than the Untermassfeld material. We wish to emphasise the distinctiveness of the Untermassfeld ground squirrel material by describing a new species.

Description of a new species:
Spermophilus kahlkei n. sp.
Etymology: The specific name honors Prof. Dr. Dr. Hans-Dietrich Kahlke (†), who successfully investigated many Quaternary faunas in Eurasia and initiated the Untermassfeld project.
Type locality and type layer: Upper Fluviatile Sands (lower part) on a high terrace of the Werra river in Southern Thuringia close to the village of Untermassfeld, Central Germany (for details of the geological setting see Ellenberg and Kahlke 1997; Kahlke 2001b; Kahlke et al. 2020)
Holotype: Articulated skeleton IQW 2012/38 656 (Mei. 35 630) with skull and mandible of the same individual IQW 2012/38 657 (Mei. 35 631) (**Fig. 1, 2g–i, 3m–r**)
Paratypes: Skull fragments IQW 2011/31 351 (Mei. 30 513), IQW 2012/38 419 (Mei. 35 393), IQW 2009/30 627 (Mei. 29 789), IQW 2012/39 320 (Mei. 36 294) (**Fig. 2**); mandible fragments IQW 2011/32 106 (Mei. 31 268), IQW 2007/29 173 (Mei. 28 335), IQW 2004/28 441 (Mei. 27 603), IQW 2012/38 582 (Mei. 35 556), IQW 2012/38 629 (Mei. 35 603), IQW 2003/27 644 (Mei. 27 149), IQW 2011/32 108 (Mei. 31 270) (**Fig. 3**)
Diagnosis: A medium-sized ground squirrel of the genus *Spermophilus* (larger than *S. citellus*, **Table 1**), in the upper molars proto-, para- and metacone are the highest tubercles, in the lower molars metaconid and paraconid. Main diagnostic character is size and proportion of the cheek teeth, which are larger than in *S. primigenius*, *S. nogaici* and early Biharian *S. polonicus* (from Kamyk) but smaller than the sample from Kozi Grzbiet, and smaller than *S. dietrichi* from Voigtstedt.
Description: The upper incisor is mesio-distally compressed and the enamel covers half of the mesial side. Buccal and lingual tips of P^3 are nearly fused, and the posterior cingulum is broad. The root is clearly grooved on the buccal side but weakly or not at all grooved on the lingual side. From all P^4-M^3 a strongly developed protocone develops an anterior parastyle that ends buccally with the paracone. On the ridge between paracone and protocone a small paraconule is present. The posterior metastyle ends buccally with the metacone. Between the paracone and metacone a small metaconule is visible, between the anterior and posterior ridge the mesostyle. The anterior margin bears a broad cingulum, and the posterior margin a narrow one. The hardly worn P^4, M^1 and M^2 display a clear distinction of metaconule and protoconule. The buccal mesostyle is more or less clearly developed. The distal margins of protocone and posterior cingulum are colinear. P^4 is molariform with an anterior cingulum narrower than in M^1 and M^2. In some specimens the inner and the anterior outer root are fused. The metastyle only weekly developed on M^3, and the metacone is only a small tubercle. The lower incisor is also laterally flattened, less curved than the upper one. On all P_4-M_3 the trigonid is a high wall, the talonid surrounds a valley. The lingual metaconid is connected with the buccal protoconid by a mesial and a distal metalophid ridge. Between these elements a trigonid basin is developed. In front of the mesial ridge is a small protoconulid. The metaconid is the highest tubercle. On the lingual part of the talonid a mesostylid is developed, and on the buccal part a mesoconid. The lingual entoconid is often strongly reduced. The buccal hypoconid is moderately developed. The posterior and the anterior margin of the tooth rarely if ever parallel (exception: P_4 and M_1) but in lingual-buccal direction they turn a bit distally, in P_4, the protoconulid protudes from the mesial margin. The enlarged hypoconid causes P_4 to be broader distally than mesially. The posterior root of P_4 is transversely broadened, mesially slightly and distally more or less clearly grooved. The lower molars usually possess one to two mesostylids. On M_3 hypo- and entoconid are hardly distinct.

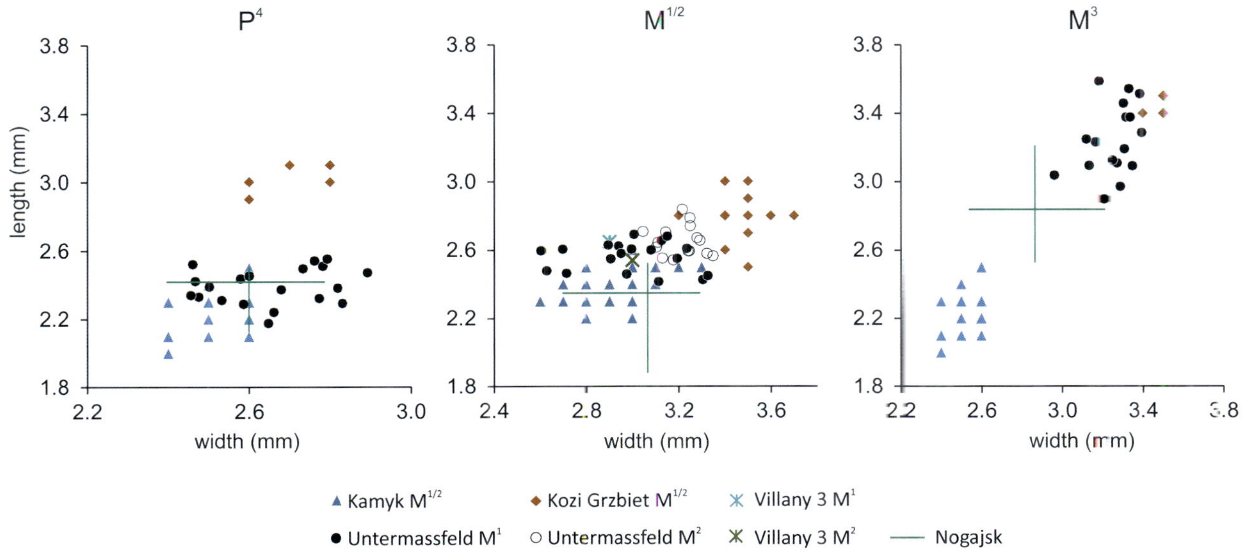

Fig. 4 *Spermophilus kahlkei* n. sp. from Untermassfeld, *S. polonicus* from Kozi Grzbiet and Kamyk, *S. primigenius* from Villány 3, and *S. nogaici* from Nogajsk. Scatter diagram of measurements of upper cheek teeth.

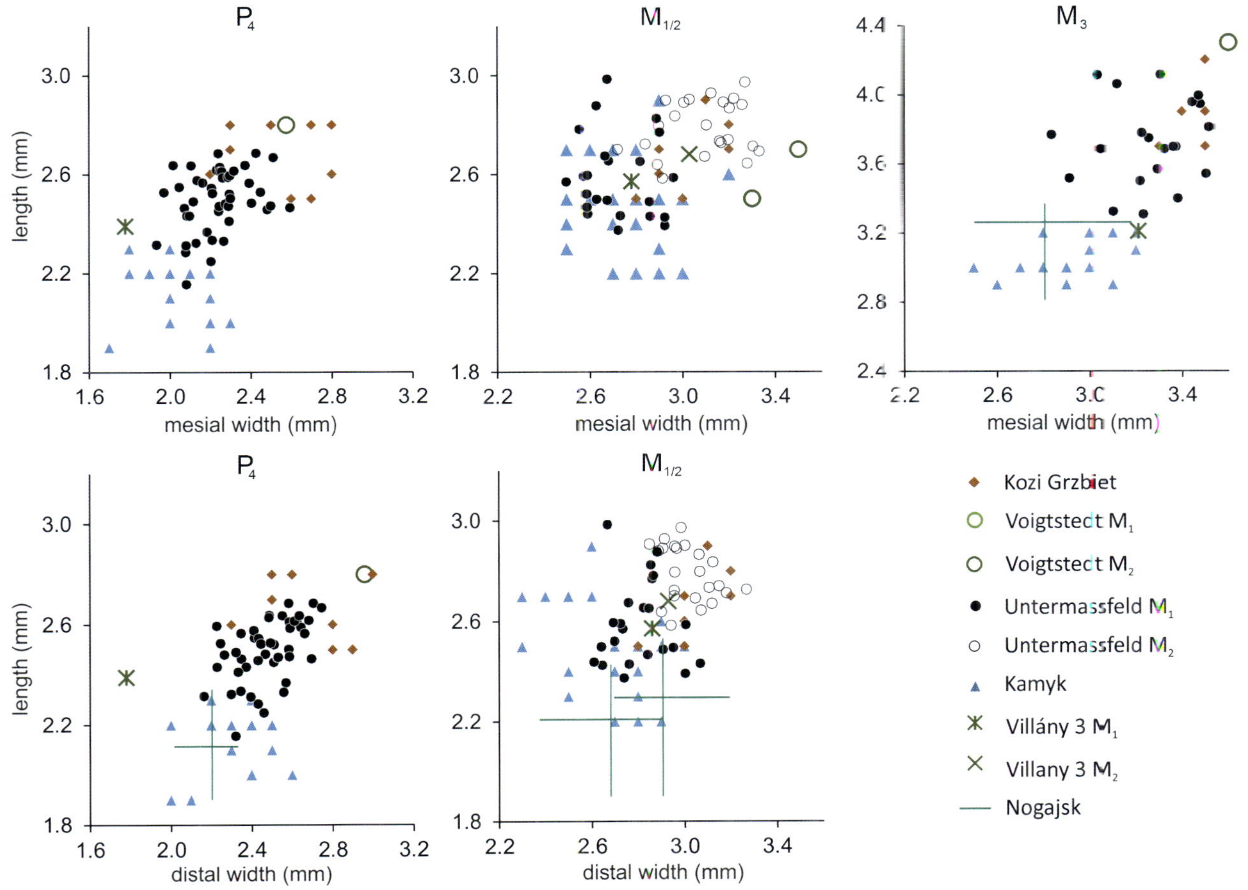

Fig. 5 *Spermophilus kahlkei* n. sp. from Untermassfeld, *S. polonicus* from Kozi Grzbiet and Kamyk, *S. dietrichi* from Voigtstedt, *S. primigenius* from Villány 3, and *S. nogaici* from Nogajsk. Scatter diagram of measurements of lower cheek teeth.

Skull

		IQW 2012/38657 (Mei. 35631)		IQW 2009/30627 (Mei. 29789)		IQW 2011/31351 (Mei. 30513)		IQW 2012/38419 (Mei. 35393)		IQW 2012/39320 (Mei. 36294)			total range			
		left	right	left	right	left	right	left	right	left	right					
		holotype		paratypes								n	min	mean	max	sd
	Greatest length	47.30														
	Infraorbital width	12.60														
	Zygomatic width	34.20														
P^3-M^3	Coronar length		12.02			12.47	12.77	11.57	11.37			12	11.03	11.77	12.77	0.479
	Alveolar length	11.94	12.07	12.32	11.50	12.88	13.09	11.61	11.99	12.66	12.63	21	11.47	12.26	13.27	0.546
P^3	Length		1.78			1.55	1.59	1.72	1.41	1.63		14	1.35	1.56	1.78	0.122
P^4	Length	2.54	2.29	2.29	2.31	2.52	2.49		2.33	2.18		20	2.18	2.39	2.55	0.106
	Width	2.76	2.59	2.83	2.53	2.46	2.73		2.48	2.65		20	2.46	2.65	2.89	0.141
M^1	Length	2.45	2.41	2.43	2.60	2.60	2.69	2.67	2.46			22	2.41	2.58	2.78	0.096
	Width	3.33	3.11	3.31	3.08	2.70	3.01	2.91	2.72			22	2.60	2.99	3.33	0.202
M^2	Length	2.58	2.54	2.55	2.67	2.71	2.74	2.36	2.70			18	2.36	2.64	2.84	0.116
	Width	3.32	3.18	3.13	3.28	3.05	3.25	3.38	3.14			18	3.05	3.24	3.38	0.099
M^3	Length	3.23	3.25		3.19	3.51	3.46	3.09	2.97			17	2.90	3.24	3.58	0.206
	Width	3.17	3.12		3.31	3.39	3.31	3.35	3.29			17	2.96	3.25	3.40	0.112

Table 1 *Spermophilus kahlkei* n. sp., Untermassfeld. Measurements (mm) of skull, mandible and postcranial bones.

Mandible		IQW 2012/38657 (Mei. 35631)		IQW 2003/27644 (Mei. 27149)	IQW 2004/28441 (Mei. 27603)	IQW 2007/29173 (Mei. 28335)	IQW 2011/32106 (Mei. 31268)	IQW 2011/32108 (Mei. 31270)	IQW 2012/38582 (Mei. 35556)	IQW 2012/38629 (Mei. 35603)					
		left	right									total range			
		holotype		paratypes							n	min	mean	max	sd
P_4-M_3	Coronar length	10.40	10.74	11.57	11.42	11.11	10.92	11.05	11.20	11.39	22	10.40	11.24	12.31	0.440
	Alveolar length	11.73	11.79	12.38	12.32	11.70	12.18	12.13	12.35	12.01	23	11.52	12.13	13.47	0.460
P_4	Length	2.28	2.53	2.64	2.47	2.35	2.15	2.32	2.33	2.31	47	2.15	2.49	2.68	0.125
	Anterior width	2.08	1.97	2.02	2.25	2.04	2.08	2.13	2.21	2.08	47	1.94	2.23	2.60	0.148
	Posterior width	2.43	2.25	2.49	2.53	2.34	2.32	2.30	2.35	2.40	47	2.17	2.47	2.75	0.144
M_1	Length	2.47	2.57	2.49	2.49	2.88	2.65	2.37	2.59	2.67	23	2.37	2.59	2.98	0.167
	Anterior width	2.59	2.50	2.68	2.86	2.63	2.68	2.72	2.59	2.67	23	2.50	2.72	2.96	0.142
	Posterior width	2.84	2.74	2.95	2.91	2.88	2.82	2.74	2.69	2.76	23	2.61	2.81	3.07	0.127
M_2	Length	2.72	2.70	2.90	2.64	2.84	2.91	2.80	2.88	2.90	23	2.58	2.79	2.97	0.11
	Anterior width	2.84	2.72	3.03	3.28	2.97	3.22	3.10	3.26	2.93	23	2.72	3.09	3.33	0.168
	Posterior width	2.96	2.96	3.01	3.07	3.12	2.85	3.07	2.89	2.96	23	2.85	3.02	3.27	0.106
M_3	Length	4.11	3.32	3.70	3.81	3.77	3.68	3.95	3.50	4.11	22	3.30	3.74	4.14	0.255
	Anterior width	3.04	3.10	3.36	3.52	2.84	3.33	3.48	3.22	3.31	22	2.84	3.20	3.60	0.208

Table 1 (continued)

Postcranial bones		IQW 2012/38656 (Mei. 35630)				total range		
		left	right					
		holotype		n	min	mean	max	sd
Humerus	Length	35.85	35.52	24	30.60	34.31	38.05	2.126
	Width of distal epiphysis	9.88	9.03	26	8.59	9.84	10.78	0.505
Ulna	Length	41.29	41.67	4	38.14	40.02	41.67	1.725
Radius	Length	34.45	33.77	8	27.71	30.44	34.45	2.373
Pelvis	Length	46.77	45.00	7	40.82	44.43	46.77	2.212
	Diameter of acetabulum	5.43	5.22	9	4.92	5.24	5.51	0.250
Femur	Length	45.38	45.72	12	39.90	43.54	46.53	2.317
	Width of distal epiphysis	8.66	8.73	9	7.58	8.19	8.73	0.422
Tibia	Length	45.38	46.28	14	39.76	43.89	47.62	2.142

Table 1 (continued)

The holotype is the most complete specimen with the entire postcranial skeleton, skull and mandible preserved (**Figs 1–3**, Kahlke in this volume, Fig. 19). It was an adult animal since the epiphyses of most long bones are fused. The posture of the animal is remarkable. In the wild, squirrels can be seen sitting upright and looking around for predators, and if they are on four legs, at least the forelegs are always straight, thus raising the front part of the body. This is not the case in this skeleton. It is neither a sleeping nor a hibernating/estivating posture.

Sexual size dimorphism is a widespread phenomenon in animals including mammals. In extant ground squirrels it occurs in some species, e.g., *Urocitellus columbianus* (formerly *Spermophilus columbianus*, see Elliott and Flinders 1991) but not in others, e.g., *Urocitellus armatus* (formerly *Spermophilus armatus*, see Eshelman and Sonnemann 2000). However, several pelves were available, encouraging us to try to determining the sex of these animals. The meagre literature on this topic in rodents refers mainly to other taxa (Arvicolinae, Murinae, see Brown and Twigg 1969), but one of the first papers (Dunmire 1955) also investigated a ground squirrel, *Otospermophilus beecheyi*, which is much larger than our animal (head and body length c. 300 mm). Dunmire found that in females the ventral region of the ischium is smaller, and the ratio of thickness to length of the pubic symphysis is higher. In most cases of the Untermassfeld material just these parts are either broken or unavailable in the in-situ block. However, in two comparative pelves of male and one female of *Callospermophilus lateralis* we recognised a considerably thicker dorsocaudal part of the ischium in the pelvis of the male than in that of the female. This 'male type' we found also in the pelvis of the holotype of *S. kahlkei* n. sp., whereas in five other specimens [IQW 2012/38 658 (Mei. 35 632), IQW 2012/38 660 (Mei. 35 634), IQW 2003/28 189 (Mei. 27 351), IQW 2003/27 643 (Mei. 27 148), IQW 2003/28 187 (Mei. 27 349)] the situation was as in the female. However, this consideration is a preliminary one and must be tested on bigger comparative samples. Unfortunately, no baculum has yet been recovered, which would clearly be indicative for male ground squirrels.

Family Gliridae Muirhead, 1819
Subfamily Glirinae Muirhead, 1819
Genus *Glis* Brisson, 1762
Glis sackdillingensis (Heller, 1930)
1 left and 3 right I^{sup}; 1 right I_{inf}; 2 left M_1; 1 left M_2; 1 right M_3.

The outline, shape, and numbers and position of ridges are typical for *Glis*. The Pliocene *Glis minor* Kowalski, 1956, the Pleistocene *Glis sackdillingensis* (Heller, 1930), and the extant *Glis glis* (Linnaeus, 1766) (Daoud 1993; Nadachowski and Daoud 1995) presumably constitute one lineage displaying two concurrent trends: increasing size and decreasing morphological complexity, i.e., a tendency to reduce the accessory ridges (Heller 1930; Koenigswald 1970; Nadachowski and Daoud 1995). This applies mainly to M^2, which is not recorded in Untermassfeld. Our M_1s resemble mostly morphotype 1 of Daoud (1993): antero- and metalophid and meso- and posterolophid lingually connected. However, one M_1 possesses two secondary extra ridges, a rare condition that does not occur in the larger sample studied by Daoud (1993). In all M_1 and M_3 from the Plio-Pleistocene of Poland morphotype I is dominant in all three *Glis* species (Daoud, 1993), so its occurrence in Untermassfeld has no special taxonomic value. However, based on size the referral of the Untermassfeld form to *Glis sackdillingensis* is certain, since they are smaller than the teeth from Sackdilling/Germany, Bayern, and Süssenborn/Germany, Thuringia, in agreement with their older stratigraphic age. Concluding from the number of right I^{sup}, at least three individuals are represented.

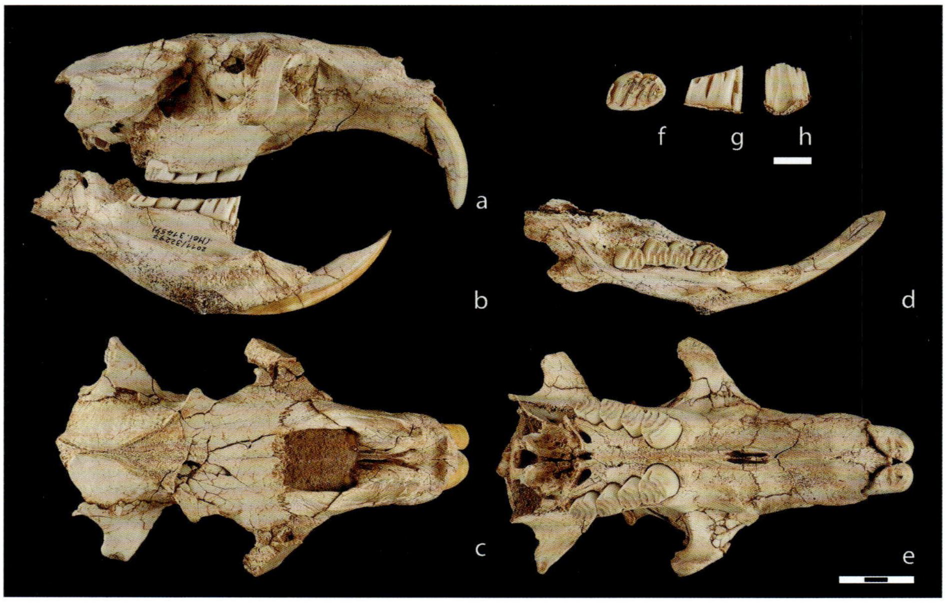

Fig. 6 *Trogontherium cuvieri* Fischer, Untermassfeld. – **a**, **c**, **e** Skull IQW 2009/30639 (Mei. 29801). – **b**, **d** Left mandible IQW 2011/32297 (Mei. 31459). – **f–h** Left P_4 IQW 1982/18014 (Mei. 17534). – **a** lateral view; **b**, **g** lingual views; **c** dorsal view; **d**, **f** occlusal views; **e** palatinal view; **h** mesial view. – Scales: **a–e** 30 mm – **f–h** 10 mm.

Suborder Castorimorpha Wood, 1955
Family Castoridae Hemprich, 1820
Subfamily Castorinae Hemprich, 1820
Genus *Castor* Linnaeus, 1758
Castor fiber Linnaeus, 1758
1 left I^{sup}; 2 left P^4; 1 right $M^{1/2}$; 1 left and 2 right M^3; 1 left and 1 right mandible with I-M_3; 1 left and 1 right P_4; 1 left $M_{1/2}$.

The extreme hypsodont, 'pillared' molars, their typical occlusal pattern, and mandible size and shape indicate the presence of *Castor fiber* Linnaeus, 1758. The evolutionary level in this lineage can be assessed by the depth of lateral striae/striidae of molars, particularly of P_4. The relative depth of lateral striae and striidae in upper and lower cheek teeth (particular suitable are P^4 and P_4) successively increases during evolution and thus correlates with the geological age of the finds (Kretzoi 1969; 1975; 1977). Heinrich (1989; 1991) modified the method and documented that the length, width and depth of the metastriid and of the mesostriid in the P_4 of *Castor* from Süssenborn, Bilzingsleben, Ehringsdorf and Taubach (all Germany, Thuringia) were inversely correlated and changed continuously with stratigraphic age. At Untermassfeld, there is only one measurable isolated P_4 available (cheek teeth in the mandibles cannot be extracted without destroying them) with unclosed crown basis. Length and width fit the size trend and are smaller than those from Süssenborn. For the striid-hypsodonty-index only a minimum value can be expressed since the crown is not yet developed. However, no new material has been recovered since 2001 which could give the index more precisely. The presence of isolated P_4 in addition to the mandibles with complete tooth rows indicates an MNI of 2.

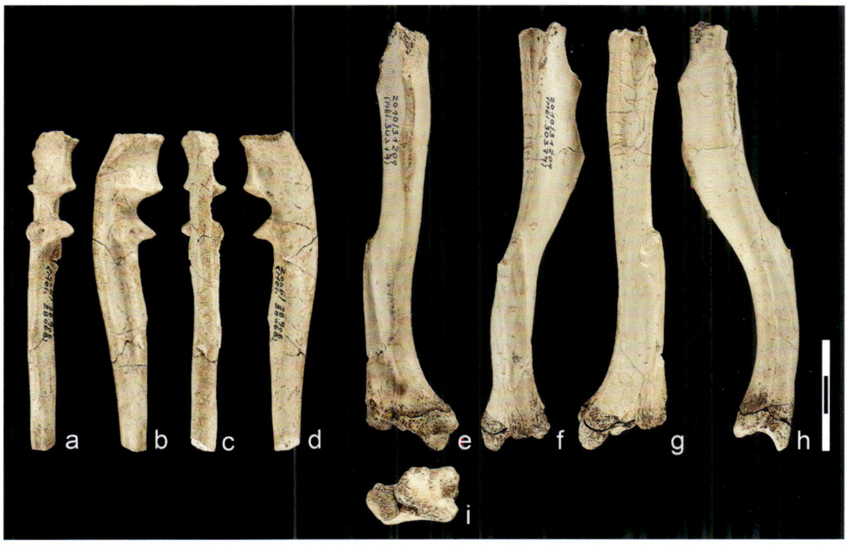

Fig. 7 *Trogontherium* Fischer, Untermassfeld. – **a–d** Right ulna IQW 2005/28 906 (Mei. 28 068). – **e–i** Right tibia IQW 2010/31 209 (Mei. 30 371). – **a**, **e** dorsal views; **b**, **f** lateral views; **c**, **g** ventral views; **d**, **h** medial views; **i** distal view. – Scale: 30 mm.

Subfamily Castoroidinae Allen, 1877
Tribe Trogontheriini Lychev, 1973
Genus *Trogontherium* Fischer, 1809
Subgenus *Trogontherium* Fischer, 1809
Trogontherium (*Trogontherium*) *cuvieri* Fischer, 1809
1 cranium; 1 left M^3; 1 left mandible with P_4-M_3; 1 left P_4; 1 right ulna fragment; 1 left pelvis fragment; 1 left and 1 right tibia fragment; 1 sacrum; 1 vertebra.

Actually, the name Trogontheriini Lychev, 1973, used by Korth (2001) in his taxonomic account, is not available, since it refers to an unpublished manuscript. However, the species name *Trogontherium cuvieri* Fischer, 1809 »is an even greater nomenclatural absurdity« (Mayhew 1978), one that is, however, widely used. Finds of *Castor* and *Trogontherium* can be distinguished by morphology and size (e. g., Owen 1869; Schreuder 1929; Mayhew 1978; Stefen 1997), for example, *Trogontherium* possesses an oval cross-section of lower incisors with a granulated enamel surface and on cheek teeth enamel folds that are not or hardly filled with cementum (Lehmann 1953; Stefen 1997). M^3 and P_4 are elongated and conical. The new material of *Trogontherium* includes a nearly complete skull and a left mandible (**Fig. 6**).
During its evolution from Miocene to Pleistocene *Trogontherium* experienced considerable metric and morphologic changes, such as an increase in the length and width of incisors and premolars and in mandible width (Mai 1979). P_4 length and width measured on the crown base show the trend better than the measures of the occlusal surface (Heinrich 1997). The size of our worn (senile) P_4 is above the range of Tegelen/The Netherlands, provincie Limburg, and below that of Maasvlakte/The Netherlands, provincie Zuid-Holland, and the size of the P_4 from the mandible is even larger than that of the Maasvlakte finds (**Table 2**).
Heinrich (1998) found that in the anterior part of the P_4 an antero-lingual flexid/striid is progressively included in the occlusal surface. In stratigraphically older finds of *Trogontherium* this element is either completely smooth or forms a shallow groove (Heinrich 1998). Therefore, Heinrich distinguished Early to Middle Pleistocene *T. boisvilletti* (Laugel, 1862) with four flexids/striids on P_4 from Middle Pleistocene *T. cuvieri* with

Skull		IQW 2009/30639 (Mei. 29801)	
		left	right
	Greatest length (without occiput)	165.0	
	Zygomatic width	> 103.0	
	Infraorbital width	43.8	
	Length of diastema	75.25	73.90
	Skull height above M^1	54.30	
I^{sup}	Mesial-distal width	10.57	10.46
	Buccal-lingual width	10.38	10.66
P^4-M^3	Coronar length	38.27	37.21
	Alveolar length	43.52	40.65
P^4	Length	11.44	11.02
	Width	12.23	13.08
M^1	Length	7.29	7.56
	Width	9.30	9.51
M^2	Length	7.40	7.68
	Width	9.74	9.57
M^3	Length	10.72	10.58
	Width	9.68	8.67
Left mandible		**IQW 2011/32297 (Mei. 31459)**	
Diastema	Length	30.61	
I_{inf}	Mesial-distal width	10.30	
	Buccal-lingual width	12.75	
P_4-M_3	Coronar length		
	Alveolar length	44.49	
P_4	Length at crown base	14.70	
	Width at crown base	11.58	
M_1	Length at crown base	7.73	
	Width at crown base	11.45	
M_2	Length at crown base	7.41	
	Width at crown base	11.48	

Table 2 *Trogontherium cuvieri*, Untermassfeld. Measurements (mm) of skull, mandible and postcranial bones.

five flexids/striids on P_4 (Heinrich 1998; Heinrich and Kolfschoten 2007). In our isolated very worn P_4 the additional striid is (if it had existed) worn away. In the premolar of the mandible the flexid forms a notch but the striid seems to reach rather deep into the tooth crown (**Fig. 6g**). Since dental morphology and size of the new finds are closer to *T. cuvieri*, we refer it to this species. The record of at least 2 individuals is concluded from the cranium with complete dentition and the isolated M^3.

Postcranial bones		IQW 2005/28906 (Mei. 28068)	IQW 2010/31209 (Mei. 30371)	IQW 2011/32048 (Mei. 31210)
Ulna	Length of semilunar notch	12.44		
	Height of olecranon	14.43		
	Length of olecranon	> 15.00		
Pelvis	Cranial-caudal diameter of acetabulum			19.94
	Dorsal-ventral diameter of acetabulum			16.45
Tibia	Width of distal epiphysis		24.80	

Table 2 (continued)

The nearly complete skull and the mandible (**Fig. 6**) presumably belong to one individual, because size and tooth abrasion are similar and they have been found in neighbouring squares. Also all castorid postcranial elements (**Fig. 7**) belong very likely to *Trogontherium*, since they are different from those of a comparative skeleton of an extant *Castor* but similar to the figures in Schreuder (1929). Although fragmentary, their spatial occurrence in the excavation area (see section 3.3.) implies they belong to the same individual as the skull and the mandible.

Suborder Myomorpha Brandt, 1855
Superfamily Dipodoidea Fischer, 1817
Family Dipodidae Fischer, 1817
Subfamily Sicistinae Allen, 1901
Genus *Sicista* Gray, 1827
Sicista sp.
1 right M_2.

The tooth morphology clearly supports referral to the genus *Sicista*, with metaconid, paraconid and protoconid forming a characteristic 'cloverleaf', an oblique ridge with ascending spur (Kowalski 1960). Beside the two extant European species *Sicista betulina* Pallas, 1779 and *S. subtilis* Pallas, 1773, a third species, *S. praeloriger* Kormos, 1930, was described from the Early Pleistocene of Betfia in Romania (Kormos 1930). According to Kowalski (1979; 2001) it is presumably the ancestor of *S. subtilis*. With regard to dental morphology it is sometimes difficult to distinguish the extant species from each other and from the Early Pleistocene one. There is a rather large overlap in size and only samples can be distinguished. Similarly, a distinction according to morphotypes requires larger samples. Therefor we refer the single M_2 to *Sicista* sp.

Superfamily Muroidea Illiger, 1811
Family Cricetidae Fischer, 1817
Subfamily Cricetinae Fischer, 1817
Genus *Cricetus* Leske, 1779
Cricetus runtonensis Newton, 1909
1 skeleton; 3 left and 2 right M^1; 2 right M^2; 1 right M^3; 1 left and 1 right M_1, 2 left M_2; 1 right M_3; 3 molar fragments; 2 left and 2 right humerus fragments; 1 radius fragment; 1 right femur fragment; 1 calcaneus; 1 phalanx of indeterminable position.

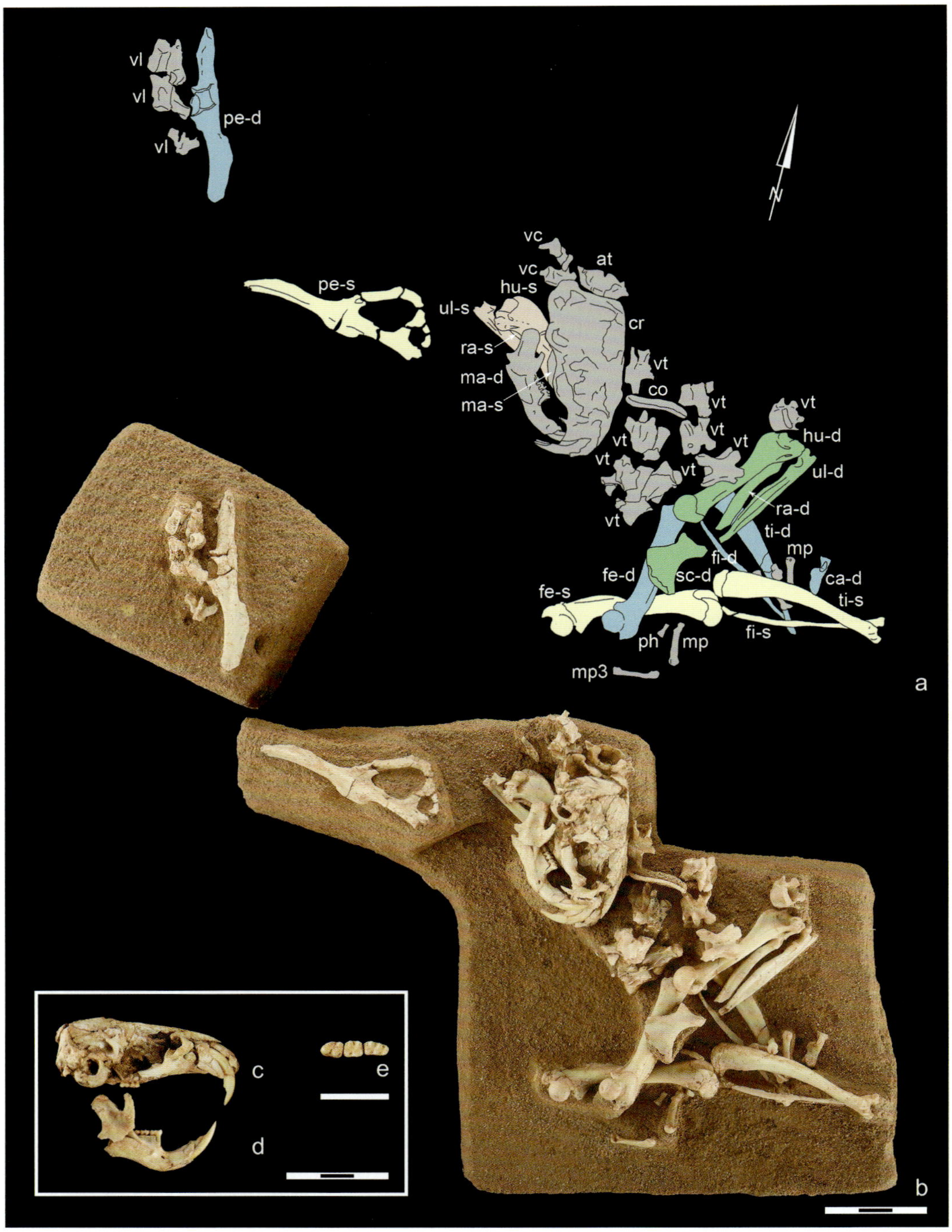

Fig. 8 *Cricetus runtonensis* Newton, Untermassfeld. Skeleton, skull and mandible IQW 2003/28 123 (Mei. 27 285). – **a** Diagrammatic positions of skeletal elements. – **b** in-situ position of skeletal elements. – **c** Skull, right lateral view. – **d** Right mandible, buccal view. – **e** Right lower tooth row, occlusal view. – Scales: **a–d** 30 mm – **e** 10 mm. – Abbreviations: at – atlas; ca – calcaneus; co – costa; cr – cranium; fe – femur; fi – fibula; hu – humerus; ma – mandible; mp – metapodial; pe – pelvis; ph – phalanx; ra – radius; sc – scapula; ti – tibia; ul – ulna; vc – vertebra cervicalis; vl – vertebra lumbalis; vt – vertebra; suffix -s indicates left specimens; suffix -d indicates right specimens.

Diagnostic features for *Cricetus* are the rhombic funnels between the tubercles and the possession of three (M^1 and M_1) and two (M^2 and M_2) pairs of tubercles. In M_1 the anterior pair is more slender than the middle one, and anterior and posterior pairs are shifted buccally. The new material includes tooth loci not represented in the pre-2001 sample and an articulated complete skeleton of a so-called 'giant hamster', by which we mean it is larger than the extant 'large hamster' *Cricetus cricetus* (Linnaeus, 1758). This skeleton is displayed as photo and a diagram in **Fig. 8**, which shows the find position of skeletal elements. The measurements (**Table 3**) confirm that our finds belong to a species slightly larger than the extant *C. cricetus* and the fossil *C. praeglacialis* (Schaub, 1930). It is within the range of specimens from the late Early Pleistocene (e.g., Zalesiaki/Poland, województwo Łodz; possibly Za Hájovnou Cave/Czech Republic, Olomoucký kraj) and early Middle Pleistocene (e.g., West Runton; Kozi Grzbiet/Poland, województwo Kielce) remains of *C. runtonensis* Newton, 1909 (Pradel 1988; Maul 2001a; Sabol and Zervanová 2014), but considerably smaller than the late Middle and Late Pleistocene of *C. major* Woldřich, 1881 finds from Výpustek/Czech Republic, Jihomoravský kraj; Petersbuch 2/Germany, Bayern; Genkingen 2/Germany, Baden-Württemberg, and Taubach/Germany, Thuringia (Woldřich 1880; Schaub 1935; Fahlbusch 1976; Ziegler 1995; Maul 2001a). However, a distinction from the late Early Pleistocene and early Middle Pleistocene populations (e.g., Villány 8/Hungary, megye Baranya, Hír 1996; 1997a; 1997b) is sometimes difficult. We do not agree with Schaub (1930) and Fahlbusch (1976), who argued that all Pleistocene giant hamsters belong to the species *C. major*, nor with Kowalski (2001), who lumped all Pleistocene large *Cricetus* into the extant species *C. cricetus*, considering them as extreme size variants.

In any case, the molars from Untermassfeld fall in the range of the Biharian finds from West Runton, Kozi Grzbiet and Zalesiaki 1 and are smaller than the Toringian remains from Petersbuch 2, Uppony/Hungary, megye Borsod, and Taubach/Germany, Thuringia. Also morphologically they differ from *Cricetus major* by the lack of accessory ridges. On the basis of size, we refer the Untermassfeld finds to *C. runtonensis*.

The complete skeleton enables us to relate dental and skeletal measurements (**Table 3**). Few measurements are published on hamster postcranial bones, which are often incomplete, lacking epiphyes. Such data are published from the Early Pleistocene of Sackdilling/Germany, Bayern (Heller 1930), and from the Late Pleistocene site Roter Berg/Germany, Thuringia (Schaub 1935). They indicate that the size of postcranial bones of Late Pleistocene *C. major* also exceed those of Early Pleistocene *C. runtonensis* from Untermassfeld and Sackdilling.

We intended to determine the sex of the hamster skeleton from Untermassfeld. As yet there is no publication on pelves of Eurasian cricetines. However, some observations in New World Cricetidae, Arvicolinae, and Murinae made by Dunmire (1955) and Brown and Twigg (1969) can be used. Assuming *Cricetus* is comparable to *Arvicola*, we can conclude from the above mentioned literature that our skeleton belongs to an adult male. The epiphyses are fused, indicating adulthood. The thickness of pubis is greater and the length of the descending ramus in relation to the length of the horizontal ramus of the ischium is smaller in males than in females (Dunmire 1955; Brown and Twigg 1969). Further, in males the obturator foramen is less elongated, the angle between the horizontal and descending rami of the ischium is lower (c. 90°), and the descending ramus of the ischium (caudal region) is thicker than in females. According to her study on pelves of extant *C. cricetus*, S. Könnicke (Dresden, pers. comm., May 2016) confirms this conclusion. 3 isolated left M^1 and the complete maxilla of the skeleton witness a record of at least 4 hamster individuals.

Subfamily Arvicolinae Gray, 1821
Genus *Myodes* Pallas, 1811
Myodes hintonianus (Kretzoi, 1958)
4 left and 2 right M^1; 8 left and 4 right M^2; 2 left and 2 right M_1; 6 left and 1 right M_2; 2 right M_3; 1 molar fragment (**Table 4**).

The possession of roots, crown cement, rounded and incompletely closed dentine fields, the shape of the anterior cap, the schmelzmuster (Koenigswald 1980) and enamel differentiation allow referral to the genus

Skull		IQW 2003/28 123 (Mei. 27 285)	IQW 1984/19 760 (Mei. 19 280)	IQW 2016/42 772 (Mei. 48 081)	IQW 2003/27 695 (Mei. 27 200)	IQW 2016/42 773 (Mei. 48 082)	IQW 1983/18 736 (Mei. 18 256)	IQW 2016/42 774 (Mei. 48 083)	IQW 2016/42 778 (Mei. 48 087)
	Greatest length	55.20							
M^1-M^3	Alveolar length	8.30							
M^1	Length		3.31	3.49	3.52	3.46			
	Anterior width		(1.67)	1.86		2.05			
	Medial width		2.12	2.02	2.34	2.25			
	Posterior width		2.02	2.11	2.43	2.34			
M^2	Length						2.83		
	Anterior width						2.27	2.50	
	Posterior width						2.13		
M^3	Length								2.72
	Anterior width								2.35
	Posterior width								1.84

Mandible		IQW 2003/28 123 (Mei. 27 285)		IQW 1984/19 761 (Mei. 19 281)	IQW 2016/42 771 (Mei. 48 080)	IQW 1985/20 425 (Mei. 19 945)	IQW 2016/42 770 (Mei. 48 079)	IQW 1983/18 780 (Mei. 18 300)
		left	right					
M_1-M_3	Coronar length	9.57	9.45					
	Alveolar length	9.76	9.79					
M_1	Length	3.44	3.53		3.44			
	Anterior width	1.33	1.36					
	Medial width	1.78	1.79		1.95			
	Posterior width	1.96	2.02	2.09	1.97			
M_2	Length	2.75	2.73			2.86	2.93	
	Anterior width	2.23	2.27			2.10	2.31	
	Posterior width	2.36	2.38			2.22	2.24	
M_3	Length	3.09	3.04					2.42
	Anterior width	2.30	2.33					1.69
	Posterior width	1.84	1.94					1.83

Table 3 *Cricetus runtonensis*, Untermassfeld. Measurements (mm) of upper cheek teeth, lower cheek teeth and postcranial bones.

Myodes. The linea sinuosa is worn in all our specimens, and the relative crown height is > 118. This does not provide further taxonomic guidance, since the assumed ancestor, *M. kretzoii* (Kowalski, 1958) (*M. sebaldi* Heller, 1963 is a synonym), also has such high crowns (Carls and Rabeder 1988). At Untermassfeld, all M_1 belong to the *hintonianus* morphotype (sensu Rabeder 1981). At Deutsch-Altenburg 2C and 4B/Austria,

Postcranial bones		IQW 2003/28 123 (Mei. 27 285)	
		left	right
Humerus	Length		45.08
	Length without distal epiphysis	15.21	41.40
Pelvis	Length	56.52	55.98
	Cranial-caudal diameter of acetabulum	7.09	8.18
Femur	Length	54.58	
	Length without distal epiphysis		46.50
Tibia	Length	52.74	54.40

Table 3 (continued)

Niederösterreich, this morphotype occurs with > 80 % frequency (Rabeder 1981) but in the stratigraphically older *M. kretzoii* the *kretzoii* type dominates (Carls and Rabeder 1988). In extant *Myodes glareolus* Schreber, 1870 the *glareolus* type dominates (Rabeder 1981; Nadachowski 1982). Our 4 M_1 agree with *Myodes hintonianus* from other sites and we refer them to this species. The MNI is 8 based on left M^2.

Genus *Pliomys* Méhely, 1914

Pliomys episcopalis Méhely, 1914

2 left and 3 right M^1; 2 left M^2; 4 left and 3 right M^3; 4 left and 7 right M_1; 1 left and 2 right M_2; 3 left and 3 right M_3; 14 molar fragments (**Table 4**).

Pliomys has roots, no crown cementum, closed dentine fields and positive enamel differentiation as in *Dinaromys* and *Dolomys*, from which it is distinguished by its smaller size and differences in the occlusal pattern. The evolutionary level in *Pliomys* is determined by the height of the linea sinuosa and the percentage of M_1 morphotypes. At Untermassfeld, the tips of sinuids are all worn, and the relative crown height is > 90. Our finds are larger than *Pliomys schernfeldensis* (relative crown height < 80, according to Carls and Rabeder 1988), where the *simplicior* morphotype occurs with a frequency > 50 %, and *Pliomys hollitzeri* Rabeder, 1981, with a smaller and broader anterior cap and a reduced T2-fold on M^3 (cf. Rabeder, 1981). Our finds also cannot belong to *P. lenki* (= *P. coronensis*) with its typical lingual incision on the anterior cap. *Pliomys bolkayi* is larger. At Untermassfeld, the *episcopalis* morphotype is dominant (5 specimens) followed by the *simplicior* (3 specimens) and *chalinei* (2 specimens) morphotypes. We therefore refer all our specimens to *P. episcopalis.*The right M_1 indicate an MNI of 7.

Genus *Mimomys* Forsyth Major, 1902

Mimomys savini Hinton, 1910

4 left and 7 right M^1; 1 right M^2; 5 left and 3 right M^3; 6 left and 9 right M_1; 1 right M_2 (**Table 4**).

		n	min	mean	max	sd
Myodes hintonianus	Length	3	2.12	2.18	2.23	
	Maximal crown height				2.65	
	Maximal relative crown height				118.73	
Pliomys episcopalis	Length	6	2.36	2.49	2.62	0.087
	Maximal crown height				2.66	
	Maximal relative crown height				101.22	
Mimomys savini	Length	12	2.99	3.25	3.62	0.168
	Maximal crown height				4.60	
	Maximal relative crown height				126.99	
Mimomys pusillus	Length	33	2.13	2.52	3.20	0.285
	Maximal crown height				3.62	
	Maximal relative crown height				113.29	
Microtus thenii	Length	50	2.20	2.60	2.95	0.173
	A/L	50	43.51	47.76	54.12	2.262

Table 4 Arvicolinae, Untermassfeld. Measurements (mm) of M_1. Maximal relative crown height (absolute crown height × 100 / M_1 length) refers to the specimen with the highest crown of the sample; A/L index after Meulen (1973).

The molars are rooted and have crown cementum, and M_1 have a simple ACC consisting of lingual salient angle (LSA) 3, LSA 4 and buccal salient angle (BSA) 3. These features indicate the genus *Mimomys*. The Untermassfeld material belongs to a derived large species (L > 2.8 mm). Sinuid tips are worn in all specimens with a crown base, thus the relative crown height (crown height / crown length) is H/L > 140.
Our finds do not belong to *Mimomys pliocaenicus* (Forsyth Major, 1889), since the type from Castelfranco di Sopra/Italy, regione Toscana, is 3.78 mm and thus much larger than any of the M_1 from Untermassfeld. The type also has more acute salient angles and less confluent dentine fields, and islet and edge occur at H/L c. 100 (Masini and Torre 1987), whereas in Untermassfeld they lack already at H/L = 140. *M. ostramosensis* Jánossy et Van der Meulen, 1975 is clearly larger, the M^1 has three unfused roots and the re-entrant angles are deeper than in our finds (Jánossy and Meulen 1975). The M_1 of the sample from Osztramos 3/Hungary, megye Borsod, have an enamel islet and a *Mimomys* edge when they are higher than H/L = 140. The M_1 length of *Mimomys tornensis* from Osztramos 3 (type locality) is similar but the morphology is different from Untermassfeld. *M. tornensis* has an extreme amount of crown cementum and the salient angles are more acute and the dentine fields completely closed. The type specimen of *M. coelodus* Kretzoi, 1954 from Kislang/Hungary, megye Fejér, resembles our finds, but at 3.0 mm the M_1 is a little below the mean of Untermassfeld. In contrast to the Untermassfeld specimens, *M. coelodus* retains an islet at the relative crown height of 130 (Kretzoi 1954; Rabeder 1981) which is lacking at Untermassfeld already in higher levels.
Our material conforms to *Mimomys savini* Hinton, 1910. Fejfar (1972) observed that for most samples the order of mean values of the length of M_1 corresponds to the stratigraphic order. There exist pros (Lozano-Fernández et al. 2013) and cons (Martin 2014; Palmqvist et al. 2014) using M_1 length of *Mimomys savini* to imply the stragigraphic order of the sites. However, in our case this order could fit, since Untermassfeld is larger than the samples from Betfia 2/Romania, județ Bihor, Deutsch-Altenburg $2C_1$/Austria, Niederösterreich etc. and it is smaller than the M_1 from the type locality West Runton and other late Biharian faunas (Voigtstedt etc.). Heinrich (1978) quantified the successive change of the relative enamel thickness of M_1

in *Arvicola* and expressed this relation with the ratio SDQ (percental relation between trailing and leading edges of the molar). The SDQ values are higher in older than in younger samples, which is a trend that begins already in *Mimomys savini*, the phylogenetic ancestor of *Arvicola* (Heinrich 1987). Values at Untermassfeld (minimum = 129.59 mm, mean = 151.00 mm, maximum = 179.18 mm, n = 11) are higher than that of Voigtstedt (minimum = 107.1 mm, mean = 133.66 mm, maximum = 159.64 mm, n = 20), which is in agreement with the greater geological age of Untermassfeld. However, *M. coelodus* might be part of the *Mimomys savini* – *Arvicola* lineage, whereas *M. pliocaenicus*, *M. ostramosensis*, and *M. tornensis* are not. The MNI 9 is inferred from right M_1.

Mimomys pusillus (Méhely, 1914)
12 left and 9 right M^1; 17 left and 12 right M^2; 5 left and 5 right M^3; 2 left and 3 right mandibles with M_1-M_2; 1 left mandible with M_1; 28 left and 23 right M_1; 3 left and 7 right M_2; 4 left and 2 right M_3; 4 molar fragments (**Table 4**).

This is a small (M_1 length < 2.8 mm), derived *Mimomys* species, based on its occlusal surface pattern, schmelzmuster, crown cementum, and hypsodonty. It differs from *M. reidi* Hinton, 1910, with all dentine fields confluent (in our specimens mainly closed), and a lower relative crown height of 100 whereas that from Untermassfeld is > 110. *M. reidi* has little crown cementum – in the Untermassfeld molars cementum fills 50 % of the re-entrants. *Pitymimomys pitymyoides* (Jánossy et Van der Meulen, 1975) differs in having T2 and T3 broadly confluent (Jánossy and Meulen 1975), whereas in our finds T1 and T2 are confluent, and cementum is more abundant.
Our finds are similar to *M. pusillus* (Méhely, 1914) and *M. blanci* Van der Meulen, 1973. Islet and a *Mimomys* edge occasionally occur in rooted specimens of both species (Méhely 1914; Meulen 1973). According to Meulen (1973), the latter is a more evolved descendant of *M. pusillus*, but every character described as being typical of *M. blanci* can also be found in *M. pusillus*. Nor can one demonstrate that *M. blanci* has a higher linea sinuosa, since specimens from the type localities of *M. pusillus* and *M. blanci*, Betfia 2 and Monte Peglia/Italy, regione Umbria, respectively, are already so hypsodont that their complete sinuids cannot be measured. Therefore we consider *M. blanci* as a synonym and refer the finds from Untermassfeld to *M. pusillus*. The left M_1 indicate at least 28 individuals.

Genus *Microtus* Schrank, 1798
Microtus (*Allophaiomys*) Kormos, 1933
Microtus (*Allophaiomys*) *thenii* Malez et Rabeder, 1984
1 skeleton; 1 maxilla fragment; 7 I^{sup}; 31 left and 33 right M^1; 30 left and 24 right M^2; 16 left and 14 right M^3; 1 left and 1 right mandible with M_1-M_2 resp. with M_1; 1 right mandible with I_{inf}, M_1-M_2; 1 left mandible with I_{inf}, M_1-M_2; 1 left mandible with I_{inf}-M_1; 1 right mandible with M_1-M_3; 5 left and 6 right mandibles with M_1-M_2; 1 right mandible with M_2; 1 left and 1 right mandible without teeth; 5 I_{inf}; 44 left and 29 right M_1; 30 left and 22 right M_2; 8 left and 10 right M_3; 38 molar fragments (**Table 4**)

The new material comprises isolated molars and also an articulated skeleton (**Fig. 9**). Rootless molars with abundant crown cementum, and relatively closed dentine fields support referral to *Microtus*. Species of that taxon are identified by the shape and size of the anteroconid complex (ACC) of M_1, based on quantification by morphotypes or morphometric data (indexes A/L, B/W, C/W according to Meulen 1973). At Untermassfeld only one species of *Microtus* occurs. M_1 morphology varies continuously, but nearly all specimens represent *pitymyoid* morphotypes (T4-T5 broadly confluent), with extreme rarely *microtid* (T4-T5

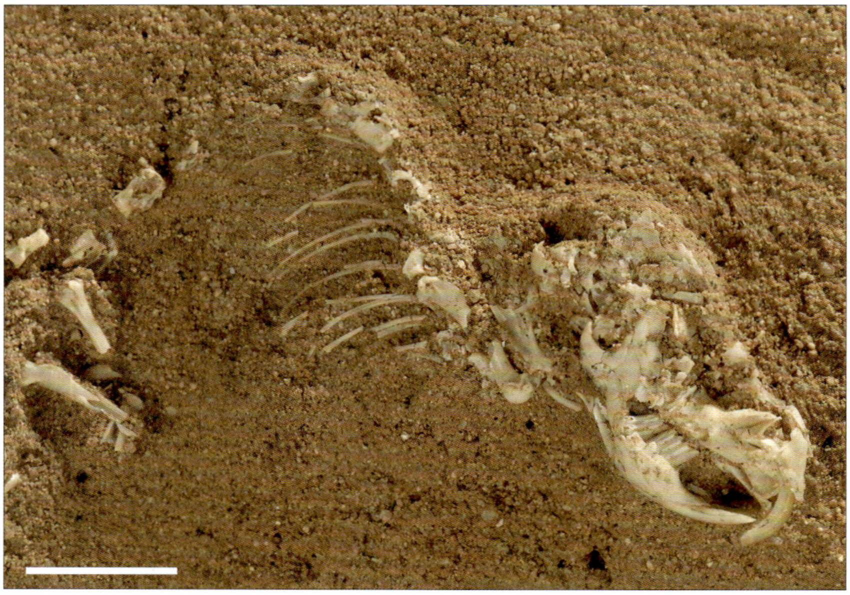

Fig. 9 *Microtus* (*Allophaiomys*) *thenii* Malez et Rabeder, Untermassfeld. – Partial skeleton IQW 2009/30628 (Mei. 29790), lateral view. – Scale: 10mm.

closed) morphotypes. The A/L is 47.76. This is similar to the conditions in Les Valerots/France, département Côte-d'Or (Heinrich 1982), with *Microtus* (*Allophaiomys*) *pliocaenicus nutiensis* Chaline, 1972 and *Microtus* (*Suranomys*) *malei burgondiae*. However, in Les Valerots the less derived *pliocaenicus* and *ratticepoides-nivaloides* morphotypes predominate (Chaline, 1972), resembling Monte Peglia *Microtus* (*Allophaiomys*) sp. A and B (Meulen 1973; Masini and Torre 1987). At Untermassfeld advanced allophaiomyid (T4-T5-anterior cap open) morphotypes occur, but in low numbers. Pitymyoid morphotypes also occur in *Microtus pitymyoides* from Bourgade (Chaline 1972), in *Microtus gregaloides* from Nagyharsányhegy 4/Hungary, megye Baranya, *Microtus arvalidens* von Villány 6/Hungary, megye Baranya – but all these samples have higher A/L values and a frequent BSA 4 (very rare at Untermassfeld). Atapuerca/TD 5 is rather similar (Cuenca-Bescós et al. 1995): several *Microtus* species, dominant pitymyoid M_1. However *Terricola arvalidens* from Atapuerca and *Microtus arvalidens* from Villány 6 (both with developed BRA4), *Stenocranius gregaloides* (Cuenca-Bescós et al. 1995), and *Microtus gregaloides* from Nagyharsányhegy 4, are all more highly evolved. The material from Včeláre 4D/Slovak Republic, Východoslvenský kraj, is similar (Fejfar and Horáček 1983).
Untermassfeld shows the greatest similarity with *Microtus thenii* Malez et Rabeder, 1984, from Podumci 1/Croatia, Šibensko-kninska županija, with dominant pitymyoid M_1 morphotypes (*hintoni* and *submalei*) (Malez and Rabeder 1984). The A/L values are similar, as is the rare occurrence of allophaiomyid M_1 (T4-T5-anterior cap are confluent: *praehintoni*- and *laguroides*-morphotype). Also, the enamel thickness is similar to *Microtus thenii*. Thus we refer the Untermassfeld population to this species. 44 left M_1 plus the skeleton indicate an MNI of 45.

Family Muridae Illiger, 1811
Subfamily Murinae Illiger, 1811
Genus *Apodemus* Kaup, 1829
Apodemus sylvaticus (Linnaeus, 1758)
2 left and 1 right I^{sup}; 1 left and 3 right M^2; 1 left and 1 right I_{inf}; 5 left and 3 right M_1; 2 left and 1 right M_2; 1 right M_3.

The finds had been assigned to *Apodemus* because of tubercle t7 in M^1, Sm in M_1, two roots on M_1 and M_2 (Niethammer 1978), the number and orientation of M^1 roots (after Herold 1956, they show the most common pattern in *A. sylvaticus*, *A. flavicollis* and *A. mystacinus*). For morphological and metric reasons, we can exclude *Rhagapodemus*, *Apodemus agrarius* (Pallas, 1771), *A. leptodus* Kretzoi, 1956, *A. mystacinus* (Danford et Alston, 1877), *A. dominans* Kretzoi, 1959, *Parapodemus coronensis* Schaub, 1938 (after Martin Suarez and Mein 1998, now *Apodemus coronensis*). Fossil and extant samples of *Apodemus flavicollis* (Melchior, 1834) are larger, and *A. maastrichtiensis* Kolfschoten, 1985 and *A. microps* Kratochvil et Rosicky, 1952 (now *A. uralensis*) are smaller. The finds best fit *Apodemus sylvaticus* (Linnaeus, 1758). Three further teeth found since 2001 are within the range (minimum = 1.42 mm, mean = 1.66 mm, maximum = 1.85 mm, n = 8) of the previously described material. However, accessory tubercles are larger in the fossils from Untermassfeld than in extant samples from the region. *A. alsomyoides* from Villány 3/Hungary, megye Baranya (Schaub 1938), and Schernfeld (Dehm 1962) also displays such strong tubercles. However, these finds are all larger than those from Untermassfeld. Although similar in size to our material, there are slight differences also to *A. atavus* Heller, 1936, which had been described by Rietschel and Storch (1974). We no longer harbor our previous doubts (*A.* cf. *sylvaticus* in Maul 2001a) and refer our murid finds to *A. sylvaticus*. The MNI is 5, based on left M_1.

Suborder Hystricomorpha Brandt, 1855
Family Hystricidae Fischer, 1817
Genus *Hystrix* Linnaeus, 1758
Hystrix refossa Gervais, 1852
1 left and 1 right I_{inf}.

The two incisors resemble those of *Castor*, but can be clearly distinguished by their cross section (an almost equilateral triangle) and the cover of the enamel (see Maul 2001a). The porcupine from Untermassfeld was previously (Maul 2001a) identified to the generic level only. In the Pleistocene of Europe two *Hystrix* species are recorded. The incisor diameter of the larger *H. refossa* Gervais, 1852 from Gundersheim/Germany, Rheinland-Pfalz (age MN 16a) (mesio-distal × linguo-buccal width: 5.7 × 7.2 mm, Heller 1936) and Erpfinger cave (MN17) (mesio-distal width: 5.5 mm and 6.5 mm, Lehmann 1953) is about the same as at Untermassfeld (mesio-distal × linguo-buccal width: 5.1 × 7.1 mm and 5.3 × 7.2 mm). Since we can exclude the smaller species *H. vinogradovi* Argyropulo, 1941, the incisors of which are considerably smaller (I_{inf} from the Late Pleistocene of Burgtonna/Germany, Thuringia, with mesio-distal × linguo-buccal width 4.2–5.2 mm × 5.1–6.1 mm, Maul 2001b), *H. refossa* is the only possible species to which to refer our material. Due to the almost identical dimensions of the two incisors from Untermassfeld, it is very likely that both finds belong to one individual.

This consideration holds true also for the gnaw marks produced by a porcupine (Maul 2001b). Similar marks from the Late Pleistocene site of Burgtonna were attributed to *H. vinogradovi* and are also smaller than the gnaw marks on the bones from Untermassfeld: the radius of the facets in Burgtonna material is c. 40 mm, whereas those from Untermassfeld are > 43 mm, mostly closer to 50 mm. Since the radius of these marks correlates with the size of the skull, we conclude that all gnawing marks in Untermassfeld were made by *H. refossa*.

In addition to the 16 bones described in Maul (2001b), further bones have been found with gnaw-marks corresponding in size to the porcupine, i. e., marks on an antler fragment IQW 2003/28 127 (Mei. 27 289), on a large mammal rib IQW 2009/30 596 (Mei. 29 758), on a pelvis of *Equus* IQW 2014/41 849 (Mei. 47 158) (Kahlke in this volume, Fig. 27), and on a tibia of *Stephanorhinus* IQW 2015/42 168 (Mei. 47 477).

3.2. Overview of the fauna

Table 5 shows the current faunal list of small mammals, including more than 7,000 remains (comprising 1,246 catalogue nos on 01 January 2016, which has been enlarged to 3,658 catalogue nos, at cut-off date 01 January 2017). See appendix for catalogue nos. of the entire material from the excavation seasons 1979–2016. In comparison with 2001 (Maul 2001a), four new taxa were recovered (indicated with +), one new species was described (*Spermophilus kahlkei* n. sp.), and in 14 cases the taxonomic referral (T) or the nomenclature (N) was changed. In addition to the specimens determined to the species or genus level, there are 300 specimens (small fragments of incisor, and maxillas and mandibulas without teeth) of arvicolines, and 253 incisor fragments of rodents. 239 very fragmented remains of teeth and more than 3,830 remains of postcranial bones are referable to micromammals only. They constitute a not inconsiderable amount of bones and teeth. The minimum number of individuals (MNI) comprising the entire small mammal material of the excavation seasons 1979–2016 is calculated by the highest number of a skeletal element or tooth of one body side with consideration of age classes (e. g., wear stage from unworn to very worn). However, this could not be realised in all cases since many of the remains are fragmentary.

3.3. Taphonomy

The Pleistocene small mammal remains from Untermassfeld represent a fluviatile taphocoenosis, since the recorded taxa originate from different habitats and were transported by water, accumulated, and ultimately buried in the excavation area. The majority of the remains are isolated finds, although articulated specimens have been found as well. After counting the various skeletal elements and calculating the completeness, we previously inferred that the majority of small mammal finds originated from owl pellets, while finds of *Spermophilus* were covered by sediment as more or less complete bodies (Maul 2001a). Since 2001 this observation has been confirmed by the excavation of complete and partial skeletons (Kahlke in this volume, 1689 ff.). Extracting articulated and more or less complete skeletons and skeletal parts (legs, vertebral column etc.) became possible due to improvements in the excavation and preparation techniques. These preparations confirmed the assumption that at least remains of *Spermophilus* were accumulated as whole animals, and not as remains of prey.
Another nearly complete skeleton belongs to *Cricetus* (**Fig. 8**). An interesting pattern emerges when we examine the orientation of diaphyses of consecutive limb bones (scapula – humerus – ulna/radius, pelvis – femur – tibia). They form parallel lines, which might reflect the direction of the water movement. Also a skeleton of *Microtus* has been excavated (**Fig. 9**), an animal that 'normally' constitutes the main part of prey animals in a pellet of a bird or in an excrement of a mammalian carnivore (Andrews 1990). The surface of most remains displays traces of weathering or digestion.

3.4. Stratigraphy

A palaeomagnetic reversal had been recorded at Untermassfeld (Wiegank 1997), which is considered – in combination with the faunal content – by most researchers as the base of the Jaramillo subchron. In response to an alternative dating of the Untermassfeld fauna to the Matuyama/Brunhes boundary (Kolfschoten and Markova 2005), we reitterated the evidence for placing Untermassfeld at the base of the Jaramillo subchron (Maul et al. 2007). The main objection of Kolfschoten and Markova (l.c.) was that we re-

Taxon	NISP	MNI
Erinaceomorpha		
Erinaceus cf. *europaeus*	3	1
Soricomorpha		
Talpa cf. *europaea*	45	6
Talpa minor (T)	2	2
Desmana thermalis (+)	1	1
Sorex (*Sorex*) *minutus* (+)	1	1
Sorex (*Sorex*) *runtonensis* (N)	16	7
Sorex (*Drepanosorex*) *margaritodon* (T)	15	5
Beremendia fissidens (T)	13	3
Lagomorpha		
Hypolagus brachygnathus (+)	1	1
Lepus sp. (T)	4	2
Hypolagus vel *Lepus*	18	
Rodentia		
Sciurus whitei (T)	4	1
Spermophilus kahlkei n. sp.	1,512	35
	+ 51 skeletal units + 27 isolated skulls (see Kahlke in this volume, Fig. 18, Table 6)	45
Glis sackdillingensis	9	3
Castor fiber	9	2
Trogontherium (*Trogontherium*) *cuvieri* (N)	10	2
Sicista sp. (+)	1	1
Cricetus runtonensis (T)	25	3
	+ 1 skeleton	1
Myodes hintonianus (NT)	32	8
Pliomys episcopalis	48	7
Mimomys savini	36	9
Mimomys pusillus (N)	137	28
Microtus (*Allophaiomys*) *thenii* (NT)	362	44
	+ 1 partial skeleton	1
Apodemus sylvaticus (T)	22	5
Hystrix refossa (T)	2	1
Arvicolinae indet. (jaws without teeth, incisors)	300	
Rodentia indet. (incisors)	253	
sum	**2,961**	**225**

Table 5 Updated faunal list of small mammals from Untermassfeld. Number of identified specimens (NISP) and minimum number of individuals (MNI) at cut-off date 01 January 2017. Compared to the faunal list of Maul (2001a): + – new taxon; N – changed nomenclature; T – changed taxonomy. In addition, the small mammal material comprises more than 4,050 indeterminable postcranial and teeth fragments.

corded in Untermassfeld a *Microtus* species, whereas in other sites referred to the Jaramillo subchron, such as Korotoyak/Russia, Voronezhskaya oblast', and Colle Curti/Italy, regione Marche, *Allophaiomys* occurs. Nomenclature can be rather suggestive. The name *Allophaiomys* gives the fauna a relativly older impression, and the name *Microtus* a younger one, since it is commonly known that all extant *Microtus* evolved from morphologically more primitive forms and passed a so-called *Allophaiomys* stage. However, as it has been shown several times previously (Maul 2001a; Maul et al. 2007) and in this paper, *Microtus thenii* recorded in Untermassfeld, is still not a 'true *Microtus*' since it does not include the typical *Microtus* morphotypes but a variety of different types, including some primitive, 'allophaiomyid' ones. On the other hand, the finds from Colle Curti could be also considered as primitive members of the *M.* (*Pallasiinus*)-lineage, thus giving them a more modern character. In other words, during the time around the Jaramillo subchron in the various sites many different morphotypes are recorded and probably different species in parallel lineages existed. The modern members of the lineages (subgenera) *Stenocranius*, *Terricola*, *Pallasiinus* and *Microtus* with clear *Microtus* morphotypes, however, occur only in the interval between Jaramillo and the Matuyama/Brunhes boundary (Maul et al. 2007).

In summary, based mainly on the evolutionary level of *Microtus* (*Allophaiomys*) *thenii*, the site of Untermassfeld must be correlated with the Jaramillo palaeomagnetic subchron and not with the onset of the Brunhes chron. This is confirmed by the new data on *Sorex* (*Drepanosorex*) *margaritodon*, and *Spermophilus kahlkei* n. sp., and the occurrence of *Mimomys pusillus* and *Hypolagus brachygnathus*, which went extinct in Central Europe before the beginning of the Middle Pleistocene.

Fig. 10 summarises the ranges of the taxa recorded at Untermassfeld. For further correlation, the following localities might be biostratigraphic equivalents of the Untermassfeld fauna, based on the evolutionary levels of *Microtus* and *Mimomys*: Neuleiningen 5 and 15/Germany, Rheinland-Pfalz (Maul 1996); Maasvlakte 1/The Netherlands, provincie Zuid-Holland (Vervoort-Kerkhoff and Kolfschoten 1988); Zuurland 6/The Netherlands, provincie Zuid-Holland (Kolfschoten 1988); Zalesiaki 1A/Poland, województwo Łodz (Nadachowski 1989); Cherevychne/upper layers/Ukraine, Odes'ka oblast' (Topachevskij et al. 1987); Karaj Dubina/Ukraine, Khersons'ka oblast'; Port Katon/Russia, Krasnodarskij kraj (Markova 1982); Petropavlovka 2/Russia, Voronezhskaya oblast'; Uryv 3/Russia, Voronezhskaya oblast' (Agadjanian 1977); Les Valerots/France, département Côte-d'Or (Chaline 1972); Le Vallonnet/France, département Alpes-Maritimes (Chaline 1971); Včeláre 4D/Slovak Republic, Vychodoslovenský kraj (Fejfar and Horáček 1983); Mladeč 1/Czech Republic, Olomoucký kraj (Horáček and Ložek 1984); Únětice/Czech Republic, okres Praha-západ; Somssich-hegy 2/Hungary, megye Baranya (Jánossy 1983); Villány 6/Hungary, megye Baranya (Kretzoi 1956; Meulen 1973); Betfia 5/Romania, judeţ Bihor (Kretzoi 1941; Terzea 1968); Atapuerca/TD 5/Spain, provincia Burgos (Cuenca-Bescós et al. 1995; Cuenca-Bescós and García 2007); Huescar 1/Spain, provincia Andalucia (Mazo et al. 1985); Pietrafitta/Italy, regione Calabria (Gentili et al. 1996); Podumci 1/Croatia, Šibensko-kninska županija (Malez and Rabeder 1984); Tourkobounia 2/Greece, periféria Attike (Meulen and Doukas 2001).

3.5. Ecology

To estimate the habitat requirements of the recorded taxa, we depend on the requirements of their nearest living relatives, often the members of the same genus. Sometimes this might work very well; in other cases it can be misleading. Typical inhabitants of open habitats are members of the genera *Spermophilus*,

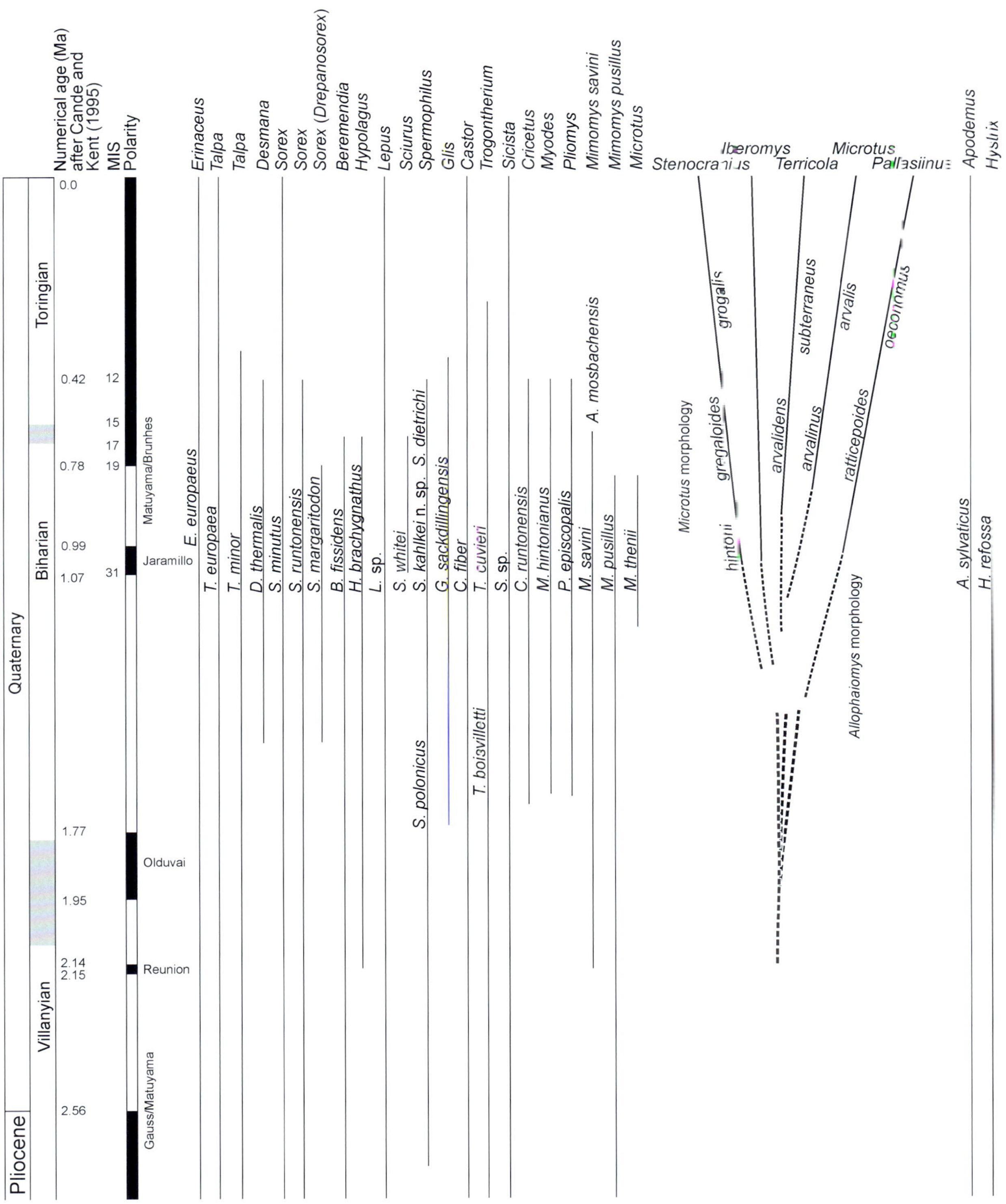

Fig. 10 Stratigraphic ranges of the taxa recorded at Untermassfeld, and related species. Numerical ages of the palaeomagnetic reversal are after Cande and Kent (1995).

Cricetus and *Microtus*, possibly also *Pliomys*, most *Mimomys* species (here *M. pusillus*), *Sicista* and *Hystrix*. *Lepus* relies on high-speed endurance running to escape from their enemies and inhabits open to semi-open areas, and also light forests. The extant *Sicista subtilis* is a decided steppe dweller, whereas *Sicista betulina* lives in northern Europe and Asia in forest and marsh zones. It makes a subterranean burrow in the summer and hibernates in the winter. *Sciurus*, *Myodes*, *Glis* and *Apodemus* are forest forms, and *Castor* is also reliant on trees. *Desmana*, *Sorex* (*Drepanosorex*), and *Mimomys savini* required flowing or standing water. *Trogontherium* is obviously aquatic, as is *Castor*. However, due to its different proportions of skull and skeleton (Schreuder 1929; Mayhew 1978; Fostowicz-Frelik 2008) *Trogontherium* was possibly better adapted for digging. Other mammals which require rather soft ground to dig burrowing systems are *Talpa*, *Sorex*, and *Hypolagus*. *Erinaceus* and most *Sorex* species (here *S. minutus*, and *S. runtonensis*) are generalists. *Beremendia* indicates temperate, humid and relatively stable climate and was an opportunist, and probably like the American *Blarina* it lived in forests, grasslands, marshes and brushy areas and is not restricted to any habitat (Rofes and Cuenca-Bescós 2009). However, this might have zoogeographic or stratigraphic reasons, since in the Early Pleistocene *Crocidura* does not occur north of Austria and Hungary and is known in Germany only from the Middle Pleistocene onwards (Rzebik-Kowalska 1998). Conversely, taxa considered as cold indicators, like Dicrostonychini, are recorded in France and Poland in nearly contemporaneous sites (Les Valerots, Zalesiaki 1) (Kowalski 2001).

4. Conclusions

The investigation of the new small mammals recovered since the previous report (Maul 2001a) provided several taxonomic additions and changes. The study resulted in the record of now 23 (morphologically distinguishable) small mammal taxa. This number – 75 % the size of the extant small mammal fauna (excluding bats) of Thuringia (30 species) and 90 % the size of local faunas (25 species) – scarcely differs from modern values. We therefore assume that the small mammal fauna is rather completely sampled. However, larger sample sizes for several taxa would be desirable. At the moment only *Spermophilus*, *Microtus* and *Mimomys pusillus* are recorded in sufficiently large numbers. Another limitation is the number and quality of comparable, well-dated faunas in the region. Only a network of faunal successions can improve the regional and transregional stratigraphic framework.

Acknowledgements

I am indebted to the colleagues from Weimar: G. Utschig for screen washing and pre-sorting of samples, S. Döring and T. Korn for photographs, D. Rössler for preparation of the articulated skeletons, E. Haase for layout, R.-D. Kahlke and H.-D. Kahlke (†) for valuable discussion, comments, and inspiration. K. T. Smith (Frankfurt/M.) improved the English and discussed various issues of the manuscript, I. Horáček (Prague) identified the humerus of *Sorex minutus*. Moreover, I am very grateful to S. Parfitt (London) and his reviewer's comments and suggestions, which improved the manuscript. For loaning comparatrive material I thank G. Storch (†), K. Krohmann, T. Lehmann (all Frankfurt/M.), R. Beutel (Jena), I. Thomas (†) (Berlin), K. Kowalski (†) (Krakow), O. Fejfar (Prague), and L. Kordos (Budapest).

References

Agadjanian, A.K., 1977. Quartäre Kleinsäuger aus der Russischen Ebene. Quartär 27/28, 111–145.

Andrews, P., 1990. Owls, Caves and Fossils. Chicago.

Brown, J.C., Twigg, G.I., 1969. Studies on the pelvis in British Muridae and Cricetidae (Rodentia). Journal of Zoology 158, 81–132.

Brunner, G., 1953. Das Extremitätenskelett der Waldspitzmaus, *Sorex araneus* Linné, 1758. Säugetierkundliche Mitteilungen 1, 59–63.

Cande, S.C., Kent, D.V., 1995. Revised calibration of the geomagnetic polarity timescale for the Late Cretaceous and Cenozoic. Journal of Geophysical Research 100, 6093–6095.

Carls, N., Rabeder, G., 1988. Die Arvicoliden (Rodentia, Mammalia) aus dem Ältest-Pleistozän von Schernfeld (Bayern). Beiträge zur Paläontologie von Österreich 14, 123–237.

Čermák, S., 2009. The Plio-Pleistocene record of *Hypolagus* (Lagomorpha, Leporidae) from the Czech and Slovak Republics with comments on systematics and classification of the genus. Bulletin of Geosciences 84, 497–524.

Čermák, S., Fladerer, F.A., Maul, L.C., Angelone, C., 2012. The Evolutionary History of European *Hypolagus* (Archaeolaginae, Leporidae): State of the Art. 4th World Lagomorph Conference, 23–27 July, 2012. Abstracts. Vienna, 35.

Chaline, J., 1971. La microfaune du Vallonnet (A.-M.) et le problème des corrélations micro-macromammifères a la limite Pléistocène inférieur-moyen. Bulletin du Musée d'Anthropologie Prèhistorique de Monaco 17, 65–69.

1972. Les rongeurs du Pléistocène moyen et supérieur de France (Systématique, Biostratigraphie, Paléoclimatologie). Paris.

Cuenca-Bescós, G., García, N., 2007. Biostratigraphic succession of the Early and Middle Pleistocene mammal faunas of the Atapuerca cave sites (Burgos, Spain). In: Kahlke, R.-D., Maul, L.C., Mazza, P.P.A. (Eds.), Late Neogene and Quaternary biodiversity and evolution: Regional developments and interregional correlations. Proceedings of the 18th International Senckenberg Conference (VI International Palaeontological Colloquium in Weimar). Volume II. Courier Forschungsinstitut Senckenberg 259, 99–110.

Cuenca-Bescós, G., Canudo, J.I., Laplana, C., 1995. Los arvicólidos (Rodentia, Mammalia) de los niveles inferiores de Gran Dolina (Pleistoceno Inferior, Atapuerca, Burgos, España). Revista Española de Paleontología 10, 202–218.

Daoud, A., 1993. Evolution of *Gliridae* (*Rodentia, Mammalia*) in the Pliocene and Quaternary of Poland. Acta zoologica cracoviensia 36, 199–231.

Dehm, R., 1962. Altpleistocäne Säuger von Schernfeld bei Eichstätt in Bayern. Mitteilungen der Bayerischen Staatssammlung für Paläontologie und historische Geologie 2, 17–61, pls. 3–6.

Dunmire, W.W., 1955. Sex Dimorphism in the Pelvis of Rodents. Journal of Mammalogy 36, 356–361.

Ellenberg, J., Kahlke, R.-D., 1997. Die quartärgeologische Entwicklung des mittleren Werratals und der Bau der unterpleistozänen Komplexfundstelle Untermaßfeld. In: Kahlke, R.-D. (Ed.), Das Pleistozän von Untermaßfeld (Thüringen). Teil 1. Monographien des Römisch-Germanischen Zentralmuseums 40, 1, 29–62, pls. 1–18.

Elliott, C.L., Flinders, J.T., 1991. *Spermophilus columbianus*. Mammalian Species 372, 1–9.

Eshelman, B.D., Sonnemann, C.S., 2000. *Spermophilus armatus*. Mammalian Species 637, 1–6.

Fahlbusch, V., 1976. *Cricetus major* Woldřich (Mammalia, Rodentia) aus der mittelpleistozänen Spaltenfüllung Petersbuch 1. Mitteilungen der Bayerischen Staatssammlung für Paläontologie und historische Geologie 16, 71–81, pls. 4–5.

Fejfar, O., 1966. Über zwei neue Säugetiere aus dem Altpleistozän von Böhmen. Neues Jahrbuch für Geologie und Paläontologie, Monatshefte 11, 680–691.

1972. Die Wühlmäuse (Microtidae, Mammalia) der älteren Sammlungen aus Stránská skála bei Brno. Anthropos 20, nova seria 12, 165–174.

Fejfar, O., Horáček, I., 1983. Zur Entwicklung der Kleinsäugerfaunen im Villányium und Alt-Biharium auf dem Gebiet der ČSSR. Schriftenreihe für geologische Wissenschaften 19/20, 111–207.

Fejfar, O., Heinrich, W.-D., Lindsay, E.H., 1998. Updating the Neogene Rodent biochronology in Europe. Mededelinger Nederlands Instituut voor Geowetenschappen TNO 60, 533–554.

Fladerer, F.A., 1987. Beitrag zur Entwicklung von *Hypolagus* und *Lepus* (Lagomorpha, Mammalia) im Pliopleistozän von Mitteleuropa. Sitzungsberichte der Österreichischen Akademie der Wissenschaften, Mathematisch-Naturwissenschaftliche Klasse, Abteilung I, 196, 123–138.

Fladerer, F.A., Reiner, G., 1995. Evolutionary shifts in the first premolar pattern of *Hypolagus beremendensis* (Petényi, 1864) (Lagomorpha, Mammalia) in the Plio-Pleistocene of Central Europe. Acta zoologica cracoviensia 39, 147–160.

Fostowicz-Frelik, Ł., 2008. First record of *Trogontherium cuvieri* (Mammalia, Rodentia) from the middle Pleistocene of Poland and review of the species. Geodiversitas 30, 765–778.

Furió, M., Agustí, J., Mouskhelishvili, A., Sanisidro, Ó., Santos-Cubedo, A., 2010. The Paleobiology of the Extinct Venomous Shrew *Beremendia* (Soricidae, Insectivora, Mammalia) in Relation to the Geology and Paleoenvironment of Dmanisi (Early Pleistocene, Georgia). Journal of Vertebrate Paleontology 30, 928–942.

Gentili, S., Abbazzi, L., Masini, F., Ambrosetti, P., Argenti, P., Torre, D., 1996. Voles from the Early Pleistocene of Pietrafitta (central Italy, Perugia). Acta zoologica cracoviensia 39, 185–199.

Gromov, I.M., Bibikov, D.I. Kalabukhov, N.I., Mejer, M.N., 1965. Nazemnye belich'i (Marmotinae). Fauna SSSR. Mlekopitayushchie III/2. Moskva, Leningrad. (in Russian)

Heinrich, W.-D., 1978. Zur biometrischen Erfassung eines Evolutionstrends bei *Arvicola* (Rodentia, Mammalia) aus dem Pleistozän Thüringens. Säugetierkundliche Informationen 2, 3–21.

1982. Eine altbiharische Kleinsäugerfauna aus Südthüringen. Zeitschrift für geologische Wissenschaften 10, 923–928.

1987. Neue Ergebnisse zur Evolution und Biostratigraphie von *Arvicola* (Rodentia, Mammalia) im Quartär Europas. Zeitschrift für geologische Wissenschaften 15, 389–406.

1989. Biometrische Untersuchungen an Fossilresten des Bibers (*Castor fiber* L.) aus thüringischen Travertinen: Ein Beitrag zur

Altersstellung der Fundstelle Bilzingsleben. Ethnographisch-Archäologische Zeitschrift 30, 394–403.

1991. Biometrische Untersuchungen an Fossilresten des Bibers (*Castor fiber* L.) aus der Fundstätte Bilzingsleben. In: Fischer, K., Guenther, E.W., Heinrich, W.-D., Mania, D., Musil, R., Nötzold, T. (Eds.), Bilzingsleben IV. *Homo erectus* – seine Kultur und seine Umwelt. Veröffentlichungen des Landesmuseums für Vorgeschichte Halle 44. Berlin, 35–62, pls. 35–47.

1997. Über *Trogontherium cuvieri* (Mammalia, Rodentia: Castoridae) aus dem mittelpleistozänen Travertinkomplex Bilzingsleben II in Thüringen. In: Bilzingsleben V. *Homo erectus* – seine Kultur und seine Umwelt. Bad Homburg, Leipzig, 135–182.

1998. Evolutionary trends in *Trogontherium cuvieri* (Mammalia, Rodentia, Castoridae) and their implications. Mededelingen Nederlands Instituut voor Toegepaste Geowetenschappen TNO 60, 573–578.

Heinrich, W.-D., Kolfschoten, T.v., 2007. Erster Skelettfund von *Trogontherium cuvieri* (Altbiber). In: Thieme, H. (Ed.), Die Schöninger Speere: Mensch und Jagd vor 400 000 Jahren. Stuttgart, 118–123.

Helgen, K.M., Cole, R.F., Helgen, L.E., Wilson, D.E., 2009. Generic Revision in the Holarctic Ground Squirrel Genus *Spermophilus*. Journal of Mammalogy 90, 270–305.

Heller, F., 1930. Eine Forest-Bed-Fauna aus der Sackdillinger Höhle (Oberpfalz). Neues Jahrbuch für Mineralogie, Geologie und Paläontologie, Abhandlungen, Abteilung B, Beilagen-Band 63, 247–298, pls. XV–XVIII.

1936. Eine oberpliocäne Wirbeltierfauna aus Rheinhessen. Neues Jahrbuch für Mineralogie, Geologie und Paläontologie, Abhandlungen, Abteilung B, Beilagen-Band 76, 99–160, pls. VII–XI.

1954. Neue Fundstellen altdiluvialer *Desmana*-Reste in Südwestdeutschland. Neues Jahrbuch für Geologie und Paläontologie, Monatshefte 10, 465–475.

Hellmund, M., Ziegler, R., 2012. A Ruscinian mammalian microfauna from a fissure filling near Sondershausen (Thuringia, Central Germany). Swiss Journal of Palaeontology 131, 77–94.

Herold, W., 1956. Über die Variabilität der Molaren-Wurzeln des Oberkiefers bei einigen *Apodemus*-Arten. Wissenschaftliche Zeitschrift der Humboldt-Universität zu Berlin, Mathematisch-Naturwissenschaftliche Reihe 6, 237–244.

Hír, J., 1996. New results in the study of Hungarian Plio-Pleistocene cricetids. Acta zoologica cracoviensia 39, 213–218.

1997a. A Comparative study on the dental morphology of the Early Pleistocene *Cricetus praeglacialis* Schaub, 1930 and recent Hungarian *Cricetus cricetus* L. Folia Historico Naturalia Musei Matraensis 22, 51–79.

1997b. *Cricetus runtonensis solymarenis* ssp. n. (Mammalia, Rodentia) from the late Middle Pleistocene of Solymár. Annales Historico-naturales Musei Nationalis Hungarici, pars Mineralogica et Palaeontologica 89, 23–42.

Horáček, I., Ložek, V., 1984. Z výzkumu výplně Mladečské jeskyně u Litovle. Česky kras 35, 98–100.

Jánossy, D., 1963. Die altpleistozäne Wirbeltierfauna von Kővesvárad bei Répáshuta (Bükk-Gebirge). Annales historico-naturales Musei Nationalis Hungarici, pars Mineralogica et Palaeontologica 55, 109–141, pl. I.

1965. Die Insectivoren-Reste aus dem Altpleistozän von Voigtstedt in Thüringen. Paläontologische Abhandlungen, Abteilung A Paläozoologie 2, 2/3. Berlin, 663–679, pl. XL.

1972. Ein kleiner Hystrix aus dem Altpleistozän der Fundstelle Osztramos 8 (Nordungarn). Vertebrata Hungarica 13, 163–182, pl. I.

1983. Lemming-remain from the Older Pleistocene of Southern Hungary (Villány, Somssich-hegy 2). Fragmenta Mineralogica et Palaeontologica 11, 55–60.

Jánossy, D., Meulen, A.J.v.d., 1975. On *Mimomys* (Rodentia) from Osztramos-3, North Hungary. Proceedings of the Koninklijke Nederlandse Akademie van Wetenschappen, Series B Palaeontology, geology, physics, chemistry, anthropology 78, 381–391.

Kahlke, R.-D., 1997 (Ed.). Das Pleistozän von Untermaßfeld bei Meiningen (Thüringen), Teil 1. Monographien des Römisch-Germanischen Zentralmuseums 40, 1, I–VII, 1–418, pls. 1–67.

2001a (Ed.). Das Pleistozän von Untermaßfeld bei Meiningen (Thüringen), Teil 2. Monographien des Römisch-Germanischen Zentralmuseums 40, 2, I–VIII, 419–698, pls. 68–131, suppl. I.

2001b. Verbesserte geologische Standardprofile zur unterpleistozänen Komplexfundstelle Untermaßfeld. In: Kahlke, R.-D. (Ed.), Das Pleistozän von Untermaßfeld bei Meiningen (Thüringen). Teil 2. Monographien des Römisch-Germanischen Zentralmuseums 40, 2, 433–445, suppl. I.

2001c (Ed.). Das Pleistozän von Untermaßfeld bei Meiningen (Thüringen), Teil 3. Monographien des Römisch-Germanischen Zentralmuseums 40, 3, I–VI, 699–1030, pls. 132–151, suppl. I–XV.

2001d. Die unterpleistozäne Komplexfundstelle Untermaßfeld – Zusammenfassung des Kenntnisstandes sowie synthetische Betrachtung zu Genesemodell, Paläoökologie und Stratigraphie. In: Kahlke, R.-D. (Ed.), Das Pleistozän von Untermaßfeld bei Meiningen (Thüringen). Teil 3. Monographien des Römisch-Germanischen Zentralmuseums 40, 3, 931–1030, suppl. I–XV.

2006. Untermassfeld. A late Early Pleistocene (Epivillafranchian) fossil site near Meiningen (Thuringia, Germany) and its position in the development of the European mammal fauna. British Archaeological Reports, International Series 1578. Oxford.

2007. Late Early Pleistocene large mammals and the concept of an Epivillafranchian biochron. In: Kahlke, R.-D., Maul, L.C., Mazza, P.P.A. (Eds.), Late Neogene and Quaternary biodiversity and evolution: Regional developments and interregional correlations. Proceedings of the 18th International Senckenberg Conference (VI International Palaeontological Colloquium in Weimar), Volume II. Courier Forschungsinstitut Senckenberg 259, 265–278.

2020. Excavation progress and research at the Early Pleistocene site of Untermassfeld during the years 1997–2015. In: Kahlke, R.-D. (Ed.), The Pleistocene of Untermassfeld near Meiningen (Thüringen, Germany). Part 4. Monographien des Römisch-Germanischen Zentralmuseums 40, 4, 1031–1078.

in this volume. The Pleistocene of Untermassfeld – A synopsis on site origin, palaeobiodiversity, taphonomic characteristics, palaeoenvironment, chronostratigraphy, and significance in western Palaearctic faunal history. 1635–1734.

Kahlke, R.-D., Linnemann, U., Gärtner, A., 2020. New results on the origin and geological history of the Early Pleistocene site of Untermassfeld. In: Kahlke, R.-D. (Ed.), The Pleistocene of Untermassfeld near Meiningen (Thüringen, Germany). Part 4. Monographien des Römisch-Germanischen Zentralmuseums 40, 4, 1079–1104.

Koenigswald, W. v., 1970. Mittelpleistozäne Kleinsäugerfauna aus der Spaltenfüllung Petersbuch bei Eichstätt. Mitteilungen der Bayerischen Staatssammlung für Paläontologie und historische Geologie 10, 407–432.

1980. Schmelzstruktur und Morphologie in den Molaren der Arvicolidae (Rodentia). Abhandlungen der Senckenbergischen Naturforschenden Gesellschaft 539, 1–129.

Kolfschoten, T.v., 1988. The Pleistocene mammalian faunas from the Zuurland boreholes at Brielle, The Netherlands. Mededelingen van de Werkgroep voor Tertiaire en Kwartaire Geologie 25, 73–86.

Kolfschoten, T.v., Markova, A.K., 2005. Response of the European mammalian fauna to the mid-Pleistocene transition. In: Head, M.J., Gibbard, P.L. (Eds.), Early-Middle Pleistocene Transitions: The Land-Ocean Evidence. London, 221–229.

Kormos, T., 1930. Diagnosen neuer Säugetiere aus der oberpliozänen Fauna des Somlyóberges bei Püspökfürdő. Annales historico-naturales Muse Nationalis Hungarici 27, 237–246.

1934. Neue Insektenfresser, Fledermäuse und Nager aus dem Oberpliozän der Villányer Gegend. Földtani Közleny 64, 296–321.

Korth, W.W., 2001. Comments on the Systematics and Classification of the Beavers (Rodentia, Castoridae). Journal of Mammalian Evolution 8, 279–296.

Kowalski, K., 1960. Pliocene Insectivores and Rodents from Rębielice Królewskie (Poland). Acta zoologica cracoviensia 5, 155–201, pls. XIX–XXI.

1979. Fossil *Zapodidae (Rodentia, Mammalia)* from the Pliocene and Quaternary of Poland. Acta zoologica cracoviensia 23, 199–210, pls. 123–124.

2001. Pleistocene Rodents of Europe. Folia Quaternaria 72, 1–389.

Kretzoi, M., 1941. Die unterpleistozäne Säugetierfauna von Betfia bei Nagyvárad. Földtani Közlöny 71, 308–355.

1954. Bericht über die Calabrische (Villafranchische) Fauna von Kisláng, Kom. Fejér. Földtani Intezet Évi Jelentes 1953, 1, 239–264.

1956. Die altpleistozänen Wirbeltierfaunen des Villányer Gebirges. Geologica Hungarica, Series Palaeontologia 27, 1–264.

1965. *Drepanosorex* – neu definiert. Vertebrata Hungarica 7, 117–129.

1969. Die *Castor*- und *Trogontherium*-Reste aus den Kiesen von Süßenborn bei Weimar. Paläontologische Abhandlungen, Abteilung A Paläozoologie 3, 3/4. Berlin, 771–784, pl. XLVI.

1975. Die *Castor*-Funde aus dem Travertinkomplex von Weimar-Ehringsdorf. Abhandlungen des Zentralen Geologischen Instituts, Paläontologische Abhandlungen 23, 513–532.

1977. Die *Castor*-Reste aus den Travertinen von Taubach bei Weimar. Quartärpaläontologie 2, 389–400, pls. XLV–XLVI.

Lehmann, U., 1953. Eine Villafranchiano-Fauna von der Erpfinger Höhle (Schwäbische Alb). Neues Jahrbuch für Geologie und Paläontologie, Monatshefte 10, 437–464.

López-Martínez, N., 2008. The Lagomorph Fossil Record and the Origin of the European Rabbit. In: Alves, P.C., Ferrand, N., Hackländer, K. (Eds.), Lagomorph Biology: Evolution, Ecology, and Conservation. Berlin, Heidelberg, 27–46.

Lozano-Fernández, I., Agustí, J., Cuenca-Bescós, G., Blain, H.-A., López-García, J.M., Vallverdú, J., 2013. Pleistocene evolutionary trends in dental morphology of *Mimomys savini* (Rodentia, Mammalia) from Iberian Peninsula and discussion about the origin of the genus *Arvicola*. Quaternaire 24, 179–190.

Mai, H., 1979. Die Biberarten *Castor* und *Trogontherium* aus den altpleistozänen Schichten von Mauer an der Elsenz. Schriften des Naturwissenschaftlichen Vereins von Schleswig-Holstein 49, 35–46.

Malez, M., Rabeder, G., 1984. Neues Fundmaterial von Kleinsäugern aus der altpleistozänen Spaltenfüllung Podumci 1 in Norddalmatien (Kroatien, Jugoslawien). Beiträge zur Paläontologie von Österreich 11, 439–510.

Markova, A.K., 1982. Pleistotsenovye gryzuny Russkoj ravniny. Moskva. (in Russian)

Martin, R.A., 2014. A critique of vole clocks. Quaternary Science Reviews 94, 1–6.

Martín-Suárez, E., Mein, P., 1998. Revision of the genera *Parapodemus*, *Apodemus*, *Rhagamys*, and *Rhagapodemus* (Rodentia, Mammalia). Geobios 31, 87–97.

Masini, F., Torre, D., 1987. Review of the Villafranchian Arvicolids of Italy. Geologia Romana 26, 127–133.

Maul, L., 1996. Biochronological implications of the arvicolids (Mammalia: Rodentia) from the Pliocene and Pleistocene faunas of Neuleiningen (Rheinland-Pfalz, southwest Germany). Acta zoologica cracoviensia 39, 349–356.

2001a. Die Kleinsäugerreste (Insectivora, Lagomorpha, Rodentia) aus dem Unterpleistozän von Untermaßfeld. In: Kahlke, R.-D. (Ed.), Das Pleistozän von Untermaßfeld bei Meiningen (Thüringen). Teil 3. Monographien des Römisch-Germanischen Zentralmuseums 40, 2, 783–887, pls. 144–145.

2001b. Nagespuren von Kleinsäugern an Knochenmaterial aus dem Unterpleistozän von Untermaßfeld. In: Kahlke, R.-D. (Ed.), Das Pleistozän von Untermaßfeld bei Meiningen (Thüringen). Teil 3. Monographien des Römisch-Germanischen Zentralmuseums 40, 3, 905–916, pls. 148–151.

Maul, L., Rzebik-Kowalska, B., 1998. A record of *Macroneomys brachygnathus* Fejfar, 1966 (Mammalia, Insectivora, Soricidae) in the early Middle Pleistocene (late Biharian) locality of Voigtstedt (Germany) and the history of the genus *Macroneomys*. Acta zoologica cracoviensia 41, 79–100.

Maul, L.C., 2007. Erster Nachweis von *Hypolagus* (Leporidae, Mammalia) in der unterpleistozänen Fundstelle Untermaßfeld (Thüringen, Mitteldeutschland). Beiträge zur Geologie von Thüringen, Neue Folge 14, 215–233.

2011. Implications from a semi-articulated skeleton of *Cricetus runtonensis* in the late Early Pleistocene of Thuringia (Central Germany). In: Monecke, S., Pévet, P. (Eds.), Program & Conference Proceedings. Strasbourg, 41.

Maul, L.C., Heinrich, W.-D., Parfitt, S.A., Paunescu, A.-C., 2007. Comment on the correlation between magnetostratigraphy and the evolution of *Microtus* (Arvicolidae, Rodentia, Mammalia) during the Early and early Middle Pleistocene. In: Kahlke, R.-D., Maul, L.C., Mazza, P.P.A. (Eds.), Late Neogene and Quaternary biodiversity and evolution: Regional developments and interregional correlations. Proceedings of the 18th International Senckenberg Conference (VI International Palaeontological Colloquium in Weimar), Volume II. Courier Forschungsinstitut Senckenberg 259, 243–263.

Mayhew, D.F., 1978. Reinterpretation of the extinct beaver *Trogontherium* (Mammalia, Rodentia). Philosophical Transactions of the Royal Society of London, series B Biological Sciences 281, 407–438.

Mazo, A.V., Sesé-Benito, C., Ruiz Bustos, A., Peña, J.A., 1985. Geologia y Paleontologia de los yacimientos Plio-Pleistocenos de Huescar (depresiòn de Guadix Baza, Granada). Estudios Geologicos 41, 467–493.

Méhely, L. v., 1914. Fibrinae Hungariae: Die ternären und quartären wurzelzähnigen Wühlmäuse Ungarns. Annales Musei Nationalis Hungarici 12, 155–243, pls. I–VIII.

Meulen, A.J.v.d., 1973. Middle Pleistocene Smaller Mammals from the Monte Peglia, (Orvieto, Italy) with Special Reference to the Phylogeny of *Microtus* (Arvicolidae, Rodentia). Quaternaria 17, 1–144, pls. I–XI.

Meulen, A.J.v.d., Doukas, C.S., 2001. The Early Biharian rodent fauna from Tourkoubounia 2. Lynx, nóva série 32, 255–277.

Nadachowski, A., 1982. Late Quaternary Rodents of Poland with Special Reference to Morphotype Dentition Analysis of Voles. Warszawa, Kraków.

1989. 7.5 Rodentia. Folia Quaternaria 59–60, 151–176.

Nadachowski, A., Daoud, A., 1995. Patterns of Myoxid Evolution in the Pliocene and Pleistocene of Europe. Hystrix 6, 141–149.

Niethammer, J., 1978. Familie Muridae Gray, 1821 – Echte Mäuse. In: Niethammer, J., Krapp, F. (Eds.), Handbuch der Säugetiere Europas. Bd. 1. Wiesbaden, 281–289.

Owen, R., 1869. On the Distinction between *Castor* and *Trogontherium*. Geological Magazine 6, 49–56.

Palmqvist, P., González-Donoso, J.M., De Renzi, M., 2014. Rectilinear evolution in arvicoline rodents and numerical dating of Iberian Early Pleistocene sites. Quaternary Science Reviews 98, 100–109.

Popov, V.V., 2003. Late Pliocene Soricidae (Insectivora, Mammalia) from Varshets (North Bulgaria). Acta zoologica cracoviensia 46, 43–72.

Pradel, A., 1988. Fossil hamsters (Cricetinae, Rodentia) from the Pliocene and Quaternary of Poland. Acta zoologica cracoviensia 31, 235–296.

Rabeder, G., 1972. Die Insectivoren und Chiropteren (Mammalia) aus dem Altpleistozän von Hundsheim (Niederösterreich). Annalen des Naturhistorischen Museums Wien 76, 375–474.

1981. Die Arvicoliden (Rodentia, Mammalia) aus dem Pliozän und dem älteren Pleistozän von Niederösterreich. Beiträge zur Paläontologie von Österreich 8, 1–343, pls. 1–15.

Reumer, J.W.F., 1984. Ruscinian and early Pleistocene Soricidae (Insectivora, Mammalia) from Tegelen (The Netherlands) and Hungary. Scripta Geologica 73, 1–173.

1985. The generic status and species of *Drepanosorex* reconsidered (Mammalia, Soricidae). Revue de Paléobiologie 4, 53–58.

1998. A Classification of the Fossil and Recent Shrews. In: Wójcik, J.M., Wolsan, M. (Eds.), Evolution of Shrews. Białowieża, 5–22.

Reumer, J.W.F., Hoek Ostende, L.W.v.d., 2003. Petauristidae and Sciuridae (Mammalia, Rodentia) from Tegelen, Zuurland, and the Maasvlakte (the Netherlands). Deinsea 10, 455–467.

Reumer, J.W.F., Mayhew, D.F., Veen, J.C.v., 2005. Small mammals from the Late Pliocene Oosterschelde dredgings. Deinsea 11, 103–118.

Rietschel, S., Storch, G., 1974. Außergewöhnlich erhaltene Waldmäuse (*Apodemus atavus* Heller 1936) aus dem Ober-Pliozän von Willershausen am Harz. Senckenbergiana lethaea 54, 491–519, pls. 1–5.

Rofes, J., Cuenca-Bescós, G., 2009. First record of *Beremendia fissidens* (Mammalia, Soricidae) in the Pleistocene of the Iberian Peninsula, with a review of the biostratigraphy, biogeography and palaeoecology of the species. Comptes Rendus Palevol 8, 21–37.

2013. First record of *Sorex* (*Drepanosorex*) *margaritodon* (Mammalia, Soricidae) in Western Europe: biostratigraphy, biogeography and evolution of the species. Paläontologische Zeitschrift 87, 1–13.

Rofes, J., Moya-Costa, R., Bennàsar, M., Blain, H.-A., Cuenca-Bescós, G., 2016. Biostratigraphy, palaeogeography and palaeoenvironmental significance of *Sorex runtonensis* Hinton, 1911 (Mammalia, Soricidae): First record from the Iberian Peninsula. Palaeogeography, Palaeoclimatology, Palaeoecology 459, 508–517.

Rümke, C.G., 1985a. Remarks on the European desmanine moles (Desmaninae, Talpidae, Insectivora). Proceedings of the Koninklijke Nederlandse Akademie van Wetenschappen, Series B Palaeontology, geology, physics, chemistry, anthropology 88, 415–426.

1985b. A review of fossil and recent Desmaninae (Talpidae, Insectivora). Utrecht Micropaleontological Bulletins, Special publication 4, 1–241, pls. 1–4.

Rzebik-Kowalska, B., 1971. The Pliocene and Pleistocene Insectivores (*Mammalia*) of Poland. I. *Erinaceidae* and *Desmaninae*. Acta zoologica cracoviensia 16, 435–462.

1972. The Insectivora from Stránská skála near Brno. Anthropos 20, nóva série 12, 65–70.

1976. The Neogene and Pleistocene Insectivores (*Mammalia*) of Poland. III. *Soricidae*: *Beremendia* and *Blarinoides*. Acta zoologica cracoviensia 21, 359–386.

1991. Pliocene and Pleistocene *Insectivora* (*Mammalia*) of Poland. VIII. *Soricidae*: *Sorex* Linnaeus, 1758, *Neomys* Kaup, 1829, *Macroneomys* Fejfar, 1966, *Paenelimnoecus* Baudelot, 1972 and *Soricidae* indeterminata. Acta zoologica cracoviensia 34, 323–424.

1998. Fossil History of Shrews in Europe. In: Wójcik, J.M., Wolsan, M. (Eds.), Evolution of Shrews. Białowieża, 23–92.

2000. Insectivora (Mammalia) from the Early and early Middle Pleistocene of Betfia in Romania. I. Soricidae Fischer von Waldheim, 1817. Acta zoologica cracoviensia 43, 1–53.

2013. *Sorex bifidus* n. sp. and the rich insectivore mammal fauna (Erinaceomorpha, Soricomorpha, Mammalia) from the Early Pleistocene of Żabia Cave in Poland. Palaeontologia Electronica

[Online] 16.2.12A. Available: https://palaeo-electronica.org/content/2013/428-a-new-shrew-species. [Accessed: 6 April 2020]

Sabol, M., Zervanová, J., 2014. *Cricetus* cf. *runtonensis* (Newton, 1909) (Cricetidae, Rodentia) from Za Hájovnou Cave (Moravia, the Czech Republic). Acta Musei Nationalis Pragae, Series B Historia Naturalis 70, 55–58.

Sansalone, G., Kotsakis, T., Piras, P., 2015. *Talpa fossilis* or *Talpa europaea*? Using geometric morphometrics and allometric trajectories of humeral moles remains from Hungary to answer a taxonomic debate. Palaeontologia Electronica [Online] 18.2.42A. Available: https://palaeo-electronica.org/content/2015/1293-plio-pleistocene-moles. [Accessed: 6 April 2020]

Schaub, S., 1930. Quartäre und jungtertiäre Hamster. Abhandlungen der Schweizerischen Palaeontologischen Gesellschaft 49, 1–49, pls. I–II.

1935. Die Hamster vom Roten Berg bei Saalfeld. Eclogae Geologicae Helvetiae 28, 580–590.

1938. Tertiäre und quartäre Murinae. Abhandlungen der schweizerischen paläontologischen Gesellschaft 61, 1–39, pl. I.

Schreuder, A., 1929. *Conodontes (Trogontherium)* and *Castor* from the Teglian Clay compared with the Castoridae from other localities. Archives du Musée Teyler, Serié 3, 6, 99–321, pls. I–XVI.

1940. A Revision of the Fossil Water-Moles (Desmaninae). Archives Néerlandaises de Zoologie 4, 201–333.

Stefen, C., 1997. *Steneofiber eseri* (Castoridae, Mammalia) von der Westtangente bei Ulm im Vergleich zu anderen Biberpopulationen. Stuttgarter Beiträge zur Naturkunde, Serie B Geologie und Paläontologie 255, 1–73.

Storch, G., Franzen, J.L., Malec, F., 1973. Die altpleistozäne Säugerfauna (Mammalia) von Hohensülzen bei Worms. Senckenbergiana lethaea 54, 311–343.

Suárez, E.M., Mein, P., 2004. The late Pliocene locality of Saint-Vallier (Drôme, France). Eleven micromammals. Geobios 37, Supplement 1, S115–S125.

Sulimski, A., 1964. Pliocene Lagomorpha and Rodentia from Węże 1 (Poland). Acta Palaeontologica Polonica 9, 149–244.

Terzea, E., Jurcsák, T., 1968. Bemerkungen über die mittelpleistozänen Faunen von Betfia. Berichte der deutschen Gesellschaft für geologische Wissenschaften, Reihe A Geologie und Paläontologie 13, 381–390.

Thenius, E., 1948. Fischotter und Bisamspitzmaus aus dem Altquartär von Hundsheim in Niederösterreich. Sitzungsberichte der österreichischen Akademie der Wissenschaften, Mathematisch-Naturwissenschaftliche Klasse 157, 187–202.

Topachevskij, V.A., Skorik, A.F., Rekovets, L.I., 1987. Gryzuny verkhneneogenovykh i ranneantropogenovykh otlozhenij khadzhibejskogo limana. Kiev.

Vervoort-Kerkhoff, Y., Kolfschoten, T.v., 1988. Pleistocene and Holocene mammalian faunas from the Maasvlakte near Rotterdam (The Netherlands). Mededelingen van de Werkgroep voor Tertiaire en Kwartaire Geologie 25, 87–98.

Wiegank, F., 1997. Paläomagnetische Charakteristik des Unterpleistozäns von Untermaßfeld. In: Kahlke, R.-D. (Ed.), Das Pleistozän von Untermaßfeld bei Meiningen (Thüringen). Teil 1. Monographien des Römisch-Germanischen Zentralmuseums 40, 1, 63–69.

Wilson, D.E., Reeder, D.M., 2005. Mammal Species of the World: A Taxonomic and Geographic Reference. Volume 1 and 2. Baltimore.

Woldřich, J.N., 1880. Diluviale Fauna von Zuzlawitz bei Winterberg im Böhmerwalde. Sitzungsberichte der kaiserlichen Akademie der Wissenschaften, Mathematisch-Naturwissenschaftliche Classe 82, 7–66, pls. I–IV.

Ziegler, R., 1995. Pleistozäne Säugetierfaunen von Genkingen bei Reutlingen (Baden-Württemberg). Stuttgarter Beiträge zur Naturkunde, Serie B Geologie und Paläontologie 234, 1–43.

Appendix

List of studied material

Erinaceus cf. *europaeus* Linnaeus, 1758
1 right P^3: IQW 1984/19 899 (Mei. 19 419),
1 left mandible fragment without teeth: IQW 1984/20 179 (Mei. 19 699),
1 right mandible fragment without teeth: IQW 1985/20 420 (Mei. 19 940).

Talpae cf. *europaea* Linnaeus, 1758
1 left C^{sup}: IQW 1985/20 894 (Mei. 20 413),
2 right C^{sup}: IQW 1985/20 896 (Mei. 20 415), IQW 1991/23 787 (Mei. 23 315),
1 left mandible with A_2-M_1: IQW 1985/20 895 (Mei. 20 414),
1 right mandible with A_4-M_1: IQW 2016/43 764 (Mei. 49 073),
2 left mandible fragments without teeth: IQW 2009/30 358 (Mei. 29 520), IQW 2015/42 004 (Mei. 47 313),
1 right mandible fragment without teeth: IQW 2009/30 358 (Mei. 29 520),
3 A_{inf}: IQW 1983/18 715 (Mei. 18 235), IQW 1983/18 718 (Mei. 18 238), IQW 2009/30 358 (Mei. 29 520),
1 right C_{inf}: IQW 2016/43 768 (Mei. 49 077),
2 left M_3: IQW 1983/18 705 (Mei. 18 225), IQW 1983/18 706 (Mei. 18 226),
1 right M_{inf}: IQW 1983/18 716 (Mei. 18 236),
6 left humerus fragments: IQW 2009/30 358 (Mei. 29 520), IQW 2012/38 390 (Mei. 35 364), IQW 2015/41 978 (Mei. 47 287), IQW 2015/42 003 (Mei. 47 312), IQW 2016/43 763 (Mei. 49 072), IQW 1983/18 704 (Mei. 18 224),
4 right humerus fragments: IQW 2009/30 358 (Mei. 29 520), IQW 2009/30 634 (Mei. 29 796), IQW 1983/18 702 (Mei. 18 222), IQW 1983/18 703 (Mei. 18 223),
3 left radius fragments: IQW 2016/43 765 (Mei. 49 074), IQW 1985/20 897 (Mei. 20 416), IQW 1985/20 898 (Mei. 20 417),
2 right radius fragments: IQW 1985/20 893 (Mei. 20 412), IQW 2009/30 561 (Mei. 29 723),
1 right radius and ulna: IQW 2003/27 717 (Mei. 27 222),
2 right ulna fragments: IQW 2009/30 358 (Mei. 29 520), IQW 1982/18 535 (Mei. 18 055),
6 distal phalanges: IQW 1985/20 899 (Mei. 20 418), IQW 1991/23 788 (Mei. 23 316), IQW 1991/23 789 (Mei. 23 317), IQW 1991/23 790 (Mei. 23 318), IQW 2016/43 766 (Mei. 49 075), IQW 2016/43 767 (Mei. 49 076),
6 phalanges of indeterminable position: IQW 2009/30 358 (Mei. 29 520).

Talpa minor Freudenberg, 1914
2 left humerus fragments: IQW 2005/28 814 (Mei. 27 976), IQW 1984/20 180 (Mei. 19 700).

Desmana thermalis Kormos, 1930
1 right mandible fragment with isolated M_1, P_4, and P_1: IQW 2016/43 835 (Mei. 49 144).

Sorex (*Sorex*) *minutus* Linnaeus, 1766
1 left humerus fragment: IQW 2016/43 195 (Mei. 48 504).

Sorex (*Sorex*) *runtonensis* Hinton, 1910
1 left M^1: IQW 2016/43 200 (Mei. 48 509),
1 left mandible fragment with I_{inf}-M_3: IQW 1984/19 841 (Mei. 19 361),
1 right mandible fragment with I_{inf}-M_1: IQW 1983/18 710 (Mei. 18 230),
1 left mandible fragment with I_{inf}-P_4: IQW 2016/43 199 (Mei. 48 508),
1 left mandible fragment with M_1-M_3: IQW 1983/19 535 (Mei. 19 055),
1 right mandible fragment with M_1-M_3: IQW 1983/18 717 (Mei. 18 237),
2 left mandible fragments with M_1-M_2: IQW 1983/18 707 (Mei. 18 227), IQW 1983/19 534 (Mei. 19 054),
1 right mandible fragment with M_1-M_2: IQW 1984/19 822 (Mei. 19 342),
1 left mandible fragment with M_2-M_3: IQW 1983/18 723 (Mei. 18 243),
1 right mandible fragment with M_2-M_3: IQW 2016/43 197 (Mei. 48 506),
1 right mandible fragment with M_2: IQW 1983/18 713 (Mei. 18 233),
1 left mandible fragment without teeth: IQW 2016/43 196 (Mei. 48 505),
1 left I_{inf}: IQW 1983/19 536 (Mei. 19 056),
2 right I_{inf}: IQW 1983/19 537 (Mei. 19 057), IQW 2016/43 198 (Mei. 48 507).

Sorex (*Drepanosorex*) *margaritodon* Kormos, 1930
1 left maxilla with P^4-M^2: IQW 2016/43 193 (Mei. 48 502),
1 right I^{sup}: IQW 1983/18 709 (Mei. 18 229),
1 right P^4: IQW 1983/18 719 (Mei. 18 239),
1 left M^1: IQW 2009/30 495 (Mei. 29 657),
2 left M^2: IQW 1985/20 959 (Mei. 20 478), IQW 2016/43 194 (Mei. 48 503),
1 left mandible fragment with M_1-M_3: IQW 1983/18 722 (Mei. 18 242),
2 right mandible fragments with M_1-M_3: IQW 1983/18 721 (Mei. 18 241), IQW 1984/19 840 (Mei. 19 360),
1 left mandible fragment with M_1-M_2: IQW 1983/18 714 (Mei. 18 234),
1 right mandible fragment with M_1-M_2: IQW 1984/19 839 (Mei. 19 359),
1 left mandible fragment with P_3-P_4: IQW 2016/43 192 (Mei. 48 501),
2 left mandible fragments with P_4-M_3: IQW 2011/35 480 (Mei. 32 690), IQW 2015/42 006 (Mei. 47 315),
1 femur fragment: IQW 2016/42 797 (Mei. 48 106).

Beremendia fissidens (Petényi, 1864)
2 left I^{sup}: IQW 1983/18 711 (Mei. 18 231), IQW 1984/19 818 (Mei. 19 338),
1 right I^{sup}: IQW 1983/18 708 (Mei. 18 228),
1 right P^4: IQW 1983/18 720 (Mei. 18 240),
1 right M^1: IQW 1983/18 712 (Mei. 18 232),
1 left mandible with I_{inf}-M_3: IQW 2012/38 410 (Mei. 35 384),
1 left mandible with M_1-M_3: IQW 1984/19 816 (Mei. 19 336),
1 right mandible with M_1-M_3: IQW 1984/19 813 (Mei. 19 333),
1 right mandible with M_2-M_3: IQW 1984/19 815 (Mei. 19 335),
1 left mandible with M_3: IQW 1984/19 814 (Mei. 19 334),
1 left I_{inf}: IQW 1984/19 817 (Mei. 19 337),
1 right I_{inf}: IQW 1983/18 724 (Mei. 18 244),
1 femur fragment: IQW 2016/44 777 (Mei. 50 086).

Hypolagus brachygnathus Kormos, 1930
1 left P_3: IQW 2007/29 557 (Mei. 28 719).

Lepus sp.
2 right P^2: IQW 1984/20 178 (Mei. 19 698), IQW 2014/41 830 (Mei. 47 139),
2 right humerus fragments distal: IQW 1991/23 716 (Mei. 23 245), IQW 1991/23 717 (Mei. 23 246).

Hypolagus brachygnathus vel *Lepus* sp.
1 left P^3: IQW 2015/42 007 (Mei. 47 316),
2 left P^{sup}: IQW 2010/30 897 (Mei. 30 059),
1 right M^2: IQW 2016/42 783 (Mei. 48 092),
2 M^{sup}: IQW 2016/42 782 (Mei. 48 091), IQW 2010/31 030 (Mei. 30 192),
1 dP_3: IQW 2016/42 781 (Mei. 48 090),
2 femur fragments: IQW 1994/24 582 (Mei. 24 111), IQW 2015/41 978 (Mei. 47 287),
2 tibia fragments: IQW 2016/44 014 (Mei. 49 323),
2 phalanges of indeterminable position: IQW 1986/21 496 (Mei. 21 015), IQW 2016/44 065 (Mei. 49 3749).
1 vertebra: IQW 1982/18 599 (Mei. 18 119),
4 costae: IQW 1992/23 912 (Mei. 23 441), IQW 2011/31 239 (Mei. 30 401), IQW 2011/31 259 (Mei. 30 421), IQW 2011/31 371 (Mei. 30 533).

Sciurus whitei Hinton, 1914
1 right I^{sup}: IQW 1991/23 723 (Mei. 23 252),
1 left $M^{1/2}$: IQW 1985/20 417 (Mei. 19 937),
1 right $M^{1/2}$: IQW 1983/19 420 (Mei. 18 940),
1 left $M_{1/2}$: IQW 1984/19 831 (Mei. 19 351).

Spermophilus kahlkei n. sp.
38 skeletal units: IQW 2003/27 592 (Mei. 27 097), IQW 2003/27 643 (Mei. 27 148), IQW 2003/27 652 (Mei. 27 157), IQW 2003/28 125 (Mei. 27 287), IQW 2003/28 187 (Mei. 27 349), IQW 2003/28 188 (Mei. 27 350), IQW 2003/28 189 (Mei. 27 351), IQW 2003/28 364 (Mei. 27 526), IQW 2005/28 960 (Mei. 28 122), IQW 2008/29 802 (Mei. 28 964),

IQW 2009/30 023 (Mei. 29 185), IQW 2009/30 214 (Mei. 29 376), IQW 2009/30 359 (Mei. 29 521), IQW 2009/30 544 (Mei. 29 706), IQW 2009/30 588 (Mei. 29 750), IQW 2009/30 622 (Mei. 29 784), IQW 2009/30 624 (Mei. 29 786), IQW 2009/30 636 (Mei. 29 798), IQW 2009/30 637 (Mei. 29 799), IQW 2009/30 638 (Mei. 29 800), IQW 2010/30 772 (Mei. 29 934), IQW 2010/31 104 (Mei. 30 266), IQW 2011/31 387 (Mei. 30 549), IQW 2011/32 268 (Mei. 31 431), IQW 2011/35 477 (Mei. 32 687), IQW 2012/38 429 (Mei. 35 403), IQW 2012/38 457 (Mei. 35 431), IQW 2012/38 568 (Mei. 35 542), IQW 2012/38 569 (Mei. 35 543), IQW 2012/38 602 (Mei. 35 576), IQW 2012/38 656 (Mei. 35 630), IQW 2012/38 656 (Mei. 35 630), IQW 2012/38 658 (Mei. 35 632), IQW 2012/38 659 (Mei. 35 633), IQW 2012/38 660 (Mei. 35 634), IQW 2012/38 661 (Mei. 35 635), IQW 2012/38 662 (Mei. 35 636), IQW 2013/41 665 (Mei. 47 005),
42 cranium fragments: IQW 2009/30 118 (Mei. 29 280), IQW 2011/31 351 (Mei. 30 513), IQW 2012/38 419 (Mei. 35 393), IQW 2012/38 472 (Mei. 35 446), IQW 2012/38 565 (Mei. 35 539), IQW 2012/38 600 (Mei. 35 574), IQW 2012/38 601 (Mei. 35 575), IQW 2012/38 607 (Mei. 35 581), IQW 2012/38 622 (Mei. 35 596), IQW 2012/38 757 (Mei. 35 731), IQW 2012/38 758 (Mei. 35 732), IQW 2012/39 320 (Mei. 36 294), IQW 2013/41 663 (Mei. 47 003), IQW 2013/41 664 (Mei. 47 004), IQW 2015/41 996 (Mei. 47 305), IQW 1985/20 931 (Mei. 20 450), IQW 2003/27 707 (Mei. 27 212), IQW 2003/27 711 (Mei. 27 216), IQW 2004/28 460 (Mei. 27 622), IQW 2004/28 769 (Mei. 27 931), IQW 2007/29 523 (Mei. 28 685), IQW 2008/29 811 (Mei. 28 973), IQW 2009/30 448 (Mei. 29 610), IQW 2009/30 518 (Mei. 29 680), IQW 2009/30 627 (Mei. 29 789), IQW 2010/30 933 (Mei. 30 095), IQW 2010/30 968 (Mei. 30 130), IQW 2010/31 105 (Mei. 30 267), IQW 2010/31 106 (Mei. 30 268), IQW 2011/31 373 (Mei. 30 535), IQW 2011/32 186 (Mei. 31 350), IQW 2013/41 538 (Mei. 36 884), IQW 2015/41 998 (Mei. 47 307), IQW 1982/17 979 (Mei. 17 499), IQW 2010/30 836 (Mei. 29 998), IQW 2010/30 933 (Mei. 30 095), IQW 2016/43 226 (Mei. 48 535), IQW 2016/43 232 (Mei. 48 541), IQW 2016/43 233 (Mei. 48 542), IQW 2016/43 242 (Mei. 48 551), IQW 2016/43 242 (Mei. 48 551), IQW 2016/43 246 (Mei. 48 555),
3 maxilla fragments: IQW 2011/31 370 (Mei. 30 532), IQW 2010/30 933 (Mei. 30 095), IQW 2016/43 729 (Mei. 49 038),
44 left mandible fragments: IQW 2011/32 106 (Mei. 31 268), IQW 2011/32 108 (Mei. 31 270), IQW 2012/38 416 (Mei. 35 390), IQW 2012/38 456 (Mei. 35 430), IQW 2012/38 557 (Mei. 35 531), IQW 2012/38 561 (Mei. 35 535), IQW 2012/38 564 (Mei. 35 538), IQW 2012/38 582 (Mei. 35 556), IQW 2012/38 603 (Mei. 35 577), IQW 2012/38 605 (Mei. 35 579), IQW 2012/38 609 (Mei. 35 583), IQW 2012/38 627 (Mei. 35 601), IQW 2012/38 629 (Mei. 35 603), IQW 2012/38 633 (Mei. 35 607), IQW 2013/41 537 (Mei. 36 883), IQW 2015/41 989 (Mei. 47 298), IQW 2015/41 997 (Mei. 47 306), IQW 2016/42 755 (Mei. 48 064), IQW 2010/30 933 (Mei. 30 095), IQW 2012/38 752 (Mei. 35 726), IQW 2016/43 069 (Mei. 48 378), IQW 2016/43 216 (Mei. 48 525), IQW 2016/43 225 (Mei. 48 534), IQW 2016/43 226 (Mei. 48 535), IQW 2016/43 230 (Mei. 48 539), IQW 2016/43 231 (Mei. 48 540), IQW 2016/43 237 (Mei. 48 546), IQW 2016/43 238 (Mei. 48 547), IQW 2016/43 239 (Mei. 48 548), IQW 2016/43 240 (Mei. 48 549), IQW 2016/43 243 (Mei. 48 552), IQW 2016/43 247 (Mei. 48 556), IQW 2016/43 345 (Mei. 48 654), IQW 2016/43 350 (Mei. 48 659), IQW 2016/43 475 (Mei. 48 784), IQW 1983/18 791 (Mei. 18 311), IQW 2003/27 639 (Mei. 27 144), IQW 2007/29 151 (Mei. 28 313), IQW 2011/31 334 (Mei. 30 496), IQW 2012/38 409 (Mei. 35 383), IQW 1983/19 446 (Mei. 18 966), IQW 2010/30 798 (Mei. 29 960), IQW 2012/38 766 (Mei. 35 740),
47 right mandible fragments: IQW 1982/17 896 (Mei. 17 416), IQW 1983/18 792 (Mei. 18 312), IQW 1983/18 793 (Mei. 18 313), IQW 1983/18 794 (Mei. 18 314), IQW 1983/18 795 (Mei. 18 315), IQW 1983/18 796 (Mei. 18 316), IQW 1983/19 444 (Mei. 18 964), IQW 1985/20 418 (Mei. 19 938), IQW 1985/20 931 (Mei. 20 450), IQW 1985/20 934 (Mei. 20 453), IQW 1985/20 949 (Mei. 20 468), IQW 1985/20 950 (Mei. 20 469), IQW 1985/20 952 (Mei. 20 471), IQW 1985/20 953 (Mei. 20 472), IQW 1985/20 954 (Mei. 20 473), IQW 1985/20 955 (Mei. 20 474), IQW 1985/20 956 (Mei. 20 475), IQW 1985/20 957 (Mei. 20 476), IQW 2003/27 641 (Mei. 27 146), IQW 2003/27 642 (Mei. 27 147), IQW 2003/27 644 (Mei. 27 149), IQW 2003/28 154 (Mei. 27 316), IQW 2003/28 406 (Mei. 27 568), IQW 2004/28 441 (Mei. 27 603), IQW 2007/29 173 (Mei. 28 335), IQW 2007/29 524 (Mei. 28 686), IQW 2007/29 525 (Mei. 28 687), IQW 2008/29 740 (Mei. 28 902), IQW 2008/29 741 (Mei. 28 903), IQW 2008/29 798 (Mei. 28 960), IQW 2008/29 813 (Mei. 28 975), IQW 2009/30 442 (Mei. 29 604), IQW 2009/30 540 (Mei. 29 702), IQW 2009/30 613 (Mei. 29 775), IQW 2009/30 619 (Mei. 29 781), IQW 2009/30 620 (Mei. 29 782), IQW 2009/30 629 (Mei. 29 791), IQW 2009/30 633 (Mei. 29 795), IQW 2010/30 679 (Mei. 29 841), IQW 2010/30 688 (Mei. 29 850), IQW 2010/30 922 (Mei. 30 084), IQW 2010/30 945 (Mei. 30 107), IQW 2010/31 097 (Mei. 30 259), IQW 2010/31 098 (Mei. 30 260), IQW 2010/31 127 (Mei. 30 289), IQW 2011/31 372 (Mei. 30 534), IQW 2011/32 103 (Mei. 31 265),
140 incisor fragments: IQW 2007/29 370 (Mei. 28 532), IQW 2009/30 157 (Mei. 29 319), IQW 2009/30 540 (Mei. 29 702), IQW 2010/31 056 (Mei. 30 218), IQW 2012/38 413 (Mei. 35 387), IQW 2012/38 456 (Mei. 35 430), IQW 2012/38 769 (Mei. 35 743), IQW 2016/43 513 (Mei. 48 822), IQW 2016/43 514 (Mei. 48 823), IQW 2016/43 515 (Mei. 48 824), IQW 2016/43 516 (Mei. 48 825), IQW 1985/20 958 (Mei. 20 477), IQW 2003/28 135 (Mei. 27 297), IQW 2004/28 442 (Mei. 27 604), IQW 2004/28 477 (Mei. 27 639), IQW 2008/29 815 (Mei. 28 977), IQW 2009/30 038 (Mei. 29 200), IQW 2009/30 119 (Mei. 29 281), IQW 2009/30 376 (Mei. 29 538), IQW 2010/30 836 (Mei. 29 998), IQW 2010/30 941

(Mei. 30 103), IQW 2010/30 947 (Mei. 30 109), IQW 2010/31 008 (Mei. 30 170), IQW 2010/31 012 (Mei. 30 174), IQW 2010/31 038 (Mei. 30 200), IQW 2011/31 335 (Mei. 30 497), IQW 2011/31 370 (Mei. 30 532), IQW 2011/35 480 (Mei. 32 690), IQW 2012/38 432 (Mei. 35 406), IQW 2012/38 545 (Mei. 35 519), IQW 2012/38 624 (Mei. 35 598), IQW 2012/38 636 (Mei. 35 610), IQW 2016/42 756 (Mei. 48 065), IQW 2016/42 757 (Mei. 48 066), IQW 2016/42 758 (Mei. 48 067), IQW 2016/42 759 (Mei. 48 068), IQW 2016/42 760 (Mei. 48 069), IQW 2016/42 761 (Mei. 48 070), IQW 2016/42 762 (Mei. 48 071), IQW 2016/42 763 (Mei. 48 072), IQW 2016/42 764 (Mei. 48 073), IQW 2016/42 765 (Mei. 48 074), IQW 2016/43 398 (Mei. 48 707), IQW 2016/43 402 (Mei. 48 711), IQW 2016/43 409 (Mei. 48 718), IQW 2016/43 446 (Mei. 48 755), IQW 2016/43 447 (Mei. 48 756), IQW 2016/43 448 (Mei. 48 757), IQW 2016/43 449 (Mei. 48 758), IQW 2016/43 450 (Mei. 48 759), IQW 2016/43 451 (Mei. 48 760), IQW 2016/43 453 (Mei. 48 762), IQW 2016/43 454 (Mei. 48 763), IQW 2016/43 455 (Mei. 48 764), IQW 2016/43 464 (Mei. 48 773), IQW 2016/43 466 (Mei. 48 775), IQW 2016/43 486 (Mei. 48 795), IQW 2016/43 504 (Mei. 48 813), IQW 2016/43 508 (Mei. 48 817), IQW 2016/43 519 (Mei. 48 828), IQW 2016/43 529 (Mei. 48 838), IQW 2016/43 530 (Mei. 48 839), IQW 2016/43 543 (Mei. 48 852), IQW 2016/43 554 (Mei. 48 863), IQW 2016/43 586 (Mei. 48 895), IQW 2016/43 599 (Mei. 48 908), IQW 2016/43 600 (Mei. 48 909), IQW 2016/43 603 (Mei. 48 912), IQW 2016/43 607 (Mei. 48 916), IQW 2016/43 618 (Mei. 48 927), IQW 2016/43 622 (Mei. 48 931), IQW 2016/43 637 (Mei. 48 946), IQW 2016/43 658 (Mei. 48 967), IQW 2016/43 666 (Mei. 48 975), IQW 2016/43 679 (Mei. 48 988), IQW 2016/43 705 (Mei. 49 014), IQW 2016/43 734 (Mei. 49 043), IQW 2016/43 736 (Mei. 49 045), IQW 2016/43 745 (Mei. 49 054), IQW 2016/43 753 (Mei. 49 062), IQW 2014/41 859 (Mei. 47 168), IQW 2016/43 209 (Mei. 48 518), IQW 2016/43 210 (Mei. 48 519), IQW 2016/43 214 (Mei. 48 523), IQW 2016/43 224 (Mei. 48 533), IQW 2016/43 242 (Mei. 48 551), IQW 2016/43 243 (Mei. 48 552), IQW 2016/43 272 (Mei. 48 581), IQW 2016/43 345 (Mei. 48 654), IQW 2016/43 351 (Mei. 48 660), IQW 2010/31 056 (Mei. 30 218), IQW 2012/38 413 (Mei. 35 387), IQW 2012/38 448 (Mei. 35 422), IQW 2012/38 627 (Mei. 35 601),

100 P^{sup}/M^{sup}: IQW 2016/43 589 (Mei. 48 898), IQW 2016/43 596 (Mei. 48 905), IQW 2016/43 606 (Mei. 48 915), IQW 2016/43 613 (Mei. 48 922), IQW 2016/43 463 (Mei. 48 772), IQW 2016/43 465 (Mei. 48 774), IQW 2016/43 467 (Mei. 48 776), IQW 2016/43 468 (Mei. 48 777), IQW 2016/43 608 (Mei. 48 917), IQW 2016/43 610 (Mei. 48 919), IQW 2016/43 611 (Mei. 48 920), IQW 2016/43 615 (Mei. 48 924), IQW 2016/43 616 (Mei. 48 925), IQW 2016/43 617 (Mei. 48 926), IQW 2016/43 621 (Mei. 48 930), IQW 2016/43 623 (Mei. 48 932), IQW 2016/43 624 (Mei. 48 933), IQW 2016/43 626 (Mei. 48 935), IQW 2016/43 627 (Mei. 48 936), IQW 2016/43 629 (Mei. 48 938), IQW 2016/43 630 (Mei. 48 939), IQW 2016/43 633 (Mei. 48 942), IQW 2016/43 634 (Mei. 48 943), IQW 1983/18 804 (Mei. 18 324), IQW 1983/18 810 (Mei. 18 330), IQW 1983/18 812 (Mei. 18 332), IQW 1983/18 813 (Mei. 18 333), IQW 1983/18 814 (Mei. 18 334), IQW 1983/18 816 (Mei. 18 336), IQW 1983/18 817 (Mei. 18 337), IQW 1983/18 818 (Mei. 18 338), IQW 1983/18 829 (Mei. 18 349), IQW 1983/18 833 (Mei. 18 353), IQW 1983/18 835 (Mei. 18 355), IQW 1983/18 841 (Mei. 18 361), IQW 1983/18 842 (Mei. 18 362), IQW 1983/18 921 (Mei. 18 441), IQW 1983/19 007 (Mei. 18 527), IQW 1983/19 422 (Mei. 18 942), IQW 1983/19 424 (Mei. 18 944), IQW 1983/19 427 (Mei. 18 947), IQW 1983/19 428 (Mei. 18 948), IQW 1984/19 842 (Mei. 19 362), IQW 1984/19 843 (Mei. 19 363), IQW 1984/19 844 (Mei. 19 364), IQW 1984/19 845 (Mei. 19 365), IQW 1984/19 846 (Mei. 19 366), IQW 1984/19 847 (Mei. 19 367), IQW 1984/19 848 (Mei. 19 368), IQW 1984/19 849 (Mei. 19 369), IQW 1984/19 850 (Mei. 19 370), IQW 1984/19 851 (Mei. 19 371), IQW 1984/19 852 (Mei. 19 372), IQW 1984/19 853 (Mei. 19 373), IQW 1984/19 854 (Mei. 19 374), IQW 1984/19 855 (Mei. 19 375), IQW 1984/19 856 (Mei. 19 376), IQW 1984/19 857 (Mei. 19 377), IQW 1984/19 858 (Mei. 19 378), IQW 1984/19 859 (Mei. 19 379), IQW 1984/19 860 (Mei. 19 380), IQW 1984/19 861 (Mei. 19 381), IQW 1984/19 862 (Mei. 19 382), IQW 1984/19 863 (Mei. 19 383), IQW 1984/19 867 (Mei. 19 387), IQW 1984/19 867 (Mei. 19 387), IQW 1984/19 867 (Mei. 19 387), IQW 1984/19 867 (Mei. 19 387), IQW 1984/19 868 (Mei. 19 388), IQW 1984/19 869 (Mei. 19 389), IQW 1984/19 870 (Mei. 19 390), IQW 1984/19 871 (Mei. 19 391), IQW 1984/19 872 (Mei. 19 392), IQW 1984/19 873 (Mei. 19 393), IQW 1984/19 874 (Mei. 19 394), IQW 1984/19 875 (Mei. 19 395), IQW 1984/19 876 (Mei. 19 396), IQW 1984/19 877 (Mei. 19 397), IQW 1984/19 878 (Mei. 19 398), IQW 1984/19 879 (Mei. 19 399), IQW 1984/19 880 (Mei. 19 400), IQW 1984/19 881 (Mei. 19 401), IQW 1984/19 897 (Mei. 19 417), IQW 1984/19 898 (Mei. 19 418), IQW 1985/20 932 (Mei. 20 451), IQW 1985/20 937 (Mei. 20 456), IQW 1985/20 938 (Mei. 20 457), IQW 1985/20 939 (Mei. 20 458), IQW 1985/20 940 (Mei. 20 459), IQW 1985/20 941 (Mei. 20 460), IQW 1985/20 942 (Mei. 20 461), IQW 1985/20 943 (Mei. 20 462), IQW 1985/20 944 (Mei. 20 463), IQW 1985/20 945 (Mei. 20 464), IQW 1985/20 948 (Mei. 20 467), IQW 2016/43 597 (Mei. 48 906), IQW 2016/43 598 (Mei. 48 907), IQW 2016/43 601 (Mei. 48 910), IQW 2016/43 605 (Mei. 48 914), IQW 2016/43 612 (Mei. 48 921),

105 P_{inf}/M_{inf}: IQW 2012/38 627 (Mei. 35 601), IQW 2016/43 730 (Mei. 49 039), IQW 2016/43 731 (Mei. 49 040), IQW 2016/43 732 (Mei. 49 041), IQW 2016/43 733 (Mei. 49 042), IQW 2016/43 739 (Mei. 49 048), IQW 2016/43 740 (Mei. 49 049), IQW 2016/43 744 (Mei. 49 053), IQW 2016/43 746 (Mei. 49 055), IQW 2016/43 747 (Mei. 49 056), IQW 2016/43 748 (Mei. 49 057), IQW 2016/43 749 (Mei. 49 058), IQW 2016/43 750 (Mei. 49 059), IQW 2016/43 751 (Mei. 49 060), IQW 2016/43 752 (Mei. 49 061), IQW 2016/43 754 (Mei. 49 063), IQW 2016/43 756 (Mei. 49 065),

IQW 1982/17 897 (Mei. 17 417), IQW 1982/18 369 (Mei. 17 889), IQW 1983/18 768 (Mei. 18 288), IQW 1983/18 798 (Mei. 18 318), IQW 1983/18 799 (Mei. 18 319), IQW 1983/18 800 (Mei. 18 320), IQW 1983/18 801 (Mei. 18 321), IQW 1983/18 802 (Mei. 18 322), IQW 1983/18 803 (Mei. 18 323), IQW 1983/18 806 (Mei. 18 326), IQW 1983/18 808 (Mei. 18 328), IQW 1983/18 809 (Mei. 18 329), IQW 1983/18 820 (Mei. 18 340), IQW 1983/18 822 (Mei. 18 342), IQW 1983/18 823 (Mei. 18 343), IQW 1983/18 824 (Mei. 18 344), IQW 1983/18 825 (Mei. 18 345), IQW 1983/18 826 (Mei. 18 346), IQW 1983/18 828 (Mei. 18 348), IQW 1983/18 831 (Mei. 18 351), IQW 1983/18 832 (Mei. 18 352), IQW 1983/18 834 (Mei. 18 354), IQW 1983/18 836 (Mei. 18 356), IQW 1983/18 839 (Mei. 18 359), IQW 1983/18 840 (Mei. 18 360), IQW 1983/18 843 (Mei. 18 362), IQW 1983/19 165 (Mei. 18 685), IQW 1983/19 165 (Mei. 18 685), IQW 1983/19 426 (Mei. 18 946), IQW 1984/19 829 (Mei. 19 349), IQW 1984/19 830 (Mei. 19 350), IQW 1984/19 832 (Mei. 19 352), IQW 1984/19 833 (Mei. 19 353), IQW 1984/19 834 (Mei. 19 354), IQW 1984/19 836 (Mei. 19 356), IQW 1984/19 837 (Mei. 19 357), IQW 1984/19 838 (Mei. 19 358), IQW 1984/19 848 (Mei. 19 348), IQW 1984/19 864 (Mei. 19 384), IQW 1984/19 865 (Mei. 19 385), IQW 1984/19 866 (Mei. 19 386), IQW 1984/19 882 (Mei. 19 402), IQW 1984/19 883 (Mei. 19 403), IQW 1984/19 884 (Mei. 19 404), IQW 1984/19 885 (Mei. 19 405), IQW 1984/19 886 (Mei. 19 406), IQW 1984/19 887 (Mei. 19 407), IQW 1984/19 887 (Mei. 19 407), IQW 1984/19 888 (Mei. 19 408), IQW 1984/19 889 (Mei. 19 409), IQW 1984/19 890 (Mei. 19 410), IQW 1984/19 891 (Mei. 19 411), IQW 1984/19 892 (Mei. 19 412), IQW 1984/19 893 (Mei. 19 413), IQW 1984/19 894 (Mei. 19 414), IQW 1984/19 895 (Mei. 19 415), IQW 1984/19 896 (Mei. 19 416), IQW 1984/19 900 (Mei. 19 420), IQW 1984/19 901 (Mei. 19 421), IQW 1984/19 902 (Mei. 19 422), IQW 1984/19 903 (Mei. 19 423), IQW 1984/19 904 (Mei. 19 424), IQW 1984/19 908 (Mei. 19 428), IQW 1985/20 933 (Mei. 20 452), IQW 1985/20 935 (Mei. 20 454), IQW 1985/20 936 (Mei. 20 455), IQW 1985/20 946 (Mei. 20 465), IQW 1985/20 947 (Mei. 20 466), IQW 2010/31 069 (Mei. 30 231), IQW 2016/43 443 (Mei. 48 752), IQW 2016/43 476 (Mei. 48 785), IQW 2016/43 477 (Mei. 48 786), IQW 2016/43 478 (Mei. 48 787), IQW 2016/43 479 (Mei. 48 788), IQW 2016/43 480 (Mei. 48 789), IQW 2016/43 481 (Mei. 48 790), IQW 2016/43 482 (Mei. 48 791), IQW 2016/43 483 (Mei. 48 792), IQW 2016/43 484 (Mei. 48 793), IQW 2016/43 738 (Mei. 49 047), IQW 2016/43 742 (Mei. 49 051), IQW 2016/43 743 (Mei. 49 052), IQW 2016/43 758 (Mei. 49 067), IQW 2016/43 759 (Mei. 49 068), IQW 2016/43 760 (Mei. 49 069), IQW 2016/43 761 (Mei. 49 070), IQW 2016/43 762 (Mei. 49 071), IQW 2016/44 018 (Mei. 49 327),

99 molar fragments: IQW 1985/20 951 (Mei. 20 470), IQW 1994/23 834 (Mei. 23 362), IQW 2009/30 376 (Mei. 29 538), IQW 2009/30 478 (Mei. 29 640), IQW 2009/30 615 (Mei. 29 777), IQW 2010/31 056 (Mei. 30 218), IQW 2011/32 074 (Mei. 31 236), IQW 2011/35 480 (Mei. 32 690), IQW 2011/35 506 (Mei. 32 716), IQW 2011/35 523 (Mei. 32 733), IQW 2012/38 411 (Mei. 35 385), IQW 2012/38 433 (Mei. 35 407), IQW 2012/38 557 (Mei. 35 531), IQW 2012/38 590 (Mei. 35 564), IQW 2012/38 778 (Mei. 35 752), IQW 2016/43 347 (Mei. 48 656), IQW 2016/43 456 (Mei. 48 765), IQW 2016/43 470 (Mei. 48 779), IQW 2016/43 485 (Mei. 48 794), IQW 2016/43 487 (Mei. 48 796), IQW 2016/43 489 (Mei. 48 798), IQW 2016/43 491 (Mei. 48 800), IQW 2016/43 492 (Mei. 48 801), IQW 2016/43 495 (Mei. 48 804), IQW 2016/43 496 (Mei. 48 805), IQW 2016/43 497 (Mei. 48 806), IQW 2016/43 498 (Mei. 48 807), IQW 2016/43 499 (Mei. 48 808), IQW 2016/43 500 (Mei. 48 809), IQW 2016/43 501 (Mei. 48 810), IQW 2016/43 502 (Mei. 48 811), IQW 2016/43 503 (Mei. 48 812), IQW 2016/43 505 (Mei. 48 814), IQW 2016/43 518 (Mei. 48 827), IQW 2016/43 520 (Mei. 48 829), IQW 2016/43 521 (Mei. 48 830), IQW 2016/43 523 (Mei. 48 832), IQW 2016/43 525 (Mei. 48 834), IQW 2016/43 527 (Mei. 48 836), IQW 2016/43 528 (Mei. 48 837), IQW 2016/43 531 (Mei. 48 840), IQW 2016/43 532 (Mei. 48 841), IQW 2016/43 533 (Mei. 48 842), IQW 2016/43 534 (Mei. 48 843), IQW 2016/43 535 (Mei. 48 844), IQW 2016/43 536 (Mei. 48 845), IQW 2016/43 537 (Mei. 48 846), IQW 2016/43 538 (Mei. 48 847), IQW 2016/43 539 (Mei. 48 848), IQW 2016/43 540 (Mei. 48 849), IQW 2016/43 542 (Mei. 48 851), IQW 2016/43 545 (Mei. 48 854), IQW 2016/43 546 (Mei. 48 855), IQW 2016/43 547 (Mei. 48 856), IQW 2016/43 548 (Mei. 48 857), IQW 2016/43 549 (Mei. 48 858), IQW 2016/43 550 (Mei. 48 859), IQW 2016/43 552 (Mei. 48 861), IQW 2016/43 553 (Mei. 48 862), IQW 2016/43 555 (Mei. 48 864), IQW 2016/43 556 (Mei. 48 865), IQW 2016/43 557 (Mei. 48 866), IQW 2016/43 558 (Mei. 48 867), IQW 2016/43 559 (Mei. 48 868), IQW 2016/43 561 (Mei. 48 870), IQW 2016/43 562 (Mei. 48 871), IQW 2016/43 563 (Mei. 48 872), IQW 2016/43 564 (Mei. 48 873), IQW 2016/43 565 (Mei. 48 874), IQW 2016/43 566 (Mei. 48 875), IQW 2016/43 568 (Mei. 48 877), IQW 2016/43 570 (Mei. 48 879), IQW 2016/43 574 (Mei. 48 883), IQW 2016/43 580 (Mei. 48 889), IQW 2016/43 593 (Mei. 48 902), IQW 2016/43 604 (Mei. 48 913), IQW 2016/43 625 (Mei. 48 934), IQW 2016/43 657 (Mei. 48 966), IQW 2016/43 673 (Mei. 48 982), IQW 2016/43 698 (Mei. 49 007), IQW 2016/43 726 (Mei. 49 035), IQW 2016/43 735 (Mei. 49 044), IQW 2016/43 757 (Mei. 49 066), IQW 2010/30 836 (Mei. 29 998), IQW 2016/43 345 (Mei. 48 654), IQW 2016/43 349 (Mei. 48 658),

25 scapula fragments: IQW 2007/29 159 (Mei. 28 321), IQW 2007/29 370 (Mei. 28 532), IQW 2009/30 376 (Mei. 29 538), IQW 2009/30 376 (Mei. 29 538), IQW 2010/30 910 (Mei. 30 072), IQW 2011/31 275 (Mei. 30 437), IQW 2011/31 363 (Mei. 30 525), IQW 2011/35 480 (Mei. 32 690), IQW 2012/38 414 (Mei. 35 388), IQW 2012/38 432 (Mei. 35 406), IQW 2012/38 433 (Mei. 35 407), IQW 2012/38 447 (Mei. 35 421), IQW 2012/38 448 (Mei. 35 422), IQW 2012/38 456 (Mei. 35 430), IQW 2012/38 634 (Mei. 35 608), IQW 2016/43 306 (Mei. 48 615), IQW 2016/43 258 (Mei. 48 567),

IQW 2016/43 283 (Mei. 48 592), IQW 2010/30 933 (Mei. 30 095), IQW 2016/43 255 (Mei. 48 564), IQW 2016/43 269 (Mei. 48 578), IQW 2016/43 278 (Mei. 48 587), IQW 2016/43 292 (Mei. 48 601), IQW 2016/43 294 (Mei. 48 603),

80 humerus fragments: IQW 2013/41 628 (Mei. 46 968), IQW 1982/17 898 (Mei. 17 418), IQW 1982/17 953 (Mei. 17 473), IQW 1983/19 429 (Mei. 18 949), IQW 1983/19 438 (Mei. 18 958), IQW 1983/19 443 (Mei. 18 963), IQW 1983/19 445 (Mei. 18 965), IQW 1990/23 599 (Mei. 23 128), IQW 2003/27 598 (Mei. 27 103), IQW 2004/28 643 (Mei. 27 805), IQW 2004/28 779 (Mei. 27 941), IQW 2007/29 148 (Mei. 28 310), IQW 2007/29 372 (Mei. 28 534), IQW 2008/29 815 (Mei. 28 977), IQW 2009/30 157 (Mei. 29 319), IQW 2009/30 357 (Mei. 29 519), IQW 2009/30 442 (Mei. 29 604), IQW 2009/30 444 (Mei. 29 606), IQW 2009/30 461 (Mei. 29 623), IQW 2009/30 467 (Mei. 29 629), IQW 2009/30 478 (Mei. 29 640), IQW 2009/30 478 (Mei. 29 640), IQW 2009/30 616 (Mei. 29 778), IQW 2009/30 621 (Mei. 29 783), IQW 2010/30 844 (Mei. 30 006), IQW 2010/31 093 (Mei. 30 255), IQW 2010/31 114 (Mei. 30 276), IQW 2011/32 062 (Mei. 31 224), IQW 2011/32 064 (Mei. 31 226), IQW 2011/32 265 (Mei. 31 428), IQW 2011/35 480 (Mei. 32 690), IQW 2012/38 413 (Mei. 35 387), IQW 2012/38 432 (Mei. 35 406), IQW 2012/38 438 (Mei. 35 412), IQW 2012/38 456 (Mei. 35 430), IQW 2012/38 557 (Mei. 35 531), IQW 2012/38 560 (Mei. 35 534), IQW 2012/38 566 (Mei. 35 540), IQW 2012/38 567 (Mei. 35 541), IQW 2012/38 594 (Mei. 35 568), IQW 2012/38 618 (Mei. 35 592), IQW 2012/38 627 (Mei. 35 601), IQW 2012/38 627 (Mei. 35 601), IQW 2012/38 770 (Mei. 35 744), IQW 2016/43 277 (Mei. 48 586), IQW 2016/43 308 (Mei. 48 617), IQW 2016/43 326 (Mei. 48 635), IQW 2016/43 335 (Mei. 48 644), IQW 2016/43 338 (Mei. 48 647), IQW 2016/43 327 (Mei. 48 636), IQW 2016/43 293 (Mei. 48 602), IQW 2016/43 303 (Mei. 48 612), IQW 2010/30 933 (Mei. 30 095), IQW 2016/43 211 (Mei. 48 520), IQW 2016/43 218 (Mei. 48 527), IQW 2016/43 260 (Mei. 48 569), IQW 2016/43 263 (Mei. 48 572), IQW 2016/43 264 (Mei. 48 573), IQW 2016/43 266 (Mei. 48 575), IQW 2016/43 267 (Mei. 48 576), IQW 2016/43 268 (Mei. 48 577), IQW 2016/43 275 (Mei. 48 584), IQW 2016/43 279 (Mei. 48 588), IQW 2016/43 284 (Mei. 48 593), IQW 2016/43 289 (Mei. 48 598), IQW 2016/43 301 (Mei. 48 610), IQW 2016/43 304 (Mei. 48 613), IQW 2016/43 309 (Mei. 48 618), IQW 2016/43 310 (Mei. 48 619), IQW 2016/43 313 (Mei. 48 622), IQW 2016/43 317 (Mei. 48 626), IQW 2016/43 329 (Mei. 48 638), IQW 2016/43 339 (Mei. 48 648), IQW 2016/43 340 (Mei. 48 649),

48 ulna fragments: IQW 1982/17 895 (Mei. 17 415), IQW 1983/19 430 (Mei. 18 950), IQW 1983/19 432 (Mei. 18 952), IQW 1983/19 436 (Mei. 18 956), IQW 1994/23 815 (Mei. 23 343), IQW 1994/23 817 (Mei. 23 345), IQW 1994/23 818 (Mei. 23 346), IQW 1994/23 820 (Mei. 23 348), IQW 1994/23 821 (Mei. 23 349), IQW 1994/23 822 (Mei. 23 350), IQW 1994/23 824 (Mei. 23 352), IQW 1994/23 825 (Mei. 23 353), IQW 1994/23 826 (Mei. 23 354), IQW 1994/23 827 (Mei. 23 355), IQW 1994/23 828 (Mei. 23 356), IQW 1994/23 829 (Mei. 23 357), IQW 1994/23 830 (Mei. 23 358), IQW 1994/23 831 (Mei. 23 359), IQW 1994/23 832 (Mei. 23 360), IQW 1994/23 833 (Mei. 23 361), IQW 1994/23 836 (Mei. 23 364), IQW 1994/23 837 (Mei. 23 365), IQW 1994/23 838 (Mei. 23 366), IQW 1994/23 839 (Mei. 23 367), IQW 2007/29 159 (Mei. 28 321), IQW 2009/30 157 (Mei. 29 319), IQW 2009/30 478 (Mei. 29 640), IQW 2010/30 934 (Mei. 30 096), IQW 2011/32 062 (Mei. 31 224), IQW 2011/32 104 (Mei. 31 266), IQW 2011/35 474 (Mei. 32 684), IQW 2011/35 520 (Mei. 32 730), IQW 2012/38 413 (Mei. 35 387), IQW 2012/38 414 (Mei. 35 388), IQW 2012/38 448 (Mei. 35 422), IQW 2012/38 559 (Mei. 35 533), IQW 2012/38 560 (Mei. 35 534), IQW 2012/38 634 (Mei. 35 608), IQW 2012/38 634 (Mei. 35 608), IQW 2012/38 765 (Mei. 35 739), IQW 2012/38 769 (Mei. 35 743), IQW 1983/19 165 (Mei. 18 685), IQW 2016/43 206 (Mei. 48 515), IQW 2016/43 225 (Mei. 48 534),

43 radius fragments: IQW 1983/19 433 (Mei. 18 953), IQW 1994/23 840 (Mei. 23 368), IQW 1994/23 841 (Mei. 23 369), IQW 1994/23 842 (Mei. 23 370), IQW 1994/23 843 (Mei. 23 371), IQW 1994/23 844 (Mei. 23 372), IQW 1994/23 845 (Mei. 23 373), IQW 1994/23 846 (Mei. 23 374), IQW 1994/23 847 (Mei. 23 375), IQW 1994/23 848 (Mei. 23 376), IQW 1994/23 849 (Mei. 23 377), IQW 1994/23 850 (Mei. 23 378), IQW 1994/23 851 (Mei. 23 379), IQW 2007/29 159 (Mei. 28 321), IQW 2007/29 166 (Mei. 28 328), IQW 2008/29 769 (Mei. 28 931), IQW 2009/30 478 (Mei. 29 640), IQW 2010/30 944 (Mei. 30 106), IQW 2010/30 999 (Mei. 30 161), IQW 2010/31 062 (Mei. 30 224), IQW 2011/31 244 (Mei. 30 406), IQW 2011/32 097 (Mei. 31 259), IQW 2012/38 418 (Mei. 35 392), IQW 2012/38 432 (Mei. 35 406), IQW 2012/38 433 (Mei. 35 407), IQW 2012/38 447 (Mei. 35 421), IQW 2012/38 448 (Mei. 35 422), IQW 2012/38 456 (Mei. 35 430), IQW 2012/38 557 (Mei. 35 531), IQW 2012/38 627 (Mei. 35 601), IQW 2012/38 632 (Mei. 35 606), IQW 2012/38 634 (Mei. 35 608), IQW 2012/38 768 (Mei. 35 742), IQW 2016/43 328 (Mei. 48 637), IQW 2016/43 249 (Mei. 48 558), IQW 2016/43 253 (Mei. 48 562), IQW 2016/43 302 (Mei. 48 611), IQW 2016/43 307 (Mei. 48 616), IQW 2016/43 312 (Mei. 48 621), IQW 2016/43 319 (Mei. 48 628), IQW 2016/43 320 (Mei. 48 629), IQW 2016/43 331 (Mei. 48 640),

28 pelvis fragments: IQW 1983/19 490 (Mei. 19 010), IQW 2007/29 159 (Mei. 28 321), IQW 2009/30 157 (Mei. 29 319), IQW 2009/30 376 (Mei. 29 538), IQW 2009/30 438 (Mei. 29 600), IQW 2009/30 617 (Mei. 29 779), IQW 2010/30 910 (Mei. 30 072), IQW 2010/30 943 (Mei. 30 105), IQW 2010/30 972 (Mei. 30 134), IQW 2011/32 095 (Mei. 31 257), IQW 2011/35 480 (Mei. 32 690), IQW 2011/35 480 (Mei. 32 690), IQW 2011/35 519 (Mei. 32 729), IQW 2012/38 413 (Mei. 35 387), IQW 2012/38 414 (Mei. 35 388), IQW 2012/38 432 (Mei. 35 406), IQW 2012/38 456 (Mei. 35 430), IQW 2012/38 556 (Mei. 35 530), IQW 2012/38 558 (Mei. 35 532), IQW 2012/38 767 (Mei. 35 741),

1 left pelvis: IQW 2016/44 009 (Mei. 49 318), IQW 1982/17 979 (Mei. 17 499), IQW 1994/24 581 (Mei. 24 110),
51 femur fragments: IQW 1980/16 398 (Mei. 15 909), IQW 1980/17 166 (Mei. 16 687), IQW 1983/19 165 (Mei. 18 685), IQW 1987/22 185 (Mei. 21 704), IQW 1987/22 186 (Mei. 21 705), IQW 1987/22 269 (Mei. 21 788), IQW 2003/27 640 (Mei. 27 145), IQW 2003/28 150 (Mei. 27 312), IQW 2003/28 344 (Mei. 27 506), IQW 2004/28 443 (Mei. 27 605), IQW 2009/30 157 (Mei. 29 319), IQW 2009/30 168 (Mei. 29 330), IQW 2009/30 376 (Mei. 29 538), IQW 2009/30 438 (Mei. 29 600), IQW 2009/30 442 (Mei. 29 604), IQW 2009/30 505 (Mei. 29 667), IQW 2009/30 511 (Mei. 29 673), IQW 2009/30 540 (Mei. 29 702), IQW 2009/30 608 (Mei. 29 770), IQW 2009/30 610 (Mei. 29 772), IQW 2009/30 612 (Mei. 29 774), IQW 2009/30 630 (Mei. 29 792), IQW 2010/30 934 (Mei. 30 096), IQW 2010/31 039 (Mei. 30 201), IQW 2010/31 056 (Mei. 30 218), IQW 2011/32 065 (Mei. 31 227), IQW 2011/32 100 (Mei. 31 262), IQW 2011/35 480 (Mei. 32 690), IQW 2011/35 525 (Mei. 32 735), IQW 2012/38 395 (Mei. 35 369), IQW 2012/38 414 (Mei. 35 388), IQW 2012/38 432 (Mei. 35 406), IQW 2012/38 433 (Mei. 35 407), IQW 2012/38 448 (Mei. 35 422), IQW 2012/38 456 (Mei. 35 430), IQW 2012/38 456 (Mei. 35 430), IQW 2012/38 557 (Mei. 35 531), IQW 2012/38 608 (Mei. 35 582), IQW 2012/38 627 (Mei. 35 601), IQW 2012/38 630 (Mei. 35 604), IQW 2012/38 634 (Mei. 35 608), IQW 2012/38 637 (Mei. 35 611), IQW 2016/43 305 (Mei. 48 614), IQW 2016/43 314 (Mei. 48 623), IQW 1982/17 893 (Mei. 17 413), IQW 1982/17 979 (Mei. 17 499), IQW 1982/18 437 (Mei. 17 957), IQW 2016/43 225 (Mei. 48 534),
46 tibia fragments: IQW 1980/16 559 (Mei. 16 080), IQW 1981/17 709 (Mei. 17 231), IQW 1983/19 431 (Mei. 18 951), IQW 1983/19 605 (Mei. 19 125), IQW 1994/23 816 (Mei. 23 344), IQW 1994/23 819 (Mei. 23 347), IQW 1994/23 835 (Mei. 23 363), IQW 2003/28 287 (Mei. 27 449), IQW 2007/29 107 (Mei. 28 269), IQW 2007/29 168 (Mei. 28 330), IQW 2008/29 744 (Mei. 28 906), IQW 2009/30 157 (Mei. 29 319), IQW 2009/30 158 (Mei. 29 320), IQW 2009/30 266 (Mei. 29 428), IQW 2009/30 607 (Mei. 29 769), IQW 2010/30 807 (Mei. 29 969), IQW 2010/30 889 (Mei. 30 051), IQW 2010/30 941 (Mei. 30 103), IQW 2010/31 102 (Mei. 30 264), IQW 2011/31 349 (Mei. 30 511), IQW 2011/31 370 (Mei. 30 532), IQW 2011/32 098 (Mei. 31 260), IQW 2011/35 480 (Mei. 32 690), IQW 2011/35 519 (Mei. 32 729), IQW 2012/38 394 (Mei. 35 368), IQW 2012/38 413 (Mei. 35 387), IQW 2012/38 414 (Mei. 35 388), IQW 2012/38 421 (Mei. 35 395), IQW 2012/38 432 (Mei. 35 406), IQW 2012/38 433 (Mei. 35 407), IQW 2012/38 447 (Mei. 35 421), IQW 2012/38 448 (Mei. 35 422), IQW 2012/38 456 (Mei. 35 430), IQW 2012/38 560 (Mei. 35 534), IQW 2012/38 572 (Mei. 35 546), IQW 2012/38 627 (Mei. 35 601), IQW 2012/38 770 (Mei. 35 744), IQW 2016/43 201 (Mei. 48 510), IQW 2016/43 315 (Mei. 48 624), IQW 2016/43 316 (Mei. 48 625), IQW 1982/18 428 (Mei. 17 948), IQW 1983/19 419 (Mei. 18 939), IQW 2016/43 215 (Mei. 48 524), IQW 2016/43 226 (Mei. 48 535),
2 fibula fragments: IQW 1982/17 899 (Mei. 17 419), IQW 2012/38 432 (Mei. 35 406),
24 astragali: IQW 2007/29 370 (Mei. 28 532), IQW 2009/30 608 (Mei. 29 770), IQW 2010/30 889 (Mei. 30 051), IQW 2012/38 448 (Mei. 35 422), IQW 2012/38 456 (Mei. 35 430), IQW 2012/38 560 (Mei. 35 534), IQW 2012/38 627 (Mei. 35 601), IQW 2012/38 769 (Mei. 35 743), IQW 1983/19 435 (Mei. 18 955), IQW 1987/22 270 (Mei. 21 789), IQW 2016/44 083 (Mei. 49 392), IQW 2016/44 087 (Mei. 49 396), IQW 2016/44 101 (Mei. 49 410), IQW 2016/44 107 (Mei. 49 416), IQW 1983/19 434 (Mei. 18 954), IQW 2016/44 082 (Mei. 49 391), IQW 2016/44 090 (Mei. 49 399), IQW 2016/44 092 (Mei. 49 401), IQW 2016/44 104 (Mei. 49 413), IQW 2016/44 105 (Mei. 49 414), IQW 2016/44 106 (Mei. 49 415), IQW 2016/44 108 (Mei. 49 417), IQW 2016/43 345 (Mei. 48 654), IQW 2016/44 089 (Mei. 49 398),
24 calcanei: IQW 2009/30 376 (Mei. 29 538), IQW 2009/30 520 (Mei. 29 682), IQW 2009/30 608 (Mei. 29 770), IQW 2012/38 432 (Mei. 35 406), IQW 2012/38 433 (Mei. 35 407), IQW 2012/38 447 (Mei. 35 421), IQW 2012/38 448 (Mei. 35 422), IQW 2012/38 634 (Mei. 35 608), IQW 2016/43 202 (Mei. 48 511), IQW 2016/43 452 (Mei. 48 761), IQW 2016/43 471 (Mei. 48 780), IQW 2016/43 493 (Mei. 48 802), IQW 1983/19 165 (Mei. 18 655), IQW 2016/43 472 (Mei. 48 781), IQW 2016/43 474 (Mei. 48 783), IQW 2016/44 110 (Mei. 49 419), IQW 2016/44 876 (Mei. 50 185), IQW 1983/19 422 (Mei. 18 943), IQW 1983/19 425 (Mei. 18 945), IQW 1983/19 442 (Mei. 18 962), IQW 2016/43 469 (Mei. 48 778), IQW 2016/44 111 (Mei. 49 420), IQW 2016/44 875 (Mei. 50 184),
327 metapodials/phalanx fragments of indeterminable position: IQW 2007/29 370 (Mei. 28 532), IQW 2009/30 119 (Mei. 29 281), IQW 2009/30 157 (Mei. 29 319), IQW 2009/30 376 (Mei. 29 538), IQW 2009/30 383 (Mei. 29 545), IQW 2009/30 443 (Mei. 29 605), IQW 2009/30 631 (Mei. 29 793), IQW 2010/30 934 (Mei. 30 096), IQW 2010/30 944 (Mei. 30 106), IQW 2011/31 243 (Mei. 30 405), IQW 2011/32 062 (Mei. 31 224), IQW 2011/35 480 (Mei. 32 690), IQW 2011/35 519 (Mei. 32 729), IQW 2011/35 520 (Mei. 32 730), IQW 2012/38 400 (Mei. 35 374), IQW 2012/38 413 (Mei. 35 387), IQW 2012/38 432 (Mei. 35 406), IQW 2012/38 433 (Mei. 35 407), IQW 2012/38 447 (Mei. 35 421), IQW 2012/38 448 (Mei. 35 422), IQW 2012/38 455 (Mei. 35 429), IQW 2012/38 456 (Mei. 35 430), IQW 2012/38 557 (Mei. 35 531), IQW 2012/38 559 (Mei. 35 533), IQW 2012/38 560 (Mei. 35 534), IQW 2012/38 588 (Mei. 35 562), IQW 2012/38 627 (Mei. 35 601), IQW 2012/38 634 (Mei. 35 608), IQW 2012/38 635 (Mei. 35 609), IQW 2012/38 769 (Mei. 35 743), IQW 2010/31 069 (Mei. 30 231), IQW 2016/43 207 (Mei. 48 516), IQW 2016/43 213 (Mei. 48 522), IQW 2016/43 223 (Mei. 48 532), IQW 2016/43 341 (Mei. 48 650), IQW 2012/38 628 (Mei. 35 602), IQW 2016/43 220 (Mei. 48 529), IQW 2016/43 368 (Mei. 48 677), IQW 2016/43 369 (Mei. 48 678), IQW 2016/43 370 (Mei. 48 679),

IQW 2016/43 371 (Mei. 48 680), IQW 2016/43 372 (Mei. 48 681), IQW 2016/43 461 (Mei. 48 770), IQW 2016/43 511 (Mei. 48 820), IQW 2016/43 522 (Mei. 48 831), IQW 2016/43 526 (Mei. 48 835), IQW 2016/43 569 (Mei. 48 878), IQW 2016/43 575 (Mei. 48 884), IQW 2016/43 590 (Mei. 48 899), IQW 2016/43 591 (Mei. 48 900), IQW 2016/43 609 (Mei. 48 918), IQW 2016/43 614 (Mei. 48 923), IQW 2016/43 619 (Mei. 48 928), IQW 2016/43 444 (Mei. 48 753), IQW 2016/43 445 (Mei. 48 754), IQW 2016/43 457 (Mei. 48 766), IQW 2016/43 458 (Mei. 48 767), IQW 2016/43 459 (Mei. 48 768), IQW 2016/43 460 (Mei. 48 769), IQW 2016/43 462 (Mei. 48 771), IQW 2016/43 506 (Mei. 48 815), IQW 2016/43 507 (Mei. 48 816), IQW 2016/43 509 (Mei. 48 818), IQW 2016/43 635 (Mei. 48 944), IQW 2016/43 636 (Mei. 48 945), IQW 2016/43 639 (Mei. 48 948), IQW 2016/43 640 (Mei. 48 949), IQW 2016/43 641 (Mei. 48 950), IQW 2016/43 642 (Mei. 48 951), IQW 2016/43 643 (Mei. 48 952), IQW 2016/43 644 (Mei. 48 953), IQW 2016/43 645 (Mei. 48 954), IQW 2016/43 647 (Mei. 48 956), IQW 2016/43 648 (Mei. 48 957), IQW 2016/43 649 (Mei. 48 958), IQW 2016/43 650 (Mei. 48 959), IQW 2016/43 651 (Mei. 48 960), IQW 2016/43 654 (Mei. 48 963), IQW 2016/43 655 (Mei. 48 964), IQW 2016/43 656 (Mei. 48 965), IQW 2016/43 659 (Mei. 48 968), IQW 2016/43 662 (Mei. 48 971), IQW 2016/43 663 (Mei. 48 972), IQW 2016/43 664 (Mei. 48 973), IQW 2016/43 665 (Mei. 48 974), IQW 2016/43 668 (Mei. 48 977), IQW 2016/43 669 (Mei. 48 978), IQW 2016/43 670 (Mei. 48 979), IQW 2016/43 672 (Mei. 48 981), IQW 2016/43 674 (Mei. 48 983), IQW 2016/43 675 (Mei. 48 984), IQW 2016/43 676 (Mei. 48 985), IQW 2016/43 678 (Mei. 48 987), IQW 2016/43 680 (Mei. 48 989), IQW 2016/43 681 (Mei. 48 990), IQW 2016/43 682 (Mei. 48 991), IQW 2016/43 683 (Mei. 48 992), IQW 2016/43 684 (Mei. 48 993), IQW 2016/43 685 (Mei. 48 994), IQW 2016/43 686 (Mei. 48 995), IQW 2016/43 687 (Mei. 48 996), IQW 2016/43 688 (Mei. 48 997), IQW 2016/43 689 (Mei. 48 998), IQW 2016/43 690 (Mei. 48 999), IQW 2016/43 691 (Mei. 49 000), IQW 2016/43 692 (Mei. 49 001), IQW 2016/43 693 (Mei. 49 002), IQW 2016/43 694 (Mei. 49 003), IQW 2016/43 695 (Mei. 49 004), IQW 2016/43 696 (Mei. 49 005), IQW 2016/43 697 (Mei. 49 006), IQW 2016/43 699 (Mei. 49 008), IQW 2016/43 700 (Mei. 49 009), IQW 2016/43 701 (Mei. 49 010), IQW 2016/43 702 (Mei. 49 011), IQW 2016/43 703 (Mei. 49 012), IQW 2016/43 704 (Mei. 49 013), IQW 2016/43 706 (Mei. 49 015), IQW 2016/43 708 (Mei. 49 017), IQW 2016/43 709 (Mei. 49 018), IQW 2016/43 710 (Mei. 49 019), IQW 2016/43 711 (Mei. 49 020), IQW 2016/43 712 (Mei. 49 021), IQW 2016/43 714 (Mei. 49 023), IQW 2016/43 715 (Mei. 49 024), IQW 2016/43 716 (Mei. 49 025), IQW 2016/43 717 (Mei. 49 026), IQW 2016/43 718 (Mei. 49 027), IQW 2016/43 720 (Mei. 49 029), IQW 2016/43 724 (Mei. 49 033), IQW 2016/43 728 (Mei. 49 037), IQW 2016/43 737 (Mei. 49 046), IQW 2007/29 370 (Mei. 28 532), IQW 2008/29 806 (Mei. 28 968), IQW 2009/30 157 (Mei. 29 319), IQW 2009/30 584 (Mei. 29 746), IQW 2010/30 934 (Mei. 30 096), IQW 2010/30 944 (Mei. 30 106), IQW 2011/32 062 (Mei. 31 224), IQW 2011/32 094 (Mei. 31 256), IQW 2011/35 480 (Mei. 32 690), IQW 2012/38 401 (Mei. 35 375), IQW 2012/38 414 (Mei. 35 388), IQW 2012/38 432 (Mei. 35 406), IQW 2012/38 433 (Mei. 35 407), IQW 2012/38 456 (Mei. 35 430), IQW 2012/38 557 (Mei. 35 531), IQW 2012/38 559 (Mei. 35 533), IQW 2012/38 581 (Mei. 35 555), IQW 2012/38 583 (Mei. 35 557), IQW 2012/38 586 (Mei. 35 560), IQW 2012/38 587 (Mei. 35 561), IQW 2012/38 589 (Mei. 35 563), IQW 2012/38 593 (Mei. 35 567), IQW 2012/38 627 (Mei. 35 601), IQW 2012/38 769 (Mei. 35 743), IQW 2016/43 203 (Mei. 48 512), IQW 2016/43 345 (Mei. 48 654), IQW 1983/19 165 (Mei. 18 685), IQW 1983/19 165 (Mei. 18 685), IQW 1983/19 165 (Mei. 18 685), IQW 2016/43 221 (Mei. 48 530), IQW 2016/43 229 (Mei. 48 538), IQW 2016/43 346 (Mei. 48 655), IQW 2016/43 488 (Mei. 48 797), IQW 2016/43 490 (Mei. 48 799), IQW 2016/43 517 (Mei. 48 826), IQW 2016/43 544 (Mei. 48 853), IQW 2016/43 560 (Mei. 48 869), IQW 2016/43 567 (Mei. 48 876), IQW 2016/43 571 (Mei. 48 880), IQW 2016/43 572 (Mei. 48 881), IQW 2016/43 573 (Mei. 48 882), IQW 2016/43 576 (Mei. 48 885), IQW 2016/43 577 (Mei. 48 886), IQW 2016/43 578 (Mei. 48 887), IQW 2016/43 579 (Mei. 48 888), IQW 2016/43 581 (Mei. 48 890), IQW 2016/43 582 (Mei. 48 891), IQW 2016/43 583 (Mei. 48 892), IQW 2016/43 584 (Mei. 48 893), IQW 2016/43 585 (Mei. 48 894), IQW 2016/43 587 (Mei. 48 896), IQW 2016/43 588 (Mei. 48 897), IQW 2016/43 592 (Mei. 48 901), IQW 2016/43 594 (Mei. 48 903), IQW 2016/43 595 (Mei. 48 904), IQW 2016/43 620 (Mei. 48 929), IQW 2016/43 628 (Mei. 48 937), IQW 2016/43 632 (Mei. 48 941), IQW 2016/43 638 (Mei. 48 947), IQW 2016/43 652 (Mei. 48 961), IQW 2016/43 653 (Mei. 48 962), IQW 2016/43 660 (Mei. 48 969), IQW 2016/43 661 (Mei. 48 970), IQW 2016/43 667 (Mei. 48 976), IQW 2016/43 671 (Mei. 48 980), IQW 2016/43 677 (Mei. 48 986), IQW 2016/43 719 (Mei. 49 028), IQW 2016/43 721 (Mei. 49 030), IQW 2016/43 722 (Mei. 49 031), IQW 2016/43 723 (Mei. 49 032), IQW 2016/43 725 (Mei. 49 034), IQW 2016/43 727 (Mei. 49 036), IQW 2016/43 741 (Mei. 49 050), IQW 2016/43 755 (Mei. 49 064), IQW 2016/44 369 (Mei. 49 678), IQW 2016/44 380 (Mei. 49 689), IQW 2016/44 381 (Mei. 49 690), IQW 2016/44 382 (Mei. 49 691), IQW 2016/44 394 (Mei. 49 703), IQW 2016/44 413 (Mei. 49 722), IQW 2016/44 423 (Mei. 49 732), IQW 2016/44 424 (Mei. 49 733), IQW 2016/44 425 (Mei. 49 734), IQW 2016/44 426 (Mei. 49 735), IQW 2016/44 427 (Mei. 49 736), IQW 2016/44 428 (Mei. 49 737), IQW 2016/44 429 (Mei. 49 738), IQW 2016/44 440 (Mei. 49 749), IQW 2016/44 472 (Mei. 49 781), IQW 2016/44 492 (Mei. 49 801), IQW 2016/44 495 (Mei. 49 804), IQW 2016/44 501 (Mei. 49 810), IQW 2016/44 502 (Mei. 49 811), IQW 2016/44 512 (Mei. 49 821), IQW 2016/44 518 (Mei. 49 827), IQW 2016/44 523 (Mei. 49 832) IQW 2016/44 534 (Mei. 49 843), IQW 2016/44 540 (Mei. 49 849), IQW 2016/44 566 (Mei. 49 875), IQW 2016/44 574 (Mei. 49 883), IQW 2016/44 578

(Mei. 49887), IQW 2016/44583 (Mei. 49892), IQW 2016/44584 (Mei. 49893), IQW 2016/44585 (Mei. 49894), IQW 2016/44586 (Mei. 49895), IQW 2016/44587 (Mei. 49896), IQW 2016/44596 (Mei. 49905), IQW 2016/44597 (Mei. 49906), IQW 2016/44612 (Mei. 49921), IQW 2016/44613 (Mei. 49922), IQW 2016/44614 (Mei. 49923), IQW 2016/44617 (Mei. 49926), IQW 2016/44618 (Mei. 49927), IQW 2016/44633 (Mei. 49942), IQW 2016/44635 (Mei. 49944), IQW 2016/44636 (Mei. 49945), IQW 2016/44637 (Mei. 49946), IQW 2016/44638 (Mei. 49947), IQW 2016/44639 (Mei. 49948), IQW 2016/44640 (Mei. 49949), IQW 2016/44641 (Mei. 49950), IQW 2016/44642 (Mei. 49951), IQW 2010/30944 (Mei. 30106),

84 vertebra fragments: IQW 2003/27698 (Mei. 27203), IQW 2003/28286 (Mei. 27448), IQW 2007/29159 (Mei. 28321), IQW 2007/29244 (Mei. 28406), IQW 2007/29366 (Mei. 28528), IQW 2007/29547 (Mei. 28709), IQW 2010/30667 (Mei. 29829), IQW 2010/30889 (Mei. 30051), IQW 2010/30940 (Mei. 30102), IQW 2010/30997 (Mei. 30159), IQW 2010/30998 (Mei. 30160), IQW 2010/31055 (Mei. 30217), IQW 2011/32031 (Mei. 31193), IQW 2011/32101 (Mei. 31263), IQW 2011/35519 (Mei. 32729), IQW 2012/38432 (Mei. 35406), IQW 2012/38447 (Mei. 35421), IQW 2012/38455 (Mei. 35429), IQW 2012/38557 (Mei. 35531), IQW 2012/38559 (Mei. 35533), IQW 2012/38560 (Mei. 35534), IQW 2012/38627 (Mei. 35601), IQW 2012/38635 (Mei. 35609), IQW 2012/38769 (Mei. 35743), IQW 2012/38770 (Mei. 35744), IQW 2016/43241 (Mei. 48550), IQW 2016/43352 (Mei. 48661), IQW 2016/43353 (Mei. 48662), IQW 2016/43354 (Mei. 48663), IQW 2016/43355 (Mei. 48664), IQW 2016/43356 (Mei. 48665), IQW 2016/43357 (Mei. 48666), IQW 2016/43358 (Mei. 48667), IQW 2016/43359 (Mei. 48668), IQW 2016/43360 (Mei. 48669), IQW 2016/43361 (Mei. 48670), IQW 2016/43362 (Mei. 48671), IQW 2016/43363 (Mei. 48672), IQW 2016/43364 (Mei. 48673), IQW 2016/43365 (Mei. 48674), IQW 2016/43366 (Mei. 48675), IQW 2016/43367 (Mei. 48676), IQW 2003/28081 (Mei. 27243), IQW 2010/30836 (Mei. 29998), IQW 2010/30933 (Mei. 30095), IQW 2016/43205 (Mei. 48514), IQW 2016/43212 (Mei. 48521), IQW 2016/43217 (Mei. 48526), IQW 2016/43241 (Mei. 48550), IQW 2016/43245 (Mei. 48554), IQW 2016/43342 (Mei. 48651), IQW 2016/43343 (Mei. 48652), IQW 2016/43345 (Mei. 48654), IQW 2016/43373 (Mei. 48682), IQW 2016/43374 (Mei. 48683),

3 sacrum fragments: IQW 2009/30438 (Mei. 29600), IQW 2012/38447 (Mei. 35421), IQW 2012/38769 (Mei. 35743),

26 costa fragments: IQW 2003/27708 (Mei. 27213), IQW 2003/28083 (Mei. 27245), IQW 2004/28478 (Mei. 27640), IQW 2007/29159 (Mei. 28321), IQW 2009/30376 (Mei. 29538), IQW 2009/30442 (Mei. 29604), IQW 2010/31055 (Mei. 30217), IQW 2011/32062 (Mei. 31224), IQW 2012/38399 (Mei. 35373), IQW 2012/38412 (Mei. 35386), IQW 2012/38433 (Mei. 35407), IQW 2012/38462 (Mei. 35436), IQW 2012/38557 (Mei. 35531), IQW 2012/38627 (Mei. 35601), IQW 2012/38634 (Mei. 35608), IQW 2012/38769 (Mei. 35743), IQW 2016/43351 (Mei. 48660),

161 indeterminable postcranial fragments: IQW 2008/29756 (Mei. 28918), IQW 2009/30376 (Mei. 29538), IQW 2011/35520 (Mei. 32730), IQW 2004/28476 (Mei. 27638), IQW 2004/28476 (Mei. 27638), IQW 2004/28476 (Mei. 27638), IQW 2004/28476 (Mei. 27638), IQW 2004/28476 (Mei. 27638), IQW 2004/28476 (Mei. 27638), IQW 2004/28476 (Mei. 27638), IQW 2004/28476 (Mei. 27638), IQW 2004/28476 (Mei. 27638), IQW 2004/28482 (Mei. 27644), IQW 2010/30933 (Mei. 30095), IQW 2016/43225 (Mei. 48534), IQW 2016/43345 (Mei. 48654), IQW 2016/44137 (Mei. 49446), IQW 2016/44643 (Mei. 49952), IQW 2016/44644 (Mei. 49953), IQW 2016/44659 (Mei. 49968), IQW 2016/44660 (Mei. 49969), IQW 2016/44661 (Mei. 49970), IQW 2016/44662 (Mei. 49971), IQW 2016/44663 (Mei. 49972), IQW 2016/44664 (Mei. 49973), IQW 2016/44665 (Mei. 49974), IQW 2016/44666 (Mei. 49975), IQW 2016/44667 (Mei. 49976), IQW 2016/44668 (Mei. 49977), IQW 2016/44669 (Mei. 49978), IQW 2016/44670 (Mei. 49979), IQW 2016/44671 (Mei. 49980), IQW 2016/44672 (Mei. 49981), IQW 2016/44673 (Mei. 49982), IQW 2016/44674 (Mei. 49983), IQW 2016/44675 (Mei. 49984), IQW 2016/44676 (Mei. 49985), IQW 2016/44677 (Mei. 49986), IQW 2016/44678 (Mei. 49987), IQW 2016/44679 (Mei. 49988), IQW 2016/44680 (Mei. 49989), IQW 2016/44681 (Mei. 49990), IQW 2016/44682 (Mei. 49991), IQW 2016/44683 (Mei. 49992), IQW 2016/44684 (Mei. 49993), IQW 2016/44685 (Mei. 49994), IQW 2016/44686 (Mei. 49995), IQW 2016/44687 (Mei. 49996), IQW 2016/44688 (Mei. 49997), IQW 2016/44689 (Mei. 49998), IQW 2016/44690 (Mei. 49999), IQW 2016/44691 (Mei. 50000), IQW 2016/44692 (Mei. 50001), IQW 2016/44693 (Mei. 50002), IQW 2016/44694 (Mei. 50003), IQW 2016/44695 (Mei. 50004), IQW 2016/44696 (Mei. 50005), IQW 2016/44697 (Mei. 50006), IQW 2016/44698 (Mei. 50007), IQW 2016/44699 (Mei. 50008), IQW 2016/44700 (Mei. 50009), IQW 2016/44701 (Mei. 50010), IQW 2016/44702 (Mei. 50011), IQW 2016/44703 (Mei. 50012), IQW 2016/44704 (Mei. 50013), IQW 2016/44705 (Mei. 50014), IQW 2016/44706 (Mei. 50015), IQW 2016/44707 (Mei. 50016), IQW 2016/44708 (Mei. 50017), IQW 2016/44709 (Mei. 50018), IQW 2016/44710 (Mei. 50019), IQW 2016/44711 (Mei. 50020), IQW 2016/44712 (Mei. 50021), IQW 2016/44713 (Mei. 50022), IQW 2016/44714 (Mei. 50023), IQW 2016/44715 (Mei. 50024), IQW 2016/44716 (Mei. 50025), IQW 2016/44717 (Mei. 50026), IQW 2016/44718 (Mei. 50027), IQW 2016/44721 (Mei. 50030), IQW 2016/44745 (Mei. 50054), IQW 2016/44761 (Mei. 50070), IQW 2016/44865 (Mei. 50174), IQW 2016/44866 (Mei. 50175), IQW 2016/44867 (Mei. 50176), IQW 1983/19165 (Mei. 18685), IQW 1994/23823 (Mei. 23351), IQW 2009/30072 (Mei. 29234), IQW 2010/30674 (Mei. 29836), IQW 2016/43236 (Mei. 48545).

Glis sackdillingensis (Heller, 1930)
1 left I^{sup}: IQW 1991/23721 (Mei. 23250),
3 right I^{sup}: IQW 1991/23718 (Mei. 23247), IQW 1991/23719 (Mei. 23248), IQW 1991/23720 (Mei. 23249),
1 right I_{inf}: IQW 1991/23722 (Mei. 23251),
2 left M_1: IQW 1983/18727 (Mei. 18247), IQW 1985/20960 (Mei. 20479),
1 left M_2: IQW 2016/42779 (Mei. 48088),
1 right M_3: IQW 1985/20961 (Mei. 20480).

Castor fiber Linnaeus, 1758
1 left I^{sup}: IQW 2012/38553 (Mei. 35527),
2 left P^4: IQW 1988/22462 (Mei. 21981), IQW 2015/42008 (Mei. 47317),
1 left $M^{1/2}$: IQW 1999/26580 (Mei. 26109),
1 left M^3: IQW 2009/30635 (Mei. 29797), 2 right M^3: IQW 1980/16192 (Mei. 15703), IQW 1999/26581 (Mei. 26110),
1 left mandible with I_{inf}-M_3: IQW 1980/16190 (Mei. 15701),
1 right mandible with I_{inf}-M_3: IQW 1980/16191 (Mei. 15702),
1 right P_4: IQW 1988/22463 (Mei. 21982).

Trogontherium (*Trogontherium*) *cuvieri* Fischer, 1809
1 cranium: IQW 2009/30639 (Mei. 29801),
1 left M^3: IQW 1985/20359 (Mei. 19879),
1 left mandible fragment: IQW 2011/32297 (Mei. 31459),
1 left P_4: IQW 1982/18014 (Mei. 17534),
1 right ulna fragment: IQW 2005/28906 (Mei. 28068),
1 left pelvis fragment: IQW 2011/32048 (Mei. 31210),
1 left tibia fragment: IQW 2013/41766 (Mei. 47076),
1 right tibia fragment: IQW 2010/31209 (Mei. 30371),
1 sacrum: IQW 2011/32047 (Mei. 31209),
1 vertebra: IQW 2011/32046 (Mei. 31208).

Sicista sp.
1 right M_2: IQW 2016/43191 (Mei. 48500).

Cricetus runtonensis Newton, 1909
1 skeleton: IQW 2003/28123 (Mei. 27285),
3 left M^1: IQW 1983/18788 (Mei. 18308), IQW 1984/19760 (Mei. 19280), IQW 2003/27695 (Mei. 27200),
2 right M^1: IQW 2016/42772 (Mei. 48081), IQW 2016/42773 (Mei. 48082),
2 right M^2: IQW 1983/18736 (Mei. 18256), IQW 2016/42774 (Mei. 48083),
1 right M^3: IQW 2016/42778 (Mei. 48087),
1 left M_1: IQW 1984/19761 (Mei. 19281),
1 right M_1: IQW 2016/42771 (Mei. 48080),
2 left M_2: IQW 1985/20425 (Mei. 19945), IQW 2016/42770 (Mei. 48079),
1 right M_3: IQW 1983/18780 (Mei. 18300),
3 molar fagments: IQW 1991/23791 (Mei. 23319), IQW 2016/42777 (Mei. 48086), IQW 2016/42776 (Mei. 48085),
2 left humerus fragments: IQW 2016/42767 (Mei. 48076), IQW 1982/17894 (Mei. 17414),
2 right humerus fragments: IQW 2012/38631 (Mei. 35605), IQW 2016/42766 (Mei. 48075),
1 radius fragment: IQW 2012/38631 (Mei. 35605),
1 right femur fragment distal: IQW 2016/42768 (Mei. 48077),
1 calcaneus: IQW 2012/38631 (Mei. 35605),
1 phalanx of indeterminable position: IQW 2012/38631 (Mei. 35605).

Myodes hintonianus (Kretzoi, 1958)
4 left M^1: IQW 1983/19307 (Mei. 18827), IQW 2016/43052 (Mei. 48361), IQW 2016/43045 (Mei. 48354), IQW 2016/43047 (Mei. 48356),
2 right M^1: IQW 2016/43041 (Mei. 48350), IQW 2016/43043 (Mei. 48352),
8 left M^2: IQW 1983/19341 (Mei. 18861), IQW 1983/19344 (Mei. 18864), IQW 1983/19456 (Mei. 18976), IQW 1983/19473 (Mei. 18993), IQW 1983/19512 (Mei. 19032), IQW 2016/43040 (Mei. 48349), IQW 2016/43051 (Mei. 48360), IQW 2016/43044 (Mei. 48353),

4 right M^2: IQW 1983/19302 (Mei. 18822), IQW 1983/19389 (Mei. 18909), IQW 2016/43049 (Mei. 48358), IQW 2016/43050 (Mei. 48359),
2 left M_1: IQW 2016/43039 (Mei. 48348), IQW 1983/18737 (Mei. 18257),
2 right M_1: IQW 1984/19764 (Mei. 19284), IQW 2016/43073 (Mei. 48382),
6 left M_2: IQW 1983/19278 (Mei. 18798), IQW 1983/19297 (Mei. 18817), IQW 1983/19332 (Mei. 18852), IQW 1983/19377 (Mei. 18897), IQW 2016/43042 (Mei. 48351), IQW 2016/43046 (Mei. 48355),
1 right M_2: IQW 1983/19465 (Mei. 18985);2 right M_3: IQW 1983/19350 (Mei. 18870), IQW 2016/43038 (Mei. 48347),
1 molar fragment: IQW 2016/43048 (Mei. 48357).

Pliomys episcopalis Méhely, 1914
2 left M^1: IQW 1983/19366 (Mei. 18886), IQW 2016/43082 (Mei. 48391),
3 right M^1: IQW 1983/19267 (Mei. 18787), IQW 1983/19290 (Mei. 18810), IQW 1983/19331 (Mei. 18851),
2 left M^2: IQW 1983/19321 (Mei. 18841), IQW 2016/43087 (Mei. 48396),
4 left M^3: IQW 1983/19285 (Mei. 18805), IQW 1983/19464 (Mei. 18984), IQW 1983/19497 (Mei. 19017), IQW 1985/20971 (Mei. 20490),
3 right M^3: IQW 1983/18692 (Mei. 18212), IQW 1983/18767 (Mei. 18287), IQW 1983/18790 (Mei. 18310),
4 left M_1: IQW 1983/18694 (Mei. 18214), IQW 1983/18741 (Mei. 18261), IQW 1983/18760 (Mei. 18280), IQW 2016/43074 (Mei. 48383),
7 right M_1: IQW 1983/19326 (Mei. 18846), IQW 1984/19763 (Mei. 19283), IQW 2010/30691 (Mei. 29853), IQW 2016/43071 (Mei. 48380), IQW 2016/43072 (Mei. 48381), IQW 2016/43075 (Mei. 48384), IQW 2016/43069 (Mei. 48378),
1 left M_2: IQW 1983/19379 (Mei. 18899),
2 right M_2: IQW 1983/19271 (Mei. 18791), IQW 1983/19365 (Mei. 18885),
3 left M_3: IQW 1983/19498 (Mei. 19018), IQW 1983/19520 (Mei. 19040), IQW 2016/43068 (Mei. 48377),
3 right M_3: IQW 1983/19367 (Mei. 18887), IQW 1983/19382 (Mei. 18902), IQW 1983/19491 (Mei. 19011),
14 molar fragments: IQW 1985/20974 (Mei. 20493), IQW 2016/43064 (Mei. 48373), IQW 2016/43076 (Mei. 48385), IQW 2016/43077 (Mei. 48386), IQW 2016/43078 (Mei. 48387), IQW 2016/43079 (Mei. 48388), IQW 2016/43080 (Mei. 48389), IQW 2016/43081 (Mei. 48390), IQW 2016/43083 (Mei. 48392), IQW 2016/43084 (Mei. 48393), IQW 2016/43085 (Mei. 48394), IQW 2016/43086 (Mei. 48395), IQW 2016/43088 (Mei. 48397), IQW 2016/43070 (Mei. 48379).

Mimomys savini Hinton, 1910
4 left M^1: IQW 1983/19484 (Mei. 19004), IQW 1999/26582 (Mei. 26111), IQW 1999/26584 (Mei. 26113), IQW 1999/26586 (Mei. 26115),
7 right M^1: IQW 1999/26587 (Mei. 26116), IQW 1983/19378 (Mei. 18898), IQW 1999/26583 (Mei. 26112), IQW 1999/26585 (Mei. 26114), IQW 2009/30260 (Mei. 29422), IQW 2009/30618 (Mei. 29780), IQW 2016/43035 (Mei. 48344),
1 right M^2: IQW 1999/26589 (Mei. 26118),
5 left M^3: IQW 1983/18728 (Mei. 18248), IQW 1983/19412 (Mei. 18932), IQW 1985/20972 (Mei. 20491), IQW 1985/20973 (Mei. 20492), IQW 1999/26572 (Mei. 26101),
3 right M^3: IQW 1983/18731 (Mei. 18251), IQW 1983/18789 (Mei. 18309), IQW 1983/19293 (Mei. 18813),
6 left M_1: IQW 1984/19675 (Mei. 19195), IQW 1984/19682 (Mei. 19202), IQW 1984/19755 (Mei. 19275), IQW 1984/19757 (Mei. 19277), IQW 1984/19827 (Mei. 19347), IQW 1985/20986 (Mei. 20505),
9 right M_1: IQW 2012/38455 (Mei. 35429), IQW 1983/18695 (Mei. 18215), IQW 1984/19825 (Mei. 19345), IQW 1984/19826 (Mei. 19346), IQW 1985/20985 (Mei. 20504), IQW 1985/20989 (Mei. 20508), IQW 1999/26603 (Mei. 26132), IQW 2016/43033 (Mei. 48342), IQW 2016/43036 (Mei. 48345),
1 right M_2: IQW 2012/38455 (Mei. 35429).

Mimomys pusillus (Méhely, 1914)
12 left M^1: IQW 1983/19308 (Mei. 18828), IQW 1983/19314 (Mei. 18834), IQW 1983/19351 (Mei. 18871), IQW 1983/19360 (Mei. 18880), IQW 1983/19391 (Mei. 18911), IQW 1983/19404 (Mei. 18924), IQW 2016/42986 (Mei. 48295), IQW 2016/42992 (Mei. 48301), IQW 2016/42993 (Mei. 48302), IQW 2016/43008 (Mei. 48317), IQW 2016/43022 (Mei. 48331), IQW 2016/43023 (Mei. 48332),
9 right M^1: IQW 1983/19373 (Mei. 18893), IQW 1983/19472 (Mei. 18992), IQW 1983/19493 (Mei. 19013), IQW 2016/42979 (Mei. 48288), IQW 2016/42994 (Mei. 48303), IQW 2016/42995 (Mei. 48304), IQW 2016/43006 (Mei. 48315), IQW 2016/43007 (Mei. 48316), IQW 2016/43025 (Mei. 48334),

17 left M^2: IQW 1983/19 280 (Mei. 18 800), IQW 1983/19 312 (Mei. 18 832), IQW 1983/19 320 (Mei. 18 840), IQW 1983/19 330 (Mei. 18 850), IQW 1983/19 353 (Mei. 18 873), IQW 1983/19 408 (Mei. 18 928), IQW 1983/19 521 (Mei. 19 041), IQW 1983/19 525 (Mei. 19 045), IQW 2016/42 982 (Mei. 48 291), IQW 2016/42 985 (Mei. 48 294), IQW 2016/42 991 (Mei. 48 300), IQW 2016/43 001 (Mei. 48 310), IQW 2016/43 005 (Mei. 48 314), IQW 2016/43 026 (Mei. 48 335), IQW 2016/43 028 (Mei. 48 337), IQW 2016/43 029 (Mei. 48 338), IQW 2016/43 010 (Mei. 48 319),
12 right M^2: IQW 1983/19 281 (Mei. 18 801), IQW 1983/19 388 (Mei. 18 908), IQW 1983/19 397 (Mei. 18 917), IQW 1983/19 399 (Mei. 18 919), IQW 2016/42 980 (Mei. 48 289), IQW 2016/42 981 (Mei. 48 290), IQW 2016/42 987 (Mei. 48 296), IQW 2016/42 988 (Mei. 48 297), IQW 2016/42 990 (Mei. 48 299), IQW 2016/43 027 (Mei. 48 336), IQW 2016/43 030 (Mei. 48 339), IQW 2016/43 348 (Mei. 48 657),
5 left M^3: IQW 1983/18 743 (Mei. 18 263), IQW 1985/20 975 (Mei. 20 494), IQW 1983/18 755 (Mei. 18 275), IQW 1983/19 398 (Mei. 18 918), IQW 1983/19 404 (Mei. 18 924),
5 right M^3: IQW 1983/19 500 (Mei. 19 020, IQW 1999/26 576 (Mei. 26 105), IQW 1999/26 579 (Mei. 26 108), IQW 1983/18 747 (Mei. 18 267), IQW 1983/18 754 (Mei. 18 274),
2 left mandible fragments with M_1-M_2: IQW 2014/41 831 (Mei. 47 140), IQW 2016/43 013 (Mei. 48 322),
3 right mandible fragments with M_1-M_2: IQW 2007/29 374 (Mei. 28 536), IQW 2011/32 092 (Mei. 31 254), IQW 2016/43 012 (Mei. 48 321),
1 left mandible with M_1: IQW 2016/43 020 (Mei. 48 329),
28 left M_1: IQW 1983/18 734 (Mei. 18 254), IQW 1983/18 750 (Mei. 18 270), IQW 1983/18 765 (Mei. 18 285), IQW 1983/18 772 (Mei. 18 292), IQW 1983/18 777 (Mei. 18 297), IQW 1983/19 342 (Mei. 18 862), IQW 1983/19 398 (Mei. 18 918), IQW 1984/19 676 (Mei. 19 196), IQW 1984/19 677 (Mei. 19 197), IQW 1984/19 679 (Mei. 19 199), IQW 1984/19 756 (Mei. 19 276), IQW 1984/19 758 (Mei. 19 278), IQW 1984/19 759 (Mei. 19 279), IQW 1985/20 988 (Mei. 20 507), IQW 1999/26 575 (Mei. 26 104), IQW 2010/30 684 (Mei. 29 846), IQW 2016/42 976 (Mei. 48 285), IQW 2016/42 978 (Mei. 48 287), IQW 2016/42 983 (Mei. 48 292), IQW 2016/43 011 (Mei. 48 320), IQW 2016/43 016 (Mei. 48 325), IQW 2016/43 017 (Mei. 48 326), IQW 2016/43 018 (Mei. 48 327), IQW 2016/43 032 (Mei. 48 341), IQW 1983/18 775 (Mei. 18 295), IQW 1983/19 292 (Mei. 18 812), IQW 1983/19 307 (Mei. 18 827), IQW 1985/20 983 (Mei. 20 502),
23 right M_1: IQW 2016/42 975 (Mei. 48 284), IQW 1983/18 730 (Mei. 18 250), IQW 1983/18 739 (Mei. 18 259), IQW 1983/18 753 (Mei. 18 273), IQW 1983/18 757 (Mei. 18 277), IQW 1983/18 763 (Mei. 18 283), IQW 1983/18 783 (Mei. 18 303), IQW 1983/19 283 (Mei. 18 803), IQW 1983/19 385 (Mei. 18 905), IQW 1983/19 479 (Mei. 18 999), IQW 1983/19 481 (Mei. 19 001), IQW 1984/19 678 (Mei. 19 198), IQW 1984/19 680 (Mei. 19 200), IQW 1984/19 681 (Mei. 19 201), IQW 1984/19 774 (Mei. 19 294), IQW 1985/20 986 (Mei. 20 505), IQW 1985/20 987 (Mei. 20 506), IQW 1999/26 574 (Mei. 26 103), IQW 1999/26 577 (Mei. 26 106), IQW 1999/26 578 (Mei. 26 107), IQW 2016/43 014 (Mei. 48 323), IQW 2016/43 015 (Mei. 48 324), IQW 2016/43 019 (Mei. 48 328),
3 left M_2: IQW 2016/42 984 (Mei. 48 293), IQW 2016/42 999 (Mei. 48 308), IQW 2016/43 002 (Mei. 48 311),
7 right M_2: IQW 1999/26 588 (Mei. 26 117), IQW 2010/30 855 (Mei. 30 017), IQW 2016/42 989 (Mei. 48 298), IQW 2016/42 996 (Mei. 48 305), IQW 2016/43 004 (Mei. 48 313), IQW 2016/43 021 (Mei. 48 330), IQW 2016/43 024 (Mei. 48 333),
4 left M_3: IQW 1983/19 274 (Mei. 18 794), IQW 1983/19 478 (Mei. 18 998), IQW 2016/42 998 (Mei. 48 307), IQW 2016/43 003 (Mei. 48 312),
2 right M_3: IQW 2016/42 997 (Mei. 48 306), IQW 2016/43 000 (Mei. 48 309),
4 molar fragments: IQW 1983/19 336 (Mei. 18 856), IQW 1983/19 526 (Mei. 19 046), IQW 2016/42 973 (Mei. 48 282), IQW 2016/42 974 (Mei. 48 283).

Microtus (*Allophaiomys*) *thenii* Malez et Rabeder, 1984
1 skeleton: IQW 2009/30 628 (Mei. 29 790),
1 maxilla fragment: IQW 2012/38 769 (Mei. 35 743),
7 I^{sup}: IQW 2012/38 432 (Mei. 35 406), IQW 2012/38 433 (Mei. 35 407), IQW 2012/38 627 (Mei. 35 601), IQW 2012/38 769 (Mei. 35 743),
31 left M^1: IQW 1983/19 277 (Mei. 18 797), IQW 1983/19 289 (Mei. 18 809), IQW 1983/19 310 (Mei. 18 830), IQW 1983/19 315 (Mei. 18 835), IQW 1983/19 358 (Mei. 18 878), IQW 1983/19 359 (Mei. 18 879), IQW 1983/19 363 (Mei. 18 883), IQW 1983/19 368 (Mei. 18 888), IQW 1983/19 381 (Mei. 18 901), IQW 1983/19 413 (Mei. 18 933), IQW 1983/19 452 (Mei. 18 972), IQW 1983/19 476 (Mei. 18 996), IQW 1983/19 480 (Mei. 19 000), IQW 1983/19 485 (Mei. 19 005), IQW 1983/19 508 (Mei. 19 028), IQW 1983/19 511 (Mei. 19 031), IQW 1983/19 522 (Mei. 19 042), IQW 2016/42 825 (Mei. 48 134), IQW 2016/42 828 (Mei. 48 137), IQW 2016/42 836 (Mei. 48 145), IQW 2016/42 837 (Mei. 48 146), IQW 2016/42 845 (Mei. 48 154), IQW 2016/42 871 (Mei. 48 180), IQW 2016/42 872 (Mei. 48 181),

IQW 2016/42 876 (Mei. 48 185), IQW 2016/42 890 (Mei. 48 199), IQW 2016/42 895 (Mei. 48 204), IQW 2016/42 901 (Mei. 48 210), IQW 2016/42 914 (Mei. 48 223), IQW 2016/42 915 (Mei. 48 224), IQW 2016/42 918 (Mei. 48 227),
33 right M^1: IQW 1983/19 322 (Mei. 18 842), IQW 1983/19 339 (Mei. 18 859), IQW 1983/19 364 (Mei. 18 884), IQW 1983/19 371 (Mei. 18 891), IQW 1983/19 454 (Mei. 18 974), IQW 1983/19 486 (Mei. 19 006), IQW 1983/19 509 (Mei. 19 029), IQW 1983/19 513 (Mei. 19 033), IQW 2009/30 632 (Mei. 29 794), IQW 2016/42 827 (Mei. 48 136), IQW 2016/42 829 (Mei. 48 138), IQW 2016/42 830 (Mei. 48 139), IQW 2016/42 831 (Mei. 48 140), IQW 2016/42 832 (Mei. 48 141), IQW 2016/42 833 (Mei. 48 142), IQW 2016/42 834 (Mei. 48 143), IQW 2016/42 835 (Mei. 48 144), IQW 2016/42 865 (Mei. 48 174), IQW 2016/42 866 (Mei. 48 175), IQW 2016/42 867 (Mei. 48 176), IQW 2016/42 868 (Mei. 48 177), IQW 2016/42 869 (Mei. 48 178), IQW 2016/42 870 (Mei. 48 179), IQW 2016/42 873 (Mei. 48 182), IQW 2016/42 884 (Mei. 48 193), IQW 2016/42 889 (Mei. 48 198), IQW 2016/42 893 (Mei. 48 202), IQW 2016/42 903 (Mei. 48 212), IQW 2016/42 916 (Mei. 48 225), IQW 2016/42 917 (Mei. 48 226), IQW 2016/42 919 (Mei. 48 228), IQW 2016/42 920 (Mei. 48 229), IQW 2016/42 922 (Mei. 48 231),
30 left M^2: IQW 1983/18 693 (Mei. 18 213), IQW 1983/18 749 (Mei. 18 269), IQW 1983/19 272 (Mei. 18 792), IQW 1983/19 324 (Mei. 18 844), IQW 1983/19 354 (Mei. 18 874), IQW 1983/19 402 (Mei. 18 922), IQW 1983/19 403 (Mei. 18 923), IQW 1983/19 457 (Mei. 18 977), IQW 1983/19 469 (Mei. 18 989), IQW 1983/19 482 (Mei. 19 002), IQW 1983/19 488 (Mei. 19 008), IQW 1983/19 494 (Mei. 19 014), IQW 1983/19 502 (Mei. 19 022), IQW 1983/19 504 (Mei. 19 024), IQW 1999/26 590 (Mei. 26 119), IQW 2016/42 816 (Mei. 48 125), IQW 2016/42 817 (Mei. 48 126), IQW 2016/42 820 (Mei. 48 129), IQW 2016/42 840 (Mei. 48 149), IQW 2016/42 848 (Mei. 48 157), IQW 2016/42 858 (Mei. 48 167), IQW 2016/42 859 (Mei. 48 168), IQW 2016/42 862 (Mei. 48 171), IQW 2016/42 877 (Mei. 48 186), IQW 2016/42 880 (Mei. 48 189), IQW 2016/42 887 (Mei. 48 196), IQW 2016/42 896 (Mei. 48 205), IQW 2016/42 897 (Mei. 48 206), IQW 2016/42 898 (Mei. 48 207), IQW 2016/42 904 (Mei. 48 213),
24 right M^2: IQW 1983/19 294 (Mei. 18 814), IQW 1983/19 325 (Mei. 18 845), IQW 1983/19 328 (Mei. 18 848), IQW 1983/19 396 (Mei. 18 916), IQW 1983/19 414 (Mei. 18 934), IQW 1983/19 450 (Mei. 18 970), IQW 1983/19 475 (Mei. 18 995), IQW 1983/19 516 (Mei. 19 036), IQW 1983/19 527 (Mei. 19 047), IQW 1983/19 530 (Mei. 19 050), IQW 2016/42 819 (Mei. 48 128), IQW 2016/42 826 (Mei. 48 135), IQW 2016/42 841 (Mei. 48 150), IQW 2016/42 844 (Mei. 48 153), IQW 2016/42 847 (Mei. 48 156), IQW 2016/42 854 (Mei. 48 163), IQW 2016/42 861 (Mei. 48 170), IQW 2016/42 863 (Mei. 48 172), IQW 2016/42 864 (Mei. 48 173), IQW 2016/42 885 (Mei. 48 194), IQW 2016/42 888 (Mei. 48 197), IQW 2016/42 891 (Mei. 48 200), IQW 2016/42 905 (Mei. 48 214), IQW 2016/42 839 (Mei. 48 148),
16 left M^3: IQW 1983/18 784 (Mei. 18 304), IQW 1985/20 976 (Mei. 20 495), IQW 1985/20 977 (Mei. 20 496), IQW 1985/20 979 (Mei. 20 498), IQW 1985/20 981 (Mei. 20 500), IQW 1985/20 982 (Mei. 20 501), IQW 1999/26 595 (Mei. 26 124), IQW 2016/42 911 (Mei. 48 220), IQW 2016/42 925 (Mei. 48 234), IQW 2016/42 927 (Mei. 48 236), IQW 2016/42 928 (Mei. 48 237), IQW 2016/42 930 (Mei. 48 239), IQW 2016/42 933 (Mei. 48 242), IQW 2016/42 938 (Mei. 48 247), IQW 2016/42 939 (Mei. 48 248), IQW 2016/42 940 (Mei. 48 249),
14 right M^3: IQW 1985/20 978 (Mei. 20 497), IQW 1985/20 980 (Mei. 20 499), IQW 1999/26 594 (Mei. 26 123), IQW 2016/42 875 (Mei. 48 184), IQW 2016/42 912 (Mei. 48 221), IQW 2016/42 923 (Mei. 48 232), IQW 2016/42 924 (Mei. 48 233), IQW 2016/42 926 (Mei. 48 235), IQW 2016/42 929 (Mei. 48 238), IQW 2016/42 931 (Mei. 48 240), IQW 2016/42 934 (Mei. 48 243), IQW 2016/42 935 (Mei. 48 244), IQW 2016/42 936 (Mei. 48 245), IQW 2016/42 937 (Mei. 48 246),
1 left and right mandible with M_1-M_2 resp. with M_1: IQW 2009/30 038 (Mei. 29 200),
1 left mandible with I_{inf}, M_1-M_2: IQW 2009/30 514 (Mei. 29 676),
1 right mandible with I_{inf}, M_1-M_2: IQW 2007/29 165 (Mei. 28 327),
1 left mandible with I_{inf}-M_1: IQW 1983/19 524 (Mei. 19 044),
1 right mandible with M_1-M_3: IQW 2009/30 574 (Mei. 29 736),
5 left mandibles with M_1-M_2: IQW 1984/19 907 (Mei. 19 427), IQW 1999/26 600 (Mei. 26 129), IQW 2007/29 610 (Mei. 28 772), IQW 2016/42 804 (Mei. 48 113), IQW 2016/42 805 (Mei. 48 114),
6 right mandibles with M_1-M_2: IQW 1983/18 696 (Mei. 18 216), IQW 1984/19 772 (Mei. 19 292), IQW 1984/19 773 (Mei. 19 293), IQW 1999/26 601 (Mei. 26 130), IQW 2016/42 808 (Mei. 48 117), IQW 2016/42 815 (Mei. 48 124),
3 left mandibles with M_1: IQW 1999/26 602 (Mei. 26 131), IQW 2016/42 809 (Mei. 48 118), IQW 2016/42 910 (Mei. 48 219),
1 right mandible with M_2: IQW 2009/30 468 (Mei. 29 630),
1 left mandible without teeth: IQW 2016/42 950 (Mei. 48 259),
1 right mandible without teeth: IQW 1983/18 490 (Mei. 18 010),
5 I_{inf}: IQW 2012/38 433 (Mei. 35 407), IQW 2012/38 455 (Mei. 35 429), IQW 2012/38 627 (Mei. 35 601), IQW 2011/31 370 (Mei. 30 532),
44 left M_1: IQW 1983/18 726 (Mei. 18 246), IQW 1983/18 729 (Mei. 18 249), IQW 1983/18 732 (Mei. 18 252), IQW 1983/18 738 (Mei. 18 258), IQW 1983/18 740 (Mei. 18 260), IQW 1983/18 752 (Mei. 18 272), IQW 1983/18 762

(Mei. 18282), IQW 1983/18766 (Mei. 18286), IQW 1983/18770 (Mei. 18290), IQW 1983/18773 (Mei 18293), IQW 1983/19298 (Mei. 18818), IQW 1983/19303 (Mei. 18823), IQW 1983/19453 (Mei. 18973), IQW 1984/19624 (Mei. 19144), IQW 1984/19626 (Mei. 19146), IQW 1984/19628 (Mei. 19148), IQW 1984/19630 (Mei 19150), IQW 1984/19631 (Mei. 19151), IQW 1984/19632 (Mei. 19152), IQW 1984/19633 (Mei. 19153), IQW 1984/19766 (Mei. 19286), IQW 1984/19768 (Mei. 19288), IQW 1985/20990 (Mei. 20509), IQW 1985/20991 (Mei. 20510), IQW 1985/20992 (Mei. 20511), IQW 1985/20994 (Mei. 20513), IQW 1985/20995 (Mei. 20514), IQW 1999/26593 (Mei. 26122), IQW 1999/26597 (Mei. 26126), IQW 1999/26599 (Mei. 26128), IQW 2016/42787 (Mei. 48096), IQW 2016/42789 (Mei. 48098), IQW 2016/42790 (Mei. 48099), IQW 2016/42801 (Mei. 48110), IQW 2016/42802 (Mei. 48111), IQW 2016/42803 (Mei. 48112), IQW 2016/42806 (Mei. 48115), IQW 2016/42807 (Mei. 48116), IQW 2016/42810 (Mei. 48119), IQW 2016/42811 (Mei. 48120), IQW 2016/42813 (Mei. 48122), IQW 2016/42814 (Mei. 48123), IQW 2016/42900 (Mei. 48209), IQW 2016/42908 (Mei. 48217),

29 right M_1: IQW 1983/18744 (Mei. 18264), IQW 1983/18751 (Mei. 18271), IQW 1983/18764 (Mei. 18284), IQW 1983/18786 (Mei. 18306), IQW 1983/19286 (Mei. 18806), IQW 1983/19291 (Mei. 18811), IQW 1983/19327 (Mei. 18847), IQW 1984/19623 (Mei. 19143), IQW 1984/19625 (Mei. 19145), IQW 1984/19627 (Mei. 19147), IQW 1984/19629 (Mei. 19149), IQW 1984/19765 (Mei. 19285), IQW 1984/19767 (Mei. 19287), IQW 1984/19769 (Mei. 19289), IQW 1984/19770 (Mei. 19290), IQW 1984/19775 (Mei. 19295), IQW 1985/20993 (Mei. 20512), IQW 1999/26596 (Mei. 26125), IQW 1999/26598 (Mei. 26127), IQW 2016/42788 (Mei. 48097), IQW 2016/42812 (Mei. 48121), IQW 2016/42906 (Mei. 48215), IQW 2016/42907 (Mei. 48216), IQW 2016/42932 (Mei. 48241), IQW 2016/42971 (Mei. 48280), IQW 2011/31370 (Mei. 30532), IQW 2012/38432 (Mei. 35406), IQW 2012/38433 (Mei. 35407), IQW 2012/38455 (Mei. 35429),

30 left M_2: IQW 1983/19264 (Mei. 18784), IQW 1983/19268 (Mei. 18788), IQW 1983/19287 (Mei. 18807), IQW 1983/19304 (Mei. 18824), IQW 1983/19318 (Mei. 18838), IQW 1983/19323 (Mei. 18843), IQW 1983/19345 (Mei. 18865), IQW 1983/19355 (Mei. 18875), IQW 1983/19375 (Mei. 18895), IQW 1983/19387 (Mei. 18907), IQW 1983/19393 (Mei. 18913), IQW 1983/19405 (Mei. 18925), IQW 1983/19417 (Mei. 18937), IQW 1983/19458 (Mei. 18978), IQW 1983/19470 (Mei. 18990), IQW 1983/19505 (Mei. 19025), IQW 1983/19529 (Mei. 19049), IQW 1983/19533 (Mei. 19053), IQW 2016/42838 (Mei. 48147), IQW 2016/42843 (Mei. 48152), IQW 2016/42850 (Mei. 48159), IQW 2016/42853 (Mei. 48162), IQW 2016/42856 (Mei. 48165), IQW 2016/42860 (Mei. 48169), IQW 2016/42878 (Mei. 48187), IQW 2016/42881 (Mei. 48190), IQW 2016/42886 (Mei. 48195), IQW 2016/42894 (Mei. 48203), IQW 2016/42899 (Mei. 48208), IQW 2016/42921 (Mei. 48230),

22 right M_2: IQW 1983/19284 (Mei. 18804), IQW 1983/19296 (Mei. 18816), IQW 1983/19317 (Mei. 18837), IQW 1983/19337 (Mei. 18857), IQW 1983/19369 (Mei. 18889), IQW 1983/19372 (Mei. 18892), IQW 1983/19394 (Mei. 18914), IQW 1983/19401 (Mei. 18921), IQW 1983/19410 (Mei. 18930), IQW 1983/19507 (Mei. 19027), IQW 2016/42818 (Mei. 48127), IQW 2016/42842 (Mei. 48151), IQW 2016/42846 (Mei. 48155), IQW 2016/42849 (Mei. 48158), IQW 2016/42851 (Mei. 48160), IQW 2016/42852 (Mei. 48161), IQW 2016/42855 (Mei. 48164), IQW 2016/42857 (Mei. 48166), IQW 2016/42874 (Mei. 48183), IQW 2016/42882 (Mei. 48191), IQW 2016/42892 (Mei. 48201), IQW 2016/42902 (Mei. 48211),

8 left M_3: IQW 1983/19329 (Mei. 18849), IQW 1983/19357 (Mei. 18877), IQW 1983/19370 (Mei. 18890), IQW 1983/19489 (Mei. 19009), IQW 1983/19514 (Mei. 19034), IQW 2016/42821 (Mei. 48130), IQW 2016/42823 (Mei. 48132), IQW 2016/42883 (Mei. 48192),

10 right M_3: IQW 1983/19301 (Mei. 18821), IQW 1983/19333 (Mei. 18853), IQW 1983/19392 (Mei. 18912), IQW 1983/19409 (Mei. 18929), IQW 1983/19415 (Mei. 18935), IQW 1983/19463 (Mei. 18983), IQW 2016/42822 (Mei. 48131), IQW 2016/42824 (Mei. 48133), IQW 2016/42879 (Mei. 48188), IQW 2016/42913 (Mei. 48222),

38 molar fragments: IQW 2012/38432 (Mei. 35406), IQW 2012/38455 (Mei. 35429), IQW 2012/38627 (Mei. 35601), IQW 2016/42785 (Mei. 48094), IQW 2016/42786 (Mei. 48095), IQW 2016/42943 (Mei. 48252), IQW 2016/42951 (Mei. 48260), IQW 2016/42953 (Mei. 48262), IQW 2016/42954 (Mei. 48263), IQW 2016/42955 (Mei. 48264), IQW 2016/42957 (Mei. 48266), IQW 2016/42958 (Mei. 48267), IQW 2016/42959 (Mei. 48268), IQW 2016/42960 (Mei. 48269), IQW 2016/42961 (Mei. 48270), IQW 2016/42962 (Mei. 48271), IQW 2016/42963 (Mei. 48272), IQW 2016/42964 (Mei. 48273), IQW 2016/42965 (Mei. 48274), IQW 2016/42966 (Mei. 48275), IQW 2016/42969 (Mei. 48278), IQW 2016/42970 (Mei. 48279), IQW 2016/42909 (Mei. 48218), IQW 1983/18735 (Mei. 18255), IQW 1984/19771 (Mei. 19291).

Arvicolinae indet.

10 incisor fragments: IQW 2012/38456 (Mei. 35430), IQW 2016/42967 (Mei. 48276), IQW 2016/42544 (Mei. 47853), IQW 2016/42952 (Mei. 48261), IQW 2016/42968 (Mei 48277), IQW 2016/42690 (Mei. 47999), IQW 2016/42691 (Mei. 48000), IQW 2016/42693 (Mei. 48002), IQW 2016/42694 (Mei. 48003), IQW 2016/42695 (Mei. 48004)

9 maxilla fragments without teeth: IQW 1983/19 338 (Mei. 18 858), IQW 1983/19 459 (Mei. 18 979), IQW 2016/42 799 (Mei. 48 108), IQW 2016/43 432 (Mei. 48 741), IQW 2016/44 272 (Mei. 49 581), IQW 2016/44 342 (Mei. 49 651), IQW 2016/44 343 (Mei. 49 652), IQW 2016/44 537 (Mei. 49 846), IQW 2016/44 538 (Mei. 49 847),
20 mandible fragments without teeth: IQW 2016/42 679 (Mei. 47 988), IQW 1983/18 490 (Mei. 18 010), IQW 1983/18 811 (Mei. 18 331), IQW 1983/19 400 (Mei. 18 920), IQW 1983/19 461 (Mei. 18 981), IQW 1983/19 468 (Mei. 18 988), IQW 1983/19 487 (Mei. 19 007), IQW 1983/19 496 (Mei. 19 016), IQW 1983/19 499 (Mei. 19 019), IQW 1983/19 506 (Mei. 19 026), IQW 1983/19 515 (Mei. 19 035), IQW 1983/19 523 (Mei. 19 043), IQW 2016/42 753 (Mei. 48 062), IQW 2016/44 262 (Mei. 49 571), IQW 2016/44 283 (Mei. 49 592), IQW 2016/44 293 (Mei. 49 602), IQW 2016/44 340 (Mei. 49 649), IQW 2016/44 341 (Mei. 49 650), IQW 2016/44 451 (Mei. 49 760), IQW 2016/44 626 (Mei. 49 935),
261 molar fragments: IQW 2016/42 549 (Mei. 47 858), IQW 2016/42 547 (Mei. 47 856), IQW 2016/42 548 (Mei. 47 857), IQW 2016/42 545 (Mei. 47 854), IQW 2016/42 546 (Mei. 47 855), IQW 2016/42 537 (Mei. 47 846), IQW 2016/42 538 (Mei. 47 847), IQW 2016/42 539 (Mei. 47 848), IQW 2016/42 540 (Mei. 47 849), IQW 2016/42 541 (Mei. 47 850), IQW 2016/42 542 (Mei. 47 851), IQW 2016/42 543 (Mei. 47 852), IQW 2016/42 550 (Mei. 47 859), IQW 2016/42 551 (Mei. 47 860), IQW 2016/42 552 (Mei. 47 861), IQW 2016/42 553 (Mei. 47 862), IQW 2016/42 554 (Mei. 47 863), IQW 2016/42 555 (Mei. 47 864), IQW 2016/42 556 (Mei. 47 865), IQW 2016/42 557 (Mei. 47 866), IQW 2016/42 558 (Mei. 47 867), IQW 2016/42 559 (Mei. 47 868), IQW 2016/42 560 (Mei. 47 869), IQW 2016/42 561 (Mei. 47 870), IQW 2016/42 562 (Mei. 47 871), IQW 2016/42 563 (Mei. 47 872), IQW 2016/42 564 (Mei. 47 873), IQW 2016/42 565 (Mei. 47 874), IQW 2016/42 566 (Mei. 47 875), IQW 2016/42 567 (Mei. 47 876), IQW 2016/42 568 (Mei. 47 877), IQW 2016/42 569 (Mei. 47 878), IQW 2016/42 570 (Mei. 47 879), IQW 2016/42 571 (Mei. 47 880), IQW 2016/42 572 (Mei. 47 881), IQW 2016/42 573 (Mei. 47 882), IQW 2016/42 574 (Mei. 47 883), IQW 2016/42 575 (Mei. 47 884), IQW 2016/42 576 (Mei. 47 885), IQW 2016/42 577 (Mei. 47 886), IQW 2016/42 578 (Mei. 47 887), IQW 2016/42 579 (Mei. 47 888), IQW 2016/42 580 (Mei. 47 889), IQW 2016/42 581 (Mei. 47 890), IQW 2016/42 582 (Mei. 47 891), IQW 2016/42 583 (Mei. 47 892), IQW 2016/42 584 (Mei. 47 893), IQW 2016/42 585 (Mei. 47 894), IQW 2016/42 586 (Mei. 47 895), IQW 2016/42 587 (Mei. 47 896), IQW 2016/42 588 (Mei. 47 897), IQW 2016/42 589 (Mei. 47 898), IQW 2016/42 590 (Mei. 47 899), IQW 2016/42 591 (Mei. 47 900), IQW 2016/42 592 (Mei. 47 901), IQW 2016/42 593 (Mei. 47 902), IQW 2016/42 594 (Mei. 47 903), IQW 2016/42 595 (Mei. 47 904), IQW 2016/42 596 (Mei. 47 905), IQW 2016/42 597 (Mei. 47 906), IQW 2016/42 598 (Mei. 47 907), IQW 2016/42 599 (Mei. 47 908), IQW 2016/42 600 (Mei. 47 909), IQW 2016/42 601 (Mei. 47 910), IQW 2016/42 602 (Mei. 47 911), IQW 2016/42 603 (Mei. 47 912), IQW 2016/42 604 (Mei. 47 913), IQW 2016/42 605 (Mei. 47 914), IQW 2016/42 606 (Mei. 47 915), IQW 2016/42 607 (Mei. 47 916), IQW 2016/42 608 (Mei. 47 917), IQW 2016/42 609 (Mei. 47 918), IQW 2016/42 610 (Mei. 47 919), IQW 2016/42 611 (Mei. 47 920), IQW 2016/42 612 (Mei. 47 921), IQW 2016/42 613 (Mei. 47 922), IQW 2016/42 614 (Mei. 47 923), IQW 2016/42 615 (Mei. 47 924), IQW 2016/42 616 (Mei. 47 925), IQW 2016/42 617 (Mei. 47 926), IQW 2016/42 618 (Mei. 47 927), IQW 2016/42 619 (Mei. 47 928), IQW 2016/42 620 (Mei. 47 929), IQW 2016/42 621 (Mei. 47 930), IQW 2016/42 622 (Mei. 47 931), IQW 2016/42 623 (Mei. 47 932), IQW 2016/42 624 (Mei. 47 933), IQW 2016/42 625 (Mei. 47 934), IQW 2016/42 626 (Mei. 47 935), IQW 2016/42 627 (Mei. 47 936), IQW 2016/42 628 (Mei. 47 937), IQW 2016/42 629 (Mei. 47 938), IQW 2016/42 630 (Mei. 47 939), IQW 2016/42 631 (Mei. 47 940), IQW 2016/42 632 (Mei. 47 941), IQW 2016/42 633 (Mei. 47 942), IQW 2016/42 634 (Mei. 47 943), IQW 2016/42 635 (Mei. 47 944), IQW 2016/42 636 (Mei. 47 945), IQW 2016/42 637 (Mei. 47 946), IQW 2016/42 638 (Mei. 47 947), IQW 2016/42 639 (Mei. 47 948), IQW 2016/42 640 (Mei. 47 949), IQW 2016/42 641 (Mei. 47 950), IQW 2016/42 642 (Mei. 47 951), IQW 2016/42 643 (Mei. 47 952), IQW 2016/42 644 (Mei. 47 953), IQW 2016/42 645 (Mei. 47 954), IQW 2016/42 646 (Mei. 47 955), IQW 2016/42 647 (Mei. 47 956), IQW 2016/42 648 (Mei. 47 957), IQW 2016/42 649 (Mei. 47 958), IQW 2016/42 650 (Mei. 47 959), IQW 2016/42 651 (Mei. 47 960), IQW 2016/42 652 (Mei. 47 961), IQW 2016/42 653 (Mei. 47 962), IQW 2016/42 654 (Mei. 47 963), IQW 2016/42 655 (Mei. 47 964), IQW 2016/42 656 (Mei. 47 965), IQW 2016/42 657 (Mei. 47 966), IQW 2016/42 658 (Mei. 47 967), IQW 2016/42 659 (Mei. 47 968), IQW 2016/42 660 (Mei. 47 969), IQW 2016/42 661 (Mei. 47 970), IQW 2016/42 662 (Mei. 47 971), IQW 2016/42 663 (Mei. 47 972), IQW 2016/42 664 (Mei. 47 973), IQW 2016/42 665 (Mei. 47 974), IQW 2016/42 666 (Mei. 47 975), IQW 2016/42 667 (Mei. 47 976), IQW 2016/42 668 (Mei. 47 977), IQW 2016/42 669 (Mei. 47 978), IQW 2016/42 670 (Mei. 47 979), IQW 2016/42 671 (Mei. 47 980), IQW 2016/42 672 (Mei. 47 981), IQW 2016/42 673 (Mei. 47 982), IQW 2016/42 674 (Mei. 47 983), IQW 2016/42 675 (Mei. 47 984), IQW 2016/42 676 (Mei. 47 985), IQW 2016/42 677 (Mei. 47 986), IQW 2016/42 678 (Mei. 47 987), IQW 2016/42 680 (Mei. 47 989), IQW 2016/42 681 (Mei. 47 990), IQW 2016/42 682 (Mei. 47 991), IQW 2016/42 683 (Mei. 47 992), IQW 2016/42 684 (Mei. 47 993), IQW 2016/42 685 (Mei. 47 994), IQW 2016/42 686 (Mei. 47 995), IQW 2016/42 687 (Mei. 47 996), IQW 2016/42 688 (Mei. 47 997), IQW 2016/42 689 (Mei. 47 998), IQW 2016/42 692 (Mei. 48 001), IQW 2016/42 696 (Mei. 48 005), IQW 2016/42 697 (Mei. 48 006), IQW 2016/42 698 (Mei. 48 007), IQW 2016/42 699 (Mei. 48 008), IQW 2016/42 700 (Mei. 48 009), IQW

2016/42 701 (Mei. 48 010), IQW 2016/42 702 (Mei. 48 011), IQW 2016/44 012 (Mei. 49 321), IQW 2016/44 029 (Mei. 49 338), IQW 1983/19 270 (Mei. 18 790), IQW 1983/19 276 (Mei. 18 796), IQW 1983/19 279 (Mei. 18 799), IQW 1983/19 295 (Mei. 18 815), IQW 1983/19 300 (Mei. 18 820), IQW 1983/19 305 (Mei. 18 825), IQW 1983/19 311 (Mei. 18 831), IQW 1983/19 313 (Mei. 18 833), IQW 1983/19 319 (Mei. 18 839), IQW 1983/19 334 (Mei. 18 854), IQW 1983/19 340 (Mei. 18 860), IQW 1983/19 343 (Mei. 18 863), IQW 1983/19 346 (Mei. 18 866), IQW 1983/19 347 (Mei. 18 867), IQW 1983/19 348 (Mei. 18 868), IQW 1983/19 352 (Mei. 18 872), IQW 1983/19 356 (Mei. 18 876), IQW 1983/19 361 (Mei. 18 881), IQW 1983/19 362 (Mei. 18 882), IQW 1983/19 376 (Mei. 18 896), IQW 1983/19 380 (Mei. 18 900), IQW 1983/19 383 (Mei. 18 903), IQW 1983/19 390 (Mei. 18 910), IQW 1983/19 395 (Mei. 18 915), IQW 1983/19 406 (Mei. 18 926), IQW 1983/19 407 (Mei. 18 927), IQW 1983/19 411 (Mei. 18 931), IQW 1983/19 416 (Mei. 18 936), IQW 1983/19 418 (Mei. 18 938), IQW 1983/19 447 (Mei. 18 967), IQW 1983/19 448 (Mei. 18 968), IQW 1983/19 449 (Mei. 18 969), IQW 1983/19 451 (Mei. 18 971), IQW 1983/19 460 (Mei. 18 980), IQW 1983/19 467 (Mei. 18 987), IQW 1983/19 471 (Mei. 18 491), IQW 1983/19 477 (Mei. 18 997), IQW 1983/19 483 (Mei. 19 003), IQW 1983/19 492 (Mei. 19 012), IQW 1983/19 495 (Mei. 19 015), IQW 1983/19 499 (Mei. 19 019), IQW 1983/19 501 (Mei. 19 021), IQW 1983/19 510 (Mei. 19 030), IQW 1983/19 516 (Mei. 19 033), IQW 1983/19 517 (Mei. 19 037), IQW 1983/19 528 (Mei. 19 048), IQW 1983/19 531 (Mei. 19 051), IQW 2016/42 798 (Mei. 48 107), IQW 2016/43 421 (Mei. 48 730), IQW 2016/43 422 (Mei. 48 731), IQW 2016/43 423 (Mei. 48 732), IQW 2016/43 424 (Mei. 48 733), IQW 2016/43 425 (Mei. 48 734), IQW 2016/43 426 (Mei. 48 735), IQW 2016/43 427 (Mei. 48 736), IQW 2016/43 428 (Mei. 48 737), IQW 2016/43 429 (Mei. 48 738), IQW 2016/43 430 (Mei. 48 739), IQW 2016/43 431 (Mei. 48 740), IQW 2016/44 233 (Mei. 49 542), IQW 2016/44 234 (Mei. 49 543), IQW 2016/44 235 (Mei. 49 544), IQW 2016/44 236 (Mei. 49 545), IQW 2016/44 237 (Mei. 49 546), IQW 2016/44 238 (Mei. 49 547), IQW 2016/44 239 (Mei. 49 548), IQW 2016/44 240 (Mei. 49 549), IQW 2016/44 241 (Mei. 49 550), IQW 2016/44 242 (Mei. 49 551), IQW 2016/44 243 (Mei. 49 552), IQW 2016/44 244 (Mei. 49 553), IQW 2016/44 245 (Mei. 49 554), IQW 2016/44 246 (Mei. 49 555), IQW 2016/44 247 (Mei. 49 556), IQW 2016/44 270 (Mei. 49 579), IQW 2016/44 284 (Mei. 49 593), IQW 2016/44 285 (Mei. 49 594), IQW 2016/44 301 (Mei. 49 610), IQW 2016/44 302 (Mei. 49 611), IQW 2016/44 303 (Mei. 49 612), IQW 2016/44 304 (Mei. 49 613), IQW 2016/44 395 (Mei. 49 704), IQW 2016/44 408 (Mei. 49 717), IQW 2016/44 409 (Mei. 49 718), IQW 2016/44 410 (Mei. 49 719), IQW 2016/44 411 (Mei. 49 720), IQW 2016/44 412 (Mei. 49 721), IQW 2016/44 450 (Mei. 49 759), IQW 2016/44 503 (Mei. 49 812), IQW 2016/44 531 (Mei. 49 840), IQW 2016/44 532 (Mei. 49 841), IQW 2016/44 535 (Mei. 49 844), IQW 2016/44 536 (Mei. 49 845), IQW 2016/44 738 (Mei. 50 047), IQW 2016/44 758 (Mei. 50 067), IQW 2016/44 762 (Mei. 50 071), IQW 2016/44 763 (Mei. 50 072), IQW 2016/44 764 (Mei. 50 073), IQW 2016/44 765 (Mei. 50 074), IQW 2016/44 766 (Mei. 50 075).

Apodemus sylvaticus (Linnaeus, 1758)
1 maxilla fragment without teeth: IQW 2016/44 767 (Mei. 50 076),
2 left I^{sup}: IQW 1991/23 794 (Mei. 23 322), IQW 2016/42 534 (Mei. 47 843),
1 right I^{sup}: IQW 1991/23 793 (Mei. 23 321),
3 right M^1: IQW 1983/18 746 (Mei. 18 266), IQW 1985/20 963 (Mei. 20 482), IQW 1985/20 967 (Mei. 20 486),
1 left M^2: IQW 1983/18 748 (Mei. 18 268),
1 left I_{inf}: IQW 1991/23 792 (Mei. 23 320),
1 right I_{inf}: IQW 1991/23 795 (Mei. 23 323),
5 left M_1: IQW 1983/18 725 (Mei. 18 245), IQW 1983/18 745 (Mei. 18 265), IQW 1985/20 962 (Mei. 20 481), IQW 1985/20 966 (Mei. 20 485), IQW 2016/42 535 (Mei. 47 844),
3 right M_1: IQW 1985/20 968 (Mei. 20 487), IQW 1985/20 969 (Mei. 20 488), IQW 1983/18 787 (Mei. 18 307),
2 left M_2: IQW 1985/20 964 (Mei. 20 483), IQW 2016/42 536 (Mei. 47 845),
1 right M_2: IQW 1985/20 965 (Mei. 20 484),
1 right M_3: IQW 1985/20 970 (Mei. 20 489).

Hystrix refossa Gervais, 1852
1 left I_{inf}: IQW 1994/24 539 (Mei. 24 068);1 right I_{inf}: IQW 1984/20 221 (Mei. 19 741).

Rodentia indet.
253 incisor fragments collectively catalogued under: IQW 2016/44 081 (Mei. 49 390), IQW 2016/44 096 (Mei. 49 405), IQW 1983/19 165 (Mei. 18 685), IQW 2016/43 089 (Mei. 48 398), IQW 2016/43 090 (Mei. 48 399), IQW 2016/43 091 (Mei. 48 400), IQW 2016/43 092 (Mei. 48 401), IQW 2016/43 093 (Mei. 48 402), IQW 2016/43 094 (Mei. 48 403), IQW 2016/43 095 (Mei. 48 404), IQW 2016/43 096 (Mei. 48 405), IQW 2016/43 097 (Mei. 48 406), IQW 2016/43 098 (Mei. 48 407), IQW 2016/43 099 (Mei. 48 408), IQW 2016/43 100 (Mei. 48 409), IQW 2016/43 101 (Mei. 48 410), IQW 2016/43 102 (Mei. 48 411), IQW 2016/43 103 (Mei. 48 412), IQW 2016/43 104 (Mei. 48 413), IQW 2016/43 105

(Mei. 48414), IQW 2016/43106 (Mei. 48415), IQW 2016/43107 (Mei. 48416), IQW 2016/43108 (Mei. 48417), IQW 2016/43109 (Mei. 48418), IQW 2016/43110 (Mei. 48419), IQW 2016/43111 (Mei. 48420), IQW 2016/43112 (Mei. 48421), IQW 2016/43113 (Mei. 48422), IQW 2016/43114 (Mei. 48423), IQW 2016/43115 (Mei. 48424), IQW 2016/43116 (Mei. 48425), IQW 2016/43117 (Mei. 48426), IQW 2016/43118 (Mei. 48427), IQW 2016/43119 (Mei. 48428), IQW 2016/43120 (Mei. 48429), IQW 2016/43121 (Mei. 48430), IQW 2016/43122 (Mei. 48431), IQW 2016/43123 (Mei. 48432), IQW 2016/43124 (Mei. 48433), IQW 2016/43125 (Mei. 48434), IQW 2016/43126 (Mei. 48435), IQW 2016/43127 (Mei. 48436), IQW 2016/43128 (Mei. 48437), IQW 2016/43129 (Mei. 48438), IQW 2016/43130 (Mei. 48439), IQW 2016/43131 (Mei. 48440), IQW 2016/43132 (Mei. 48441), IQW 2016/43133 (Mei. 48442), IQW 2016/43134 (Mei. 48443), IQW 2016/43135 (Mei. 48444), IQW 2016/43136 (Mei. 48445), IQW 2016/43137 (Mei. 48446), IQW 2016/43138 (Mei. 48447), IQW 2016/43139 (Mei. 48448), IQW 2016/43140 (Mei. 48449), IQW 2016/43141 (Mei. 48450), IQW 2016/43142 (Mei. 48451), IQW 2016/43143 (Mei. 48452), IQW 2016/43144 (Mei. 48453), IQW 2016/43145 (Mei. 48454), IQW 2016/43146 (Mei. 48455), IQW 2016/43147 (Mei. 48456), IQW 2016/43148 (Mei. 48457), IQW 2016/43149 (Mei. 48458), IQW 2016/43150 (Mei. 48459), IQW 2016/43151 (Mei. 48460), IQW 2016/43152 (Mei. 48461), IQW 2016/43153 (Mei. 48462), IQW 2016/43154 (Mei. 48463), IQW 2016/43155 (Mei. 48464), IQW 2016/43156 (Mei. 48465), IQW 2016/43157 (Mei. 48466), IQW 2016/43158 (Mei. 48467), IQW 2016/43159 (Mei. 48468), IQW 2016/43160 (Mei. 48469), IQW 2016/43161 (Mei. 48470), IQW 2016/43162 (Mei. 48471), IQW 2016/43163 (Mei. 48472), IQW 2016/43164 (Mei. 48473), IQW 2016/43165 (Mei. 48474), IQW 2016/43166 (Mei. 48475), IQW 2016/43167 (Mei. 48476), IQW 2016/43168 (Mei. 48477), IQW 2016/43169 (Mei. 48478), IQW 2016/43170 (Mei. 48479), IQW 2016/43171 (Mei. 48480), IQW 2016/43172 (Mei. 48481), IQW 2016/43173 (Mei. 48482), IQW 2016/43174 (Mei. 48483), IQW 2016/43175 (Mei. 48484), IQW 2016/43176 (Mei. 48485), IQW 2016/43177 (Mei. 48486), IQW 2016/43178 (Mei. 48487), IQW 2016/43179 (Mei. 48488), IQW 2016/43180 (Mei. 48489), IQW 2016/43181 (Mei. 48490), IQW 2016/43182 (Mei. 48491), IQW 2016/43183 (Mei. 48492), IQW 2016/43184 (Mei. 48493), IQW 2016/43185 (Mei. 48494), IQW 2016/43186 (Mei. 48495), IQW 2016/43187 (Mei. 48496), IQW 2016/43188 (Mei. 48497), IQW 2016/43189 (Mei. 48498), IQW 2016/43190 (Mei. 48499), IQW 2016/43973 (Mei. 49282), IQW 2016/44011 (Mei. 49320), IQW 2016/44013 (Mei. 49322), IQW 2016/44020 (Mei. 49329), IQW 2016/44035 (Mei. 49344).

Micromammalia indet.
239 indeterminable tooth fragments collectively catalogued under: IQW 2009/30382 (Mei. 29544), IQW 2010/31084 (Mei. 30246), IQW 2011/32156 (Mei. 31320), IQW 2016/44091 (Mei. 49400), IQW 2016/44091 (Mei. 49400), IQW 2016/44095 (Mei. 49404), IQW 2016/44096 (Mei. 49405), IQW 2016/44097 (Mei. 49406), IQW 2016/44099 (Mei. 49408), IQW 2016/44102 (Mei. 49411), IQW 2016/44104 (Mei. 49413), IQW 2016/44104 (Mei. 49413), IQW 2016/44109 (Mei. 49418), IQW 2016/44115 (Mei. 49424), IQW 2016/44118 (Mei. 49427), IQW 2016/44119 (Mei. 49428), IQW 2016/44123 (Mei. 49432), IQW 2016/44124 (Mei. 49433), IQW 2016/44126 (Mei. 49435), IQW 2016/44127 (Mei. 49436), IQW 2016/44132 (Mei. 49441), IQW 2016/44133 (Mei. 49442), IQW 2016/44138 (Mei. 49447), IQW 2016/44144 (Mei. 49453), IQW 2016/44148 (Mei. 49457), IQW 2016/44155 (Mei. 49464), IQW 2016/44155 (Mei. 49464), IQW 2016/44156 (Mei. 49465), IQW 2016/44157 (Mei. 49466), IQW 2007/29176 (Mei. 28338), IQW 2007/29435 (Mei. 28597), IQW 2016/44043 (Mei. 49352), IQW 2016/44050 (Mei. 49359), IQW 2016/44061 (Mei. 49370), IQW 2016/44075 (Mei. 49384), IQW 2016/44079 (Mei. 49388), IQW 2016/44162 (Mei. 49471), IQW 2016/44162 (Mei. 49471), IQW 2016/44162 (Mei. 49471), IQW 2016/44200 (Mei. 49509), IQW 2016/44201 (Mei. 49510), IQW 2016/44720 (Mei. 50029), IQW 2016/42780 (Mei. 48089), IQW 1990/23601 (Mei. 23130), IQW 2016/44041 (Mei. 49350), IQW 2016/44646 (Mei. 49955), IQW 2016/44647 (Mei. 49956), IQW 2016/44648 (Mei. 49957), IQW 2016/44649 (Mei. 49958), IQW 2016/44650 (Mei. 49959), IQW 2016/44651 (Mei. 49960), IQW 2016/44652 (Mei. 49961), IQW 2016/44653 (Mei. 49962), IQW 2016/44654 (Mei. 49963), IQW 2016/44655 (Mei. 49964), IQW 2016/44656 (Mei. 49965), IQW 2016/44657 (Mei. 49966), IQW 2016/44869 (Mei. 50178), IQW 2016/44870 (Mei. 50179), IQW 2016/44871 (Mei. 50180), IQW 2016/44872 (Mei. 50181), IQW 2013/41656 (Mei. 46996),
c. 3,830 indeterminable postcranial fragments collectively catalogued under: IQW 2012/38413 (Mei. 35387), IQW 2010/31056 (Mei. 30218), IQW 2012/38456 (Mei. 35430), IQW 2010/31056 (Mei. 30218), IQW 2012/38635 (Mei. 35609), IQW 2009/30540 (Mei. 29702), IQW 2007/29370 (Mei. 28532), IQW 2010/31056 (Mei. 30218), IQW 2010/30700 (Mei. 29862), IQW 1983/19435 (Mei. 18955), IQW 1987/22270 (Mei. 21789), IQW 2012/39374 (Mei. 36348), IQW 2016/42956 (Mei. 48265), IQW 1982/17962 (Mei. 17482), IQW 1983/19165 (Mei. 18685), IQW 1987/22186 (Mei. 21705), IQW 2003/28085 (Mei. 27247), IQW 2003/28086 (Mei. 27248), IQW 2016/42709 (Mei. 48018), IQW 2016/42944 (Mei. 48253), IQW 2016/43066 (Mei. 48375), IQW 2016/43067 (Mei. 48376), IQW 2016/43440 (Mei. 48749), IQW 2016/43441 (Mei. 48750), IQW 2016/43442 (Mei. 48751), IQW 2016/43473

(Mei. 48 782), IQW 2016/43 494 (Mei. 48 803), IQW 2016/43 524 (Mei. 48 833), IQW 2016/43 551 (Mei. 48 860), IQW 2016/43 602 (Mei. 48 911), IQW 2016/43 631 (Mei. 48 940), IQW 2016/43 707 (Mei. 49 016), IQW 2016/43 713 (Mei. 49 022), IQW 2016/44 022 (Mei. 49 331), IQW 2016/44 267 (Mei. 49 576), IQW 2016/44 274 (Mei. 49 583), IQW 2016/44 286 (Mei. 49 595), IQW 2016/44 295 (Mei. 49 604), IQW 2016/44 299 (Mei. 49 608), IQW 2016/44 311 (Mei. 49 620), IQW 2016/44 312 (Mei. 49 621), IQW 2016/44 313 (Mei. 49 622), IQW 2016/44 314 (Mei. 49 623), IQW 2016/44 315 (Mei. 49 624), IQW 2016/44 316 (Mei. 49 625), IQW 2016/44 317 (Mei. 49 626), IQW 2016/44 318 (Mei. 49 627), IQW 2016/44 319 (Mei. 49 628), IQW 2016/44 320 (Mei. 49 629), IQW 2016/44 348 (Mei. 49 657), IQW 2016/44 349 (Mei. 49 658), IQW 2016/44 350 (Mei. 49 659), IQW 2016/44 351 (Mei. 49 660), IQW 2016/44 352 (Mei. 49 661), IQW 2016/44 374 (Mei. 49 683), IQW 2016/44 375 (Mei. 49 684), IQW 2016/44 376 (Mei. 49 685), IQW 2016/44 377 (Mei. 49 686), IQW 2016/44 378 (Mei. 49 687), IQW 2016/44 391 (Mei. 49 700), IQW 2016/44 392 (Mei. 49 701), IQW 2016/44 397 (Mei. 49 706), IQW 2016/44 398 (Mei. 49 707), IQW 2016/44 399 (Mei. 49 708), IQW 2016/44 435 (Mei. 49 744), IQW 2016/44 437 (Mei. 49 746), IQW 2016/44 439 (Mei. 49 748), IQW 2016/44 444 (Mei. 49 753), IQW 2016/44 445 (Mei. 49 754), IQW 2016/44 446 (Mei. 49 755), IQW 2016/44 447 (Mei. 49 756), IQW 2016/44 448 (Mei. 49 757), IQW 2016/44 449 (Mei. 49 758), IQW 2016/44 452 (Mei. 49 761), IQW 2016/44 454 (Mei. 49 763), IQW 2016/44 459 (Mei. 49 768), IQW 2016/44 460 (Mei. 49 769), IQW 2016/44 461 (Mei. 49 770), IQW 2016/44 462 (Mei. 49 771), IQW 2016/44 463 (Mei. 49 772), IQW 2016/44 464 (Mei. 49 773), IQW 2016/44 465 (Mei. 49 774), IQW 2016/44 466 (Mei. 49 775), IQW 2016/44 467 (Mei. 49 776), IQW 2016/44 468 (Mei. 49 777), IQW 2016/44 469 (Mei. 49 778), IQW 2016/44 470 (Mei. 49 779), IQW 2016/44 487 (Mei. 49 796), IQW 2016/44 488 (Mei. 49 797), IQW 2016/44 491 (Mei. 49 800), IQW 2016/44 496 (Mei. 49 805), IQW 2016/44 497 (Mei. 49 806), IQW 2016/44 498 (Mei. 49 807), IQW 2016/44 508 (Mei. 49 817), IQW 2016/44 516 (Mei. 49 825), IQW 2016/44 520 (Mei. 49 829), IQW 2016/44 554 (Mei. 49 863), IQW 2016/44 555 (Mei. 49 864), IQW 2016/44 556 (Mei. 49 865), IQW 2016/44 557 (Mei. 49 866), IQW 2016/44 558 (Mei. 49 867), IQW 2016/44 559 (Mei. 49 868), IQW 2016/44 560 (Mei. 49 869), IQW 2016/44 561 (Mei. 49 870), IQW 2016/44 562 (Mei. 49 871), IQW 2016/44 591 (Mei. 49 900), IQW 2016/44 605 (Mei. 49 914), IQW 2016/44 607 (Mei. 49 916), IQW 2016/44 620 (Mei. 49 929), IQW 2016/44 729 (Mei. 50 038), IQW 2016/44 742 (Mei. 50 051), IQW 2016/44 756 (Mei. 50 065), IQW 2016/44 776 (Mei. 50 085), IQW 2010/31 007 (Mei. 30 169), IQW 1986/21 186 (Mei. 20 705), IQW 1984/19 879 (Mei. 19 391), IQW 2009/30 150 (Mei. 29 312), IQW 2016/42 708 (Mei. 48 017), IQW 2016/42 784 (Mei. 48 093), IQW 2016/43 034 (Mei. 48 343), IQW 2016/43 037 (Mei. 48 346), IQW 2016/43 063 (Mei. 48 372), IQW 2016/43 065 (Mei. 48 374), IQW 2016/43 438 (Mei. 48 747), IQW 2016/43 439 (Mei. 48 748), IQW 2016/44 264 (Mei. 49 573), IQW 2016/44 273 (Mei. 49 582), IQW 2016/44 298 (Mei. 49 607), IQW 2016/44 310 (Mei. 49 619), IQW 2016/44 347 (Mei. 49 656), IQW 2016/44 372 (Mei. 49 681), IQW 2016/44 373 (Mei. 49 682), IQW 2016/44 396 (Mei. 49 705), IQW 2016/44 422 (Mei. 49 731), IQW 2016/44 458 (Mei. 49 767), IQW 2016/44 507 (Mei. 49 816), IQW 2016/44 539 (Mei. 49 848), IQW 2016/44 565 (Mei. 49 874), IQW 2016/44 577 (Mei. 49 886), IQW 2016/44 590 (Mei. 49 899), IQW 2016/44 595 (Mei. 49 904), IQW 2016/44 629 (Mei. 49 938), IQW 2016/44 631 (Mei. 49 940), IQW 2016/44 741 (Mei. 50 050), IQW 2016/44 750 (Mei. 50 059), IQW 2016/44 769 (Mei. 50 078), IQW 2016/44 770 (Mei. 50 079), IQW 2016/44 771 (Mei. 50 080), IQW 2016/44 772 (Mei. 50 081), IQW 2016/44 773 (Mei. 50 082), IQW 2016/44 774 (Mei. 50 083), IQW 2016/44 775 (Mei. 50 084), IQW 2012/39 317 (Mei. 36 291), IQW 1983/19 434 (Mei. 18 954), IQW 1987/22 183 (Mei. 21 705), IQW 2016/42 704 (Mei. 48 013), IQW 2016/42 791 (Mei. 48 100), IQW 2016/42 942 (Mei. 48 251), IQW 2016/43 204 (Mei. 48 513), IQW 2016/43 208 (Mei. 48 517), IQW 2016/43 219 (Mei. 48 528), IQW 2016/43 228 (Mei. 48 537), IQW 2016/43 256 (Mei. 48 565), IQW 2016/43 259 (Mei. 48 568), IQW 2016/44 251 (Mei. 49 560), IQW 2016/44 268 (Mei. 49 577), IQW 2016/44 287 (Mei. 49 596), IQW 2016/44 415 (Mei. 49 724), IQW 2016/44 499 (Mei. 49 808), IQW 2016/44 625 (Mei. 49 934), IQW 2016/44 726 (Mei. 50 035), IQW 2003/28 205 (Mei. 27 367), IQW 2016/42 947 (Mei. 48 256), IQW 2016/43 433 (Mei. 48 742), IQW 2016/43 434 (Mei. 48 743), IQW 2016/43 435 (Mei. 48 744), IQW 2016/44 294 (Mei. 49 603), IQW 2016/44 306 (Mei. 49 615), IQW 2016/44 307 (Mei. 49 616), IQW 2016/44 308 (Mei. 49 617), IQW 2016/44 344 (Mei. 49 653), IQW 2016/44 345 (Mei. 49 654), IQW 2016/44 390 (Mei. 49 699), IQW 2016/44 434 (Mei. 49 743), IQW 2016/44 563 (Mei. 49 872), IQW 2016/44 564 (Mei. 49 873), IQW 2016/44 588 (Mei. 49 897), IQW 2016/44 627 (Mei. 49 936), IQW 2016/44 630 (Mei. 49 939), IQW 2016/44 737 (Mei. 50 046), IQW 2016/44 749 (Mei. 50 058), IQW 2016/44 768 (Mei. 50 077), IQW 2010/30 930 (Mei. 30 092), IQW 2016/42 775 (Mei. 48 084), IQW 2016/44 719 (Mei. 50 028), IQW 2016/43 510 (Mei. 48 819), IQW 2016/43 512 (Mei. 48 821), IQW 2016/43 235 (Mei. 48 544), IQW 2016/44 014 (Mei. 49 323), IQW 2016/44 063 (Mei. 49 372), IQW 2016/44 379 (Mei. 49 688), IQW 2016/44 739 (Mei. 50 048), IQW 2008/29 763 (Mei. 28 925), IQW 2008/29 764 (Mei. 28 926), IQW 2010/31 065 (Mei. 30 227), IQW 2010/30 700 (Mei. 29 862), IQW 2003/28 080 (Mei. 27 242), IQW 2003/28 226 (Mei. 27 388), IQW 2005/28 928 (Mei. 28 090), IQW 2016/44 041 (Mei. 49 350), IQW 2016/44 041 (Mei. 49 350), IQW 2007/29 204 (Mei. 28 366), IQW 2007/29 358 (Mei. 28 520), IQW 2008/29 796 (Mei. 28 958), IQW 2009/29 926 (Mei. 29 088), IQW 2009/30 040 (Mei. 29 202), IQW 2009/30 044 (Mei. 29 206), IQW 2009/30 120 (Mei. 29 282), IQW

2009/30121 (Mei. 29283), IQW 2009/30201 (Mei. 29363), IQW 2009/30359 (Mei. 29521), IQW 2009/30401 (Mei. 29563), IQW 2009/30415 (Mei. 29577), IQW 2009/30482 (Mei. 29644), IQW 2009/30559 (Mei. 29721), IQW 2010/30679 (Mei. 29841), IQW 2010/30827 (Mei. 29989), IQW 2010/30874 (Mei. 30036), IQW 2010/30931 (Mei. 30093), IQW 2010/31029 (Mei. 30191), IQW 2011/32099 (Mei. 31261), IQW 2012/39320 (Mei. 36294), IQW 2012/39342 (Mei. 36316), IQW 2013/41657 (Mei. 46997), IQW 2013/41700 (Mei. 47040), IQW 2015/41982 (Mei. 47291), IQW 2016/44053 (Mei. 49362), IQW 2016/44080 (Mei. 49389), IQW 2016/44081 (Mei. 49390), IQW 2016/44081 (Mei. 49390), IQW 2016/44081 (Mei. 49390), IQW 2016/44084 (Mei. 49393), IQW 2016/44085 (Mei. 49394), IQW 2016/44086 (Mei. 49395), IQW 2016/44088 (Mei. 49397), IQW 2016/44088 (Mei. 49397), IQW 2016/44091 (Mei. 49400), IQW 2016/44091 (Mei. 49400), IQW 2016/44091 (Mei. 49400), IQW 2016/44093 (Mei. 49402), IQW 2016/44095 (Mei. 49404), IQW 2016/44096 (Mei. 49405), IQW 2016/44096 (Mei. 49405), IQW 2016/44096 (Mei. 49405), IQW 2016/44097 (Mei. 49406), IQW 2016/44099 (Mei. 49408), IQW 2016/44100 (Mei. 49409), IQW 2016/44100 (Mei. 49409), IQW 2016/44102 (Mei. 49411), IQW 2016/44102 (Mei. 49411), IQW 2016/44103 (Mei. 49412), IQW 2016/44104 (Mei. 49413), IQW 2016/44109 (Mei. 49418), IQW 2016/44112 (Mei. 49421), IQW 2016/44112 (Mei. 49421), IQW 2016/44112 (Mei. 49421), IQW 2016/44113 (Mei. 49422), IQW 2016/44114 (Mei. 49423), IQW 2016/44115 (Mei. 49424), IQW 2016/44116 (Mei. 49425), IQW 2016/44117 (Mei. 49426), IQW 2016/44119 (Mei. 49428), IQW 2016/44120 (Mei. 49429), IQW 2016/44121 (Mei. 49430), IQW 2016/44122 (Mei. 49431), IQW 2016/44123 (Mei. 49432), IQW 2016/44125 (Mei. 49434), IQW 2016/44127 (Mei. 49436), IQW 2016/44128 (Mei. 49437), IQW 2016/44129 (Mei. 49438), IQW 2016/44130 (Mei. 49439), IQW 2016/44131 (Mei. 49440), IQW 2016/44132 (Mei. 49441), IQW 2016/44133 (Mei. 49442), IQW 2016/44134 (Mei. 49443), IQW 2016/44135 (Mei. 49444), IQW 2016/44136 (Mei. 49445), IQW 2016/44138 (Mei. 49447), IQW 2016/44143 (Mei. 49452), IQW 2016/44144 (Mei. 49453), IQW 2016/44146 (Mei. 49455), IQW 2016/44147 (Mei. 49456), IQW 2016/44148 (Mei. 49457), IQW 2016/44152 (Mei. 49461), IQW 2016/44154 (Mei. 49463), IQW 2016/44156 (Mei. 49465), IQW 2016/44157 (Mei. 49466), IQW 2016/44166 (Mei. 49475), IQW 2016/44166 (Mei. 49475), IQW 2016/44166 (Mei. 49475), IQW 2016/44166 (Mei. 49475), IQW 1982/18027 (Mei. 17547), IQW 1982/18027 (Mei. 17547), IQW 1983/19165 (Mei. 18685), IQW 1980/16194 (Mei. 15705), IQW 1980/17245 (Mei. 16767), IQW 1980/17270 (Mei. 16792), IQW 1982/17958 (Mei. 17478), IQW 1982/18160 (Mei. 17680), IQW 1982/18534 (Mei. 18054), IQW 1982/18536 (Mei. 18056), IQW 1982/18570 (Mei. 18090), IQW 1983/18687 (Mei. 18207), IQW 1983/18688 (Mei. 18208), IQW 1983/18689 (Mei. 18209), IQW 1983/18724 (Mei. 18244), IQW 1983/18742 (Mei. 18262), IQW 1983/19462 (Mei. 18982), IQW 1983/19499 (Mei. 19019), IQW 1983/19499 (Mei. 19019), IQW 1983/19604 (Mei. 19124), IQW 1983/19607 (Mei. 19127), IQW 1984/20279 (Mei. 19799), IQW 1987/21821 (Mei. 21340), IQW 1987/22186 (Mei. 21705), IQW 1987/22273(Mei. 21792), IQW 1987/22327 (Mei. 21846), IQW 1988/22629 (Mei. 22148), IQW 1988/22638 (Mei. 22157), IQW 2003/27593 (Mei. 27098), IQW 2003/27616 (Mei. 27121), IQW 2003/27695 (Mei. 27200), IQW 2003/28082 (Mei. 27244), IQW 2003/28135 (Mei. 27297), IQW 2003/28317 (Mei. 27479), IQW 2004/28476 (Mei. 27638), IQW 2004/28482 (Mei. 27644), IQW 2004/28602 (Mei. 27764), IQW 2004/28642 (Mei. 27804), IQW 2004/28663 (Mei. 27825), IQW 2007/29609 (Mei. 28771), IQW 2008/29812 (Mei. 28974), IQW 2008/29842 (Mei. 29004), IQW 2009/29924 (Mei. 29086), IQW 2009/30016 (Mei. 29178), IQW 2009/30021 (Mei. 29183), IQW 2010/31030 (Mei. 30192), IQW 2013/41767 (Mei. 47077), IQW 2015/42034 (Mei. 47343), IQW 2016/42710 (Mei. 48019), IQW 2016/42711 (Mei. 48020), IQW 2016/42712 (Mei. 48021), IQW 2016/42713 (Mei. 48022), IQW 2016/42714 (Mei. 48023), IQW 2016/42715 (Mei. 48024), IQW 2016/42716 (Mei. 48025), IQW 2016/42717 (Mei. 48026), IQW 2016/42718 (Mei. 48027), IQW 2016/42719 (Mei. 48028), IQW 2016/42720 (Mei. 48029), IQW 2016/42726 (Mei. 48035), IQW 2016/42727 (Mei. 48036), IQW 2016/42728 (Mei. 48037), IQW 2016/42730 (Mei. 48039), IQW 2016/42731 (Mei. 48040), IQW 2016/42732 (Mei. 48041), IQW 2016/42733 (Mei. 48042), IQW 2016/42734 (Mei. 48043), IQW 2016/42735 (Mei. 48044), IQW 2016/42736 (Mei. 48045), IQW 2016/42737 (Mei. 48046), IQW 2016/42738 (Mei. 48047), IQW 2016/42739 (Mei. 48048), IQW 2016/42740 (Mei. 48049), IQW 2016/42741 (Mei. 48050), IQW 2016/42742 (Mei. 48051), IQW 2016/42743 (Mei. 48052), IQW 2016/42744 (Mei. 48053), IQW 2016/42745 (Mei. 48054), IQW 2016/42746 (Mei. 48055), IQW 2016/42747 (Mei. 48056), IQW 2016/42748 (Mei. 48057), IQW 2016/42749 (Mei. 48058), IQW 2016/42750 (Mei. 48059), IQW 2016/42751 (Mei. 48060), IQW 2016/42752 (Mei. 48061), IQW 2016/42792 (Mei. 48101), IQW 2016/42793 (Mei. 48102), IQW 2016/42794 (Mei. 48103), IQW 2016/42795 (Mei. 48104), IQW 2016/42796 (Mei. 48105), IQW 2016/42800 (Mei. 48109), IQW 2016/42945 (Mei. 48254), IQW 2016/43009 (Mei. 48318), IQW 2016/43053 (Mei. 48362), IQW 2016/43222 (Mei. 48531), IQW 2016/43232 (Mei. 48541), IQW 2016/43233 (Mei. 48542), IQW 2016/43234 (Mei. 48543), IQW 2016/43244 (Mei. 48553), IQW 2016/43248 (Mei. 48557), IQW 2016/43250 (Mei. 48559), IQW 2016/43251 (Mei. 48560), IQW 2016/43252 (Mei. 48561), IQW 2016/43254 (Mei. 48563), IQW 2016/43257 (Mei. 48566), IQW 2016/43261 (Mei. 48570), IQW 2016/43262 (Mei. 48571), IQW 2016/43265 (Mei. 48574), IQW 2016/43270 (Mei. 48579), IQW 2016/43271

(Mei. 48 580), IQW 2016/43 273 (Mei. 48 582), IQW 2016/43 274 (Mei. 48 583), IQW 2016/43 276 (Mei. 48 585), IQW 2016/43 280 (Mei. 48 589), IQW 2016/43 281 (Mei. 48 590), IQW 2016/43 282 (Mei. 48 591), IQW 2016/43 285 (Mei. 48 594), IQW 2016/43 286 (Mei. 48 595), IQW 2016/43 287 (Mei. 48 596), IQW 2016/43 288 (Mei. 48 597), IQW 2016/43 290 (Mei. 48 599), IQW 2016/43 291 (Mei. 48 600), IQW 2016/43 295 (Mei. 48 604), IQW 2016/43 296 (Mei. 48 605), IQW 2016/43 297 (Mei. 48 606), IQW 2016/43 298 (Mei. 48 607), IQW 2016/43 299 (Mei. 48 608), IQW 2016/43 300 (Mei. 48 609), IQW 2016/43 311 (Mei. 48 620), IQW 2016/43 318 (Mei. 48 627), IQW 2016/43 321 (Mei. 48 630), IQW 2016/43 322 (Mei. 48 631), IQW 2016/43 323 (Mei. 48 632), IQW 2016/43 324 (Mei. 48 633), IQW 2016/43 325 (Mei. 48 634), IQW 2016/43 330 (Mei. 48 639), IQW 2016/43 332 (Mei. 48 641), IQW 2016/43 333 (Mei. 48 642), IQW 2016/43 334 (Mei. 48 643), IQW 2016/43 336 (Mei. 48 645), IQW 2016/43 337 (Mei. 48 646), IQW 2016/43 344 (Mei. 48 653), IQW 2016/43 375 (Mei. 48 684), IQW 2016/43 376 (Mei. 48 685), IQW 2016/43 377 (Mei. 48 686), IQW 2016/43 378 (Mei. 48 687), IQW 2016/43 379 (Mei. 48 688), IQW 2016/43 380 (Mei. 48 689), IQW 2016/43 381 (Mei. 48 690), IQW 2016/43 382 (Mei. 48 691), IQW 2016/43 383 (Mei. 48 692), IQW 2016/43 384 (Mei. 48 693), IQW 2016/43 385 (Mei. 48 694), IQW 2016/43 386 (Mei. 48 695), IQW 2016/43 387 (Mei. 48 696), IQW 2016/43 388 (Mei. 48 697), IQW 2016/43 391 (Mei. 48 700), IQW 2016/43 392 (Mei. 48 701), IQW 2016/43 393 (Mei. 48 702), IQW 2016/43 394 (Mei. 48 703), IQW 2016/43 395 (Mei. 48 704), IQW 2016/43 396 (Mei. 48 705), IQW 2016/43 397 (Mei. 48 706), IQW 2016/43 399 (Mei. 48 708), IQW 2016/43 400 (Mei. 48 709), IQW 2016/43 403 (Mei. 48 712), IQW 2016/43 404 (Mei. 48 713), IQW 2016/43 405 (Mei. 48 714), IQW 2016/43 406 (Mei. 48 715), IQW 2016/43 407 (Mei. 48 716), IQW 2016/43 408 (Mei. 48 717), IQW 2016/43 410 (Mei. 48 719), IQW 2016/43 411 (Mei. 48 720), IQW 2016/43 412 (Mei. 48 721), IQW 2016/43 413 (Mei. 48 722), IQW 2016/43 414 (Mei. 48 723), IQW 2016/43 415 (Mei. 48 724), IQW 2016/43 416 (Mei. 48 725), IQW 2016/43 417 (Mei. 48 726), IQW 2016/43 418 (Mei. 48 727), IQW 2016/43 419 (Mei. 48 728), IQW 2016/43 420 (Mei. 48 729), IQW 2016/43 646 (Mei. 48 955), IQW 2016/43 954 (Mei. 49 263), IQW 2016/43 956 (Mei. 49 265), IQW 2016/43 957 (Mei. 49 266), IQW 2016/43 958 (Mei. 49 267), IQW 2016/43 959 (Mei. 49 268), IQW 2016/43 960 (Mei. 49 269), IQW 2016/43 962 (Mei. 49 271), IQW 2016/43 963 (Mei. 49 272), IQW 2016/43 964 (Mei. 49 273), IQW 2016/43 965 (Mei. 49 274), IQW 2016/43 966 (Mei. 49 275), IQW 2016/43 969 (Mei. 49 278), IQW 2016/43 970 (Mei. 49 279), IQW 2016/43 971 (Mei. 49 280), IQW 2016/43 972 (Mei. 49 281), IQW 2016/43 974 (Mei. 49 283), IQW 2016/43 976 (Mei. 49 285), IQW 2016/43 977 (Mei. 49 286), IQW 2016/43 978 (Mei. 49 287), IQW 2016/43 979 (Mei. 49 288), IQW 2016/43 980 (Mei. 49 289), IQW 2016/43 981 (Mei. 49 290), IQW 2016/43 982 (Mei. 49 291), IQW 2016/43 983 (Mei. 49 292), IQW 2016/43 984 (Mei. 49 293), IQW 2016/43 986 (Mei. 49 295), IQW 2016/43 987 (Mei. 49 296), IQW 2016/43 988 (Mei. 49 297), IQW 2016/43 989 (Mei. 49 298), IQW 2016/43 990 (Mei. 49 299), IQW 2016/43 991 (Mei. 49 300), IQW 2016/43 992 (Mei. 49 301), IQW 2016/43 993 (Mei. 49 302), IQW 2016/43 994 (Mei. 49 303), IQW 2016/43 995 (Mei. 49 304), IQW 2016/43 996 (Mei. 49 305), IQW 2016/43 998 (Mei. 49 307), IQW 2016/44 000 (Mei. 49 309), IQW 2016/44 001 (Mei. 49 310), IQW 2016/44 002 (Mei. 49 311), IQW 2016/44 003 (Mei. 49 312), IQW 2016/44 005 (Mei. 49 314), IQW 2016/44 006 (Mei. 49 315), IQW 2016/44 008 (Mei. 49 317), IQW 2016/44 010 (Mei. 49 319), IQW 2016/44 014 (Mei. 49 323), IQW 2016/44 019 (Mei. 49 328), IQW 2016/44 021 (Mei. 49 330), IQW 2016/44 023 (Mei. 49 332), IQW 2016/44 024 (Mei. 49 333), IQW 2016/44 025 (Mei. 49 334), IQW 2016/44 026 (Mei. 49 335), IQW 2016/44 027 (Mei. 49 336), IQW 2016/44 028 (Mei. 49 337), IQW 2016/44 030 (Mei. 49 339), IQW 2016/44 030 (Mei. 49 339), IQW 2016/44 031 (Mei. 49 340), IQW 2016/44 032 (Mei. 49 341), IQW 2016/44 033 (Mei. 49 342), IQW 2016/44 033 (Mei. 49 342), IQW 2016/44 034 (Mei. 49 343), IQW 2016/44 035 (Mei. 49 344), IQW 2016/44 035 (Mei. 49 344), IQW 2016/44 035 (Mei. 49 344), IQW 2016/44 035 (Mei. 49 344), IQW 2016/44 036 (Mei. 49 345), IQW 2016/44 037 (Mei. 49 346), IQW 2016/44 037 (Mei. 49 346), IQW 2016/44 037 (Mei. 49 346), IQW 2016/44 039 (Mei. 49 348), IQW 2016/44 040 (Mei. 49 349), IQW 2016/44 040 (Mei. 49 349), IQW 2016/44 041 (Mei. 49 350), IQW 2016/44 041 (Mei. 49 350), IQW 2016/44 041 (Mei. 49 350), IQW 2016/44 042 (Mei. 49 351), IQW 2016/44 042 (Mei. 49 351), IQW 2016/44 042 (Mei. 49 351), IQW 2016/44 042 (Mei. 49 351), IQW 2016/44 042 (Mei. 49 351), IQW 2016/44 042 (Mei. 49 351), IQW 2016/44 042 (Mei. 49 351), IQW 2016/44 043 (Mei. 49 352), IQW 2016/44 043 (Mei. 49 352), IQW 2016/44 043 (Mei. 49 352), IQW 2016/44 044 (Mei. 49 353), IQW 2016/44 045 (Mei. 49 354), IQW 2016/44 046 (Mei. 49 355), IQW 2016/44 047 (Mei. 49 356), IQW 2016/44 048 (Mei. 49 357), IQW 2016/44 049 (Mei. 49 358), IQW 2016/44 050 (Mei. 49 359), IQW 2016/44 051 (Mei. 49 360), IQW 2016/44 052 (Mei. 49 361), IQW 2016/44 052 (Mei. 49 361), IQW 2016/44 054 (Mei. 49 363), IQW 2016/44 055 (Mei. 49 364), IQW 2016/44 055 (Mei. 49 364), IQW 2016/44 056 (Mei. 49 365), IQW 2016/44 057 (Mei. 49 366), IQW 2016/44 058 (Mei. 49 367), IQW 2016/44 059 (Mei. 49 368), IQW 2016/44 060 (Mei. 49 369), IQW 2016/44 061 (Mei. 49 370), IQW 2016/44 062 (Mei. 49 371), IQW 2016/44 066 (Mei. 49 375), IQW 2016/44 067 (Mei. 49 376), IQW 2016/44 068 (Mei. 49 377), IQW 2016/44 069 (Mei. 49 378), IQW 2016/44 070 (Mei. 49 379), IQW 2016/44 070 (Mei. 49 379), IQW 2016/44 072 (Mei. 49 381), IQW 2016/44 073 (Mei. 49 382), IQW 2016/44 074 (Mei. 49 383), IQW 2016/44 076 (Mei. 49 385), IQW 2016/44 077 (Mei. 49 386), IQW 2016/44 077 (Mei. 49 386), IQW 2016/44 077 (Mei. 49 386), IQW 2016/44 078 (Mei. 49 387), IQW

2016/44079 (Mei. 49388), IQW 2016/44094 (Mei. 49403), IQW 2016/44098 (Mei. 49407), IQW 2016/44139 (Mei. 49448), IQW 2016/44139 (Mei. 49448), IQW 2016/44140 (Mei. 49449), IQW 2016/44142 (Mei. 49451), IQW 2016/44158 (Mei. 49467), IQW 2016/44159 (Mei. 49468), IQW 2016/44160 (Mei. 49469), IQW 2016/44163 (Mei. 49472), IQW 2016/44164 (Mei. 49473), IQW 2016/44165 (Mei. 49474), IQW 2016/44195 (Mei. 49504), IQW 2016/44197 (Mei. 49506), IQW 2016/44198 (Mei. 49507), IQW 2016/44203 (Mei. 49512), IQW 2016/44204 (Mei. 49513), IQW 2016/44205 (Mei. 49514), IQW 2016/44206 (Mei. 49515), IQW 2016/44207 (Mei. 49516), IQW 2016/44208 (Mei. 49517), IQW 2016/44209 (Mei. 49518), IQW 2016/44210 (Mei. 49519), IQW 2016/44211 (Mei. 49520), IQW 2016/44212 (Mei. 49521), IQW 2016/44213 (Mei. 49522), IQW 2016/44214 (Mei. 49523), IQW 2016/44215 (Mei. 49524), IQW 2016/44216 (Mei. 49525), IQW 2016/44217 (Mei. 49526), IQW 2016/44218 (Mei. 49527), IQW 2016/44219 (Mei. 49528), IQW 2016/44220 (Mei. 49529), IQW 2016/44221 (Mei. 49530), IQW 2016/44222 (Mei. 49531), IQW 2016/44223 (Mei. 49532), IQW 2016/44224 (Mei. 49533), IQW 2016/44225 (Mei. 49534), IQW 2016/44226 (Mei. 49535), IQW 2016/44227 (Mei. 49536), IQW 2016/44228 (Mei. 49537), IQW 2016/44229 (Mei. 49538), IQW 2016/44230 (Mei. 49539), IQW 2016/44231 (Mei. 49540), IQW 2016/44232 (Mei. 49541), IQW 2016/44248 (Mei. 49557), IQW 2016/44249 (Mei. 49558), IQW 2016/44250 (Mei. 49559), IQW 2016/44253 (Mei. 49562), IQW 2016/44254 (Mei. 49563), IQW 2016/44255 (Mei. 49564), IQW 2016/44256 (Mei. 49565), IQW 2016/44257 (Mei. 49566), IQW 2016/44258 (Mei. 49567), IQW 2016/44259 (Mei. 49568), IQW 2016/44261 (Mei. 49570), IQW 2016/44263 (Mei. 49572), IQW 2016/44265 (Mei. 49574), IQW 2016/44269 (Mei. 49578), IQW 2016/44275 (Mei. 49584), IQW 2016/44276 (Mei. 49585), IQW 2016/44277 (Mei. 49586), IQW 2016/44278 (Mei. 49587), IQW 2016/44279 (Mei. 49588), IQW 2016/44280 (Mei. 49589), IQW 2016/44281 (Mei. 49590), IQW 2016/44282 (Mei. 49591), IQW 2016/44288 (Mei. 49597), IQW 2016/44289 (Mei. 49598), IQW 2016/44290 (Mei. 49599), IQW 2016/44291 (Mei. 49600), IQW 2016/44292 (Mei. 49601), IQW 2016/44300 (Mei. 49609), IQW 2016/44305 (Mei. 49614), IQW 2016/44321 (Mei. 49630), IQW 2016/44322 (Mei. 49631), IQW 2016/44324 (Mei. 49633), IQW 2016/44325 (Mei. 49634), IQW 2016/44326 (Mei. 49635), IQW 2016/44327 (Mei. 49636), IQW 2016/44328 (Mei. 49637), IQW 2016/44329 (Mei. 49638), IQW 2016/44330 (Mei. 49639), IQW 2016/44331 (Mei. 49640), IQW 2016/44332 (Mei. 49641), IQW 2016/44333 (Mei. 49642), IQW 2016/44334 (Mei. 49643), IQW 2016/44335 (Mei. 49644), IQW 2016/44336 (Mei. 49645), IQW 2016/44337 (Mei. 49646), IQW 2016/44338 (Mei. 49647), IQW 2016/44339 (Mei. 49648), IQW 2016/44353 (Mei. 49662), IQW 2016/44354 (Mei. 49663), IQW 2016/44355 (Mei. 49664), IQW 2016/44356 (Mei. 49665), IQW 2016/44357 (Mei. 49666), IQW 2016/44358 (Mei. 49667), IQW 2016/44359 (Mei. 49668), IQW 2016/44360 (Mei. 49669), IQW 2016/44361 (Mei. 49670), IQW 2016/44362 (Mei. 49671), IQW 2016/44363 (Mei. 49672), IQW 2016/44364 (Mei. 49673), IQW 2016/44365 (Mei. 49674), IQW 2016/44366 (Mei. 49675), IQW 2016/44367 (Mei. 49676), IQW 2016/44370 (Mei. 49679), IQW 2016/44371 (Mei. 49680), IQW 2016/44383 (Mei. 49692), IQW 2016/44384 (Mei. 49693), IQW 2016/44385 (Mei. 49694), IQW 2016/44386 (Mei. 49695), IQW 2016/44387 (Mei. 49696), IQW 2016/44388 (Mei. 49697), IQW 2016/44389 (Mei. 49698), IQW 2016/44393 (Mei. 49702), IQW 2016/44400 (Mei. 49709), IQW 2016/44401 (Mei. 49710), IQW 2016/44402 (Mei. 49711), IQW 2016/44403 (Mei. 49712), IQW 2016/44404 (Mei. 49713), IQW 2016/44405 (Mei. 49714), IQW 2016/44406 (Mei. 49715), IQW 2016/44407 (Mei. 49716), IQW 2016/44414 (Mei. 49723), IQW 2016/44416 (Mei. 49725), IQW 2016/44417 (Mei. 49726), IQW 2016/44418 (Mei. 49727), IQW 2016/44430 (Mei. 49739), IQW 2016/44431 (Mei. 49740), IQW 2016/44432 (Mei. 49741), IQW 2016/44433 (Mei. 49742), IQW 2016/44436 (Mei. 49745), IQW 2016/44438 (Mei. 49747), IQW 2016/44441 (Mei. 49750), IQW 2016/44443 (Mei. 49752), IQW 2016/44453 (Mei. 49762), IQW 2016/44455 (Mei. 49764), IQW 2016/44471 (Mei. 49780), IQW 2016/44473 (Mei. 49782), IQW 2016/44474 (Mei. 49783), IQW 2016/44475 (Mei. 49784), IQW 2016/44476 (Mei. 49785), IQW 2016/44477 (Mei. 49786), IQW 2016/44478 (Mei. 49787), IQW 2016/44479 (Mei. 49788), IQW 2016/44480 (Mei. 49789), IQW 2016/44481 (Mei. 49790), IQW 2016/44482 (Mei. 49791), IQW 2016/44483 (Mei. 49792), IQW 2016/44484 (Mei. 49793), IQW 2016/44485 (Mei. 49794), IQW 2016/44486 (Mei. 49795), IQW 2016/44489 (Mei. 49798), IQW 2016/44490 (Mei. 49799), IQW 2016/44493 (Mei. 49802), IQW 2016/44494 (Mei. 49803), IQW 2016/44500 (Mei. 49809), IQW 2016/44509 (Mei. 49818), IQW 2016/44510 (Mei. 49819), IQW 2016/44511 (Mei. 49820), IQW 2016/44513 (Mei. 49822), IQW 2016/44514 (Mei. 49823), IQW 2016/44515 (Mei. 49824), IQW 2016/44517 (Mei. 49826), IQW 2016/44521 (Mei. 49830), IQW 2016/44522 (Mei. 49831), IQW 2016/44524 (Mei. 49833), IQW 2016/44525 (Mei. 49834), IQW 2016/44526 (Mei. 49835), IQW 2016/44527 (Mei. 49836), IQW 2016/44528 (Mei. 49837), IQW 2016/44529 (Mei. 49838), IQW 2016/44530 (Mei. 49839), IQW 2016/44533 (Mei. 49842), IQW 2016/44541 (Mei. 49850), IQW 2016/44542 (Mei. 49851), IQW 2016/44543 (Mei. 49852), IQW 2016/44544 (Mei. 49853), IQW 2016/44545 (Mei. 49854), IQW 2016/44546 (Mei. 49855), IQW 2016/44547 (Mei. 49856), IQW 2016/44548 (Mei. 49857), IQW 2016/44549 (Mei. 49858), IQW 2016/44550 (Mei. 49859), IQW 2016/44551 (Mei. 49860), IQW 2016/44552 (Mei. 49861), IQW 2016/44553 (Mei. 49862), IQW 2016/44567 (Mei. 49876), IQW 2016/44568 (Mei. 49877), IQW 2016/44569

(Mei. 49 878), IQW 2016/44 570 (Mei. 49 879), IQW 2016/44 571 (Mei. 49 880), IQW 2016/44 572 (Mei. 49 881), IQW 2016/44 573 (Mei. 49 882), IQW 2016/44 575 (Mei. 49 884), IQW 2016/44 579 (Mei. 49 888), IQW 2016/44 580 (Mei. 49 889), IQW 2016/44 581 (Mei. 49 890), IQW 2016/44 582 (Mei. 49 891), IQW 2016/44 592 (Mei. 49 901), IQW 2016/44 593 (Mei. 49 902), IQW 2016/44 598 (Mei. 49 907), IQW 2016/44 599 (Mei. 49 908), IQW 2016/44 600 (Mei. 49 909), IQW 2016/44 601 (Mei. 49 910), IQW 2016/44 602 (Mei. 49 911), IQW 2016/44 603 (Mei. 49 912), IQW 2016/44 604 (Mei. 49 913), IQW 2016/44 606 (Mei. 49 915), IQW 2016/44 608 (Mei. 49 917), IQW 2016/44 609 (Mei. 49 918), IQW 2016/44 610 (Mei. 49 919), IQW 2016/44 611 (Mei. 49 920), IQW 2016/44 615 (Mei. 49 924), IQW 2016/44 616 (Mei. 49 925), IQW 2016/44 619 (Mei. 49 928), IQW 2016/44 623 (Mei. 49 932), IQW 2016/44 624 (Mei. 49 933), IQW 2016/44 632 (Mei. 49 941), IQW 2016/44 634 (Mei. 49 943), IQW 2016/44 645 (Mei. 49 954), IQW 2016/44 727 (Mei. 50 036), IQW 2016/44 728 (Mei. 50 037), IQW 2016/44 731 (Mei. 50 040), IQW 2016/44 732 (Mei. 50 041), IQW 2016/44 733 (Mei. 50 042), IQW 2016/44 734 (Mei. 50 043), IQW 2016/44 735 (Mei. 50 044), IQW 2016/44 736 (Mei. 50 045), IQW 2016/44 740 (Mei. 50 049), IQW 2016/44 743 (Mei. 50 052), IQW 2016/44 746 (Mei. 50 055), IQW 2016/44 747 (Mei. 50 056), IQW 2016/44 748 (Mei. 50 057), IQW 2016/44 751 (Mei. 50 060), IQW 2016/44 752 (Mei. 50 061), IQW 2016/44 753 (Mei. 50 062), IQW 2016/44 754 (Mei. 50 063), IQW 2016/44 755 (Mei. 50 064), IQW 2016/44 757 (Mei. 50 066), IQW 2016/44 759 (Mei. 50 068), IQW 2016/44 760 (Mei. 50 069), IQW 2016/44 778 (Mei. 50 087), IQW 2016/44 779 (Mei. 50 088), IQW 2016/44 780 (Mei. 50 089), IQW 2016/44 781 (Mei. 50 090), IQW 2016/44 782 (Mei. 50 091), IQW 2016/44 783 (Mei. 50 092), IQW 2016/44 784 (Mei. 50 093), IQW 2016/44 785 (Mei. 50 094), IQW 2016/44 786 (Mei. 50 095), IQW 2016/44 787 (Mei. 50 096), IQW 2016/44 788 (Mei. 50 097), IQW 2016/44 789 (Mei. 50 098), IQW 2016/44 790 (Mei. 50 099), IQW 2016/44 791 (Mei. 50 100), IQW 2016/44 792 (Mei. 50 101), IQW 2016/44 793 (Mei. 50 102), IQW 2016/44 794 (Mei. 50 103), IQW 2016/44 795 (Mei. 50 104), IQW 2016/44 796 (Mei. 50 105), IQW 2016/44 797 (Mei. 50 106), IQW 2016/44 798 (Mei. 50 107), IQW 2016/44 799 (Mei. 50 108), IQW 2016/44 800 (Mei. 50 109), IQW 2016/44 801 (Mei. 50 110), IQW 2016/44 802 (Mei. 50 111), IQW 2016/44 803 (Mei. 50 112), IQW 2016/44 804 (Mei. 50 113), IQW 2016/44 805 (Mei. 50 114), IQW 2016/44 806 (Mei. 50 115), IQW 2016/44 807 (Mei. 50 116), IQW 2016/44 808 (Mei. 50 117), IQW 2016/44 809 (Mei. 50 118), IQW 2016/44 810 (Mei. 50 119), IQW 2016/44 811 (Mei. 50 120), IQW 2016/44 812 (Mei. 50 121), IQW 2016/44 813 (Mei. 50 122), IQW 2016/44 814 (Mei. 50 123), IQW 2016/44 815 (Mei. 50 124), IQW 2016/44 816 (Mei. 50 125), IQW 2016/44 817 (Mei. 50 126), IQW 2016/44 818 (Mei. 50 127), IQW 2016/44 819 (Mei. 50 128), IQW 2016/44 820 (Mei. 50 129), IQW 2016/44 821 (Mei. 50 130), IQW 2016/44 822 (Mei. 50 131), IQW 2016/44 823 (Mei. 50 132), IQW 2016/44 824 (Mei. 50 133), IQW 2016/44 825 (Mei. 50 134), IQW 2016/44 826 (Mei. 50 135), IQW 2016/44 827 (Mei. 50 136), IQW 2016/44 828 (Mei. 50 137), IQW 2016/44 829 (Mei. 50 138), IQW 2016/44 830 (Mei. 50 139), IQW 2016/44 831 (Mei. 50 140), IQW 2016/44 832 (Mei. 50 141), IQW 2016/44 833 (Mei. 50 142), IQW 2016/44 834 (Mei. 50 143), IQW 2016/44 835 (Mei. 50 144), IQW 2016/44 836 (Mei. 50 145), IQW 2016/44 837 (Mei. 50 146), IQW 2016/44 838 (Mei. 50 147), IQW 2016/44 839 (Mei. 50 148), IQW 2016/44 840 (Mei. 50 149), IQW 2016/44 841 (Mei. 50 150), IQW 2016/44 842 (Mei. 50 151), IQW 2016/44 843 (Mei. 50 152), IQW 2016/44 844 (Mei. 50 153), IQW 2016/44 845 (Mei. 50 154), IQW 2016/44 846 (Mei. 50 155), IQW 2016/44 847 (Mei. 50 156), IQW 2016/44 848 (Mei. 50 157), IQW 2016/44 849 (Mei. 50 158), IQW 2016/44 850 (Mei. 50 159), IQW 2016/44 851 (Mei. 50 160), IQW 2016/44 852 (Mei. 50 161), IQW 2016/44 853 (Mei. 50 162), IQW 2016/44 854 (Mei. 50 163), IQW 2016/44 855 (Mei. 50 164), IQW 2016/44 856 (Mei. 50 165), IQW 2016/44 857 (Mei. 50 166), IQW 2016/44 858 (Mei. 50 167), IQW 2016/44 859 (Mei. 50 168), IQW 2016/44 860 (Mei. 50 169), IQW 2016/44 861 (Mei. 50 170), IQW 2016/44 864 (Mei. 50 173), IQW 2016/44 868 (Mei. 50 177), IQW 2016/44 877 (Mei. 50 186), IQW 2016/44 879 (Mei. 50 188), IQW 2003/28 087 (Mei. 27 249), IQW 2016/44 071 (Mei. 49 380), IQW 2016/44 519 (Mei. 49 828), IQW 2016/44 628 (Mei. 49 937), IQW 1988/22 551 (Mei. 22 070), IQW 2016/42 946 (Mei. 48 255), IQW 2016/44 504 (Mei. 49 813), IQW 2010/30 715 (Mei. 29 877), IQW 2012/38 412 (Mei. 35 386), IQW 2009/30 084 (Mei. 29 246), IQW 2016/43 968 (Mei. 49 277), IQW 2016/44 723 (Mei. 50 032), IQW 2016/44 724 (Mei. 50 033), IQW 2016/44 873 (Mei. 50 182), IQW 2016/44 874 (Mei. 50 183), IQW 2004/28 698 (Mei. 27 860), IQW 1983/18 710 (Mei. 18 230), IQW 1983/18 713 (Mei. 18 233), IQW 1983/18 722 (Mei. 18 242), IQW 1983/19 165 (Mei. 18 685), IQW 2016/42 706 (Mei. 48 015), IQW 2016/42 948 (Mei. 48 257), IQW 2016/42 949 (Mei. 48 258), IQW 2016/44 064 (Mei. 49 373), IQW 2016/44 252 (Mei. 49 561), IQW 2016/44 266 (Mei. 49 575), IQW 2016/44 346 (Mei. 49 655), IQW 2016/44 442 (Mei. 49 751), IQW 2016/44 621 (Mei. 49 930), IQW 2009/29 976 (Mei. 29 138), IQW 2010/30 700 (Mei. 29 862), IQW 2009/29 927 (Mei. 29 089), IQW 2009/29 973 (Mei. 29 135), IQW 2009/30 074 (Mei. 29 236), IQW 2011/31 313 (Mei. 30 475), IQW 2011/32 152 (Mei. 31 316), IQW 2013/41 676 (Mei. 47 016), IQW 2013/41 701 (Mei. 47 041), IQW 1990/23 600 (Mei. 23 129), IQW 2004/28 481 (Mei. 27 643), IQW 2016/44 863 (Mei. 50 172), IQW 1983/19 165 (Mei. 18 685), IQW 1983/19 422 (Mei. 18 942), IQW 1983/19 425 (Mei. 18 945), IQW 1983/19 442 (Mei. 18 962), IQW 2016/42 703 (Mei. 48 012), IQW 2016/42 705 (Mei. 48 014), IQW 2016/42 707 (Mei. 48 016), IQW 2016/42 754 (Mei. 48 063), IQW 2016/42 769 (Mei. 48 078), IQW 2016/42 941 (Mei. 48 250), IQW 2016/42 972 (Mei. 48 281), IQW

2016/42 977 (Mei. 48 286), IQW 2016/43 031 (Mei. 48 340), IQW 2016/43 389 (Mei. 48 698), IQW 2016/43 436 (Mei. 48 745), IQW 2016/43 437 (Mei. 48 746), IQW 2016/44 309 (Mei. 49 618), IQW 2016/44 419 (Mei. 49 728), IQW 2016/44 420 (Mei. 49 729), IQW 2016/44 421 (Mei. 49 730), IQW 2016/44 456 (Mei. 49 765), IQW 2016/44 457 (Mei. 49 766), IQW 2016/44 505 (Mei. 49 814), IQW 2016/44 506 (Mei. 49 815), IQW 2016/44 576 (Mei. 49 885), IQW 2016/44 594 (Mei. 49 903), IQW 2016/44 725 (Mei. 50 034).

JELLE W. F. REUMER · RALF-DIETRICH KAHLKE

NEW RESULTS ON CERCOPITHECIDS FROM THE EARLY PLEISTOCENE SITE OF UNTERMASSFELD

Abstract

The scant sample of *Macaca sylvanus* from the Early Pleistocene locality of Untermassfeld (Thuringia, Germany) is increased by the recent finds of a deciduous right upper incisor and a left maxillary fragment bearing a partial tooth row with P^3-M^2 preserved. The material is described; it brings the number of recorded individuals to at least three. We refrain from giving a subspecific name for the Untermassfeld macaques due to the large overlap in size among European fossil records. *Macaca sylvanus* is among the least recorded large mammal species of the Untermassfeld fauna; their tree- and rock-climbing behaviour probably let them escape from occasional flooding.

1. Introduction

Barbary apes or macaques (*Macaca sylvanus*) are a rare but not uncommon element of many Early, Middle and even Late Pleistocene mammal faunas of the western Palaearctic (e.g., Kahlke 1961, 503; Franzen 1973; Delson 1974; 1980; Adam 1975; Fladerer 1991; Alba et al. 2008; 2011; 2016; Castaños et al. 2011; Sardella et al. 2015; Reumer et al. 2018). In all of the corresponding sites, macaque remains occurred in small numbers only and usually represent not more than one or a few single individuals.
After Thenius (1965) announced the identification of a *Macaca* phalanx from the early Middle Pleistocene of Voigtstedt as the first proof of a fossil macaque in Central Germany, Kahlke (1982, 943) reported the occurrence of this genus at the newly discovered site of Untermassfeld. A third, stratigraphically younger find from Central Germany was mentioned by Mania (1983, 159) from the late Middle Pleistocene of Bilzingsleben. Zapfe (2001) was the first to describe the *Macaca* excavated at Untermassfeld during the late 1970s and 1984.

2. Previous finds

After the discovery of the site of Untermassfeld in 1978, systematic excavations have been carried out in extended annual campaigns (Kahlke 1997; 2001a; 2001b; 2020). The unusually high concentration of vertebrate carcasses and other necromass incorporated in fluviatile sediments was induced by the accumulating effect of a clastic mudflow fan influencing the riverine water flow (Ellenberg and Kahlke 1997; Kahlke et al. 2020). The huge amount of vertebrate remains from Untermassfeld represent one single contemporaneous fauna, assigned to Marine Isotope Stage (MIS) 31 (Kahlke 2001c; 2006; Maul et al. 2007; Kahlke et al. 2020; in this volume).

The first *Macaca* finds from Untermassfeld occurred during the 1979 field season. These are five elements [C inf. dex. IQW 1980/16 566 (Mei. 16 087)b, I_1 dex. IQW 1980/16 566 (Mei. 16 087)a, P_4 dex. IQW 1980/16 566 (Mei. 16 087)c, M_1 dex. IQW 1980/16 566 (Mei. 16 087)d, M_3 dex. IQW 1980/16 566 (Mei. 16 087)e, Zapfe 2001], which were recovered during rescue excavations (Kahlke 1997; 2001c, Fig. 8; 2006, Fig. 41). Their origin was most probably from the main bone bed of the site, the lower part of the Upper Fluviatile Sands [UFS(l); Kahlke 2006; Kahlke et al. 2020], that form the infill of an erosional channel. Another early macaque find, a relatively large M_3 dex. [IQW 1984/20 021 (Mei 16 541), Zapfe 2001], was found on 16 August 1984 in grid square Q 501, 0.08 m above the site's 0-level (Kahlke 2001c, supplement IX). This fossil certainly originates from the UFS(l). Following Zapfe (2001, 890) the hitherto discovered dental elements represent at least one male and one female individual.

3. New finds

3.1. Discoveries

During recent excavations, two additional *Macaca* specimens came to light. A deciduous right upper incisor IQW 2011/32 185 (Mei. 31 349) was found on 08 July 2010 in grid square Q 38, 2.42 m below the site's 0-level. A left maxillary fragment bearing a partial tooth row with P^3-M^2 preserved IQW 2013/41 540 (Mei. 36 886) appeared on 12 June 2012 in grid square Q 609, 1.99 m below the site's 0-level. Like the previous finds, both specimens originate from the channel filling of the site, i. e., from the fossil bearing lower part of the Upper Fluviatile Sands [UFS(l)].

For comparison we studied and measured material of extant *Macaca sylvanus* from the collections of the Institute of Special Zoology and Evolutionary Biology (Phyletisches Museum) of the Friedrich-Schiller-Universität Jena (PMJ) and a fossil from the private collection of Mr. and Mrs. Verhulsdonk, Nijmegen, The Netherlands (described in Reumer et al. 2018).

3.2. Description

3.2.1. Right upper milk incisor

This is a rootless deciduous incisor (**Fig. 1**), which was most probably naturally shed during tooth replacement. The incisal edge is sharp and bears a minute central boss (crenulation) that is slightly worn. The crown has a maximum mesio-distal width of 5.1 mm, lingual height 6.2 mm, buccal height 6.1 mm, maximum lingual-buccal thickness 4.1 mm. After the affiliation to a cervid was excluded (pers. comm. M. Breda to R.-D. K., November 2016), we compared the specimen with deciduous incisors of several neonate bovids and with that of extant macaques. The resemblance in both shape and size allows the identification of the tooth as a *Macaca sylvanus* right dI^1.

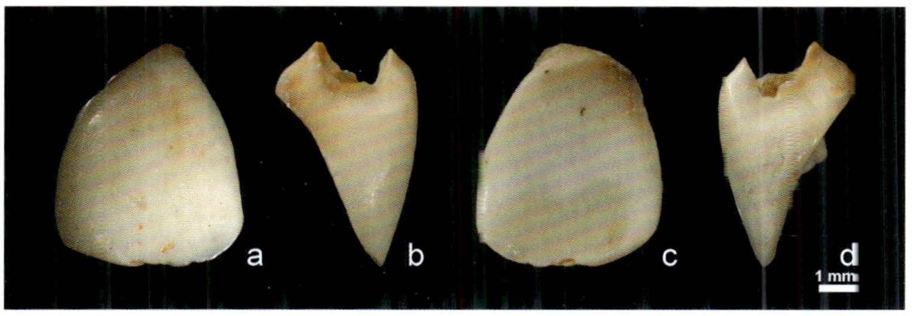

Fig. 1 *Macaca sylvanus* (Linnaeus), Untermassfeld. – **a–d** Right deciduous upper incisor (dI[1] dex.) IQW 2011/32 185 (Mei. 31 349), buccal, distal, lingual, mesial views.

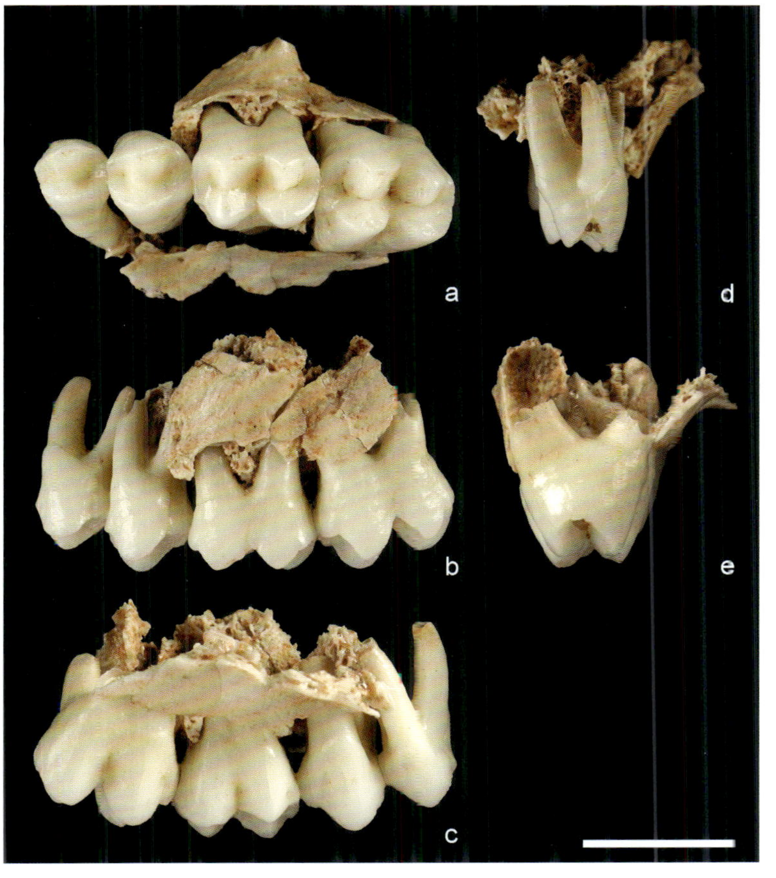

Fig. 2 *Macaca sylvanus* (Linnaeus), Untermassfeld. – **a–e** Left maxillary fragment IQW 2013/41 540 (Mei. 36 886) with P^3-M^2, occlusal, buccal, lingual, mesial, distal views. – Scale: 10 mm.

3.2.2. Left maxillary fragment with P^3-M^2

This maxilla (**Fig. 2**) is of a subadult individual; its permanent premolars are entirely erupted, but the roots of M^2 are still in the process of being formed and not yet fully developed. There is no detectable wear. All preserved elements, including P^3 and P^4, have three roots, one lingual and two buccal. Slight cingular structures can be discerned on the mesio-lingual side of the molars, most clearly so in M^2. Both M^1 and M^2 have a mesial (anterior) fovea that is slightly more pronounced than the distal (posterior) one. Measurements are given in **Tables 1–4** [L – length; W – maximum width; MW – mesial (anterior) width; DW – distal (posterior) width]; comparative measurements of the Quibas sample follow Alba et al. (2011).

specimen	L	W
IQW 2013/41 540 (Mei. 36 886), sin.	4.9	6.7
PMJ, Mam. 8020, sin.	5.4	6.7
PMJ, Mam. 8020, dex.	5.5	6.4
PMJ, Mam. 8021, sin.	5.3	6.5
PMJ, Mam. 8021, dex.	5.0	6.5
PMJ, Mam. 7387, sin.	5.5	6.4
PMJ, Mam. 7387, dex.	5.3	6.2
Quibas, GCP-CV4055, dex.	5.5	6.3
Quibas, Q05-Ec-87, dex.	5.2	6.3
Quibas, Q05-Ec-88, dex.	5.6	6.9

Table 1 P^3 measurements (mm) of *Macaca sylvanus*.

specimen	L	W
IQW 2013/41 540 (Mei. 36 886), sin.	5.8	6.9
PMJ, Mam. 8020, sin.	5.7	7.0
PMJ, Mam. 8020, dex.	5.4	7.2
PMJ, Mam. 8021, sin.	5.4	7.2
PMJ, Mam. 8021, dex.	5.2	7.0
PMJ, Mam. 7387, sin.	5.8	6.7
PMJ, Mam. 7387, dex.	5.4	6.7
Quibas, GCP-CV4055, dex.	5.3	6.8
Quibas, Q05-Ec-87, dex.	5.1	6.7

Table 2 P^4 measurements (mm) of *Macaca sylvanus*.

specimen	L	MW	DW
IQW 2013/41 540 (Mei. 36 886), sin.	8.3	7.7	7.3
PMJ, Mam. 8020, sin.	8.0	7.0	6.4
PMJ, Mam. 8020, dex.	8.2	7.0	6.5
PMJ, Mam. 8021, sin.	7.7	7.0	6.4
PMJ, Mam. 8021, dex.	7.5	7.3	6.5
PMJ, Mam. 7387, sin.	8.6		
PMJ, Mam. 7387, dex.	8.2	7.5	6.7
Quibas, GCP-CV4054, dex.	8.9	8.5	8.3
Quibas, Q05-Ec-87, dex.	8.4	8.0	
Quibas, Q05-Ec-88, dex.	8.0	7.9	7.8

Table 3 M^1 measurements (mm) of *Macaca sylvanus*.

specimen	L	MW	DW
IQW 2013/41 540 (Mei. 36 886), sin.	9.6	8.8	8.2
Coll. Verhulsdonk, 00177 HvH, dex.	8.9	8.3	7.8
PMJ, Mam. 8020, sin.	9.7	8.5	7.5
PMJ, Mam. 8020, dex.	9.6	8.7	7.6
PMJ, Mam. 8021, sin.	9.1	8.8	7.3
PMJ, Mam. 8021, dex.	8.9	8.6	7.4
PMJ, Mam. 7387, sin.	9.4	8.7	7.7
PMJ, Mam. 7387, dex.	9.2	8.6	7.9
Quibas, GCP-CV4054, dex.	9.1	8.9	7.7
Quibas, Q03-E-155b, sin.	9.8	9.6	8.5
Quibas, Q03-E-155a, dex.	9.6	9.3	8.3
Quibas, Q05-Ec-87, dex.	9.5	9.1	8.5
Quibas, Q05-Ec-88, dex.	9.2	9.5	8.6

Table 4 M^2 measurements (mm) of *Macaca sylvanus*.

4. Comparison and taxonomy

Order Primates Linnaeus, 1758
Suborder Haplorrhini Pocock, 1918
Parvorder Catarrhini É. Geoffroy Saint-Hilaire, 1812
Family Cercopithecidae Gray, 1821
Genus *Macaca* Lacépède, 1799
Macaca sylvanus (Linnaeus, 1758)

The literature provides ample discussions about the taxonomy of Pliocene to extant European macaques (e.g., Delson 1980; Fa 1989). Three fossil subspecies are commonly recognised from continental Europe (Szalay and Delson 1979; Sardella et al. 2015). There is a large overlap in sizes and no diagnostically clear-cut morphological difference between Pliocene *Macaca sylvanus prisca* Gervais, 1859, Late Plio- to Early Pleistocene *M. sylvanus florentina* (Cocchi, 1872), Middle-Late Pleistocene *M. sylvanus pliocena* Owen, 1846 and recent *M. sylvanus sylvanus*. In this framework Castaños et al. (2011, 819) noted the following about their Late Pleistocene *M. sylvanus* from Lezetxiki II cave: »The comparative morphometric study of the teeth with those of Plio-Pleistocene fossils and a sample of modern female macaques confirms the great size variability of this species and the absence of reliable criteria for the attribution of the specimen to either of the two subspecies recorded from the Pleistocene«. Similar remarks about the taxonomy were made earlier by Fladerer (1991), Rook et al. (2001), Zapfe (2001), and Alba et al. (2008; 2011). A complete revision of Plio-Pleistocene European *Macaca* is, however, well beyond the scope of the present paper. Here, we adhere with the near common opinion that European macaques, including the Untermassfeld material, belong to *Macaca sylvanus*; we wish to refrain from giving or suggesting any closer subspecific identification (for a discussion on the matter see also Reumer et al. 2018).

5. Discussion

5.1. Abundance of *Macaca* in the Untermassfeld fossil record

The fine scale documented field work at the site has produced a total of 14,291 large mammal remains (as of 01 July 2020). From this material altogether eight macaque fossils were identified (see sections 2. and 3.), which represent less than 0.06 % of the total. According to the discovered number of specimens, macaques are among the least recorded large mammal species of the Untermassfeld fauna [for minimum numbers of individuals (MNI) see Kahlke 2006, Table 9; in this volume, Table 8]. Most probably all finds originate from the same sedimentological unit, the UFS(l) (see sections 2. and 3.1.). This means the individual elements were deposited at the site during one single inundation event (on the origin of the Untermassfeld thanatocoenosis see Kahlke 2001c; 2006; Kahlke et al. 2020; in this volume). However, this does not prove a common date of death for the detected individuals with permanent dentition.
The isolated dI^1 (see section 3.2.1.) is evidently of a juvenile monkey. Furthermore, we have two right lower third molars, of two separate individuals of both sexes. These two molars are unworn, while the available M_1 IQW 1980/16566 (Mei. 16087)d is well worn. The other elements (upper tooth row, lower incisor, canine and premolar) can theoretically be combined with any of the mentioned third molars to form the remains of two individuals. The MNI is therefore three, including the juvenile. The actual number of individuals might be four if we consider the worn M_1 to originate from a different individual than the two third molars.

5.2. Habitat

Present-day Barbary macaques live preferably in mountainous areas with diverse vegetation, such as Mediterranean shrub (maquis), oak forests or conifer stands. The majority of animals nowadays live in stands of Atlas cedar (*Cedrus atlantica*), or in dryer ecosystems with dwarf palms (*Chamaerops humilis*) and oak (*Quercus*) (Blanco 1998). Blanco (l.c.) also mentioned the fact that the present habitat preference of macaques is the result of human persecution. This effect was absent during the Pleistocene, allowing for a broader habitat range. Haltenorth and Diller (1977) mentioned their preference for a rocky environment and forest with little undergrowth and noted that they prefer to sit in elevated spots as it »makes them feel safer«.

Near the site of Untermassfeld, along the flanks of the Werra river valley, there were rocky outcrops of Middle Triassic limestone (see geological map in Ellenberg and Kahlke 1997, Fig. 4) that could have been used by the animals. In addition, trees of a riparian forest (palaeolandscape reconstruction in Kahlke 2001c, 983; 2006, 87f.; in this volume, 1719f.) could have served the same purpose. Because of their climbing abilities, macaques were only marginally endangered by the inundation of the river valley and therefore only rarely represented in the thanatocoenosis.

6. Conclusions

Macaca sylvanus was a rare but palaeoecologically interesting element of the Untermassfeld fauna. Here we describe the latest finds, a right upper milk incisor, dI^1, and a left upper partial tooth row with P^3-M^2. The species found suitable conditions to inhabit the valley flanks and the riparian forest habitats along the Werra river. The macaques' tree- and rock-climbing behavior led to a low risk during riverine floods and thus to a relatively low representation in the preserved necromass. The new finds raise the number of recorded individuals to at least three.

Acknowledgements

We thank the Senckenberg teams for excavation and preparation of the Untermassfeld fossils. M. Breda (Ferrara) recognised the described milk incisor from within the cervid remains of the Untermassfeld collection during her stay in 2013 and first proposed its assignment to *Macaca*. Photographs were made by L. C. Maul (**Fig. 1**) and S. Döring (**Fig. 2**), E. Haase arranged the figures (all Weimar). J. Arnold (Weimar) and M. Krüger (Jena) arranged the access to the mammal collection of the Institute of Special Zoology and Evolutionary Biology of the Friedrich-Schiller-Universität Jena for morphometric comparisons. H. and K. Verhulsdonk (Nijmegen) allowed us to study the M^2 in their collection. E. Delson (New York) reviewed the paper and supported us by valuable hints and suggestions.

References

Adam, K.D., 1975. Die mittelpleistozäne Säugetier-Fauna aus dem Heppenloch bei Gutenberg (Württemberg). Abhandlungen zur Karst- und Höhlenkunde, Reihe D Paläontologie, Zoologie 1, 1–247.

Alba, D.M., Calero, J.A.C., Mancheño, M.A., Montoya, P., 2011. Fossil remains of *Macaca sylvanus florentina* (Cocchi, 1872) (Primates, Cercopithecidae) from the Early Pleistocene of Quibas (Murcia, Spain). Journal of Human Evolution 61, 703–718.

Alba, D.M., Moyà-Solà, S., Madurell, J., Aurell, P., 2008. Dento-gnathic remains of *Macaca* (Primates, Cercopithecidae) from the late early Pleistocene of Terrassa (Catalonia, Spain). Journal of Human Evolution 55, 1160–1163.

Alba, D.M., Madurell-Malapeira, J., Delson, E., Vinuesa, V. Susanna, I., Patrocinio Espigares, M., Ros-Montoya, S., Martínez-Navarro, B., 2015. First record of macaques from the Early Pleistocene of Incarcal (NE Iberian Peninsula). Journal of Human Evolution 96, 139–144.

Blanco, J.C., 1998. Mamíferos de España. I. Insectívoros, Quirópteros, Primates y Carnívoros de la península Ibérica, Baleares y Canarias. Barcelona.

Castaños, P., Murelaga, X., Arrizabalaga, A., Iriarte, M.-J., 2011. First evidence of *Macaca sylvanus* (Primates, Cercopithecidae) from the Late Pleistocene of Lezetxiki II cave (Basque Country, Spain). Journal of Human Evolution 60, 816–820.

Delson, E., 1974. Preliminary review of Cercopithecid distribution in the Circum Mediterranean Region. Mémoires du Bureau des Recherches Géologiques et Minières (France) 78, 131–135.

1980. Fossil macaques, phyletic relationships and a scenario of deployment. In: Lindburg, D.G. (Ed.), The Macaques: Studies in Ecology, Behavior and Evolution. Van Nostrand Reinhold primate behavior and development series. New York, 10–30.

Ellenberg, J., Kahlke, R.-D., 1997. Die quartärgeologische Entwicklung des mittleren Werratals und der Bau der unterpleistozänen Komplexfundstelle Untermaßfeld. In: Kahlke, R.-D. (Ed.), Das Pleistozän von Untermaßfeld bei Meiningen (Thüringen). Teil 1. Monographien des Römisch-Germanischen Zentralmuseums 40, 1, 29–62, pls. 1–18.

Fa, J.E., 1989. The genus *Macaca*: a review of taxonomy and evolution. Mammal Review 19, 45–81.

Fladerer, F., 1991. Der erste Fund von *Macaca* (Cercopithecidae, Primates) im Jungpleistozän von Mitteleuropa. Zeitschrift für Säugetierkunde 56, 272–283.

Franzen, J.L., 1973. Ein Primate aus den altpleistozänen Schneckenmergeln von Hohensülzen (Rheinhessen). Senckenbergiana lethaea 54, 345–358.

Haltenorth, T., Diller, H., 1977. Säugetiere Afrikas und Madagaskars. München, Bern, Wien.

Kahlke, H.-D., 1961. Revision der Säugetierfaunen der klassischen deutschen Pleistozän-Fundstellen von Süßenborn, Mosbach und Taubach. Geologie 10, 493–532.

1982. *Hippopotamus antiquus* Desmarest, 1822 aus dem Pleistozän von Meiningen in Südthüringen (Bezirk Suhl). Zeitschrift für geologische Wissenschaften 10, 943–949.

Kahlke, R.-D. (Ed.), 1997. Das Pleistozän von Untermaßfeld bei Meiningen (Thüringen). Teil 1. Monographien des Römisch-Germanischen Zentralmuseums 40, 1, I–VI, 1–418, pls. 1–67.

(Ed.), 2001a. Das Pleistozän von Untermaßfeld bei Meiningen (Thüringen). Teil 2. Monographien des Römisch-Germanischen Zentralmuseums 40, 2, I–VIII, 419–698, pls. 68–131, suppl. I.

(Ed.), 2001b. Das Pleistozän von Untermaßfeld bei Meiningen (Thüringen). Teil 3. Monographien des Römisch-Germanischen Zentralmuseums 40, 3, I–VI, 699–1030, pls. 132–151, suppl. I–XV.

2001c. Die unterpleistozäne Komplexfundstelle Untermaßfeld – Zusammenfassung des Kenntnisstandes sowie synthetische Betrachtungen zu Genesemodell, Paläoökologie und Stratigraphie. In: Kahlke, R.-D. (Ed.), Das Pleistozän von Untermaßfeld bei Meiningen (Thüringen). Teil 3. Monographien des Römisch-Germanischen Zentralmuseums 40, 3, 931–1030, suppl. I–XV.

2006. Untermassfeld. A late Early Pleistocene (Epivillafranchian) fossil site near Meiningen (Thuringia, Germany) and its position in the development of the European mammal fauna. British Archaeological Reports, International Series 1578. Oxford.

2020. Excavation progress and research at the Early Pleistocene site of Untermassfeld during the years 1997–2015. In: Kahlke, R.-D. (Ed.), The Pleistocene of Untermassfeld near Meiningen (Thüringen, Germany). Part 4. Monographien des Römisch-Germanischen Zentralmuseums 40, 4, 1031–1178; Foldout I.

in this volume. The Pleistocene of Untermassfeld – A synopsis of site origin, palaeobiodiversity, taphonomic characteristics, palaeoenvironment, chronostratigraphy, and significance in western Palaearctic faunal history. 1635–1734.

Kahlke, R.-D., Linnemann, U., Gärtner, A., 2020. New results on the origin and geological history of the Early Pleistocene site of Untermassfeld. In: Kahlke, R.-D. (Ed.), The Pleistocene of Untermassfeld near Meiningen (Thüringen, Germany). Part 4. Monographien des Römisch-Germanischen Zentralmuseums 40, 4, 1079–1104.

Mania, D., 1983. Zum Stand der Untersuchung weiterer Fossilgruppen aus dem Travertin von Bilzingsleben. In: Mai, D.H., Mania, D., Nötzold, T., Toepfer, V., Vlček, E., Heinrich, W.-D., Bilzingsleben II. Homo erectus – seine Kultur und seine Umwelt. Veröffentlichungen des Landesmuseums für Vorgeschichte in Halle 36, 157–162.

Maul, L.C., Heinrich, W.-D., Parfitt, S.A., Paunescu, A.-C., 2007. Comment on the correlation between magnetostratigraphy and the evolution of *Microtus* (Arvicolidae, Rodentia, Mammalia) during the Early and early Middle Pleistocene. In Kahlke, R.-D., Maul, L.C., Mazza, P.P.A. (Eds.), Late Neogene and Quaternary biodiversity and evolution: Regional developments and interregional correlations. Proceedings of the 18th International Senckenberg Conference (VI International Palaeontological Colloquium in Weimar). Volume II. Courier Forschungsinstitut Senckenberg 259, 243–263.

Reumer, J.W.F., Mol, D., Kahlke, R.-D., 2018. First finds of Pleistocene *Macaca sylvanus* (Cercopithecidae, Primates) from the North Sea. Revue de Paléobiologie 37, 555–560.

Rook, L., Mottura, A., Gentili, S, 2001. Fossil *Macaca* remains from RDB quarry (Villafranca d'Asti, Italy): new data and overview. Journal of Human Evolution 40, 187–202.

Szalay, F.A., Delson, E., 1979. Evolutionary history of the primates. New York.

Sardella, R., Bona, F., Iurino, D.A., Rook, L., Bellucci, L., 2015. The Middle and Late Pleistocene *Macaca sylvanus* Fossil Record from Italy. Folia Primatologica 86, 351.

Thenius, E., 1965. Ein Primaten-Rest aus dem Altpleistozän von Voigtstedt in Thüringen. Paläontologische Abhandlungen, Abteilung A Paläozoologie 2. Berlin, 681–686.

Zapfe, H., 2001. Zähne von *Macaca* aus dem Unterpleistozän von Untermaßfeld. In: Kahlke, R.-D. (Ed.), Das Pleistozän von Untermaßfeld bei Meiningen (Thüringen). Teil 3. Monographien des Römisch-Germanischen Zentralmuseums 40, 3, 889–895, pl. 146.

RALF-DIETRICH KAHLKE

THE PLEISTOCENE OF UNTERMASSFELD – A SYNOPSIS OF SITE ORIGIN, PALAEOBIODIVERSITY, TAPHONOMIC CHARACTERISTICS, PALAEOENVIRONMENT, CHRONOSTRATIGRAPHY, AND SIGNIFICANCE IN WESTERN PALAEARCTIC FAUNAL HISTORY

Abstract

Since 1979, the Institute of Quaternary Palaeontology Weimar and its successor institutions have been carrying out field work near Untermassfeld (Southern Thuringia, central Germany) to document and recover an Early Pleistocene fossil vertebrate assemblage that is unique in its completeness and preservation quality. A total of 37 annual field seasons, adding up to 127 months of active excavations, were carried out to dig an area of 667 connected grid squares (1 m² each) up to a maximum depth of 5.5 m below ground level, paralleled by detailed recording of the geological sections. For future research, a highly promising fossil-bearing reserve area was left untouched. Altogether, a total of almost 175 t of sediment was processed from Untermassfeld for the extraction of small fossils. During field work and associated laboratory activities, samples were taken for 23 analytical techniques. Preparations as well as drawings and photographic documentation of the Untermassfeld finds was carried out, without interruption, for 42 years. For the Untermassfeld vertebrate finds a new conservation method was developed. The entire excavation material is stored at the Senckenberg Research Station of Quaternary Palaeontology in Weimar as a separate Untermassfeld collection comprising 18,250 catalogued specimens or series. A total of 49 project topics were assigned to Untermassfeld, which have led to significant results.
The origin and history of the fossil site of Untermassfeld are due to specific interactions between palaeoclimatological, palaeobiological, and geomorphological as well as sedimentological, geochemical, and neotec-

tonic processes. In the wake of repeated catastrophic flood events, in a zone of reduced transport energy on the lee side of a clastic mudflow fan (CMF), large amounts of necromass accumulated, composed of parts of larger mammal carcasses, more or less disarticulated micromammal skeletal units, single bones, excrement, invertebrate remains, and vegetational residues. This fossil accumulation was preserved until the present day because its erosion was prevented by neotectonic subsidence movements during the Pleistocene valley deepening.

The fossil site of Untermassfeld yielded a vertebrate fauna that is exceedingly rich in species and individuals. In contrast, evidence for invertebrates is comparatively scarce. A complete oxidation of the fossil-bearing sediments hindered or prevented the preservation of micro- and macro-botanical records. According to current knowledge, the range of the fossil fauna from Untermassfeld comprises remains and/or marks of 115 taxa of gastropods, insects, fishes, amphibians, reptiles, birds, and mammals. The latter are represented by bovids, cervids, hippopotamids, suids, rhinocerotids, equids, elephantids, felids, mustelids, ursids, canids, sciurids, glirids, castorids, dipodids, cricetids, murids, hystricids, leporids, cercopithecids, erinaceids, talpids, and soricids. The individual animal groups are described and documented in parts 1 to 5 of this monograph, as well as being taxonomically classified and discussed in terms of their palaeoecological and biostratigraphical significance. Pathological findings on skeletal and dental elements of several large mammal species have enabled conclusions to be drawn about the life history of the individuals concerned.

Different geological and hydrological processes have led to the formation of the diverse depositional environments which make up the area of the present-day site. While the Lower Fluviatile Sands (LFS) deposited in the floodplain areas contain only moderately high numbers of finds, the channel filling with the lower part of the Upper Fluviatile Sands [UFS(l)] has in some areas extremely dense bone accumulations, especially in its deeper section. Two areas with different bone distribution patterns can be distinguished northwest of the mudflow fan, namely a turbulence zone immediately leeward of the fan, and an aligment zone corresponding to the direction of flow. Analyses of the smaller-scale distribution of finds led to the differentiation of four essential distribution patterns of large mammal remains, namely anatomically connected skeletal units, isolated finds, polyspecific bone concentrations, and bone accumulations. About 90 % of the identifiable large mammal material was found in an anatomically isolated state. Within the LFS, and also in the channel infill, isolated skeletal elements were concentrated in dense piles containing remains of different species and individuals. Obstruction bones acted as mechanical barriers and thus caused the agglomeration of further necromass transported by the riverine water flow. Extended bone accumulations resulted from the amalgamation of several polyspecific bone concentrations.

The majority of vertebrate remains from the site show hardly any signs of weathering or transportation. Only in rare cases were these kinds of modifications identified, such as closely spaced cracking caused by strong insolation, abrasion due to rolling, and etching traces of a delicately-veined root system. As insect marks, pupation chambers of an ideterminate type of dermestid could be detected on some bone and antler surfaces. *Pachycrocuta brevirostris* left numerous bite and feeding marks on the find material. Several skeletal elements of different large mammal species exhibit distinct gnaw marks of rodents, and one single find shows the results of herbivore osteophagy. The remains of *Salmo* cf. *salar* often show digestive corrosion. Increased carbonate and reduced phosphate contents of the bones from Untermassfeld indicate diagenetic remineralisation processes. In the deepest part of the channel fill, skeletal elements were demineralised. The complete fossil-bearing stratigraphic sequence of Untermassfeld was overprinted by a neotectonic stress regime.

Minimum numbers of individuals (MNI), each subdivided into juvenile/subadult and adult individual ages, have been determined for all vertebrate species recorded from Untermassfeld. Thereby, a total of 443 individuals of larger mammals were documented. Among the herbivores, *Eucladoceros giulii*, *Pseudodama vallonnetensis*, and *Bison menneri* are the most numerous. The most common carnivores according to their MNI are *Pachycrocuta brevirostris*, *Canis mosbachensis*, and *Ursus* ex gr. *U. dolinensis*. The total MNI of fishes, amphibians, reptiles, birds and small mammals from Untermassfeld amounts to 298.

In addition to the depositional regime of the site and the weight and transportability of the corresponding carcasses or parts thereof, the range, number of individuals, and mortality patterns of the recorded vertebrate species are determined by their population densities, behavioural patterns and locomotory skills, especially swimming, climbing, and digging capabilities, as well as by their specific biotope preferences. Taphonomic features of the fossil contents of the LFS point to a predominantly calm deposition of the sedimentary material and corresponding necromass. The deepest part of the channel initially received extraordinarily high quantities of cadaver and skeletal parts from the river water flowing off immediately after the peak of a flood. The majority of individually connected skeletal elements of several microvertebrate species are likely to have come from eroded burrows from which hibernating animals were washed away. The deposition of cervid remains can be restricted to periods of winter half-years. A similar assessment results from the occurrence of the anadromous Atlantic salmon and from the insectogenic modifications of skeletal elements.

The habitat types assigned to the animal species detected combined with the geological and geomorphological conditions of the region enabled as detailed a picture of the landscape as possible at the time of the site's formation to be derived. The floodplain valley of the Early Pleistocene Werra was traversed by a dynamic watercourse all year round. Wettish sections of the floodplain would have been partly marshy, whilst permanently waterlogged areas would have been mainly covered with softwood, so that thickets were probably present. In drier locations, tall mixed forests with less undergrowth would have spread with lush, herbaceous vegetation thriving on the often loose soils. The limestone slopes close to the site were mainly comprised of open areas alongside habitats with species-poor tree and shrub vegetation. The plateaus flanking the Werra valley have loamy-clay and sandy-loam soils that would have given rise to park-like landscapes and areas with few trees.

While at least two flood events with accumulations of vertebrate remains are verifiable for the LFS, the subsequent incising, and refilling of the erosion channel occurred in the course of only one particularly destructive event. In the river valleys of the Central European low mountain regions, extreme high water levels occur during the winter months when temperatures above freezing lead to a combination of snowmelt in high altitudes with persistent rainfall in lower regions. Such a constellation is likely to have been the cause of the flood events leading to the introduction of necromass at Untermassfeld. The maximum flooded area of the middle and upper Werra, including its tributaries upstream of the site, and thus the potential area of origin of the necromass covered c. 94 km². An S-shaped narrowing of the valley 12 km upstream of the Untermassfeld fossil deposits had a particular influence on the course of the floods in the Werra valley. Regionally, a warm climate characterised by Atlantic-Mediterranean influences prevailed with significantly higher temperatures than are observed today in the region of Southern Thuringia. Compared to current conditions, the lower amounts of precipitation were concentrated in the winter half-year.

The chronostratigraphical classification of the formation of the Untermassfeld fossil site and its fossil content is based on a combination of litho- and magnetostratigraphy, U/Pb dating of redeposited carbonate, and the biostratigraphic data of all relevant taxa of the recorded fauna. In all likelihood, the Untermassfeld fossil deposit was formed at the beginning of the Early Pleistocene Jaramillo polarity subchron. Given the position of the palaeomagnetic reversal within the Untermassfeld Pleistocene sequence as well as the reconstructed process sequence, the fossil assemblage can be assigned to MIS 31, which corresponds to a pronounced warm interval that overlaps the onset of the Jaramillo. In light of the recorded data, the Untermassfeld fauna lived during the early Bavelian and can be dated to about 1.07–1.06 Ma BP.

The Untermassfeld fossil assemblage contains one of the most complete vertebrate find complexes in the western Palaearctic from the period between 1.2 and 0.9 Ma BP. In particular, the recorded large mammal taxa support the idea of inserting a separate biochron, the Epivillafranchian, between the Villafranchian and Galerian biostratigraphic units defined for this area. The Epivillafranchian appears as a climatically unstable transitional period between the preceding 41 ka and the following 100 ka periodicity.

1. Introduction and framework data

The first cut into the Pleistocene sediments in the area known today as the fossil deposits of Untermassfeld took place around 1900; but it did not yield any fossil vertebrate remains. It was not until January 1978, during local sand mining, 2 km south of the town of Meiningen on the eastern slope of the Werra (**Fig. 1**), that teeth and skeletal elements of Pleistocene large mammals were discovered and promptly handed over to the then Institute of Quaternary Palaeontology in Weimar (R.-D. Kahlke 1997a, 6 ff.). Beginning the following year, documented rescue excavations were carried out due to the sand mining operations that were still sporadically taking place at the site. Given the unusual quantity of fossils found in exceptionally good conditions, planned excavations were then carried out at Untermassfeld (50.542657 N, 10.419383 E) from 1983 onwards. Owing to several serious structural changes at the aforementioned Weimar research institution, new concepts for the financing of field work and the preparation, documentation, and evaluation of the finds needed to be repeatedly introduced from 1990 onwards.

Legal regulations applying to recovered finds and fossil material still in the ground fall under the Thuringian Monument Protection Act (Thüringer Denkmalschutzgesetz, ThürDSchG). Corresponding finds are the property of the federal state of Thuringia regardless of the time of their discovery and the way they were recovered (see also Keiler and Wenzel 2019). On 05 July 1994, the site of Untermassfeld was officially entered into the Thuringian register of monuments. Based on current geological and taphonomic research results from the site, the Thuringian State Office for Heritage and Archaeology has placed further areas under protection with effect from 19 August 2021 (Protection number, SM 190). This means that a total of c. 9,300 m² around the fossil deposit at Untermassfeld is now registered as a monument.

Fieldwork near Untermassfeld, which began on 18 June 1979 (Kahlke 1997a), was carried out by members of the Weimar Quaternary Palaeontology division with the occasional participation of interns from various training institutes, students, and guest researchers. Three to four excavators typically worked on the fossil deposit at the same time. Until the end of the 1980s, the annual field seasons lasted over five months. Due to the often immense quantities of finds from active excavation sections and increasingly complex conservation techniques (see section 3.), the field seasons were successively reduced in the following years to 10 weeks per year, including one working week each for opening and covering the fossil-bearing areas (for details see R.-D. Kahlke 1997a; 2001a; 2016; 2020; Kahlke and Keiler 2017; 2018; 2019). A total of 37 annual excavation seasons, adding up to 127 months of active field work were carried out at Untermassfeld. In the process, an area of 667 grid squares (Q, 1 m² each) was dug (status up to and including 2015 in R.-D. Kahlke 2020, Fig. 8: Excavated grid square incompletely recorded by mistake, as well as with reversed colour symbols; for the configuration of the complete excavation area see **Fig. 13** of this paper) and in most cases completely excavated (see section 2.1.). In the southern area of the site, depths of up to 5.50 m below ground level were reached (Kahlke et al. 2020, Fig. 11). In some areas, subsections of fossil-free squares (17) were cut to prepare geological sections. Final excavation work was carried out during the last field season which took place from 06 May to 03 July 2019 (Kahlke and Keiler 2019).

Fig. 1 View from the south looking out over the bend of the Werra south of the city of Meiningen (background) and the Untermassfeld research excavation (arrowhead) on the eastern slope of the Werra. Photo: O. Braasch, 01 July 1995; published with kind permission of the Thuringian State Office for Heritage and Archaeology (Archive inventory number 5528/018-1).

Preparations as well as drawings and photographic documentation of the Untermassfeld finds was carried out, without interruption, for over 42 years (see section 2.2.). As a rule, conservation of all new finds from the respective preceding excavation campaign could be carried out during excavation-free periods. Exceptions to this were large in-situ block preparations, whose time-consuming and material-intensive preparation usually took several years (e.g., R.-D. Kahlke 2020, 1049 ff.). Preparation of the excavation plans (Kahlke 2001e, foldouts I–XV; **Foldouts** II and III of this paper) took place parallel to the cataloguing of the finds throughout the year. The entire excavation material is stored at the Senckenberg Research Station of Quaternary Palaeontology in Weimar as a separate Untermassfeld collection and housed in specially manufactured cabinets in climate-controlled storerooms (Kahlke 2020, 1051; see section 3. of this paper).
The cataloguing of the Untermassfeld finds was completed on 01 July 2020. A total of 18,250 specimens or series have been recorded (in brackets, number of individual items included). Among them are 37 geological items, 116 specimens or series of molluscs (c. 560 individuals), and a small series of ostracods consisting of 3 specimens. In addition to 25 individually catalogued, hitherto indeterminable microvertebrate-remains, the following have all been documented: 81 (94 specimens) fish, 88 (158) amphibians, 30 (33) reptiles, 43 (103) birds, and 7 (70) as yet unspecified eggshell fragments (counting of microvertebrates by L. C. Maul). The main part of the Untermassfeld collection consists of 14,291 catalogued large mammal finds. This number does not include indeterminable bone fragments that are stored according to the respective grid squares of their discovery. 3,531 further catalogue numbers refer to small mammal remains, representing a total of about 7,040 individual specimens. For large mammal finds, 48 in-situ preparations were made (see section 3.) or are in progress, and a further 99 were made for microvertebrate finds (see sections 3., 6.1.2.5.).

Due to the rich diversity of species and large quantity of fossil material, as well as the geological and taphonomic features documented in detail during the long-term excavations, the analysis of Untermassfeld was entrusted to an interdisciplinary working group. With a large team of scientists that has changed over the years, the research results on the Pleistocene of Untermassfeld forms part of the Weimar Quaternary Palaeontology division's series of monographic treatises on transregionally important Pleistocene fossil sites (Bibliography: Fiedler 2012; Collections: Kahlke et al. 2014). A total of 49 project topics (R.-D. Kahlke 2020, Table 9) were assigned to Untermassfeld, which have led to significant results and were each concluded with corresponding publications. Bibliographies can be found in the accounts of the research history from 1978–1989 (R.-D. Kahlke 1997a), 1990–1996 (R.-D. Kahlke 2001a), and 1997–2015 (R.-D. Kahlke 2020). The results of the individual projects mentioned above form the basis for the present summary of current knowledge on the site and its fossil material, as well as the conclusions that can be drawn from it concerning the faunal history of the western Palaearctic. This contribution follows on from – and expands on – earlier analyses (Kahlke 1997c; 2001d; 2006).

2. Excavation and documentation

2.1. Excavation strategy and limiting factors

The Pleistocene sediment sequence in the area of Untermassfeld on the eastern slope of the Werra begins at 359 m above sea level, i.e., at a relative height of 69 m above the present floodplain of the Werra river (Ellenberg and Kahlke 1997, 37; R.-D. Kahlke 2001b, 433) (**Fig. 1**). The fossil material, often found in high concentrations, was deposited in a sequence of up to 8 m thick fluviatile sands (Kahlke et al. 2020 and references therein). While the Lower Fluviatile Sands (LFS) were deposited as high flood sediments, the lower part of the Upper Fluviatile Sands [UFS(l)], which is particularly rich in fossils, represents the infill of an erosion channel (see section 4.; frontispiece of this volume).
After emergency rescue excavations had been carried out at the site during the summer of 1978 (Kahlke 1997a, 8), documented rescue excavations began the following year, during which the grid network and corresponding grid squares, which were used and expanded over the following four decades, were established. From then on, areas with finds were excavated in horizontal sectional planes, mostly at intervals of 0.25–0.40 m (Kahlke 1997a, Fig. 12; 2001a, Figs. 1–3; 2001b, Figs. 1, 5; 2020, frontispiece, Figs. 2b, 4a, 6a, 15a). Occasional deviations from this pattern were due to the size, shape, and deposition of the fossil material. From 1983 onwards, an excavation-spanning site 0-level was defined, 1.00 m above the main planum (= Planum I), and about 2.00 m below the slightly sloping ground level. With the definition of the site 0-level, the three-dimensional recording of the individual finds became possible, regardless of their depth below the ground surface (see below).
The discovery and sampling of the first vertebrate finds from Untermassfeld took place near the boundary between LFS and UFS(l) (Kahlke 1997a, Fig. 5), identifiable in the overall excavation plan by single-digit and low two-digit grid square numbers (Kahlke 2001b, Fig. 8; **Foldout II**). Until 1984, the excavation progressed in the eastern direction up to the edge of the LFS (Ellenberg and Kahlke 1997, Figs. 5, 9, 20, Pl. 3/2; Kahlke 1997a, Fig. 6), and up to the interlocking of the LFS and UFS(l) with a coarse clastic mudflow fan (CMF) in the south (Ellenberg and Kahlke 1997, Figs. 5, 10, 18, Pls. 3/2, 8/2; Kahlke et al. 2020, Foldout I). The corresponding areas were each dug down to the level of Planum I (i.e., 3.00 m below ground level). In 1984, an exploration section measuring 8.00 m × 16.00 m was cut, as Planum II, into the sediments at the south-eastern edge of the site 1.00 m above Planum I, i.e., at site 0-level, in order to check the CMF for

possible fossils (Ellenberg and Kahlke 1997, Figs. 5, 20, Pl. 16/2; Kahlke et al. 2020, Foldout I). No significant fossil material was found. In order to clarify the geological situation, an exploration trench was also dug up to 4.50 m below Planum I, i.e., up to 5.50 m below site 0-level, which intersected the floodplain deposits (FPD) and reached the underlying Younger Coarse Gravels (YCG) (Ellenberg and Kahlke 1997, Pl. 3/1; Kahlke et al. 2020, Foldout I).

In a westerly and north-westerly direction, i.e., towards the valley, the fossil-bearing UFS(l) were also initially dug to the level of Planum I during the years 1980 to 1986 (Ellenberg and Kahlke 1997, Figs. 5, 11, 13–16, 20, Pls. 7/1–2, 8/1, 9/1, 10/1–2; Kahlke 1997a, Figs. 9, 16; Kahlke et al. 2020, Foldout I). This was preceded by drilling to determine the thickness of the fossil-bearing sands (Ellenberg and Kahlke 1997, 42; Kahlke et al. 2020, Foldout I). A limiting factor in the completion of the work in this area was the transition to the fossil-free UFS(u) above the silt band (SB), with a SW dip of up to 20° (Ellenberg and Kahlke 1997, 54).

In view of the expected high density of finds within the fluviatile trench mainly filled with UFS(l) (see section 4.), the north-western area of Planum I was lowered by 1.00–1.50 m, as Planum III, over an area of 26.5 m² during the years 1989–1992, in accordance with the general SW dip of the Pleistocene layer sequence (2.50–2.00 m below site 0-level, i.e., 4.50–4.00 m below ground level) (R.-D. Kahlke 2001a, Figs. 1–3; 2001b, Figs. 5–6). In the process, the floodplain deposits (FPD) under the fluviatile sands as well as the reworked fossil-bearing zone of the Younger Corse Gravels (YCG) were cut (see section 4.).

After completion of the excavations in the north-western section of the site, work continued northwards from 1993 to 1996 above Planum I. In addition to remaining parts of the UFS(l), mainly fossil-bearing LFS were excavated at this time (Kahlke 1997a, Fig. 10; 2001a, Figs. 5–6; 2001b, Figs. 1–4; 2020, Fig. 9; Kahlke et al. 2020, Foldout I). By 1995, a total of 584 connected grid squares had been opened at Untermassfeld; with the 1996 field season that area was increased to 612 squares. In the process, further digging was carried out beyond the northern boundary of the fossil deposit in order to rule out any loss of finds. In 2000, a field season without recovery of any large mammal finds was conceded (Kahlke 2020, Table 4) in order to ensure the complete examination of what were ultimately fossil-free grid squares lying immediately north of an extensive bone concentration in the north-western section of the excavation area above Planum I (see section 6.1.1.1., **Foldout II**). The absence of fossil vertebrate remains in areas up to 100 m north of the excavation site can be confirmed by the Untermassfeld sand quarry, which was operated for several decades, as well as by the digging of exploration trenches up to 4–5 m deep (Ellenberg and Kahlke 1997, 42, Figs. 5, 8).

In 2001, excavation activities were again moved to the southern part of the site, this time below Planum I. Based on the experience in the NW area of the site (see above), it was assumed that the channel filling northwest of the front of the coarse clastic mudflow fan (CMF; Kahlke et al. 2020, Foldout I) also contained high concentrations of finds (see book cover). In total, 61 grid squares (Kahlke 2020, Fig. 1) were excavated up to a maximum depth of 3.50 m below site 0-level, i.e., 2.50 m below Planum I (about 5.50 m below ground level) until completion of the field work at Untermassfeld. Between grid squares Q 234 and 154 in the southeast and the connecting square Q 574 with finds from the 1987–1990 field seasons in the northwest, the basal channel filling was documented and its rich fossil content, consisting of 4,449 individual finds, was completely recovered (Kahlke 2016; 2020, frontispiece, Figs. 1–7, 12a, 15; Kahlke and Keiler 2017; 2018; 2019; 2020; Kahlke et al. 2020, Figs. 1–8, 10–14; **Foldout II** of this paper). For future research, a highly promising fossil-bearing reserve area was left untouched between the west-facing front of the CMF and the south-western boundary of the channel fill (Kahlke 2020, 1039 f.). For potential finds, SW dipping of the corresponding layer sequence results in depths of, predominantly, more than 5.00 m below ground level.

The dimensions and spatial configuration of the sedimentary bodies at Untermassfeld, investigated in 667 connected grid squares, were mainly determined by geological factors. The lower boundary of the fossil deposit is defined by the accumulation of LFS and the depth of the refilled erosion channel (see section 4.2., phases II and III) and the upper boundary by the loss of the accumulating effect of the CMF due to successive sediment cover (see section 4.2., phase IV). Upstream, to the south, the northern boundary of the CMF formed the boundary of the fossil deposit. On the slope, i.e., to the east, the river bank limited the accumulation of finds. On the valley side, in a westerly direction, the fossil accumulation was shaped by the influence of the CMF on the flow regime of the Werra at the time the deposit was formed. In addition, in a northerly direction, i.e., downstream, the range of the accumulating effect of the CMF determined the fossil deposit both in the horizontally deposited LFS and in the erosion channel, which was mainly filled with UFS(I).

2.2. Field documentation and cataloguing

In order to document the fossil site of Untermassfeld, detailed recording of the geological sections was carried out in parallel alongside the respective excavations from 1983 onwards (Ellenberg and Kahlke 1997; R.-D. Kahlke 2001b; Kahlke et al. 2020). This served as a basis for both diagrammatic section of the excavated areas and a perspective block diagram of the exposed stratigraphic sequence (Kahlke et al. 2020). From this, the process sequence of the formation and preservation of the fossil deposit was derived (see section 4.2.).

From 1983, three-dimensional measurements were obtained for all identifiable large mammal remains and microvertebrate finds that could be represented on a scale of 1:10. The necessary transfer of reference points for the grid squares and site 0-level was carried out during the 1980s with a BRT 006 opto-mechanical theodolite (Carl Zeiss, Jena). Until 1996, the find heights relative to the site 0-level were measured with the Ni 060 levelling device (Carl Zeiss Jena) and from the year 2000 a Spectra Physics Laser Plane L-220 was used. The further transfer of reference points for the site 0-level and grid squares, which was required due to the progress of the excavation, was carried out in 2011 with the TCR407 Total Station (Leica) using the corresponding TachyCAD software.

From 1979 onwards, all finds were documented in field drawings. The documentation of the field work was supplemented by a weekly excavation diary, kept from 1989 onwards. From 1992, copies of the finds' labels with the respective entries on date, grid square, find depth, find circumstances, existing photographs, object number, and number of the corresponding field drawing were added. Later, information on the primary state of preservation of the fossils at the time of their discovery as well as on the type of sediment they were embedded in was also included. After completion of preparation and conservation, final drawings were created for each of the field seasons, up to and including the year 2004, which also detailed the complete contours of finds recovered in sediment blocks. These sets of drawings formed the basis for the manual preparation of, initially, 15 detailed excavation plans that were published as combinable foldouts (Kahlke 2001e, Fig. 9, foldout I–XV). In these plans, find depth intervals are indicated by various symbols. In 2008, these individual plans were digitised and merged.

Since 2007, the complete excavation plan of Untermassfeld has been digitally created using AutoCAD. Prepared and conserved finds were recorded with a flatbed scanner or, in the case of large-sized objects, photographed. Based on these recordings the complete contour of each fossil was drawn individually on a scale of 1:10. The finds were then added to the AutoCAD plan using the 3D coordinates determined in the field. A corresponding object thickness (extrusion height) was assigned to each recorded fossil, so that a spatial image of the distribution of the finds is also available. In two-dimensional representations (**Fold-**

outs II and **III**), the overlapping of skeletal elements in zones of high find concentration means that for some of the items only a partial contour can be identified. In the digital excavation plan, each individual piece can be selectively accessed and any hidden parts exposed.
During the excavation work at Untermassfeld, the deepest finds in the southern part of the site were recovered from grid square Q 41 (**Foldouts II** and **III**), 3.50 m below site 0-level. These were postcranial micromammal remains [IQW 2016/44 152 (Mei. 49 461)] recovered from basal sandy loam (BSL) in the deepest part of the channel, near the front of the CMF (Kahlke 2020, Fig. 10; Kahlke et al. 2020, 1092 f., Figs. 8, 11). The vertebrate remains lying closest to the surface were found in the northern part of the site outside of the erosion channel. In the LFS of grid squares Q 716 and 717, two individually associated left antler fragments [IQW 1995/25 122 (Mei. 24 651), IQW 1995/25 119 (Mei. 24 648)] were found within a polyspecific bone concentration (see section 6.1.1.1., **Foldout II**) at depths of 1.30 m above site 0-level. Combined with five other, slightly deeper, embedded pieces [IQW 1995/25 108 (Mei. 24 637), IQW 1995/25 113 (Mei. 24 642), IQW 1995/25 117 (Mei. 24 646), IQW 1995/25 118 (Mei. 24 647), IQW 1995/25 121 (Mei. 24 650)] they form the basal section of an antler of *Pseudodama vallonnetensis*. The vertebrate remains documented at Untermassfeld were found above and below the site 0-level at a total height interval of 4.80 m. This range reflects the depth of the erosion channel filled with fossil-bearing sediment as well as the SW dipping of the site's Pleistocene stratigraphic sequence (Kahlke et al. 2020, Foldout I).
Around 88 % of the large mammal remains from Untermassfeld stored at the Weimar Quaternary Palaeontology division have their associated context documented. Areas seeming to be fossil-free in the central part of the excavation site (**Foldout II**), roughly between grid squares Q 462, 113, 489 and 525, represent an area where find losses occurred before the deposit was reported in February 1978, as well as undocumented areas from the time of both emergency and initial rescue excavations (see section 2.1.). About 1,700 catalogued specimens from these areas are available, most of them in very good condition.
The catalogue of the Untermassfeld collection (see section 1.), which was compiled as a database for future research, was kept in handwritten form until the end of 2006 and then digitised. From 2011 onwards, the information was transferred to the FileMaker database management system. The recorded data usually include taxon, skeletal element, grid square, find depth, field object number, field drawing number, find date, remarks on the discovery as well as the method of preparation/new preparation of the individual finds. In addition, the Senckenberg Research Station of Quaternary Palaeontology Weimar stores the excavation diaries, all field drawings of the geological sections, field and final drawings of the excavated squares as well as the hand drawn final excavation plans. Original graphics are also housed at Weimar, especially those relating to osteological finds from previous publications.
What remains undocumented is a not inconsiderable number of osteological finds that were broken and stolen from the active excavation areas over a number of years by trespassers (R.-D. Kahlke 2001a; Beales 2014; Baar 2017a; 2017b; Callaway 2017; Kleinhubbert 2017; Roebroeks et al. 2018; Keiler and Wenzel 2019; Gaudzinski-Windheuser 2020).

2.3. Screen washing and sampling

From 1980 onwards, and increasingly so from 1983, fossil-bearing sediment from Untermassfeld was washed to extract molluscs, ostracods and small vertebrate remains. Whenever possible, this material was taken close to large mammal remains, as skeletal elements of smaller animals were often found there in relatively large numbers. By 1989 an amount totalling around 120 t had been processed down to a smallest sieve mesh size of 0.71 mm (Kahlke 1997a, 11). During the 1990 to 1996 excavation years, another 50 t of

Analysis / research object	Number of samples	Sampling data and/or results	
Granulometric analysis of fluviatile sands	25	Ellenberg and Kahlke 1997	
Granulometric and geochemical analyses of gravels, sands, clays, and carbonates	38	Kahlke 2020; Kahlke et al. 2020	
Geochemical analysis of re-deposited pisolithic ore	2	Lange in Ellenberg and Kahlke 1997	
Thin sections for lithological analyses of travertine, dolomite, clay pebble and mudflow matrix	4	Kahlke et al. 2020	
Sediment transfer preparations	2	Kahlke 2001f; 2020	
Palaeomagnetic analysis of the Pleistocene succession	85	Wiegank 1997	
Thermoluminescence dating of fluviatile sands	1	Ellenberg and Kahlke 1997	**
Combined ESR/U-series dating of cervid teeth	6	Kahlke 2020	**
Amino acid racemisation dating of gastropod slug plates	2	Kahlke 2020	*
U-Pb small-scale isochrone dating of re-deposited travertine	57 spots	Gerdes et al. 2020	
U-Pb dating of detrital zircon and provenance determination	15	Kahlke 2020	*
Palynological record of fluviatile sands and coprolites	16	Kahlke 2020	**
Phytoliths	10	Kahlke 2020	**
Impressions of macrobotanical remains in travertine and coprolites	20	Kahlke 2020	**
Molluscs (parallel to micromammal sampling)	160	Krolopp 1997	
Chemical analysis of fossil bone	2	Bleck in Keiler 1995	
$\delta^{18}O$ analysis for palaeotemperature determination on equid remains	7	Stephan et al. 2001	
Hibernation marks in *Spermophilus* incisors	2	Trunova 2001	**
Tooth root cementum analysis on cervid remains for determination of season of death	4	Kahlke 2001e	**
Tooth root cementum analysis on ursid molar for determination of season of death	1	Kahlke 2020	**
Stable isotope analysis of large herbivore and carnivore tooth enamel	51	Kahlke 2020	**
Matrix-Assisted Laser Desorption Ionisation Mass Spectrometry for ancient osteocalcin analysis on bovid remains	1	Nielsen-Marsh et al. 2003; Kahlke 2020	**
Ancient DNA analysis on cervid remains	2	Kahlke 2020	**

Table 1 Samples taken at the excavation site and on material recovered from there. – * Study in progress. – ** Available material did not lead to results.

additional sediment was washed (Maul 2001, 786). Between 1997–1999 no excavations took place (Kahlke 2020), however, the washing and sieving resumed in parallel with the continuation of the excavations in the southern part of the site below the main planum (see section 2.1.). In addition to the samples taken in the field (Kahlke 2020, Fig. 10), residues resulting from the removal of finds from sediment blocks were also washed and processed. Thus, by 2013 another 4.3 t of sediment had been analysed and sorted down to a smallest sieve mesh size of 0.8 mm. Altogether, a total of almost 175 t of sediment was processed from Untermassfeld for the extraction of small fossils.

During field work and associated laboratory activities, samples were taken for a wide range of different analytical techniques (**Table 1**). The work did not always lead to usable results, and in several cases the samples have not yet been evaluated.

During the sampling, anatomically fully preserved dentition and skeletal elements as well as other particularly significant finds were strictly avoided. In justified cases, both fragmentary fossil and sediment material is available in sufficient quantities for future sampling. Chemically untreated (see section 3.) bone material has also been stored for this purpose.

3. Preparation and conservation of vertebrate remains

In order to protect the sensitive bone surfaces as much as possible, fossil material was, for the most part, only roughly exposed at the excavation site. Palette knives, spatulas with rounded edges and brushes were typically used for this purpose. Individual finds and fossil assemblages at risk of fracture were secured with partly metal-reinforced plaster jackets before being removed as a block (Kahlke 2020, Figs. 2a, 4, 5, 12a). Cranes were often required for this purpose. Final extraction of the finds was carried out in the laboratories of the Senckenberg Research Station of Quaternary Palaeontology, mostly by manual-mechanical means (Kahlke 2020, Fig. 12) using dental instruments (Keiler 1995; 2009, 180–182). Devices with ultrasonic handpieces were used to remove carbonate-cemented sandy matrix from bone surfaces (Keiler 2001a, 456f.) and gentle cleaning of the fossil surfaces was carried out with water or acetone.
During the years 1978 to 1991, surface impregnation was carried out using varnish, so called Geiseltallack, to consolidate the dried vertebrate material. This process was developed for palaeontological finds in the late 1920s (Brüning 1962). Depending on the manufacturer, the varnish mentioned above was produced according to different recipes. These were always mixtures of cellulose nitrates (CN) and plasticisers dissolved in various organic solvents. Since the varnish reacts to inherent moisture with a strong white bloom formation, it was not possible to conserve the finds in the field. The fragility of the varnish over time also proved to be a particular problem (Keiler 1995, 13–14; 2001a, 451–453), as it became brittle over a period of 10 to 30 years following application, so that correspondingly treated preparations increasingly lost stability.
Due to the disadvantages mentioned above, a new conservation method was developed for the Untermassfeld specimens, which has been in use since 1992. In this method, the osteological material is impregnated with durable and plasticiser-free polyvinyl butyral (PVB) for consolidation (Keiler 2001c). PVB dissolved in alcohol or acetone (Mowital B 30 H from Hoechst AG) retains a residual elasticity after curing which is beneficial for the preservation of the fossils. It is also waterproof and is resistant to ageing, as demonstrated by industrial tests (pers. comm. J.-A. Keiler, July 2000). The PVB impregnation is a reversible stabilisation method. If necessary, cured polyvinyl butyral can be completely removed from the fossils by solvent washing. In addition, due to its good moisture compatibility, PVB dissolved in acetone can be used without complications for stabilising finds in the field.
After consolidation, fragments of shattered finds that belonged together were glued and deformations caused by diagenetic or tectonic factors (see section 6.2.5.) were corrected as much as possible in order to achieve anatomically correct final preparations that adhered to palaeontological requirements. Duosan rapid, a cellulose nitrate-based adhesive, was used until 1989. Unfortunately, however, this adhesive became brittle in the same way as the varnish (see above), so that treated finds lost their stability and eventually broke apart in places (Keiler 1991; 2017, 81). Since 1990, only adhesives based on cyanoacrylates, epoxy resin or polyvinyl butyral at different viscosities have been used, all of which have better resistance to ageing than cellulose nitrate adhesives (Keiler 1995, 17–18; 2001a, 454–456).
For the preparation and conservation of larger finds with anatomical articulation or complicated attachment of individual elements, as well as for more or less completely preserved skulls, the so-called in-situ preparation technique has proven to be successful (frontispiece of this volume). In this method, the original

position of the fossils in the embedding sediment is preserved (Keiler 1995, Pls. 12–15; 2001a, Pls. 68, 70, 71; Kahlke 2000, Fig. 1; 2001d, Figs. 1, 4, 6, 13; 2006, Figs. 4, 18, 31, 35, 45, 46a; 2007, Figs. 3, 6; 2009, Figs. 7, 9; 2020, Figs. 5, 12; Breda et al. 2020, Figs. 11, 15; Bukhsianidze 2020, Figs. 2, 8; Kahlke et al. 2020, Fig. 4; Hemmer and Kahlke in this volume, Fig. 9; Van Essen in this volume, Fig. 1). Moreover, even the smallest fragments of deformed and/or severely shattered skeletal elements remain in their discovery position (Keiler 1995, 19–20; 2001a, Pl. 69/2; 2008). The occasionally high sediment content in these specimens is no obstacle to computer tomographic investigations (Keiler 2009), as shown by corresponding images of sand-filled skulls of *Bison menneri* (Hufeland-Klinikum Weimar) and *Ursus* ex gr. *U. dolinensis* (Max Planck Institute for Evolutionary Anthropology Leipzig).

Since 2002, Untermassfeld has yielded skeletons and partial skeletons of microvertebrates in different states of disarticulation (see section 6.1.2.5.), all of which were recovered in sediment blocks and preserved as in-situ preparations (Kahlke 2020, Fig. 11a). After being brought into the laboratory, the respective skeletal units were partially freed from the surrounding sediment using preparation needles and fine brushes, and together with the remaining block were impregnated for consolidation (Rössler and Stache 2009). In some cases, particularly unstable flat finds, such as large mammal scapulae and hyaena coprolite concentrations, were embedded in an artificial matrix of synthetic resin or plaster using the transfer technique and thus permanently preserved in their original position (Keiler 1994; 1995, pls. 17–19; 2001a, 457; see section 6.1.2.6. of this paper).

The entire Untermassfeld collection is kept under climate control at about 50 % relative humidity and a constant room temperature of around 20 °C in several storerooms of the Senckenberg Research Station of Quaternary Palaeontology Weimar. In detail, the finds are housed as follows (as of 01 January 2021):

Building A:

- Collection storeroom IV with 8 cupboards and 4 cabinets containing 185 occupied drawers, large specimens in top sections;
- Collection storeroom V with 40 cupboards and 4 cabinets containing 699 drawers, large specimens in top sections;
- Type collection storeroom, including 8 transport boxes with a plasticiser-free, fitted inner lining made of expanded polypropylene (EEP): bovids, cervids (2), equids, felids (2), ursids, rodents;
- Micromammal laboratory, temporarily containing 15 flat drawers.

Building B:

- Collection storeroom VI with 5 cupboards, containing 96 drawers, upper sections with large preparations.

The fossil-bearing areas of the site itself were protected as far as possible from weathering, collapse and, increasingly, from the intervention of looters. During the 1980s, open profiles between the field seasons were covered with square straw bales and sheets. However, the instability and rapid deterioration of these ineffective protective structures called for better solutions. Later, in order to protect the profiles from collapse, sandbag barriers and sediment-filled big bags were inserted into the active excavation sections (Kahlke 2020, Fig. 6). Assemblages that had not yet been exposed to the point where they could be salvaged, were protected against frost until the next field season by sturdy polystyrene foam boxes. Excavation sections sunk into the main planum (see section 2.1.) were covered with wooden beams and steel plates weighing several tonnes (Kahlke 2020, Fig. 7).

4. Geological history and site origin

The origin and history of the fossil site of Untermassfeld are due to a very specific series of interactions between palaeoclimatological, palaeobiological and geomorphological as well as sedimentological, geochemical and neotectonic processes. Its reconstruction is based on systematically conducted geological profiles (Ellenberg and Kahlke 1997; R.-D. Kahlke 2001b; Kahlke et al. 2020, Foldout I) and corresponding documentation of the spatial distribution of finds (Kahlke 2001e, foldouts I–XV; see section 2.2. and **Foldouts II** and **III** of this paper).

4.1. Pleistocene succession

The Pleistocene profile of Untermassfeld overlies Middle Triassic marine carbonates on the eastern slope of the Werra valley. In the following description of the sedimentary sequence, including the reconstruction of the site development based on it as well as corresponding taphonomic processes, the following abbreviations have been used for the individual lithological units (in stratigraphic order from top to bottom), analogous to the presentation in Kahlke et al. (2020):

YCG	Younger Coarse Gravels
FPD	Floodplain deposits with remains of fossil soil (i) on top
LFS	Lower Fluviatile Sands
CMF	Coarse clastic mudflow fan
SCB	Sandy carbonate breccia with large sized boulders
TRA	Travertine
BSL	Basal sandy loam
UFS(l)	Upper Fluviatile Sands (lower part)
SB	Silt band, part of palaeosoil (ii)
UFS(u)	Upper Fluviatile Sands (upper part)
CS	Cover sediments

The Pleistocene sequence of Untermassfeld starts with 7.0–8.0 m thick deposits of riverine YCG, likely Eburonian (Ellenberg and Kahlke 1997) or – less likely – Menapian (Ellenberg 2002) in age (see section 8.1.). The uppermost section contains a fluviatile reworked zone up to 0.2 m thick with mixed sand and occasional large amounts of stratigraphically younger fossil vertebrate remains (see section 4.2.). The YCG are overlain by up to 2.0 m thick, fossil-free FPD of sandy loam consistency covered by a palaeosoil (i) (Kahlke et al. 2020, 1086 f.)
The main part of the sandy river deposits in the area of the site and beyond consist of LFS, a 2.5–8.0 m thick series of multi-coloured, fine to medium grained sands (Ellenberg and Kahlke 1997, 37 ff.; R.-D. Kahlke 2001b, 434 ff.; Kahlke et al. 2020). Fossils recovered from several sand strata of the LFS were found to be extremely well preserved (see section 4.2.).
In the southern and south-eastern areas of the site a 1.0–2.0 m thick CMF is wedged into the LFS (Ellenberg and Kahlke 1997, 37, Fig. 10, Pl. 17; Kahlke et al. 2020, 1087 ff.). It consists of calcareous to silty erosion products of the Triassic limestone slope. On the western, downhill facing front of the CMF its uppermost 4.0–5.0 cm are solidly calcified and have been termed SCB. Parts of the SCB are covered by a 5.0–10.0 cm thick TRA (Kahlke et al. 2020, 1091 f., Fig. 7), from which U-Pb dates were obtained (Gerdes et al. 2020; see section 8.2. of this paper).

The most characteristic structure of the Untermassfeld fossil deposit is an up to 5.0 m deep erosion channel that surrounds the CMF and extends deeper into the LFS in a SE-NW direction over the entire site area (R.-D. Kahlke 2001b, 423 ff., Fig. 8; Kahlke et al. 2020, 1095 f.). Occasionally it intersects the FPD and, at its deepest part, reaches into the YCG. The coarse-grained gravels halted any further expansion downwards, as they resist fluviatile excavation more than sands and clays. The erosion channel filling contained the majority of the recovered vertebrate remains with the largest amount of finds in the deepest part of the channel. Here, the vertebrate remains were embedded in different reworked sediments (Kahlke et al. 2020, 1083 ff.). In several places, the uppermost decimetres of the YCG were interspersed with fluviatile sands and freshly introduced necromass (Kahlke et al. 2020, Fig. 4). Directly west of the CMF, a BSL was excavated in thicknesses of up to 0.6 m. In addition to the discharge of sediment particles from the YCG, FPD and CMF, it also contained vertebrate remains of the Untermassfeld fauna that became incorporated during the initial filling of the channel (Kahlke et al. 2020, Fig. 13a).
The majority of the vertebrate finds were recovered from the UFS(l), i.e., reddish brown, fine to medium sands in a mostly flat horizontal stratification that formed the main part of the channel filling (Kahlke 2001e; 2006; Kahlke et al. 2020, 1094 f., Fig. 13b; see section 4.2. of this paper and frontispiece of this volume). Within the channel, the UFS(l) overlay both YCG and FPD, including their reworked zones, as well as the various components of the CMF and its immediate surroundings. The front of the mudflow fan itself was partially underwashed, resulting in the break-up or lateral intrusion of sedimentary bodies, including a large block of SCB covered with travertine. Postgenetic movements of this kind can be demonstrated by the presence of clear fracture edges and the spatial overlapping of SBC plates (Kahlke et al. 2020, 1089 ff., Fig. 12). Numerous vertebrate remains were found both above and below the erratic blocks in the UFS(l). The palaeosoil (ii) which formed in the uppermost 0.3 m of the UFS(l) indicates an interruption in the sedimentation after the channel was filled (Kahlke et al. 2020, Figs. 8, 14d).
The upper border of the palaeosoil (ii) is the SB, a silty layer of < 1–15 cm in thickness that divides the UFS into a lower (l) and an upper (u) part (Ellenberg and Kahlke 1997, 42 ff., Figs. 13–14, Pls. 7–9.1; Kahlke et al. 2020, Figs. 8, 14a). According to its grain size spectrum, the SB acted as an impermeable water barrier. The non-fossiliferous UFS(u) overlie the UFS(l) and SB uncomformably, their primary texture being strongly influenced by solifluid expansion and compression movements as well as gravitational compensatory movements (Ellenberg and Kahlke 1997, 40 ff., pls. 7–11; Kahlke et al. 2020, 1095 f., Fig. 8). 1.0–2.0 m thick layers of CS formed by hillside loam and intercalated layers of limestone clasts, and a recent soil complete the profile (Ellenberg and Kahlke 1997, Figs. 13, 15, 16).

4.2. Four-phase model of site evolution

Using knowledge gained from the Pleistocene profile of Untermassfeld along with the distribution of finds in its layers, a detailed model of the development of the site has been derived (Kahlke et al. 2020). Four phases can be distinguished:

(I) Processes prior to the downslide of the mudflow fan (**Fig. 2, I**):
During an Early Pleistocene cold period, most probably the Eburonian, large areas of gravel (YCG) were accumulated in the still flat central Werra valley, consisting mainly of erosion material from the neighbouring Permo-Carboniferous ridge of the Thuringian Forest and the South-west Thuringian Triassic area (Ellenberg and Kahlke 1997, Fig. 1). Profound interglacial weathering of the YCG, documented by a distinct bleaching of pebbles as well as by an incipient kaolinisation of feldspars, excludes the possibility of any original, i.e.,

isochronous, vertebrate remains being present. When undisturbed, the YCG are covered by sandy-clayey FPD, which have not yet yielded any fossil remains so far. Pedogenetic processes in their uppermost section indicate a phase lasting several hundred to a few thousand years of interrupted sedimentation with no erosion processes. In the area of the site as well as in the northern vicinity of the excavated area (Ellenberg and Kahlke 1997, Fig. 5) the FPD are overlain by LFS, a sequence of sandy high flood sediments up to 8.0m thick.
Due to opposing neotectonic movements (Ellenberg and Kahlke 1997, 53–56ff.), no extended erosion took place in the area of the present site, so that the Pleistocene profile including its fossil content (see below, phases II and III) became party inclined, but still remained largely intact.

(II) Downslide of the mudflow fan and onset of necromass accumulation (**Fig. 2, II**):
The repeated accumulation of individual strata of sandy high flood sediments, which vary in colour depending on the source material, are grouped together as the LFS. This accumulation indicates seasonally increased precipitation (Kahlke 2001e, 984; 2006, 89; see section 7.3. of this paper) and thus correspondingly periodic moisture penetration of the surrounding landscape surface. The loamy weathered limestone detritus of the adjacent valley slope, as well as the FPD in the river valley itself, were ideal slides for the mudflow that formed the CMF, which in turn caused the fossil accumulation. The CMF interfingers with both the LFS and the subsequently deposited UFS(I) (Kahlke et. al. 2020, 1100 and references therein). The significant decrease in the thickness of both of the aforementioned sand bodies towards the periphery of the CMF shows that the latter was already moving down from the adjacent slope into the river valley during the deposition of the LFS, where it established its effect on the fluviatile sedimentation processes and remained there during the deposition of the UFS(I). After the fan's movement came to a halt, exposed parts began to consolidate due to the effect of carbonate-saturated surface and spring waters.
The LFS originated as floodplain deposits of a rapid and laterally shifting watercourse with flow rates of 0.4–0.5m/s (Ellenberg and Kahlke 1997, 57ff.). The CMF intercalated into the sands, changing the sedimentation process of the affected river area. On the lee side of the fan, a zone of reduced transport energy developed in the flowing water, which led to the deposition of some parts of the debris it carried, including animal and plant necromass. Several strata of the LFS, which clearly represent individual flooding events, contained remarkable amounts of fossil vertebrate remains. Their distribution patterns are described in sections 6.1.1.1., 6.1.2.1. and 6.1.2.3. Because the individual strata of the LFS do not extend over large areas, their microstratigraphical correlation is difficult to assess and the number of flood events that led to fossil deposition cannot yet be stated with any certainty. That said, at least two fossiliferous accumulations, separated by a continuous silt deposit with a maximum thickness of 0.3m (R.-D. Kahlke 2001b, 434ff., Fig. 2), can be detected within them. It is very likely, however, that more than two flood events occurring very close together, one after another, led to the fossil record of the LFS leeside of the CMF

(III) Incising of the channel and refill by fossiliferous sediments (**Fig. 2, III**):
Following the deposition of the LFS, a particularly strong flood event, apparently caused by a special meteorological constellation in the catchment area of the river (see section 7.3.), led to the formation of an erosion channel up to 5.0m deep in the immediate vicinity of the CMF. The bedded, but unconsolidated LFS and in some sections the underlying FPD and the uppermost decimetres of the YCG became intersected. The carbonic consolidation of lateral areas of the CMF resulted in a certain resistance to weathering. The fan was therefore only marginally undercut, so that individual sedimentary bodies, namely primary CMF components, SCB, large boulders and TRA, collapsed into the adjacent channel.
Immediately following its incising, the filling of the channel started in certain areas with the sedimentation of reworked YCG and redeposited FPD accumulated as BSL, followed by the rapid deposition of UFS(I). The

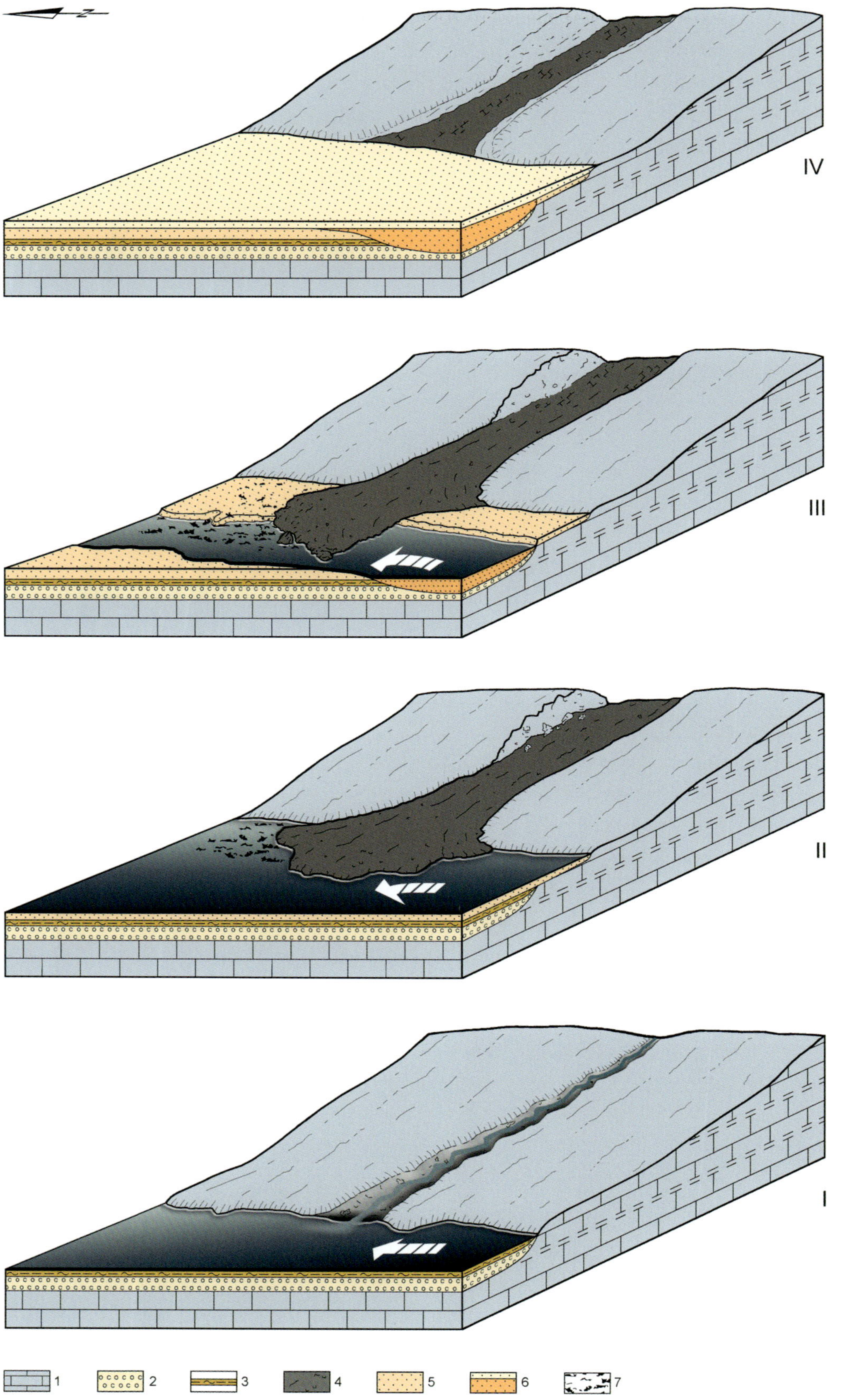
N
IV
III
II
I
1
2
3
4
5
6
7

CMF, again acting as a barrier, created a leeward zone of reduced transport energy in the flowing water. The sediments deposited there contained the main mass of vertebrate finds recovered at Untermassfeld (Ellenberg and Kahlke 1997, 57; R.-D. Kahlke 2001b, 443f.; 2001d, 966ff.; **Foldout II** of this paper, frontispiece of this volume). Significant amounts of necromass had already been deposited in the reworked zone of the YCG and in the BSL (Kahlke et al. 2020, Figs. 4, 13a; **Fig. 3** of this paper). The channel lag contained vertebrate remains in extremely high concentrations, which in some areas numbered more than 200 specimens/m² (**Foldouts II** and **III**). Their large- and small-scale spatial distribution is described in sections 6.1.1.2. and 6.1.2.

The necromass deposited immediately after the flood peak consisted of the parts of larger mammal carcasses, more or less disarticulated micromammal skeletal units, single and often gnawed bones, excrement, invertebrate remains, and vegetational residues, which the flood collected from areas upstream of the site (Kahlke 2001e, Fig. 26; 2006, Fig. 78) and also from its immediate vicinity (see section 6.1.2.6.). The exceptionally powerful and destructive flooding of the river valley, which caused the erosion and refilling of the fossiliferous channel, extended over large areas upstream of the site (see section 7.3.). Judging by the sedimentological characteristics and the distribution of the vertebrate material deposited, this event was more likely to have occurred over a few hours or days than over weeks or months. The filling of the erosion channel, with upwardly decreasing amounts of incorporated large vertebrate remains, ended with the deposition of a silty layer, the SB. A palaeosoil (ii) within the uppermost UFS(l) and the SB itself indicates sedimentation calming with no renewed flooding.

(IV) Further flooding, solifluction and deposition of cover sediments (**Fig. 2, IV**):

After the filling of the prominent erosion channel with UFS(l) and sediments of the SB, the fluviatile flooding events continued in the middle Werra valley and a new sequence of riverine sands, the UFS(u), were deposited. As the accumulating effect of the CMF, which only slightly protruded from the sandy subsoil and was ultimately completely covered by sediment, came to a standstill, no further vertebrate remains were accumulated in the area of the site.

A complete dissolution of the primary texture of the UFS(u) indicates intensive post-sedimentary solifluction processes above the SB which acted as an aquifuge. The sediment movements were facilitated by the neotectonic SW dip of the above-mentioned layers (Kahlke et al. 2020, Fouldout I). At the same time, the SB acted like a shield for the fossil content of the underlying fossiliferous UFS(l). The vertebrate remains contained therein were protected from solifluidal movements and thus from further displacement and destruction.

The CS, with recent top soil, reflect a normal development of the Werra valley after the genesis of the fluviatile site.

From the reconstructed development phases of the fossil site it becomes clear that the high concentration of the Untermassfeld vertebrate material and its mostly very good quality of preservation was caused by

Fig. 2 Untermassfeld, development phases before, during and after site formation, arrow marks flow direction of the Werra river. – 1 Middle Triassic limestone. – 2 Younger Coarse Gravels. – 3 Floodplain deposits. – 4 Coarse clastic mudflow fan. – 5 Lower Fluviatile Sands. – 6 Upper Fluviatile Sands, lower and upper parts. – 7 Necromass.

I Situation prior to the downslide of the mudflow fan: Eastern slope of the Werra valley from Triassic carbonates, floodplain with accumulated Younger Coarse Gravels (YCG) and floodplain deposits (FPD), beginning of sedimentation of Lower Fluviatile Sands (LFS) during a flood.

II Situation immediately after the downslide of the mudflow fan: Change of the flow regime and onset of necromass accumulation in Lower Fluviatile Sands (LFS) during a flood.

III Situation after incising of an erosion channel and its refilling by fossiliferous sediments: Incision of Younger Coarse Gravels (YCG) and floodplain deposits (FPD) in the deepest part of the channel, undercutting of the edges of the clastic mudflow fan (CMF) and erosion of fossil-bearing Lower Fluviatile Sands (LFS) as well as refilling of the channel with fossil-bearing reworked sediments and the lower part of the Upper Fluviatile Sands [UFS(l)].

IV Situation after filling of the erosion channel and covering of the fans valley side with sand: The missing accumulating effect of the buried clastic mudflow fan (CMF) causes fossil-free deposition of the upper part of the Upper Fluviatile Sands [UFS(u)].

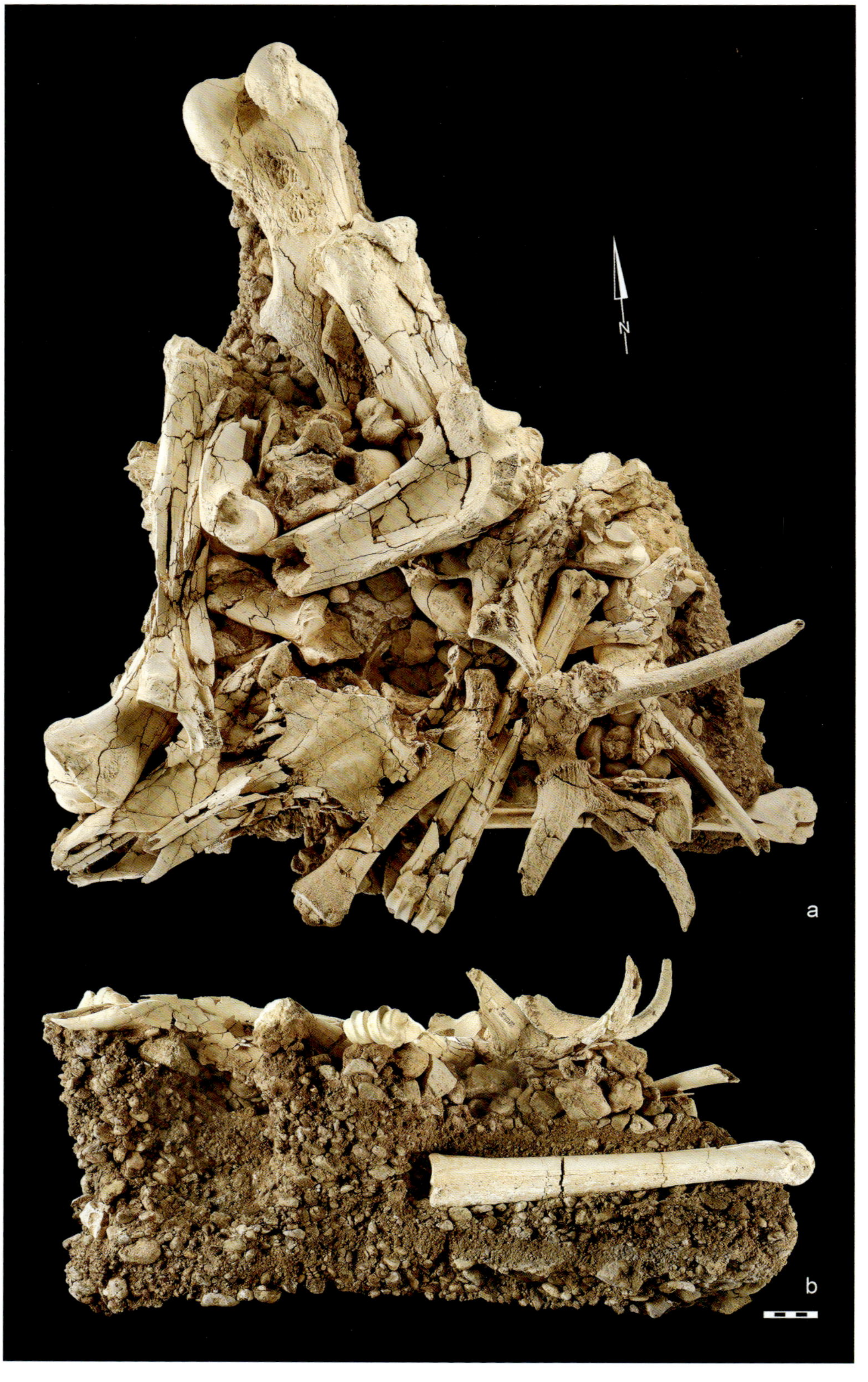

Fig. 3 Untermassfeld research excavation, block IQW 2018/45431 (Mei. 50740) from grid square Q 528, see **Foldout III**. – **a–b** In-situ preparation of reworked Younger Coarse Gravels (YCG) mixed with Upper Fluviatile Sands (lower part) [UFS(l)] and a polyspecific bone concentration 2.05–1.94 m below site 0-level, with deformed skull of a female *Eucladoceros giulii* Kahlke. – **b** Left metatarsal of *Eucladoceros giulii* Kahlke deposited 8.0–6.0 cm below the bone concentration in reworked YCG. – Scale: 50 mm.

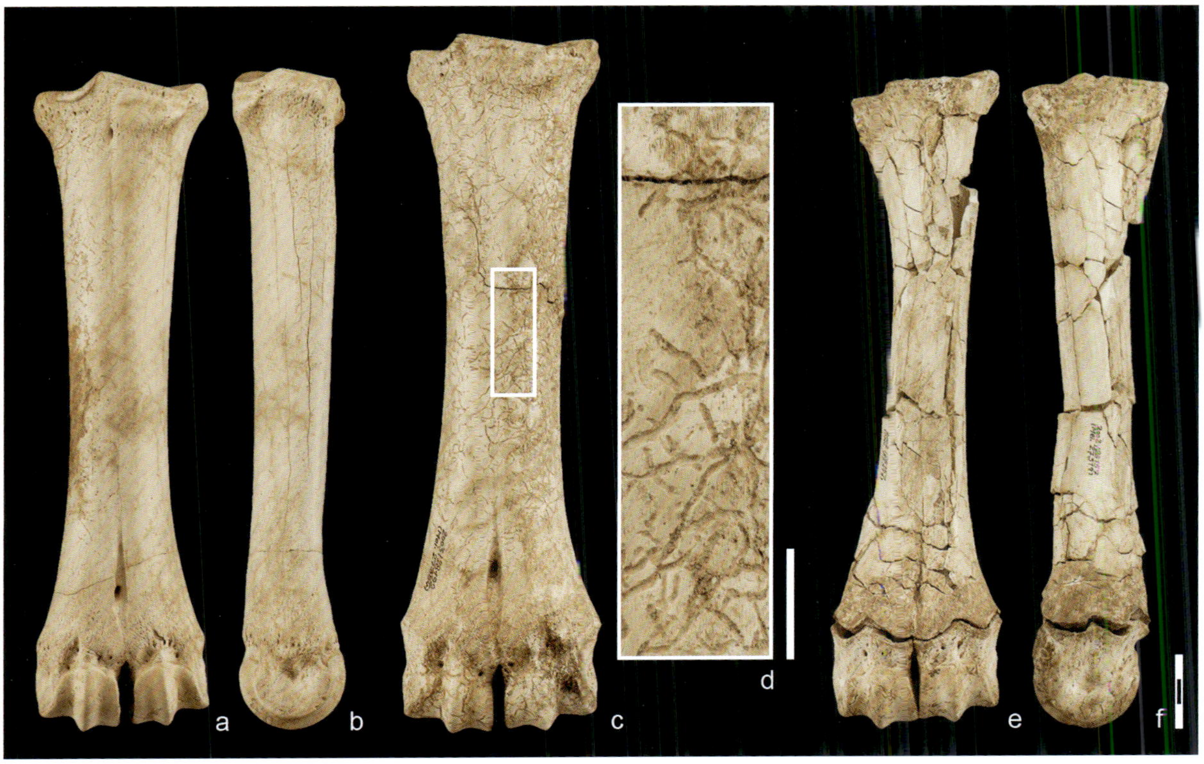

Fig. 4 *Bison menneri* Sher, Untermassfeld. – **a–b** Right metacarpal III–IV IQW 2011/31 290 (Mei. 30452) in an undeformed, very good preservation state, anterior, lateral views. – **c–d** Right metacarpal III–IV IQW 2003/28 192 (Mei. 27354) in an undeformed condition with root etchings on the bone surface, general anterior view, close anterior view. – **e–f** Right metacarpal III–IV IQW 2002/28 157 (Mei. 27319), juvenile, neotectonically deformed and shattered, anterior, lateral views. – Scales: **a–c**, **e–f** 30 mm – **d** 10 mm.

the interaction of several factors and processes. Warm climatic conditions with seasonally high precipitation enabled the development of a diverse landscape (see section 7.2.), which offered habitats for a rich and varied fauna (see sections 5.2. and 6.3.). Massive precipitation in combination with the geomorphology of the middle Werra valley repeatedly led to flooding events (see section 6.4.) and a corresponding threat to numerous vertebrate species. A mudflow fan that slid into the Werra valley acted as an accumulating factor for necromass transported by the river. In its immediate leeward side, vertebrate remains in various decomposition states (see section 6.) were deposited and rapidly became embedded. Nothing even approximately comparable to this configuration of geological events could be detected a second time outside of the excavation site, despite intensive sand mining and surveying which took place around the area for decades. To date, no fossil vertebrate remains have ever been found in the corresponding sediments.

The unusually rich fossil site of Untermassfeld was preserved until the present day because its erosion was prevented by neotectonic subsidence movements during the Pleistocene valley deepening. The accumulation of impermeable silt on the downhill inclined fossil-bearing sands protected the original distribution patterns of the buried vertebrate remains from solifluidal displacement. A slight carbonate impregnation by stratum water from the Triassic limestones of the valley slope also contributed to the very good condition of the finds (**Fig. 4a–b**), including manifold traces of life preserved on them (see section 6.2., **Fig. 4c–d**). The acidic nature of the YCG that were cut in the deepest part of the channel had the opposite effect. Skeletal elements embedded in the reworked zone had undergone chemical dissolution, thus becoming unstable and deformed by the load of the covering sediments (e.g., **Fig. 3**, skull of *Eucladoceros giulii*). Depending on the orientation of the skeletal elements, deformation and/or fragmentation also occurred in the tectonically stressed sections of the fluviatile sands (**Fig. 4e–f**).

5. Recorded palaeobiodiversity

The fossil site of Untermassfeld yielded a vertebrate fauna that is exceedingly rich in species and individuals. In contrast, evidence for invertebrates is comparatively scarce. Geochemical causes, namely a complete oxidation of the fossil-bearing sediments, hindered or prevented the preservation of micro- and macro-botanical remains.

5.1. Flora

In order to extract possible pollen spectra (M. Stebich, Weimar), the loamy FPD below the fossil-bearing sediments (see section 4.1.) was sampled along with the LFS and UFS(l) (details in Kahlke 2020, 1044). An attempt was also made to extract pollen from the coprolites of *Pachycrocuta brevirostris*, which were preserved in abundant amounts at the site (see section 6.1.2.6.). In all cases only a few, barely significant, amounts of pollen were found from the following taxa (status as of 01 December 2021):

Class: Coniferopsida
Order: Coniferales
Pinus sp.

Class: Magnoliopsida
Order: Fagales
Betula sp.
Corylus sp.
Quercus sp.

Order: Poales
Poaceae indet.

Order: Asterales
Artemisia sp.

Phytoliths were found in the LFS as well as in the UFS(l) (details in Kahlke 2020, 1044 f.). Unfortunately, although a corresponding study was initiated, it wasn't completed.
No precisely determinable macro-botanical remains have been recovered from the fossiliferous horizons of the site. The only evidence for such remains are the impressions of plants that are occasionally visible inside and on the surface of some coprolites. In one case the fragment of a grass leaf was identified (Keiler 2001b, pl. 131, 1–2).
Within the fossil-bearing LFS, occasional streaks or lentil-like accumulations of xylitic particles were detected. With a maximum thickness of 10 mm, these inclusions cover areas of up to several dm^2 and are most likely the macro remains of plants, such as leaves, needles, bark and twigs that got washed up together and were then relatively quickly covered by sediment over the course of the flood events that led to the site's formation (see section 4.2.). In the UFS(l), mm-sized xylite particles occur regularly, increasing in quantity both within the bone concentrations as well as in their immediate vicinity. Again, these were probably the remains of former floodplain vegetation contained in the necromass.

Finally, the mechanical separation and screening of a travertine slab (TRA: Gerdes et al. 2020; Kahlke 2020, 1045; Kahlke et al. 2020, 1091 f., Figs. 7, 12; see section 4.1. of this paper) deposited in the deepest part of the LFS (Q 610) also produced absolutely no evidence of fossil vegetation (pers. comm. F. Kienast, September 2020).

5.2. Fauna

According to current knowledge, the range of the fossil faunal remains from Untermassfeld comprises the following taxa (status as of 01 December 2021):

Class: Gastropoda

Order: Mesogastropoda

Valvata cristata O. F. Müller, 1774
Bithynia leachi (Sheppard, 1823)
Belgrandia germanica (Clessin, 1882)
? *Pomatias* sp.
Acicula polita (Hartmann, 1870)

Order: Basommatophora

Lymnaea peregra (O. F. Müller, 1774)
Lymnaea truncatula (O. F. Mülller, 1774)
Planorbis planorbis (Linnaeus, 1758)
Anisus leucostoma (Millet, 1813)
Bathyomphalus contortus (Linnaeus, 1758)
Carychium minimum O. F. Müller, 1774

Order: Stylommatophora

Succinea oblonga Draparnaud, 1801
Catinella arenaria (Bouchard-Chantereaux, 1837)
Cochlicopa lubricella (Porro, 1837)
Granaria cf. *frumentum* (Draparnaud, 1801)
Pupilla muscorum (Linnaeus, 1758)
Vallonia pulchella (O. F. Müller, 1774)
Vallonia costata (O. F. Müller, 1774)
Clausilia dubia Draparnaud, 1805
Clausiliidae indet.
Cecilioides acicula (O. F. Müller, 1774)
Discus ruderatus (Férussac, 1821)
Nesovitrea hammonis (Ström, 1756)
Zonitidae indet.
Limacidae indet.
Euconulus fulvus (O. F. Müller, 1774)
Bradybaena cf. *fruticum* (O. F. Müller, 1774)
Helicella sp. indet.

Helicopsis sp. indet.
Trichia hispida (Linnaeus, 1758)
Trichia sp. indet.
Helicigona lapicida (Linnaeus, 1758)
Cepaea cf. *nemoralis* (Linnaeus, 1758)
Helicidae indet. (1)
Helicidae indet. (2)
Helicidae indet. (3)

Class: I n s e c t a
Order: Coleoptera
Dermestidae indet. (feeding traces) ***

Class: O s t e i c h t h y e s
Order: Salmoniformes
Salmo cf. *salar* Linnaeus, 1758 ***

Order: Cypriniformes
Gobio sp. ***

Class: A m p h i b i a
Order: Urodela
Lissotriton vulgaris (Linnaeus, 1758) * **
Triturus cf. *cristatus* (Laurenti, 1768)

Order: Anura
Rana temporaria Linnaeus, 1758 ***
Rana arvalis Nilsson, 1842 **
Rana dalmatina Fitzinger in Bonaparte, 1839 **
Pelophylax ridibundus (Pallas, 1771) vel *esculentus* (Linnaeus, 1758) ***
Bufo bufo (Linnaeus, 1758) **
Bufotes viridis Laurenti, 1768 **
Hyla arborea (Linnaeus, 1758) **

Class: R e p t i l i a
Order: Squamata
Anguis fragilis Linnaeus, 1758 ***
aff. *Ophisops elegans* Ménétriés, 1832 ***
Lacerta sp. ***
Eryx sp. ***
Elaphe sp. ***
cf. *Coronella* sp. ***
Natrix sp. ***

Class: A v e s	
Order: Anseriformes	
Cygnus sp. (cf. *olor* Gmelin, 1788)	
Anser sp. (aff. *subanser* Jánossy, 1983)	
Anatidae gen. et sp. indet.	***
Order: Galliformes	
Lyrurus sp.	***
Phasianidae gen. et sp. indet.	***
Francolinus (*Lambrechtia*) *capeki* (Lambrecht, 1933)	
Order: Accipitriformes	
Haliaeetus sp. (aff. *brevipes* Jánossy, 1983)	
Accipiter nisus (Linnaeus, 1758)	***
Order: Passeriformes	
Corvus sp.	
Turdus sp. (aff. *philomelos* Brehm, 1831)	
Garrulus sp. [aff. *glandarius* (Linnaeus, 1758)]	
Passeriformes indet.	
Class: M a m m a l i a	
Order: Artiodactyla	
Bison (*Poephagus*) *menneri* Sher, 1997	* **
Bison sp.	***
Capreolus cusanoides Kahlke, 2001	
Cervalces carnutorum (Laugel, 1862)	*
Eucladoceros giulii Kahlke, 1997	
Pseudodama vallonnetensis (de Lumley, Kahlke, Moigne et Moullé, 1988)	* **
Hippopotamus antiquus Desmarest, 1822	**
Sus strozzii Forsyth-Major, 1881	**
Order: Perissodactyla	
Stephanorhinus hundsheimensis (Toula, 1902)	*
Equus (*Sussemionus*) *wuesti* Musil, 2001	* **
Equus sp.	***
Order: Proboscidea	
Mammuthus meridionalis (Nesti, 1825) incl. *M.* cf. *meridionalis*	**
Order: Carnivora	
Panthera onca gombaszoegensis (Kretzoi, 1938)	
Acinonyx pardinensis pleistocaenicus (Zdansky, 1925)	
Puma pardoides brevipes, Hemmer and Kahlke in this volume	****
Lynx valdarnensis spelaeoides, Hemmer and Kahlke in this volume	****
Megantereon adroveri Pons Moyà, 1987	**
Homotherium latidens ultimum (Teilhard de Chardin, 1936)	**

Pachycrocuta brevirostris (Gervais, 1850)	*
Meles meles atavus Kormos, 1914	**
Ursus ex gr. *U. dolinensis* García et Arsuaga, 2001	**
Xenocyon lycaonoides Kretzoi, 1938	**
Canis mosbachensis Soergel, 1925	
Order: Rodentia	
Sciurus whitei Hinton, 1914	**
Spermophilus kahlkei, Maul in this volume	****
Glis sackdillingensis (Heller, 1930)	
Castor fiber Linnaeus, 1758	
Trogontherium (*Trogontherium*) *cuvieri* Fischer, 1809	*
Sicista sp.	***
Cricetus runtonensis Newton, 1909	*
Myodes hintonianus (Kretzoi, 1958)	* **
Pliomys episcopalis Méhely, 1914	
Mimomys savini Hinton, 1910	
Mimomys pusillus (Méhely, 1914)	*
Microtus (*Allophaiomys*) *thenii* Malez et Rabeder, 1984	* **
Apodemus sylvaticus (Linnaeus, 1758)	**
Hystrix refossa Gervais, 1852	**
Order: Lagomorpha	
Hypolagus brachygnathus Kormos, 1930	***
Lepus sp.	**
Order: Primates	
Macaca sylvanus (Linnaeus, 1758)	**
Order: Erinaceomorpha	
Erinaceus cf. *europaeus* Linnaeus, 1758	
Order: Soricomorpha	
Talpa cf. *europaea* Linnaeus, 1758	
Talpa minor Freudenberg, 1914	**
Desmana thermalis Kormos, 1930	***
Sorex (*Sorex*) *minutus* Linnaeus, 1766	***
Sorex (*Sorex*) *runtonensis* Hinton, 1910	*
Sorex (*Drepanosorex*) *margaritodon* Kormos, 1930	**
Beremendia fissidens (Petényi, 1864)	**

Additions and changes compared to the faunal list by Kahlke (2006):

*	Changed nomenclature.
**	Changed taxonomic classification.
***	Newly discovered in the Untermassfeld fauna.
****	New described species/subspecies.

5.2.1. Gastropods

Gastropod remains were recovered from both the Lower (LFS) and the Upper Fluviatile Sands [UFS(l)] (see section 4.1.), although for both finds and individuals the numbers are small. The preservation is uniform, so the fossils were not relocated from older stratigraphic units. Occasional new remains recovered from the UFS(l) of the channel infill (determination R. Meyrick, Weimar and Toronto; G. Utschig, Weimar) produced no changes to the list of 36 taxa published by Krolopp (1997). Eight aquatic and 28 terrestrial species represent a rather broad, but by no means complete gastropod spectrum of the former locality. The most commonly found species were the small amber snail (*Succinea oblonga*), and the blind or needle snail (*Cecilioides acicula*).

5.2.2. Insects

The high oxygen content of the fossiliferous sediments of Untermassfeld hindered the preservation of insect remains. However, feeding traces of various configurations caused by dermestid larvae (Coleoptera) were found on several osteological specimens (Keiler et al. 2020; see section 6.2.2.). As yet, no traces of limnetic insects have been identified.

5.2.3. Fishes

Despite extensive sampling of the fossil-bearing sands over the past decades (Kahlke 2020, 1045, Fig. 10), fish remains were recovered only very infrequently. Based on the total material obtained during the excavations, the fish fauna has been revised by M. Böhme (2020a). As a result, only two species have been conclusively verified. Evidence for the anadromous Atlantic salmon (*Salmo* cf. *salar*) has been established by the remains of several individuals with body lengths of around 1 m, and, although rare, remains of the cyprinid *Gobio* sp. have also been documented. On the basis of the available material, the rudd *Scardinius* sp., mentioned by G. Böhme (1997b), could not be confirmed.

5.2.4. Amphibians

As expected, the amphibian and reptile finds, recovered mainly from mud samples, remained comparatively rare. However, the present findings represent a gratifyingly high species diversity (G. Böhme 1997a; M. Böhme 2020b). In addition to the smooth newt (*Lissotriton vulgaris*) and northern crested newt (*Triturus* cf. *cristatus*), a total of seven frog species were detected, namely the common frog (*Rana temporaria*), moor frog (*Rana arvalis*), agile frog (*Rana dalmatina*), green frog (*Pelophylax ridibundus* vel *esculentus*), common toad (*Bufo bufo*), European green toad (*Bufotes viridis*), and the European tree frog (*Hyla arborea*).

5.2.5. Reptiles

Recent research has significantly advanced our knowledge relating to the Untermassfeld reptile fauna (M. Böhme 2020b), and, as a result, corrections to previous determinations have been made. Three species of lizards, namely the slow worm (*Anguis fragilis*), snake-eyed lizard (aff. *Ophisops elegans*), and a larger-sized lizard (*Lacerta* sp.), as well as four snake species, i.e., a sand boa (*Eryx* sp.), a rat snake (*Elaphe* sp.), another small-sized colubrine (cf. *Coronella* sp.), and a grass snake (*Natrix* sp.), have now been identified. The purported occurrence of pond turtles (Emydini gen. et sp. indet.) (Maul 1997), however, is based on a misidentification. Several borrowed Untermassfeld finds that were initially presumed lost (Kahlke 2006, 19), have since been recovered and returned to Weimar. Corresponding records of skink (*Chalcides* sp.) and gecko (*Eublepharis* sp.) by Tatarinov (details in Kahlke 2001e, 939) could not be confirmed on the basis of this material.

5.2.6. Birds

As expected for a fluviatile site, bird remains are among the rarer fossils found at Untermassfeld (Jánossy 1997; Manegold 2020). The finds are distributed among three species of waterfowl (Anseriformes), three species of gamefowl (Galliformes), two species of birds of prey (Accipitriformes), and four species of songbirds (Passeriformes).

In addition to a species of waterfowl (*Anser* sp.), a species of swan (*Cygnus* sp.) the size of the mute swan (*C. olor*), and possibly a duck the size of the common teal (*Anas crecca*) have been verified. Two fossils (*Lyrurus* sp.) correspond morphometrically to the extant black grouse (*L. tetrix*), but could also belong to a separate fossil form. Two fragmentary phasianid finds are within the size range of the grey partridge (*Perdix perdix*). Unfortunately, no further finds could be assigned to the *Francolinus* (*Lambrechtia*) *capeki* identified by Jánossy (1997). In addition to a previously recorded sea eagle, *Haliaeetus* sp. (aff. *brevipes*), with the discovery of the Eurasian sparrowhawk *Accipiter nisus* (Linnaeus, 1758) another bird of prey could be added to the list of Untermassfeld fauna. The majority of the bird finds can be attributed to several individuals of a rather large species of crow or corvid (*Corvus* sp.) (see section 6.3.), but thrush *Turdus* sp. (aff. *philomelos*), jay *Garrulus* sp. (aff. *glandarius*), and the remains of one or two as yet indeterminable smaller songbirds, similar in size to the Eurasian blackcap (*Sylvia atricapilla*), were also recovered.

5.2.7. Mammals

Artiodactyls are strongly represented in the Untermassfeld fauna by abundant finds of bovids, cervids, and hippopotamids, and an occasional suid. Since the initial study of the Untermassfeld b o v i d s by Sher (1997), the corresponding material has more than doubled to well over 2,300 finds. The majority of the bison remains belong to a single morphologically homogeneous species with a mostly clearly recognisable sexual dimorphism in the postcranial skeleton. This long-legged, very large-bodied animal with its high hump, possessed a relatively small skull as well as graceful horn cones (Kahlke 2001e, Fig. 1; Bukhsianidze 2020, Figs. 4, 5; **Fig. 5** of this paper).

While Sher (l.c.) classified the common Untermassfeld bison as the earliest representative of the subgenus *Bison*, Bukhsianidze (2020) has concluded that the Untermassfeld population represents a primitive yak – *B.* (*Poephagus*) *menneri* – based on the configuration of the meanwhile more frequently found cranial bones. According to current knowledge, this represents the oldest and most western fossil evidence for yaks. Consequently, Bukhsianidze (l.c.) views bisons in the strict sense [*Bison* (*Bison*)] and yaks [*B.* (*Poephagus*)] as parallel evolutionary lineages within the Holarctic region. In addition to the above-mentioned bovid, which is extremely abundant in the fossil record of Untermassfeld, another bison species has been identified on the basis of some cranial fragments and at least one metacarpal bone. This species has been provisionally listed as *Bison* sp. (Bukhsianidze l.c.).

Based on the number of finds and individuals (see section 6.3.), the c e r v i d s are the most extensively recorded taxonomic family of the Untermassfeld fauna and a total of four species have been identified (H.-D. Kahlke 1995; 1997; 2001a; Breda et al. 2020). As a relatively rare faunal element, an early type of roe deer, represented by some 30 individual finds, was described as a new species, *Capreolus cusanoides* (H.-D. Kahlke 2001a), differing in both dental and antler characters from the similarly sized early Middle Pleistocene species *C. suessenbornensis* (Kahlke 1956a; 1956b). The phylogenetic relationships of *C. cusanoides* to the Ruscinian / Early Villafranchian species *Procapreolus cusanus* are yet to be resolved.

Based on the size of its postcranial skeleton, and quite sparsely documented with a total of only 23 specimens, the Untermassfeld elk *Cervalces carnutorum* (H.-D. Kahlke 1997; 2001a; Breda et al. 2020) lies between the Villafranchian *C. gallicus* and the more massive *C. latifrons*. The latter is well documented in the Thuringian region by the early Middle Pleistocene finds from Voigtstedt (Kahlke 1965) and Süßenborn (Kahlke 1969).

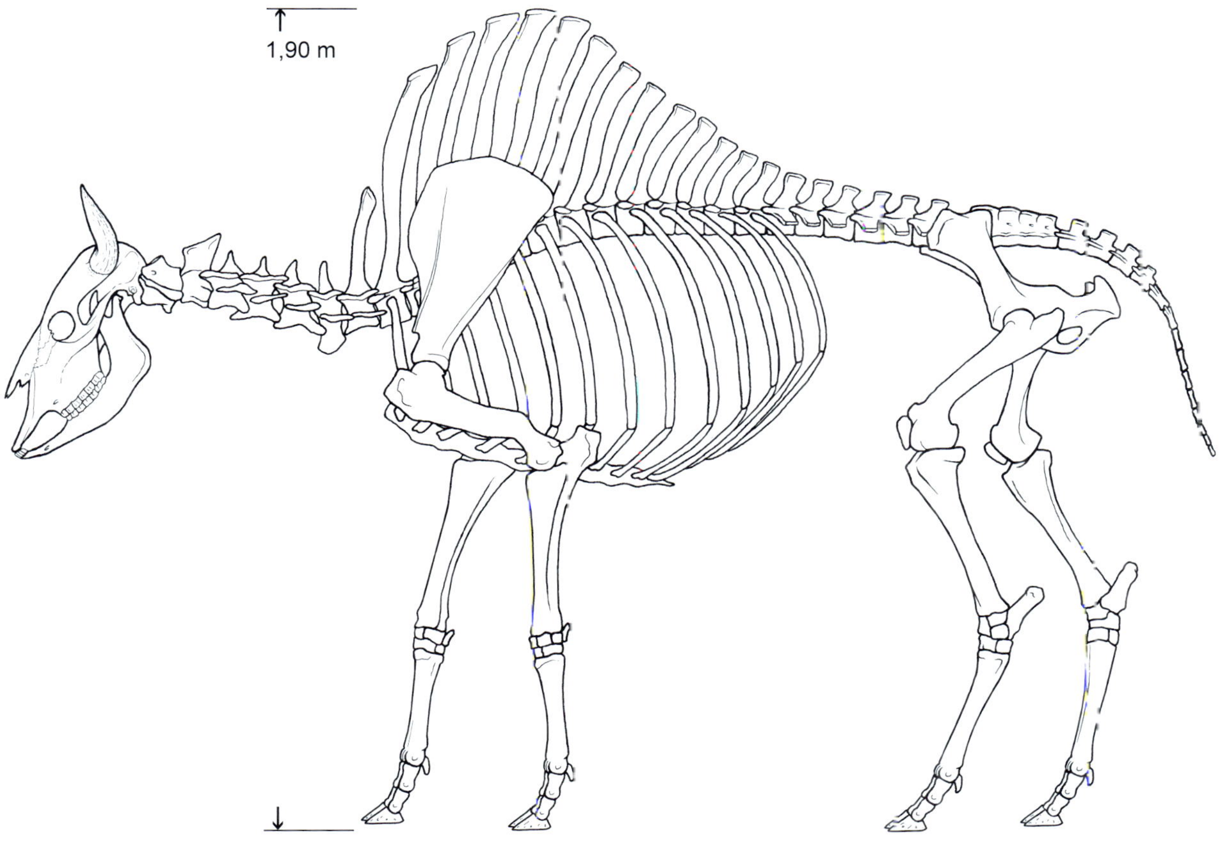

Fig. 5 *Bison menneri* Sher, Untermassfeld. – Skeletal reconstruction based on individually related (male) finds of skeletal elements (implementation: R.-D. Kahlke, E. Haase).

With about 2,800 individual finds and additional dentition and skeletal elements within in-situ block preparations, the large, long-legged deer *Eucladoceros giulii* is the most abundantly represented large mammal species in the Untermassfeld (type site) fossil material. As a result of the taphonomic conditions at the site (see section 6.4.), its skulls are underrepresented in the fossil material and complete antler finds are missing. Corresponding fragments, however, show the comb-like design typical of the genus. The morphometry of nearly all of the other skeletal elements has been documented in detail (H.-D. Kahlke 1997; 2001a; Breda et al. 2020).

The remains of the fallow-like deer as described by H.-D. Kahlke (1997; 2001a) have been augmented by a number of new skull and antler finds (e.g., frontispiece of this volume). With about 1,530 individual elements, this species is frequently represented in the find material, and has now been provisionally listed as *Pseudodama vallonnetensis* by Breda et al. (2020, 1226 ff.). The antlers differ from those of later *Dama* species by the complete lack of a palmation. The phylogenetic position of *Pseudodama* is as yet unresolved.

Iconic for the Untermassfeld fossil assemblage are the unusually numerous remains of large-bodied hippopotamids recovered from the site. Nearly every skeletal element is represented in the total *Hippopotamus* material of around 830 finds, most of which are present in smaller or larger series (H.-D. Kahlke 1982; R.-D. Kahlke 1985a; 1985b; 1987; 1997b; 2001c). For taphonomic reasons, bulky cranial and dental elements are underrepresented. According to recent research, the Early to early Middle Pleistocene hippopotamus forms of the European mainland are no longer considered in direct phylogenetic proximity to the extant African *H. amphibius*, but are now recognised as a separate species, *Hippopotamus antiquus* (Kierdorf and Kahlke 2020, 1252). The nearly complete skull of an adult individual from Untermassfeld has been placed morphologically between the Upper Villafranchian lectotype of the species from Figline (Upper Valdarno) and stratigraphically younger finds from Southern Europe (Kahlke 2001c).

Remains of suids are rare in the Untermassfeld material. Originally, 49 skeletal elements of several juvenile and adult individuals (Guérin and Faure 1997; Kahlke 2001e, 941) were assigned to *Sus scrofa priscus*, a form typical for the late Early Pleistocene of Europe. Based on tooth morphology, however, Cherin et al. (2020) and Iannucci (in this volume) have now reassigned the material to *Sus strozzii*. Correspondingly, either this species survived longer in Europe than previously recorded, or it re-immigrated from Asia during the Epivillafranchian.

Perissodactyla are represented by rhinocerotids and equids. The Untermassfeld material includes a total of around 1,070 remains of rhinocerotids (H.-D. Kahlke 2001b; R.-D. Kahlke 2001e; Kahlke and Kaiser 2011; Kotowski et al. 2020), including most notably rich postcranial material. Dentition elements, however, are underrepresented. Two skulls have verified the presence of a juvenile as well as a distinctly gracile adult female (**Fig. 6**). The finds have been assigned to *Stephanorhinus hundsheimensis* and belong to the older representatives of the species (sensu Lacombat 2006) in the western Palaearctic. Based on the ratio of the length of the nasal bones to the distance between the narial notch and the rostral rim of the orbit, combined with the degree of ossification of the nasal septum, *S. hundsheimensis* is morphologically closely related to *S. hemitoechus* and *S. kirchbergensis* (Kotowski et al. 2020).

Equids are among the less frequently represented large mammals in the find material. About 280 finds, including dentition remains and metapodials, mostly belong to a form of medium body size that was described by Musil (2001a) from Untermassfeld as a new species, *Equus wuesti*. Eisenmann and Boulbes (2020) place the species in the sussemiones, a group of predominantly Early and Middle Pleistocene equids (Eisenmann 2006) that were distributed around the northern hemisphere. They have been defined as a separate subgenus due to their peculiar dental characters (Eisenmann 2010). *E.* (*Sussemionus*) *wuesti* lies geographically and stratigraphically between the Early Pleistocene sussemione finds from Venta Micena (Spain) and those from the early Middle Pleistocene of Akhalkalaki (Georgia). The single find of a relatively large radius provides evidence for a second, apparently much rarer, horse species in the area of the Untermassfeld site, but this specimen has so far only been classified as *Equus* sp. (Eisenmann and Boulbes 2020, 1319).

A first description of the available Untermassfeld elephantid material was presented by Dubrovo (2001). Notes on subsequent new finds then followed (Kahlke 2001e; 2006). A total of 123 specimens are now available, which is indicative of more individuals than have so far been recorded (see section 6.3.). As expected, bulky and heavy skeletal elements are missing, especially skulls and mandibles of adult individuals. A small series of dentition elements is comprised of around 75 % deciduous teeth, whose taxonomic classification is much more difficult to make than that of permanent teeth. A detailed analysis by Van Essen (2011, 155 ff.; in this volume) confirmed the assignment of the material to *Mammuthus meridionalis*. On some finds, in particular a fragmentary juvenile skull, a mosaic character of features was detected, which led to the classification as *M.* cf. *meridionalis*. Full *Mammuthus trogontherii* morphology was not found on any of the material. An examination of the new finds of postcranial skeletal elements of *M. meridionalis* (**Fig. 7**) by D. Mol (Hoofddorp), which began in August 2015 on the original material, could unfortunately not be completed (pers. comm. D. Mol, December 2020).

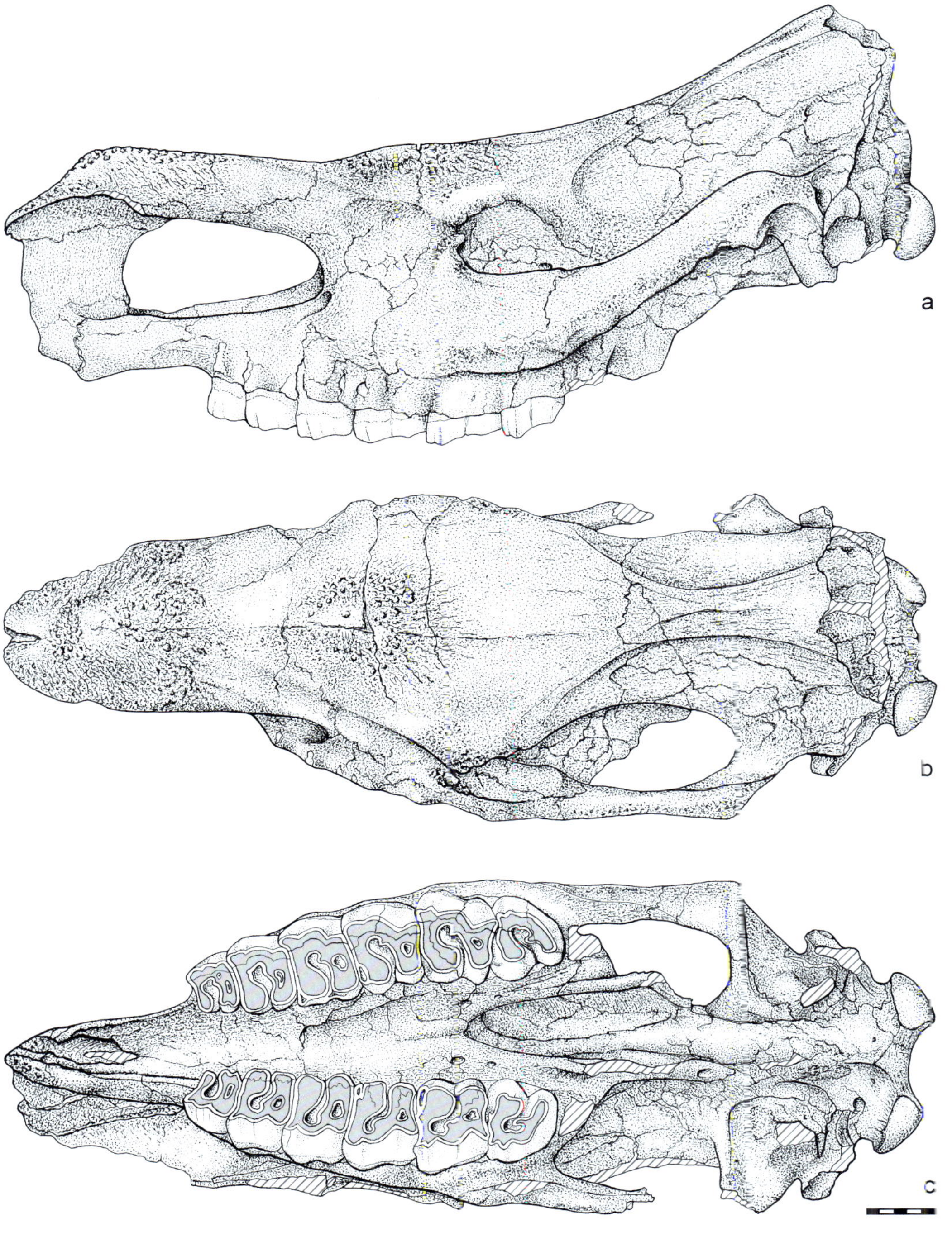

Fig. 6 *Stephanorhinus hundsheimensis* (Toula), Untermassfeld. – **a–c** Skull IQW 2009/30 270 (Mei. 29 432), female, left lateral, dorsal, ventral views. – Scale: 50 mm.

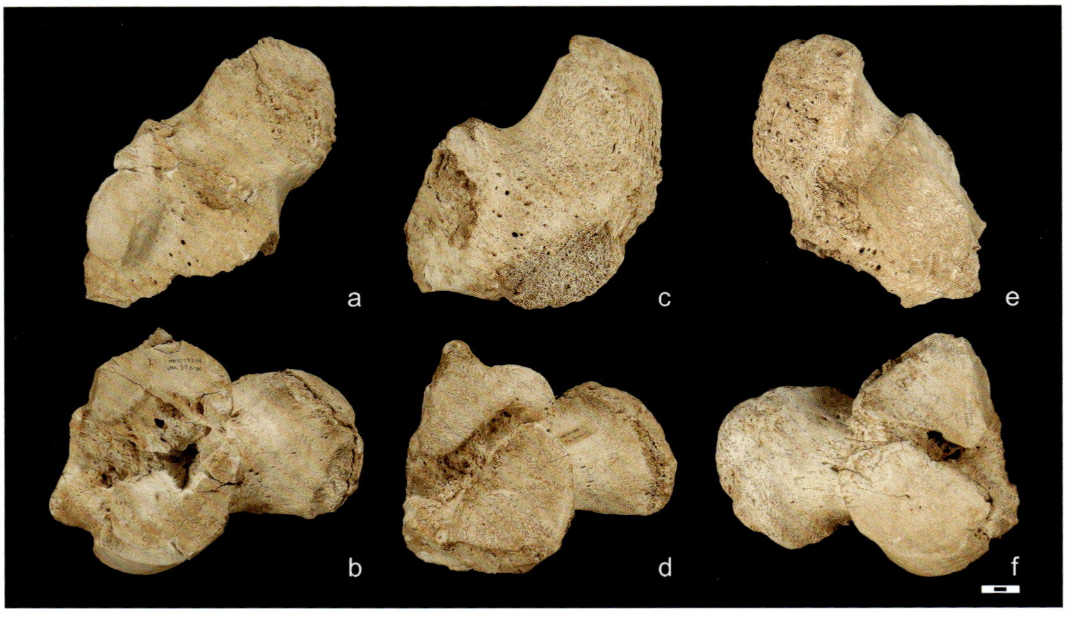

Fig. 7 *Mammuthus meridionalis* (Nesti), Untermassfeld. – **a–b** Right calcaneus IQW 1989/23299 (Mei. 22818). – **c–d** Right calcaneus IQW 1988/22488 (Mei. 22007). – **e–f** Left calcaneus IQW 2009/30418 (Mei. 29580). – **a**, **c**, **e** internal views; **b**, **d**, **f** dorsal views. – Scale: 30mm.

The extensive amount of Untermassfeld carnivore finds were distributed among felids, hyaenids, mustelids, ursids, and canids. A total of 291 felid specimens were identified and included in the investigations (Hemmer 2001; Hemmer and Kahlke in this volume), among them a partial skeleton each of cheetah, puma and lynx. According to the MNI, jaguars are the most frequently recorded cats in the Untermassfeld fauna (see section 6.3.): 47 isolated specimens were found, including the mandible of a young cub. The size variability of the dentition and skeletal elements does not correspond to that of a full population and smaller individuals in particular are obviously missing from the recovered material (Hemmer and Kahlke l.c.). The finds have been assigned to the Epivillafranchian to Middle Pleistocene *Panthera onca gombaszoegensis*.
With a total of 136 articulated elements partially found in situ, the cheetah *Acinonyx pardinensis pleistocaenicus* is the best anatomically documented felid species from Untermassfeld. The material includes two skulls with at least parts of their respective mandibles present (Hemmer 2001, pl. 135; Hemmer and Kahlke in this volume, Figs. 6–8). One of the skulls was found in articulation with the vertebral column of a forequarter skeleton (see frontispiece of this volume). In spite of their gigantic size, the physique of the Untermassfeld cheetahs is fundamentally similar to that of the extant species, so that a concordant hunting and killing mode may be inferred (Hemmer and Kahlke l.c.). *Acinonyx pardinensis pleistocaenicus* occurs in the East Asian Late Villafranchian as well as in the European Epivillafranchian and according to current knowledge differs from the older nominate form (Hemmer 2001), in particular in the design of the mandible.
48 anatomical elements belonging to an Epivillafranchian Old-World puma were also recovered at Untermassfeld. Significantly different in its morphology from both the Middle Villafranchian nominate form and the fossil cheetah-like pumas of the Americas, as well as from the recent species, this new subspecies *Puma pardoides brevipes* was a morphologically generalised cat without a far-reaching specialisation, but well adapted to life in wooded habitats (Hemmer and Kahlke in this volume).
In the 17 individual specimens of the Untermassfeld *Lynx* material there are several well preserved cervical and thoracic vertebrae with the same relative vertebral length as that recorded in the extant *L. pardinus* and *L. lynx*, in contrast to the long-necked *L. issiodorensis*. This crucial difference in functional morphology

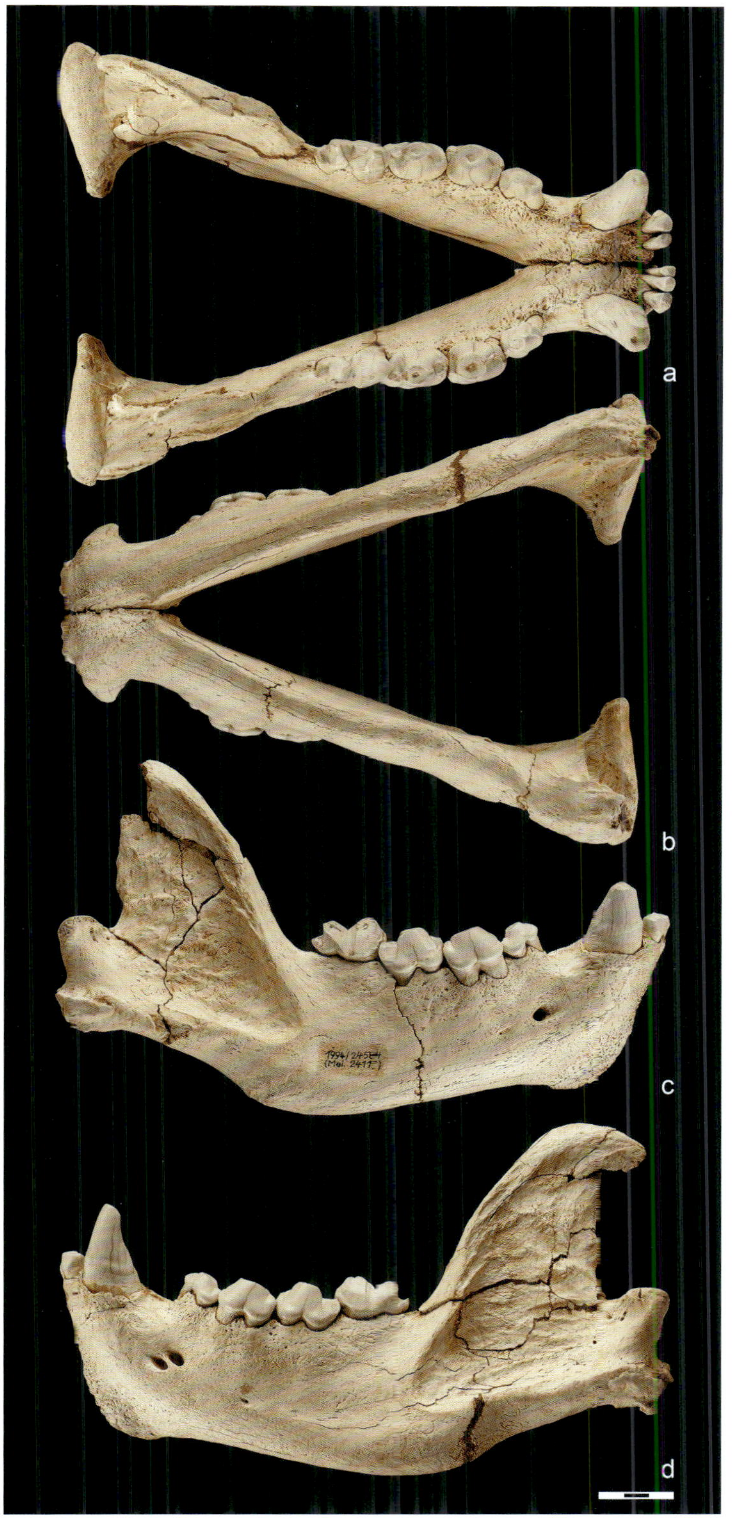

Fig. 8 *Pachycrocuta brevirostris* (Gervais), Untermassfeld. – **a–d** Left hemimandible IQW 1993/24369 (Mei. 23898) with I_2-M_1 and right hemimandible IQW 1994/24584 (Mei. 24113) with I_1-M_1, occlusal, ventral, right and left lateral views. – Scale: 30 mm.

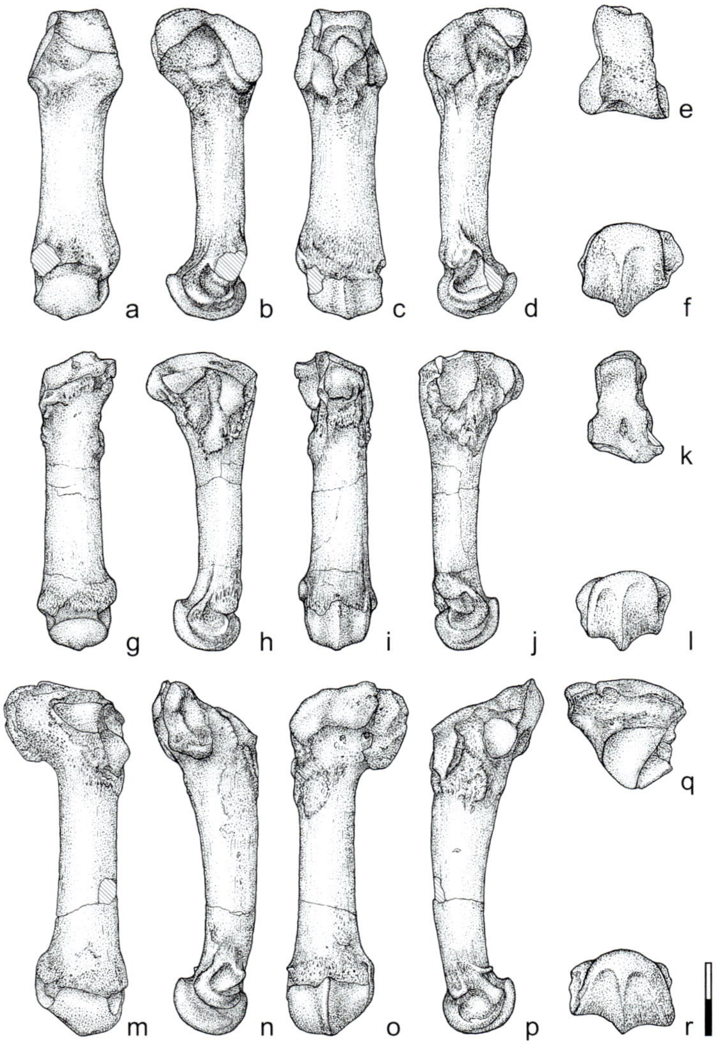

Fig. 9 *Ursus* ex gr. *U. dolinensis* García et Arsuaga, Untermassfeld. – **a–f** Left metacarpal 4 IQW 1980/16307 (Mei. 15818). – **g–l** Left metatarsal 3 IQW 1987/22001 (Mei. 21520). – **m–r** Right metatarsal 5 IQW 1989/23118 (Mei. 22637). – **a**, **g**, **m** anterior views; **b**, **h**, **p** internal views; **c**, **i**, **o** posterior views; **d**, **j**, **n** external views; **e**, **k**, **q** proximal views; **f**, **l**, **r** distal views. – Scale: 20mm.

supports an upgrade, together with the closely related Villafranchian South European lynxes, to that of a separate species, *L. valdarnensis*. The Epivillafranchian lynx from Untermassfeld has been allocated to a new chrono-subspecies, *L. valdarnensis spelaeoides* (Hemmer and Kahlke l.c.).

The dirk-toothed cat *Megantereon* is represented by 12 individual elements. The finds, which belong to two individuals, demonstrate an extremely wide size variability, indicating strong sexual dimorphism. They have been assigned to the evolutionarily advanced *Megantereon* of the European Late Villafranchian and Epivillafranchian and are listed as separate species *M. adroveri* (Hemmer and Kahlke l.c.).

The presence of scimitar cats at Untermassfeld has also been established with some notable finds. A total of 31 identified specimens include a hemimandible of a juvenile and two complete mandibles of adult individuals. For the taxonomy of Pleistocene *Homotherium*, a two species concept is advocated (Hemmer and Kahlke l.c.), which separates the more robustly built Middle to early Late Villafranchian populations of Eurasian *H. crenatidens* from the more gracile species *H. latidens*. This geochronologically younger species

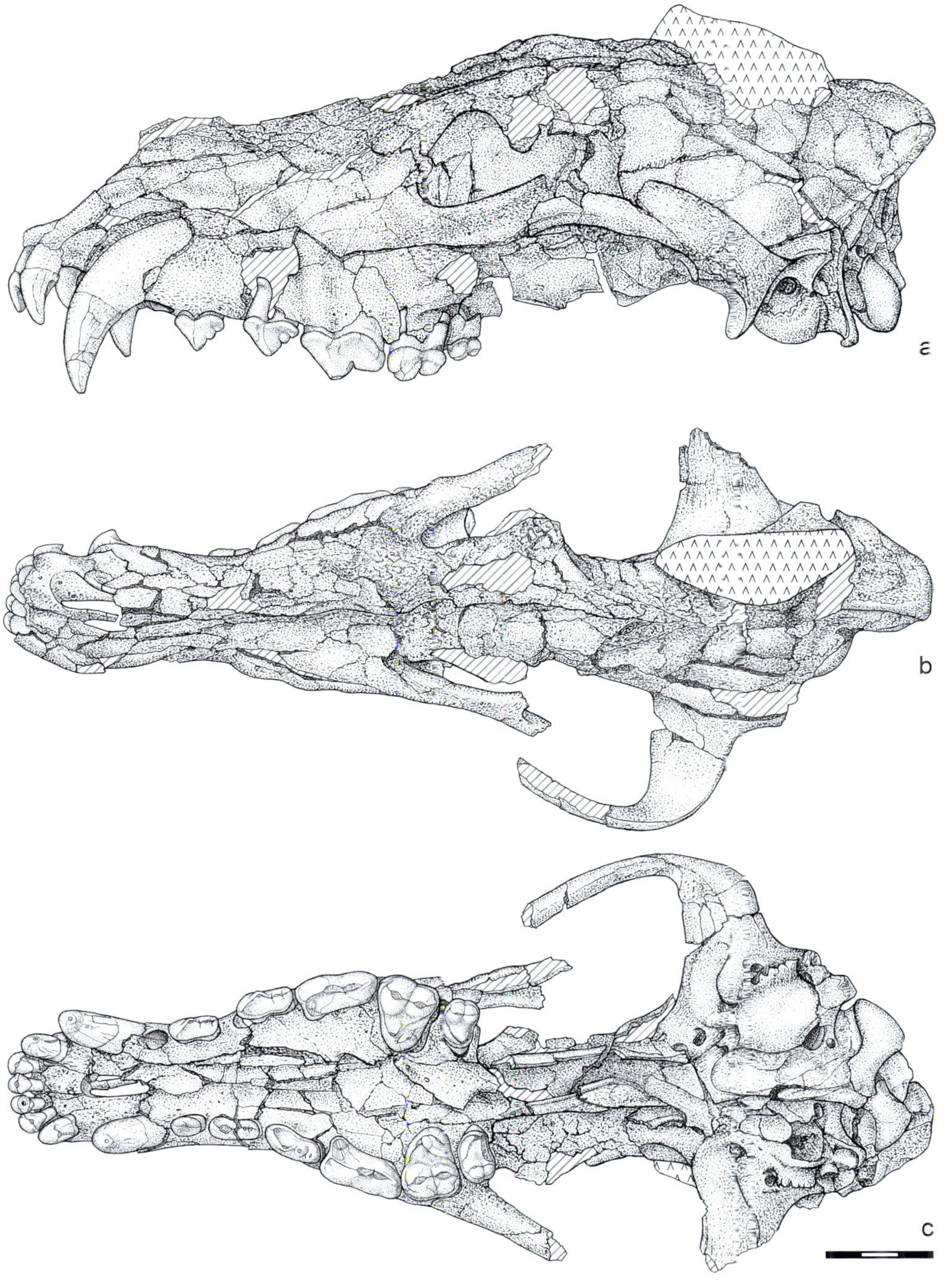

Fig. 10 *Xenocyon lycaonoides* Kretzoi, Untermassfeld. – **a–c** Skull IQW 2016/ 44170 (Mei. 49479) with a bleached porphyrite pebble from reworked Younger Coarse Gravels (YCG) pressed into the brain case, probably female, left lateral, dorsal, ventral views. – Scale: 30mm.

is subdivided into the latest Early to Middle Pleistocene *H. latidens ultimum*, the Middle to Late Pleistocene European to East Beringian *H. latidens latidens*, and *H. latidens serum* of southern North America. The Untermassfeld scimitar cat has been assigned to *H. latidens ultimum*, which roamed across Europe and into East Asia, surviving there up to the Middle Pleistocene.

A rich amount of hyaenid material of the large-bodied *Pachycrocuta brevirostris* was recovered from the site (**Fig. 8**). A total of 230 dental and skeletal elements (Turner 2001; Kahlke 2001e; Iannucci et al. in this volume), 435 corresponding coprolites (Keiler 2001b; section 6.1.2.6.), and numerous bite and other marks on skeletal elements of various species attest to the presence of the species. Territorial marking in the area of the site by faeces is likely (Keiler 2001b, 696). Detailed results from a study on the bone modifications by *Pachycrocuta* are available for the osteological material of *Bison menneri* (Arnold in this volume). Biometric analysis suggests a slight increase in the body size of *P. brevirostris* during the Early Pleistocene in Europe (Iannucci et al. in this volume).

M u s t e l i d remains are among the rarest of the Untermassfeld finds. Only three specimens of *Meles*, including a skull with dentition and a left mandible, have been described or reported on so far (Wolsan 2001; R.-D. Kahlke 2001e). Six additional finds come from the 1979–2011 field seasons: P_4 dex. IQW 2012/38 580 (Mei. 35 554), M_1 dex. fragment IQW 1997/26 273 (Mei. 25 802), ramus mandibulae dex. fragment IQW 2019/45 556 (Mei. 50 865), metacarpal 3 dex. (juv.) IQW 1980/15 544 (Mei. 15 056), astragalus dex. IQW 2019/45 557 (Mei. 50 866) and metatarsal 3 dex. proximal fragment IQW 1980/16 040 (Mei. 15 551) (det. G. Utschig). Madurell-Malapeira et al. (2011) have assigned the material to the late Early Pleistocene subspecies *M. meles atavus*.

127 identified remains of u r s i d s provide evidence for an unusually high MNI given the fluviatile conditions (see section 6.4., **Fig. 9**). Based on primitive dentition characters, as also exhibited by arctoid ursids, Musil (2001b) places the Untermassfeld population in the evolutionary lineage of brown bears as the newly described species *Ursus rodei*. According to García (2004) and García and Mazza (in this volume), the combination of characters of the medium to large-sized Untermassfeld bears shows numerous similarities with the Le Vallonnet (France) forms and those from the Early Pleistocene levels of Atapuerca Gran Dolina (Spain). The Untermassfeld material is listed accordingly as *Ursus* ex gr. *U. dolinensis* (García and Mazza l.c.).

Based on both the MNI and the number of skeletal elements recovered, the c a n i d s are the most abundantly represented carnivore family in the Untermassfeld material. A total of 106 identified finds of *Xenocyon lycaonoides* (as of 01 December 2021) were recovered from both the LFS and the channel fill [UFS(l)]. In 2015, the shattered but nearly complete skull of an adult individual was recovered from grid squares Q 528 and 538, 2.02 m below site 0-level (Kahlke et al. 2020, Fig. 13a; **Fig. 10** and **Foldout III** of this paper). Judging by the size of an associated hemimandible IQW 2021/46 226 (Mei. 51 534) that was found in the immediate vicinity (Q 528, 2.06–2.03 m below site 0-level), the skull is believed to be from an adult female. In total, eight hemimandibles from at least six individuals of both sexes were recovered from Untermassfeld. Besides isolated dentition elements, anatomically connected vertebrae (see section 6.1.2.1.) and pelvic remains, all long bones, carpals and metacarpals, tarsals and metatarsals, as well as phalanges are recorded from the site. The first finds were described by Sotnikova (2001), subsequent remains were reported later (Kahlke 2006, 31 ff., Fig. 34). Recent comparative studies by M. Sotnikova (Moscow), which could not be included in the present volume due to time constraints, support the view of *Xenocyon* as a distinct genus (pers. comm. M. S. 25 October 2021).

Evidence for the presence of the medium-sized *Canis mosbachensis* at Untermassfeld is given by an unusually large number of fossil finds. With around 790 specimens, nearly all cranial and postcranial skeletal elements are present. Six nearly complete skulls from individuals of both sexes were recovered [IQW 1982/18 052 (Mei. 17 572), IQW 1984/20 177 (Mei. 19 697), IQW 1996/25 895 (Mei. 25 424), IQW 2000/27 001 (Mei. 26 506), IQW 2008/29 757 (Mei. 28 919); **Fig. 11**] (Sotnikova 2001, pl. 102–104), one of them with the articulated cervical vertebrae I–V [IQW 1997/26 017 (Mei. 25 546); **Fig. 12**] (Kahlke 2006, Fig. 35). The in-situ block preparation IQW 2016/43 953 (Mei. 49 262) contains two incomplete skulls (Kahlke et al. 2020, Fig. 4). Substantial parts of the facial skull of three other individuals have also survived [IQW 1982/18 249 (Mei. 17 769), IQW 2011/31 339 (Mei. 30 501), IQW 2012/38 639 (Mei. 35 613)]. An even higher number of individuals is supported by a series of 32 nearly complete hemimandibles (16 sin., 16 dex.) (see section

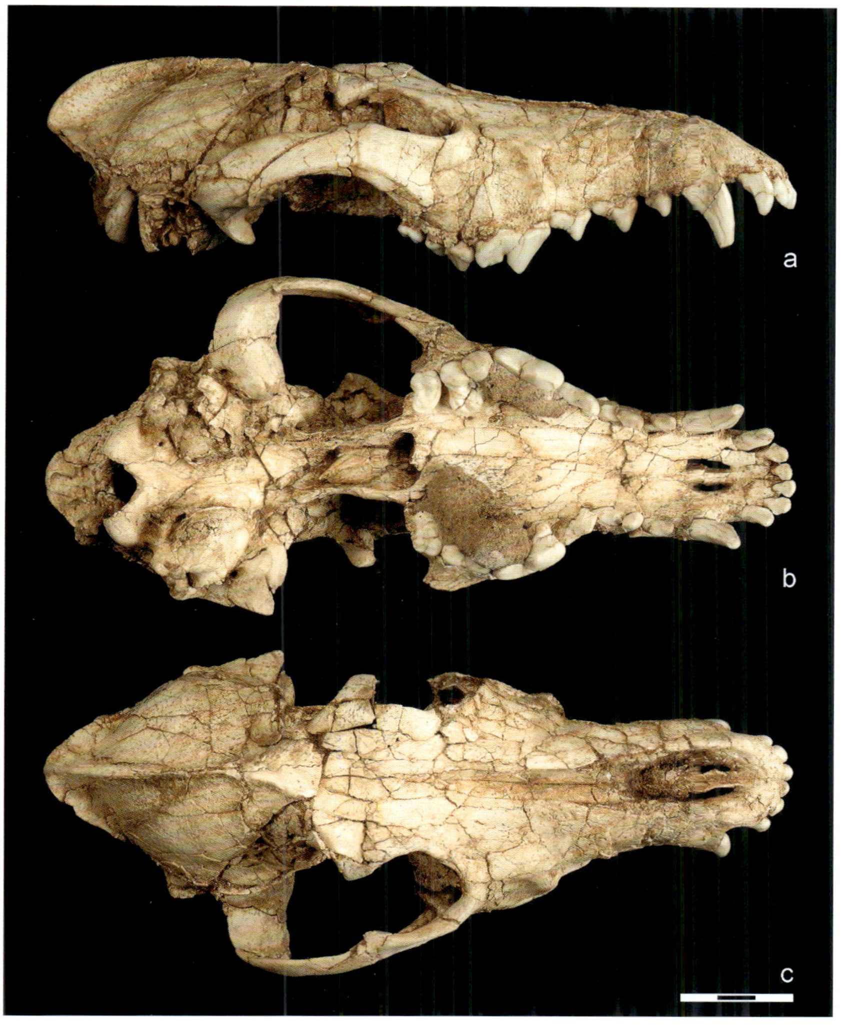

Fig. 11 *Canis mosbachensis* Soergel, Untermassfeld. – **a–c** Skull IQW 2008/29757 (Mei. 28919), probably male, right lateral ventral, dorsal views. – Scale: 30 cm.

6.3.). The total amount of material recovered offers the future possibility of a detailed population analysis of the Untermassfeld *C. mosbachensis*.

Extensive and systematic screen washing (Kahlke 2020, Fig. 19; see section 2.3.) and the recovery of numerous skeletal assemblages (Kahlke 2020, Fig. 11a; see section 6.1.2.5.) resulted in a considerable collection of small mammal remains. Rodents were most frequently recorded with a total of 14 species (Maul 2001a; in this volume). A few finds attest to the presence of a s c i u r i d and the relatively small size of the surviving teeth justifies its assignment to *Sciurus whitei*. More than 1,000 finds, including numerous skeletal units and crania (see section 6.1.2.5.), have led to the discovery of a new species of medium-sized ground squirrel, *Spermophilus kahlkei* (Maul in this volume). The presence of g l i r i d s at the site has been established by the dentition remains of some individuals of the dormouse *Glis sackdillingensis*.

The rarity of c a s t o r i d remains at Untermassfeld are most certainly due to the fact that it was repeated flooding events in the Werra valley that led to the formation of the fossil site (see section 4.2.). In addition to *Castor fiber*, several finds, including a mostly complete skull and an individually associated hemimandible, suggest the presence of *Trogontherium cuvieri* (Maul l.c.).

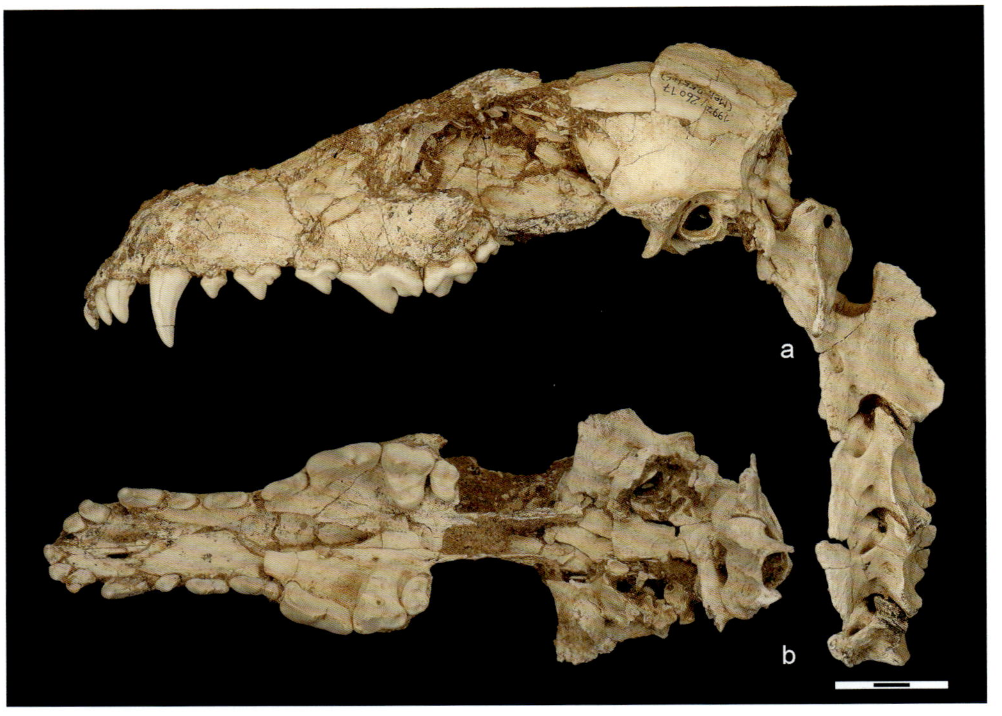

Fig. 12 *Canis mosbachensis* Soergel, Untermassfeld. – **a–b** Skull with cervical vertebrae I–V IQW 1997/26 017 (Mei. 25 546), probably male, left lateral, ventral views. – Scale: 30 cm.

Dipodids are represented by the single find of a birch mouse, *Sicista* sp. A largely disarticulated cricetid skeleton and several single specimens have been assigned to the large hamster *Cricetus runtonensis* (Maul l.c.). The bank vole *Myodes hintonianus* and the vole species *Pliomys episcopalis* and *Mimomys savini* are only sparsely represented, in contrast, *Mimomys pusillus* and *Microtus thenii* were found more frequently (Maul l.c.); a partial skeleton of the latter species is available. The presence of murids is documented by some dentition elements of the wood mouse *Apodemus sylvaticus* and the hystricid *Hystrix refossa* is represented by a pair of lower incisors (Maul 2001a; in this volume). The surfaces of several skeletal elements of larger mammals show gnawing marks by porcupines (see section 6.2.3.).

Only one leporid find, a left P_3, can be definitively assigned to *Hypolagus brachygnathus* (Maul 2007; in this volume). Several finds are evidence for the presence of a hare, *Lepus* sp. However, the vast majority of identified leporid remains cannot be assigned with any certainty to either of the mentioned genera.

A total of eight cercopithecid finds belonging to three individuals of both sexes of the macaque *Macaca sylvanus* were recovered (Zapfe 2001; Reumer and Kahlke in this volume). Given this species is an excellent climber it would not have been very susceptible to flooding.

Single finds indicate the existence of erinaceids and talpids. Besides the hedgehog *Erinaceus* cf. *europaeus* Linnaeus, 1758, two mole species of differing body size, *Talpa* cf. *europaea* and *T. minor*, have been verified. The presence of a desman, *Desmana thermalis*, has also been established by the individually associated parts of a mandible. Soricids are represented by the shrews *Sorex minutus* (single find), *S. runtonensis*, *S. margaritodon*, and *Beremendia fissidens* (Maul 2001; in this volume).

– 5.2.7.1. Pathological findings

Pathological findings on skeletal and dental elements of several large mammal species of the Untermassfeld fauna have enabled conclusions to be drawn about the life history of the individuals concerned and their interrelationships with various environmental parameters. For example, diffuse idiopathic skeletal hyperostosis (DISH) was recorded in various skeletal elements of an adult individual of *Ursus* ex gr. *U. dolinensis*, which could indicate a possible relationship to type-2 diabetes (Kahlke and Kierdorf 1997; 2005). Degenerative joint disease (osteoarthritis) was observed in several cases in the postcranial skeleton of adult *Hippopotamus antiquus* (Kahlke 1997b, 356; Kierdorf and Kahlke 2004; 2020). Here, a causal relationship with the topography of the site environment might exist (see section 7.2.). Skeletal traumata, partly associated with a post-traumatic osteomyelitis, i.e., an infectious bone disease, were found on long bones of *Bison menneri* (Kierdorf et al. 2012; in this volume) and the right front paw of a *Puma pardoides brevipes* (Hemmer and Kahlke in this volume). On one metacarpal and one metatarsal bone of *Acinonyx pardinensis pleistocaenicus* significant reactive periosteal bone lesions were found (Hemmer and Kahlke in this volume), which could have been caused by an inflammation of the periosteum (pers. comm. U. Kierdorf, June 2021). In addition, exostoses were recorded on the bodies of thoracic vertebrae of a *Lynx valdarnensis spelaeoides* (Hemmer and Kahlke l.c.).

Several individuals of *Hippopotamus antiquus* exhibit dental and jaw pathologies (Kahlke 1997b; Kierdorf and Kahlke 2004; 2020). As a result, post-eruptive enamel lesions in a dP_4 and associated bone lesions in the mandible have been verified, along with developmental defects in the root of a dP_3, pathological root resorption in a dP_2, and enamel hypoplasia in a permanent mandibular canine and in an M_1.

Since not all Untermassfeld mammalian groups have been examined for pathological conditions, further discoveries are expected.

6. Taphonomy

6.1. Spatial distribution of vertebrate remains

Different geological conditions and processes have led to the formation of the diverse depositional environments which make up the area of the present-day site (Kahlke et al. 2020; see section 4.2. of this paper). The sedimentological characteristics shaped the spatial arrangement of the vertebrate remains, and the effect of various geological and hydrological factors have enabled a distinction to be made between larger and smaller scale distribution patterns (see sections 6.1.1., 6.1.2.).

In the north-western and south-western areas of the site, the increasing dip of the fossiliferous layers towards the SW is made apparent by the tendency of the fossils to be found at increasingly deeper levels in that direction [e.g., Kahlke 2001e: compare the find depths on Foldouts VIII and XI (W) with those on IX and XII (E); Kahlke et al. 2020, Foldout I). This neotectonically caused tilt (Ellenberg and Kahlke 1997; see section 4.2. of this paper) occurred after the formation of the site and is therefore not included in the following taphonomic discussion.

6.1.1. Large-scale distribution patterns

A total of 237 grid squares was excavated within the Lower Fluviatile Sands (LFS) that accumulated leeside of the clastic mudflow fan (CMF) (see section 4.1., **Fig. 2**). These sands represent stage II of the site's evolution model (Kahlke et al. 2020; see section 4.2. of this paper). A further 376 squares, representing stage III of the site's evolution model, were examined within the channel infill. Along the north-eastern flank of the erosion channel, an additional 54 squares contained material from both of the aforementioned sediment bodies (**Fig. 13, Foldout II**). The excavation progress of the 2001–2019 field seasons followed the configuration of the SE/NW-oriented channel infill, > 1 m below site 0-level (**Foldout III**).

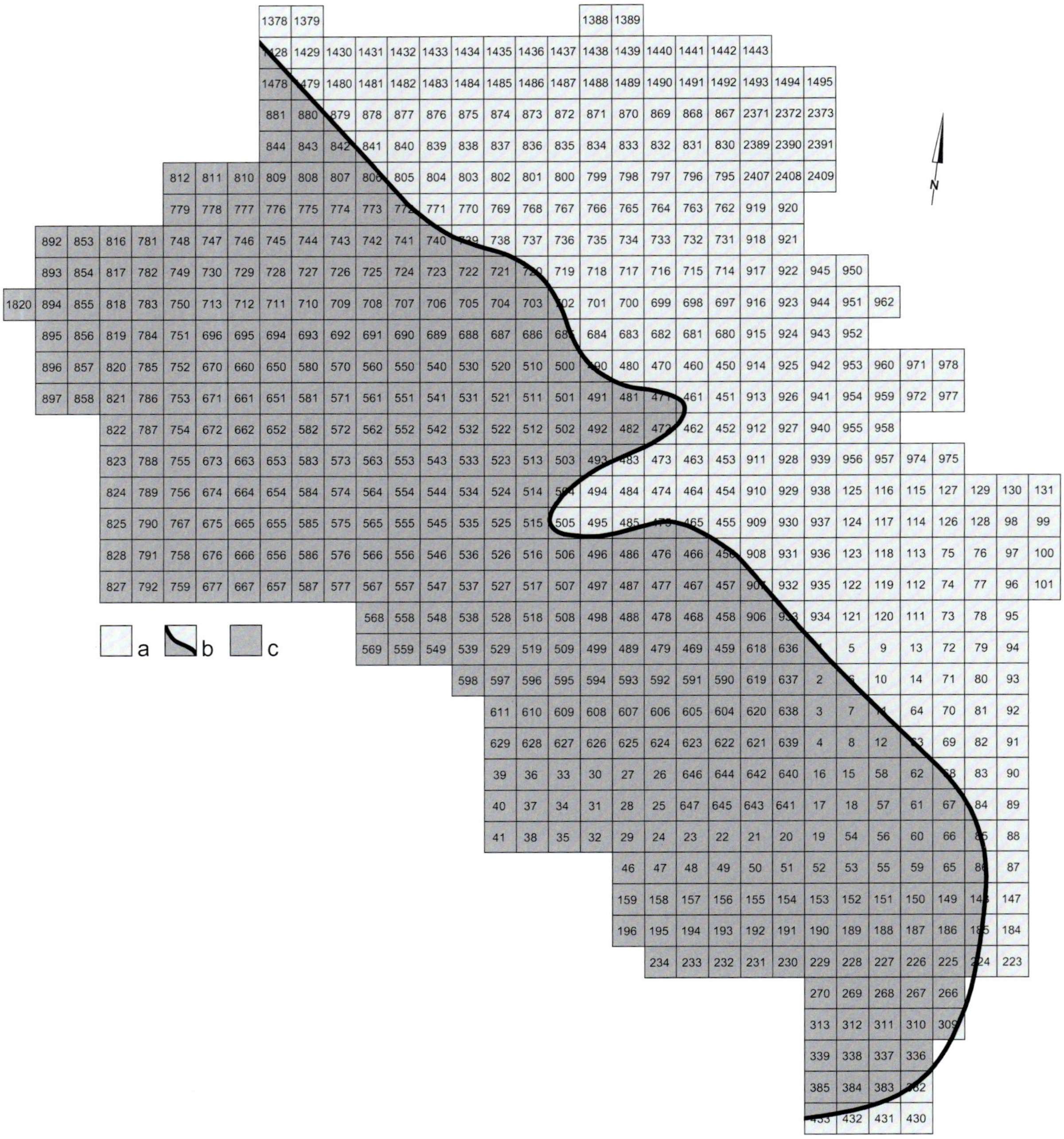

Fig. 13 Untermassfeld research excavation, total area with numbered grid squares and representation of the fossiliferous sediment bodies at site 0-level (1 m above main planum = Planum I). – **a** Flooded surfaces formed by Lower Fluviatile Sands (LFS). – **b** Flank of the erosional channel (after R.-D. Kahlke 2001b). – **c** Channel infill formed by Upper Fluviatile Sands (lower part) [UFS(l)] and other types of sediments. For distribution patterns of the vertebrate finds see **Foldouts II** and **III**.

– 6.1.1.1. Flooded surfaces

The fossiliferous LFS outcrop is situated mainly in the north-eastern part of the site (Kahlke 2001e, north-eastern area of foldout VII, eastern area of foldout XII, and entire area of foldout XIII; **Fig. 13**, **Foldout II** of this paper). The number of finds contained in the LFS was usually lower than that in the south-western adjacent sediments of the channel infill (see section 6.1.1.2.). As a rule, bone material recovered from the LFS is largely disarticulated and in some rare cases strongly fragmented, although articulated skeletal units were also found (Kahlke 2001e, 952; 2006, 48 ff.; see section 6.1.2.1. of this paper).

Conspicuous within the flooded surfaces, however, are three S/N- to SW/NE-oriented bone concentrations, each extending over several grid squares. The most pronounced concentration between grid squares Q 915/924 and 919/920 (**Foldout II**) has already been described in detail (Kahlke 2001e, 952–953; 2006, 47) and yielded about 800 separate finds. Grid square Q 921, for example, contained 118 catalogued specimens, whereas Q 922 and 923 had 110 and 223 skeletal elements, respectively. The bones of this concentration were dispersed through several sand layers within a height interval of 1.85 m, i.e., 0.65 m below (Q 924) to 1.20 m above site 0-level (Q 922), with the majority of finds deposited between 0 and 1.00 m above 0-level. Because the bone concentration was close to the riverside and nearly all long bones and mandibular rami were aligned parallel to its bank, a repeated deposition of skeletal elements in a wrack zone is suspected as being the main agent in the formation of this deposit.

Two more similarly elongated bone concentrations, but with significantly lower numbers of finds and less pronounced long bone alignment, were found between grid squares Q 490 and 763 as well as between Q 165/166 and 188/189 and contained 411 and 154 catalogued specimens, respectively (**Foldout II**). The finds of the Q 490 to 763 concentration were distributed over a depth interval of 1.00 m below to 1.30 m above site 0-level, with the highest number of finds at depths from 0.60 below to 1.30 m above 0-level. The somewhat less pronounced bone concentration between Q 165/166 and 188/189 spanned depths of 0.25 m below to 1.00 m above 0-level, with the bulk of elements found between 0-level and up to 1.0 m above it. Both structures were aligned at an acute angle or even perpendicular to the erosion channel, which became deeper in the immediate south-western vicinity. These bone deposits most probably traced the contour of temporary shoals that had formed on the flooded surfaces during the deposition of the LFS.

– 6.1.1.2. Channel filling

The higher sections of the SE/NW oriented channel infill above the main planum were excavated before the year 2001 (see section 2.1.). During the years 1989–1992, the north-western area of the site was explored. This area was where the deepest part of the channel infill was located and possessed large amounts of vertebrate finds (Kahlke 2001e, from Q 564 at foldout VIII to Q 660 at foldout XI and Q 783 at foldout X). Starting with the 2001 field season, the remaining deeper section of the erosion channel below the main planum was excavated from the immediate edge of the CMF onwards in a northwesterly direction (Kahlke 2020, Fig. 1, Table 3), until the previously studied deeper sections in the north-western area of the site were reached (**Fig. 13**, **Foldout II**, frontispiece of this volume). The fossil content of this channel filling has therefore been completely recorded.

The distribution pattern of the vertebrate finds within the sediments of the channel filling was clearly different from that of the LFS (see section 6.1.1.1.). Thousands of finds reflect turbulent, but quite regular accumulation events during the filling of the channel. Starting in the immediate vicinity of the CMF, a succession of bone concentrations and more extended bone accumulations (see sections 6.1.2.3., 6.1.2.4.) clearly marked out the contours of the former erosion channel over a distance of 26 m (**Foldout II**). Leeside of the CMF, acting as an obstacle to the flowing river water (see section 4.2.) two zones with different bone distribution patterns were formed, each shaped by the respective flow regime (**Foldout III**).

Adjoining the CMF to the northwest, between grid squares Q 21, 46, 155 and 195, is an irregularly arranged fossil accumulation of about 14 m² that is comprised of a high proportion of relatively heavy skeletal elements or anatomically connected large mammal skeletal units (see section 6.1.2.1., **Foldout III**, frontispiece of this volume). Interestingly, these bodies usually have no anchor-like prominent parts, such as antlers, protruding canines or bulky articulated ribs. **Fig. 14** shows parts of a *Bison menneri* spine (Q 157/194; **Table 2**, No 25, 26) whose transportability into the deposition area was made possible by the systematic separation of the ribs from the vertebral bodies by *Pachycrocuta brevirostris*.

Thus, in this zone of, at first glance, randomly arranged bones, a certain separation of accumulated skeletal elements becomes apparent, which is most likely a result of the transportability of different body parts in flowing water. The irregular bone configuration within the aforementioned grid squares was evidently caused by violent turbulence within the flowing water immediately leeward of the CMF. The correspondingly configured excavation area should therefore be referred to as a turbulence zone (**Fig. 15**).

To the immediate northwest of the turbulence zone, between grid squares Q 24/25 and 627/628, was an approximately 8 m² zone of clearly adjusted skeletal elements whose longitudinal axes were oriented more or less parallel or at right angles to the direction of flow (see section 6.1.2.4., **Foldout III**). In accordance with the turbulence zone, this very characteristic large mammal bone accumulation, which is essential in understanding the fluviatile character of the site, should therefore be called an alignment zone (**Fig. 16**).

A further bone accumulation with a similar SE/NW orientation but shifted by 2 m to the north joined the channel filling. This accumulation covered an area of approximately 3 m² between grid squares Q 519 and 528 (**Foldout II** and **III**) and contained a lower number of aligned skeletal elements. Following after an area with polyspecific bone concentrations (see section 6.1.2.3.) was a third bone accumulation spanning 11 m² between grid squares Q 660 and 818 (**Foldout II**). This accumulation contained an even lower proportion of aligned bones (see section 6.1.2.4.).

The arrangement of finds in the higher section of the channel filling, < 1 m below site 0-level, i.e., above the main planum of the excavation, is documented between grid squares Q 533, 582, 711, and 843 (**Foldout II**). This area was located between the recorded bone accumulations of the deepest part of the fossiliferous channel sediments and the north-eastern flank of the channel. In the higher part of the channel sediments [UFS(l)], vertebrate finds mostly occurred at lesser concentrations than those recorded in the deeper sections. The large mammal skeletal units (see section 6.1.2.1.) involved in the respective bone concentrations or accumulations of the channel filling match, in both quantity and arrangement (**Foldout II**), the described distribution pattern of the remaining finds (see section 6.1.2.2.). Overall, it becomes clear that the highly energetic flow regime that prevailed immediately after the culmination of the flood event subsided in the course of replenishing the freshly eroded channel.

On the valley side, i.e., on the south-western edge of the bone accumulation deposit in the deepest part of the channel infill, numerous micromammal skeletal units and hyaena coprolite concentrations were found in a zonal arrangement over many grid squares (see sections 6.1.2.5., 6.1.2.6.). The analysis of these smaller-scaled fossil distribution patterns has also contributed to the understanding of the site's formation and the origin of the accumulated necromass.

6.1.2. Smaller-scale distribution patterns

Analyses of the spatial distribution within the areas of the site excavated before 2001 led to the differentiation of four essential smaller-scale distribution patterns of large mammal remains, namely anatomically connected skeletal units, isolated finds, polyspecific bone concentrations, and bone accumulations (R.-D.

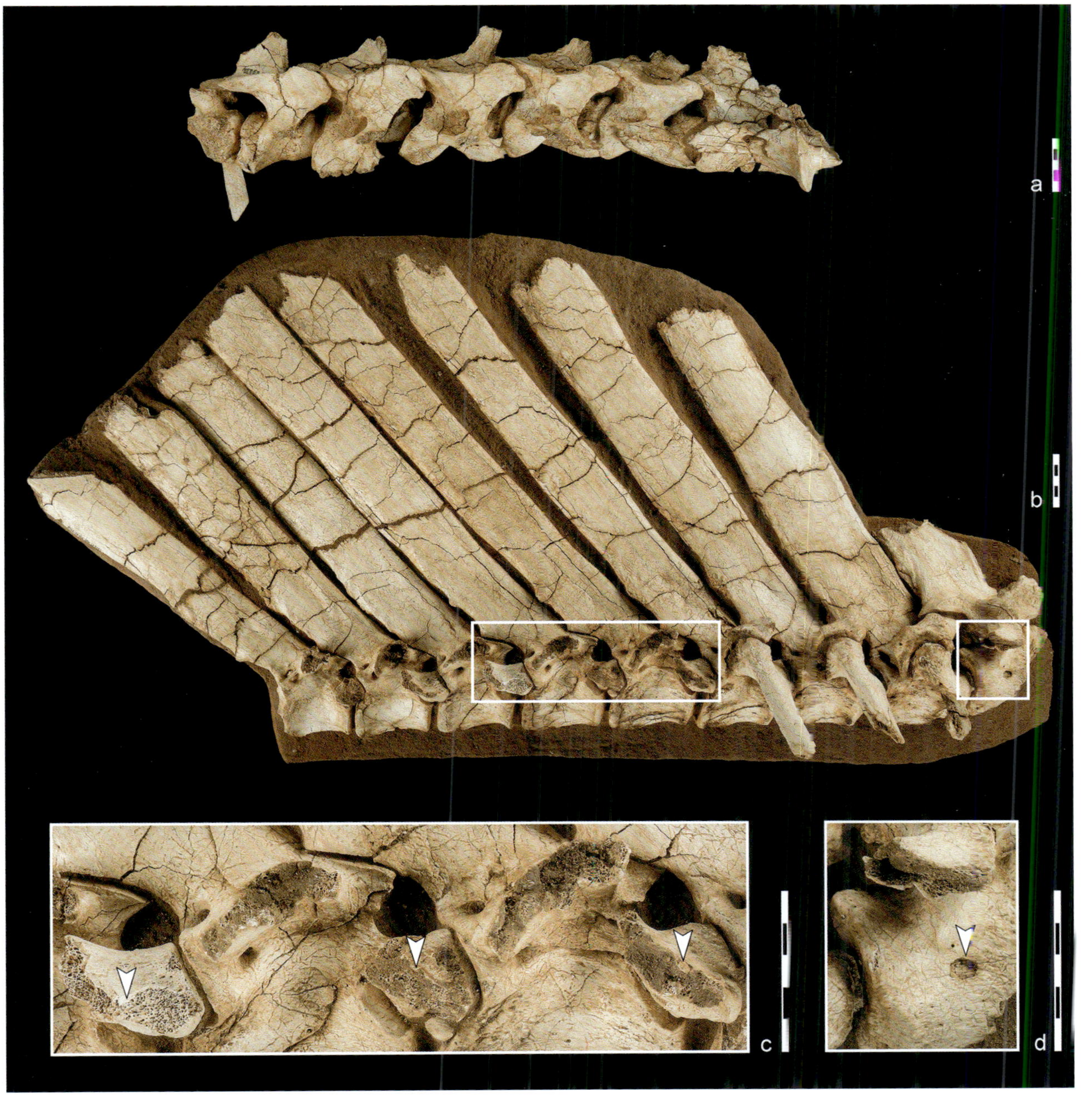

Fig. 14 Untermassfeld research excavation, articulated spine sections of *Bison menneri*. – **a** Cervical spine IQW 2004/28 696 (Mei. 27 858) consisting of CII–VII, partial TI and proximal fragment of first left costa from grid squares Q 157/194, 1.77–1.64 m below site 0-level, right lateral view (top view of the embedding status). – **b–d** Thoracic spine IQW 2003/28 190 (Mei. 27 352) consisting of TIV–XII and corresponding proximal rib fragments with traces of hyaena (*Pachycrocuta brevirostris*) gnawing (arrowheads) and biting from grid squares Q 47/48 and 157/158, 1.82–1.68 m below site 0-level, right lateral views (top views of the embedding status). – Scales: 50 mm.

Kahlke 1997c; 2001d; 2006). The existence of these distribution patterns was confirmed by subsequent excavations from 2001 onwards (**Foldout III**). Information concerning the large-scale arrangement of the finds excavated from the channel filling can be found in the overview given above (see section 5.1.1.2.).

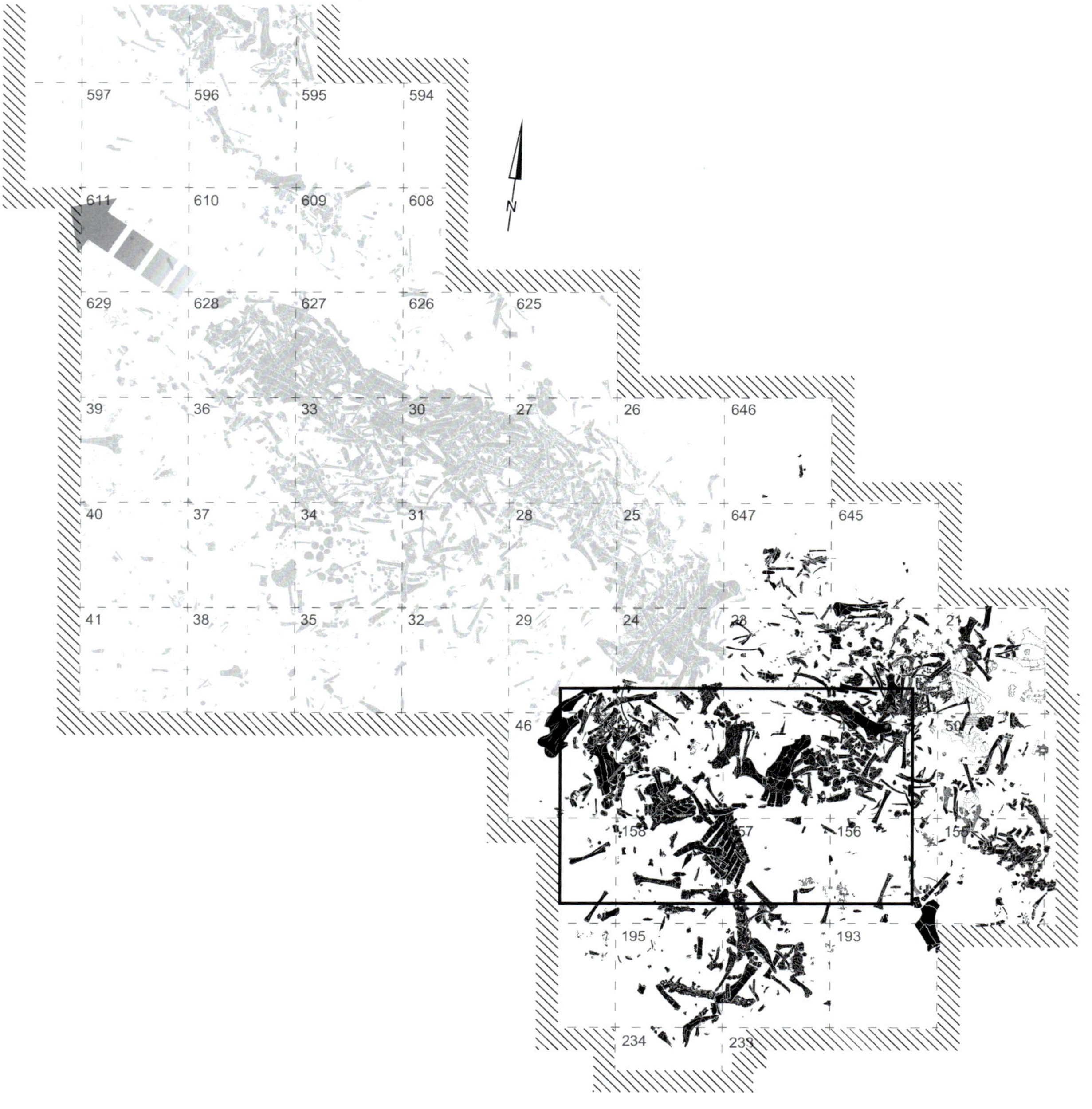

Fig. 15 Untermassfeld research excavation (2001–2015 field seasons), southern area immediately north of the coarse clastic mudflow fan (CMF). – **Above** Distribution pattern of the vertebrate finds within the channel filling. – **Below** Detail of the turbulence zone with a high proportion of heavy skeletal elements and anatomically connected skeletal units. – Black: irregular bone configuration of the turbulence zone. – Arrow: main flow direction of the river water leeward of the CMF.

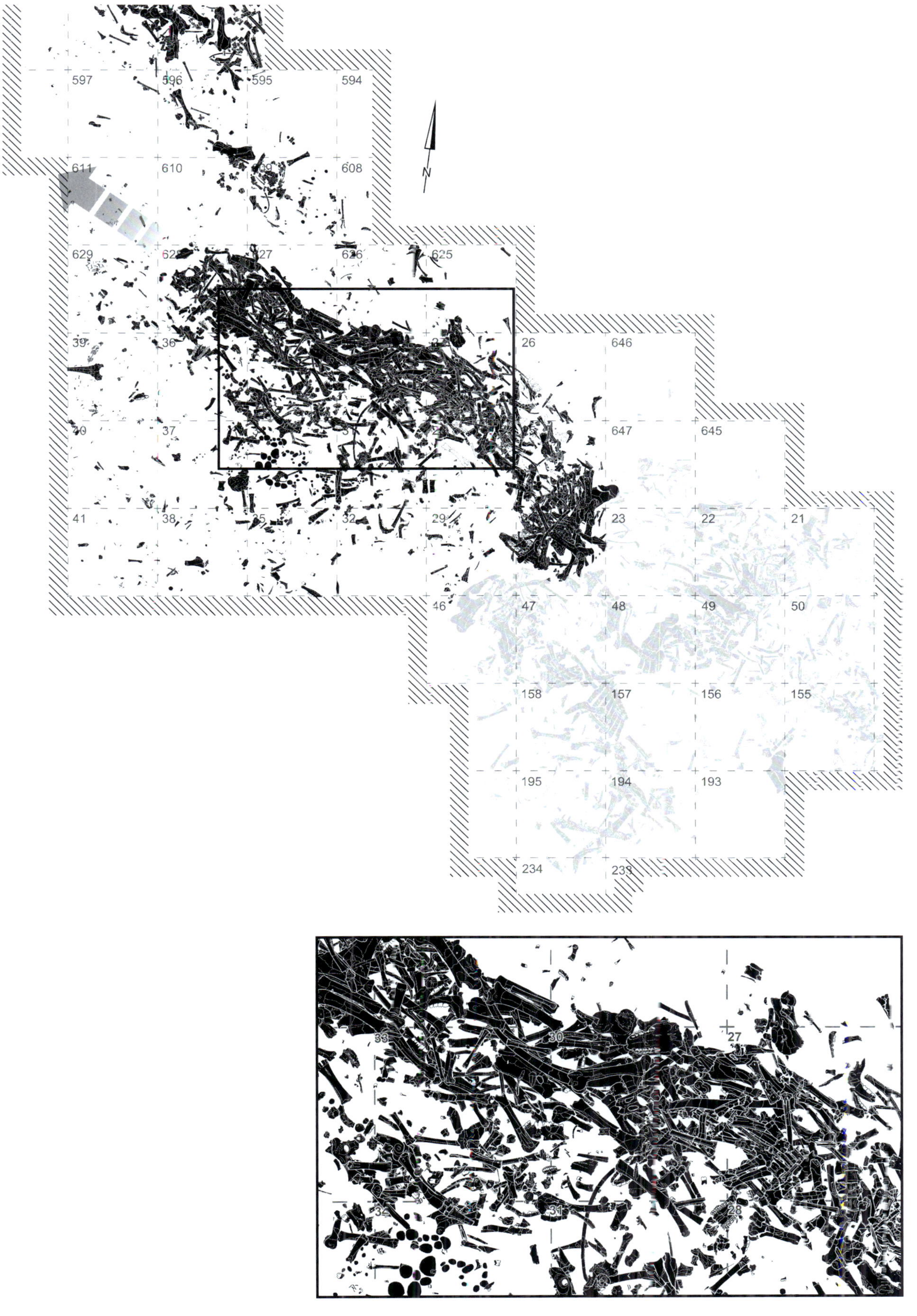

Fig. 16 Untermassfeld research excavation (2001–2015 field seasons), southern area immediately north of the coarse clastic mudflow fan (CMF). – **Above** Distribution pattern of the vertebrate finds within the channel filling. – **Below** Detail of the alignment zone with a high proportion of skeletal elements whose longitudinal axes are oriented more or less parallel with the direction of water flow. – Black: adjusted bone configuration of the alignment zone. – Arrow: main flow direction of the river water leeward of the CMF

– 6.1.2.1. Connected large mammal skeletal units

Complete, or nearly complete, skeletal finds of large mammals were not recorded at the site. Anatomically connected units are characterised by articulated skeletal elements or by disarticulations with minor displacements of individual elements that immediately reveal the anatomical context. In most cases, the connected large mammal skeletal units recovered are completely or partially preserved extremities that became dislodged from the shoulder or pelvic girdle. Occasionally, sections of vertebral columns of differing lengths and anatomical positions were deposited (listings of previous finds in Kahlke 2001e, 953 ff., Figs. 10–12; 2006, 48 ff.). Extreme torsion of skeletal units resulting from tendon contraction due to rapid dehydration has never been observed.

During the excavations which took place before 2001, 48 anatomical units of large herbivores as well as three units of *Canis mosbachensis* (Kahlke 1997c; 2001d; 2006; Sotnikova 2001) were recovered from the flooded surfaces of the LFS (Kahlke 2001e, 952 f.; 2006, 47) and, in most cases, the higher levels of the channel infill [UFS(l)] (Kahlke 2001e; 2006, 43 ff.; Kahlke et al. 2020). Altogether, these units comprise 503 individual skeletal elements representing about 5.0 % of the total amount of identifiable large mammal finds. In addition, 112 units were reconstructed from isolated specimens (Kahlke 2001e, Table 1; 2006, Table 8).

A significantly larger number of anatomically connected skeletal units were recovered from the deepest part of the channel filling during the 2001–2015 field seasons (**Fig. 17**), i. e., from the so-called channel lag deposits (Kahlke 2001e, 951 f.; 2006, 46 f.) in the southern area of the site, immediately north of the coarse clastic mudflow fan (CMF). 85 anatomically connected large mammal units consisting of 549 individual elements were recovered from this area. The units were distributed over the following large mammal species: *Bison menneri* (24), *Pseudodama vallonnetensis* (24), *Eucladoceros giulii* (21), *Hippopotamus antiquus* (5), *Stephanorhinus hundsheimensis* (4), *Acinonyx pardinensis* (4), *Xenocyon lycaonoides* (2) and *Homotherium latidens* (1) (**Table 2**). Out of a total of 4,147 catalogued large mammal remains (Kahlke 2020), comprising 4,319 single skeletal elements excavated from the aforementioned area, anatomically connected units represent about 12.7 % of the finds. Apparently, the more turbulent flow regime immediately following the peak of the flooding, when the channel lag sediments were accumulating, favoured the deposition of anatomically connected parts of large mammal cadavers.

Units from isolated elements have not yet been systematically reconstructed for the 2001–2015 excavations. Analysis of the cat finds from the deeper areas of the channel filling shows, however, that corresponding units can be identified. For example, *Puma pardoides brevipes* has a well preserved partial skeleton (Hemmer and Kahlke in this volume).

The number of anatomically connected skeletal units of large herbivores is higher than that of carnivores because of the generally larger NISP and MNI of the former represented in the 2001–2015 fossil record. However, neither the robustness of the specific skeletal structure, nor the absolute frequency of skeletal elements within the total material is directly reflected within the preserved herbivore skeletal units. The arrangement and frequency of the recorded skeletal units are primarily determined by their mechanical characteristics. The spatial arrangement corresponds to size and weight ranges, basic forms, and orientation patterns of adjacent finds (**Fig. 17**; see section 6.1.1.).

The transportability of skeletal elements and connected skeletal units by flowing river water is an essential requirement for the accumulation of such finds in the channel filling at Untermassfeld (Kahlke et al. 2020; see section 4.2. of this paper). The fact that obstructive skeletal elements, e. g., complete antlers attached to skulls or shed beams, bovid skulls with horn cores, proboscidean tusks or mandibles, rhinocerotid skulls with originally attached keratin horns, and hippopotamus mandibles with exposed canines and incisors, were rare or completely absent from the site had already been identified through analysis of the fossil mate-

No	(Main) grid square (Q) Skeletal elements [side of the body]	Species	Find depth below site 0-level	Number of elements (n)
1	Q 155	*Pseudodama vallonnetensis*	1.97–1.95 m	7
	Hum. IQW 2007/29 233 (Mei. 28 395), Rad. et Ulna IQW 2007/29 234 (Mei. 28 396), Ci IQW 2007/29 235 (Mei. 28 397), Cu IQW 2007/29 235 (Mei. 28 397), C2/C3 IQW 2007/29 235 (Mei. 28 397), Mc3/4 QW 2007/29 235 (Mei. 28 397) [sin.]			
2	Q 155	*Hippopotamus antiquus*	2.04–1.99 m	4
	Mt2 IQW 2007/29 303 (Mei. 28 465), Mt3 IQW 2007/29 569 (Mei. 28 731), Mt4 IQW 2007/29 570 (Mei. 28 732), Mt5 IQW 2007/29 313 (Mei. 28 475) [sin.]			
3	Q 50	*Pseudodama vallonnetensis*	1.88 m	6
	Tt IQW 2005/28 832 (Mei. 27 994), Tf IQW 2005/28 833 (Mei. 27 995), Tc/T4 IQW 2005/28 829 (Mei. 27 991), T1 IQW 2005/28 831 (Mei. 27 993), T2/T3 IQW 2005/28 830 (Mei. 27 992), Mt3/4 IQW 2005/28 828 (Mei. 27 990) [dex.]			
4	Q 50	*Eucladoceros giulii*	2.49–2.29 m	11
	Hum. IQW 2008/29 648 (Mei. 28 810), Rad. et Ulna IQW 2008/29 647 (Mei. 28 809), Cr IQW 2008/29 654 (Mei. 28 816), Ci IQW 2008/29 654 (Mei. 28 816), Cu IQW 2008/29 653 (Mei. 28 815), Ca IQW 2008/29 654 (Mei. 28 816), C2/C3 IQW 2008/29 654 (Mei. 28 816), C4 IQW 2008/29 653 (Mei. 28 815), Mc3/4 IQW 2008/29 651 (Mei. 28 813), Metatarsal sesamoid prox. IQW 2008/29 653 (Mei. 28 815) [sin.]			
5	Q 50	*Pseudodama vallonnetensis*	1.81 m	6
	Tib. IQW 2003/28 378 (Mei. 27 540), Tt IQW 2003/28 379 (Mei. 27 541), Tf IQW 2003/28 383 (Mei. 27 545), Tc/T4 IQW 2003/28 380 (Mei. 27 542), T1 IQW 2003/28 382 (Mei. 27 544), T2/T3 IQW 2003/28 381 (Mei. 27 543) [sin.]			
6	Q 21	*Acinonyx pardinensis*	2.08–2.03 m	15
	Vert. cerv. II–VII, Vert. thor. I–IV, 5× Costae, all skeletal elements included in block preparation IQW 2014/41 900 (Mei. 47 209) (frontispiece of this volume; Hemmer and Kahlke in this volume, Fig. 9c)			
7	Q 21	*Acinonyx pardinensis*	2.07–2.04 m	6
	Mc1, Mc2, Mc3, Mc4, PhI(1), PhIII(1), all skeletal elements included in block preparation IQW 2014/41 900 (Mei. 47 209) [dex.] (frontispiece of this volume; Hemmer and Kahlke in this volume, Fig. 9b)			
8	Q 21	*Eucladoceros giulii*	2.09–2.08 m	7
	PhI IQW 2003/28 337 (Mei. 27 499), PhI IQW 2004/28 586 (Mei. 27 748), PhII IQW 2004/28 583 (Mei. 27 745), PhII IQW 2004/28 582 (Mei. 27 744), Os sesam. dist. IQW 2004/28 587 (Mei. 27 749), PhIII IQW 2004/28 584 (Mei. 27 746), PhIII IQW 2004/28 585 (Mei. 27 747)			
9	Q 21	*Acinonyx pardinensis*	2.05 m	2
	Calva. IQW 2004/28 616 (Mei. 27 778), Vert. cerv. I IQW 2004/28 618 (Mei. 27 780) (Hemmer and Kahlke in this volume, Fig. 6a–c)			
10	Q 21	*Pseudodama vallonnetensis*	2.06–2.02 m	12
	Fem., Tib., Os malleo., Tt, Tf, Tc/T4, T2/T3, Mt3/4, 4× Ossa sesam. prox., all skeletal elements included in block preparation IQW 2014/41 900 (Mei. 47 209) [sin.] (frontispiece of this volume)			
11	Q 49	*Eucladoceros giulii*	1.39–1.36 m	3
	Vert. cerv. II–IV, all skeletal elements included in block preparation IQW 2015/41 933 (Mei. 47 242)			
12	Q 49	*Pseudodama vallonnetensis*	1.87 m	2
	Tc/T4 IQW 2003/28 409 (Mei. 27 571), T2/T3 IQW 2003/28 410 (Mei. 27 572) [dex.]			
13	Q 22	*Pseudodama vallonnetensis*	2.26 m	5
	Ci IQW 2003/28 358 (Mei. 27 520), Cu IQW 2003/28 359 (Mei. 27 521), Ca IQW 2003/28 361 (Mei. 27 523), C4 IQW 2003/28 357 (Mei. 27 519), Mc3/4 IQW 2003/28 356 (Mei. 27 518) [dex.]			

Table 2 Anatomically connected skeletal units of large mammals recovered from the southern area of the site (2001–2015 field seasons; Fig. 17).

No	(Main) grid square (Q) Skeletal elements [side of the body]	Species	Find depth below site 0-level	Number of elements (n)
14	Q 22	*Acinonyx pardinensis*	2.04 m	23
	Cr+Ci, Cu, Ca, C1, C2, C3, C4, Mc2, Mc3, Mc4, Mc5, PhI(1), PhI(2), PhI(3), PhI(4), PhI(5), PhII(2), PhII(3), PhII(5), PhIII(1), PhIII(2), PhIII(3), PhIII(5), all skeletal elements included in block preparation IQW 2014/41 900 (Mei. 47 209) [sin.] (frontispiece of this volume; Hemmer and Kahlke in this volume, Fig. 9a)			
15	Q 22	*Pseudodama vallonnetensis*	2.06–2.03 m	2
	2× Vert. cerv., both skeletal elements included in block preparation IQW 2014/41 900 (Mei. 47 209) (frontispiece of this volume)			
16	Q 22	*Pseudodama vallonnetensis*	2.05–2.01 m	7
	Vert. thor. XI–XIII, Vert. lumb. I–IV, all skeletal elements included in block preparation IQW 2014/41 900 (Mei. 47 209) (frontispiece of this volume)			
17	Q 22	*Pseudodama vallonnetensis*	2.04–1.99 m	5
	Vert. lumb. V–VII, Os sacr., Pelv., all skeletal elements included in block preparation IQW 2014/41 900 (Mei. 47 209) (frontispiece of this volume)			
18	Q 22	*Pseudodama vallonnetensis*	2.01–1.99 m	8
	Fem., Tib., Os malleo., Tt, Tf, Tc/T4, T2/T3, Mt3/4, all skeletal elements included in block preparation IQW 2014/41 900 (Mei. 47 209) [dex.] (frontispiece of this volume)			
19	Q 22	*Pseudodama vallonnetensis*	2.01 m	4
	2× PhII, 2× PhIII, all skeletal elements included in block preparation IQW 2014/41 900 (Mei. 47 209) (frontispiece of this volume)			
20	Q 22	*Pseudodama vallonnetensis*	2.01 m	9
	Rad. et Ulna, Ci, Cu, Ca, C4, Mc3/4, 2× PhI, all skeletal elements included in block preparation IQW 2014/41 900 (Mei. 47 209) [sin.] (frontispiece of this volume)			
21	Q 22	*Pseudodama vallonnetensis*	2.01 m	12
	C2/C3, Mc3/4, 4× Ossa sesam. prox., 2× PhI, 2× PhII, 2× PhIII, all skeletal elements included in block preparation IQW 2014/41 900 (Mei. 47 209) [dex.] (frontispiece of this volume)			
22	Q 22	*Pseudodama vallonnetensis*	2.01 m	13
	Rad., Cr, Ci, Cu, Ca, C2/C3, C4, Mc3/4, PhI(3), PhI(4), PhII(3), PhII(4), PhIII(4), all skeletal elements included in block preparation IQW 2014/41 900 (Mei. 47 209) [dex.] (frontispiece of this volume)			
23	Q 194	*Eucladoceros giulii*	1.82–1.72 m	9
	Vert. cerv. II–VII, Vert. thor. I–III, all skeletal elements included in a series of block preparations catalogued under IQW 2004/28 774 (Mei. 27 936)			
24	Q 194	*Bison menneri*	1.75–1.70 m	8
	Rad. et Ulna IQW 2003/28 352 (Mei. 27 514), Cr IQW 2003/28 351 (Mei. 27 513), Ci IQW 2003/28 350 (Mei. 27 512), Cu IQW 2003/28 349 (Mei. 27 511), C2/C3 IQW 2003/28 353 (Mei. 27 515), C4 IQW 2003/28 348 (Mei. 27 510), Mc3/4 IQW 2004/28 697 (Mei. 27 589) [sin.]			
25	Q 194	*Bison menneri*	1.77–1.64 m	8
	Vert. cerv. II–VII, Vert. thor. I, fragm. costa sin., all skeletal elements catalogued under IQW 2004/28 696 (Mei. 27 858) (**Fig. 14a**)			
26	Q 157	*Bison menneri*	1.82–1.68 m	18
	Vert. thor. IV–XII, 9× fragm. costae dex., all skeletal elements included in block preparation IQW 2003/28 190 (Mei. 27 352) (**Fig. 14b–d**)			

Table 2 (continued)

No	(Main) grid square (Q) Skeletal elements [side of the body]	Species	Find depth below site 0-level	Number of elements (n)
27	Q 48	*Eucladoceros giulii*	1.41–1.36 m	8
	Rad., Cr, Ci, Cu, Ca, C2/C3, C4, Mc3/4, all skeletal elements included in block preparation IQW 2015/41 933 (Mei. 47 242) [sin.] (Kahlke 2020, Fig. 12)			
28	Q 48	*Hippopotamus antiquus*	1.28–1.17 m	7
	Rad. et Ulna IQW 2015/41 933 (Mei. 47 242), Cr IQW 2015/41 933 (Mei. 47 242), Ci IQW 2015/41 933 (Mei. 47 242), Mc2 IQW 2002/27 372 (Mei. 26 877), Mc3 IQW 2002/27 374 (Mei. 26 879), Mc4 IQW 2002/27 373 (Mei. 26 878) [dex.] (partially in Kahlke 2020, Fig. 12)			
29	Q 48	*Hippopotamus antiquus*	1.30–1.20 m	2
	Fem., Pat., both skeletal elements included in block preparation IQW 2015/41 933 (Mei. 47 242) [dex.] (Kahlke 2020, Fig. 12)			
30	Q 48	*Hippopotamus antiquus*	1.29–1.17 m	9
	Rad. IQW 2003/28 323 (Mei. 27 485), Ulna IQW 2003/28 324 (Mei. 27 486), Ci IQW 2003/28 326 (Mei. 27 488), Cu IQW 2003/28 325 (Mei. 27 487), Ca IQW 2003/28 322 (Mei. 27 484), Mc4 IQW 2003/28 318 (Mei. 27 480), Mc5 IQW 2003/28 321 (Mei. 27 483), Os sesam. prox. IQW 2003/28 319 (Mei. 27 481), Os sesam. prox. IQW 2003/28 320 (Mei. 27 482) [dex.]			
31	Q 47	*Bison menneri*	1.69–1.64 m	3
	Hum. IQW 2003/27 706 (Mei. 27 211), Rad. et Ulna IQW 2003/27 705 (Mei. 27 210) [dex.]			
32	Q 47	*Eucladoceros giulii*	1.68–1.60 m	5
	Vert. thor. II–VI, all skeletal elements included in block preparation IQW 2003/28 139 (Mei. 27 301)			
33	Q 47	*Eucladoceros giulii*	1.55–1.53 m	2
	Vert. cerv. II IQW 2003/28 251 (Mei. 27 413), Vert. cerv. III IQW 2003/28 252 (Mei. 27 414)			
34	Q 47	*Bison menneri*	1.51 m	4
	PhI IQW 2002/27 571 (Mei. 27 076), PhI IQW 2002/27 569 (Mei. 27 074), PhII IQW 2002/27 572 (Mei. 27 077), PhII IQW 2002/27 570 (Mei. 27 075)			
35	Q 24	*Pseudodama vallonnetensis*	1.22–1.17 m	2
	Tc/T4 IQW 2002/27 547 (Mei. 27 052), T2/T3 IQW 2002/27 573 (Mei. 27 078) [sin.]			
36	Q 24	*Bison menneri*	1.46–1.43 m	3
	Hum. IQW 2003/27 618 (Mei. 27 123), Rad. et Ulna IQW 2003/27 617 (Mei. 27 122) [dex.]			
37	Q 24	*Bison menneri*	1.67–1.57 m	3
	Hum. IQW 2005/28 824 (Mei. 27 986), Rad. IQW 2005/28 823 (Mei. 27 985), Ulna IQW 2005/28 822 (Mei. 27 984) [dex.]			
38	Q 24	*Eucladoceros giulii*	1.45 m	3
	Vert. cerv. V–VII, all skeletal elements included in block preparation IQW 2005/28 881 (Mei. 28 043)			
39	Q 24	*Eucladoceros giulii*	1.51–1.48 m	6
	Tib. IQW 2005/28 945 (Mei. 28 107), Os malleo. IQW 2005/28 947 (Mei. 28 109), Tt IQW 2005/28 946 (Mei. 28 108), Tf IQW 2005/28 948 (Mei. 28 110), Tc/T4 IQW 2005/28 949 (Mei. 28 111), T2/T3 IQW 2005/28 950 (Mei. 28 112) [dex.]			
40	Q 24	*Eucladoceros giulii*	1.40 m	3
	Vert. lumb. II–IV, all skeletal elements included in block preparation IQW 2005/28 909 (Mei. 28 071)			
41	Q 24	*Pseudodama vallonnetensis*	1.86–1.82 m	3

Table 2 (continued)

No	(Main) grid square (Q)	Species	Find depth below site 0-level	Number of elements (n)
	Skeletal elements [side of the body]			
	Hum. IQW 2007/29 540 (Mei. 28 702), Rad. IQW 2007/29 541 (Mei. 28 703), Ulna IQW 2007/29 542 (Mei. 28 704) [sin.]			
42	Q 25	*Stephanorhinus hundsheimensis*	1.62–1.59 m	2
	Tt IQW 2005/28 810 (Mei. 27 972), Tf IQW 2005/28 808 (Mei. 27 970) [dex.]			
43	Q 25	*Pseudodama vallonnetensis*	1.98 m	14
	Tt IQW 2007/29 309 (Mei. 28 471), Tf IQW 2007/29 308 (Mei. 28 470), Tc/T4 IQW 2007/29 310 (Mei. 28 472), T1 IQW 2007/29 312 (Mei. 28 474), T2/T3 IQW 2007/29 311 (Mei. 28 473), Mt3/4 IQW 2007/29 558 (Mei. 28 720), 2× PhI IQW 2007/29 283 (Mei. 28 445), 2× PhII IQW 2007/29 283 (Mei. 28 445), 2× Ossa sesam. dist. IQW 2007/29 283 (Mei. 28 445), 2× PhIII IQW 2007/29 283 (Mei. 28 445) [sin.]			
44	Q 25	*Eucladoceros giulii*	1.99–1.86 m	16
	Tib. IQW 2009/30 125 (Mei. 29 287), Os malleo. IQW 2009/30 302 (Mei. 29 464), Tt IQW 2009/30 220 (Mei. 29 382), Tf IQW 2009/30 301 (Mei. 29 463), Tc/T4 IQW 2009/30 172 (Mei. 29 334), T2/T3 IQW 2009/29 930 (Mei. 29 092), Mt 3/4 IQW 2009/30 171 (Mei. 29 333), Os sesam. prox. IQW 2009/30 174 (Mei. 29 336), PhI(3) IQW 2010/30 724 (Mei. 29 886), PhI(4) IQW 2010/30 723 (Mei. 29 885), PhII(3) IQW 2009/30 188 (Mei. 29 350), PhII(4) IQW 2010/30 725 (Mei. 29 887), Os sesam. dist. IQW 2009/30 726 (Mei. 29 888), Os sesam. dist. IQW 2009/30 727 (Mei. 29 889), PhIII(3) IQW 2009/30 189 (Mei. 29 351), PhIII(4) IQW 2009/30 300 (Mei. 29 462) [sin.]			
45	Q 28	*Hippopotamus antiquus*	1.33 m	2
	Mc3 IQW 2008/29 775 (Mei. 28 937), Mc4 IQW 2008/29 839 (Mei. 29 001) [sin.]			
46	Q 28	*Bison menneri*	1.69–1.59 m	7
	Tib. IQW 2009/30 281 (Mei. 29 443), Os malleo. IQW 2009/30 284 (Mei. 29 446), Tt IQW 2009/30 286 (Mei. 29 448), Tf IQW 2009/30 282 (Mei. 29 444), Tc/T4 IQW 2009/30 286 (Mei. 29 448), T1 IQW 2009/30 285 (Mei. 29 447), T2/T3 IQW 2009/30 283 (Mei. 29 445) [sin.] (Kierdorf et al. 2012)			
47	Q 28	*Bison menneri*	1.62–1.61 m	13
	Mt3/4 IQW 2009/30 126 (Mei. 29 288), 2× Ossa sesam. prox. (3) IQW 2009/30 127 (Mei. 29 289), 2× Ossa sesam. prox. (4) IQW 2009/30 128 (Mei. 29 290), PhI(3) IQW 2009/30 129 (Mei. 29 291), PhI(4) IQW 2009/30 130 (Mei. 29 292), PhII(3) IQW 2009/30 131 (Mei. 29 293), PhII(4) IQW 2009/30 132 (Mei. 29 294), Os sesam. dist. (3) IQW 2009/30 133 (Mei. 29 295), Os sesam. dist. (4) IQW 2009/30 134 (Mei. 29 296), PhIII(3) IQW 2009/30 135 (Mei. 29 297), PhIII(4) IQW 2009/30 136 (Mei. 29 298) [sin.]			
48	Q 28	*Eucladoceros giulii*	1.69–1.65 m	7
	Tc/T4 IQW 2009/30 027 (Mei. 29 189), T1 IQW 2009/30 027 (Mei. 29 189), T2/T3 IQW 2009/30 027 (Mei. 29 189), Mt3/4 IQW 2009/30 028 (Mei. 29 190), PhI(4) IQW 2009/30 029 (Mei. 29 191), PhII(4) IQW 2009/30 030 (Mei. 29 192), PhIII(4) IQW 2009/30 031 (Mei. 29 193) [sin.]			
49	Q 28	*Bison menneri*	2.40–2.39 m	3
	Vert. cerv. IV IQW 2009/30 547 (Mei. 29 709), Vert. cerv. V IQW 2009/30 546 (Mei. 29 708), Vert. cerv. VI IQW 2009/30 545 (Mei. 29 707)			
50	Q 28	*Bison menneri*	1.51–1.50 m	2
	Vert. thor. IQW 2009/29 921 (Mei. 29 083), Vert. thor. IQW 2009/29 922 (Mei. 29 084)			
51	Q 27	*Bison menneri*	1.64–1.52 m	22
	Rad. et Ulna IQW 2009/29 977 (Mei. 29 139), Cr IQW 2009/ 29 979 (Mei. 29 141), Ci IQW 2009/29 980 (Mei. 29 142), Cu IQW 2009/29 981 (Mei. 29 143), Ca IQW 2009/29 982 (Mei. 29 144), C2/C3 IQW 2009/29 983 (Mei. 29 145), C4 IQW 2009/29 984 (Mei. 29 146), Mc3/4 IQW 2009/29 978 (Mei. 29 140), Mc5 IQW 2009/29 985 (Mei. 29 147), 2× Ossa sesam. prox (3) IQW 2009/29 986 (Mei. 29 148), 2× Ossa sesam. prox (4) IQW 2009/29 987 (Mei. 29 149), PhI(3) IQW 2009/29 988 (Mei. 29 150), PhI(4) IQW 2009/29 989 (Mei. 29 151), PhII(3) IQW 2009/29 990 (Mei. 29 152), PhII(4) IQW 2009/29 991 (Mei. 29 153), Os sesam. dist. (3) IQW 2009/29 992 (Mei. 29 154), Os sesam. dist. (4) IQW 2009/29 993 (Mei. 29 155), PhIII(3) IQW 2009/29 994 (Mei. 29 156), PhIII(4) IQW 2009/29 995 (Mei. 29 157) [dex.]			

Table 2 (continued)

No	(Main) grid square (Q) Skeletal elements [side of the body]	Species	Find depth below site 0-level	Number of elements (n)
52	Q 27	*Bison menneri*	1.47–1.43 m	5
	Vert. lumb. IV IQW 2009/30 010 (Mei. 29 172), Vert. lumb. V IQW 2009/30 011 (Mei. 29 173), Vert. lumb. VI IQW 2009/30 012 (Mei. 29 174), Vert. lumb. VII IQW 2009/30 013 (Mei. 29 175), Os sacrum IQW 2009/30 014 (Mei. 29 176)			
53	Q 27	*Bison menneri*	1.65–1.59 m	5
	Os malleo. IQW 2009/29 948 (Mei. 29 110), Tf IQW 2009/29 946 (Mei. 29 108), Tc/T4 IQW 2009/29 947 (Mei. 29 109), T1 IQW 2009/29 951 (Mei. 29 113), T2/T3 IQW 2009/29 950 (Mei. 29 112) [dex.]			
54	Q 27	*Bison menneri*	1.59–1.58 m	2
	Vert. thor. IQW 2009/30 232 (Mei. 29 394), Vert. thor. IQW 2009/30 233 (Mei. 29 395)			
55	Q 27	*Eucladoceros giulii*	1.67 m	2
	Tt IQW 2009/30 324 (Mei. 29 486), Tf IQW 2009/30 323 (Mei. 29 485) [sin.]			
56	Q 27	*Pseudodama vallonnetensis*	1.49 m	4
	Tc/T4 IQW 2009/29 963 (Mei. 29 125), T1 IQW 2009/29 963 (Mei. 29 125), T2/T3 IQW 2009/29 963 (Mei. 29 125), Mt3/4 IQW 2009/29 952 (Mei. 29 114) [sin.]			
57	Q 27	*Bison menneri*	1.60–1.57 m	19
	Cr, Ci, Cu, Ca, C2/C3, C4, Mc3/4, 4× Ossa sesam. prox., PhI–III(3), PhI–III(4), 2× Ossa sesam. dist., all skeletal elements included in block preparation IQW 2009/30 230 (Mei. 29 392) [sin.]			
58	Q 27	*Bison menneri*	1.49–1.47 m	4
	Tc/T4 IQW 2009/30 061 (Mei. 29 223), T1 IQW 2009/30 056 (Mei. 29 218), T2/T3 IQW 2009/30 055 (Mei. 29 217), Mt3/4 IQW 2009/30 070 (Mei. 29 232) [dex.]			
59	Q 27	*Eucladoceros giulii*	2.06–2.03 m	5
	Tf IQW 2009/30 569 (Mei. 29 731), Tc/T4 IQW 2009/30 570 (Mei. 29 732), T1 IQW 2009/30 571 (Mei. 29 733), T2/T3 IQW 2009/30 572 (Mei. 29 734), Mt3/4 IQW 2009/30 573 (Mei. 29 735) [sin.]			
60	Q 31	*Bison menneri*	1.58–1.54 m	4
	Tf IQW 2009/30 367 (Mei. 29 529), Tc/T4 IQW 2009/30 366 (Mei. 29 528), T2/T3 IQW 2009/30 365 (Mei. 29 527), Mt3/4 IQW 2009/30 362 (Mei. 29 524) [sin.]			
61	Q 31	*Bison menneri*	1.53–1.37 m	5
	Tib. IQW 2010/30 695 (Mei. 29 857), Os malleo. IQW 2010/30 698 (Mei. 29 860), Tt IQW 2010/30 696 (Mei. 29 858), Tf IQW 2010/30 699 (Mei. 29 861), Tc/T4 IQW 2010/30 697 (Mei. 29 859) [dex.]			
62	Q 30	*Homotherium latidens*	1.53–1.52 m	2
	Mand. sin. IQW 2009/30 238 (Mei. 29 400), Mand. dex. IQW 2009/30 239 (Mei. 29 401) (Kahlke et al. 2020, Fig. 13b)			
63	Q 30	*Eucladoceros giulii*	1.55–1.53 m	2
	Vert. thor. VI IQW 2009/30 142 (Mei. 29 304), Vert. thor. VII IQW 2009/30 141 (Mei. 29 303)			
64	Q 626	*Pseudodama vallonnetensis*	1.52–1.50 m	3
	Hum., Rad. et Ulna, all skeletal elements included in block preparation 2010/31 043 (Mei. 30 205) [sin.]			
65	Q 608	*Pseudodama vallonnetensis*	1.66 m	17
	Mc3/4, 4× Ossa sesam. prox., PhI–III(2), PhI–III(3), PhI–III(4), PhI–III(5), all skeletal elements included in block preparation IQW 2013/41 625 (Mei. 46 965) [sin.]			
66	Q 33	*Bison menneri*	1.75–1.50 m	11
	Hum., Rad. et Ulna, Cr, Ci, Cu, Ca, C2/C3, C4, Mc3/4, Mc5, all skeletal elements included in block preparation IQW 2010/31 064 (Mei. 30 226) [sin.]			

Table 2 (continued)

No	(Main) grid square (Q) Skeletal elements [side of the body]	Species	Find depth below site 0-level	Number of elements (n)
67	Q 33	*Stephanorhinus hundsheimensis*	1.90–1.81 m	9
	Ulna IQW 2011/31 236 (Mei. 30 398), Rad. IQW 2011/31 247 (Mei. 30 409), Cr IQW 2011/31 296 (Mei. 30 458), Ci IQW 2011/31 295 (Mei. 30 457), Cu IQW 2011/31 294 (Mei. 30 456), C2 IQW 2011/31 299 (Mei. 30 461), C3 IQW 2011/31 297 (Mei. 30 459), C4 IQW 2011/31 298 (Mei. 30 460), Mc5 IQW 2011/31 299 (Mei. 30 461) [sin.]			
68	Q 33	*Bison menneri*	1.85–1.81 m	5
	Cr, Ci, Cu, C2/C3, C4, all skeletal elements catalogued under IQW 2011/31 291 (Mei. 30 453) [sin.]			
69	Q 627	*Eucladoceros giulii*	1.70–1.63 m	2
	Vert. cerv. II–III, both skeletal elements included in block preparation IQW 2011/32 045 (Mei. 31 207)			
70	Q 627	*Eucladoceros giulii*	1.68 m	3
	Vert. lumb. VI–VII, Os sacrum, all skeletal elements included in block preparation IQW 2010/31 184 (Mei. 30 346)			
71	Q 627	*Eucladoceros giulii*	1.90–1.82 m	5
	Os malleo. IQW 2010/31 141 (Mei. 30 303), Tt IQW 2010/31 138 (Mei. 30 300), Tf IQW 2010/31 137 (Mei. 30 299), Tc/T4 IQW 2010/31 139 (Mei. 30 301), T2/T3 IQW 2010/31 140 (Mei. 30 302) [sin.]			
72	Q 609	*Eucladoceros giulii*	1.30 m	2
	Tt IQW 2013/41 532 (Mei. 36 878), Tf IQW 2013/41 525 (Mei. 36 871) [sin.]			
73	Q 628	*Bison menneri*	1.72–1.70 m	2
	Vert. lumb. IQW 2012/38 362 (Mei. 35 336), Vert. lumb. IQW 2012/38 363 (Mei. 35 337)			
74	Q 628	*Bison menneri*	1.95 m	12
	Rad, Cr, Ci, Cu, Ca, C2/C3, C4, Mc3/4, 2× PhI, 2× Ph II, all skeletal elements included in block preparation IQW 2011/38 089 (Mei. 35 063) [dex.]			
75	Q 628	*Pseudodama vallonnetensis*	2.06 m	8
	Tib. IQW 2011/38 077 (Mei. 35 051), Os malleo. IQW 2011/38 074 (Mei. 35 048), Tt IQW 2011/38 072 (Mei. 35 046), Tf IQW 2011/38 071 (Mei. 35 045), Tc/T4 IQW 2011/38 073 (Mei. 35 047), T1 IQW 2011/38 076 (Mei. 35 050), T2/T3 IQW 2011/38 075 (Mei. 35 049), Mt3/4 IQW 2011/38 078 (Mei. 35 052) [sin.]			
76	Q 628	*Pseudodama vallonnetensis*	1.77–1.76 m	2
	Vert. cerv. II IQW 2011/31 314 (Mei. 30 476), Vert. cerv. III IQW 2011/31 315 (Mei. 30 477)			
77	Q 519	*Stephanorhinus hundsheimensis*	2.02–1.81 m	4
	Hum. IQW 2016/43 910 (Mei. 49 219), Rad. IQW 2016/43 912 (Mei. 49 221), Ulna IQW 2016/43 911 (Mei. 49 220), C3 IQW 2015/42 176 (Mei. 47 485) [dex.]			
78	Q 519	*Stephanorhinus hundsheimensis*	2.02–1.80 m	7
	Vert. thor. XVI IQW 2016/43 913 (Mei. 49 222), Vert. thor. XVII IQW 2016/43 914 (Mei. 49 223), Vert. thor. XVIII IQW 2016/43 915 (Mei. 49 224), Vert. thor. IXX IQW 2016/43 916 (Mei. 49 225), Vert. lumb. I IQW 2016/43 917 (Mei. 49 226), Vert. lumb. II IQW 2016/43 918 (Mei. 49 227), Vert. lumb. III IQW 2016/43 919 (Mei. 49 228)			
79	Q 519	*Xenocyon lycaonoides*	1.93 m	5
	Vert. lumb. IV IQW 2015/42 154 (Mei. 47 463), Vert. lumb. V IQW 2015/42 153 (Mei. 47 462), Vert. lumb. VI IQW 2015/42 152 (Mei. 47 461), Vert. lumb. VII IQW 2015/42 151 (Mei. 47 460), Os sacrum IQW 2015/42 150 (Mei. 47 459)			
80	Q 518	*Pseudodama vallonnetensis*	2.00–1.98 m	3
	Tc/T4 IQW 2015/42 022 (Mei. 47 331), T2/T3 IQW 2016/43 836 (Mei. 49 145), Mt3/4 IQW 2015/42 025 (Mei. 47 334) [dex.]			

Table 2 (continued)

No	(Main) grid square (Q) Skeletal elements [side of the body]	Species	Find depth below site 0-level	Number of elements (n)
81	Q 518	*Xenocyon lycaonoides*	2.05–2.02 m	2
	Rad. IQW 2015/42 013 (Mei. 47 322), Ulna IQW 2015/42 014 (Mei. 47 323) [sin.]			
82	Q 597	*Pseudodama vallonnetensis*	2.32–2.30 m	4
	Tc/T4 IQW 2016/44 178 (Mei. 49 487), T1 IQW 2016/44 180 (Mei. 49 489), T2/T3 IQW 2016/44 179 (Mei. 49 488), Mt3/4 IQW 2016/44 177 (Mei. 49 486) [dex.]			
83	Q 597	*Eucladoceros giulii*	2.31 m	2
	Cr, Ci, both skeletal elements catalogued under IQW 2016/44 184 (Mei. 49 493) [sin.]			
84	Q 528	*Eucladoceros giulii*	2.02–1.98 m	7
	Rad. et Ulna IQW 2016/43 890 (Mei. 49 199), Cr IQW 2016/43 772 (Mei. 49 031), C2/C3 IQW 2015/42 500 (Mei. 47 809), Mc3/4 IQW 2016/43 783 (Mei. 49 092), PhI(3) IQW 2016/43 846 (Mei. 49 155), Os sesam. prox. IQW 2016/43 867 (Mei. 49 176) [dex.]			
85	Q 528	*Bison menneri*	2.05–1.98 m	15
	Cr IQW 2016/43 950 (Mei. 49 259), Ca IQW 2015/42 052 (Mei. 47 361), C4 IQW 2015/42 050 (Mei. 47 359), Mc3/4 IQW 2016/43 945 (Mei. 49 254), Mc5 IQW 2016/43 948 (Mei. 49 257), Os sesam. prox. (3) IQW 2016/43 949 (Mei. 49 258), Os sesam. prox. (3) IQW 2015/42 064 (Mei. 47 373), Os sesam. prox. (4) IQW 2015/42 063 (Mei. 47 372), Os sesam. prox. (4) IQW 2015/42 062 (Mei. 47 371), PhI(3) IQW 2016/42 514 (Mei. 47 823), PhI(4) IQW 2016/43 883 (Mei. 49 192), PhII(3) IQW 2016/43 839 (Mei. 49 148), PhII(4) IQW 2016/43 946 (Mei. 49 255), PhIII(3) IQW 2016/43 895 (Mei. 49 204), PhIII(4) IQW 2016/43 894 (Mei. 49 203) [sin.]			
			Total n	549

Table 2 (continued)

rial recovered before 2001 (Kahlke 2001e, 959 ff.; 2006, 55 f.; Kotowski et al. 2020, 1274). In many cases, these kinds of parts of the carcasses or skeletons had probably already become embedded or destroyed upstream of the present site.

Clear examples of skeletal unit modifications by *Pachycrocuta* as a condition for later transportation into the channel filling can be seen on two articulated segments of *Bison menneri* vertebra columns, namely a section comprising cervical vertebrae CII–VII and thoracic vertebra TI with the proximal fragment of the first left rib [IQW 2004/28 696 (Mei. 27 858); **Table 2**, No 25, **Fig. 14a**], as well as a »buffalo hump« consisting of TIV–XII with a complete set of articulated right costal fragments – mostly rib heads only [IQW 2003/28 190 (Mei. 27 352); **Table 2**, No 26, **Fig. 14b–d**]. Two more modified spinal units belong to *Eucladoceros giulii*: CII–TIII [IQW 2004/28 774 (Mei. 27 936); **Table 2**, No 23], and TII–VI [IQW 2003/28 139 (Mei. 27 301); **Table 2**, No 32]. According to the vertebrae represented, as well as their respective individual age at death, the described spinal segments belonged to different individuals. Both bison neck and hump as well as the cervid spinal sections show clear hyaena gnaw marks, especially on the processes and the remaining proximal rib fragments (**Fig. 14**). The almost complete removal of protruding processes and rib bodies allowed the corresponding spinal column sections to be transported in the flowing water. Spinal columns, or even parts thereof, with more or less complete ribs in articulation were not found.

Some of the skeletal units found can be combined into individual units, such as the remains of the cheetah *Acinonyx pardinensis* recovered from squares Q 21, 22 and 49 (see frontispiece). In this case, the skull, the cervical spine, parts of the thoracic spine, inclusive of some costal fragments, and elements of both front

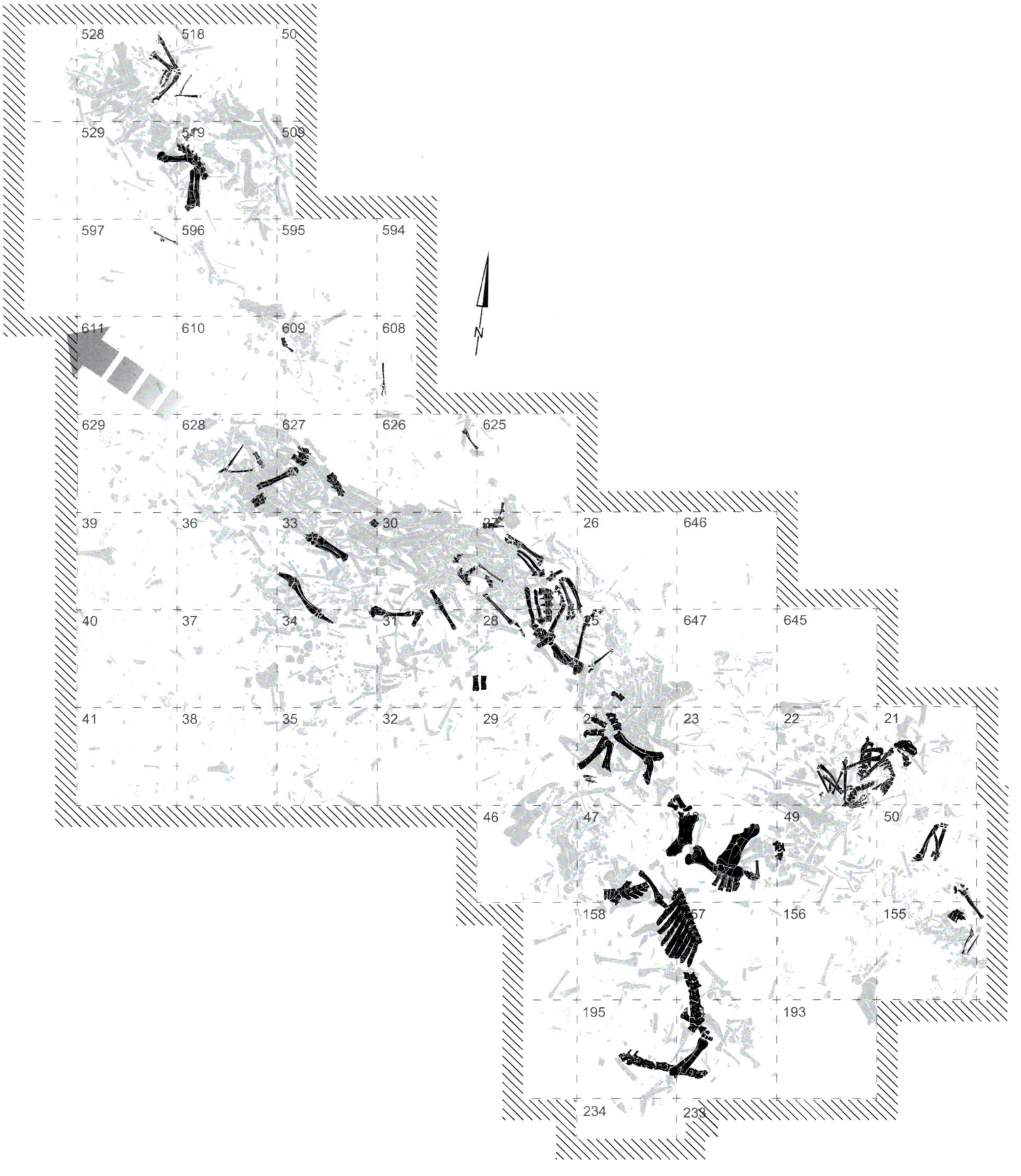

Fig. 17 Untermassfeld research excavation (2001–2015 field seasons), southern area immediately north of the coarse clastic mudflow fan (CMF), with highlighted finds of anatomically connected skeletal units of large mammals. Black – connected skeletal units, in some cases, units are concealed by overlaying finds (complete listing in **Table 2**). Grey – disarticulated remains of large mammals. Arrow – main flow direction of the river water leeward of the CMF.

paws of the individual – 46 elements in total – were found (**Table 2**, No 6–9, 13). Additional isolated finds have also been assigned to this individual cat (Hemmer and Kahlke in this volume).
An exemplary sequence of disarticulation events largely unaffected by carnivore activities was determined by linking a number of selected find complexes involving *Hippopotamus* extremities (R.-D. Kahlke 1997c, Figs. 1–6; 1999a, Figs. 3–8). Similarly, the reconstructed decomposition process of a bear cadaver illustrated its disintegration in flowing water (Kahlke and Kierdorf 1997, 162 ff.).

– 6.1.2.2. Isolated skeletal elements
The partial or complete disintegration of carcasses and skeletons resulted in a large number of disarticulated skeletal elements. About 95.0 % of the identifiable large mammal material from the excavations before 2001 was found in an anatomically isolated state. For the 2001–2015 seasons, the proportion of disarticulated large mammal elements was approximately 87.3 % (for data sources see section 6.1.2.1). The vast majority of small vertebrate remains were also found as isolated skeletal elements (Maul 2001a, 858 ff.; in this volume). In addition, however, a remarkable number of microvertebrate skeletons and disarticulated, but recognisably related, individual bone units were found. These were regularly arranged in a particular sector of the channel filling (see section 6.1.2.5.). The arrangement of the isolated large mammal skeletal elements recovered from the individual parts of the site is documented in **Foldouts II** and **III**.

– 6.1.2.3. Polyspecific large mammal bone concentrations
As described earlier (Kahlke 2001e; 2006), in the flooded surfaces represented by the Lower Fluviatile Sands (LFS; see section 4.) and also in the channel infill, i. e., the Upper Fluviatile Sands [UFS (l)], isolated skeletal elements were concentrated in dense piles containing remains of different species and individuals. In many cases, relatively massive and heavy partial cadavers were preserved as articulated skeletal units, or large sized skeletal elements. So-called obstruction bones acted as mechanical barriers and thus caused the agglomeration of further necromass transported by the riverine water flow (examples in Kahlke 2001e, 955 f.; 2006, 50 and corresponding foldouts). Additional polyspecific bone concentrations were uncovered during the 2001–2015 campaigns in squares Q 194, 155, 49/21, 22 and 46/47 of the channel filling, immediately northwest of the CMF (**Foldout III**).
Both the extent of the polyspecific bone concentrations themselves and the size of the skeletal elements involved vary from case to case. Now and then the bone concentrations appear to be imbricated, with their component parts inclined in a more or less upstream direction (Kahlke 2006, 50) and, in some cases, anatomically connected skeletal units are also entangled in the piles (Kahlke 2001e, 955; 2006, 50, Figs. 17, 45; **Fig. 17** of this paper). Parts of plants from the former floodplain vegetation were also caught up in the bone concentrations, as evidenced by irregularly distributed, mm sized, and mostly flattened xylite particles (see section 5.1.).

– 6.1.2.4. Polyspecific large mammal bone accumulations
During the 1988–1995 field seasons (R.-D. Kahlke 2001a, 420 ff.), a bone accumulation with constantly high concentrations of large mammal skeletal elements that covered almost 11 m² was recovered from the deepest section of the channel filling between grid squares Q 660 and 818 (Kahlke 2001e, foldouts X–XI; **Foldout II** of this paper). In its central section the accumulation was up to 1.85 m thick, and, following the channel morphology, the thickness decreased in the peripheral grid squares (**Table 3**). From the arrangement of the skeletal elements present, it is clear that this accumulation resulted from the amalgamation of several polyspecific bone concentrations. The respective obstruction bones (see section 6.1.2.3.) can be easily recognised in grid squares Q 660, 670, 750, 751, and 784.

Grid square (Q)	NCSP	Depth below site 0-level
660	124	2.26–1.50
670	82	2.35–1.20
750	81	2.50–1.00
751	171	2.85–1.00
752	44	2.35–1.00
783	134	2.58–1.40
818	74	2.50–2.30

Table 3 Number of catalogued specimens (NCSP) from the bone accumulation between grid squares Q 660 and 818 (for location see **Foldout III**).

During later field seasons, a further impressive mammal bone accumulation spanning approximately 7.5 m² was exposed between grid squares Q 24/25 and 627/628 (**Fig. 16**, **Foldout III**). Here, extremely high bone concentrations formed with thicknesses between 1.75 and 1.03 m (**Table 4**). In contrast to the large mammal bone accumulation mentioned in the paragraph above, just one obstruction bone was found, which had been deposited at the south-eastern edge of the accumulation (**Foldout III**, Q 25), near to the south-eastern turbulence zone (see section 6.1.1.2.). This obstruction bone was the right femur of a *Mammuthus meridionalis* (Kahlke 2020, Fig. 2a) that was oriented perpendicularly to the flow direction of the river water. However, in contrast, within the bone accumulation itself the vast majority of the longitudinal axes of the skeletal elements were clearly aligned parallel to the flow direction. The presence of the bulky femur undoubtedly caused the formation of the immediately adjacent bone accumulation (**Fig. 16**).

A third remarkable bone accumulation containing > 203 specimens per m² (Q 528, block preparation with additional bones covered by sediment) and covering an area of approximately 3 m² was discovered between grid squares Q 519 and 528 (see section 6.1.1.2., **Foldout III**). In these two grid squares, as well as in Q 518, several bulky long bones may have acted as obstruction bones. Compared to the bone accumulation excavated between squares Q 24/25 and 627/628 (see above), this new accumulation to the north was characterised by thicknesses of between only 1.04 and 0.46 m (**Table 5**), lower bone concentrations within the single squares, and a smaller number of skeletal elements set parallel or perpendicular to the flow direction. These corresponding features suggest a slightly less energetic flow regime.

In the north-western continuation of the channel filling, following on from an area with polyspecific bone concentrations situated between grid squares Q 553, 565 and 581 (Kahlke 2001e, foldout VIII; **Foldout II** of this paper), is the previously described 11 m² bone accumulation characterised by clearly recognisable obstruction bones and a relatively low degree of alignment between the individual skeletal elements (Kahlke 2001e, foldout X–XI; **Foldout II** of this paper).

Grid square (Q)	NCSP	Depth below site 0-level
24	193	2.88–1.18
25	213	2.33–1.18
27	251	2.85–1.17
28	202	2.88–1.13
30	260	2.15–1.12
33	267	2.87–1.26
627	124	2.69–1.20
628	172	2.78–1.08

Table 4 Number of catalogued specimens (NCSP) from the bone accumulation between grid squares Q 24/25 and 627/628 (for location see **Foldout III**).

Grid square (Q)	NCSP	Depth below site 0-level
518	64	2.05–1.01
519	131	2.21–1.20
528	203	2.20–1.74
529	72	2.55–1.61

Table 5 Number of catalogued specimens (NCSP) from the bone accumulation between grid squares Q 519 and 528 (for location see **Foldout III**).

To summarise, all the described evidence points to a flow regime on the lee side of the CMF which led to repeated formations of extended bone accumulations. Within these structures, both the concentration of skeletal elements and the degree of their alignment decreased along the course of water flow, i.e., to the NW. In other words, the influence of the mudflow fan on the arrangement of the deposited fossils decreased with increasing distance.

During the filling of the deepest part of the erosion channel, a limited area on the edge of the adjacent mudflow fan was washed out and subsequently collapsed (**Fig. 2, III**). As a result, two solidly calcified sheets of sandy carbonate breccia (SCB) from the upper part of the CMF were deposited in grid squares Q 595–597, 609–611, 627 and 628 (Kahlke et al. 2020, Figs. 7a–c, 12; see section 4.2. of this paper). Vertebrate finds both below and above the breccia sheets testify to their synchronous embedding along with the bone accumulations (Kahlke et al. 2020, Fig. 12). The position of the SCB bodies in the erosion channel caused changes in the hydrodynamic situation directly leeward of the mudflow fan and thus modified the spatial arrangement of the bone elements buried in the channel filling. The bone accumulation of grid squares Q 628 to 596 is offset by 2 m to the north compared to the accumulation excavated immediately southeast of the breccia sheets (**Foldout III**).

– 6.1.2.5. Microvertebrate skeletal units

On the south-western periphery of the zone of bone accumulations deposited in the deepest part of the channel infill, between grid squares Q 234 and 156 in the southeast, and Q 611 and 597 in the northwest, skeletons and partial skeletons, as well as individual skulls of microvertebrates, mostly micromammals, were found in different states of disarticulation (**Fig. 18**). Such finds appeared exclusively in the lower part of the Upper Fluviatile Sands [ULS(1)], but not in any other sediment, such as reworked Younger Coarse Gravels (YCG) or basal sandy loam (BSL) (Kahlke et al. 2020).

At depths between 2.46 m (Q 39) and 1.10 m (Q 29) below site 0-level and spread over an area of 32 m^2, four predominantly articulated skeletons, 47 partial or disarticulated skeletons, and 27 isolated skulls of the ground squirrel *Spermophilus kahlkei* (species description by Maul in this volume) were discovered (**Table 6**; stored as a separate part of the Untermassfeld collection). Based on the number of skull finds from the aforementioned area, a minimum number (MNI) of 45 *Spermophilus* individuals has been recorded, each represented by skeletal associations and/or largely intact skulls. In addition, more or less articulated partial skeletons of an agile frog [*Rana dalmatina* IQW 2010/30 892 (Mei. 30 054), grid square Q 34, 1.75–1.71 below site 0-level; Böhme 2020b, Fig. 3], a hamster [*Cricetus runtonensis* IQW 2003/28 123 (Mei. 27 285), grid square Q 156, 1.65–1.60 m below site 0-level; Maul in this volume, Fig. 8] and a vole [*Microtus thenii* IQW 2009/30 628 (Mei. 29 790), grid square Q 29, 1.50 below site 0-level; Maul in this volume, Fig. 9] were found within this area. The articulated finds were taken from the fossiliferous layer as in-situ preparations and stored under glass lids (Kahlke 2020, 1047 ff., Fig. 11a).

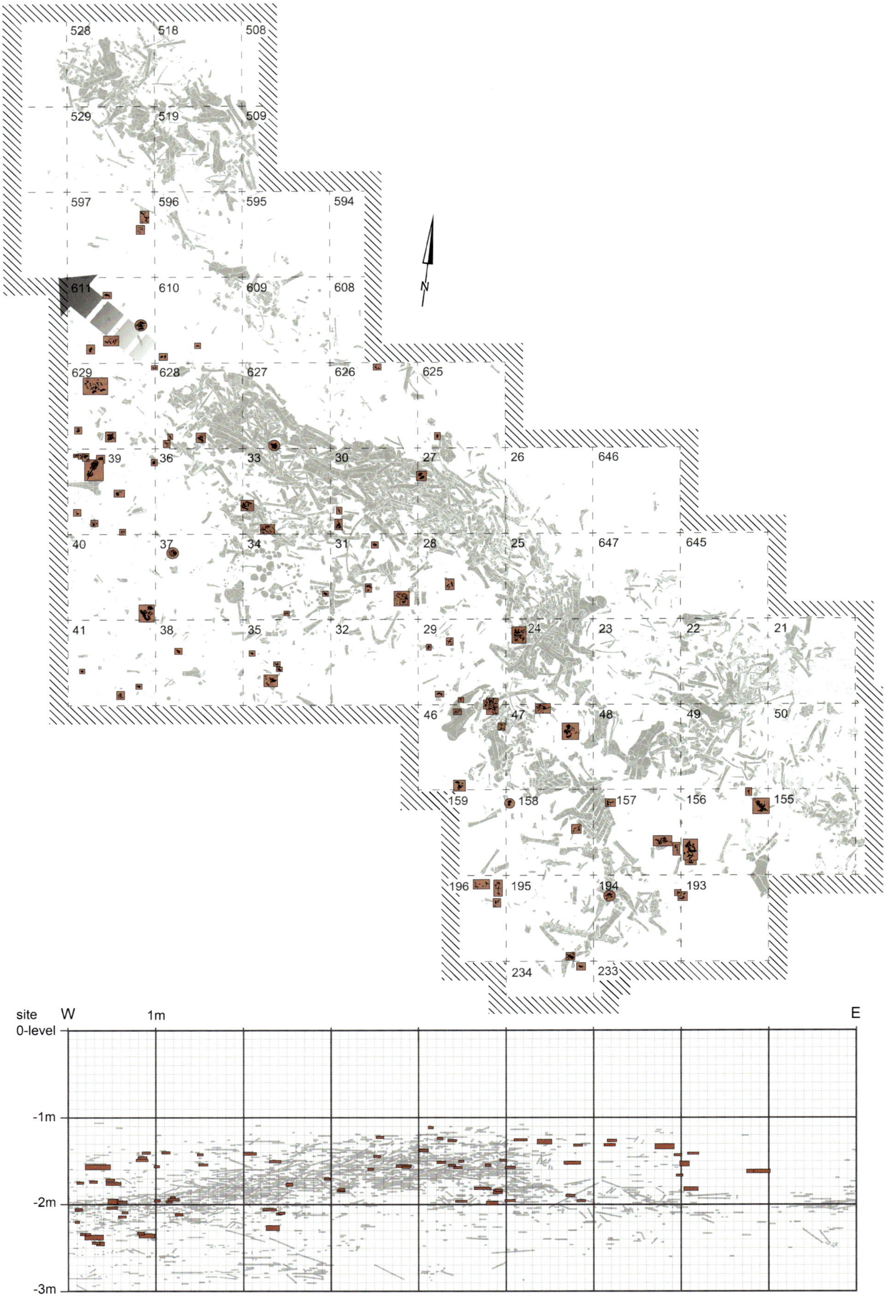
528
518
508
529
519
509
597
596
595
594
N
611
610
609
608
629
628
627
626
625
39
36
33
30
27
26
646
40
37
34
31
28
25
647
645
41
38
35
32
29
24
23
22
21
46
47
48
49
50
159
158
157
156
155
196
195
194
193
234
233
site
0-level
W
1m
E
-1m
-2m
-3m

Grid square (Q)	Predominantly articulated skeletons (n)	Predominantly disarticulated/partial skeletons (n)	Isolated skulls (n)	MNI (based on skulls)
193		1		
156	1 (**Fig. 20**)	3 (incl. 1 *Cricetus*)		2
194		1	1	2
157		3		2
234			1	1
195			1	1
158		1	1	1
47		2		2
24	1			1
196		3 (**Fig. 23**)		1
46		3 (**Fig. 22**)	2	3
29		4 (incl. 1 *Microtus*)	1	3
28		1		
27			2	2
625		1		
31		3	1	2
30		2		1
626		1		
35		1	3	3
34	1 (*Rana*)		1	2
33		2		1
627			1	1
38			1	1
37			1	1
628		3		
610		1	1	1
41		2	1	1
40	1 (**Fig. 21**)			1
39	1 (**Fig. 19**)	5	4	5
629		3	1	4
611		1	3	3
597		2		
			Total MNI	48

Table 6 Numbers (n) of microvertebrate skeletons, partial skeletons and isolated skulls (*Spermophilus* unless otherwise noted) recovered from the southern area of the site (2001–2015 field seasons; see **Fig. 18**).

←

Fig. 18 Untermassfeld research excavation (2001–2015 field seasons), southern area, immediately north of the coarse clastic mudflow fan (CMF), with finds of microvertebrate skeletal units and skulls highlighted in red. – **Above** Distribution pattern of the vertebrate finds within the channel filling, excavation plan. – **Below** Find depths below site 0-level (2006–2015), note neotectonically caused SW dipping of the fossiliferous channel infill. – Grey: remains of large mammals. – Red rectangles with black elements: in-situ preparations of skeletal units. – Red circles with black elements: preparations without sediment matrix. – Arrow: main flow direction of the river water leeward of the CMF.

The previously mentioned, more complete, microvertebrate finds were evenly distributed both horizontally and vertically in a south-western (left, valley facing) portion of the channel filling (**Fig. 18**). Clear frequency maxima are not recognisable. Within one of the large mammal bone accumulations (see section 6.1.2.4.), as well as on its north-eastern (right, slope facing) margin, hardly any correspondingly complete microvertebrate finds were recorded. The schematic representation of the respective find depths below site 0-level (**Fig. 18 below**) reflects the neotectonically caused SW dipping of the fossiliferous channel infill (Ellenberg and Kahlke 1997, 53ff.; Kahlke et al. 2020, Foldout I). Nevertheless, it becomes clear that the few single finds of microvertebrates in the slope facing side (grid squares Q 27, 625 and 626) all appeared around the upper boundary of the large mammal bone accumulations.

While most of the isolated dental and skeletal remains of micromammals excavated from the site before 2001 originated from disintegrated pellets or droppings of the respective hunters (Maul 2001a, 858ff.), the hypothesis that *Spermophilus* material discovered in close vicinity to the mudflow fan (CMF) was derived from destroyed animal burrows has previously been posited (Kahlke 2001e; 2006; Maul 2001a). Ground squirrels use burrows for nocturnal sleep and hibernation. Similarly, all of the other microvertebrates found represented by complete or partial skeletons (**Table 6**) also dig burrows in the ground or live seasonally in such structures: *Cricetus* hibernates in burrows; *Microtus* is winter active, but also tunnels and digs burrows; extant *Rana dalmatina* hibernate terrestrially under moss cushions, clods, stones or piles of leaves, or they dig hiding places in the ground.

The listed microvertebrate skeletons and individually related bone units of the 2001–2015 field seasons (**Table 6**), however, do not represent the contents of burrows dug into the fossiliferous lower part of the Upper Fluviatile Sands [UFS(l)] in the immediate site area. The corresponding sediment structures were not detected, nor could the animals have successfully penetrated through the densely packed accumulations of large mammal bones that were deposited there. In addition, the many hyaena coprolites embedded next to individually related ground squirrel finds (see section 6.1.2.6.) would have been damaged or destroyed by the branching systems of *Spermophilus* tunnels, which is not the case.

The regularity of the arrangement of the microvertebrate skeletal units which occur almost exclusively on the valley facing periphery of the large mammal bone accumulations within the deepest part of the channel filling (**Fig. 18 above**) can only be explained by their being sorted by running water. The individual carcasses, or parts thereof, would have been successively transported from upstream areas and deposited parallel to and overlapping with the large mammal bone accumulations. As with the bulk of large mammal bones embedded directly northeast of them (**Fig. 18 below**), the even distribution of the microvertebrate remains over the same depth interval suggests a simultaneous and successive supply of both types of necromass. Whereas turbulent sedimentation conditions prevailed when filling the deepest part of the erosion channel (Kahlke et al. 2020; see section 4. of this paper), the *Spermophilus* skeletons were embedded in varying states of completeness, therefore it must be assumed that they underwent different transportation processes. The degrees of decomposition of the transported microvertebrate cadavers most probably depended on the duration of the transport within the river valley, which in turn would have roughly correlated with the individually covered distances, as well as on small-scale flow conditions at the place where the remains became embedded.

The *Spermophilus* axial skeleton IQW 2012/38656 (Mei. 35630) is preserved in almost every detail up to the presence of complete sets of phalanges (**Fig. 19**). It belonged to an individual that found death in the immediate vicinity of its final embedding place. Only the corresponding cranium IQW 2012/38657 (Mei. 35631) was separated post mortem and drifted over just 3 cm to the north, almost parallel to the flow direction of the river water.

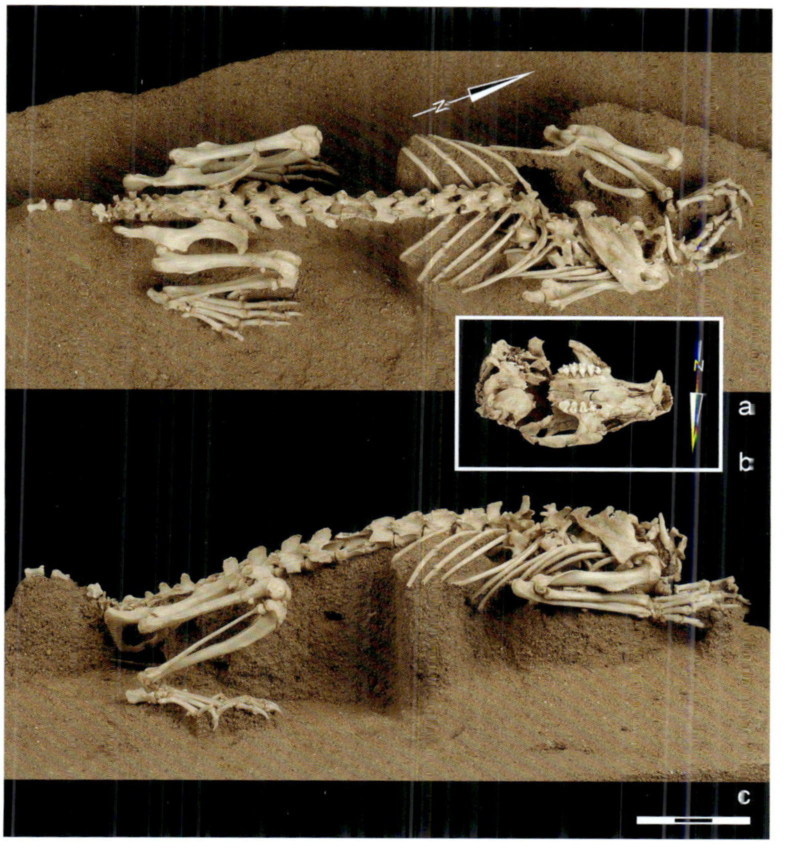

Fig. 19 Untermassfeld research excavation, in-situ preparation of an articulated skeleton of *Spermophilus kahlkei* Maul (in this volume; holotype). – **a**, **c** Axial skeleton IQW 2012/38656 (Mei. 35630) from grid square Q 39, 2.43–2.35 m below site 0-level, dorsal and right lateral views. – **b** Individually associated skull IQW 2012/38657 (Mei. 35631) from grid square Q 39, 2.46–2.43 m below site 0-level, embedded 3 cm north of the axial skeleton, palatal view. – Scale: 30 mm.

The predominantly articulated, partial ground squirrel skeletons show only a few (**Fig. 20**) or a moderate number of shifted skeletal elements (**Fig. 21**). These are mostly skulls, hemi-mandibles, and long bones with relatively large surfaces. At this stage, the drifting of skeletal elements more or less often follows the river water flow direction, as, for example, in the case of the *Cricetus* skeleton (Maul in this volume, Fig. 8). However, in view of the powerful flow regime reconstructed for the origin of the fossil bearing Upper Fluviatile Sands (Kahlke et al. 2020; see section 4. of this paper), mechanical damage and loss to cadavers during transport cannot be excluded.

In a situation where advanced decomposition of partial cadavers has occurred, in addition to increased horizontal displacement of skeletal elements (**Fig. 22**), massive vertical drifting leads to complete disintegration of the individual bone units (**Fig. 23**). Separated skeletal elements can sink up to 115 mm into the surrounding sediment matrix. This involves the embedding of isolated skeletal elements at different inclination angles up to the vertical position of long bones [IQW 2012/38660 (Mei. 35634)]. Such embedding scenarios indicate the presence of water-saturated sand or sand-water mixtures with corresponding hydraulic properties at the bottom of the erosion channel.

The predominantly *Spermophilus* microvertebrate skeletons deposited in the erosion channel represent freshly deceased individuals, i. e., those who fell victim to the ongoing flood event (see section 6.4.2.).

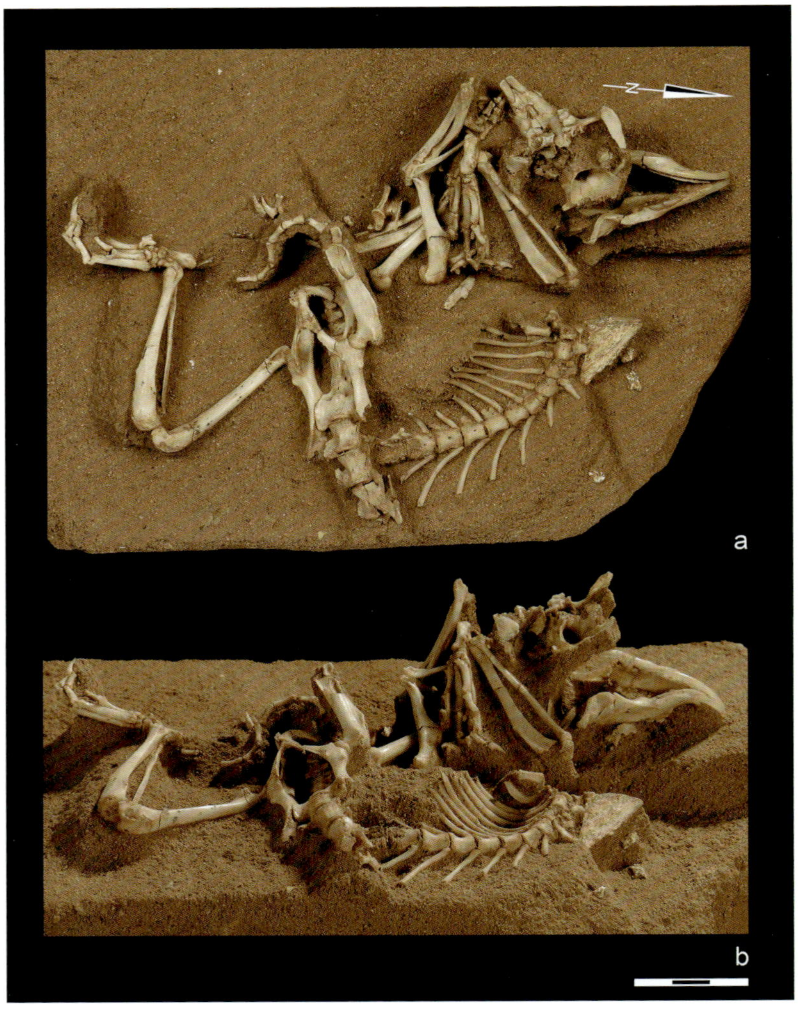

Fig. 20 Untermassfeld research excavation, in-situ preparation of a predominantly articulated partial skeleton of *Spermophilus kahlkei* Maul (in this volume) IQW 2003/28189 (Mei. 27351) from grid square Q 156, 1.84–1.80 m below site 0-level. – **a–b** Top and side views of the embedding status. – Scale: 30 mm.

– 6.1.2.6. Hyaena coprolite concentrations

Concentrations of hyaena (*Pachycrocuta brevirostris*) coprolites represent further examples of smaller-scale distribution patterns of fossils arranged by transport sorting. Generally, *Pachycrocuta* coprolites occur in all fossil-bearing sediments of the site and are usually spheroid or ovoid bodies that are occasionally compressed but, in most cases, undeformed (detailed description in Keiler 2001b). During the 2001–2017 excavations in the southern part of the site, below the main planum (= Planum I), i. e., >1 m below site 0-level (see section 2.1.), a total of 215 coprolites were recovered, bringing the total number of corresponding finds from Untermassfeld to 435. Within the channel infill, coprolites (**Table 7**) were found at depths of 2.49–1.27 m below site 0-level. Higher concentrations of identifiable finds ($n \geq 10/m^2$) were recorded in grid squares Q 30, 155, 547 and 609 (**Foldout III**). An unusually high number, of more than 100 densely packed specimens, was found in grid square Q 34 (in part **Fig. 24a**). Q 154 contained a flat mass of compacted *Pachycrocuta* faeces (**Fig. 24b–c**).

The differently sized coprolites found individually or in small groups were usually more or less intact, however occasional deformations had occurred. Accumulations of large amounts of mostly deformed and/or

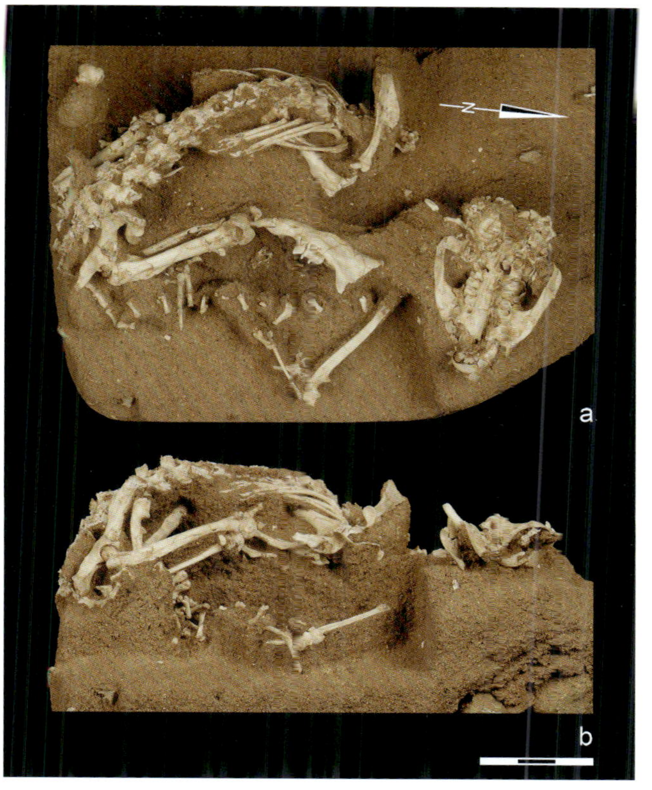

Fig. 21 Untermassfeld research excavation, in-situ preparation of a predominantly articulated partial skeleton of *Spermophilus kahlkei* Maul (in this volume) IQW 2012/38659 (Mei. 35633) from grid square Q 40, 2.39–2.34 m below site 0-level. – **a–b** Top and side views of the embedding status. – Scale: 30 mm.

abraded coprolites, or completely disintegrated faeces, were discovered in the deepest part of the erosion channel. In contrast to all other finds (Keiler 2001b and above-mentioned material), fossil faeces from grid squares Q 34 and 154 showed varying levels of decomposition. Nearly intact, more or less spheroid shaped or slightly deformed bodies have dry cracks of various widths and depths (**Fig. 25a–b**). Rapid flushing along these cracks, as well as on the surface of the coprolites, led to an increasing separation of partial bodies (**Fig. 25c**), up to and including the complete disintegration of the spherical droppings (**Fig. 25d**). This probably led to a complete flattening of the faeces (**Fig. 25e**). The resulting bodies, 0.3–2.0 cm in height, cover surfaces of up to 320 cm² (**Fig. 24b–c**). Corresponding preparations were possible through application of the transfer method (see section 3.). The underside of the flattened faeces mirrors the contours of the relatively smooth bottom of the channel filling (**Fig. 24c**). Within the coprolite mass, bone fragments of up to 2 cm in length are visible which have rounded edges caused by dissolution during digestion (**Fig. 25e**), part of the hyaena's defecation process. Finally, sandy sediments and larger-sized fresh bone fragments with sharp broken edges became attached to the upper surface of the flat and probably sticky faecal bodies (**Figs. 24b, 25f**).

The origin and transportation of the coprolites to the fossil site of Untermassfeld can be reconstructed from the sequence and intensity of the decomposition stages described above. While the numerous, largely undamaged, and often individually found specimens were transported into the fluviatile sedimentation area over very short distances (Keiler 2001b) and embedded rapidly, the unusually large coprolite conglomeration of grid square Q 34 and the flattened faecal body from square Q 154 point to the occurrence of very different taphonomic processes.

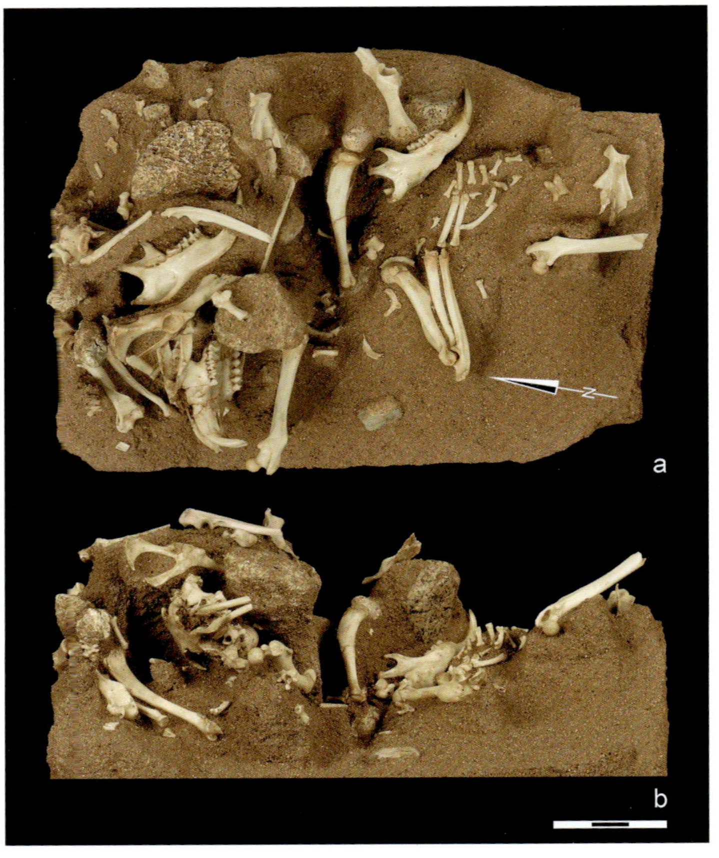

Fig. 22 Untermassfeld research excavation, in-situ preparation of a predominantly disarticulated partial skeleton of *Spermophilus kahlkei* Maul (in this volume) IQW 2009/30637 (Mei. 29799) from grid square Q 46, 2.08–1.97 m below site 0-level. – **a–b** Top and side views of the embedding status. – Scale: 30 mm.

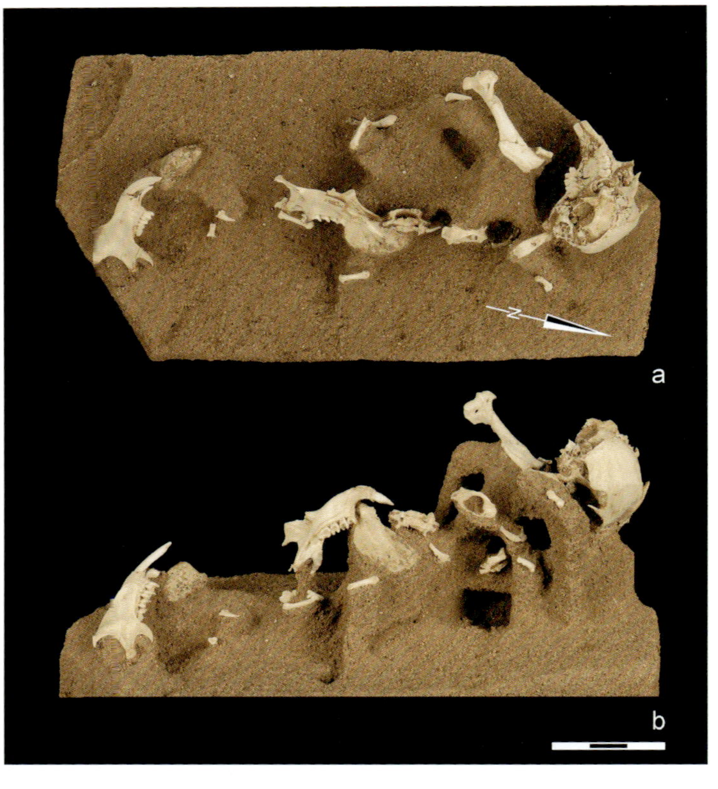

Fig. 23 Untermassfeld research excavation, in-situ preparation of a completely disarticulated partial skeleton of *Spermophilus kahlkei* Maul (in this volume) IQW 2009/30636 (Mei. 29798) from grid square Q 196, 1.90–1.83 m below site 0-level. – **a–b** Top and side views of the embedding status. – Scale: 30 mm.

Grid square (Q)	n	Grid square (Q)	n	Grid square (Q)	n
21	2	38	5	528	5
23	1	39	1	529	2
24	1	41	1	547	14
28	1	50	2	595	2
29	1	154	1*	596	3
30	10	155	10	609	11
31	2	156	1	627	2
33	7	158	1	628	3
34	103	159	1	645	2
35	3	519	3	646	1
36	5	527	1	647	1
37	6				
				Total n	215

Table 7 Numbers (n) of coprolites recovered from the southern part of the site of Untermassfeld (2001–2017 field seasons). * – flat mass of compacted faeces.

The shrinkage cracks observed on the largely intact coprolites indicate volume loss via dehydration (Vogeltanz 1965; image sample https://www.flickr.com/photos/dave_irving/4312816256, accessed 14 September 2017). This in turn implies that the faeces were deposited on land, i.e., along the river bank or in the adjacent floodplains. The existence of hyaena latrines marking boundaries of clan ranges (Kruuk 1972, 258ff.) in the immediate vicinity of the fossil site has already been postulated by Keiler (2001b). During dehydration, the high proportion of mineral bone substance (hydroxyapatite) causes a slight hardening of the hyaena droppings. In dry environments, therefore individual droppings of modern spotted hyaena (*Crocuta crocuta*) lying on the ground surface may be preserved for a year or more (Larkin et al. 2000, 21). In the case of flood events, saturated faeces are transportable over short distances of possibly up to tens of metres. Due to identical transport properties in flowing water, the droppings would have been concentrated at common deposition sites and then modified in the manner described above (on diminution of consolidated coprolites in moving water see Häntzschel et al. 1968, 4ff.). The post-mortem discharge of rectal content from freshly deceased hyaena individuals cannot be ruled out as being involved in the formation of the coprolite accumulations (see also Keiler 2001b, 696).

The *Pachycrocuta* coprolite concentrations were found at distances of approximately 5–10 m NW from the northern edge of the original (i.e., undisturbed) coarse clastic mudflow fan (Ellenberg and Kahlke 1997; see section 4. of this paper). The copious accumulation of coprolites in Q 34 was deposited slightly more than 1 m above the surface of the CMF's lateral extrusion (Kahlke et al. 2020, Fig. 11). The majority of coprolites found during the 2001–2017 excavations were embedded in the south-western periphery of the zone of large mammal bone accumulations within the channel infill (**Table 7**, **Foldout III**). The same applies

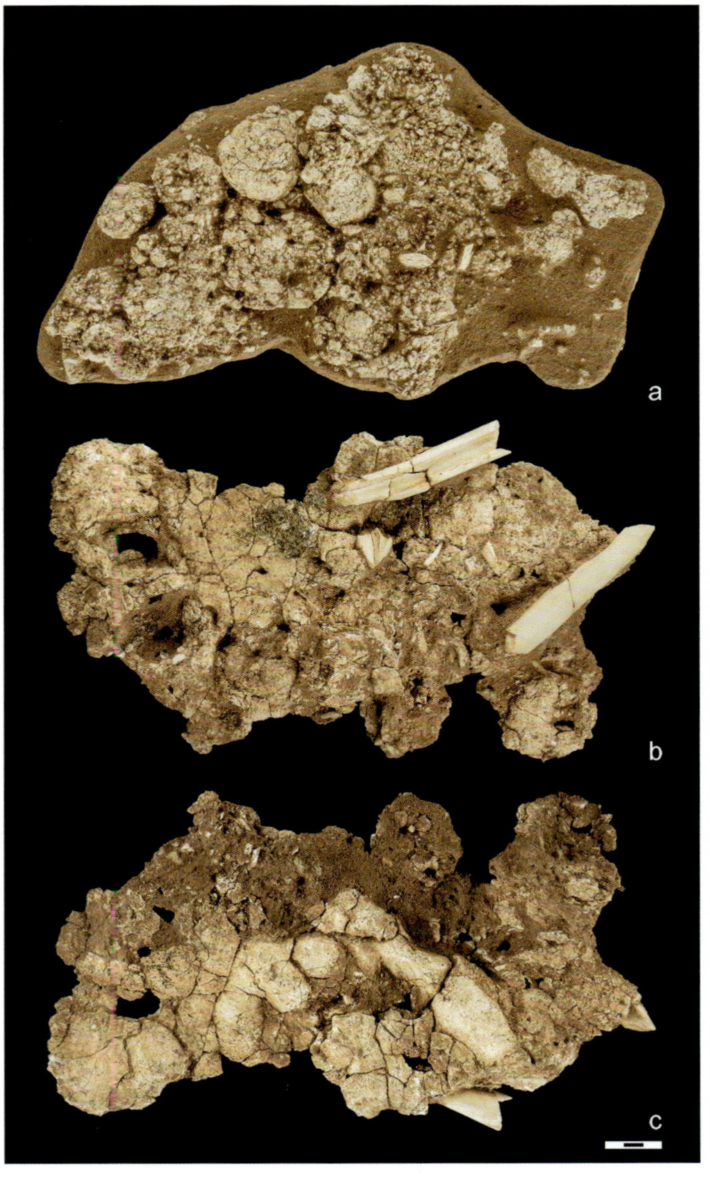

Fig. 24 Untermassfeld research excavation, *Pachycrocuta* coprolites. – **a** In-situ preparation of accumulated coprolites IQW 2017/44990 (Mei. 50299) in different stages of decomposition from grid square Q 34, 1.77–1.70 m below site 0-level, top view. – **b–c** Flat mass of compacted faeces IQW 2017/44993 (Mei. 50303) from grid square Q 154, 2.05–2.03 m below site 0-level, top and bottom views. – Scale: 30 mm.

to almost all of the recovered microvertebrate skeletal units (see section 6.1.2.5.). The in-situ preparation IQW 2010/30892 (Mei. 30054), originating from grid square Q 34, 1.75–1.71 m below site 0-level, is such an example and is comprised of coprolites and the partial skeleton of an agile frog (Böhme 2020b, Fig. 3) embedded immediately next to each other. Certain similar transport properties between microvertebrate cadavers, or parts thereof, and hyaena droppings in flowing water are likely to provide an explanation for such an arrangement pattern.

Fig. 25 Untermassfeld research excavation, in-situ preparation of *Pachycrocuta* coprolites, detailed views. – **a** Nearly intact specimen IQW 2017/44990 (Mei. 50299) with narrow cracks from grid square Q 34, 1.73–1.70 m below site 0-level. – **b** Specimen IQW 2017/44991 (Mei. 50300) with wide cracks from grid square Q 34, 1.75–1.74 m below site 0-level. – **c** Specimen IQW 2017/44991 (Mei. 50300) with separated partial bodies from grid square Q 34, 1.74 m below site 0-level. – **d** Disintegrated spherical dropping IQW 2017/44991 (Mei. 50300) from grid square Q 34, 1.74 m below site 0-level. – **e** Etched bone fragment within flattened faecal mass IQW 2017/44993 (Mei. 50303) from square Q 154, 2.05 m below site 0-level. – **f** Fluviatile sediment and fresh bone fragments attached to the upper surface of flat faecal bodies IQW 2017/44993 (Mei. 50303) from grid square Q 154, 2.03 m below site 0-level. – **a–d**, **f** top views; **e** bottom view. – Scales: 10 mm.

6.2. Postmortem bone modifications

6.2.1. Weathering

The majority of the remains of both large mammals and small vertebrates recovered from the site show hardly any signs of transport or weathering. Most of the osteological material was embedded in a relatively fresh condition, sometimes in the form of articulated specimens or as small vertebrate skeletons (see section 6.1.2.1., 6.1.2.5.). Even the exposed parts of isolated skeletal elements survived in a largely intact state. Traces of grinding and polishing are absent. In several instances, bones recovered in a completely intact state had sediment-free cavities. In such cases, the skeletal element was obviously embedded with an intact periosteum (Arribas and Palmqvist 1998). The necromass introduced into the sedimentation area of the site was fragmented by fluvial transport (see section 4.2., 6.1.) and in some cases possibly by trampling.

Occasionally, skeletal elements exhibit closely spaced parallel cracks within the compact bone. Such modifications can occur during strong insolation as a result of rapid degreasing. A comparison of the respective position in relation to the compass direction has not yet been carried out, but such an approach could lead to insights regarding the sun exposure of the bones before or after their introduction into the site.

A few bone fragments from various large mammal species which had experienced a significant amount of abrasion due to rolling were obviously transported as single elements over longer distances. These pieces were probably already exposed to weathering and fluviatile transport before the flood events that led to the formation of the fossil accumulation.

Numerous osteological finds show the etching traces of a delicately-veined root system on their surfaces (e. g., Guérin and Faure 1997, Pl. 66/3–4; H.-D. Kahlke 1997, Pls. 37/7, 44/1–2; 2001b, Pls. 79, 80, 83, 89/1–4; R.-D. Kahlke 1997b, 282, Pls. 51/10, 61/9; Sher 1997, Pls. 22/1–3, 25/1–3, 27/2, 29/1–12, 31/ 5, 7, 11, 16, 17, 32/2, 33/3–5, 8, 11, 12; Musil 2001b, Pl. 120/2; Sotnikova 2001, Pls. 103/3, 105/2, 110, 111/3, 112/1, 113; Turner 2001, Pls. 122/2, 124, 1; Breda et al. 2020, Figs. 4, 6, 8d; Kierdorf and Kahlke 2020, Fig. 1; Arnold in this volume, Fig. 27; Hemmer and Kahlke in this volume, Figs. 4b, 4d, 5c, 6d–f, 14g–l, 29; Iannucci et al. in this volume, Fig. 7; Kierdorf et al. in this volume, Fig. 1; **Figs. 4c–d, 14b** of this paper). In addition, numerous coprolites are traversed by root cavities (Keiler 2001b, 695f., Pl. 131/3–4). Following sediment deposition, the well-fertilised areas of the necromass accumulation were then apparently covered by predominantly herbaceous vegetation. Although the age of this vegetation cannot be exactly determined, the arrangement of the surviving traces reflects the status of the separation of the skeletal elements and other tissues at the time of root penetration. For example, on a skull of *Stephanorhinus hundsheimensis* the frontal horn had already been separated or destroyed before root penetration occurred (Kotowski et al. 2020, 1274f.).

6.2.2. Insect marks

Three morphologically well-differentiated types of insect bone modifications were found on the osteological material from Untermassfeld (Keiler et al. 2020). The traces on an antler of *Pseudodama vallonnetensis*, described as branched tubes, are the pupation chambers of an indeterminate type of dermestid, probably of the genus *Dermestes*. Judging from the individual antler development, the infested cervid died sometime around the middle of August to the beginning of September. Hemispherical pits of varying size and intensity are another type of bone modification observed on the *Pseudodama* antler fragments and may be a combination of the traces of tenebrionid and dermestid larvae. In addition, clustered shallow pits are present on the bone surfaces of several cervid finds, which were probably also caused by dermestid larvae (Keiler et al. l.c.).

An accurate determination of the exact colonisation period for the Untermassfeld cervid carcasses cannot be ascertained from the aforementioned insect marks as the incubation periods of the various dermestid species vary depending on temperature and habitat moisture levels. The pupal structures of the branched tubes and the hemispherical pits are, however, fully formed. Given a development period for dermestid larvae of at least 26 days, it can be assumed that the carcasses, or cadaver parts, from Untermassfeld must have remained on the surface and stayed relatively dry for at least a month before becoming embedded in the fluvial sediment. This may explain why only about 0.1 % of the Untermassfeld large mammal finds show insectogenic modifications (Keiler et al. l.c.). Considering the dynamic fluviatile regime in the area around where the fossil deposit originated (see section 4.2.), as well as the presence of numerous carnivores (see section 5.2.7.), accessible carcass remains that survived for several weeks on the surface would have only occurred in rare cases.

Traces of limnic insects have not yet been found on the Untermassfeld material. High sedimentation rates during the formation of the fossil deposit seem to have prevented their activities on the deposited necromass.

For the sake of completeness, it should be mentioned that several coprolites of *Pachycrocuta brevirostris* show traces of parasitic activities. The causative agents are most likely to be coprophagous beetles and their larvae (Keiler 2001b, 694 f.).

6.2.3. Chew and gnaw marks

The impact of most carnivore action recorded in the Untermassfeld faunal assemblage (see section 5.2.7.) facilitated and modified the decomposition of the carcass or skeletal units before and after they had been transported into the site. Corresponding traces are described and/or illustrated in most of the papers on the individual large mammal groups and in other publications (e.g., H.-D. Kahlke 1997; 2001b; R.-D. Kahlke 1997b; 2001d; 2006; Dubrovo 2001; Musil 2001a; 2001b; Turner 2001; Kahlke and Gaudzinski 2005; Steguweit and Kahlke 2014; Kotowski et al. 2020). *Pachycrocuta brevirostris* in particular left numerous bite and feeding marks on the find material (**Fig. 26**). Secondary to this, corresponding marks of the abundantly preserved canids might also be present. However, for felids strong penetration of the teeth into the skeleton of any respective prey would have been rather rare.

An example of the regular arrangement of *Pachycrocuta* feeding marks was demonstrated at Untermassfeld on remains of *Hippopotamus antiquus* (R.-D. Kahlke 1997b, 282, Tabs. 2, 3). Reconstructed activities served primarily to extract fat-rich tissue. The exposure of large areas of cancellous bone by *Pachycrocuta* was mainly carried out when the respective skeletal element was too big to be bitten through completely.

A separate study refers to hyaenid modifications of *Bison menneri* bones (Arnold 2018; in this volume). Using this extensive material, the feeding strategy of *Pachycrocuta* was characterised and compared with that of extant species. Naturally, the activities of the Untermassfeld hyaenas concentrated on skeletal elements with high meat and marrow content and/or a low bone mineral density. Given the abundant supply of fresh flood victims, or parts of them (see section 4.2., 6.1.), it was mainly the nutrient-rich limb bones that were eaten. Clear preferences for particular feeding tendencies can be observed. While humeri and tibiae were consumed from the proximal to the distal ends, the opposite direction was preferred for ulnae and radii. The reconstructed activities of *P. brevirostris* took place in the immediate vicinity of the embedding area of the fossil finds as well as upstream of the present site (see section 6.4.).

All types of hyaena chew and gnaw marks, as defined by Maguire et al. (1980), could be detected in varying frequency on the bison material from Untermassfeld. There are clear quantitative similarities between the findings from Untermassfeld and those of an Early Pleistocene *Pachycrocuta* den in the Haro River Quarry (Pakistan, Saunders and Dawson 1998). Compared to extant hyaenids, the distribution of the categories produced by *P. brevirostris* from Untermassfeld resembles that of *Parahyaena brunnea* and *Hyaena hyaena*. The most striking differences are found when comparing with *Crocuta crocuta* (Arnold l.c.).

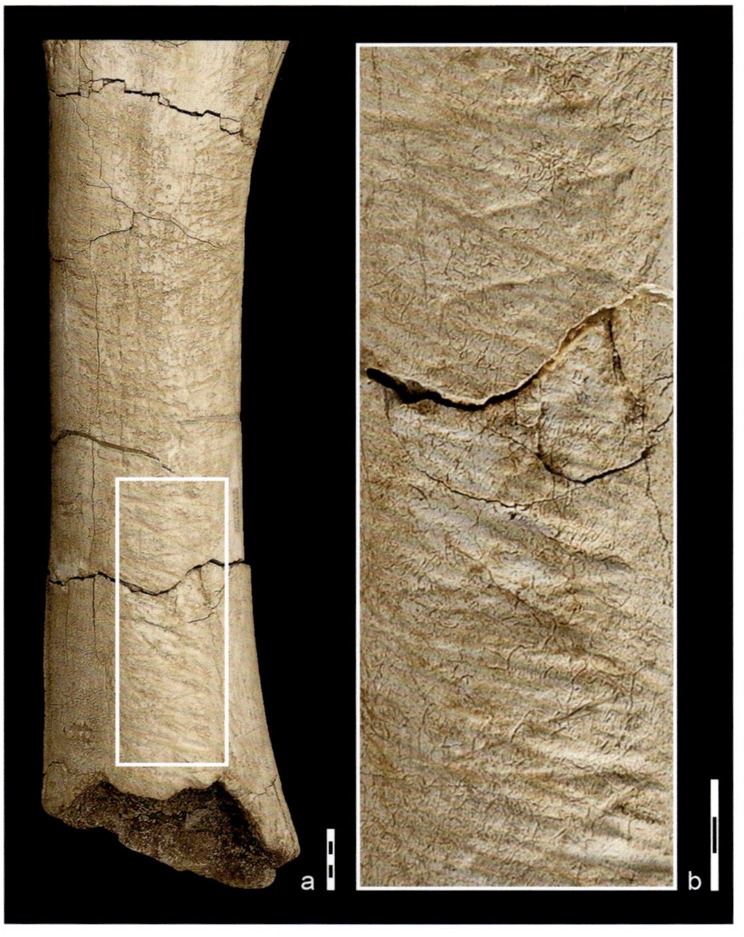

Fig. 26 *Mammuthus meridionalis* (Nesti), Untermassfeld. – **a–b** Right femur IQW 2007/29518 (Mei. 28680) with extensive gnaw marks by *Pachycrocuta brevirostris*, anterior views. – Scales: **a** 50 mm – **b** 30 mm.

Several skeletal elements of different large mammal species show distinct gnaw marks of rodents (Maul 2001b; **Fig. 27** of this paper). Based on the width of the corresponding grooves and the width of the facet-segments and facet-semicircles produced by gnawing, the porcupine *Hystrix refossa* was identified as the causative agent for the majority of the gnaw marks. Smaller gnawing traces may have been made by the frequently documented ground squirrel *Spermophilus kahlkei*. Some of the skeletal elements gnawed by rodents also show feeding marks of carnivores. This indicates that defleshed bones were being utilised. The concentration of skeletal elements with rodent gnaw marks in several immediately adjacent grid squares (Maul 2001b, Fig. 2) could indicate that at least some skeletal elements were gnawed in the immediate vicinity of the embedding site. For other finds, an origin upstream of the floodplain areas can be assumed.

The single find of a metatarsal diaphyseal fragment from *Eucladoceros giulii* shows clear traces of herbivore osteophagy (Kahlke 2001d), for which *Bison menneri* or *E. giulii* are most likely to be responsible. This osteophagic behaviour was probably triggered by occasional, possibly seasonal, mineral deficits in the diet of these ruminants.

Based on deliberate misinformation and misidentification, the presence of alleged anthropogenic cut marks on bone fragments from Untermassfeld was published. These claims have been corrected by Roebroeks et al. (2018).

Fig. 27 *Equus wuesti* Musil, Untermassfeld. – **a–b** Right pelvis fragment IQW 2014/41849 (Mei. 47158) with extensive gnaw marks by *Hystrix refossa*, dorsal views. – Scales: **a** 30mm – **b** 10mm.

6.2.4. Digestive etching and corrosion

Various digestion traces are visible on the osteological material. The remains of the migratory Atlantic salmon *Salmo* cf. *salar* often show etching or corrosion (Böhme 2020a) and some of the specimens are also crushed or deformed. Four of the six recovered individuals are affected (see section 6.3.). Judging from the clear evidence of bone digestion, the fossils must originate from carnivore faeces. The salmon from Untermassfeld reached body lengths of up to about 1m, resulting in body weights of up to 10kg. The bears, *Ursus* ex gr. *U. dolinensis*, and the large canids *Xenocyon lycaonoides* must be considered as prime hunters for these large fish (Böhme l.c.), given that both were found at the site in unusually high MNI (see section 6.3.). The occurrence of Atlantic salmon in the middle Werra valley and the evidence of their utilisation as prey in the fossil material helps to provide a seasonal time frame for the necromass accumulations at the site (see section 6.5.).

Occasionally, bone fragments of up to 2 cm in length are also found in the material, which, in addition to rounded edges, have highly polished surface areas. These are corroded bone splinters that have passed through the stomach of a hyaena (see section 6.1.2.6., **Fig. 25e**) and probably originate from completely reworked coprolites. The regurgitation of larger, already acid-etched bone fragments, as described by Zapfe (1939) from the extant *Crocuta crocuta*, is also plausible for the Untermassfeld *P. brevirostris*.

6.2.5. Diagenetic and neotectonic modifications

Following sediment deposition, hydrological, geochemical, and neotectonic processes all modified the embedded necromass of the Untermassfeld fossil deposit. Oxidation and leaching prevented preservation of the available biomolecules in the osteological material. Thus, neither extraction of the bone protein osteocalcin for analysis by Matrix-Assisted Laser Desorption Ionisation Mass Spectrometry (MALDI-MS), nor the recovery of sequenceable DNA were successful (Kahlke 2020, 1047).

The constant supply of carbonate-rich water from the Triassic Muschelkalk of the adjacent slope of the river valley (Ellenberg and Kahlke 1997, 36 f.) caused a slight impregnation of the Pleistocene stratigraphic sequence and its fossil content. Increased carbonate and reduced phosphate contents indicate diagenetic de- and remineralisation processes within the bones (Keiler 1995, 7; 2001a, 449). Due to the complete recrystallisation of the root cementum of isolated cervid teeth, no seasonal time period for the individual age of death of the animals could be determined (Kahlke 2001e, 977).

In the deepest part of the channel fill, numerous highly demineralised, and thus unstable, skeletal elements were found. This is especially true in and immediately above the reworked fossil-bearing zone of the underlying YCG, where pore waters from the acidic, secondary carbonate-free gravels had often significantly dissolved the embedded fossil material. The resulting reduced stability of the bones led to deformations in some skeletal elements (e. g., skulls in **Figs. 3**, **10**).

In addition, the fossil-bearing stratigraphic sequence of Untermassfeld was overprinted by a neotectonic stress regime. In the western and south-western areas of the site in particular, the fossil-bearing layers increasingly dip to the SW as a result of tilting movements south of a fault zone (Ellenberg and Kahlke 1997, 54 ff.; measurements in Kahlke 2001e, 934; perspective block diagram in Kahlke et al. 2020, Foldout I). The resulting sedimentary plastic deformations mainly affected larger skeletal elements (e. g., a skull of *Hippopotamus antiquus*, Kahlke 2001c). Corresponding deformities and fractures occurred mainly on finds from the channel fill near the north-eastern, i. e., slope-facing, flank.

6.3. Bone spectra, minimum numbers of individuals and mortality patterns

In earlier reviews of the Untermassfeld fossil assemblage (Kahlke 1997c; 2001d; 2006), the spectra of previously recovered bones were qualitatively characterised. Quantitative data were determined for the most frequently occurring species: *Bison menneri*, *Eucladoceros giulii*, *Pseudodama vallonnetensis*, *Hippopotamus antiquus*, *Stephanorhinus hundsheimensis*, and *Equus wuesti*. The bone spectrum of *B. menneri* remains recovered up to and including 2015 and consisting of 2,169 catalogued specimens has already been examined in detail (Arnold 2018; in this volume). Find spectra of less common species are presented in the individual studies of the Untermassfeld monographs.

Within the entire volume of large mammal material recovered from Untermassfeld, complete or near-complete trunk skeletons have not been recorded. Even undamaged vertebrae are underrepresented. Complete ribs of all species are also counted among the rarer finds. Their fragments, however, occur in larger numbers. Obviously, the trunk skeletons of the large mammals would have been exposed to the most rapid processes of decomposition. Beside the swelling of the carcasses through gases and the resulting mechanical instability of, especially, the thoracic and abdominal cavities, the activities of carnivores would have been the main cause of their disintegration. The relatively thin layer of skin covering the abdomen and the anal region would have offered easily accessible entrance points to carnivores looking to utilise the carcass.
Complete or nearly complete skulls of larger mammals were also rare at Untermassfeld. Of those that were recovered, some were highly fragmented or plastically deformed according to the neotectonic stress patterns acting at the site (see section 6.2.5.). Fortunately, however, they represent a rather broad spectrum of species: *Bison menneri* (Kahlke 2001e, Fig. 1; 2006, Fig. 4; Keiler 2008, Figs. 12–32; Bukhsianidze 2020, Figs. 2–4), *Eucladoceros giulii* (H.-D. Kahlke 2001a, Pl. 74; R.-D. Kahlke 2001e, Fig. 2; 2006, Fig. 9, 11; 2009, Fig. 10–11; Keiler 2008, Figs. 11, 33; Breda et al. 2020, Figs. 10–11; **Fig. 3** of this paper), *Pseudodama vallonnetensis* (Breda et al. 2020, Figs. 15–16), *Hippopotamus antiquus* (Kahlke 2001c, Fig. 5, Pls. 77–78 2006, Fig. 12; Keiler 2008, Figs. 1–3), *Sus strozzii* (Guérin and Faure 1997, Pl. 63; Iannucci in this volume Fig. 1), *Stephanorhinus hundsheimensis* (H.-D. Kahlke 2001b, Pls. 79–80; R.-D. Kahlke 2009, Fig. 12; Kotowski et al. 2020), *Mammuthus meridionalis* (Kahlke 2001e, Fig. 4; 2006, Fig. 18), *Acinonyx pardinensis pleistocaenicus* (Hemmer 2001, Pl. 135; Kahlke 2006, Fig. 20; 2009, Fig. 16; Hemmer and Kahlke in this volume, Fig. 6), *Pachycrocuta brevirostris* (Kahlke 2001e, Fig. 5; 2006, Fig. 28; Keiler 2008, Figs. 10–11; Iannucci et al. in this volume, Fig. 1), *Meles meles atavus* (Wolsan 2001, Fig. 1; Madurell-Malapeira et al. 2011), *Ursus* ex gr. *U. dolinensis* (Kahlke 2001e, Fig. 6; 2006, Fig. 31; Musil 2001b, Pl. 114; Keiler 2008, Fig. 34), *Xenocyon lycaonoides* (Kahlke et al. 2020, Fig. 13a; **Fig. 10** of this paper), *Canis mosbachensis* (Sotnikova 2001, Pls. 102–104; Kahlke 2006, Fig. 35; 2009, Fig. 20; Kahlke et al. 2020, Fig. 4b; **Figs. 11–12** of this paper) and *Trogontherium cuvieri* (Maul in this volume, Fig. 6).
Obstructive elements of the cranial region (e.g., complete antlers attached to skulls, shed beams, skulls with wide sweeping horn cores, tusks or mandibles of elephants, complete hippopotamus mandibles with exposed canines and incisors), and complete shoulder blades or pelves of larger mammals are rare or completely absent. These parts of carcasses or skeletons probably became frequently entangled further upstream and were embedded or destroyed outside the limits of the present site.
Small and easily transportable bones, such as patellas, sesamoids and phalanges, are also significantly underrepresented in the Untermassfeld material. This also indicates that a certain amount of transport sorting took place. Even with a relatively small but sufficient transport energy, the river's current would have enabled such elements to be carried away in significant quantities.
Minimum numbers of individuals (MNI) have been determined for all mammal species recorded from Untermassfeld based on the most frequently recovered dental and skeletal elements and their properties, such as respective body side, body size, sex, and age at death. The previous compilation of the available material (Kahlke 2006) indicated an MNI of 313 larger-sized mammals. The current material (status as of 01 December 2021) represents at least 443 individuals of large mammal species (**Table 8**). Fishes, amphibians, reptiles, birds and small mammals with body weights of less than 10 kg belong to at least 298 more individuals (**Table 9**). Thus, the vertebrate total MNI of the site of Untermassfeld amounts to 741.
Among the larger mammals, individuals of the following herbivores have been detected most frequently: *Eucladoceros giulii* (20.6 %), *Pseudodama vallonnetensis* (19.5 %), *Bison menneri* (15.6 %), *Stephanorhinus hundsheimensis* (8.1 %) and *Hippopotamus antiquus* (5.2 %). The most common carnivores according to their MNI are *Pachycrocuta brevirostris* (5.4 %), *Canis mosbachensis* (4.7 %), *Ursus* ex gr. *U. dolinensis*

Species	MNI juveniles/ subadults	MNI adults	MNI total	data
Bison menneri	18	51	69	Based on Arnold 2018
Bison sp.	0	1	1	Bukhsianidze 2020
Capreolus cusanoides	1	4	5	Breda et al. 2020
Cervalces carnutorum	2	5	7	Breda et al. 2020
Eucladoceros giulii	38	53	91	Breda et al. 2020
Pseudodama vallonnetensis	32	54	86	Breda et al. 2020
Hippopotamus antiquus	13	10	23	Kierdorf and Kahlke 2020
Sus strozzii	3	3	6	R.-D. Kahlke
Stephanorhinus hundsheimensis	13	23	36	Kahlke 2006; Arnold in Kotowski et al. 2020
Equus wuesti	5	8	13	Kahlke 2006; Eisenmann and Boulbes 2020
Equus sp.	0	1	1	Eisenmann and Boulbes 2020
Mammuthus meridionalis (incl. *M.* cf. *meridionalis*)	4	6	10	pers. comm. D. Mol, August 2015
Panthera onca gombaszoegensis	1	6	7	Hemmer and Kahlke in this volume
Acinonyx pardinensis pleistocaenicus	0	3	3	Hemmer and Kahlke in this volume
Puma pardoides brevipes	0	2	2	Hemmer and Kahlke in this volume
Lynx valdarnensis spelaeoides	0	3	3	Hemmer and Kahlke in this volume
Megantereon adroveri	0	2	2	Hemmer and Kahlke in this volume
Homotherium latidens ultimum	1	4	5	Hemmer and Kahlke in this volume
Pachycrocuta brevirostris	18	6	24	Kahlke 2006; G. Utschig
Meles meles atavus	1	2	3	G. Utschig, R.-D. Kahlke
Ursus ex gr. *U. dolinensis*	2	9	11	R.-D. Kahlke
Xenocyon lycaonoides	0	6	6	R.-D. Kahlke
Canis mosbachensis	2	19	21	R.-D. Kahlke
Castor fiber	0	2	2	Kahlke 2006; Maul in this volume
Trogontherium cuvieri	0	2	2	Maul in this volume
Hystrix refossa	0	1	1	Kahlke 2006; Maul in this volume
Macaca sylvanus	1	2	3	Reumer and Kahlke in this volume
Total MNI	**155**	**288**	**443**	

Table 8 Summary of the minimum number of individuals (MNI) and individual age composition of large mammals (including castorids and hystricids) from Untermassfeld.

(2.5 %), *Panthera onca gombaszoegensis* (1.6 %) and *Xenocyon lycaonoides* (1.6 %). Among the vertebrate species with body weights below 10 kg, the rodents *Spermophilus kahlkei* (26.8 %), *Microtus thenii* (15.1 %) and *Mimomys pusillus* (9.4 %), crows/ravens *Corvus* sp. (6.4 %), and the moor frog *Rana arvalis* (2.7 %) are most frequently represented according to their respective MNI.

For species with high numbers of individuals, namely *Bison menneri*, *Eucladoceros giulii*, *Pseudodama vallonnetensis*, *Hippopotamus antiquus*, *Stephanorhinus hundsheimensis*, *Equus wuesti*, *Mammuthus meri-*

Species (fishes, amphibians, reptiles, birds)	MNI	Species (small mammals)	MNI
Salmo cf. *salar*	6	*Sciurus whitei*	1
Gobio sp.	1	*Spermophilus kahlkei*	80
Lissotriton vulgaris	3	*Glis sackdillingensis*	3
Triturus cf. *cristatus*	1	*Sicista* sp.	1
Rana temporaria	3	*Cricetus runtonensis*	4
Rana arvalis	8	*Myodes hintonianus*	8
Rana dalmatina	4	*Pliomys episcopalis*	7
Pelophylax ridibundus vel *esculentus*	1	*Mimomys savini*	9
Bufo bufo	2	*Mimomys pusillus*	28
Bufotes viridis	2	*Microtus thenii*	45
Hyla arborea	4	*Apodemus sylvaticus*	5
Anguis fragilis	1	*Hypolagus brachygnathus*	1
aff. *Ophisops elegans*	1	*Lepus* sp.	2
Lacerta sp.	1	*Erinaceus* cf. *europaeus*	1
Eryx sp.	2	*Talpa* cf. *europaea*	6
Elaphe sp.	2	*Talpa minor*	2
cf. *Coronella* sp.	1	*Desmana thermalis*	1
Natrix sp.	2	*Sorex minutus*	1
Cygnus sp.	1	*Sorex runtonensis*	7
Anser sp.	1	*Sorex margaritodon*	5
Anatidae gen. et sp. indet.	1	*Beremendia fissidens*	3
Lyrurus sp.	2		
Phasianidae gen. et sp. indet.	2		
Francolinus capeki	1		
Haliaeetus sp.	1		
Accipiter nisus	1		
Corvus sp.	19		
Turdus sp.	1		
Garrulus sp.	1		
Passeriformes indet.	2		
Sum of MNI	78		220
Total MNI			**298**

Table 9 Summary of the minimum number of individuals (MNI) of fishes, amphibians, reptiles, birds and small mammals weighing less than 10 kg from Untermassfeld; calculation based on data by Böhme (2020a; 2020b), Manegold (2020), Kahlke (see section 6.1.2.5.), and Maul (in this volume).

dionalis, and *Pachycrocuta brevirostris*, high proportions of juvenile to young adult individuals were recorded.

Sher (1997) demonstrated that the age and sex composition of the previously recovered *Bison menneri* material corresponded to that of a natural population. The currently determined MNI of juveniles/subadults

and adults (**Table 8**), which has been almost doubled by numerous new finds (Kahlke 2006; Van Asperen and Kahlke 2017; Arnold in this volume), does not contradict Sher's findings.
In comparison to the other cervid species, both *Capreolus cusanoides* and *Alces carnutorum* are represented by only a few individuals, including one and two juveniles, respectively (**Table 8**). Based on the wear stages of mandibular tooth rows of the most commonly recorded *Eucladoceros giulii*, H.-D. Kahlke (1997, 241–243) reconstructed the respective proportions of the age classes represented in the available material. The resulting mortality patterns showed a high number of juvenile individuals in their first year as well as old individuals aged between 8 and 11 years, while individuals in their prime were underrepresented. These patterns, which are also evident in the bone assemblages recovered at later dates (Breda et al. 2020; **Table 8** of this paper), reflect a U-shaped attritional mortality (discussion in Kahlke and Gaudzinski 2005). Based on the material available to date (H.-D. Kahlke 1997, 203–205), the individual age distribution of *Pseudodama vallonnetensis*, also frequently found, differs from that of the larger *E. giulii*. However, numerous new finds, especially of mandibular remains with D_4 (Breda et al. 2020; **Table 8** of this paper), slightly shift the completed mortality spectrum. The latest evidence puts the number of individuals in the first three years of life at a higher proportion in relation to the material discussed by H.-D. Kahlke (1997).
The fossil record of *Hippopotamus antiquus*, which has grown by at least three individuals since previous surveys were conducted (Kahlke 2006), comprises 56.5 % juveniles (**Table 8**). In extant populations, the proportion of calves of up to one year in age was found to be 5.9–13.8 % (Laws 1968, 46). The quite sparse remains of *Sus strozzii* belong to at least three juveniles and three fully adult animals.
In the cases of both *Stephanorhinus hundsheimensis* and *Equus wuesti*, the contribution of juveniles to the established MNI is rather high at 36.1 % and 38.58 %, respectively (**Table 8**). Where *Mammuthus meridionalis* is concerned, 60 % of the, at least, 10 recorded animals were found to be adult individuals (**Table 8**), two of them over 50 AEY (African elephant years) old (pers. comm. D. Mol, August 2015).
From the total Untermassfeld material, at least 22 individuals of the usually rare felids have been identified (Hemmer 2001; Hemmer and Kahlke in this volume; **Table 8** of this paper). However, since the findings are distributed over six species (**Table 8**) age profiles cannot be established. The most frequently detected of the felids, *Panthera onca gombaszoegensis,* is represented by at least seven individuals, including evidence for one juvenile with milk dentition. A cub has also been identified among the five recorded individuals of the saber-toothed cat *Homotherium latidens ultimum*.
For *Pachycrocuta brevirostris* the composition of the fossil material is more meaningful. In contrast to all other carnivores, young animals predominate with 75 % of the recorded MNI (**Table 8**), indicating a particular threat to juveniles. On the basis of dental development, Turner (2001, 686 f.) distinguishes five juvenile age classes, all of which relate to ages below one year, suggesting that the animals were non-seasonal breeders. The abundant species *Ursus* ex gr. *U. dolinensis* and *Canis mosbachensis* are mostly represented by adult individuals. Juveniles of *Xenocyon lycaonoides* were not detected.
Burrow-tunneling species are also seldom detected at Untermassfeld. For *Meles,* only three individuals were recorded and *Hystrix* is documented by only one individual. It is not surprising that the beavers *Castor* and *Trogontherium*, both species with pronounced water affinity, are each represented by the elements of only two individuals. In a similar way, remains of the good climber *Macaca* were also rare, with only two adult individuals of both sexes and one juvenile being detected.
Data on the mortality spectra of the more commonly recorded micromammals, i. e. *Spermophilus kahlkei*, *Microtus thenii* and *Mimomys pusillus*, are not available. In the case of rodents with rootless teeth (*Microtus*), it is impossible to distinguish age class based on the molars that were exclusively used for taxonomic determination (pers. comm. L. C. Maul, August 2020). For *M. pusillus*, and *S. kahlkei*, which both have rooted molars, well-tested individual ageing methods based on dentition do not yet exist.

6.4. Origin, composition and accumulation of the necromass

Although it can be assumed that components of the former floodplain vegetation were contained in the necromass accumulated at Untermassfeld, corresponding remains have only been preserved in indeterminable traces (see section 5.1.). Plant necromass can therefore not be considered in the following.

6.4.1. Large mammals

In addition to the depositional regime of the site (see section 4.2.) and the weight and transportability of the corresponding carcasses or parts thereof (see section 6.1.), the range, number of individuals, and mortality patterns of the recorded vertebrate species (see section 6.3.) are determined by their population densities, behavioural patterns and locomotory skills, especially swimming, climbing and even digging capabilities, as well as by their specific biotope preferences (see section 7.1.).

According to their high MNI (see section 6.3.), not only *Eucladoceros giulii*, *Pseudodama vallonnetensis* and *Bison menneri*, but also *Stephanorhinus hundsheimensis* and even *Hippopotamus antiquus* attained especially high population densities in the region around the site. All these species found sufficient habitats in the riverine floodplain and in the landscapes of the plateaus flanking the Werra valley (see section 7.2.). According to the palaeoenvironmental and taphonomic conditions at the time of the site's formation, the carnivore species *Pachycrocuta brevirostris*, *Canis mosbachensis* and *Ursus* ex gr. *U. dolinensis* are represented in particularly high numbers of individuals. Their strong representation is based on the seasonal availability of corresponding food resources (Hemmer 2001; Böhme 2020a; Hemmer and Kahlke in this volume). The same applies to the relatively frequent large cats *Panthera onca gombaszoegensis,* which has a pronounced affinity to water-rich habitats, and *Homotherium latidens ultimum*, with prey take over and scavenging as essential parts of its dietary traits (Hemmer 2001). Equids, *Mammuthus meridionalis*, *Cervalces carnutorum*, *Capreolus cusanoides* and *Sus strozzii* were represented in the site's catchment area with apparently lower population densities than the herbivores mentioned above. The same naturally applies to the predators *Xenocyon lycaonoides*, *Acinonyx pardinensis pleistocaenicus*, *Lynx valdarnensis spelaeoides*, *Puma pardoides brevipes,* and *Megantereon adroveri*.

In view of the physiological abilities and ethology of the respective species present in the Untermassfeld fauna, the individual causes of death were probably very diverse (Kahlke 2001e, 971). Against the background of a carnivore fauna that was rich in species and included many active hunters (for predator spectra see Hemmer 2001; Hemmer and Kahlke in this volume), it can be presumed that carcasses of prey in successive stages of decomposition were continuously present in the floodplain valley. Hemmer (l.c., 770) estimates that about 46 % of the detected ungulate remains were related to carnivore kills upstream. At the same time, this means that probably more than half of the large mammal individuals documented for Untermassfeld are flood victims; other possible causes of death, of quantitatively minor importance, are mentioned by Kahlke (2001e, 971).

The flood situation in the Werra valley upstream of the site thus had significant effects on the mortality spectra of the individual species. Clear evidence for this is the age composition of the total material of *Bison menneri* (**Table 8**), which corresponds to that of a natural population. Similar evidence to that of Untermassfeld is reported by Haynes (1988) from a modern mass drowning assemblage of American bison (*Bison bison*) from Wood Buffalo National Park (Alberta, Canada). As highlighted by Roth (1969), due to a strong social instinct many herd animals ignore their individual motivation to cross a body of water and instead follow the general movement of the herd. In the Untermassfeld assemblage, the mortality spectrum of *B. menneri* is obviously dominated by a herd related drowning scenario.

Among the cervids that possessed quite good swimming abilities, the two more frequently recorded species *Eucladoceros giulii* and *Pseudodama vallonnetensis* show differences in their mortality patterns (see section 6.3., **Table 8**). Prime aged individuals of the robust *E. giulii*, comparable in size with the American elk *Cervus canadensis*, have been recorded in relatively smaller numbers than those of the more delicate *P. vallonnetensis*. In the latter species, similar in size to the extant Persian fallow deer *Dama mesopotamica*, adult individuals were apparently more vulnerable to flooding events.
The mortality rate recorded for young *Hippopotamus antiquus* with deciduous dentition is slightly above the values of recent populations (see section 6.3., **Table 8**). Under today's non-catastrophic conditions, the mortality rate during the first year of life is between 45 and 54 % (Laws 1968, 46). Recent observations have shown that separation of calves from their mothers due to the action of water currents is the most common cause of death. If there were flood victims among the adult *Hippopotamus* individuals from Untermassfeld, then they most probably died in the turbulence of the flowing water near where the valley narrowed, 12 km upstream of the site (Kahlke 2001e, Fig. 26; see section 7.3. of this paper).
Extant European wild boars are only occasionally willing to swim across wider and more turbulent bodies of water. However, the comparatively low evidence for the presence of suids at Untermassfeld (**Table 8**) does not allow for more accurate conclusions on population dynamics to be made for these species.
The proportion of juveniles among the recorded individuals is very high for both *Stephanorhinus hundsheimensis* and *Equus wuesti* (**Table 8**). Both perissodactyl species were certainly not powerful swimmers. For rhinos it is generally difficult to keep their heavily built heads above the water level for longer periods of time, and the swimming ability of *S. hundsheimensis* was probably somewhere between the capabilities of the African black rhinoceros (*Diceros bicornis*) and those of recent Asian rhinocerotids (Child 1968; Roth and Child 1968; Roth 1969; Goddard 1970; Anonymous 1998; Kahlke 2001e and references therein). It is therefore probably safe to assume that all age stages of the Untermassfeld rhinoceroses were equally at risk from flooding events. This also applies to *E. wuesti*.
Elephants are known to be able to survive by actively swimming or drifting in water for extended periods of time and can cover long distances (compilation in Kahlke 2001e, 973). Because of their well-developed swimming abilities, supported by their trunk, which is used like a snorkel, and the pneumatised inner construction of their skull (Sondaar 1986) as well as their large body mass, the animals would have been less at risk from high floods. The majority of adult individuals from Untermassfeld (**Table 8**) probably died because they failed to reach a safe bank due to their poor climbing abilities.
For the species-rich felid spectrum, Hemmer and Kahlke (in this volume) have developed a flood vulnerability ratio (FVR). Based on the number of recovered skeletal elements of a species and its recorded MNI, the resistance to flood events of relatively common species (**Table 8**) can be determined within certain limits. According to the FVR, the cheetah *Acinonyx pardinensis pleistocaenicus* was generally not able to withstand the destructive effects of exceptional water flows, whereas the situation with *Panthera onca gombaszoegensis* was obviously quite different. Due to its preference for habitats close to rivers, the European jaguar was accustomed to flood events in a similar way as the extant jaguar and thus able to routinely escape dangerous situations (Hemmer and Kahlke l.c.). According to the FVR, something similar could have been the case with *Homotherium latidens ultimum*. Here, too, some familiarity with possible dangers emanating from flood events cannot be ruled out.
As noted earlier, among the numerous remains of the hyaena *Pachycrocuta brevirostris*, juveniles with life spans of less than one year clearly dominate (Turner 2001; Kahlke 2001e, 973; **Table 8**). Although mortality rates are difficult to quantify for recent hyaena populations, the observations by Kruuk (1972, 35 f.) and Van Jaarsveld (1993) in particular on extant spotted hyaenas (*Crocuta crocuta*) show that under non-catastrophic conditions the mortality of individuals during the first year of life is lower than documented in

the Untermassfeld material. Lower mortality rates of corresponding juveniles were also found in the Early Pleistocene assemblage of Venta Micena (Spain) and in numerous Middle to Late Pleistocene hyaena caves in Europe (compilation in Kahlke and Gaudzinski 2005, 1212). In the flooded area upstream of the Untermassfeld site, there must have been hyaena dens with very young and thus highly vulnerable animals. It is most likely that the sandy-clayey loose sediments that made up the floodplain encouraged the construction of these burrows. Adult *Pachycrocuta* individuals may have been capable swimmers due to their physical strength and are represented by impressive finds of only a few individuals (Iannucci et al. in this volume) in the Untermassfeld material.

Ursus ex gr. *U. dolinensis* is represented in large numbers in the material by mainly adult individuals (**Table 8**) and their presence might be related to the annual migrations of the Atlantic salmon *Salmo* cf. *salar* into the Werra (Böhme 2020a) (see section 6.5.). It is very likely that the bears were surprised by high tides while fishing in shallow river waters. Digestion marks on the majority of salmon finds support this idea. Böhme (l.c.) considers a similar relationship possible for *Xenocyon lycaonoides*.

In marked contrast to the hyaenas (see above), the wolf *Canis mosbachensis* (**Table 8**), which was recovered in large numbers, is dominated by adult individuals of both sexes. This does not reflect the age ratio of a natural population. As noted earlier (Kahlke 2001e, 973), the dens as well as the so-called »rendezvous sites« of the young of *C. mosbachensis* were located outside the flood-prone valley areas.

Both species of beaver, *Castor fiber* und *Trogontherium cuvieri*, which, according to their body masses, fit into the group of mammals dealt with here, along with the macaques, *Macaca sylvanus*, left evidence of only a few individuals (**Table 8**). Either water affinity or climbing ability minimised the risk posed by flooding in these cases. Even remnants of species that burrow into the ground, for example *Meles meles atavus* and *Hystrix refossa*, are rarely detected (**Table 8**). If their burrows were rapidly flooded, then they may have died and become embedded there. In addition, the tendency of extant *M. meles* to build their burrows on southern and western slopes (Lüps and Wandeler 1993, 876f. and references therein) could offer an explanation for the scarcity of badger remains in the find material. Drowning in burrows and subsequent embedding would also be conceivable for some of the smaller mustelid species as well as for the foxes, which are also absent from the find material.

The remains of other species, which would be expected from an ecological point of view in the Untermassfeld fauna, e.g., ermine, least weasel, and wildcat, are completely lacking in the fossil record. These animals were obviously able to survive by reaching branches of bushes or trees above the rising water level. If otters were also present in the early Pleistocene Werra, they would also not have been at great risk from the flooding events.

The hyaena gnawing frequently observed on the Untermassfeld large mammal material (Arnold in this volume; section 6.2.3. of this paper) took place at different locations within the Werra floodplain. Numerous more or less well preserved coprolites in the area of the site indicate that hyaenas were feeding in the immediate vicinity and were probably also marking their territory with feces. However, it is not very likely that the hyaenas were systematically and persistently dissecting complete or partial carcasses during periods of turbulent water flow, as reconstructed for the area leeward of the Untermassfeld clastic mudflow fan (CMF) (see section 4.2.). Evidently, a not insignificant proportion of gnawed skeletal elements must have originated from floodplains located upstream. In several cases, the transportability of parts of bulky carcasses into the deposition area was only made possible by the separation of anchor-like prominent sections created by hyaena predation, as shown in the examples of a *Bison menneri* spine (see section 6.1.1.2.) and the skull of a *Stephanorhinus hundsheimensis* (Kotowski et al. 2020, 1274).

The carcasses and their remains carried from the upstream sections of the flooded Werra valley into the area of the present site were subject to gradual decomposition during transport. In addition to hydrodynamic traction and pressure forces exerted by the turbulently flowing river water, collisions with natural obstacles and flotsam would have also contributed to their disintegration. On the lee side of the clastic mudflow fan, cut by the excavation area (CMF, see section 4.1.; **Fig. 2**), necromass was accumulated on a large scale during both the formation period of the Lower Fluviatile Sands (LFS, stage II of the site's evolution model) as well as the filling of the channel with Upper Fluviatile Sands [UFS(l), stage III] over the course of several flood events (see section 4.2.; **Fig. 2**).

Differences in the taphonomic features (see section 6.1.) are consistent with the characteristics of the respective sedimentation processes concluded from the geological evidence. Compared to the channel fill, the mostly lower find concentrations within the LFS and often somewhat lower degrees of disarticulation as well as less fragmented skeletal elements (see section 6.1.1.1., **Foldouts II** and **III**) point to a predominantly calm deposition of the sedimentary material and corresponding necromass carried by the flowing flood waters. Along the eastern river bank, skeletal elements were repeatedly washed ashore in a wrack zone.

In contrast, the arrangement of the find material recovered from the channel fill, is completely different (see section 4.1., 6.1.1.2., **Foldouts II** and **III**). Vertically, the spatial distribution of finds within the channel fill reflects a process of varying intensity of necromass input. With the onset of sediment filling, the deepest part of the channel initially received extraordinarily high quantities of cadaver and skeletal parts from the river water flowing off immediately after the peak of a flood. With progressive filling of the erosion channel, the concentration of deposited vertebrate remains reduced with decreasing transport energy, until in the uppermost levels only a few isolated body parts were embedded.

Generally, the decomposition and rearrangement processes undergone by the cadavers and skeletal remains in the area of the site, which were controlled by flowing water and their specific body construction, follow the described sequence: large mammal skeletal unit – isolated skeletal elements – polyspecific bone concentration – polyspecific bone accumulation (see sections 6.1.2.1.–4.). Horizontally, the arrangement of the large mammal remains deposited in the channel fill on the lee side of the CMF indicates two spatially separated zones cause by different flow regimes (see section 6.1.1.2.). In the immediate vicinity of the CMF, turbulently flowing water formed a largely irregular distribution of often relatively heavy skeletal elements and anatomically connected skeletal units (**Fig. 15**). A zone of clearly aligned elements follows to the northwest. The longitudinal axes of most skeletal elements here were oriented more or less parallel with the direction of the water flow (**Fig. 16**), thus indicating a directed current at a corresponding distance from the CMF.

6.4.2. Microvertebrates

The quantity and mortality of the microvertebrates recovered from Untermassfeld (Maul 2001a, 859; Kahlke 2020, 1045, Fig. 10; see section 2.3.) have been established, as with the large mammals, by the biotope preferences of the individual species (see section 7.1.), the population densities determined from them, as well as by their specific behavioural patterns and respective physical capabilities. Survivability in flowing water and climbing ability are crucial here. The influence of predators is a further decisive factor.

The fish fauna of Untermassfeld (see section 5.2.3., **Table 9**), from which only scant fragments have survived, were of course not flood victims. The Atlantic salmon *Salmo* cf. *salar*, which dominates the finds, is an anadromous species that migrates from the North Atlantic into European river systems to reach spawning grounds in the hypo- and metarhithral headwaters during late autumn and winter (Böhme 2020a). The occurrence of salmon may explain the relatively high MNI of *Ursus* ex gr. *U. dolinensis* and possibly also of *Xenocyon lycaonoides* (Böhme l.c.; see section 6.4.1., **Table 8**).

The amphibians (10 species) and reptiles (7 species) (G. Böhme 1997a; M. Böhme 2020b, see sections 5.2.4., 5.2.5, **Table 9**), found at high levels of diversity but only in small numbers, probably represent in most cases prey remains of birds of prey from the immediate and wider vicinity of the site (M. Böhme l.c.).

The skeleton of an agile frog *Rana dalmatina* (**Table 6**) can be attributed to an individual that was flushed out from a cavity located upstream during hibernation (see below in this section).

The 12 bird species now recorded at Untermassfeld (Jánossy 1997; Manegold 2020) by no means reflect the entire avifauna expected for the area around the site and only a small proportion of them are likely to be flood victims. Bird corpses can be transported by water for quite a long time, according to Schäfer (1976) up to 40 days, due to their plumage, the pneumatised spaces in their bodies, and their large relative surface areas. The transport of bird remains from large distances to the site is therefore entirely conceivable. Skeletal elements that belong to a single crow or corvid (*Corvus* sp.) described by Jánossy (1997, 98–99) could, however, belong to an individual that died near the site. The remarkable amount of large corvid remains in the Untermassfeld fauna (**Table 9**) can be explained by the abundant availability of carrion in the flood-prone Werra valley.

The relatively high MNI of the rodents *Spermophilus kahlkei*, *Mimomys pusillus* and *Microtus thenii* among the recorded small vertebrates (**Table 9**) point to high population densities for these species. According to current knowledge, all three species were terrestrial. Ground squirrels with their limited swimming abilities drown quickly in large and fast flowing bodies of water. The *Spermophilus kahlkei* individuals represented at Untermassfeld by complete or partial skeletons (see section 6.1.2.5., **Table 6**) were transported by river water to the site over varying spans of time and from different distances, where they further disintegrated before finally becoming embedded. However, since the origin of the Untermassfeld necromass accumulations occurred during the winter period (Kahlke 2001e; 2006; see section 6.5. of this paper), most, if not all, of the ground squirrels represented by individually connected skeletal elements are likely to have come from eroded burrows from which the hibernating animals were washed away. The same applies to a recorded *Cricetus* cadaver and probably also for a *Microtus* individual (**Table 6**). Even the aforementioned *Rana dalmatina* individual, found embedded lying on its back (Böhme 2020b, Fig. 3) between *Pachycrocuta* coprolites (see section 6.1.2.6.), might have been washed out from a cavity in fluviolimnic deposits eroded upstream. The retreat of the described microvertebrate individuals to burrows or other underground shelters should have facilitated the preservation of unusual amounts of microvertebrate skeletal units at Untermassfeld. Most of the represented individuals were probably flushed out while still alive. A while later, after enduring various fates of transportation and disarticulation, their remnants then became embedded in the area of the site (see section 6.1.2.5.).

Desmana thermalis almost certainly had a semiaquatic lifestyle and this is also likely for *Mimomys savini* and *Sorex margaritodon* (pers. comm. L. C. Maul, July 2021). Good climbing ability, which enabled at least some of the individuals present in the flooded area to survive, can be assumed for *Sciurus whitei*, *Glis sackdillingensis*, *Sicista* sp., *Myodes hintonianus,* and *Apodemus sylvaticus* (for extant comparisons see, e.g., Niethammer 1978, 356; Storch 1978, 251; Pucek 1982, 514; Viro and Niethammer 1982, 136f.). As expected, the recorded MNI for both the swimming and climbing species of the Untermassfeld micromammals is low (**Table 9**).

The majority of recovered microvertebrate remains originates from disintegrated pellets or droppings of predators. Rodents, especially voles, are the preferred prey of diurnal and nocturnal birds of prey and carnivores (Andrews 1990). Various diurnal birds of prey feed on the extant *Spermophilus citellus* (Ramos-Lara et al. 2014, 79); of these, the Eurasian sparrowhawk *Accipiter nisus* has been recorded at Untermassfeld along with another predator, a large species of crow or corvid, *Corvus* sp. Arvicolines are also the regular prey of crows and corvids (Haffer 1993, 1849, 1929, 2019). In contrast, the sea eagle *Haliaeetus* sp., also recovered from the site, probably only consumed small mammals on rare occasions (for extant comparison see Glutz von Blotzheim et al. 1971). However, a large proportion of extant voles are part of the prey of owls (e.g., Niethammer 1982, 368; Niethammer and Krapp 1982, 306–307, 412; Tast 1982, 389). Although owls have not yet been recorded in the fauna of Untermassfeld, it may be assumed that the majority of the recovered vole remains entered the find layer via their pellets (Maul 2001a, 859). Owls do not, however, play a role as predators of *Spermophilus*, all species of which are diurnal.

The leporids, which are only sparsely documented in the Untermassfeld fossil assemblage, might have occasionally been hunted by *Lynx valdarnensis spelaeoides* and *Puma pardoides brevipes* as well as by *Canis mosbachensis* (Hemmer 2001; Hemmer and Kahlke in this volume). The diet of *Meles meles atavus* certainly included members of the arvicolines. However, these relationships probably did not have any discernible influence on the recorded thanatocoenosis.

The calmer sedimentation conditions of the LFS (see section 4.2., 6.4.1.) led to an enrichment of micromammal remains within the polyspecific large mammal bone concentrations acting as transport obstacles (see section 6.1.2.3.; Maul 2001a, 858f.). More turbulent conditions during the filling of the channel, but with a decreasing tendency, resulted in a different distribution pattern of disarticulated microvertebrate fossils. In contrast to decreasing concentrations of large mammal remains with the progressive filling of the channel (see section 6.4.1.), the amount of micromammalian finds obtained from corresponding samples (Kahlke 2020, Fig. 10) increases. While 1.5 specimens/kg sediment were found in the deepest part of the channel, the amount increases to 4.6 specimens/kg in the uppermost part (calculation by G. Utschig and L. C. Maul, Weimar). Decreasing flow velocities thus caused an increased deposition of microvertebrate fossils in the UFS(l).

The numerous microvertebrate skeletal units recovered from the UFS(l) (see section 6.1.2.5.) were embedded along the valley facing periphery of large mammal bone accumulations (**Fig. 18, Foldout III**). This zonal arrangement is the result of fluvial transport segregation.

6.5. Seasonality of necromass deposition

On the basis of the fossil material recovered from Untermassfeld, the seasonal origin of the site can be narrowed down. The numerous antler finds in particular provide clear clues as to the timing of its formation. Among the beam remains of *Pseudodama vallonnetensis* examined by H.-D. Kahlke (1997; 2001a), 89.7 % of the finds were found to be attached to their pedicles (skull fragments with pedicles included). According to Breda et al. (2020, 1234), only one single shed antler and, in addition, the bare pedicles of five more individuals of *P. vallonnetensis* were found in the total material of this species (MNI = 86, **Table 8**). No evidence of shed antlers from *Eucladoceros giulii* (MNI = 91, **Table 8**) and *Capreolus cusanoides* (MNI = 5, **Table 8**) was found. The deposition of cervid remains during several high flood events can thus be restricted to periods between October and the end of March/beginning of April, i.e., to the respective winter half-years.

A similar assessment results from the occurrence of the anadromous *Salmo* cf. *salar* (Böhme 2020a). Atlantic salmon reach their spawning grounds in European river systems with Atlantic discharge during late autumn and winter. Locating the Untermassfeld flood events during the winter months is also supported by insectogenic modifications of skeletal elements (Keiler et al. 2020). Here, too, antlers are particularly suitable for deriving seasonal information. For example, the basal antler section of an individual of *P. vallonnetensis* that died during velvet antler growth, i.e., according to today's fallow deer antler cycle between mid-August and the beginning of September, shows clearly identifiable pupation chambers of dermestid larvae (Keiler et al. l.c.). The time required to create such structures is at least four weeks. This means that the embedding of the above-mentioned fossil must have taken place at the earliest in mid-September to early October, or even later in the year. Finally, the complete absence of bat remains may also be an indication that the site formed during the winter months, i.e., during the animals' hibernation period.

In order to determine the season at the time of death of individuals from one herbivore and one carnivore species, analyses of the tooth root cementum were undertaken (Kahlke 2001a, 424; 2020, 1045f.). Unfortunately, the anticipated structures were destroyed by diagenetic recrystallisation processes in the samples of *Pseudodama vallonnetensis* as well as in an M_3 of *Ursus* ex gr. *U. dolinensis* (pers. comm. H. Kierdorf, November 2004; N. García, February 2007). The attempt to detect hibernation marks in the dentin of two incisors of *Spermophilus kahlkei* was also unsuccessful (pers. comm. Y. E. Selkova to L. C. Maul, April 2004). However, since corresponding marks are only detectable over a period of 1.5–2.0 months after the end of hibernation (Trunova and Klevezal 1999), this observation does not contradict the seasonal assignment of the site's genesis.

The taphonomic information presented in this chapter seeks to serve as a general characterisation of the find situation encountered at Untermassfeld as well as to provide essential palaeobiological facts that can be deduced from the findings. More detailed taphonomic analyses must be reserved for later studies.

7. Palaeoenvironment

Riverine landscapes with their attractive wetlands, riparian zones and different types of water bodies are not only colonised by animal species with a corresponding habitat association. Faunal elements from the wider environment seek out flowing waterways for drinking and, in some cases, for feeding. It can be assumed, therefore, that the floodplain landscapes of the Early Pleistocene Werra valley were also regularly frequented by species from the adjacent areas. The flood events reconstructed from the sedimentological and taphonomic findings (see sections 4. and 6.) thus focused on evidence of fauna from a wider catchment area in the region around the site. Skeletal elements of smaller vertebrates were also recorded as digestive remains, mainly from birds of prey.

As far as possible, the individual species found at Untermassfeld have been assigned to their respective habitat types (see section 7.1.). These are combined with the geological and geomorphological conditions of the region (see section 4.2.) in order to derive as detailed a picture of the landscape as possible at the time of the site's formation (see section 7.2.). In doing so, we have mainly relied upon the descriptions by Kahlke (2001e; 2006). A graphic reconstruction of the immediate and wider site environment was omitted, as neither meaningful macrobotanical fossil evidence nor corresponding pollen spectra are available from Untermassfeld (see section 5.1.).

7.1. Palaeohabitat affiliation

In the Untermassfeld gastropod fauna already described by Krolopp (1997), only 4.9 % of the individuals recorded represent aquatic species. Among the 28 terrestrial species, there are notably many inhabitants of open areas, which, at most, would have only had sporadic brushwood cover. In contrast, species inhabiting woodland are almost completely absent. While the frequently found small amber snail *Succinea oblonga* is considered to have an affinity for moisture, the hypogeal blind awl snail *Cecilioides acicula*, which was also found in large numbers, is restricted to warm dry sites, especially dry grasslands and scree fields. The soil-dwelling cecilioids are rarely found on the surface; soil-burrowing animals or floods, however, can unearth the shells of dead individuals en masse (Wächtler 1929; Zeissler 1981, 41). The remains of the moss snail *Pupilla muscorum* and the smooth grass snail *Vallonia pulchella*, which also occur in remarkable quantities at the site, are mainly found in open areas, but are also widespread in shrubbery and sparse forests (Zeissler 1981, 21f.).

According to current knowledge, the fish fauna of Untermassfeld (Böhme 2020a) is comprised solely of the anadromous Atlantic salmon *Salmo* cf. *salar*, with at least six recorded individuals, and the rare cyprinid *Gobio* sp. (comparative data on more species-rich Pleistocene occurrences in Kahlke 2001e, 977). Spawning grounds of the Atlantic salmon are shallow, oxygen-rich fast-flowing waters with stony bedload.

The quite diverse amphibian fauna is evidence of the presence of several types of water bodies in the Werra valley around the site (Kahlke 2001e, 978; Böhme 2020b, 1157). Calmly flowing waters were colonised by the common frog *Rana temporaria*, whereas the majority of amphibian species spawn in still water habitats outside the main discharge. While some of the water bodies may have only been seasonal, the northern crested newt *Triturus* cf. *cristatus* and the green frog *Pelophylax* indicate the existence of some perennial ponds or pools (Böhme l.c.). The presence of an articulated skeleton of the common frog *Rana dalmatina* (see 6.1.2.5.), meanwhile, suggests that the banks of the Werra river were part of its land habitat. Extensive dry, open habitats with sandy to rocky soils are indicated by the occurrence of the European green toad *Bufotes viridis*, a sand boa (*Eryx* sp.), and a snake-eyed lizard (aff. *Ophisops elegans*).

Evidence demonstrating the presence of several bird species with a clear affinity to water (Jánossy 1997; Manegold 2020) is not surprising. Swans (*Cygnus*), geese (*Anser*), and ducks (Anatidae) as well as sea eagles (*Haliaeetus*) require the existence of lakes or a river. The sea eagle, which nests in rocky cliffs or on old trees, feeds by catching fish or hunting water birds but also by taking carrion, which would have often been available in the carnivore-rich and flood-prone Werra valley. At least three of the recorded species indicate the existence of forest vegetation. The Eurasian sparrowhawk *Accipiter nisus* breeds in diverse woodlands but avoids pure deciduous forests. Eurasian jays (*Garrulus glandarius*) are also forest dwellers, and the black grouse *Lyrurus tetrix* prefers a combination of open forests, moors, and bogs (Manegold 2020 and references therein).

The uncertainties that existed in the ecological classification of *Bison menneri*, a species that is extensively documented in the Untermassfeld fossil assemblage (Sher 1997; Bukhsianidze 2020), have not been resolved by the inclusion of the species in the subgenus *Poephagus* (Bukhsianidze l.c.). This is due to the fact that neither the Middle to Late Pleistocene yaks so far recorded (compilation in Kahlke 1999b, 55ff.) nor corresponding extant populations can be considered as analogues of the Untermassfeld bovids with regard to their habitat requirements. The mesowear signature of *B. menneri* is indicative of a variable diet that contained both high proportions of browse and more abrasive foodstuffs (Van Asperen and Kahlke 2017). As mixed feeders, the Untermassfeld bison herds would have been able to utilise the resources provided by both open landscapes and more densely vegetated areas.

Interestingly, a second bison species (*Bison* sp.) has been identified in the Untermassfeld fossil material on the basis of a few finds (Bukhsianidze 2020, 1190ff.). The rarity of corresponding evidence could be an indication of singular or occasional seasonal immigration. Since the fossil accumulations of Untermassfeld were formed during winter half-years (see section 6.5.), these bovids could have been seasonal migrants from north-eastern regions. A parallel of the evidence of infrequent winter visitors is found in the Early Middle Pleistocene large mammal fauna of Süssenborn in Central Germany (Soergel 1939, 833f.; Kahlke 1999b, 94).

The morphological analysis of cranial material from the cervid *Eucladoceros giulii*, frequently represented in the find material, indicates a predominantly browsing lifestyle (Breda et al. 2020): the presence of corresponding forested areas is therefore assumed. Like its closely related fossil species *Cervalces gallicus* and *C. latifrons*, the moose *C. carnutorum*, must also be regarded as a specialised browser (Breda 2008; Breda et al. l.c.). Similar to present-day moose, *C. carnutorum* might have migrated along riparian corridors in order to utilise a wide variety of forested environments. *P. vallonnetensis*, which is also common in Untermassfeld, was more of a mixed feeder judging by its skull and dentition morphology (Breda et al. l.c.). This classification is supported by the fact that *Pseudodama* can be regarded as a prey animal of the open land

hunter *Acinonyx pardinensis pleistocaenicus* (Hemmer 2001, 727; Hemmer and Kahlke in this volume). Therefore, open and park-like landscapes must have been essential components of the habitat of *P. vallonnetensis*. More difficult to assess is the habitat preference of the roe deer *Capreolus cusanoides*, which was only recorded in small numbers of individuals. Similar to today's representatives of the genus, the animals probably preferred parts of the landscape that were rich in cover. Mesowear analyses to reconstruct the respective dietary traits are in preparation for *Eucladoceros* and *Pseudodama* (E. van Asperen, Durham).
The abundant remains of the semiaquatic hippos *Hippopotamus antiquus* (R.-D. Kahlke 1997b; 2001c; Kierdorf and Kahlke 2020) are of particular importance for the reconstruction of the palaeoecological conditions at the time of the formation of the Untermassfeld fossil record. Similar to the extant species *H. amphibius*, the animals, which stay true to their chosen location, probably occupied shallow and open waters surrounded by semi-open woodland during the day. Stony locations with strong currents would have been avoided if possible. At night, they would have sought out grazing areas rich in grasses, but food resources would also have been found at greater distances from their aquatic territories (Eltringham 1999). According to the geomorphology of the middle Werra valley, the larger portion of the grazing areas of the Untermassfeld hippos were located in the upland regions beyond the valley of the Werra river. The height difference between the floodplain and the surrounding uplands reached up to 60 meters. In addition, the Lower Triassic limestones tended to outcrop along the flanks of the river valley to form escarpments. For the Untermassfeld hippos, this resulted in the necessity to climb relatively steep slopes each day to reach their grazing areas. Corresponding signs of stress in the form of osteoarthritic lesions can be found on the joint surfaces of the large, and therefore heavy, animals (Kierdorf and Kahlke 2020).
The massive structure and body proportions of the wild boar *Sus strozzii* probably hindered any serious attempts at fast locomotion. Significantly wetter habitats were more likely to be the preferred habitats of this species than those of the extant *S. scrofa* (Cherin et al. 2020; Iannucci et al. 2020).
With the presence of *Mammuthus meridionalis* and *Stephanorhinus hundsheimensis*, the Untermassfeld fauna has two large mammals that were able to live in savannah-like landscapes as well as in open woodland. Being tall with a relatively short body length, the aforementioned elephant species was able to obtain leaves, twigs and bark even from the upper parts of the tree vegetation (Garutt and Bajguševa 1981, 13 f.). It is very likely that their diet, and with it the more frequently visited parts of the landscape, underwent seasonal changes. While fresh open vegetation would have been preferred in spring and at the beginning of summer, food obtained from trees probably gained in importance later in the year. Similar preferences, by exploiting the lowest levels of tree vegetation, are likely also applicable to *S. hundsheimensis*. Dietary variability, ranging from grazing to predominantly browsing regimes, characterises the species as the most ecologically tolerant rhinoceros of the Eurasian Plio-Pleistocene (Kahlke and Kaiser 2011).
The feeding traits of *Equus wuesti* are still largely unclear (Boulbes and Van Asperen 2019; Eisenmann and Boulbes 2020). According to Musil (2001a), who places the species in the ancestry of *E. altidens*, the animals were probably grazers who found suitable living conditions in more or less open areas and sought out the Werra floodplain for water.
Among the felids represented in the Untermassfeld fauna, clear habitat preferences can be detected for the jaguar *Panthera onca gombaszoegensis* and for the giant cheetah *Acinonyx pardinensis pleistocaenicus*. To a lesser extent, this is also true for the scimitar cat *Homotherium latidens ultimum*. The jaguar is the most common big cat recorded at the site. Its central habitat was riparian landscapes, as is also the case for extant jaguars (Hemmer and Kahlke in this volume, 1478; see section 6.4.1. of this paper). *A. pardinensis*, on the other hand, should be regarded as an open land hunter that developed superior speeds over short and medium distances according to its functional morphology (Hemmer 2001, 725 ff.) and the flood vulnerability ratio calculated for the Untermassfeld fossil material (FVR; Hemmer and Kahlke l.c., 1502). The species

avoided dense wooded areas and shrubbery as well as completely open spaces without cover; its preferred habitat being parkland landscapes. Judging by its body structure and paw development, *H. latidens* was specialised in swiftly roaming extensive areas (Hemmer 2001, 758). Its FVR, calculated from the Untermassfeld material, is similar to that calculated for the jaguar. For the homotheres, this could indicate at least some familiarity with the destructive effect of running water in the Werra valley (Hemmer and Kahlke l.c., 1553). For *Puma pardoides brevipes*, *Lynx valdarnensis spelaeoides* and *Megantereon adroveri* no preferred habitat types can be identified from their diet (Hemmer 2001, Fig. 20; Hemmer and Kahlke l.c.), but presumably both open and covered parts of the landscape formed part of their home range territory.

Pachycrocuta brevirostris should be regarded as an opportunistic scavenger that due to its impressive body and dentition size was quite capable of forcing the takeover of other carnivores' prey (Turner and Antón 1996, 463f.). The highly probable behaviour of territorial marking by means of faeces (Keiler 2001b; see section 6.1.2.6. of this paper), indicates that the immediate site environment was a preferred part of the hyaena's territory. Extant spotted hyaenas (*Crocuta crocuta*) arrange their »latrines« along frequently used paths as well as near water (Kruuk 1972, 258ff., fig. 62). The animals are quite familiar with water bodies and can feed on prey at depths of up to 1.5m (Kruuk 1975, 21, 73; Larkin et al. 2000, 20). A similar situation may be assumed for the Untermassfeld *P. brevirostris*. Hyaena dens were almost certainly established on the valley floor along the Werra (see section 6.4.1.).

Similar to the situation assumed for the lynx *L. valdarnensis*, the cover-rich, mainly south-facing slopes of the river valley could also have been utilised by the badger *Meles meles atavus* for the construction of its dens and the rearing of young.

The habitat preferences of extant brown bears might in principle be transferable to the morphologically less-specialised Untermassfeld *Ursus* ex gr. *U. dolinensis*. If this is the case, then the omnivorous diet of the animals would have varied seasonally. During their active periods, the bears would have certainly roamed over considerable areas without any close connection to particular habitats. The relatively high MNI of *Ursus* ex gr. *U. dolinensis* in the Untermassfeld material can be explained by annual migrations of Atlantic salmon (*Salmo* cf. *salar*), which were seasonally hunted by the bears in the Werra (Böhme 2020a, see sections 6.2.4., 6.4.2. of this paper).

It can also be assumed that *Xenocyon lycaonoides*, whose body size roughly corresponds to that of today's grey wolves (**Fig. 10**), hunted the approximately 1m-long Atlantic salmon (Böhme l.c.). A link to any specific habitats, however, is not likely. Similar to the prey spectra of extant (smaller) southern wolves, *Canis mosbachensis*, the second canid of the Untermassfeld fauna, was able to kill prey up to a maximum size of *Pseudodama vallonnetensis* (Hemmer 2001, 768). Strict habitat preferences can also be ruled out for this wolf. The high population density inferred from the rich fossil material of *C. mosbachensis* (see section 5.2.7.) could have led to a continuous displacement of the foxes, which remain unrecorded in the fossil material during the entire formation period of the Untermassfeld fossil accumulation.

Several rodent species, some of them with high MNI (**Tables 6, 9**), indicate widespread open habitats. This is especially true for the ground squirrel *Spermophilus kahlkei*, the large hamster *Cricetus runtonensis*, and the vole *Microtus thenii*, and maybe also for *Pliomys episcopalis*, *Mimomys pusillus*, the birch mouse *Sicista* sp., and the porcupine *Hystrix refossa* (Maul in this volume). The squirrel *Sciurus whitei*, the dormouse *Glis sackdillingensis*, the red-backed vole *Myodes hintonianus* and the wood mouse *Apodemus sylvaticus*, however, indicate woodland (Maul l.c.). The semi-aquatic beaver *Castor fiber* also needs wooded areas situated near water. Whereas, according to Mayhew (1978), the second Untermassfeld beaver species, *Trogontherium cuvieri*, should be regarded as an inhabitant of stagnant to slow-flowing waters, *Castor*, due to corresponding sympatry, could have been at least partially excluded from such biotopes. The morphology of the incisors of *T. cuvieri* could indicate digging activities (Fostowicz-Frelik 2008). The species apparently preferred largely non-forested enclaves along the water bodies it inhabited, as it mostly subsists during vegetation regeneration phases following flood events (Mayhew l.c., 433). *Mimomys savini* was an aquatic species.

Leporids are only documented by small amounts of find material. The presence of the rabbit *Hypolagus brachygnathus* is indicative of loose soils that allow the creation of suitable burrows. Hares of the genus *Lepus* live in open to semi-open areas and also in open woodland, where the animals can escape predators by high-speed endurance running.

The Barbary macaque, *Macaca sylvanus* (Reumer and Kahlke in this volume; see section 6.4.1. of this paper), which is also rarely found because of its climbing ability, preferred light-filled forests and – if available – landscapes with rocky components in a similar way to extant members of the species. The animals' preference for water bodies has also been observed (Haltenorth and Diller 1977, 270).

For the hedgehog *Erinaceus* cf. *europaeus* (Maul in this volume), of which only a few specimens were found, dry habitats in forests as well as in more open landscapes are typical. Moist, loose soils are indicated by the presence of two mole species *Talpa* cf. *europaea* and *T. minor*. The desman *Desmana thermalis* and the shrew *Sorex margaritodon* are semi-aquatic. *Beremendia fissidens*, according to Maul (l.c.), indicates temperate and humid climatic conditions, but can live in a variety of landscape types.

7.2. Palaeolandscape reconstruction

The floodplain valley of the Early Pleistocene Werra, which in accordance with its geological age was still shallow and quite wide in places, was traversed by a dynamic watercourse all year round during periods of normal water levels. The riverbed, as well as the spatially variable bank sections, consisted mainly of reddish sands of different grain size, crushed rock stone and gravels, varying degrees of reworked clastic mudflow fans, and clayey-silty sediments. In places, the riverbed was covered by dense vegetation and in sheltered river sections there were open and clear water areas with little to no currents. In addition to the main stream and its more or less active side channels, oxbow lakes with stagnant waters and siltation zones were found in waterlogged parts of the floodplain. Due to special meteorological-hydrological constellations (see section 7.3.), the outflowing water volume could swell, mostly in winter, and turn the Werra into a raging torrent with considerable erosive power and, when the valley was narrowed, into a highly turbulent watercourse. Available floodplain areas would have been regularly flooded. Slopes of the river valley drained in smaller, temporarily swelling rivulets and springs with occasional, and mostly seasonal, travertine formation.

Due to their increased water supply, floodplains are usually comprised of habitats whose character deviates from the zonal ecological conditions of the region (Walter and Breckle 1986, 124 ff.). At the time of the site's formation, ponds and pools existed in the Werra valley that followed the changing relief of the river. Depending on insolation and depth, these small bodies of water would have had different temperatures. Wettish sections of the floodplain would have been partly marshy, whilst permanently waterlogged areas were most likely to have been mainly covered with softwood, so that thickets would probably have been present, at least near the river banks. In particularly affected areas, woodland would have suffered some level of damage depending on the duration and periodic sequence of flooding and thus tended to be low-growing. In more stable, drier locations, tall mixed forests with less undergrowth would have spread with lush, herbaceous vegetation thriving on the often loose soils.

The Muschelkalk slopes close to the site were mainly comprised of open areas alongside habitats with species-poor tree and shrub vegetation. Dry grasslands and sparse areas of scree were particularly widespread on the south-facing slopes of the valley on the right, i.e., on the side where the site was located. Free-standing, massive limestone benches (Werksteinbänke in the local geological terminology) and crum-

bling cliffs derived from the respective properties of the outcropping rocks. Clayey-limestone hill-wash with corresponding vegetation cover formed at the foot of the slope. On suitable, less steep slopes, there might have been well-trodden hippopotamus paths, which the animals would have regularly used in order to reach their grazing areas.
The plateaus flanking the Werra valley south of the Thuringian Forest Hills were covered by limestone-rich, loamy-clayey soils over Middle Triassic Muschelkalk and sandy-loamy soils over Lower Triassic Bunter sandstone. The areas above the Lower Muschelkalk in particular would have tended to dry out quickly. Depending on the soil conditions and the local relief, park-like landscapes and areas with few trees extended outside the Werra valley.

7.3. Palaeohydrological conditions

The sedimentological characteristics of the fossil-bearing sediments of Untermassfeld prove that the region was regularly hit by fluvial floods during the depositional period of the Lower Fluviatile Sands (LFS) and the channel fill [UFS(l) etc.], which reached thicknesses of up to 8.0m and 5.0m, respectively (Ellenberg and Kahlke 1997, 37–40; Kahlke 2001b, 434ff.; Kahlke et al. 2020; section 4.2. of this paper). A singular meteorological constellation can thus be excluded as a cause of flooding. Rather, the number and sequence of sand layers indicate constantly recurring events that were largely controlled by the hydrological conditions of the region. Only some of the floods documented in the stratigraphic sequence of Untermassfeld led to a mass influx of animal necromass. While at least two (possibly a few more) flooding events with their respective accumulations of vertebrate remains are verifiable for the LFS (see section 4.2., phase II of site development), the subsequent incising and refilling of the erosion channel occurred in the course of only one particularly destructive event in the Werra valley (phase III of site development). Upstream of the site, such flood events were characterised by an unusually high flow velocity and considerable erosive force. This led to the drowning of numerous animals (see section 6.4.1.) as well as to the discharge of sediment and necromass from the flooded floodplains [Kehraus-(= clear-out)effect after Kahlke 2001e, 967; 2006, 74].
In the river valleys of the Central European low mountain regions, particularly high water levels occur during the winter months when temperatures above freezing lead to a combination of snowmelt in high altitudes with persistent rainfall in lower regions. Such a constellation is very likely to have been the cause of the flood events at Untermassfeld. According to heavy mineral analyses of Pleistocene Werra gravels (Ellenberg 1975), the water catchment area comprised the mountain range of the Thuringian Forest southeast of Schmalkalden, the headwaters of the Werra in the Thuringian Slate Mountains, and parts of the South-west Thuringian Triassic area. The maximum flooded area of the middle and upper Werra, including its tributaries upstream of the site, and thus the potential area of origin of the necromass covered c. 94km² (Kahlke 2001e, 967, Fig. 26).
An S-shaped narrowing of the valley 12km upstream of the Untermassfeld fossil deposits (Kahlke 2001e, Fig. 26) had a particular influence on the catastrophic course of the floods for life in the Werra valley. The so-called Eye of the Needle, which can still be seen today, is formed by steeply rising limestone walls (Lower Muschelkalk, Middle Triassic). The present-day height difference between the floodplain and the adjacent plateau is almost 100m. Present-day valley widths in the geomorphological level of the Younger Coarse Gravels (YCG, see section 4.1.) show that the Early Pleistocene floodplain valley of the Werra in the area of the Eye of the Needle narrowed from a width of greater than 1.5km to markedly less than 1.0km. Thus, a geomorphological bottleneck was present that led to turbulent flow at the narrowest point, which then caused a subsequent back damming effect in the upstream floodplains.

7.4. Palaeoclimate parameters

The habitat affiliation of the faunal elements (Kahlke 2001e, 983 f.; see section 7.1. of this paper) and the reconstructed landscape (see section 7.2.) clearly indicate warm climatic conditions. The fossil evidence mainly consists of temperate and thermophilic elements from different biotopes at varying degrees of attachment. Characteristic species of cool- to cold-temperate conditions are missing. Stephan et al. (2001) determined mean annual temperatures 1 °C above current values of the region from $\delta^{18}O$ isotope values obtained from bioapatite phosphate of equid metapodials. This figure should be regarded as a minimum value (Kahlke 2001e, 984).
The winter temperatures of the Werra valley at the time of the site's formation can be substantiated by the abundant evidence of *Hippopotamus antiquus* (see section 5.2.7.). Since the hippos could not have survived if their water sites were covered with ice, mild winters with minimum air temperatures of not more than a few degrees below zero may be inferred. Regular and consistent winter snow cover is therefore unlikely in the landscapes in the immediate vicinity of the site. At one point, the purported occurrence of pond turtles (Maul 1997) seemed to open up a possibility for estimating summer temperatures (Kahlke 2001e 984). However, the corresponding evidence was based on a misidentification, as Böhme (2020b, 1153, Fig. 14) has since shown.
More exact palaeoclimatic parameters can be derived from the recorded herpetofauna (Böhme 2020b). A combination of temperature conditions for the two most thermophilous reptiles from Untermassfeld, namely sand boa *Eryx* sp. and snake-eyed lizard aff. *Ophisops elegans*, indicate a distinctly warm climate. At the time of the formation of the Untermassfeld fossil site, mean annual temperatures (MAT) were 13 °C. Winters were mild, as expected, with mean cold month temperatures (CMT) of 2–3 °C, while summers were hot, with warm month temperatures (WMT) of 24 °C. Compared to today's conditions, the reconstructed palaeoclimate was significantly warmer. The MAT was 5 °C and the WMT as much as 8 °C higher than today (Böhme l.c., 1155 ff.). The bio-climatic estimate of the mean annual precipitation (MAP) results in a value of 622 mm, i.e., about 6 % less than is recorded in this area today. This value should be regarded as an upper limit, the actual precipitation amounts were probably lower (Böhme l.c., 1156).
Different local climates can be reconstructed for river floodplains, slopes and plateaus in the immediate and wider vicinity of the site according to different solar radiation and moisture-holding capacities. Annual precipitation in the Werra valley was sufficient to sustain open water areas and mostly waterlogged floodplains with a variety of vegetation all year round. During the winter months, in particular, there was more extensive precipitation (see section 6.5.) that regularly led to flooding, occasionally with catastrophic effects for life in the valley. Underground and surface drainage of the adjacent uplands contributed to the year-round floodplain moisture levels, so that generally drier conditions prevailed with similar temperature conditions as in the river valley. As the Muschelkalk slopes would have retained almost no water, in the south-facing locations in particular higher levels of aridity and above-average soil and air warming would have been expected during the summer months.
Regionally, a warm climate characterised by Atlantic-Mediterranean influences prevailed with significantly higher temperatures than are observed today in the region of Southern Thuringia. Compared to current conditions, the lower amounts of precipitation were concentrated in the winter half-year. During this season, characteristic meteorological constellations repeatedly led to sometimes severe flooding of the valley floors of the Werra and its tributaries.

8. Chronostratigraphy

The geochronological classification of the formation of the Untermassfeld fossil site and the fossil assemblage it contains is based on the combination of several applicable chronostratigraphical methods (for further dating methods see Kahlke 2020, 1042 f.).

8.1. Litho- and magnetostratigraphy

The basis for lithostratigraphic dating of the fossiliferous sediments of Untermassfeld is the chronostratigraphy of the fluviatile gravel terraces of the Werra valley in Thuringia (Ellenberg 1975; 2002; Unger in Unger and Kahlke 1995; Ellenberg and Kahlke 1997). The Pleistocene sequence of Untermassfeld starts with riverine Younger Coarse Gravels (YCG, see section 4.1.), which were most probably deposited during the Eburonian (Ellenberg and Kahlke l.c.) or possibly also during the Menapian (Ellenberg 2002, 241 f.). During the warm period of the Waalian and/or the Early Bavelian a deep chemical weathering of the YCG took place, which rules out the preservation of any potential vertebrate remains that might have been present. Fossil finds from the reworked uppermost section of the YCG can be attributed to a subsequent flood event at the time of the site's necromass accumulation (see section 4.2.). The non-weathered sequence of strata from Untermassfeld, i. e., the floodplain deposits (FPD) to Upper Fluviatile Sands (UFS), is younger than the weathering process documented in the underlying gravels (Kahlke et al. 2020; see setion 4.). According to the geological data framework mentioned above, the earliest deposition date for the fossiliferous Untermassfeld sediments may be Late Waalian, but is more likely to be Bavelian or an early period of the Cromerian complex (Kahlke 2001e, 985; for global correlation see Cohen and Gibbard 2019). The next youngest gravels occurring at lower levels of the Thuringian Werra valley are terraces of Cromerian and Elsterian age (Ellenberg and Kahlke 1997, Fig. 3, Tab. 2).
The Quaternary profile of Untermassfeld (see section 4.1.) is reversely magnetised upwards from the YCG to the middle section of the Lower Fluviatile Sands (LFS) (R.-D. Kahlke 1995, Fig. 19, based on Wiegank 1997). The upper section of the LFS and the Upper Fluviatile Sands (UFS), however, show normal magnetisation. Thus, the Untermassfeld fauna can be assigned to either the Cobb Mountain cryptochron (1.208–1.187 Ma BP), an early period of the Jaramillo polarity subchron (1.071–0.990 Ma BP), or to the beginning of the Brunhes magnetochron (since 0.773 Ma BP) (absolute data after Cohen and Gibbard 2019). Short-term palaeomagnetic excursion events immediately before and after the Jaramillo (Cohen and Gibbard l.c.) are excluded as having been the depositional periods of the Lower and Upper Fluviatile Sands due to their short duration. In addition, the magnetic polarity reversal detected in the stratigraphic sequence of Untermassfeld cannot correspond to the onset of the Cobb Mountain cryptochron or to the Matuyama/Brunhes (M/B) boundary for biostratigraphical reasons (see section 8.3).

8.2. U/Pb dating of redeposited carbonate

Within the fossil-bearing channel fill from Untermassfeld, redeposited sandy carbonate breccia (SCB) covered by a 5.0–10.0 cm thick travertine (TRA, Kahlke et al. 2020, 1091 f., Figs. 6, 7a–d, 9a; see section 4.1. of this paper) was found. Its precipitation is associated with a water discharge at the clastic mudflow fan (CMF), which caused the deposition of large amounts of necromass at the site (see section 4.2.). Over the course of the downslide of parts of the CMF, the travertine entered the fossiliferous erosion channel (Kahlke et al. l.c., 1088 f.). This travertine originated before the fossiliferous refilling of the erosion channel [UFS(I)] and most likely also prior to the deposition of the fossil bearing part of the LFS (Kahlke et al. l.c.). In situ U-Pb isotope analyses date the redeposited carbonate to 1.14 ± 0.48 Ma BP (Gerdes et al. 2020). This age predates both the onset of the Jaramillo polarity subchron (see section 8.1.) and the accumulation of the necromass at the site.

8.3. Biostratigraphy

The evaluation of the fossil material supports the chronostratigraphic classification of the faunal assemblage of Untermassfeld obtained by means of geoscientific data. The biostratigraphic significance of the represented animal groups is quite different. As expected, malaco- (Krolopp 1997), ichthyo- (Böhme 2020a), herpeto- (Böhme 2020b), and avifauna (Jànossy 1997, Manegold 2020), but also some mammal species, enable statements to be made on biodiversity, palaeoenvironment, biogeography and taphonomic processes (see sections 5.–7.) rather than on the age of the fossil deposit.
The conclusion reached following the geological findings, that the age difference between the fossil assemblages of the LFS and the channel filling [UFS(I) etc.] is not of biostratigraphical relevance (Kahlke et al. 2020, 1103), is confirmed by the fine stratigraphical distribution of the detected mammal taxa. Both fossil-bearing sedimentary units contain identical faunal elements. In the LFS, which are characterised by lower find concentrations, some very rare large mammal taxa could either not be determined or detected with any certainty (*Sus stozzii*, *Acinonyx pardinensis pleistocaenicus*, *Megantereon adroveri*, *Homotherium latidens ultimum*, *Meles meles atavus*, *Macaca sylvanus*). The distribution of small mammal taxa is similar. While the entire species spectrum of the site can be found in the UFS(I), the LFS often lack evidence of rarer species (*Sciurus whitei*, *Castor fiber*, *Trogontherium cuvieri*, *Cricetus runtonensis*, *Apodemus sylvaticus*, *Hystrix refossa*, *Hypolagus brachygnathus*, *Lepus* sp., *Erinaceus* cf. *europaeus*, *Talpa minor*, *Desmana thermalis*, *Sorex minutus*, *S. runtonensis*, *Beremendia fissidens*; pers. comm. L. C. Maul, September 2021). The mass of the sediments sampled for small vertebrates may also play a role. According to the sections exposed during the field seasons, and due to the abundance of fossils, significantly more sample material was washed and sorted from the UFS(I) than from the LFS (Kahlke 2001a, 420; 2020, 1045 f.). A detailed analysis of biostratigraphically informative mammal taxa, including graphical representations of their biostratigraphical ranges, can be found in Kahlke (2001e, 985–992; 2006, 90–96). The statements made there are summarised and updated here.

The large-bodied Untermassfeld *Bison* (*Poephagus*) *menneri* replaced the smaller and more gracile *Leptobos* species of the Villafranchian as well as representatives of the more lightly built *Eobison* group (Sher 1997, 168 f.; Kahlke 2001e, 985; Maniakas and Kostopoulos 2017, 161 ff.; Kostopoulos et al. 2018). *B.* (*Poephagus*) possibly represents a sister group of *B.* (*Bison*) (Bukhsianidze 2020, 1192). In temporal proximity to the M/B boundary, more advanced representatives of the genus *Bison* already occurred in the western Palaearctic (see compilation in Kahlke 2001e, 988 f.).
Fossil finds of *Capreolus cusanoides* have not yet been recorded from faunal associations older than those of Untermassfeld. The next youngest roe deer of the western Palaearctic belong to early Middle Pleistocene populations of *C. suessenbornensis* (compilation in Kahlke 2001e, 989; on taxonomy see Breda et al. 2020, 1204). The Untermassfeld elk *Cervalces carnutorum* lies between the Middle to Late Villafranchian *C. gallicus* and the Middle Pleistocene *C. latifrons* in terms of its body size and length of antler beams (compilation in Breda et al. 2020, 1207). Representatives of the *carnutorum* elk occur up to and above the M/B

boundary (Kahlke 1990, 86). The tendency of increasing body size within the genus *Eucladoceros* from the late Villafranchian onwards culminates in *E. giulii*, which is equipped with particularly long metapodials (H.-D. Kahlke 1997, 225 ff.). The species occurs approximately between 1.4/1.3 and 0.8 Ma BP in Europe (Breda et al. 2020, 1213). No records after the M/B transition have been recovered. *Pseudodama vallonnetensis* continues a trend towards relative shortening of the premolar series that has been evident within the *Pseudodama* group since the late Middle Villafranchian (H.-D. Kahlke 1997, 206, Fig. 15; Breda et al. 2020, 1238). The species can be traced to within temporal proximity of the M/B boundary (Van der Made et al. 2017, 316 ff.).

In its cranial morphology, in particular, the Untermassfeld *Hippopotamus antiquus* embodies a clearly more advanced form compared to the Late Villafranchian finds of Southern Europe, but does not reach the developmental stage of the Galerian finds from Italy (Mazza 1991; 1995; Kahlke 2001c). The Untermassfeld hippos differ from younger finds of the Central European early Middle Pleistocene due to their particular robustness and size (Kahlke 1997b). Recent studies show that, contrary to earlier assumptions, *Sus strozzii*, detected in the European Late Villafranchian, occurred in Europe at least until the end of the Early Pleistocene (Cherin et al. 2020). Its occurrence in the Untermassfeld fauna is therefore quite plausible.

The *Stephanorhinus hundsheimensis* rhinoceroses from the site overlap morphometrically with the Late Villafranchian species of Southern Europe (for stratigraphy and increase in size see Lacombat 2006). Their body sizes and proportions differ from those of younger Early Middle Pleistocene populations from Voigtstedt, Süßenborn and Hundsheim (H.-D. Kahlke 2001b, 549 ff.).

Mammuthus meridionalis found in the Untermassfeld fossil material does not contradict the chronostratigraphic classification of the fauna in the late Early Pleistocene period, but it cannot contribute to a more precise dating (Van Essen in this volume, 1383).

The Untermassfeld jaguar differs morphometrically from the Late Villafranchian *Panthera onca toscana* and is assigned to the post-Villafranchian to Middle Pleistocene *P. onca gombaszoegensis* (Hemmer 2001, 699 ff.; Hemmer et al. 2010; Hemmer and Kahlke in this volume, 1468 ff.). Due to the stratigraphic range of the latter chrono-subspecies (Kahlke 2001e, 990, Fig. 27), a more exact stratigraphic delimitation of the Untermassfeld fauna cannot be deduced from the corresponding finds. The Untermassfeld cheetah *Acinonyx pardinensis pleistocaenicus* diverges in morphology and body size from the Middle to Late Villafranchian *A. pardinensis pardinensis* of Europe and Western Asia (Hemmer 2001, 712 ff.). Based on current knowledge, the nominate subspecies had already disappeared before Europe was repopulated from East Asia by the very large *pleistocaenicus* cheetahs (Hemmer and Kahlke in this volume, 1499).

European pumas can be classified by their dental morphology into a group of forms recorded in the Middle Villafranchian and a younger group which occurred from the Late Early Pleistocene onward (Hemmer and Kahlke l.c., 1519). Certain differences in the dentition of early and later pumas had already been noted by Madurell-Malapeira et al. (2010) for Spanish finds. The description of the chronostratigraphically younger subspecies as *P. pardoides brevipes* is based on new Untermassfeld finds (Hemmer and Kahlke l.c., 1504 ff.). It is possible that late pumas were only replaced in Europe by the leopard *Panthera pardus* during the early Middle Pleistocene (Hemmer 2001, 733). Recent finds also led to new insights regarding the evolution of early lynxes in Europe. Italian Late Villafranchian lynxes are contrasted as *Lynx valdarnensis valdarnensis* with the younger chrono-subspecies *L. valdarnensis spelaeoides* with the type locality Untermassfeld (Hemmer and Kahlke l.c., 1524 ff.). The early Middle Pleistocene lynxes of the German sites of Mauer and Mosbach may represent late forms of *L. valdarnensis spelaeoides*.

The Untermassfeld finds of the dirk-toothed cat belong to the most evolved stage of *Megantereon* with extremely enlarged upper canines and strongly reduced upper and lower premolars. This cat was widespread from the Italian Late Villafranchian Farneta Faunal Unit into the post-Villafranchian (Hemmer and Kahlke l.c., 1534). According to current knowledge, the Untermassfeld finds, upgraded to species level as *M. adroveri*, represent the most recent occurrence of dirk-toothed cats in Europe. The scimitar cat *Homotherium crenatidens*, characteristic of the Eurasian Middle and Late Villafranchian, was replaced during the late Early Pleistocene by the more evolved *H. latidens ultimum*, which spread rapidly from Europe to East Asia and survived there until the Middle Pleistocene (Hemmer and Kahlke l.c., 1551).

The *Pachycrocuta brevirostris* hyaenas, which were widespread in Europe from the Late Villafranchian to early Middle Pleistocene, show a weak but significant increase in body size during their evolutionary history (Iannucci et al. in this volume). Since Untermassfeld has provided the largest representative for the species so far recorded, they can be assigned to a post-Villafranchian population. *Ursus* ex gr. *U. dolinensis* from Untermassfeld differs from the Villafranchian *U. etruscus* by an incipient reduction of premolars (García 2004; García and Mazza in this volume). Based on morphological similarities with the late Early Pleistocene *U. dolinensis*, the Untermassfeld ursids may represent an Epivillafranchian metapopulation that led to the Middle and Late Pleistocene cave bears of the western Palaearctic (García l.c.; García and Mazza l.c.).

As Sotnikova (2001, 623 ff., Fig. 4) demonstrated, the Untermassfeld large canids *Xenocyon lycaonoides* differ from the Late Villafranchian forms (*X.* ex gr. *falconeri*) by a reduced or absent entoconid on the M1. This tendency, of lower carnassial entoconid reduction, continued during the Middle Pleistocene. Several other dentition features link the lower carnassials of *X. lycaonoides* with older post-Villafranchian forms, but distinguish it from early Middle Pleistocene ones (Sotnikova l.c., 629). The Untermassfeld wolf *Canis mosbachensis* embodies a distinct evolutionary stage of medium-sized canids. Its cranial and dentition proportions, as well as some features of dentition morphology, separate it from late Middle to early post-Villafranchian representatives of the genus (Sotnikova l.c., 608 ff.). The length ratios of P^4 and M^{1-2}, as well as the ratio between the radius and humerus, show that the Untermassfeld wolf is morphologically more primitive than early Middle Pleistocene European representatives.

The small mammal fauna recovered from the excavation site is assigned to the Jaramillo subchron period according to the evolutionary level of *Microtus* (*Allophaiomys*) *thenii*, the dates of *Sorex* (*Drepanosorex*) *margaritodon* and *Spermophilus kahlkei*, and the presence of *Mimomys pusillus* and *Hypolagus brachygnatus* (Maul et al. 2007, 250; in this volume, 1596, Fig. 10). Both a chronostratigraphically earlier classification and a temporal proximity to the M/B boundary are excluded. Thus, the finds belong to the younger section of the Early Biharian (*Mimomys pusillus*-*Mimomys savini*-zone) of the micromammal biochronology sensu Fejfar et al. (1998).

According to the biostratigraphic data, the synchronous occurrence of the Untermassfeld mammals can be narrowed down to the period between 1.1 and 0.9 Ma BP (see also Kahlke 2007, Fig. 8). A combination of this interval with the available litho- and magnetostratigraphic evidence (see section 8.1.) clarifies the chronostratigraphic classification of the site. In all likelihood, the Untermassfeld fossil deposit was formed at the beginning of the Early Pleistocene Jaramillo polarity subchron (1.071–0.990 Ma BP after Cohen and Gibbard 2019). Shackleton (1995) correlates the Jaramillo with marine oxygen isotope stages (MIS) 31–27. Given the position of the palaeomagnetic reversal within the Untermassfeld Pleistocene sequence (see section 8.1.) as well as the reconstructed process sequence (see section 4.2.), the fossil assemblage can be assigned to MIS 31, which corresponds to a pronounced warm interval (»super interglacial«; see, e.g., Melles et al. 2012; Aubry et al. 2016; Oliveira et al. 2017) that overlaps the onset of the Jaramillo (Cohen and Gibbard 1999). In light of the recorded data, the Untermassfeld fauna lived during the early Bavelian (Bavel Interglacial sensu Zagwijn and Jong 1984) and can be dated to about 1.07–1.06 Ma BP.

8.4. The Epivillafranchian biochron of the western Palaearctic

Fossil evidence on the evolutionary history of the Early and early Middle Pleistocene mammal faunas of the western Palaearctic is available in very different quantities for the individual regions and time intervals. In earlier studies, both the next oldest and next youngest local mammal faunas compared to Untermassfeld were compiled by country (Kahlke 2001e, 992–1004; 2006, 97–107). A transregional overview of western Palearctic large mammal sites from the entire time span between 2.6 and 0.4 Ma BP was presented by Kahlke et al. (2011). In the above-mentioned compilations it becomes clear that the amount of analysable fossil mammal faunas from the intervals 1.8–1.2 Ma BP and 0.8–0.4 Ma BP has until now been significantly higher than those from between 1.2 and 0.9 Ma BP, which is particularly relevant for faunal comparisons with Untermassfeld (**Fig. 28**). Accordingly, the faunal history of this period remained uncertain for a long time (Kahlke 2007, 265 f.).

The Untermassfeld fossil assemblage contains one of the most complete vertebrate find complexes in Europe and the western Palaearctic (sensu Holt et al. 2013), respectively, from the period between 1.2 and 0.9 Ma BP. In particular, the recorded large mammal taxa support the idea of inserting a separate biochron (sensu Lindsay 1990, 5–6; Steininger and Piller 1999), the Epivillafranchian, between the Villafranchian and Galerian (Kahlke 2007). As a basis for definition, the association *Bison menneri / Eucladoceros giulii / Pseudodama vallonnetensis / Capreolus cusanoides / Stephanorhinus hundsheimensis* (»*Bison menneri-Eucladoceros giulii* assemblage chron«) is used with Untermassfeld as the reference site. In addition, Bellucci et al. (2015) propose two supraregional bioevents as chronostratigraphic markers of the Epivillafranchian. The *Bison menneri* first occurrence (FO) is named as the lower boundary and the *Crocuta crocuta* FO as the upper time marker. Both occurrences coincide with the Untermassfeld findings. The FO of *Praemegaceros verticornis*, also discussed in this context by Bellucci (l.c.), could, however, be younger in Central Europe than in the Mediterranean region based on previous investigations (Mäuser 1987, 27 ff.; H.-D. Kahlke 1997, 271). Other characteristic faunal elements of the western Palaearctic Epivillafranchian biochron are *Cervalces carnutorum*, *Panthera onca gombaszoegensis*, *Acinonyx pardinensis pleistocaenicus*, *Puma pardoides brevipes*, *Lynx valdarnensis spelaeoides*, *Megantereon adroveri*, *Homotherium latidens ultimum*, *Pachycrocuta brevirostris*, *Ursus* ex gr. *U. dolinensis*, *Xenocyon lycaonoides* and *Canis mosbachensis* in mostly distinct evolutionary stages (Kahlke 2007; Kahlke et al. 2011; Bellucci et al. 2015, Tab. 1; sections 5.2.7., 8.3. of this paper). The term Epivillafranchian biochron is now used for numerous regions (e.g., García et al. 2008; Madurell-Malapeira et al. 2009; 2011; 2015; Martínez-Navarro et al. 2015; Iannucci et al. 2020; 2021; Konidaris et al. 2021).

Within the global temperature evolution, the time interval between 1.2 and 0.9 Ma BP is generally regarded as the final phase of the 41 ka periodicity. However, it cannot be overlooked that after more than 600 ka of largely consistent temperature variations, a new pattern of global climate cycles emerged (see Raymo and Nisancioglu 2003, Fig. 1; Lisiecki and Raymo 2005, Fig. 4). The Epivillafranchian period appears rather as a climatically unstable transitional period between the preceding 41 ka and the following 100 ka periodicity. For about 300 ka, there was a temporally variable succession of increasingly cold stages, with intermediary stages that were particularly warm (Snyder 2016). One such period was MIS 31, to which the Untermassfeld fauna is assigned (see section 8.3.).

The interplay of Atlantic-Mediterranean and Continental-Eurasian climatic influences, which was pronounced in Epivillafranchian Europe, led to the emergence or immigration of numerous new animal species. These formed the main foundation of the western Palearctic faunal development during the time of the subsequent and, in its global effectiveness, completely different 100 ka climate regime.

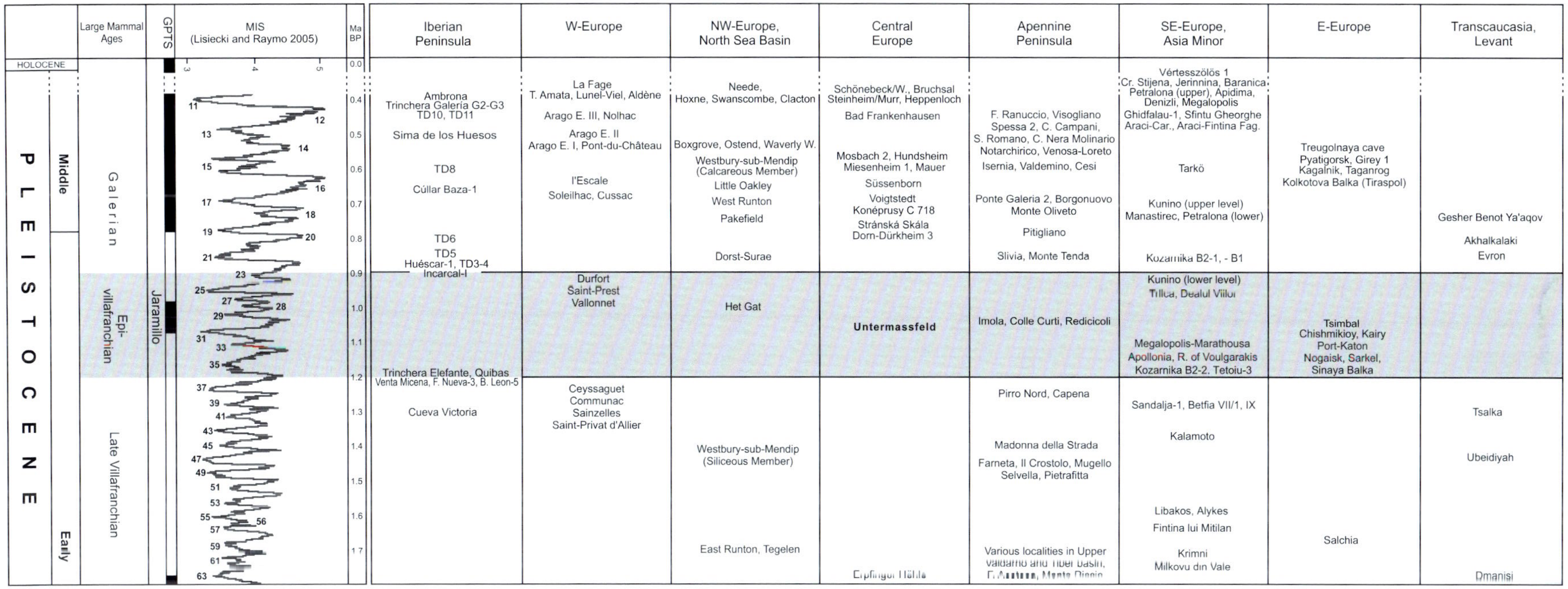

Fig. 28 Chronostratigraphic position of principal fossil mammal sites of the western Palaearctic from the period 1.8–0.4 Ma BP (based on Kahlke et al. 2011). Highlighted in grey – Epivillafranchian biochron.

Acknowledgements

My special thanks go to G. Utschig (Weimar) who supported the preparation of the present study over many years, in particular through the provision of numerous field and collection data, the recording of taxonomic groups of finds and anatomical units, and the determination of minimum numbers of individuals. Overviews of the excavated areas were compiled by G. Utschig in collaboration with E. Haase and J.-A. Keiler (both Weimar). The latter kindly contributed excavation and collection data as well as information on conservation work. My thanks also go to J. Arnold (Weimar), who determined current find numbers for several large mammal species, and D. Rössler (Weimar), who compiled find quantities and facilitated the damage-free movement of large preparations. The photographic work on the osteological finds was done by S. Döring and T. Korn (both Weimar). I am very grateful to E. Haase for the compilation of find data from selected excavation areas as well as for the mounting of the photo illustrations, for the production of all the graphics used here, and above all for the excavation plans (attached here as foldouts), which have been continuously compiled over more than a decade. My thanks go to M. Sotnikova (Moscow) for her assistance in the recovery and repatriation of borrowed Untermassfeld fossils. I am indebted to U. Kierdorf (Hildesheim) for information on palaeopathological findings. L. C. Maul (Weimar) contributed information on micromammals along with corresponding literature references. D. Mol (Hoofddorp) determined the MNI of the proboscideans represented in the Untermassfeld material. M. Böhme (Tübingen) supported the work with discussions and notes on the herpetofauna. C. Nielsen-Marsh (Leipzig) translated and edited the entire text. On the occasion of the review of an earlier version of this contribution, U. Kierdorf gave valuable advice, for which I am also grateful.

References

Andrews, P., 1990. Owls, Caves and Fossils. Predation, Preservation, and Accumulation of Small Mammal Bones in Caves, with an analysis of the Pleistocene Cave Faunas from Westbury-sub-Mendip, Somerset, UK. Chicago, IL.

Anonymous (U. Rai), 1998. Major Indian Reserve Devastated by Monsoon Floods. Cat News 29, 22–23.

Arnold, J., 2018. Modifikation frühpleistozäner Bisonreste von Untermaßfeld (Südthüringen, Deutschland) durch *Pachycrocuta*. Master thesis, Friedrich-Schiller-Universität Jena, Germany.

in this volume. Bovid bone modifications by *Pachycrocuta* from the Early Pleistocene site of Untermassfeld. 1423–1464.

Arribas, A., Palmqvist, P., 1998. Taphonomy and palaeoecology of an assemblage of large mammals: Hyaenid activity in the Lower Pleistocene site at Venta Micena (Orce, Guadix-Baza Basin, Granada, Spain). Geobios 31, supplement 3, 3–47.

Aubry, A.M.R., Vernal, A. de, Hillaire-Marcel, C., 2016. The »warm« Marine Isotope Stage 31 in the Labrador Sea: Low surface salinities and cold subsurface waters prevented winter convection. Paleoceanography and Paleoclimatology 31, 1206–1224.

Baales, M., 2014. Untermassfeld – Or the struggle for finding the earliest traces of human occupation in Central Europe: A comment on: »Hominin dispersals from the Jaramillo subchron in central and south-western Europe: Untermassfeld (Germany) and Vallparadís (Spain)« by J. García et al., Quaternary International 316 (2013), 73–93. Quaternary International 337, 254–256.

Baar, M., 2017a. Thüringen fordert Knochen zurück. Thüringer Allgemeine (Erfurt) 291, 1.

2017b. Eiszeitforscher: Scheinsensation war Diebstahl. Thüringer Allgemeine 291, 17.

Bellucci, L., Sardella, R., Rook, L., 2015. Large mammal biochronology framework in Europe at Jaramillo: The Epivillafranchian as a formal biochron. Quaternary International 389, 84–89.

Böhme, G., 1997a. Die Amphibienreste aus dem Unterpleistozän von Untermaßfeld. In: Kahlke, R.-D. (Ed.), Das Pleistozän von Untermaßfeld bei Meiningen (Thüringen). Teil 1. Monographien des Römisch-Germanischen Zentralmuseums 40, 1, 81–87.

1997b. Fossile Fischfaunen aus dem jüngeren Känozoikum Deutschlands. Quartär 47/48, 113–138.

Böhme, M., 2020a. New results on fishes from the Early Pleistocene site of Untermassfeld. In: Kahlke, R.-D. (Ed.), The Pleistocene of Untermassfeld near Meiningen (Thüringen, Germany). Part 4. Monographien des Römisch-Germanischen Zentralmuseums 40, 4, 1133–1140.

2020b. New results on amphibians and reptiles from the Early Pleistocene site of Untermassfeld - bioclimatic analysis of the herpetofauna. In: Kahlke, R.-D. (Ed.), The Pleistocene of Untermassfeld near Meiningen (Thüringen, Germany). Part 4. Monographien des Römisch-Germanischen Zentralmuseums 40, 4, 1141–1158.

Boulbes, N., van Asperen, E.N., 2019. Biostratigraphy and Palaeoecology of European *Equus*. Frontiers in Ecology and Evolution 7, 301.

Breda, M., 2008. Palaeoecology and Palaeoethology of the Plio-Pleistocene Genus *Cervalces* (Cervidae, Mammalia) in Eurasia. Journal of Vertebrate Paleontology 28, 886–899.

Breda, M., Kahlke, R.-D., Lister, A., 2020. New results on cervids from the Early Pleistocene site of Untermassfeld. In: Kahlke, R.-D. (Ed.), The Pleistocene of Untermassfeld near Meiningen (Thüringen, Germany). Part 4. Monographien des Römisch-Germanischen Zentralmuseums 40, 4, 1197–1249.

Brüning, H., 1962: Geiseltallack-Tauchverfahren. Eine einfache Methode zur Härtung fossilen Knochenmaterials. Der Präparator 8, 28.

Bukhsianidze, M., 2020. New results on bovids from the Early Pleistocene site of Untermassfeld. In: Kahlke, R.-D. (Ed.), The Pleistocene of Untermassfeld near Meiningen (Thüringen, Germany). Part 4. Monographien des Römisch-Germanischen Zentralmuseums 40, 4, 1169–1195.

Callaway, E., 2017. Archaeologists say human-evolution study used stolen bone. Nature 551, 279–280.

Cherin, M., Alba, D.M., Crotti, M., Menconero, S., Moullé, P.-É., Sorbelli, L., Madurell-Malapeira, J., 2020. The post-Jaramillo persistence of *Sus strozzii* (Suidae, Mammalia) in Europe: New evidence from the Vallparadís Section (NE Iberian Peninsula) and other coeval sites. Quaternary Science Reviews 233, 106234.

Child, G., 1968. Behaviour of large mammals during the formation of Lake Kariba. Salisbury.

Cohen, K.M., Gibbard, P.L., 2019. Global chronostratigraphical correlation table for the last 2.7 million years, version 2019 QI-500. Quaternary International 500, 20–31.

Dubrovo, I.A., 2001. Remains of Elephantidae from the Lower Pleistocene site of Untermassfeld. In: Kahlke, R.-D. (Ed.), Das Pleistozän von Untermaßfeld bei Meiningen (Thüringen). Teil 2. Monographien des Römisch-Germanischen Zentralmuseums 40, 2, 589–605, pls. 95–101.

Eisenmann, V., 2006. Pliocene and Pleistocene Equids: palaeontology versus molecular biology. In: Kahlke, R.-D., Maul, L.C., Mazza, P.P.A. (Eds.), Late Neogene and Quaternary biodiversity and evolution: Regional developments and interregional correlations. Proceedings of the 18th International Senckenberg Conference (VI International Palaeontological Colloquium in Weimar). Volume I. Courier Forschungsinstitut Senckenberg 256, 71–89.

2010. *Sussemionus*, a new subgenus of *Equus* (Perissodactyla, Mammalia). Comptes Rendus Biologies 333, 235–240.

Eisenmann, V., Boulbes, N., 2020. New results on equids from the Early Pleistocene site of Untermassfeld. In: Kahlke, R.-D. (Ed.), The Pleistocene of Untermassfeld near Meiningen (Thüringen, Germany). Part 4. Monographien des Römisch-Germanischen Zentralmuseums 40, 4, 1295–1321.

Ellenberg, J., 1975. Schwermineralverteilung, Sedimentation und Stratigraphie im Quartär der mittleren Werra. Zeitschrift für geologische Wissenschaften 3, 1389–1404.

2002. Der thüringische Zersatzgrobschotter des Unterpleistozäns. In: Kahlke, R.-D., Wunderlich, J. (Eds.), Tertiär und Quartär in Thüringen. Beiträge zur Geologie von Thüringen, Neue Folge 9, 233–246

Ellenberg, J., Kahlke, R.-D., 1997. Die quartärgeologische Entwicklung des mittleren Werratals und der Bau der unterpleistozänen Komplexfundstelle Untermaßfeld. In: Kahlke, R.-D. (Ed.), Das Pleistozän von Untermaßfeld bei Meiningen (Thüringen). Teil 1. Monographien des Römisch-Germanischen Zentralmuseums 40, 1, 29–62, pls. 1–18.

Eltringham, S.K., 1999. The hippos. Natural history and conservation. London.

Fejfar, O., Heinrich, W.-D., Lindsay, E.H., 1998. Updating the Neogene Rodent biochronology in Europe. In: van Kolfschoten, T., Gibbard, P.L. (Eds.), The Dawn of the Quaternary. Proceedings of the SEQS-EuroMam symposium 1996. Mededelingen Nederlands Instituut voor Toegepaste Geowetenschappen TNO 60, 533–553.

Fiedler, B., 2012. Quartärpaläontologische Forschung in Weimar 1962–2011. Eine Bibliographie. Frankfurt am Main.

Fostowicz-Frelik, Ł, 2008. First record of *Trogontherium cuvieri* (Mammalia: Rodentia) from Poland and review of the species. Geodiversitas 30, 765–778.

García, N., 2004. New results on the remains of Ursidae from Untermaßfeld: comparisons with *Ursus dolinensis* from Atapuerca and other Early and Middle Pleistocene sites. In: Maul, L.C., Kahlke, R.-D. (Eds.), Late Neogene and Quaternary biodiversity and evolution: Regional developments and interregional correlations. 18th International Senckenberg Conference, VI International Palaeontological Colloquium in Weimar, Weimar (Germany), 25th–30th April, 2004, Conference Volume. Terra Nostra, Schriften der Alfred-Wegener-Stiftung 2004/2, 112–113.

García, N., Mazza, P.P.A, in this volume. New results on ursids from the Early Pleistocene site of Untermassfeld. 1385–1401.

García, N., Arsuaga, J.L., Bermúdez de Castro, J.M., Carbonell, E., Rosas, A., Huguet, R., 2008. The Epivillafranchian carnivore *Pannonictis* (Mammalia, Mustelidae) from Sima del Elefante (Sierra de Atapuerca, Spain) and a revision of the Eurasian occurrences from a taxonomic perspective. Quaternary International 179, 42–52.

Garutt, V.E., Bajguševa, V.S., 1981. *Archidiskodon gromovi* Garutt et Alexeeva – der älteste Elefant der Mammutlinie in Eurasien. Quartärpaläontologie 4, 7–18.

Gaudzinski-Windheuser, S., 2020. Untermassfeld – A scientific treasure trove for generations to come. In: Kahlke, R.-D. (Ed.), The Pleistocene of Untermassfeld near Meiningen (Thüringen, Germany). Part 4. Monographien des Römisch-Germanischen Zentralmuseums 40, 4, VII–VIII.

Gerdes, A., Gärtner, A., Linnemann, U., Kahlke, R.-D., 2020. In situ U-Pb geochronology of re-deposited travertine from the Early Pleistocene site of Untermassfeld. In: Kahlke, R.-D. (Ed.), The Pleistocene of Untermassfeld near Meiningen (Thüringen, Germany). Part 4. Monographien des Römisch-Germanischen Zentralmuseums 40, 4, 1105–1115.

Glutz von Blotzheim, U.N., Bauer, K.M., Bezzel, E., 1971. Handbuch der Vögel Mitteleuropas. Band 4. Falconiformes. Frankfurt am Main.

Goddard, J., 1970. Age criteria and vital statistics of a black rhinoceros population. East African Wildlife Journal 8, 105–121.

Guérin, C., Faure, M., 1997. The Wild Boar (*Sus scrofa priscus*) from the Post-Villafranchian Lower Pleistocene of Untermassfeld. In: Kahlke, R.-D. (Ed.), Das Pleistozän von Untermaßfeld bei Meiningen (Thüringen). Teil 1. Monographien des Römisch-Germanischen Zentralmuseums 40, 1, 375–383, pls. 63–67

Haffer, J., 1993. *Corvus*. In: Glutz von Blotzheim, U.N. (Ed.), Handbuch der Vögel Mitteleuropas. Band 13/III. Passeriformes (4. Teil). Corvidae – Sturnidae. Wiesbaden, 1653–2022.

Haltenorth, T., Diller, H., 1977. Säugetiere Afrikas und Madagaskars. München, Bern, Wien.

Häntzschel, W., El-Baz, F., Amstutz, G.C., 1968. Coprolites. An Annotated Bibliography. The Geological Society of America, Memoir 108, 1–132.

Haynes, G., 1988. Mass Deaths and Serial Predation: Comparative Taphonomic Studies of Modern Large Mammal Death Sites. Journal of Archaeological Science 15, 219–235.

Hemmer, H., 2001. Die Feliden aus dem Epivillafranchium von Untermaßfeld. In: Kahlke, R.-D. (Ed.), Das Pleistozän von Untermaßfeld bei Meiningen (Thüringen). Teil 3. Monographien des Römisch-Germanischen Zentralmuseums 40, 3, 699–782, pls. 132–142.

Hemmer, H., Kahlke, R.-D., in this volume. New results on felids from the Early Pleistocene site of Untermassfeld. 1465–1566.

Hemmer, H., Kahlke, R.-D., Vekua, A.K., 2010. *Panthera onca georgica* ssp. nov. from the Early Pleistocene of Dmanisi (Republic of Georgia) and the phylogeography of jaguars (Mammalia, Carnivora, Felidae). Neues Jahrbuch für Geologie und Paläontologie, Abhandlungen 257, 115–127.

Holt, B.G., Lessard, J.-P., Borregaard, M.K., Fritz, S.A., Araújo, M.B., Dimitrov, D., Fabre, P.-H., Graham, C.H., Graves, G.R., Jønsson, K.A., Nogués-Bravo, D., Wang, Z., Whittaker, R.J., Fjeldsåand, J., Rahbek, 2013. An Update of Wallace's Zoogeographic Regions of the World. Science 339, 74–78.

Iannucci. A., in this volume. New results on suids from the Early Pleistocene site of Untermassfeld. 1339–1354.

Iannucci. A., Gasparik, M., Sardella, R., 2020. First report of *Sus strozzii* (Suidae, Mammalia) from the Early Pleistocene of Hungary (Dunaalmás) and species distinction based on deciduous teeth. The Science of Nature 107, 5.

Iannucci, A., Mecozzi, B., Kahlke, R.-D., Sardella, R., in this volume. New results on hyaenids from the Early Pleistocene site of Untermassfeld. 1403–1422.

Iannucci, A., Mecozzi, B., Sardella, R., Iurino, A.A., 2021. The extinction of the giant hyena *Pachycrocuta brevirostris* and a reappraisal of the Epivillafranchian and Galerian Hyaenidae in Europe: Faunal turnover during the Early–Middle Pleistocene Transition. Quaternary Science Reviews 272, 107240.

Jánossy, D., 1997. Die Vogelreste aus dem Unterpleistozän von Untermaßfeld. In: Kahlke, R.-D. (Ed.), Das Pleistozän von Untermaßfeld bei Meiningen (Thüringen). Teil 1. Monographien des Römisch-Germanischen Zentralmuseums 40, 1, 95–100, pl. 21.

Kahlke, H.-D., 1956a. Die Cervidenreste aus den altpleistozänen Ilmkiesen von Süßenborn bei Weimar. Teil I. Die Geweihe und Gehörne. Berlin I–VIII, 1–62, pls. 1–31, 1 foldout.

1956b. Die Cervidenreste aus den altpleistozänen Ilmkiesen von Süßenborn bei Weimar. Teil II. Schädel und Gebisse. Berlin I–VIII, 1–44, figs. 1–70, pls. 1–37.

1965. Die Cerviden-Reste aus den Tonen von Voigtstedt in Thüringen. Paläontologische Abhandlungen, Abteilung A Paläozoologie 2. Berlin, 379–426, pls. XIII–XXII.

1969. Die Cerviden-Reste aus den Kiesen von Süßenborn bei Weimar. Paläontologische Abhandlungen, Abteilung A Paläozoologie 3. Berlin, 547–610, pls. XXVII–XXXVI.

1982. *Hippopotamus antiquus* Desmarest, 1822 aus dem Pleistozän von Meiningen in Südthüringen (Bezirk Suhl). Zeitschrift für geologische Wissenschaften 10, 943–949.

1990. On the Evolution, Distribution and Taxonomy of Fossil Elk/Moose. Quartärpaläontologie 8, 83–106.

1995. Ein fossiler Elch-Fund aus dem Unterpleistozän von Untermassfeld in Thüringen (Mitteldeutschland). Quartär 45/46, 227–235.

1997. Die Cerviden-Reste aus dem Unterpleistozän von Untermaßfeld. In: Kahlke, R.-D. (Ed.), Das Pleistozän von Untermaßfeld bei Meiningen (Thüringen). Teil 1. Monographien des Römisch-Germanischen Zentralmuseums 40, 1, 181–275, pls. 34–44.

2001a. Neufunde von Cerviden-Resten aus dem Unterpleistozän von Untermaßfeld. In: Kahlke, R.-D. (Ed.), Das Pleistozän von Untermaßfeld bei Meiningen (Thüringen). Teil 2. Monographien des Römisch-Germanischen Zentralmuseums 40, 2, 461–482, pls. 72–76.

2001b. Die Rhinocerotiden-Reste aus dem Unterpleistozän von Untermaßfeld. Untermaßfeld. In: Kahlke, R.-D. (Ed.), Das Pleistozän von Untermaßfeld bei Meiningen (Thüringen). Teil 2. Monographien des Römisch-Germanischen Zentralmuseums 40, 2, 501–555, pls. 79–91.

Kahlke, R.-D., 1985a. Altpleistozäne *Hippopotamus*-Reste von Untermaßfeld bei Meiningen (Bezirk Suhl, DDR) (Vorläufige Mitteilung). Säugetierkundliche Informationen 2, 227–233.

1985b. Untersuchungen zur Incisivenreduktion an altpleistozänen *Hippopotamus*-Mandibeln von Untermaßfeld bei Meiningen (Bezirk Suhl, DDR). Biologische Rundschau 23, 315–321.

1987. Die unterpleistozänen *Hippopotamus*-Reste von Untermaßfeld bei Meiningen (Bezirk Suhl, DDR) – Ein Beitrag zur Forschungs-, Entwicklungs- und Verbreitungsgeschichte fossiler Hippopotamiden in Europa. PhD thesis, Ernst-Moritz-Arndt-Universität Greifswald, Germany.

1995. Stop 15: Untermassfeld near Meiningen – Early Pleistocene fauna. In: Schirmer, W. (Ed.), Quaternary field trips in Central Europe. Volume 2. Field trips on special topics. München, 730–732.

1997a. Zur Entdeckungs- und Erforschungsgeschichte der unterpleistozänen Komplexfundstelle Untermaßfeld. In: Kahlke, R.-D. (Ed.), Das Pleistozän von Untermaßfeld bei Meiningen (Thüringen). Teil 1. Monographien des Römisch-Germanischen Zentralmuseums 40, 1, 1–28.

1997b. Die *Hippopotamus*-Reste aus dem Unterpleistozän von Untermaßfeld. In: Kahlke, R.-D. (Ed.), Das Pleistozän von Untermaßfeld bei Meiningen (Thüringen). Teil 1. Monographien des Römisch-Germanischen Zentralmuseums 40, 1, 277–374, pls. 45–62.

1997c. Bisheriger Gesamtbefund zur Geologie, Paläozoologie, Taphonomie, Ökologie und Stratigraphie der unterpleistozänen Komplexfundstelle Untermaßfeld. In: Kahlke, R.-D. (Ed.), Das Pleistozän von Untermaßfeld bei Meiningen (Thüringen). Teil 1. Monographien des Römisch-Germanischen Zentralmuseums 40, 1, 385–418.

1999a. Overview and first quantitative data on the taphonomy of the Lower Pleistocene fossil site of Untermassfeld (Thüringen, Germany). In: Gaudzinski, S., Turner, E. (Eds.), The role of early humans in the accumulation of European Lower and Middle Palaeolithic bone assemblages. Ergebnisse eines Kolloquiums. Monographien des Römisch-Germanischen Zentralmuseums 42, 7–19.

1999b. The History of the Origin, Evolution and Dispersal of the Late Pleistocene *Mammuthus-Coelodonta* Faunal Complex in Eurasia (Large Mammals). Rapid City, SD.

2000. Beuterest: Vorderbein-Skelett eines »Etruskischen Nashorns« aus Südthüringen. Natur und Museum 130, 197–199.

2001a. Die Fortführung der Forschungsarbeiten zur unterpleistozänen Komplexfundstelle Untermaßfeld in den Jahren 1989–1996. In: Kahlke, R.-D. (Ed.), Das Pleistozän von Untermaßfeld bei Meiningen (Thüringen). Teil 2. Monographien des Römisch-Germanischen Zentralmuseums 40, 2, 419–432.

2001b. Verbesserte geologische Standardprofile zur unterpleistozänen Komplexfundstelle Untermaßfeld. In: Kahlke, R.-D. (Ed.), Das Pleistozän von Untermaßfeld bei Meiningen (Thüringen). Teil 2. Monographien des Römisch-Germanischen Zentralmuseums 40, 2, 433–445, suppl. I.

2001c. Schädelreste von *Hippopotamus* aus dem Unterpleistozän von Untermaßfeld. In: Kahlke, R.-D. (Ed.), Das Pleistozän von Untermaßfeld bei Meiningen (Thüringen). Teil 2. Monographien des Römisch-Germanischen Zentralmuseums 40, 2, 483–500, pls. 77–78.

2001d. Ein Hinweis auf Herbivoren-Osteophagie im Unterpleistozän von Untermaßfeld sowie Bemerkungen zur Gabelbildung an Knochenenden. In: Kahlke, R.-D. (Ed.), Das Pleistozän von Untermaßfeld bei Meiningen (Thüringen). Teil 3. Monographien des Römisch-Germanischen Zentralmuseums 40, 3, 897–903, pl. 147.

2001e. Die unterpleistozäne Komplexfundstelle Untermaßfeld – Zusammenfassung des Kenntnisstandes sowie synthetische Betrachtungen zu Genesemodell, Paläoökologie und Stratigraphie. In: Kahlke, R.-D. (Ed.), Das Pleistozän von Untermaßfeld bei Meiningen (Thüringen). Teil 3. Monographien des Römisch-Germanischen Zentralmuseums 40, 3, 931–1030, suppl. I–XV.

2001f. Eis und heiß – Konservierte Sedimentstrukturen des Quartärs. Natur und Museum 131, 361–362.

2006. Untermassfeld – A late Early Pleistocene (Epivillafranchian) fossil site near Meiningen (Thuringia, Germany) and its position in the development of the European mammal fauna. BAR British Archaeological Reports, International Series 1578. Oxford.

2007. Late Early Pleistocene European large mammals and the concept of an Epivillafranchian biochron. In: Kahlke, R.-D., Maul, L.C., Mazza, P. (Eds.), Late Neogene and Quaternary biodiversity and evolution: Regional developments and interregional correlations. Proceedings of the 18th International Senckenberg Conference (VI International Palaeontological Colloquium in Weimar). Volume II. Courier Forschungsinstitut Senckenberg 259, 265–278.

2009. Das Leichenfeld im Werratal. Europäischer Faunenwandel vor einer Million Jahren. Natur und Museum 139, 146–159.

2016. Bericht zu den bodendenkmalpflegerischen Maßnahmen an der unterpleistozänen Komplexfundstelle Untermaßfeld im Jahr 2016. Unpublished excavation report, Weimar.

2020. Excavation progress and research at the Early Pleistocene site of Untermassfeld during the years 1997–2015. In: Kahlke, R.-D. (Ed.), The Pleistocene of Untermassfeld near Meiningen (Thüringen, Germany). Part 4. Monographien des Römisch-Germanischen Zentralmuseums 40, 4, 1031–1078.

Kahlke, R.-D., Gaudzinski, S., 2005. The blessing of a great flood: differentiation of mortality patterns in the large mammal record of the Lower Pleistocene fluvial site of Untermassfeld (Germany) and its relevance for the interpretation of faunal assemblages from archaeological sites. Journal of Archaeological Science 32, 1202–1222.

Kahlke, R.-D., Kaiser, T.M., 2011. Generalism as a subsistence strategy: advantages and limitations of the highly flexible feeding traits of Pleistocene *Stephanorhinus hundsheimensis* (Rhinocerotidae, Mammalia). Quaternary Science Reviews 30, 2250–2261.

Kahlke, R.-D., Keiler, J.-A., 2017. Bericht zu den bodendenkmalpflegerischen Maßnahmen an der unterpleistozänen Komplexfundstelle Untermaßfeld im Jahr 2017. Unpublished excavation report, Weimar.

2018. Bericht zu den bodendenkmalpflegerischen Maßnahmen an der unterpleistozänen Komplexfundstelle Untermaßfeld im Jahr 2018. Unpublished excavation report, Weimar.

2019. Bericht zu den bodendenkmalpflegerischen Maßnahmen an der unterpleistozänen Komplexfundstelle Untermaßfeld im Jahr 2019. Unpublished excavation report, Weimar.

2020. Bericht zu den bodendenkmalpflegerischen Maßnahmen an der unterpleistozänen Komplexfundstelle Untermaßfeld im Jahr 2020. Unpublished excavation report, Weimar.

Kahlke, R.-D., Kierdorf, U., 1997. Diffuse idiopathische skelettale Hyperostose bei einem Bären (*Ursus* sp.) aus dem Unterpleistozän von Untermaßfeld bei Meiningen (Südthüringen, Deutschland). Quartär 47/48, 159–175.

2005. Diabetes im Fossilbefund? Natur und Museum 135, 280–281.

Kahlke, R.-D., Keiler, J.-A., Kienast, F., 2014. Senckenberg Forschungsstation für Quartärpaläontologie Weimar. In: Werneburg, R., Mey, E. (Eds.), Thüringer Natur-Schätze. Naturkundliche Museen und Sammlungen im Freistaat Thüringen, Regensburg, 138–147.

Kahlke, R.-D., Linnemann, U., Gärtner, A., 2020. New results on the origin and geological history of the Early Pleistocene site of Untermassfeld. In: Kahlke, R.-D. (Ed.), The Pleistocene of Untermassfeld near Meiningen (Thüringen, Germany). Part 4. Monographien des Römisch-Germanischen Zentralmuseums 40, 4, 1079–1104, 1 foldout.

Kahlke, R.-D., García, N., Kostopoulos, D.S., Lacombat, F., Lister, A.M., Mazza, P.P.A., Spassov, N., Titov, V.V., 2011. Western Palaearctic palaeoenvironmental conditions during the Early and early Middle Pleistocene inferred from large mammal communities, and implications for hominin dispersal in Europe. Quaternary Science Reviews 30, 1363–1395.

Keiler, J.-A., 1991. Klebstoffe kritisch betrachtet – ein Beitrag zur paläontologischen Präparationstechnik. Der Präparator 38, 81–84.

1994. Nutzung der Transfer-Methode zur Stabilisierung pleistozäner Wirbeltierreste. Der Präparator 40, 115–118.

1995. Bergung und Präparation pleistozäner Wirbeltierreste unter Berücksichtigung des Fossilmaterials der Komplexfundstelle Untermaßfeld / Südthüringen. Restaurierung und Museumstechnik 12, 1–31, pls. 1–24.

2001a. Die Präparation und Konservierung des Fossilmaterials aus dem Unterpleistozän von Untermaßfeld. In: Kahlke, R.-D (Ed.), Das Pleistozän von Untermaßfeld bei Meiningen (Thüringen). Teil 2. Monographien des Römisch-Germanischen Zentralmuseums 40, 2, 447–459, pls. 68–71.

2001b. Die Koprolithen aus dem Unterpleistozän von Untermaßfeld. In: Kahlke, R.-D. (Ed.), Das Pleistozän von Untermaßfeld bei Meiningen (Thüringen). Teil 2. Monographien des Römisch-Germanischen Zentralmuseums 40, 2, 691–698, pls. 128–131.

2001c. Die Polyvinylbutyral-Tränkungsmethode für fossile Knochen. Der Präparator 47, 49–56.

2008. Bergung, Präparation und Konservierung fossiler Säugetierschädel aus dem Unterpleistozän von Untermaßfeld/Thüringen. Eine methodologische Betrachtung. Der Präparator 54, 62–83.

2009. Präparation fossiler Säugetierschädel im virtuellen Zeitalter. Natur und Museum 139, 178–183.

2017. Konzepte der Präventiven Konservierung für die Sammlungen der Senckenberg Forschungsstation für Quartärpaläontologie Weimar. Der Präparator 63, 66–85.

Keiler, J.-A., Wenzel, T., 2019. Raubgräber in Untermassfeld. Natur Forschung Museum 149, 184–191.

Keiler, J.-A., Benecke, M., Keiler, J., 2020. Bone modifications by insects from the Early Pleistocene site of Untermassfeld. In: Kahlke, R.-D. (Ed.), The Pleistocene of Untermassfeld near Meiningen (Thüringen, Germany). Part 4. Monographien des Römisch-Germanischen Zentralmuseums 40, 4, 1117–1132.

Kierdorf, U., Kahlke, R.-D., 2004. Pathological findings on remains of *Hippopotamus amphibius antiquus* Desmarest, 1822 (Hippopotamidae, Mammalia) from the late Lower Pleistocene site of Untermaßfeld (Thuringia, Central Germany). In: Maul, L.C., Kahlke, R.-D. (Eds.), Late Neogene and Quaternary biodiversity and evolution: Regional developments and interregional correlations. 18th International Senckenberg Conference, VI International Palaeontological Colloquium in Weimar, Weimar (Germany), 25th–30th April, 2004, Conference Volume. Terra Nostra, Schriften der Alfred-Wegener-Stiftung 2004/2, 137–138.

2020. Pathological findings on remains of hippopotamids from the Early Pleistocene site of Untermassfeld. In: Kahlke, R.-D. (Ed.), The Pleistocene of Untermassfeld near Meiningen (Thüringen, Germany). Part 4. Monographien des Römisch-Germanischen Zentralmuseums 40, 4, 1251–1272.

Kierdorf, U., Flohr, S., Kahlke, R.-D., in this volume. Post-traumatic osteomyelitis in an immature *Bison menneri* from the Early Pleistocene site of Untermassfeld. 1325–1337.

Kierdorf, U., Kahlke, R.-D., Flohr, S., 2012. Healed fracture of the tibia in a bison (*Bison menneri* Sher, 1997) from the late Early Pleistocene site of Untermassfeld (Thuringia, Germany). International Journal of Paleopathology 2, 19–24.

Kleinhubbert, G., 2017. Knochenjäger in der Fotofalle. Der Spiegel 50, 128–129.

Konidaris, G.E., Kostopoulos, D.S., Maron, M., Schaller, M., Ehlers, T.A., Aidona, E., Marini, M., Tourloukis, V., Muttoni, G., Koufos, G.D., Harvati, K., 2021. Dating of the Lower Pleistocene Vertebrate Site of Tsiotra Vryssi (Mygdonia Basin, Greece): Biochronology, Magnetostratigraphy, and Cosmogenic Radionuclides. Quaternary 4. https://doi.org/10.3390/quat4010001.

Kostopoulos, D.S., Maniakas, I., Tsoukala, E., 2018. Early bison remains from Mygdonia Basin (Northern Greece). Geodiversitas 40, 283–319.

Kotowski, A., Stefaniak, K., Kahlke, R.-D., 2020. A rhinocerotid skull from the Early Pleistocene site of Untermassfeld. In: Kahlke, R.-D. (Ed.), The Pleistocene of Untermassfeld near Meiningen (Thüringen, Germany). Part 4. Monographien des Römisch-Germanischen Zentralmuseums 40, 4, 1273–1293.

Kroloop, E., 1997. Die Molluskenreste aus dem Unterpleistozän von Untermaßfeld. In: Kahlke, R.-D. (Ed.), Das Pleistozän von Untermaßfeld bei Meiningen (Thüringen). Teil 1. Monographien des Römisch-Germanischen Zentralmuseums 40, 1, 71–75, pl. 19.

Kruuk, H., 1972. The Spotted Hyena. A Study of Predation and Social Behavior. Chicago, IL, London.

1975. Hyaena. London.

Lacombat, F., 2006. Pleistocene Rhinoceroses in Mediterranean Europe and in Massif Central (France). In: Kahlke, R.-D., Maul, L.C., Mazza, P.P.A. (Eds.), Late Neogene and Quaternary biodiversity and evolution: Regional developments and interregional correlations. Proceedings of the 18th International Senckenberg Conference (VI International Palaeontological Colloquium in Weimar). Volume I. Courier Forschungsinstitut Senckenberg 256, 57–69.

Larkin, N.R., Alexander, J., Lewis M.D., 2000. Using Experimental Studies of Recent Faecal Material to Examine Hyaena Coprolites from the West Runton Freshwater Bed, Norfolk, U.K. Journal of Archaeological Science 27, 19–31.

Laws, R.M., 1968. Dentition and ageing of the Hippopotamus. East African Wildlife Journal 6, 19–52.

Lindsay, E.H., 1990. The setting. In: Lindsay, E.H., Fahlbusch, V., Mein, P. (Eds.), European Neogene Mammal Chronology. NATO ASI Series A, Life Sciences 40, 1–14.

Lisiecki, L.E., Raymo, M.E., 2005. A Pliocene-Pleistocene stack of 57 globally distributed benthic δ18O records. Paleoceanography and Paleoclimatology 20, PA1003.

Lüps, P., Wandeler, A.I., 1993. *Meles meles* (Linnaeus, 1758) – Dachs. In: Stubbe, M., Krapp, F. (Eds.), Handbuch der Säugetiere Europas, Band 5: Raubsäuger – Carnivora (Fissipedia), Teil II: Mustelidae 2, Viverridae, Herspetidae, Felidae. Wiesbaden, 856–906.

Madurell-Malapeira, J., Alba, D.M., Moyà-Solà, S., 2009. Carnivora from the Late Early Pleistocene of Cal Guardiola (Terrassa, Vallès-Penedès Basin, Catalonia, Spain). Journal of Paleontology 83, 969–974.

Madurell-Malapeira, J., Alba, D.M., Moyà-Solà, S., Aurell-Garrido, J., 2010. The Iberian record of the puma-like cat *Puma pardoides* (Owen, 1846) (Carnivora, Felidae). Comptes Rendus Palevol 9, 55–62.

Madurell-Malapeira, J., Alba, D.M., Marmi, J., Aurell, J., Moyà-Solà, S., 2011. The taxonomic status of European Plio-Pleistocene badgers. Journal of Vertebrate Paleontology 31, 885–894.

Madurell-Malapeira, J., Alba, D.M., Espigares, M.-P., Vinuesa, V., Palmqvist, P., Martínez-Navarro, B., Moyà-Solà, S., 2015. Were large carnivorans and great climatic shifts limiting factors for hominin dispersals? Evidence of the activity of *Pachycrocuta brevirostris* during the Mid-Pleistocene Revolution in the Vallparadis Section (Vallès-Penedès Basin, Iberian Peninsula). Quaternary International 431, 42–52.

Maguire, J.M., Pemberton, D., Collett, M.H., 1980. The Makapansgat Limeworks grey breccia: hominids, hyaenas, hystricids or hillwash? Palaeontologia Africana 23, 75–98.

Manegold, A., 2020. New results on birds from the Early Pleistocene site of Untermassfeld. In: Kahlke, R.-D. (Ed.), The Pleistocene of Untermassfeld near Meiningen (Thüringen, Germany). Part 4. Monographien des Römisch-Germanischen Zentralmuseums 40, 4, 1159–1167.

Maniakas, I., Kostopoulos, D.S., 2017. Morphometric-palaeoecological dicrimination between *Bison* populations of the western Palaearctic. Geobios 50, 155–171.

Martínez-Navarro, B., Madurell-Malapeira, J., Ros-Montoya, R., Espigares, M.-P., Medin, T., Hortolà, P., Palmqvist, P., 2015. The Epivillafranchian and the arrival of pigs into Europe. Quaternary International 389, 131–138.

Maul, L., 1997. Nachweis eines Zungenbeinfragmentes einer Schildkröte (Emydini, Emydidae, Testudinata) aus dem Unterpleistozän von Untermaßfeld. In: Kahlke, R.-D. (Ed.), Das Pleistozän von Untermaßfeld bei Meiningen (Thüringen). Teil 1. Monographien des Römisch-Germanischen Zentralmuseums 40, 1, 89–94.

2001a. Die Kleinsäugerreste (Insectivora, Lagomorpha, Rodentia) aus dem Unterpleistozän von Untermaßfeld. In: Kahlke, R.-D. (Ed.), Das Pleistozän von Untermaßfeld bei Meiningen (Thüringen). Teil 3. Monographien des Römisch-Germanischen Zentralmuseums 40, 3, 783–888, pls. 144–145.

2001b. Nagespuren von Kleinsäugern an Knochenmaterial aus dem Unterpleistozän von Untermaßfeld. In: Kahlke, R.-D. (Ed.), Das Pleistozän von Untermaßfeld bei Meiningen (Thüringen). Teil 3. Monographien des Römisch-Germanischen Zentralmuseums 40, 3, 905–916, pls. 148–151.

Maul, L.C., 2007. Erster Nachweis von *Hypolagus* (Leporidae, Mammalia) in der unterpleistozänen Fundstelle Untermaßfeld (Thüringen, Mitteldeutschland). Beiträge zur Geologie von Thüringen, Neue Folge 14, 215–233.

in this volume. New results on micromammals from the Early Pleistocene site of Untermassfeld. 1567–1626.

Maul, L.C., Heinrich, W.-D., Parfitt, S.A., Paunescu, A.-C., 2007. Comment on the correlation between magnetostratigraphy and the evolution of *Microtus* (Arvicolidae, Rodentia, Mammalia) during the Early and early Middle Pleistocene. In: Kahlke, R.-D., Maul, L.C., Mazza, P.P.A. (Eds.), Late Neogene and Quaternary biodiversity and evolution: Regional developments and interregional correlations. Proceedings of the 18th International Senckenberg Conference (VI International Palaeontological Colloquium in Weimar). Volume II. Courier Forschungsinstitut Senckenberg 259, 243–263.

Mäuser, M., 1987. Geologische und paläontologische Untersuchungen an der altpleistozänen Säugetier-Fundstelle Würzburg-Schalksberg. Münchner Geowissenschaftliche Abhandlungen A, Geologie und Paläontologie 11, 1–77.

Mayhew, D.F., 1978. Reinterpretation of the extinct beaver *Trogontherium* (Mammalia, Rodentia). Philosphical Transactions of the Royal Society of London 281, B 983, 407–438.

Mazza, P., 1991. Interrelations between Pleistocene hippopotami of Europe and Africa. Bolletino della Società Paleontologica Italiana 30, 153–186.

1995. New evidence on the Pleistocene hippopotamuses of Western Europe. Geologica Romana 31, 61–241.

Melles, M., Brigham-Grette, J., Minyuk, P.S., Nowaczyk, N.R., Wennrich, V., DeConto, R.M., Anderson, P.M., Andreev, A.A., Coletti, A., Cook, T.L., Haltia-Hovi, E., Kukkonen, M., Lozhkin, A.V., Rosén, P., Tarasov, P., Vogel, H., Wagner, B., 2012. 2.8 Million Years of Arctic Climate Change from Lake El'gygytgyn, NE Russia. Science 337, 315–320.

Musil, R., 2001a. Die Equiden-Reste aus dem Unterpleistozän von Untermaßfeld. In: Kahlke, R.-D. (Ed.), Das Pleistozän von Untermaßfeld bei Meiningen (Thüringen). Teil 2. Monographien des Römisch-Germanischen Zentralmuseums 40, 2, 557–587, pls. 92–94.

2001b. Die Ursiden-Reste aus dem Unterpleistozän von Untermaßfeld. In: Kahlke, R.-D. (Ed.), Das Pleistozän von Untermaßfeld bei Meiningen (Thüringen). Teil 2. Monographien des Römisch-Germanischen Zentralmuseums 40, 2, 633–658, pls. 114–120.

Nielsen-Marsh, C.M., Ostrom, P.H., Gandhi, H., Shapiro, B., Cooper, A., Hauschka, P.V., Kahlke, R.-D., Collins, M.J., 2003. Ancient protein sequences: A new direction for molecular palaeontology. Abstracts of Papers of the American Chemical Society 225, 111.

Niethammer, J., 1978. *Apodemus sylvaticus* (Linnaeus, 1758) – Waldmaus. In: Niethammer, J., Krapp, F. (Eds.), Handbuch der Säugetiere Europas. Band 1. Rodentia I (Sciuridae, Castoridae, Gliridae, Muridae). Wiesbaden, 337–358.

1982. *Microtus subterraneus* (de Sélys-Longchamps 1836) – Kurzohrmaus. In: Niethammer, J., Krapp, F. (Eds.), Handbuch der Säugetiere Europas. Band 2/I, Rodentia II (Cricetidae, Arvicolidae, Zapodidae, Spalacidae, Hystricidae, Capromyidae). Wiesbaden, 397–418.

Niethammer, J., Krapp, F., 1982. *Microtus arvalis* (Pallas, 1779) – Feldmaus. In: Niethammer, J., Krapp, F. (Eds.), Handbuch der Säugetiere Europas. Band 2/I, Rodentia II (Cricetidae, Arvicolidae, Zapodidae, Spalacidae, Hystricidae, Capromyidae). Wiesbaden, 284–318.

Oliveira, D., Sánchez Goñi, M.F., Naughton, F., Polanco-Martínez, J.M., Jimenez-Espejo, F.J., Grimalt, J.O., Martrat, B., Voelker A.H.L., Trigo, R., Hodell, D., Abrantes, F., Desprat, S., 2017. Unexpected weak seasonal climate in the western Mediterranean region during MIS 31, a high-insolation forced interglacial. Quaternary Science Reviews 161, 1–17.

Pucek, Z., 1982. *Sicista subtilis* (Pallas, 1773) – Steppenbirkenmaus In: Niethammer, J., Krapp, F. (Eds.), Handbuch der Säugetiere Europas. Band 2/I, Rodentia II (Cricetidae, Arvicolidae, Zapodidae, Spalacidae, Hystricidae, Capromyidae). Wiesbaden, 501–515.

Ramos-Lara, N., Koprowski, J.L. Kryštufek, B., Hoffmann, I.E., 2014. *Spermophilus citellus* (Rodentia: Sciuridae). Mammalian Species 46, 913, 71–87.

Raymo, M.E., Nisancioglu, K., 2003. The 41 kyr world: Milankovitch's other unsolved mystery. Paleoceanography 18, 1011

Reumer, J.W.F., Kahlke, R.-D., in this volume. New results on cercopithecids from the Early Pleistocene site of Untermassfeld. 1627–1633.

Roebroeks, W., Gaudzinski-Windheuser, S., Baales, M., Kahlke, R.-D., 2018. Uneven Data Quality and the Earliest Occupation of Europe – the Case of Untermassfeld (Germany). Journal of Paleolithic Archaeology 1, 5–31.

Rössler, D., Stache, M., 2009. Präparation von Kleinsäugerfunden. Natur und Museum 139, 202–205.

Roth, H.H., 1969. Über das Schwimmen afrikanischer Landsäugetiere im Kariba-Stauseegebiet und ihr Verhalten gegenüber dem Flutwasser. Der Zoologische Garten 37, 12–29.

Roth, H.H., Child, G., 1968. Distribution and Population Structure of Black Rhinoceros (*Diceros bicornis* L.) in the Lake Kariba basin. Zeitschrift für Säugetierkunde 33, 214–226.

Saunders, J.J., Dawson, B.K., 1998. Bone damage patterns produced by extinct hyena, *Pachycrocuta brevirostris* (Mammalia: Carnivora), at the Haro River Quarry, Northwestern Pakistan. In: Tomida, Y., Flynn, L.J., Jacobs, L.L. (Eds.), Advances in Vertebrate Palaeontology and Geochronology. National Science Museum Monographs 14, 215–242.

Schäfer, W., 1976. Aktuopaläontologische Beobachtungen. 10. Zur Fossilisation von Vögeln. Natur und Museum 106, 275–279.

Shackleton, N.J., 1995. New Data on the Evolution of Pliocene Climatic Variability. In: Vrba, E.S., Denton, G.H., Partridge, T.C., Burckle, L.H. (Eds.), Paleoclima and Evolution. With Emphasis on Human Origins. New Haven, CT, London, 242–248.

Sher, A.V., 1997. An Early Quaternary Bison population from Untermaßfeld: *Bison menneri* sp. nov. In: Kahlke, R.-D. (Ed.) Das Pleistozän von Untermaßfeld bei Meiningen (Thüringen). Teil 1.

Monographien des Römisch-Germanischen Zentralmuseums 40, 1, 101–180, pls. 22–33.

Snyder, C.W., 2016. Evolution of global temperature over the past two million years. Nature 538, 226–228.

Soergel, W., 1939. Unter welchen klimatischen Verhältnissen lebten zur Bildungszeit der altdiluvialen Kiese von Süßenborn *Rangifer*, *Ovibos* und *Elephas trogontherii* in Mittel- und Norddeutschland? Zeitschrift der Deutschen Geologischen Gesellschaft 91, 828–835.

Sondaar, P.Y., 1986. The Island Sweepstakes. Why did pygmy elephants, dwarf deer, and large mice once populate the Mediterranean? Natural History 95, 50–57.

Sotnikova, M.V., 2001. Remains of Canidae from the Lower Pleistocene site of Untermassfeld. In: Kahlke, R.-D. (Ed.), Das Pleistozän von Untermaßfeld bei Meiningen (Thüringen). Teil 2. Monographien des Römisch-Germanischen Zentralmuseums 40, 2, 607–632, pls. 102–113.

Steguweit, L., Kahlke, R.-D., 2014. Cut marks and other indication: Archaeologist's imagination, experimental approach and the Early Pleistocene site of Untermassfeld (Central Germany). In: Steguweit, L. (Red.), Hugo Obermaier-Gesellschaft für Erforschung des Eiszeitalters und der Steinzeit e.V., 56. Jahrestagung in Braunschweig und Schöningen, 22.–26. April 2014. Erlangen, 47.

Steininger, F.F., Piller, W.E. (Eds.), unter Mitarbeit der Stratigraphischen Kommission in Deutschland, unter Mitarbeit der Stratigraphischen Kommission in Österreich, 1999. Empfehlungen (Richtlinien) zur Handhabung der stratigrapghischen Nomenklatur. Courier Forschungsinstitut Senckenberg 209, 1–19.

Stephan, E., Uerpmann, H.-P., Cramer, B., 2001. Paläotemperaturbestimmungen nach $\delta^{18}O$-Werten von Equiden-Resten aus dem Unterpleistozän von Untermaßfeld. In: Kahlke, R.-D. (Ed.), Das Pleistozän von Untermaßfeld bei Meiningen (Thüringen). Teil 3. Monographien des Römisch-Germanischen Zentralmuseums 40, 3, 917–930.

Storch, G., 1978. *Glis glis* (Linnaeus, 1766) – Siebenschläfer. In: Niethammer, J., Krapp, F. (Eds.), Handbuch der Säugetiere Europas. Band 1. Rodentia I (Sciuridae, Castoridae, Gliridae, Muridae). Wiesbaden, 243–258.

Tast, J., 1982. *Microtus oeconomus* (Pallas, 1776) – Nordische Wühlmaus, Sumpfmaus. In: Niethammer, J., Krapp, F. (Eds.), Handbuch der Säugetiere Europas. Band 2/I, Rodentia II (Cricetidae, Arvicolidae, Zapodidae, Spalacidae, Hystricidae, Capromyidae). Wiesbaden, 374–396.

Trunova, Y.E., 2001. Hibernation Mark in the Incisors of Fossil Rodents. Doklady Biological Sciences 377, 163–165.

Trunova, Y.E., Klevezal, G.A., 1999. Intraspecific Variation in the Records of Winter Hibernation in Incisor Dentin of Rodents. Russian Journal of Zoology 3, 637–646.

Turner, A., 2001. Remains of *Pachycrocuta brevirostris* (Carnivora, Hyaenidae) from the Lower Pleistocene site of Untermaßfeld. In: Kahlke, R.-D. (Ed.), Das Pleistozän von Untermaßfeld bei Meiningen (Thüringen). Teil 2. Monographien des Römisch-Germanischen Zentralmuseums 40, 2, 673–690, pls. 121–127.

Turner, A., Antón, M., 1996. The Giant Hyaena, *Pachycrocuta brevirostris* (Mammaalia, Carvivora, Hyaenidae). Geobios 29, 455–468.

Unger, K.P., Kahlke, R.-D., 1995. IX. Thüringen. In: Benda, L. (Ed.), Das Quartär Deutschlands. Berlin, Stuttgart, 199–219.

Van Asperen, E.N., Kahlke, R.-D., 2017. Dietary traits of the late Early Pleistocene *Bison menneri* (Bovidae, Mammalia) from its type site Untermassfeld (Central Germany) and the problem of Pleistocene 'wood bison'. Quaternary Science Reviews 177, 299–313.

Van der Made, J., Rosell, J., Blasco, R., 2017. Faunas from Atapuerca at the Early–Middle Pleistocene limit: The ungulates from level TD8 in the context of climatic change. Quaternary International 433, 296–346.

Van Essen, H., 2011. Discontinuitiy in dental morphology between late Early to early Middle Pleistocene mammoths (Mammalia, Proboscidea) in Europe, with special reference to Voigstedt and Edersleben. In: Van Essen, H., Tracing Transitions. An overview of the evolution and migrations of the genus *Mammuthus* Brookes, 1828 (Mammalia, Proboscidea). PhD thesis, Leiden University, The Netherlands, 121–251.

in this volume. New results on elephantids from the Early Pleistocene site of Untermassfeld. Cranial and dental remains. 1355–1384.

Van Jaarsveld, A.S., 1993. A comparative investigation of hyaenid and aardwolf life-histories, with notes on spotted hyaena mortality patterns. Transaction of the Royal Society of South Africa 48, 219–231.

Viro, P., Niethammer, J., 1982. *Clethrionomys glareolus* (Schreber, 1780). In: Niethammer, J., Krapp, F. (Eds.), Handbuch der Säugetiere Europas. Band 2/I, Rodentia II (Cricetidae, Arvicolidae, Zapodidae, Spalacidae, Hystricidae, Capromyidae). Wiesbaden, 109–146.

Vogeltanz, R., 1965. Austrocknungsstrukturen bei Koprolithen. Neues Jahrbuch für Geologie und Paläontologie, Monatshefte 6, 362–371.

Wächtler, W., 1929. Zur Lebensweise der *Caecilioides acicula* Müll. Archiv für Molluskenkunde 61, 1–14.

Walter, H., Breckle, S.-W., 1986. Ökologie der Erde. Geo-Biosphäre. Band 3: Spezielle Ökologie der Gemäßigten und Arktischen Zonen Euro-Nordasiens. Zonobiom VI–IX. Stuttgart.

Wiegank, F., 1997. Paläomagnetische Charakteristik des Unterpleistozäns von Untermaßfeld. In: Kahlke, R.-D. (Ed.), Das Pleistozän von Untermaßfeld bei Meiningen (Thüringen). Teil. 1. Monographien des Römisch-Germanischen Zentralmuseums 40, 1, 63–69.

Wolsan, M., 2001. Remains of *Meles hollitzeri* (Carnivora, Mustelidae) from the Lower Pleistocene site of Untermassfeld. In: Kahlke, R.-D. (Ed.), Das Pleistozän von Untermaßfeld bei Meiningen (Thüringen). Teil 2. Monographien des Römisch-Germanischen Zentralmuseums 40, 2, 659–671.

Zagwijn, W.H., Jong, J. de, 1984. Die Interglaziale von Bavel und Leerdam und ihre stratigraphische Stellung im niederländischen Früh-Pleistozän. Mededelingen Rijks Geologische Dienst 37 (S. J. Dijkstra Memorial Volume), 155–169.

Zapfe, H., 1939. Lebensspuren der eiszeitlichen Höhlenhyäne. Die urgeschichtliche Bedeutung der Lebensspuren knochenfressender Raubtiere. Palaeobiologica 7, 111–146.

2001. Zähne von *Macaca* aus dem Unterpleistozän von Untermaßfeld. In: Kahlke, R.-D. (Ed.), Das Pleistozän von Untermaßfeld bei Meiningen (Thüringen). Teil 3. Monographien des Römisch-Germanischen Zentralmuseums 40, 3, 889–895, pl. 146.

Zeissler, H., 1981. Schnecken und Muscheln in und um Weimar. Eine Molluskenfauna des Gebietes Weimar. Weimarer Schriften zur Heimatgeschichte und Naturkunde 44, 1–111.

LIST OF CONTRIBUTORS

Jessica Arnold
Senckenberg Forschungsstation
für Quartärpaläontologie
Am Jakobskirchhof 4
D - 99423 Weimar
jessica.arnold@senckenberg.de

Stefan Flohr
Universität Hildesheim
Institut für Biologie und Chemie
Abteilung Biologie
Universitätsplatz 1
D - 31141 Hildesheim
flohrs@uni-hildesheim.de

Nuria García
Universidad Complutense de Madrid
Facultad de Ciencias Geológicas
Departamento de Geodinámica,
Estratigrafía y Paleontología
Grupo de Investigación Ecosistemas
Cuaternarios e Centro Mixto de Evolución
y Comportamiento Humanos
C/ José Antonio Novais, 2
Ciudad Universitaria
E - 28040 Madrid
nuria.garcia.ucm@gmail.com

Helmut Hemmer
h.hemmer-mainz@gmx.de

Alessio Iannucci
Sapienza Università di Roma
PaleoFactory
Dipartimento di Scienze della Terra
Piazzale Aldo Moro 5
I - 00185 Roma
alessio.iannucci@uniroma1.it

Ralf-Dietrich Kahlke
Senckenberg Forschungsstation
für Quartärpaläontologie
Am Jakobskirchhof 4
D - 99423 Weimar
rdkahlke@senckenberg.de

Uwe Kierdorf
Universität Hildesheim
Institut für Biologie und Chemie
Abteilung Biologie
Universitätsplatz 1
D - 31141 Hildesheim
kierdo@uni-hildesheim.de

Lutz Christian Maul
Senckenberg Forschungsstation
für Quartärpaläontologie
Am Jakobskirchhof 4
D - 99423 Weimar
lmaul@senckenberg.de

Paul P. A. Mazza
Università degli Studi di Firenze
Dipartimento di Scienze della Terra
Via La Pira 4
I - 50121 Firenze
paul.mazza@unifi.it

Beniamino Mecozzi
Sapienza Università di Roma
PaleoFactory
Dipartimento di Scienze della Terra
Piazzale Aldo Moro 5
I - 00185 Roma
beniamino.mecozzi@uniroma1.it

Jelle W. F. Reumer
Universiteit Utrecht
Faculteit Geowetenschappen
Departement Aardwetenschappen
Vening Meinesz Building A
Princetonlaan 8A
NL - 3508 TC Utrecht
and
Natuurhistorisch Museum
Westzeedijk 345 (Museumpark)
NL - 3015 AA Rotterdam
and
Naturalis Biodiversity Center
Postbus 9517
NL - 2300 RA Leiden
j.w.f.reumer@uu.nl

Raffaele Sardella
Sapienza Università di Roma
PaleoFactory
Dipartimento di Scienze della Terra
Piazzale Aldo Moro 5
I - 00185 Roma
raffaele.sardella@uniroma1.it

Hans van Essen
Universiteit Leiden
Faculteit Archeologie
Van Steenis
Einsteinweg 2
NL - 2333 CC Leiden
ejavanes@xs4all.nl